Problematic Wildlife

Francesco M. Angelici
Editor

Problematic Wildlife

A Cross-Disciplinary Approach

Springer

Editor
Francesco M. Angelici
FIZV, Rome
Italy

ISBN 978-3-319-22245-5 ISBN 978-3-319-22246-2 (eBook)
DOI 10.1007/978-3-319-22246-2

Library of Congress Control Number: 2015958782

Springer Cham Heidelberg New York Dordrecht London

Printed on acid-free paper

Springer International Publishing AG Switzerland is part of Springer Science+Business Media (www.springer.com)

Preface

Arbore deiecta, quivis ligna colligit

(Latin proverb)

Go there, where nature leads

(Francis Bacon)

It is with great trepidation and honour that I write this introduction after a long work concluded in close cooperation with all authors of this book. I am very happy to have finally finished this work!

The theme is very composite, very intriguing and, at the same time, extremely hot and current. The 'Problematic Wildlife' is certainly an argument really great and varied.

I fought a lot for this to come true, we have worked hard with all the authors, and I hope that the common effort had generated a useful product. It all began with the organisation of two international conferences held in Italy; in 2007 and in 2011, they succeeded very well. But although some authors of this book have participated in these events, specifically now this book is not a collection of conference proceedings, at all! Most of the contributions were born in a total independence; those congresses have only increased my interest for these issues of wildlife management, which already existed for a long time.

Problematic species? What does it mean? In fact there is no single definition. In this volume, in which many experts of various fields of ecology, zoology and animal biology and conservation biology have worked actively, with cross-cutting and interdisciplinary approaches, we will try to clarify what is meant by this definition or, better, what can be understood. Specific cases as examples and other contributions will be considered; by contrast, they will address more general and wide-ranging reviews on the subject.

The reason for this book is precisely to grope to make a little sum of the various cases in which wildlife can cause problems to humans, problems of various kinds, of course. And bear always in mind that the man is the main cause, if not unique, of issues related to wildlife. Following the altering intervention of man over nature, the wildlife can sometimes create 'problems' that often requires human intervention again to ensure that it can again be sustainable and in harmony with human activity and habitats remain now irrevocably altered.

It is clear that without the impact of human activity, there would not be 'Problematic Wildlife'.

Of course it has not been possible to treat all existing cases, but the text is divided into sections that cover the most important case studies, with some examples that perhaps make the idea to entire classes of events. But the purpose of this book is also to try to highlight how the various issues should be addressed in wide range, with the participation and synergy of different professionals, who face various cases from different points of view of course, but that all should be most possibly coordinated for obtaining the common objective, namely, the resolution, or at least mitigation, of problems.

Each section is preceded by a brief presentation that makes continuity and connection between different themes easier. I'll be glad to get advice or reports of individual and explanatory cases, in the event of a future updated edition.

The book has been compiled for a broad audience. It is, of course, intended for industry insiders, i.e. anybody who is interested in the profession of wildlife and wildlife management, animal biology and ecology. Yet it is also a book that addresses some issues related to trade, trafficking in endangered species and bushmeat. Problems of human-wildlife conflict will also be dealt with, from the species that are 'managed' to those that become invasive pests, including problems related to species that are in decline, already extinct or endangered. Therefore, this book is also of use to university students in disciplines such as ecology, animal biology, wildlife conservation and so on, as well as to those involved in legislation related to the protection of fauna, engineers and teachers of natural science or ecology. It is also of interest to people working in protected areas, national parks, reserves, etc., in short, to all those who have to manage the land optimally so that it is preserved in the best possible way.

The issue of cryptic species (do they exist or not?) has also been added simply because this argument often creates some rather unusual situations, attracting the attention of the general public, in addition to that of scientific authorities and wildlife management institutions.

Naturally, I would like to thank all the authors, who were the true architects of this book, because they were the ones who, from the outset, have believed in its significance and realisation and have waited a long time to see its birth. I wish to thank Springer Verlag International for its adherence to this initiative and for giving value to our work and making it concrete. Thanks is also owed to all those listed below for their unpaid efforts to foster the creation of a high-quality product, for their great professionalism and for their comments and suggestions such as those published in high- profile or scientific journals.

The book is dedicated to the memory of Marco Flavio Angelici. Thinking of him gave me the strength to continue to work hard, especially when I was not in very good health and in those moments when the project seemed to be at a stalemate.

The written contributions (including four rejected manuscripts) have been reviewed and improved by the critical referee processes of the following people:

Brian R. Amman, Marc Ancrenaz, Francesco M. Angelici, Marco Apollonio, Chad Arment, Simon J. Baker, Natale E. Baldaccini, Luigi Boitani, Scott M. Brainerd, Paul R. F. Chanin, Ernest W. T. Cooper, Alayne O. Cotterill, Cheli Cresswell, J. Fernando Del Moral Sacchetti, Emmanuel Do Linh San, Paul E. Eschenfelder, Alan H. Fielding, Shandra N. Frey, Birutė M. F. Galdikas, Jaime Gongora, Simon J. Goodman, Colin P. Groves, Matt W. Hayward, Juan Herrero Cortes, Edwin E. Herricks, Tim L. Hiller, Gregg R. Howald, Robert E. Kenward, András Kovács, Sabrina Kumschick, Jan E. Janecka, Bruce D. Leopold, Laly L. Lichtenfeld, Rurik H. List, Scott R. Loss, Sandro Lovari, Christopher F. Mason, L. David Mech, Terry A. Messmer, Susan M. Miller, Solomon Mombeshora, Alessandro Montemaggiori, Darren Naish, Wolfgang Nentwig, John F. Oates, Craig Packer, Lorenzo Rossi, Holger Schielzeth, Jeffrey J. Short, K. Shawn Smallwood, Anna Songhurst, Keith Springer, Caroline J. Stem, Jeffrey B. Stetz, Diederik Strubbe, Benoît de Thoisy, Brian E. Washburn, Mark E. Welch, David S. Wilkie and Tanya Wyatt

Rome, Italy

Francesco M. Angelici
May 20, 2015

Contents

Contributors

Julien Alleau Norwegian Institute for Nature Research, Trondheim, Norway

Giovanni Amori CNR-ISE, Institute for Ecosystem Studies, Rome, Italy

Francesco Maria Angelici Fondazione Italiana per la Zoologia dei Vertebrati (FIZV), Rome, Italy

Nicola Baccetti Istituto Superiore per la Protezione e la Ricerca Ambientale (ISPRA), Ozzano Emilia (Bologna), Italy

Alessandro Balestrieri Department of Biosciences, University of Milan, Milan, Italy

Corrado Battisti "Torre Flavia" LTER (Long Term Ecological Research) Station, Protected Areas Service, Rome, Italy

Sébastien Le Bel Cirad Department ES UPR BSEF, TA C-22/D101, Campus International de Baillarguet, Montpellier Cedex 5, France

Anna Brangi Province of Pavia, Pavia, Italy

Dario Capizzi Latium Region—Regional Park Agency, Rome, Italy

David Chavernac ES Department, BSEF Research Unit, CIRAD, Montpellier, Cedex 5, France

BIOS Department, CMAE Research Unit, Elephant Research and Conservation Unit, CIRAD, Chiredzi, Zimbabwe

René Czudek Food and Agriculture Organization of the United Nations, Subregional Office for Southern Africa, Harare, Zimbabwe

Katarzyna Daleszczyk Research and Education Unit, Białowieża National Park, Białowieża, Poland

Arie Dekker Royal Netherlands Air Force (Ret), Breda, The Netherlands

Const. Castricum, The Netherlands

Amy J. Dickman Wildlife Conservation Research Unit, Department of Zoology, The Recanati-Kaplan Centre, Tubney House, University of Oxford, Oxford, UK

Andrew Dixon International Wildlife Consultants Ltd, Carmarthen, UK

Imran Ejotre Department of Biology, Bucknell University, Lewisburg, PA, USA

Amy E. Eycott Department of Biology, University of Bergen, Bergen, Norway

Philippe Gaubert Institut des Sciences de l'Evolution de Montpellier (ISEM)—UM2-CNRS-IRD, Université de Montpellier, Montpellier Cedex 05, France

Gabriele Gentile Department of Biology, Tor Vergata University, Rome, Italy

Francesca Giannini Tuscan Archipelago National Park, Portoferraio, Livorno, Italy

Mike La Grange AWMC (Pvt) Ltd, Harare, Zimbabwe

Scott R. Groepper Wildlife Biologist, Mitchell, SD, USA

Leela Hazzah Lion Guardians, Nairobi, Kenya

Kimberley J. Hockings Anthropology Centre for Conservation, Environment and Development, Oxford Brookes University, Oxford, UK

Departmento de Antropologia, Universidade Nova de Lisboa, Lisbon, Portugal

Centre for Research in Anthropology (CRIA), Lisbon, Portugal

Arturo Izurieta Galápagos National Park Directorate, Puerto Ayora, Santa Cruz, Galápagos, Ecuador

Jaana Kekkonen University of Helsinki, Helsinki, Finland

Laura A. Kurpiers Department of Biology, Bucknell University, Lewisburg, PA, USA

Paolo Lamberti Department of Earth and Environmental Sciences, University of Pavia, Pavia, Italy

John D.C. Linnell Norwegian Institute for Nature Research, Trondheim, Norway

Marco Lombardini Department of Earth and Environmental Sciences, University of Pavia, Pavia, Italy

Albert M. Manville II Advanced Academic Programs, Krieger School of Arts and Sciences, Johns Hopkins University, Washington, DC, USA

Wildlife and Habitat Conservation Solutions LLC, Falls Church, VA, USA

Division of Migratory Bird Management, U.S. Fish and Wildlife Service, Arlington, VA, USA

Cruz Marquez Charles Darwin Foundation, Puerto Ayora, Santa Cruz, Galápagos, Ecuador

Jeffrey McKee Avisure Pty Ltd, West Burleigh, QLD, Australia

Ailsa J. McKenzie School of Biology, Newcastle University, Newcastle upon Tyne, UK

Matthew R. McLennan Anthropology Centre for Conservation, Environment and Development, Oxford Brookes University, Oxford, UK

Mattia Menchetti Department of Biology, University of Florence, Sesto Fiorentino, Firenze, Italy

Alberto Meriggi Department of Earth and Environmental Sciences, University of Pavia, Pavia, Italy

Pietro Milanesi Department of Earth and Environmental Sciences, University of Pavia, Pavia, Italy

Jeroen Minderman School of Biology, University of St Andrews, Dyers Brae House, Greenside Place, St Andrews, Fife, UK

Emiliano Mori, Di.S.A.F.A., Entomology and Zoology, University of Turin, Grugliasco (Turin), Italy

Kylie Patrick Avisure Pty Ltd, Brasilia, Brazil

Claudio Prigioni Department of Earth and Environmental Sciences, University of Pavia, Pavia, Italy

Rajanathan Rajaratnam Department of Geography and Planning, University of New England, Armidale, NSW, Australia

DeeAnn M. Reeder Department of Biology, Bucknell University, Lewisburg, PA, USA

Luigi Remonti Department of Earth and Environmental Sciences, University of Pavia, Pavia, Italy

Institut Agricole Régional, Regione La Rochère 1/A, Aosta 11100, Italy

Lorenzo Rossi Associazione Orango, Via San Cristoforo, Cesena (FC), Italy

Tiger Sangay Ecosystem Management, University of New England, Armidale, NSW 2351, Australia

Ugyen Wangchuk Institute for Conservation and Environment, Bumthang, Bhutan

Björn Schulte-Herbrüggen Stockholm Resilience Centre, Stockholm University, Stockholm, Sweden

Phillip Shaw Avisure Pty Ltd, Burleigh Heads, QLD, Australia

Howard L. Snell Department of Biology, University of New Mexico, Albuquerque, NM, USA

Ragnhild Sollund Department of Criminology and Sociology of Law, University of Oslo, Oslo, Norway

Paolo Sposimo NEMO Srl, Florence, Italy

Fiona Stansfield ES Department, BSEF Research Unit, CIRAD, Montpellier, Cedex 5, France

BIOS Department, CMAE Research Unit, Elephant Research and Conservation Unit, CIRAD, Chiredzi, Zimbabwe

Washington Tapia Galápagos Conservancy, Puerto Ayora, Santa Cruz, Galápagos, Ecuador

Galápagos National Park Directorate, Puerto Ayora, Santa Cruz, Galápagos, Ecuador

Department of Biology, University of Malaga, Malaga, Spain

John Thorpe F.R.Ae.S., International Bird Strike Committee, Surrey, UK

Jörg E. Tillmann DBU Natural Heritage GmbH, Osnabrück, Germany

Martina Trinkel School of Life Sciences, University of KwaZulu-Natal, Durban, South Africa

Stephen M. Vantassel Wildlife Control Consultant, LLC, Lewistown, MT, USA

Karl Vernes Ecosystem Management, University of New England, Armidale, NSW, Australia

Mark J. Whittingham School of Biology, Newcastle University, Newcastle upon Tyne, UK

Mieke C. Zwart School of Biology, Newcastle University, Newcastle upon Tyne, UK

Part I
General Introduction: The Reasons Behind and Significance of the Book. Problematic Wildlife: Definitions and Concepts. When and Why a Wild Species May Become Problematic

Chapter 1
Problematic Wildlife at the Beginning of the Twenty-First Century: Introduction

Francesco Maria Angelici

It is not man who has to fight against a hostile nature, but nature is helpless and for generations has been the victim of humanity

(Jacques-Yves Cousteau)

Maybe this world is another planet's hell

(Aldous Leonard Huxley)

What Is the Inspiration Behind This Book?

The inspiration for this book emerged from two international congresses held in Italy on human–wildlife conflicts (Angelici et al. 2007; Angelici and Petrozzi 2011). The number and diversity of topics convinced me that a book describing in detail the various situations in which wildlife comes into "conflict" with human activities and conversely when human activities may also create a situation which exacerbates conflicts to create an uneasy coexistence was needed. Normally, the term "wildlife" has been used to describe "animals and plants that grow independently of people, usually under natural conditions" (Yarrow 2009; IDM 2013).

In this book we cover only homeothermic vertebrates wildlife species (mammals and birds) with one exception. That exception is a large reptile, i.e., the pink iguana of the Galapagos Islands (*Conolophus marthae*) as it is a special case which is fitting to the topics covered in its own section. Regarding the specific case of the wildlife trade and bushmeat (treated in this volume), also other animal species, as

F.M. Angelici (✉)
Fondazione Italiana per la Zoologia dei Vertebrati (FIZV),
Via Cleonia 30, Scala C, Rome 00152, Italy
e-mail: francescomariaangelici@gmail.com

F.M. Angelici (ed.), *Problematic Wildlife*, DOI 10.1007/978-3-319-22246-2_1

well as mammals and birds, can be interested in the phenomenon (see below). Regardless of what definition used, animal species defined as wildlife can interfere with human presence and its activities at different levels, very often creating a variety of problems that may warrant precautionary measures or otherwise human intervention.

Problematic Wildlife: Exploring Contemporary Definitions

A problematic wildlife species is an animal whose presence in a given geographical area comprising one or more habitats is no longer compatible, in either a partial or absolute sense, with one or more components of the ecosystem. Therefore, a problematic species is not necessarily one that conflicts, either directly or indirectly, with human activities, but it may also be a species that conflicts with other animal or plant species, creating problems (Angelici 2009).

According to Messmer (2000), the conflict arising between man and wildlife species may be real or only perceived, economic or aesthetic, social or political. We must consider, however, that from the technical and scientific point of view, only actual cases of a truly ecological or even social or economic nature should be taken into account, which then excludes cases that tend to be purely aesthetic, political or otherwise. Clearly, if humans did not "alter" these various environments with his presence and activities, theoretically "problematic species" would not exist. Therefore, it may be argued that without human associations or activities, the category of "problematic wildlife" would not exist. On the other hand, it is also true that situations including the incompatibility between species (competition with subsequent competitive exclusion, or niche share-out) can be created without human "intervention," i.e., simply ecological situations that are changing in time and space, precisely as a result of the change in environmental conditions in a broad sense (Park 1954, 1962). However, these cases are not included in the topics covered by this book.

In most cases, it is human society that determines, often without any consistent or special criteria, when a species becomes "problematic." A wildlife species is perceived as problematic (due to disturbance and/or damage), for example, when it becomes particularly abundant, or if it is "out of place" (Marion 1988). The problem is then connected to its density and concentration, and to the consequent damages, and not to the presence of the species itself. There are areas and situations that intensify or attenuate these hardships, such as the presence of substantial flocks of gulls on the runway and in the sky above airports, or the pigeon droppings that can accumulate in town centers. Regarding mammals, noteworthy are the cases of rat invasions of towns and villages, or the intolerable increase in foxes, wolves, and coyotes (e.g., Farrar 2007) preying on domestic animals, which in some cases even come to attack humans (e.g., Baker and Timm 1998).

A wildlife species, according to a broader definition, can be defined as problematic, because of its conservation status. Species that are no longer able to sustain themselves without human intervention or are in serious danger of extinction and require legal protection to be saved may be considered problematic (e.g., Ralls and Ballou 2004). This of course represents another aspect of human–wildlife interactions, but it is equally important and noteworthy to those affected. Concomitantly, in many of these cases, anthropogenic factors are the source of the threats jeopardizing the species existence. Thus, humans attempt to intervene again, in a timely and appropriate manner, to try to solve the "problem," in this case, to save the species, or at least the local population, from extinction (Ebenhard 1995).

Human–wildlife conflict as defined by the World Wide Fund for Nature (WWF 2005) is "any interaction between humans and wildlife that results in negative impacts on human social, economic or cultural life, on the conservation of wildlife populations, or, more in general, on the environment." Madden (2004) elaborated human–wildlife conflict occurs when the needs and behavior of wildlife impact negatively the goals of humans or when the goals of humans negatively impact the needs of wildlife. The United States Geological Survey further defined human–wildlife conflict in two ways: firstly, as when actions by wildlife conflict with human goals, i.e., life, livelihood, and life-style; and, secondly, as when human activities threaten the safety and survival of wildlife (Cline et al. 2007). However, in both cases, outcomes are dictated by human's responses to these interactions. A good discussion of the various cases of "problematic wildlife" and of "human–wildlife" interaction, including possible intervention methods, are discussed in the work of Conover (2002), one of the few rather exhaustive treatises on this discipline in general.

In virtually every part of the earth, humans have impacted wildlife (and therefore also vice versa). Thus, both the depth and breath of the situations that could be defined as "problematic wildlife" are extensive. Because standardized process for recording and validating the extent of the cost of these interactions is lacking, we frequently rely on public surveys to estimate the magnitude.

To give just one example of these surveys, Conover (1998), based on a survey conducted of farmers and ranchers, estimated the annual damages caused by wildlife to US agriculture exceeded $2 billion (USD), despite almost $2 billion (USD) spent annually on preventive measures. Else, globally, wildlife collisions with aircraft cost an estimated $2 billion (USD) annually only in the USA (International Civil Aviation Organization 2009; Biondi et al. 2014).

Conover et al. (1995) estimated the annual costs in the U.S. only related to damage from human losses, accidents, injuries, etc. because of human–wildlife interaction may approach $3 billion (USD).

Or, another example, again in the USA, only the damages and losses caused by invasive alien species would amount to almost $120 billion (USD) per year (Pimentel et al. 2005).

And this is only one aspect of the damage to wildlife, in a single nation, while there are hundreds of other different cases, and at global level. Definitely, the global cost of wildlife "damages" will be equivalent to several hundred or thousands of billions of dollars (USD) per year (see also Messmer 2009).

Beginning in prehistoric times, humans and his predecessors have had interactions with wildlife, mainly interactions with predators, for whom hominins were possible prey. So, e.g., big cats (*Panthera leo spelaea*, *P. atrox*), bears (*Ursus spelaeus*), wolves (*Canis lupus*), and hyenas (*Crocuta crocuta spelaea*) have certainly represented a direct problem, which very often influenced the very survival of the human species and its ancestors (Hart and Sussman 2009).

However, in my opinion some animal species became problematic in their own right with the advent of agriculture and, pastoralism. The wild ancestors of many pets likely had a "problematic" relationship with man, such as the wild horse (*Equus ferus*), African wild ass (*E. africanus*), aurochs (*Bos primigenius*), grey wolf (*Canis lupus*), African wildcat (*Felis silvestris lybica*), wild goat (*Capra aegagrus*), mouflon (*Ovis orientalis*), wild boar (*Sus scrofa*), red jungle fowl (*Gallus gallus*), etc. Subsequently, some of these wild species became extinct, such as aurochs, which were the predecessors of various breeds of the domestic ox; or became endangered, such as equines, the wild ancestors of the horse and donkey; while others, such as the grey wolf, reduced their range, etc. (Clutton-Brock 1999). Hence, for some species, in this case including but not limited to mammals, highly "problematic" has meant that they have had an impact on the human species (both positive as prey-food, or negative, as a competitor or predator), which has "manipulated" these species through the process of domestication, often giving rise to domestic forms which also originated from multiple species (e.g., Eriksson et al. 2008).

In this book, however, the term "problematic wildlife" refers to a broader concept following the matter of fact that at various levels of distress, in all situations or conditions, where wildlife can be seen as a problem to be solved, this term finds his proper use. In particular, it should be borne in mind that the basic problems, with the exception of some very rare cases (e.g., extraordinary climatic-meteorological events, global warming, extreme weather), are due to wrongful human actions or the impact of man and his activities, and that man is also responsible for sudden changes in the weather and/or climate (cf. Oreskes 2004). Therefore, the species that are considered "problematic species" are all those wildlife species that create problems, either directly or indirectly, e.g., species that become invasive and must be controlled, such as mice and rats (Stenseth et al. 2003), because they create problems and damage agricultural crops and domestic animals, become carriers of zoonoses, infest urban environments, etc. (Brown et al. 2007). This also includes many other species, e.g., domestic cats (*Felis silvestris catus*), rabbits (*Oryctolagus cuniculus*), parrots (e.g., *Psittacula krameri*, *Myiopsitta monachus*), (Courchamp et al. 2003; Woods et al. 2003; Parkes et al. 2014; Menchetti et al. 2016) which create a variety of problems, such as directly impacting fauna (competition, predation) or vegetation, etc. On islands, murine populations of mice (mainly *Mus musculus*) and rats (*Rattus* spp.) frequently create significant problems of various kinds (Howald et al. 2007) for seabirds by preying on their eggs and nestlings, and also indirectly causing other alterations in the insular animal communities (Towns et al. 2009). Therefore, especially in recent years, it has increasingly considered appropriate to complete projects to eradicate rats and mice (e.g., Capizzi et al. 2016). These efforts have occasionally been debated and considered controversial, especially because there is still insufficient data on the long-term effects on the island environment and animal species subsequent to eradication (Towns et al. 2006). In other situations, unexpected effects have generated additional problems which must be solved, e.g., regarding the eradication of feral cats and the subsequent expansion of the rabbit population (Bergstrom et al. 2009). However, it is undeniable that some introduced species, especially ship rats (*Rattus rattus*) and brown rats (*R. norvegicus*), have contributed significantly to the reduction or extinction of many species of fauna which are native or endemic to the islands, such as small rodents, bats, shrews, lizards, birds, etc. (Towns et al. 2006), including very large islands (Parkes and Murphy 2003) where different levels of the containment, eradication, and management of these invasive species have often been carried out.

More generally, another current problem is the introduction of alien or nonnative species into new territories (Schmitz and Simberloff 1997; Mooney 2005; Vilà et al. 2010), which often causes the elimination or substantial reduction of native fauna species. However, even in these cases, the facts must be analyzed individually, and often data are insufficient, so we can only provide a general overview (Gurevitch and Padilla 2004). Some cases, e.g., in Europe, have already been successfully addressed at the local level and have resulted in the eradication of alien species (Genovesi 2005; Baker 2006). Other times, containment, and not the complete eradication of the alien species, has been discussed as a positive and convenient, although not definitive, solution, such as in the case of the coypu (*Myocastor coypus*, Panzacchi et al. 2007). Yet it was clear that these efforts to contain have not had any overall effect, considering the short duration of their effectiveness. In the end, they are considered unproductive overall, despite their generally being very expensive (cf. Fernández Orueta et al. 2001; Gherardi and Angiolini 2004). Of course, each case must be evaluated in and of itself, and it is impossible to generalize as the various applicable options and the chances of success depend on the introduced species and on the area covered by eradication/control.

It goes without saying that all those species which, because of man and his impact, have had their geographic range and numbers dramatically reduced, to the point of being considered rare, endangered, or even extinct, may be considered "problematic wildlife." A wildlife species can also become "problematic" from the human point of view, just because, as mentioned, it collides with some activity or human interest. There are countless examples of extinct wildlife species (Diamond et al. 1989) or species that are in danger of becoming extinct because of man. According to the 2014 IUCN red list, of the 5513 known mammal species, 22 % (1199 species) are threatened and 77 species (1.4 %) are already extinct; or, for birds, out of a total of 10,425 species, 13.35 % (1373 species) are threatened and 140 species (1.3 %) are already extinct (IUCN 2014). Needless to say, almost all of these species have become extinct or are in danger of extinction because of man.

Another possibility, which has a very direct effect, regarding the problematic nature of wildlife species, is when the species becomes dangerous for the life of man (Sukumar 1991; Conover et al. 1995; Löe and Röskaft 2004; Hart and Sussman 2009). This includes not only species that occasionally present a direct danger to humans, such as the African savannah elephant (*Loxodonta africana*), African forest elephant (*L. cyclotis*), Asian elephant (*Elephas maximus*), hippo (*Hippopotamus amphibious*), and African buffalo (*Syncerus caffer*), which despite their not being predators cause hundreds of deaths each year in both Africa and Asia (Thouless 1994; Durrheim and

Leggat 1999; Dunham et al. 2010; Chomba et al. 2012), but also certain species that occasionally or in some cases repeatedly prey on man, such as the tiger (*Panthera tigris*) (e.g., Corbett 1944; Montgomery 2009; Inskip and Zimmermann 2009), lion (*P. leo*) (Packer et al. 2005; Inskip and Zimmermann 2009), leopard (*P. pardus*) (Corbett 1947), grey wolf (Linnell et al. 2002), brown bear (*Ursus arctos*), American black bear (*U. americanus*) (Herrero 2002), crocodiles (*Crocodilus* spp.) (e.g., Caldicott et al. 2005; de Silva 2008) and, to a lesser extent, other species (Kruuk 2002; Linnell et al. 2002).

Of course, the impact all these species, which are considered precisely "man eaters," have on man and his activities can be devastating (see Linnell 2016). "Man-eaters" are also very often predators of livestock and are therefore considered very "harmful," hence humans' retaliatory reaction to them is just as devastating (e.g., Corbett 1944; Woodroffe and Frank 2005) and can often lead to the extinction of the species, both locally and totally (e.g., Seidensticker 1987; Yalden 1999; Black et al. 2013). Finding patterns of coexistence, or at least ways to mitigate conflict, is a big challenge and, in my opinion, one of the aspects that will become more prevalent and pressing in the future with the steadily growing human population which is therefore increasingly in contact and overlapping with the areas affected by the presence of potentially "man-eating" or in any case very hazardous predators (Treves and Naughton-Treves 1999; Decker et al. 2002; Woodroffe et al. 2007).

Problematic Wildlife: Consequences of Translocations

Moreover, many wildlife species are hunted, especially resident species, and are managed in some way and manipulated to sustain high population levels or expand distributions. They and their management come fully under the definition of "problematic wildlife" (e.g., Reynolds and Tapper 1996; Santilli and Bagliacca 2008), such as various types of faunal input operations or the translocation of a wildlife species from one territory to another (Souty-Grosset and Grandjean 2007). The term "introduction" means the input of a species that has never been present in an area and that is a nonnative species (e.g., Kauhala 1996). This practice may lead to profound alterations in the ecological order, such as competition with native species from a nutritional and territorial point of view, and even to the extinction of native species, as well as to the spread of new diseases (Cunningham 1996).

"Reintroduction" is the input of locally extinct species which were, however, present in the past, in order to reconstitute some stable and vital populations. This practice requires very thorough measures and preliminary analysis. First, in addition to ascertaining the real presence of the species in historical time through accurate bibliographic research or direct testimony, the causes that led to its extinction must be analyzed to then proceed to the removal of these causes; otherwise reintroduction would be ineffective, if not even dangerous for other wildlife species (Seddon et al. 2007; Soorae 2011).

"Repopulation" or "restocking," on the other hand, are interventions that tend to increase, by means of artificial intake, the number of subjects in a numerically reduced population already present in the area that is the object of the operation. These are conducted with the aim of annually restoring the levels of the wild populations that have suffered a decline in number generated by habitat upheaval or, more frequently, by hunting pressure (Grahn 1993; Angelici et al. 2000). "Restocking" does not always generate positive effects, especially in the case of hunting, and can actually be not only a failure, causing high mortality that affects the results and renders costs unjustifiable (e.g., Angelici et al. 2000; Parish and Sotherton 2007), but it can also provoke the introduction of different *taxa*, i.e., resulting in hybridization between different subspecies (Blanco-Aguiar et al. 2008; Champagnon et al. 2009) or even the real introduction of different species (Angelici and Luiselli 2007). Even the introduction of pathogens (parasites, epidemics, etc.) can cause significant problems in the area subject to introduction, with heavy consequences for the wildlife populations already present in the area (see Champagnon et al. 2012).

Problematic Wildlife: Consequences from Interactions with Humans

Another aspect of "problematic wildlife" relates to interactions where human activities may present a potential or actual danger to wildlife (e.g., Arroyo and Razin 2006). For instance, for birds, electrocutions may occur when they perch on powerlines. Thus, it may be particularly problematic for larger birds such as diurnal or nocturnal raptors, but also Ciconiiformes, Gruiformes, etc. (Bevanger 1994, 1998; Bayle 1999; Lehman et al. 2007), having negative consequences for some sensitive species (Sergio et al. 2004; Rollan et al. 2010). In these cases, it is essential to study methods and strategies to prevent or limit impact (Tintó et al. 2010; Avian Power Line Interaction Committee 2012).

Even bird collision, both of single individuals or groups of more individuals or flocks, with planes and other aircraft (birdstrikes) can cause rather significant problems (Sodhi 2002; Dolbeer 2013). These incidents can be very dangerous and even cause plane crashes (Marra et al. 2009), while at the same time they may harm many bird species (Dolbeer and Wright 2009), particularly some sensitive species (e.g., Blackwell and Wright 2006).

Another major problem concerns the impact on bird and bat populations of onshore and offshore wind farms (Barclay et al. 2007; Masden et al. 2009; Hull et al. 2015), a widespread form of renewable energy that is continuously being developed and applied. The challenge is to ensure that accidents, i.e., collisions, are reduced to a minimum to avoid impacting migrating populations and in particular rare and endangered species (Kunz et al. 2007). Hence, a strict ecological assessment protocol in the areas covered must be followed before siting new wind farms and bird and bat populations must be monitored extensively before (for a time period that sufficiently represents all seasons), during, and after the installation of the structures (e.g., Masden et al. 2010; Pearce-Higgins et al. 2012; Millon et al. 2015).

Problematic Wildlife: Wildlife Trafficking and Consumption

The term "problematic wildlife" may also be applied to issues relating to trafficking in animal species (Roe et al. 2002). The "wildlife trade" is an extremely widespread problem as it represents one of the most common examples of illegal trade worldwide, after narcotics and weapons trafficking (Izzo 2010). International wildlife trade is a serious conservation problem, addressed by the United Nations' Convention on International Trade in Endangered Species of Wild Fauna and Flora (CITES), which currently has 170 member countries referred to as parties (CITES 2013). Furthermore, it represents a serious threat to a number of endangered and vulnerable species. The removal of species from regions which are part of the illegal wildlife trade may cause severe problems for the local ecosystem.

Not only that, but with the wildlife trade, the introduction of alien species in other areas is often facilitated (e.g., Carrete and Tella 2008; Menchetti et al. 2016) as is the spread of major epidemics or parasitic species related to the transported species (Smith et al. 2009). Although this book focuses mostly on birds and mammals, in a more general sense this phenomenon is linked both to live animals used as pets for personal enjoyment or collection (fish and other aquarium organisms, amphibians, reptiles, birds, mammals) (e.g., Herrera and Hennessey 2007; Carrete and Tella 2008; Gong et al. 2009) and to dead animals, as in the trade of insects, shells, skins, furs, taxidermy animals, trophies, bones, and other parts of the remains, which are also used to manufacture various objects (Rosen and Smith 2010). One of the most economically profitable trades, which is also for the most part illegal, is that of the ivory of elephant tusks, the vast majority of which are African elephants (Stiles 2004). Wildlife trafficking also includes when animals or the remains of animals are used to prepare local traditional medicine, or alternative remedies in empirical and magical-superstitious traditions, especially in medical or paramedical practice (Nóbrega Alves and Rosa 2013), and have a very significant impact on wildlife, generally in Chinese or Eastern traditional medicine (e.g., Ellis 2005).

Another unique and illegal kind of trafficking is in the meat (fresh, dried, smoked, etc.) of wild animals, referred to as "bushmeat" (Fa et al. 2002; Brashares et al. 2004). This widespread trade, mainly in sub-Saharan Africa (e.g., Angelici et al. 1999; Schenck et al. 2006), also concerns Asia and South America to a lesser extent. This kind of trafficking began at the local level, but subsequently moved to the international level (Davies 2002) as it is linked to the displacement and migration of human populations. One very important feature of bushmeat trafficking, besides its strong impact on wild species, many of which are in serious decline or in danger of extinction, is the possibility, which has already been verified many times, of the spread of severe epidemics, to varying degrees, and zoonoses, through contact with and consumption of bushmeat (Karesh et al. 2005; Wolfe et al. 2005). Because there is no sanitary monitoring of the meat, even very serious diseases have been spread through the consumption of bushmeat, for instance, epidemics caused by the Ebola virus (Rizkalla et al. 2007). Consequently, it is evident that this kind of trafficking should be subject to checks, and possibly stopped, using complex, multidisciplinary strategies such as, among others, environmental education, health education, and the fight against poverty (De Merode et al. 2004).

Problematic Wildlife: Species Discovery

Another issue that may also be considered under the heading of "problematic wildlife" is related to the study and treatment of species whose existence is suspected, according to testimonies from various sources, but has yet to be discovered in a more standard way (e.g., Van Roosmalen 2008). Of course, many of these species are subsequently discovered and formally described (Shuker 2012), while others are declared extinct (Fisher and Blomberg 2010) or nonexistent (Rossi 2012) while others are rediscovered, often accidentally and unexpectedly, after being declared extinct for a long time (Fisher 2011a). This peculiar issue falls fully under the category of "problematic wildlife" because much research is organized and carried out every year to define such situations, substantial funds are used for this purpose (e.g., Hill et al. 2006; Fisher 2011b), and these situations often affect the quality of human life (Turner 2009) in various ways, and often significantly. These issues should be addressed by the field of "cryptozoology," according to Bernard Heuvelmans, the founder of this discipline (Heuvelmans 1982). However, it must be mentioned that issues and findings which have not emerged through any scientific means whatsoever have often been grouped under this heading (Simpson 1984; Loxton and Prothero 2013). In this book, only issues understood as being branches of zoology will be addressed, i.e., in their most legitimate scientific significance, with the aim of clarifying and separating science from imagination and speculation (Arment 2004), including some relevant cases (see Shuker 2012) which have led to the discovery of unknown species and have started with circumstantial evidence and witness testimony (Geissmann et al. 2011).

Problematic Wildlife: A Redefinition

Therefore, in this book the term "problematic wildlife" will be used according to the aforementioned definition: a broad definition that can be interpreted in various ways. I believe that this way of understanding and dealing with the various issues and cases can allow for even greater interconnection between disciplines and help to remove what are often thought of as "barriers" between specialists from different disciplines, which in reality are merely fictitious and illusory. Of course, another objective is to spread awareness of some specific, exemplary cases and provoke greater thought about problems that are underestimated and not often addressed in a timely manner, i.e., when these issues are more difficult to deal with and manage, and therefore higher costs are incurred for their containment or resolution. Predictive models or approaches may sometimes be applied to predict and possibly act quickly to mitigate and/or resolve the problem (Treves et al. 2004).

How the Book Is Structured and a Short Description of Its Content

Each sector, or chapter, will examine a general theme or concept, for example by presenting an overview, and/or will examine an emblematic case, keeping in mind, of course, that the actual number of cases may vary. Each topic/chapter will be preceded by a brief summary illustrating the contents and fixing the basic theme of the book.

Who May Find This Work Useful?

Several examples of monographs on human–wildlife conflict exist that also take into account the possible ways of coexistence and methods of conflict prevention and mitigation (e.g., Conover 2002; Woodroffe et al. 2005; Reidinger and Miller 2013). This book is not designed to be a comprehensive review but rather to present case studies where wildlife has become a problem requiring human intervention. Because of the level and extent of the human footprint, the existence of many wildlife populations will increasingly depend on the level and type of interventions implemented.

Acknowledgments I wish to thank Natale E. Baldaccini and Lorenzo Rossi for improving the preliminary draft of this chapter, and in particular Terry A. Messmer for his continued support, advice, and accurate comments on the various drafts of the manuscript. Heather Milligan for correcting and improving the English text.

References

Angelici FM (2009) Mammiferi: le specie problematiche. In: Amori G, Battisti C, De Felici S (eds) I mammiferi della Provincia di Roma. Dallo stato delle conoscenze alla gestione e conservazione delle specie. Provincia di Roma, Assessorato alle Politiche dell'Agricoltura, Stilgrafica, Roma, pp 271–280

Angelici FM, Luiselli L (2007) Body size and altitude partitioning of the hares *Lepus europaeus* and *L. corsicanus* living in sympatry and allopatry in Italy. Wildl Biol 13:251–257

Angelici FM, Petrozzi F (eds) (2011) Abstracts of II international congress: problematic wildlife: conservation and management. Castello Colonna, Genazzano, Rome, 3–5 Feb 2011

Angelici FM, Luiselli L, Politano E, Akani GC (1999) Bushmen and mammal-fauna: a survey of the mammals traded in bush-meat markets of local people in the rainforests of south-eastern Nigeria. Anthropozoologica 30:51–58

Angelici FM, Riga F, Boitani L, Luiselli L (2000) Fate of captive-reared brown hares *Lepus europaeus* released at a mountain site in central Italy. Wildl Biol 6:173–178

Angelici FM, Petrozzi F, Galli A (eds) (2007) Proceedings of the first international conference: problematic wildlife: conservation and management. Abstract book of oral relations and posters, Rocca dei Papi, Montefiascone, Viterbo, 8–9 June 2007

Arment C (2004) Criptozoology: science and speculation. Coachwhip, Landisville

Arroyo B, Razin M (2006) Effect of human activities on bearded vulture behaviour and breeding success in the French Pyrenees. Biol Conserv 128:276–284

Avian Power Line Interaction Committee (APLIC) (2012) Reducing avian collisions with power lines: the state of the art in 2012. Edison Electric Institute and APLIC, Washington, DC

Baker S (2006) The eradication of coypus (*Myocastor coypus*) from Britain: the elements required for a successful campaign. In: Koike F, Clout MN, Kawamichi M, De Poorter M, Iwatsuki K (eds) Assessment and control of biological invasion risks. Shoukadoh Book Sellers, Kyoto, and IUCN, Gland, pp 142–147

Baker RO, Timm RM (1998) Management of conflicts between urban coyotes and humans in Southern California. In: Baker RO, Crabb AC (eds) Proceedings 18th vertebrate pest conference. University of California, Davis, pp 299–312

Barclay RMR, Baerwald EF, Gruver JC (2007) Variation in bat and bird fatalities at wind energy facilities: assessing the effects of rotor size and tower height. Can J Zool 85:381–387

Bayle P (1999) Preventing birds of prey problems at transmission lines in Western Europe. J Raptors Res 33:43–48

Bergstrom DM, Lucieer A, Kiefer K, Wasley J, Belbin L, Pedersen TK, Chown SL (2009) Indirect effects of invasive species removal devastate World Heritage Island. J Appl Ecol 46:73–81

Bevanger K (1994) Bird interactions with utility structures: collision and electrocution, causes and mitigating measures. Ibis 136:412–425

Bevanger K (1998) Biological and conservation aspects of birds mortality caused by electricity power lines: a review. Biol Conserv 86:67–76

Biondi KM, Belant JL, Martin JA, DeVault TL, Wang G (2014) Integrating mammalian hazards with management at U.S. civil airports: a case study. Hum Wildl Interact 8:31–38

Black SA, Fellous A, Yamaguchi N, Roberts DL (2013) Examining the extinction of the Barbary lion and its implications for felid conservation. PLoS One 8(4), e60174. doi:10.1371/journal.pone.0060174

Blackwell BF, Wright SE (2006) Collisions of Red-tailed hawks (*Buteo jamaicensis*), Turkey vultures (*Cathartes aura*), and Black vultures (*Coragyps atratus*) with aircraft: implications for bird strike. J Raptor Res 40:76–80

Blanco-Aguiar JA, González-Jara P, Ferrero ME, Sánchez-Barbudo I, Virgó E, Villafuerte R, Dávilaet JA (2008) Assessment of restocking contributions to anthropogenic hybridization: the case of the Iberian red-legged partridge. Anim Conserv 11:535–545

Brashares JS, Arcese P, Sam MK, Coppolillo PB, Sinclair ARE, Balmford A (2004) Bushmeat hunting, wildlife declines, and fish supply in West Africa. Science 306:1180–1183

Brown PR, Huth NI, Banks PB, Singleton GR (2007) Relationship between abundance of rodents and damage to agricultural crops. Agr Ecosyst Environ 120:405–415

Caldicott DGE, Croser D, Manolis C, Webb G, Britton A (2005) Crocodile attack in Australia: an analysis of its incidence and review of the pathology and management of crocodilian attacks in general. Wilderness Environ Med 16:143–159

Capizzi D, Baccetti N, Sposimo P (2016) Fifteen years of black rat eradications on Italian islands. In: Angelici FM (ed) Problematic wildlife. A cross-disciplinary approach. Springer, New York, p 205–227

Carrete M, Tella JL (2008) Wild-bird trade and exotic invasions: a new link of conservation concern? Front Ecol Environ 6:207–211

Champagnon J, Guillemain M, Gauthier-Clerc M, Lebreton J-D, Elmberg J (2009) Consequences of massive bird releases for hunting purposes: mallard *Anas platyrhynchos* in the Camargue, southern France. Wildfowl 2(Special issue):192–201

Champagnon J, Elmberg J, Guillemain M, Michel Gauthier-Clerc M, Lebretonc J-D (2012) Conspecifics can be aliens too: a review of effects of restocking practices in vertebrates. J Nat Conserv 20:231–241
Chomba C, Senzota R, Chabwela H, Mwitwa J, Nyirenda V (2012) Patterns of human–wildlife conflicts in Zambia: causes, consequences and management responses. J Ecol Nat Environ 4:303–313
CITES (2013) Member countries. CITES Secretariat, Geneva. http://www.cites.org/eng/disc/parties/index.php. Accessed 1 Dec 2014
Cline R, Sexton N, Stewart SC (2007) A human-dimensions review of human-wildlife disturbance: a literature review of impacts, frameworks, and management solutions (Report). Open-File Report 2007-1111, p 88
Clutton-Brock J (1999) A Natural history of domesticated mammals, 2nd edn. Cambridge University Press, Cambridge
Conover MR (1998) Perceptions of American agricultural producers about wildlife on their farms and ranches. Wildl Soc Bull 26:597–604
Conover MR (2002) Resolving human-wildlife conflicts: the science of wildlife damage management. CRC Press, Boca Raton
Conover MR, Pitt WC, Kessler KK, DuBow TJ, Sanborn WA (1995) Review of human injuries, illnesses, and economic losses caused by wildlife in the United States. Wildl Soc Bull 23:407–414
Corbett J (1944) Man-eaters of Kumaon. Oxford University Press, Delhi
Corbett J (1947) The man-eating leopard of Rudraprayag. Oxford University Press, Oxford
Courchamp F, Chapuis J-L, Pascal M (2003) Mammal invaders on islands: impact, control and control impact. Biol Rev 78:347–383
Cunningham AA (1996) Disease risks of wildlife translocations. Conserv Biol 10:349–353
Davies G (2002) Bushmeat and international development. Conserv Biol 16:587–589
De Merode E, Homewood K, Cowlishaw G (2004) The value of bushmeat and other wild foods to rural households living in extreme poverty in Democratic Republic of Congo. Biol Conserv 118:573–581
Decker D, Lauber T, Siemer W (2002) Human-wildlife conflict management: a practitioners' guide. Cornell University Press, Ithaca
Diamond JM, Ashmole NP, Purves PE (1989) The present, past and future of human-caused extinctions [and discussion]. Philos Trans R Soc Lond B 325:469–477
Dolbeer RA (2013) The history of wildlife strikes and management at airports. In: DeVault TL, Blackwell BF, Belant JL (eds) Wildlife in airport environments: preventing animal–aircraft collisions through science-based management. John Hopkins University Press, Baltimore, pp 1–6
Dolbeer RA, Wright SE (2009) Safety management systems: how useful will the FAA National Wildlife Strike Database be? Hum Wildl Conflicts 3:167–178
Dunham KM, Ghiurghi A, Cumbi R, Urbano F (2010) Human–wildlife conflict in Mozambique: a national perspective, with emphasis on wildlife attacks on humans. Oryx 44:185–193
Durrheim DM, Leggat PA (1999) Risk to tourists posed by wild mammals in South Africa. J Travel Med 6:172–179
Ebenhard T (1995) Conservation breeding as a tool for saving animal species from extinction. Trends Ecol Evol 10:438–443
Ellis R (2005) Tiger bone & rhino horn: the destruction of wildlife for traditional Chinese medicine. Island Press, Washington, DC
Eriksson J, Larson G, Gunnarsson U, Bed'hom B, Tixier-Boichard M, Strömstedt L, Wright D, Jungerius A, Vereijken A, Randi E, Jensen P, Andersson L, Georges M (2008) Identification of the yellow skin gene reveals a hybrid origin of the domestic chicken. PLoS Genet 4(2), e1000010. doi:10.1371/journal.pgen.1000010
Fa JE, Peres CA, Meeuwig J (2002) Bushmeat exploitation in tropical forests: an intercontinental comparison. Conserv Biol 16:232–237
Farrar RO (2007) Assessing the impact of urban coyote on people ad pets in Austin, Travis County, Texas. In: Nolte DL, Arjo WM, Stalman DH (eds) Proceedings of the 12th wildlife damage

management conference. http://digitalcommons.unl.edu/cgi/viewcontent.cgi?article=1061&context=icwdm_wdmconfproc. Accessed 20 Dec 2014

Fernández Orueta J, Ramos Aranda Y, Aranda Ramos Y (2001) Methods to control and eradicate non-native terrestrial vertebrate species. Council of Europe, Strasbourg

Fisher DO (2011a) Trajectories from extinction: where are missing mammals rediscovered ? Glob Ecol Biogeogr 20:415–425

Fisher DO (2011b) Cost, effort and outcome in mammal rediscovery: neglect of small species. Biol Conserv 144:1712–1718

Fisher DO, Blomberg SP (2010) Inferring extinction of mammals from sighting records, threats, and biological traits. Conserv Biol 26:57–67

Geissmann T, Lwin N, Aung SS, Aung TN, Aung ZM, Hla TH, Grindley M, Momberg F (2011) A new species of snub-nosed monkey, genus *Rhinopithecus* Milne-Edwards, 1872 (Primates, Colobinae), from Northern Kachin State, northeastern Myanmar. Am J Primatol 73:96–107

Genovesi P (2005) Eradications of invasive alien species in Europe: a review. Biol Invasion 7:127–133

Gherardi F, Angiolini C (2004) Eradication and control of invasive species. In: Gherardi F, Gualtieri M, Corti C (eds) Biodiversity conservation and habitat management. Encyclopedia of Life Support Systems (EOLSS), Developed under the Auspices of the UNESCO. Eolss Publishers, Oxford. http://www.eolss.net/ebooks/sample%20chapters/c12/e1-67-08.pdf. Accessed 23 Nov 2014

Gong S-P, Chow AT, Fong JJ, Shi H-T (2009) The chelonian trade in the largest pet market in China: scale, scope and impact on turtle conservation. Oryx 43:213–216

Grahn M (1993) Mortality in the pheasant *Phasianus colchicus* during the breeding season. Behav Ecol Sociobiol 32:95–101

Gurevitch J, Padilla DK (2004) Are invasive species a major cause of extinctions? Trends Ecol Evol 19:470–474

Hart D, Sussman RW (2009) Man the hunted: primates, predators, and human evolution, 2nd edn. Westview, Boulder

Herrera M, Hennessey B (2007) Quantifying the illegal parrot trade in Santa Cruz de la Sierra, Bolivia, with emphasis on threatened species. Bird Conserv Int 17:295–300

Herrero S (2002) Bear attacks: their causes and avoidance. Globe Pequot Press, Guilford

Heuvelmans B (1982) What is cryptozoology? Cryptozoology 1:1–12

Hill GE, Mennill DJ, Rolek BW, Hicks TL, Swiston KA (2006) Evidence suggesting that ivory-billed woodpeckers (*Campephilus principalis*) exist in Florida. Avian Conservation and Ecology—Écologie et conservation des oiseaux 1(3):2. http://www.ace-eco.org/vol1/iss3/art2/

Howald G, Donlan CJ, Galván JP, Russell JC, Parkes J, Samaniego A, WangY Y, Veitch D, Genovesi P, Pascal M, Saunders A, Tershy B (2007) Invasive rodent eradication on islands. Conserv Biol 21:1258–1268

Hull C, Bennett E, Stark E, Smales I, Lau J, Venosta M (2015) Wind and wildlife. In: Proceedings from the conference on wind energy and wildlife impacts, Melbourne, Australia, October 2012. Springer, Dordrecht

IDM (ed) (2013) Cambridge advanced learner's dictionary with CD rom, 4th edn. Cambridge University Press, Cambridge

Inskip C, Zimmermann A (2009) Human-felid conflict: a review of patterns and priorities Worldwide. Oryx 43:18–34

International Civil Aviation Organization (2009) Managing wildlife hazards to aircraft. Meeting of directors of Civil Aviation of the Central Caribbean (C/CAR/DCA/10), Grand Cayman, Cayman Islands

IUCN (2014) The IUCN red list of threatened species. Version 2014.2. http://www.iucnredlist.org. Accessed 10 Nov 2014

Izzo JB (2010) PC pets for a price: combating online and traditional wildlife crime through international harmonization and authoritative policies. William Mary Environ Law Policy J 34:965–998. http://scholarship.law.wm.edu/wmelpr/vol34/iss3/6. Accessed 11 Jan 2015

Karesh WB, Cook RA, Bennett EL, Newcomb J (2005) Wildlife trade and global disease emergence. Emerg Infect Dis 11:1000–1002

Kauhala K (1996) Introduced carnivores in Europe with special reference to central and northern Europe. Wildl Biol 2:197–204

Kruuk H (2002) Hunter and hunted. Relationships between carnivores and people. Cambridge University Press, Cambridge

Kunz TH, Arnett EB, Cooper BM, Erikson WP, Larkin RP, Mabee T, Morrison ML, Strickland MD, Szewczak JM (2007) Assessing impacts of wind-energy development on nocturnally active birds and bats: a guidance document. J Wildl Manag 71:2449–2486

Lehman RN, Kennedy PL, Savidge JA (2007) The state of the art in raptor electrocution research: a global review. Biol Conserv 135:459–474

Linnell JDC (2016) Predators that kill humans: myth, reality, context and politics In: Angelici FM (ed) Problematic wildlife. A cross-disciplinary approach. Springer, New York, p 357–371

Linnell JDC, Andersen R, Andersone Z, Balciauskas L, Blanco JC, Boitani L, Brainerd S, Breitenmoser U, Kojola I, Liberg O, Löe J, Okarma H, Pedersen HC, Promberger C, Sand H, Solberg EJ, Valdmann H, Wabakken P (2002) The fear of wolves: a review of wolf attacks on humans. NINA (Norwegian Institute for Nature Research), Trondheim. Oppdragsmelding 731:1–65

Löe J, Röskaft E (2004) Large carnivores and human safety: a review. Ambio 33:283–288

Loxton D, Prothero DR (2013) Abominable science!: origins of the Yeti, Nessie, and other famous cryptids. Columbia University Press, New York

Madden F (2004) Creating coexistence between humans and wildlife: global perspectives on local efforts to address human–wildlife conflict. Hum Dimens Wildl 9:247–257

Marion WR (1988) Urban wildlife: can we live with them? In: Proceedings of thirteenth vertebrate pest conference, vol 13. University of California, Davis, pp 34–38

Marra PP, Dove CJ, Dolbeer R, Dahlan NF, Heacker M, Whatton JF, Diggs NE, France C, Henkes GA (2009) Migratory Canada geese cause crash of US Airways Flight 1549. Front Ecol Environ 7:297–301

Masden EA, Haydon DT, Fox AD, Furness RW, Bullman R, Desholm M (2009) Barriers to movement: impacts of wind farms on migrating birds. ICES J Mar Sci 66:746–753

Masden EA, Fox AD, Furness RW, Bullman R, Haydon DT (2010) Cumulative impact assessments and bird/wind farm interactions: developing a conceptual framework. Environ Impact Assess Rev 30:1–7

Menchetti M, Mori E, Angelici FM (2016) Effects of the recent world invasion by ring-necked parakeets *Psittacula krameri*. In: Angelici FM (ed) Problematic wildlife. A cross-disciplinary approach. Springer, New York, p 253–266

Messmer TA (2000) The emergence of human-wildlife conflict management: turning challenges into opportunities. Int Biodeterior Biodegrad 45:97–102

Messmer TA (2009) Human–wildlife conflicts: emerging challenges and opportunities. Hum Wildlife Conflicts 3:10–17

Millon L, Juliena J-F, Julliarda R, Kerbirioua C (2015) Bat activity in intensively farmed landscapes with wind turbines and offset measures. Ecol Eng 75:250–257

Montgomery S (2009) Spell of the tiger: the man-eaters of Sundarbans. Chelsea Green, White River Junction

Mooney HA (2005) Invasive alien species: a new synthesis. Island Press, Washington, DC

Nóbrega Alves RR, Rosa IL (eds) (2013) Animals in traditional folk medicine: implications for conservation. Springer, Berlin

Oreskes N (2004) Beyond the ivory tower. The scientific consensus on climate change. Science 306:1686. doi:10.1126/science.1103618

Packer C, Ikanda D, Kissui B, Kushnir H (2005) Conservation biology: lion attacks on humans in Tanzania. Nature 436:927–928

Panzacchi M, Bertolino S, Cocchi R, Genovesi P (2007) Population control of coypu *Myocastor coypus* in Italy compared to eradication in UK: a cost-benefit analysis. Wildl Biol 13:159–171

Parish DMB, Sotherton NW (2007) The fate of released captivereared grey partridges *Perdix perdix*: implications for reintroduction programmes. Wildl Biol 13:140–149

Park T (1954) Experimental studies of interspecific competition, II: temperature, humidity and competition in two species of *Tribolium*. Physiol Zool 27:177–238
Park T (1962) Beetles, competition, and populations. Science 138:1369–1375
Parkes J, Murphy E (2003) Management of introduced mammals in New Zealand. N Z J Zool 30:335–359
Parkes J, Fisher P, Robinson S, Aguirre-Muñoz A (2014) Eradication of feral cats from large islands: an assessment of the effort required for success. N Z J Ecol 38:307–314
Pearce-Higgins JW, Stephen L, Douse A, Langston RHW (2012) Greater impacts of wind farms on bird populations during construction than subsequent operation: results of a multi-site and multi-species analysis. J Appl Ecol 49:386–394
Pimentel D, Zuniga R, Morrison D (2005) Update on the environmental and economic costs associated with alien-invasive species in the United States. Ecol Econ 52:273–288
Ralls K, Ballou JD (2004) Genetic status and management of California condors. Condor 106:215–228
Reidinger RF, Miller JE (2013) Wildlife damage management: prevention, problem solving, and conflict resolution. The John Hopkins University Press, Baltimore
Reynolds JC, Tapper SC (1996) Control of mammalian predators in game management and conservation. Mammal Rev 26:127–155
Rizkalla C, Blanco-Silva F, Gruver S (2007) Modeling the impact of Ebola and bushmeat hunting on Western lowland gorillas. Ecohealth 4:151–155
Roe D, Mulliken T, Milledge S, Mremi J, Mosha S, Grieg-Gran M (2002) Making a killing or making a living: wildlife trade, trade controls, and rural livelihoods. Biodiversity and Livelihoods Issues No. 6. IIED, London
Rollan À, Real J, Bosch R, Tintó A, Hernández-Matias A (2010) Modelling the risk of collision with power lines in Bonelli's Eagle *Hieraaetus fasciatus* and its conservation implications. Bird Conserv Int 20:279–294
Rosen GE, Smith KF (2010) Summarizing the evidence on the international trade in illegal wildlife. Ecohealth 7:24–32
Rossi L (2012) Criptozoologia, animali misteriosi tra scienza e leggenda. Photocity, Pozzuoli
Santilli F, Bagliacca M (2008) Factors influencing pheasant *Phasianus colchicus* harvesting in Tuscany, Italy. Wildl Biol 14:281–287
Schenck M, Nsame Effa E, Starkey M, Wilkie DS, Abernethy K, Telfer P, Godoy R, Treves A (2006) Why people eat bushmeat: results from two-choice, taste tests in Gabon, Central Africa. Hum Ecol 34:433–445
Schmitz DC, Simberloff D (1997) Biological invasions: a growing threat. Issues Sci Technol 13:33–40
Seddon PJ, Armstrong DP, Maloney RF (2007) Developing the science of reintroduction biology. Conserv Biol 21:303–312
Seidensticker J (1987) Bearing witness: observations on the extinction of *Panthera tigris balica* and *Panthera tigris sondaica*. In: Tilson RL, Seal US (eds) Tigers of the world. Noyes, Park Ridge, pp 1–8
Sergio F, Marchesi L, Pedrini P, Ferrer M, Penteriani V (2004) Electrocution alters the distribution and density of a top predator, the eagle owl *Bubo bubo*. J Appl Ecol 41:836–845
Shuker KPN (2012) The encyclopedia of new and rediscovered animals. Coachwhip, Landisville
Silva de A (2008) The status of the Saltwater crocodile (*Crocodylus porosus*) inhabiting the Nilwala River, Matara District and its impact on the community. Project report, p 34 http://www.iucncsg.org/365_docs/attachments/protarea/cp_s-c6223ecb.pdf. Accessed 19 Nov 2014
Simpson GG (1984) Mammals and criptozoology. Proc Am Philos Soc 128:1–19
Smith KF, Behrens M, Schloegel LM, Marano N, Burgiel S, Daszak P (2009) Reducing the risks of the wildlife trade. Science 324:594–595
Sodhi NS (2002) Competition in the air: birds versus aircraft. Auk 119:587–595
Soorae PS (ed) (2011) Global re-introduction perspectives: 2011. More case studies from around the globe. IUCN/SSC Re-introduction Specialist Group, Gland, and Environment Agency-Abu Dhabi

Souty-Grosset C, Grandjean F (2007) Translocation as a means of balancing populations and impact on natural habitats. In: Gherardi F, Gualtieri M, Corti C (eds) Biodiversity conservation and habitat management, vol II. Encyclopedia of Life Support Systems (EOLSS), Developed under the Auspices of the UNESCO. Eolss Publishers, Oxford

Stenseth NC, Leirs H, Skonhoft A, Davis SA, Pech RP, Andreassen HP, Singleton GR, Lima M, Machang'u RS, Makundi RH, Zhang Z, Brown PR, Shi D, Wan X (2003) Mice, rats, and people: the bio-economics of agricultural rodent pests. Front Ecol Environ 1:367–375

Stiles D (2004) The ivory trade and elephant conservation. Environ Conserv 31:309–321

Sukumar R (1991) The management of large mammals in relation to male strategies and conflict with people. Biol Conserv 55:93–102

Thouless CS (1994) Conflict between humans and elephants on private land in northern Kenya. Oryx 28:119–127

Tintó A, Real J, Manyosa S (2010) Predicting and correcting electrocution of birds in Mediterranean areas. J Wildl Manag 74:1852–1862

Towns DR, Atkinson IAE, Daugherty CH (2006) Have the harmful effects of introduced rats on islands been exaggerated ? Biol Invasions 8:863–891

Towns DR, Wardle DA, Mulder CPH, Yeates GW, Fitzgerald BM, Parrish GR, Bellingham PJ, Bonner KI (2009) Predation of seabirds by invasive rats: multiple indirect consequences for invertebrate communities. Oikos 118:420–430

Treves A, Naughton-Treves L (1999) Risk and opportunity for humans coexisting with large carnivores. J Hum Evol 36:275–282

Treves A, Naughton-Treves L, Harper EK, Mladenoff DJ, Rose RA, Sickley TA, Wydeven A (2004) Predicting human-carnivore conflict: a spatial model derived from 25 years of data on wolf predation on livestock. Conserv Biol 18:114–125

Turner SS (2009) Negotiating nostalgia: the rhetoricity of thylacine representation in Tasmanian Tourism. Soc Anim 17:97–114

Van Roosmalen MGM (2008) Blootsvoets door de Amazone—de evolutie op het spoor. Bert Bakker, Amsterdam

Vilà M, Basnou C, Pyšek P, Josefsson M, Genovesi P, Gollasch S, Nentwig W, Olenin S, Roques A, Roy D, Hulme PE, and DAISIE Partners (2010) How well do we understand the impacts of alien species on ecosystem services? A pan-European, cross-taxa assessment. Front Ecol Environ 8:135–144

Wolfe ND, Daszak P, Kilpatrick AM, Burke DS (2005) Bushmeat hunting, deforestation, and prediction of zoonotic disease emergence. Emerg Infect Dis 11:1822–1827

Woodroffe R, Frank LG (2005) Lethal control of African lions (*Panthera leo*): local and regional population impacts. Anim Conserv 8:91–98

Woodroffe R, Thirgood S, Rabinowitz A (2005) People and wildlife, conflict or co-existence? Cambridge University Press, Cambridge

Woodroffe R, Frank LG, Lindsey PA, ole Ranah SMK, Romañach S (2007) Livestock husbandry as a tool for carnivore conservation in Africa's community rangelands: a case–control study. Biodivers Conserv 16:1245–1260

Woods M, McDonald RA, Harris S (2003) Predation of wildlife by domestic cats *Felis catus* in Great Britain. Mammal Rev 33:174–188

WWF (World Wide Fund for Nature) (2005) Human wildlife conflict manual. Wildlife management series. WWF, Southern African Regional Programme Office (SARPO), Harare

Yalden D (1999) The history of British mammals. Poyser, London

Yarrow GK (2009) Wildlife and wildlife management. Clemson Extension. Forestry and Natural Resources, Fact Sheet 36. Revised May 2009, Clemson University, Clemson, p 4

Part II
Extinct Species, Species at Risk of Extinction and Declining Species: Some Current and Past Case Studies. Land Fragmentation and Habitat Degradation

This part examines all the cases, which can be considered very classic cases, in which man has caused, and still is causing, serious damage, albeit to varying degrees, to wildlife through environmental degradation. Examples of this are deforestation, habitat fragmentation, pollution (broadly understood), but also direct persecution, i.e. hunting, poaching and the like. As shall be illustrated, man has already brought about the extinction of many thousands of species (of mammals and birds alone) as well as many more thousands of species which are still endangered, according to a threat gradient summarised by the IUCN Red List (IUCN 2014).

In biology, extinction, i.e. the ultimate demise of a species, may be due to one or more concomitant causes, such as species' inability to adapt to environmental changes, epidemics, drastic climate changes, excessive predation, anthropogenic factors such as hunting, the destruction of the natural environment and the introduction of allochthonous competitive species. Although the term also refers to episodes of mass extinction that have occurred in past geological eras, it should not be forgotten that the loss of species is a phenomenon that occurs all the time, even in the present. In addition to the loss of biological heritage and biodiversity, the disappearance of a species leads to alterations in the food chain of which it is a link and, therefore, even has a negative impact on other organisms (Sodhy et al. 2009).

As a result of man's nefarious actions, which are often preventable, he is forced to invest a great deal of money and come up with serious, deep interventions, which are often inconclusive or late, to try to save many species (McCarthy et al. 2012).

In this section, we have specifically included a chapter adopting a general approach regarding past and present cases of the decline and extinction of mammals in the world (Daleszczyk et al. 2016), a chapter on the emblematic case of the rapid decline of the lion population (*Panthera leo*) in Africa (Trinkel and Angelici 2016) and the possible yet complex methods that could be adopted to attempt to counter this trend, a chapter on the decline of the saker falcon (*Falco cherrug*) in Asia due to direct and indirect human impact (Dixon 2016) and a chapter addressing a specific case of rapid decline at the local level, i.e. that of the European otter (*Lutra lutra*) and its subsequent, albeit unusual, recovery (Balestrieri et al. 2016).

References

Balestrieri A, Remonti L, Prigioni C (2016) Toward extinction and back: the decline and recovery of otter populations in Italy. In: Angelici FM (ed) Problematic wildlife: a cross-disciplinary approach. Springer, New York, p 91–105

Daleszczyk K, Eycott A, Tillmann JE (2016) Mammal species extinction and decline: some current and past case studies of the detrimental influence of man. In: Angelici FM (ed) Problematic wildlife: a cross-disciplinary approach. Springer, New York, p 21–44

Dixon A (2016) Direct and indirect human impact on the globally endangered Saker falcon *Falco cherrug* in Asia. In: Angelici FM (ed) Problematic wildlife: a cross-disciplinary approach. Springer, New York, p 69–89

IUCN (2014) The IUCN Red List of Threatened Species. Version 2014.3. http://www.iucnredlist.org. Downloaded 21 May 2015

McCarthy DP et al (2012) Financial costs of meeting global biodiversity conservation targets: current spending and unmet needs. Science 338:946–949. doi:10.1126/science.1229803

Sodhy NS, Brook BW, Bradshaw CJA (2009) V. 1 Causes and consequences of species extinctions. In: Levin SA, Carpenter SR, Godfray HCJ, Kinzig AP, Loreau M, Losos JB, Walker B, Wilcove DS (eds) The Princeton guide to ecology. Princeton University Press, Princeton, pp 514–520

Trinkel M, Angelici FM (2016) The decline in the lion population in Africa, and possible mitigation measures. In: Angelici FM (ed) Problematic wildlife: a cross-disciplinary approach. Springer, New York, p 45–68

Chapter 2
Mammal Species Extinction and Decline: Some Current and Past Case Studies of the Detrimental Influence of Man

Katarzyna Daleszczyk, Amy E. Eycott, and Jörg E. Tillmann

Introduction

The majority of animal species in the world are significantly and negatively influenced by human activities. The effects of human activity extend to all continents and from the high mountains to the bottom of the ocean. Regarding mammals, the latest data from the International Union for Conservation of Nature (IUCN) show that 1.4 % of assessed species are already extinct (EX) or Extinct in the Wild (EW) and a further 21 % belong to threatened categories, that is, Critically Endangered (CR), Endangered (EN) or Vulnerable (VU). Species in these categories thus comprise almost one-quarter of the 5513 mammalian species assessed by the IUCN (IUCN Red List version 2014.3). As it may take many years to demonstrate that a species really is extinct, the number of species listed as extinct in the Red List may be lower than in reality (Hilton-Taylor et al. 2009). Moreover, due to insufficient information 14 % of known species were categorised as 'Data Deficient' (IUCN 2014) so the proportion of threatened species may be even higher.

K. Daleszczyk (✉)
Research and Education Unit, Białowieża National Park,
Park Palacowy 11, Białowieża 17-230, Poland
e-mail: daleszczyk@bpn.com.pl

A.E. Eycott
Department of Biology, University of Bergen,
Thormøhlensgate 53a, Postboks 7803, Bergen 5008, Norway
e-mail: Amy.Eycott@uib.no

J.E. Tillmann
DBU Natural Heritage GmbH, An der Bornau 2, Osnabrück 49090, Germany
e-mail: Joerg.E.Tillmann@web.de

F.M. Angelici (ed.), *Problematic Wildlife*, DOI 10.1007/978-3-319-22246-2_2

Although extinctions are a part of evolution, anthropogenic environmental change has greatly accelerated the rate at which extinctions occur, especially in the past few centuries (Sodhi et al. 2009). At present naturally caused extinctions are thought to be an irrelevantly small fraction of the total of modern extinctions (Hogan et al. 2010).

In this chapter we concentrate on terrestrial mammals and present case studies of species that have been categorised in the past or are at present extinct (EX), at risk of extinction (EW or CR), or declining (EN and with a decreasing population trend), and that were or are negatively influenced by habitat loss, fragmentation or degradation—the most frequent anthropogenic threats.

Causes and Patterns of Decline and Extinction

Causes

Any phenomenon that can cause mortality rates to exceed reproductive replacement over a sustained period can cause a species to become extinct (Sodhi et al. 2009). Small populations are more susceptible to extinction because of demographic and environmental variability (Pullin 2002), and also because of reduced gene exchange which, for example, decreases the likelihood of surviving a novel pathogen.

The most common anthropogenic threats are summarised in Table 2.1 along with some examples of species suffering from them. Habitat loss, fragmentation and degradation seem to be the most frequent threats to wildlife. These threats result from continuous human population growth, intensification of agriculture and forestry, and development of industry. Habitat loss and destruction are often exacerbated by other anthropogenic pressures such as uncontrolled hunting or poaching, introduction of new species, and war and civil unrest (Table 2.1), with climate change also now included in the list of threats. Eighty-one percent of terrestrial mammal species with an EN/CR/EW/EX classification are threatened by more than one of the 11 broad categories of threat used by the IUCN, with 10 % threatened by five or more, according to our analysis of the IUCN Red List (version 2014.3).

It seems that deforestation is currently and probably will remain the principal direct and indirect cause of local extirpations of species (Sodhi et al. 2009) as forests are the broad habitat with the highest mammal species richness, and also the highest number of threatened mammal species (around one-fifth of all threatened species). Legal logging or illegal timber extraction and forest destruction may be aimed at acquiring wood or at preparing areas for settlements and agriculture. Over a third of the tropical forest biome is now covered by croplands or herbaceous plant cover (Mace et al. 2005) and deforestation remains the most frequently recorded form of land-use change (Lepers et al. 2005). Deforestation by region from 2000 to 2010 was highest in Africa and South America (0.56 and 0.46 % lost per year, FAO 2010), with the greatest loss by area in Brazil and Indonesia. South-east Asia in particular has an unfortunate combination of deforestation, many islands, and high population; many of our example species in Table 2.1 and in the case studies are from the region. Up to 21 % of south-east Asian forest species are predicted to be lost by 2100 because of past and ongoing deforestation (Sodhi et al. 2009).

Besides deforestation, agriculture is an important driver of local (and therefore potentially global) species extinction. Some of these effects are related to habitat availability and quality. For example, the development or intensification of agriculture and animal breeding (e.g. cattle ranching) are connected with loss of habitat resulting from drainage, cultivation, and use of fertilisers and pesticides (e.g. Gibbs et al. 2009). However, agricultural abandonment may also result in unfavourable changes to habitat. For example, land abandonment and lack of cattle grazing in some areas within the range of the saiga *Saiga tatarica* resulted in a decline in grassy species and encroachment of species inedible for the saiga (Larionov et al. 2008). In addition there are agricultural effects not related to habitat alteration; for example, domestic animals may transmit diseases or parasites, or compete with wild species for food. Other threats to habitat are the development of industry (e.g. salt extraction or various branches of mining) and the expansion of human settlements and associated infrastructure such as roads or dams, which leads to the isolation of small populations and restriction of migration routes.

Human activities cause direct declines in species numbers through uncontrolled hunting, poaching or persecution but also because of warfare, capture of animals as pets and trade in animal body parts. The IUCN Red List assessments cite biological resource harvesting as a threat to almost a quarter of threatened terrestrial mammal species (IUCN Red List version 2014.3). Direct conflicts with people often result in the killing of animals as a reaction to crop depredation or livestock predation by feral carnivores, both of which are frequently an outcome of earlier habitat encroachment by humans. Human hunting activities may also influence animal species indirectly, for example, by exploitation of the prey of a given carnivore. Exotic species intentionally or unintentionally introduced by people may become serious competitors or predators for indigenous species; this threat includes also transmitting new diseases or parasites against which autochthonic species may be unable to defend themselves (Sodhi et al. 2009). In particular, the rates of extinctions occurring on islands have been considerably increased by the introduction of novel predators (Sodhi et al. 2009).

Geographic Patterns

The regions with the highest overall density of mammal species are undoubtedly tropical: Mesoamerica, tropical South America, Sub-Saharan tropical Africa, south and south-east Asia. Half of the top 20 countries for numbers of threatened species are in Asia (Vié et al. 2009); the areas of highest total number of threatened mammal species are Indonesia (184 species), Mexico (100), India, Brazil and China in ranked order (Vié et al. 2009). The ranges of many threatened species are too small to be

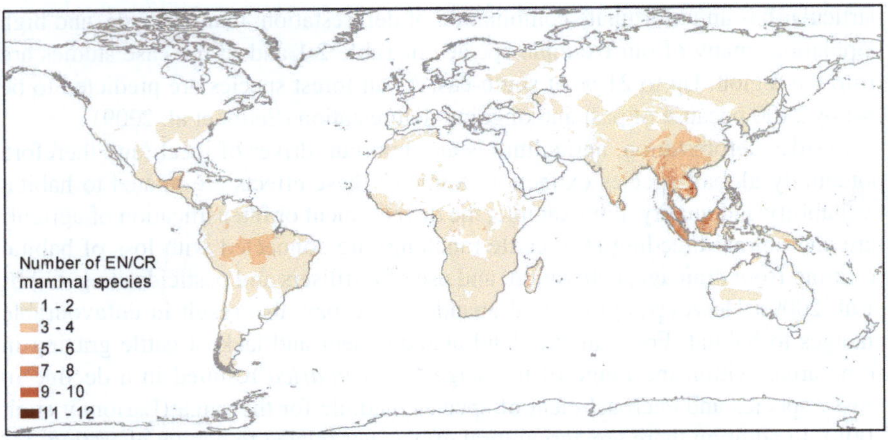

Fig. 2.1 Distribution and density of endangered or critically endangered terrestrial mammal species threatened by habitat loss and fragmentation. Distributions for several species are not visible at this scale, including two of our case study species (mountain pygmy possum and black-footed ferret). Based on data from IUCN website http://www.iucnredlist.org

visible on a map at a global scale (Fig. 2.1). The highest density of extant terrestrial mammal species threatened by habitat loss occurs in south-east Asia (Fig. 2.1). Specifically, the area where Laos borders Vietnam along the Annamite Chain, eastern Cambodia, central continental Malaysia and the southern coast of Sumatra have particularly high numbers of mammal species threatened by fragmentation.

The areas with the greatest amount of threatened animals as a percentage of their total mammal fauna are mostly island countries in the Indian Ocean, the south Pacific, and the Caribbean (see the IUCN analysis by Vié et al. 2009, p. 31). Bhutan, Bangladesh and India are the only non-island countries in the 'top 20' countries for percentage. However, the Millennium Ecosystem Assessment states that the trend is now moving away from islands with a more balanced share of extinctions being continental. Historical island extinctions were often based on species over-exploitation and the introduction of competitor species, whereas continental extinctions are more likely to be caused by habitat loss (Mace et al. 2005).

Few analyses of global patterns of mammal extinction take account of the period of human history between the advent of farming and the beginning of useful historical records of species occurrence. For example, few or no globally threatened mammals are found across most of Europe, but archaeological and historical evidence shows that wolf *Canis lupus* and lynx *Lynx lynx* were present in the British Isles (Hetherington et al. 2005; Buczacki 2002), and on mainland Europe tarpan *Equus gmelini* or tarpan-like equids (Sommer et al. 2011) and the European bison *Bison bonasus* (Benecke 2005) were all widely distributed after the most recent glaciation. IUCN data generally refers to data since 1500, but much of the land clearance for agriculture in Europe happened before then.

Taxonomic and Trait Patterns

Some mammalian orders contain a greater proportion of threatened species than others. For example, almost half of the extinct species in the IUCN terrestrial mammals dataset are from the order Rodentia. Rodents also account for the greatest number of threatened species, followed by primates. However, they are not the most at-risk order by percentage of species—63 % of the even-toed ungulates (Perissodactyla) are endangered or critically endangered, as are 60 % of the echidna/platypus grouping (Monotremata) and half of the pangolins (Pholidota) and elephants (Proboscidea). Elephants and even-toed ungulates are generally large-bodied in comparison to mammals as a whole, meaning they require a greater habitat area and many of them are exploited by humans for food. As for the largest mammalian orders, 7 % of bats and 9 % of rodents are endangered/critically endangered, as are 12 % of the Eulipotyphla (an order containing mostly small insectivores), and 10 % of the Carnivora (IUCN Red List version 2014.3).

It is complicated and difficult to assess and characterise the response of taxonomic groups to anthropogenic change because of interactions between different threats (Stork et al. 2009). Extinction risk is clumped phylogenetically, particularly for those species who are threatened by hunting rather than by habitat loss or invasive species (Fritz and Purvis 2010). However, meta-analytical studies suggest that carnivores are more sensitive to habitat area reduction than omnivores, and arboreal species are more sensitive to habitat area reduction than terrestrial ones (Prugh et al. 2008). Specialist species with narrow ecological niches are more susceptible to habitat loss and degradation, as are species with limited distributional ranges (Cardillo et al. 2005). Large bodied mammals are more vulnerable to extinction as they are disproportionately selected by human hunters (Jerozolimski and Peres 2003) and also have larger home ranges, thus in fragmented habitats are more likely to come into contact with humans (Peres 2001). Indeed, threatened mammals are an order of magnitude heavier than non-threatened ones (Sodhi et al. 2009). Generally, the generation time of a species (interval from birth to reproductive age) is related to its body mass, so larger, longer-lived, and slower-reproducing species are usually unable to compensate for high rates of harvesting and their potential for population recovery in the short term is low (Webb et al. 2002; Sodhi et al. 2009). Species which have a high intrinsic rate of population increase (e.g. high birth rates or rapid generation time) are less vulnerable, as they can compensate for greater mortality (Bodmer et al. 1997).

Present and Past Case Studies

In Table 2.1 we list examples of terrestrial mammals suffering from habitat loss, destruction and fragmentation. We find them on almost all continents and in almost every mammalian order. Other anthropogenic threats are frequently relevant (Table 2.1). To show the variety and omnipresence of detrimental man-caused factors influencing mammals we chose several threatened species from various orders and different continents to analyse their biological characteristics and reasons for decline more thoroughly.

Asian Elephant Elephas maximus *(EN)*

The Asian elephant *Elephas maximus* (order Proboscidea, family Elephantidae) is the largest terrestrial Asian animal. An adult female weighs 3000 kg and an adult male weighs 4500 kg. The Asian elephant is a generalist which browses and grazes on a variety of plants and thus occurs in grassland, various types of forest such as tropical and subtropical forests, moist and dry deciduous forests and dry thorn forests as well as in cultivated areas, from sea level to 3000 m (Sukumar 2000; Choudhury 2009). An elephant's home range varies from a few tens to a few hundreds of square kilometres (Fernando et al. 2005). The Asian elephant can consume as much as 150 kg of wet-weight biomass a day (Choudhury et al. 2008). Elephants live in matriarchal family herds of 6–10 individuals; males 10–15 years old disperse and establish their own home ranges (Sukumar 2000). Their lifespan is 60–70 years (Shoshani and Eisenberg 1982). Age of sexual maturity is typically 11–14 years in females and at least 15 years in males (Vidya and Sukumar 2005). A female produces one calf every 4.5–5 years (Sukumar 2000). Average annual mortality rates over 5 years of age are 1–3 % in females and up to 6 % in males under natural conditions. However, ivory poaching can cause male mortality rates to rise up to 20 %. Annual population growth rate of the species under natural conditions does not exceed 1.5 % (Sukumar 2000).

The former range of the Asian elephant encompassed about 9 million km^2 while the present total area inhabited by the now isolated elephant populations covers less than half a million square kilometres (Choudhury et al. 2008). The Asian elephant population estimates of 30,000–50,000 individuals are thought to be only a crude guess (Hedges 2006). Over 50 % of the remaining Asian elephants live in India (Sukumar 2000).

Reasons for decline. The IUCN has estimated a decline of at least 50 % in the population size of the Asian elephant in the last three generations. Major threats to the species are habitat loss, fragmentation and degradation. The Asian elephant lives in the region of the world with the densest human population, which has an annual growth rate of 1–3 %, resulting in competition between humans and elephants for space. Settlement, agriculture including slash-and-burn shifting cultivation, coal mining, logging, and other activities have significantly encroached on natural habitat. Combined with land fragmentation from roads, railway lines, canals, and dams, human settlement has reduced the elephants' habitat and resulted in severe conflicts between humans and elephants (e.g. Leimgruber et al. 2003; Hedges et al. 2005; Hedges 2006). Because elephants require much larger areas of natural habitat than most other Asian terrestrial mammals, they are also among species most vulnerable to habitat fragmentation and degradation. Moreover, their great size and large food requirements predispose them to cause large-scale crop and property damage. Over 25 years, more than 1150 people and 370 elephants were killed as a result of human–elephant conflicts in northeastern India (Choudhury 2004). People suffering from damage by elephants to their property and crops demand protection or compensation from government authorities. Lack of such reparation results in the killing of elephants in retaliation, and local antagonism toward the species and its conservation (Hill et al. 2002). In addition elephants are poached to obtain ivory, meat and other body parts (Hedges 2006). Elephants injured by poachers often retaliate by killing humans and damaging their property (Choudhury 2004) and so a vicious circle develops. By removing tusker males, poaching limits the species genetic variability and may also lead to highly skewed age and sex ratios, which can have a significant impact on population dynamics (Hedges 2006). For instance, in Periyar Tiger Reserve in southern India, the adult male:female sex ratio changed from 1:6 to 1:122 over a 20 year period (Hedges 2006).

Tiger Panthera tigris *(EN)*

The tiger *Panthera tigris* (order Carnivora, family Felidae) is one of the most highly ranked felids of tropical Asia in terms of the vulnerability index and, moreover, is actively threatened (Nowell and Jackson 1996). Of the nine subspecies three are already EX, two CR and the remaining four, as well as the species in general, are classified as EN.

Depending on subspecies adult females can reach up to 110–167 kg and males over 300 kg in weight (Mazák 1981). The tiger is most often found in tropical and subtropical moist deciduous forests, followed by temperate and deciduous mixed forest, tropical and subtropical dry deciduous forest, and also occasionally in coniferous forest, mangrove forest, tropical grassland and shrubland (Sanderson et al. 2006). While tigers eat a variety of animals from large ungulates or even young elephants to rodents, fish and insects, the tiger needs a healthy prey base to form a viable population (Biswas and Sankar 2002). Tigers are territorial; the size of their home range depends on the abundance of prey and varies from a tens to a few hundreds of square kilometres (Matjuschkin et al. 1980; Sunquist 1981; Barlow et al. 2011). Average age at first reproduction is 3.4 years for females and 4.8 years for males (Smith and McDougal 1991). The litter usually consists of 2–3 cubs (Sankhala 1978), and interbirth interval is about 2 years (Smith and McDougal 1991). For both males and females the mean number of offspring incorporated into the breeding population is two (Smith and McDougal 1991). In the wild the tiger may live for up to 15 years (McDougal 1991).

The historic range of the tiger extended across Asia from Turkey in the west to the eastern coast of Russia (Nowell and Jackson 1996). However, according to Sanderson et al. (2006), the tiger has lost 93 % of its range and now lives in highly fragmented populations (Fig. 2.2). The global population numbers are assessed at

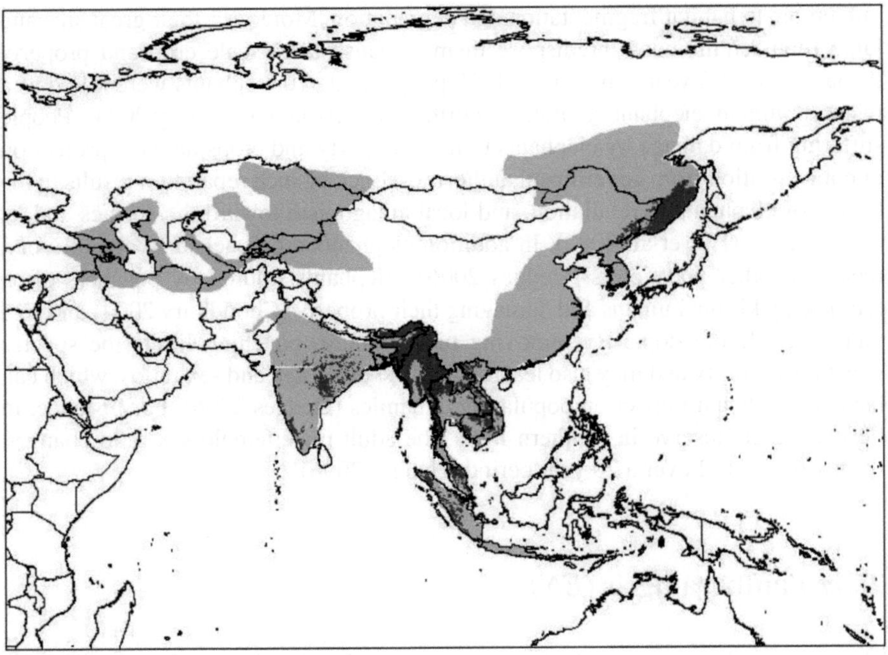

Fig. 2.2 Historical (*light grey*) and present (*dark grey*) distribution of the tiger *Panthera tigris*, redrawn from Luo et al. (2004) (with modifications) by Kate West

3200–3600 individuals, approximately half of the size estimated a decade ago (Seidensticker 2010) and the effective population size (number of reproductively successful adults) is estimated to be less than 2500 tigers (Chundawat et al. 2011).

Reasons for decline. Habitat loss, poaching of tigers for various body parts and the exotic pet trade, and overhunting of tigers' ungulate prey resulting in prey scarcity are all major threats for the tiger (Dinerstein et al. 2007; Tian et al. 2011). Like the Asian elephant, the tiger's range is a region of high density and rapid growth of human population, resulting in rapid conversion of forest habitat to agriculture and settlements combined with degradation of forests by commercial logging. Habitat fragmentation has caused isolation of tiger populations and added the threat of genetic deterioration (Wikramanayake et al. 2011).

Hunting of tigers for sport was once among the causes of population decline. Moreover, the main prey of the tiger—large ungulates—are also heavily hunted by people and have to compete with domestic animals. It is suggested that tigers may be unable to reproduce successfully at prey densities below 2–5 ungulates per km^2 (Nowell and Jackson 1996).

Lucrative illegal markets now exist for tiger products such as skins, teeth and claws (Nowell and Jackson 1996). Another problem is poaching of tigers for the illegal trade of their bones and other body parts used in traditional Asian medicine;

this issue includes intensive breeding, so-called farming, of tigers in China. Some authors consider this trade to be an even greater threat for the tiger than loss of habitat (Pullin 2002).

The bigger the felid, the more severe the conflict with people (Inskip and Zimmermann 2009). Tigers are thus also killed because of conflicts resulting from tiger attacks on people and livestock. In the earlier part of the twentieth century tigers were officially considered pests with bounties paid for their annihilation (Nowell and Jackson 1996). The majority of attacks on people were by unhealthy tigers, either those with physical deformities (Gurung et al. 2008) or which had been wounded (mainly by people); almost half of attacks were provoked (Goodrich et al. 2011). Removal of unhealthy tigers from the wild seems to reduce the number of human deaths (Goodrich et al. 2011), which is important as human–tiger conflicts create negative attitudes towards tiger conservation in local communities (Goodrich 2010).

Mountain Pygmy-Possum **Burramys parvus** *(CR)*

The mountain pygmy-possum is a small rodent-like Australian marsupial belonging to the order Diprotodontia, family Burramyidae. The average adult weight is 40 g but varies from 30 g in spring up to 82 g in autumn, prior to hibernation (NSW National Parks and Wildlife Service 2002). The species was originally described from fossils (Broom 1896) and presumed to be extinct, but in 1966 was discovered in the Australian Alps. It is the only Australian mammal confined to alpine environments. The mountain pygmy-possum lives in patches of periglacial boulders with associated shrubby heathland (NSW National Parks and Wildlife Service 2002). The climate is harsh but the boulders provide the possum with deep, insulated hibernacula and protected nesting sites (NSW National Parks and Wildlife Service 2002). The species' diet changes seasonally and consists of invertebrates, seeds and fruit (Menkhorst et al. 2008). Survival from birth to breeding is significantly higher in females (Mansergh and Scotts 1990). Females are sedentary and live longer than dispersing males with over 11 years recorded as the maximum longevity for a wild female and only 3 years for a male (Mansergh and Scotts 1990). Habitat fragmentation is likely to decrease food availability and increase predator risk in suboptimal areas and thus skew the sex ratio (Broome 2001; Mitrovski et al. 2008). Females have only one litter of up to 4 young per year (Mansergh and Scotts 1990). They are capable of mating at 1 year of age, but only half of them will survive due to their inability to maintain fat reserves for hibernation. Juveniles spend 5 months in hibernation and adults 7 months (Geiser and Broome 1991). The mountain pygmy-possum is the only Australian mammal that depends on winter snow cover for its survival. When snow depth exceeds 80–100 cm, snow temperature at the ground level remains at the optimal 2 °C for the hibernation of the species (NSW National Parks and Wildlife Service 2002).

The former distribution of the species comprised only 2250 km^2, but the confirmed current range seems to cover less than 7 km^2 (Broome 2008). There are less than 1800 adult mountain pygmy-possums and the population is declining (Heinze et al. 2004).

Reasons for decline. The habitat of the mountain pygmy-possum is highly restricted and thus each environmental modification may have a deleterious effect. Unfortunately for the species its habitat is within the range of ski resorts and so has been destroyed and fragmented by development of roads and infrastructure for the downhill skiing industry, as well as by rock extraction for dam construction. Levelling and grooming of land for ski trails has greatly affected the amount of available habitat; in some parts as much as 80 % of the habitat has been disturbed or eliminated (Mitrovski et al. 2007). The remaining patches of good habitat are small and separated from each other by large areas of unsuitable habitat (Caughley 1986). Mitrovski et al. (2007) found limited dispersal across barriers and claim that each separate area containing the mountain pygmy-possum should be considered as an independent gene pool. In one of the Mt Butler populations, within the area of a developing resort, levels of genetic variation fell to approximately one-third of the initial level within 10 years—the most rapid loss of genetic diversity in a mammal ever recorded (Mitrovski et al. 2008). McCarthy and Broome (2000) demonstrated that mountain pygmy-possums are very sensitive to reductions in population growth rate. With a 15 % reduction in current survival and recruitment rates, the probability of decline of equilibrium populations of 20 females to 5 females or fewer within 100 years is about 90 %.

Habitat is also reduced in quality by diminishing depth and duration of snow cover. Low snow cover together with high skier and snowboarder use result in visible damage to vegetation. Winter snow grooming machinery can damage vegetation used by mountain pygmy-possum for food (Broome 2001) and may also affect hibernation of the animals by causing energetically costly arousals due to noise, vibration and changes in hibernacula temperatures and thus may decrease winter survival (NSW National Parks and Wildlife Service 2002). Another threat is predation from introduced foxes *Vulpes vulpes* and feral cats *Felis catus* (NSW National Parks and Wildlife Service 2002).

Black-Footed Ferret **Mustela nigripes** *(in the Past EW, Presently EN) and Prairie Dogs* **Cynomys** *spp.: Joint Case Study of a Carnivore and Its Prey*

The black-footed ferret *Mustela nigripes* is a North American carnivore from the family Mustelidae. It weighs 650–1400 g (Biggins and Schroeder 1988). Female black-footed ferrets reach sexual maturity at 1 year of age (Miller et al. 1996).

The species has low reproductive rates; the mean litter size at emergence of young found in the last free-living population was 3.3 (Forrest et al. 1988).

The black-footed ferret is limited to open habitat used by prairie dogs (*Cynomys* spp.) and its historical range included the grasslands and mountain basins of interior North America from southern Canada to northern Mexico overlapping the combined ranges of *Cynomys ludovicianus*, *C. leucurus* and *C. gunnisoni* (Hillman and Clark 1980). The black-footed ferret depends nearly completely on prairie dogs (Biggins and Schroeder 1988); they constitute about 90 % of its diet (Clark 1986) and the ferret also uses their burrows for shelter and litter rearing (Hillman and Clark 1980). Stromberg et al. (1983) estimated that populations of about 400–1400 *C. ludovicianus* or *C. leucurus* are needed to support the annual requirements of one reproductive female ferret and her young. However, black-footed ferrets do not considerably reduce prairie dog populations as they kill only what they can eat and prairie dogs have comparatively high breeding potential which counterbalances predation by ferrets (Hillman and Clark 1980).

Reasons for decline. This extreme dependence on prairie dogs made the black-footed ferret especially vulnerable to extinction when the numbers of its prey were greatly reduced. Conversion of prairie dog habitat to agricultural cropland was extensive and, moreover, farmers and ranchers considered prairie dogs to be pests and competitors of domestic livestock which resulted in organised state poisoning programmes to control prairie dogs. Eradication of prairie dogs began in the early 1900s or even earlier (Knowles et al. 2002). Furthermore, sylvatic plague, a non-native disease, was brought to the west coast of North America from eastern Asia in the beginning of the twentieth century, presumably on ships (Biggins et al. 2011), and became another major threat. The plague spread rapidly and caused high mortality among prairie dogs, further reducing their numbers. It was estimated that in 1960 prairie dogs occupied only 2 % of their original range. Unfortunately, ferrets also turned out to be susceptible to sylvatic plague (Miller et al. 1996). As a result of all these factors, the ferret population declined to near extinction by the late 1970s.

In 1987 the black-footed ferret was considered EW. Prior to this, 18 black-footed ferrets had been captured from the last-known population in Wyoming, in order to start captive breeding (Miller et al. 1996). Seven of those ferrets contributed unique genetic material and are considered founders of the captive population. At present there are 18 reintroduced populations but only three of them are self-sustaining (Belant et al. 2008). In 2008 there were approximately 300 captive ferrets and about 500 breeding adults in the wild (Belant et al. 2008). That genetic bottleneck means that the genetic variability of the black-footed ferrets is a major concern (Biggins and Schroeder 1988).

The future of the black-footed ferret depends on availability of prairie dogs (Jachowski and Lockhart 2009); so to conserve the ferret, prairie dogs must also be conserved at appropriate densities and distribution (Miller et al. 2007).

Lesser Bilby Macrotis leucura *(EX)*

A frequent problem with extinct species is that even very basic data on them are lacking or sparse. The lesser bilby *Macrotis leucura* is an example. It was an Australian marsupial (order Peramelemorphia, family Thylacomyidae) of medium size (310–435 g, Johnson 1989). The species was confined to dry habitats in central Australia, such as dunes, sandplains with grassy hummocks, and sparse low trees and shrub (Burbidge et al. 1988). It dug out its shelters in the form of deep burrows which were important refuges from heat (Johnson 1989). The lesser bilby fed on insects, grass seeds and bulbs (Johnson 1989). Its litter consisted of one or two, rarely three, offspring (Johnson 1989).

Being previously common, the lesser bilby was drastically reduced in the early 1900s (Nowak 1991). The last specimen was collected in 1931 (Johnson 2008). According to information from Aborigines some populations possibly survived longer (Burbidge et al. 1988), maybe into the 1960s, but there is no evidence that the species still persists.

Reasons for decline. To explain numerous declines in mammals from arid lands of Australia, three main hypotheses have been proposed: introduction of exotic predators, competition from exotic herbivores and changes in fire regimes (Maxwell et al. 1996). All those factors seem to have played a part in the extinction of the lesser bilby. Predation from introduced feral cats and foxes, competition with introduced rabbits *Oryctolagus cuniculus* for burrows, trapping of bilbies for their pelt, and degradation of habitat due to changes in fire regimes are all thought to have contributed to the extinction of the species (Johnson 1989). Feral cats are supposed to have entered Australia from seventeenth-century shipwrecks on the west coast of the continent and foxes appeared later, reaching central Australia by the 1930s; native mammals seem to be unable to cope with them (Burbidge et al. 1988). The first rabbits were introduced in the 1890s and quickly became widespread (Burbidge et al. 1988).

Changes in fire regimes resulted from the depopulation of the deserts and could have caused degradation of habitat. Aborigines used fire for a variety of purposes, such as hunting and regeneration of food plants. This resulted in a mosaic of habitat types in different stages of regeneration and such environmental diversity benefited mammals. Traditional burning patterns of Aborigines also prevented the occurrence of extensive wildfires in summer. When the Aborigines left their lands, the fire regime changed to one consisting of rare but extensive summer wildfires, usually caused by lightning. This increased habitat homogeneity and restricted food and shelter availability, causing a rapid population decrease and even extinction of some species (Burbidge et al. 1988; Johnson 1989).

Tasmanian Tiger Thylacinus cynocephalus *(EX)*

The Tasmanian tiger *Thylacinus cynocephalus* (order Dasyuromorphia, family Thylacinidae) is another extinct species where our knowledge contains many gaps. It was the last member of the family of carnivorous marsupials that lived in Australia.

The Tasmanian tiger was also the largest carnivorous marsupial in historic times, weighing between 15 and 35 kg (Mooney and Rounsevell 2008). Information on its prey type and size is limited (Wroe et al. 2007), but kangaroos and wallabies are suggested to have formed its main source of food (Mooney and Rounsevell 2008). The Tasmanian tiger was most often encountered in open forest and grassland, with shelters found in caves, hollow logs or dense vegetation (Mooney and Rounsevell 2008). The social structure seemed to include both stable family groups with fixed home ranges (probably territories) and nomadic solitary individuals (Paddle 2002). Breeding took place in winter or spring with two to three offspring weaned (Mooney and Rounsevell 2008). Record longevity was about 8.5 years (Collins 1973).

Reasons for decline. Several thousand years ago the Tasmanian tiger was widespread on the Australian mainland and its decline there was suggested to be caused by competition with and predation by the introduced dingo *Canis lupus dingo*. The Tasmanian tiger became extinct on the Australian mainland about 2000 years ago but survived in Tasmania (Mooney and Rounsevell 2008). However, based on anatomical details of the Tasmanian tiger, conclusions about the range of the species' prey and therefore on the possibility of competition with the dingo are ambiguous (Jones and Stoddart 1998; Wroe et al. 2007). Spontaneous predation by dingoes may be doubtful as dogs seemed to be afraid of Tasmanian tigers and almost all historical data on dogs killing Tasmanian tigers included the company of people with some kind of weapon (Paddle 2002). The Tasmanian tiger seemed to be under considerable stress long before the arrival of the dingo because of its competition with Aborigines, who also used the species as food, and the appearance of the dingo in Australia probably only sped up the process of its extinction there (Paddle 2002). There could also have been a disease introduced with alien species that affected the Tasmanian tiger (Mooney and Rounsevell 2008).

Tasmanian tigers that survived in Tasmania were probably not very numerous as Tasmania did not provide a good habitat for them (Mooney and Rounsevell 2008). After the introduction of sheep by Europeans, Tasmanian tigers were considered pests. Although the reported number of sheep killed by Tasmanian tigers was much exaggerated, extensive persecution of the species started in the first half of the nineteenth century and was reinforced by bounties (Paddle 2002). In 1936 the species was probably already extinct (Mooney and Rounsevell 2008).

European Bison Bison bonasus *(in the Past EW, Presently VU): An Expanded Case Study and Promising Example*

The European bison *Bison bonasus* (order Cetartiodactyla, family Bovidae) is the largest terrestrial mammal in Europe. Adult males weigh 440–840 kg and adult females 340–540 kg (Krasińska and Krasiński 2002). The European bison is predominantly a grazer, but its diet is supplemented with some browse and bark. It prefers deciduous and mixed forests in moderate climate zones and preferentially feeds in glades and riverside meadows (Daleszczyk et al. 2007). It is a social species: cows with calves and sub-adults form mixed groups of a dozen or so individuals while adult males live solitarily or in bull groups of 2–3 males, joining mixed groups during the rutting season (Krasińska and Krasiński 2007). Home ranges of groups or solitary bulls range from a few tens to 150 km^2 (Krasińska and Krasiński 2007). Cows start calving from 4 years of age, giving birth to one calf every second year. Bulls reach sexual maturity at 3–4 years of age, but begin to participate in reproduction 2–3 years later (Krasińska and Krasiński 2007).

Extinction in the Wild, Restoration, and Current and Future Challenges for Its Conservation

The fate of the European bison is one example of how humans brought a species to the brink of extinction in a few centuries but then was able to save it through great efforts.

The European bison was historically distributed throughout western, central, and south-eastern Europe and the Caucasus. Along with the aurochs *Bos primigenius* that became extinct in 1627 it was the largest terrestrial mammal within historical times in Europe. Morphophysiologically the European bison is adapted to graze (Hofmann 1989) and its diet is dominated by grass and herbaceous plants (Gębczyńska et al. 1991; Kowalczyk et al. 2011). This is one aspect among others in its ecology that argue for the European bison being fundamentally an open habitat species rather than a forest species; a combination of increasing replacement of open steppe habitats by forest cover in the Holocene and increasing human pressure forced bison into remote forests for refuge (Kerley et al. 2012)—this beginning the gradual extinction process.

The bison was hunted as a considerable food resource but also because it competed with domestic stock and arable farming. Overharvesting and increasing habitat loss since settlement of the human population led to a continuous reduction of the range of the European bison. By the early Middle Ages the range of the bison was already dramatically fragmented to small relict populations.

The European bison was one of the first species to be protected by law, mainly to serve the rulers of the area as a challenging and majestic game species. The bison as charismatic game was so popular that early attempts were undertaken to reintroduce bison to habitats where it had already become extinct, for example, in Mecklenburg in 1689 and in Saxony in 1733 (Pucek et al. 2004). These early reintroduction projects failed due to poaching (Tillmann 2008).

In the early twentieth century only two populations survived in the wild: in Białowieża Forest (over 700 individuals of *B. b. bonasus*, Wróblewski 1927) and in the western Caucasus mountains (between 400 and 500 *B. bn caucasicus*, Heptner et al. 1966). These remaining two populations were finally exterminated within a very short time in 1919 and 1927, respectively, as a result of warfare and extensive poaching. Only 54 individuals survived in a few European zoological gardens.

This precarious situation of the European bison and the urgent need to rescue it from extirpation was reported to a broader audience in 1923 by the Polish zoologist Jan Sztolcman at the first International Conservation Congress in Paris. In consequence, 16 participating nations founded the International Society for the Conservation of the European Bison in order to coordinate a breeding programme to increase the population size and to maintain the remaining genetic variability, with the aim of re-establishing free-ranging herds in the European bison's former range. The most important tool in achieving this goal has been the European Bison Pedigree Book (EBPB) which was first published in 1932 and continues today. It is the first documentation of the breeding stock of an endangered species used as a basis for its conservation programme. The greatest challenge of the EBPB is the conservation of the remaining genetic variability as the bison went through a severe genetic bottleneck—the entire contemporary genetic variability derives from only 12 founders—and inbreeding remains as a major threat to the world population (Tokarska et al. 2011).

The breeding programme started in 1929 in Białowieża with a slow but steady increase in the captive world population, succeeding such that in 1952 the first reintroduction to the wild was launched in Białowieża Forest. Since then, additional captive and free-ranging populations have been founded across Europe. By 2014 the global population was a little over 5200 bison, of which 69 % were free ranging but distributed across 35 isolated and generally small populations in Poland, Belarus, Lithuania, Russia, Ukraine, Slovakia, Romania and Germany (Raczyński 2014). Although the population of the European bison has grown during the twentieth century, the species still faces an uncertain future. Consequently in Europe, the bison is included in Appendix III (protected fauna species) of the Bern Convention on the conservation of European wildlife and natural habitats and is classified as an endangered species (VU: D1) by the 2014 IUCN Red List of Threatened Species. The European Union recognised this special responsibility and listed the European bison as a strictly protected priority species of community interest, whose conservation requires the designation of special areas of conservation (EU Habitats Directive, annex II and IV).

With its current status the world population of European bison is still not regarded as saved from extinction (Pucek et al. 2004). A great danger is the further loss of genetic variability. Most free-ranging bison populations have less than 50 individuals, with only ten of the 35 free-ranging populations numbering over 100 individuals (Raczyński 2014), additionally the populations are geographically isolated. For this reason most free-ranging populations are prone to catastrophic events such as epidemics and extinction and thus their protection should include the establishment of metapopulations and interventions to provide gene exchange (e.g. Daleszczyk and Bunevich 2009; Kuemmerle et al. 2011).

The range of the free-ranging populations is usually limited by habitat suitability but much more by acceptance (Lawreszuk 2012). The know-how and the will to coexist with European bison and large mammals in general have been lost over the centuries. In order to integrate this species into today's anthropogenic landscapes, conservation faces manifold challenges.

Two recent projects—one in Poland and one in Germany—are pilot schemes to develop new sustainable conservation programmes for free-ranging bison populations.

The EU-Life project 'BISON LAND—European bison conservation in the Białowieża Forest, Poland' is a sensitive approach integrating local to regional land-use interests, to improve habitat suitability and habitat size for the biggest free-ranging European bison population. This should form a basis for further dispersal and a sustainable population increase as part of a wider ecological network (Kowalczyk et al. 2010).

The BISON LAND project has implemented a set of public awareness-raising activities and published attractive dissemination materials. Various environmental activities were accompanied by public consultations and information-education campaigns that enhanced knowledge and awareness of the region's population of this species. The bison was promoted as a tourist attraction in the region (Lawreszuk 2012). Conflict management, e.g. the protection of agricultural areas affected by bison or managing damage-causing individuals, was an important tool in increasing acceptance by the local population. Building or equipping farmers with fences and catching troublesome individuals to transport them deep into the forest were the most effective measures (Kowalczyk et al. 2010). Compensating damages to crops and meadows but moreover contracting meadows from local farmers for the needs of bison were also found to be very effective tools for improving tolerance, as this actively involved farmers in bison management as part of their income and their local environment.

Beside the optimisation of the human dimension in bison management, various other actions have been implemented focusing on the improvement of its living conditions. For example, feeding sites, forest meadows and watering places have been created as well as existing forest meadows managed specifically as ideal grazing sites for bison. These practical management tools were implemented in a spatially strategic way in order to guide bison to ecological corridors connecting their current home range with further suitable habitat.

As a result of the comprehensive conservation measures, the bison population increased in size and range, damages and conflicts with land-use were reduced and empathy for free-ranging bison among the local community was raised. The concept of extending the bison range by the designation of ecological corridors could, on this basis, then be implemented in the Provincial Land Development Plan. This allowed free migrations so that gene flow between bison groups became possible. Many problems of the 'refugee species', as the European bison was described by Kerley et al. (2012), could be mitigated by the management model applied in BISON LAND.

In today's reintroduction attempts of European bison the focus lies not only on the conservation of this endangered species per se but also on the reintroduction of its ecological role and its interactions with the environment that have gone extinct alongside local extinction of the bison. However, until now, there had never been a serious attempt to reintroduce a population of free-ranging bison to its former range in western Europe. The initiative to reintroduce the European bison in an intensively managed commercial forest in North-Rhine Westphalia, Germany, was formulated by various stakeholders in 2003 and finally happened in 2013 (Tillmann et al. 2013). The reintroduction of a small population of European bison is at this stage unique for western Europe and aims on the one hand to contribute to the conservation of this highly endangered species and on the other hand to fill again its abandoned ecological niche in a central European forest landscape. This innovative reintroduction project relies on bison management experiences gathered in the free-ranging populations in Eastern Europe, particularly the modern management approaches as developed within BISON LAND. Beyond that, this attempt faces challenges that are different or additional compared to those faced by the eastern European populations.

One very important pillar of this project was to inform and involve the public and relevant stakeholders from the very beginning. In this context, the initial consultations and information events for local to regional stakeholders, authorities and the wider public revealed several potential points of conflict, particularly arising in the fields of agriculture, forestry and tourism. To counter these concerns, a feasibility study was conducted to evaluate habitat suitability and habitat capacity for the European bison. Experts on various free-ranging bison populations in eastern Europe were consulted in order to incorporate existing knowledge and the project area was found to be suitable for carrying a free-ranging herd of European bison (Lindner et al. 2010; Tillmann et al. 2012). After these comprehensive information campaigns, questionnaire surveys and interviews among the local community revealed a broad acceptance of the idea to reintroduce European bison in this region (Decker et al. 2010).

The reintroduction itself began in 2009 with a thorough Environmental Impact Assessment and programmes of stakeholder participation and conflict management. During the initial captivity phase, European bison were found to be manageable in the Rothaargebirge mountain range and therefore permission to release the herd was given by the state of North Rhine-Westphalia in the winter of 2012/2013. This was a milestone in efforts to conserve this species and their ecological role in human-dominated landscapes and can be taken as an exemplar for reintroduction projects elsewhere.

Both the BISON LAND and the Rothaargebirge project have shown that bison not only shape their natural environment but moreover their presence in the wild has an outstanding human dimension. The paradigm of 'conservation with development' has attracted increasing support from conservationists as conservation efforts often lack money and the economic value of wildlife is being taken into account more and more. The integration of human development needs with bison conservation objectives can result in the establishment of mutually beneficial relationships as can be observed in almost all free-ranging bison populations. Bison based ecotourism ventures can earn direct revenues for local and regional communities. Besides the economic value of this charismatic species it has an additional educational value.

Summing up, the European bison plays not only a significant role as a flagship species in the ecosystem but moreover has an aesthetic, cultural and economic value that should be considered and integrated into regional development programmes and marketing concepts. The conservation needs of the European bison can be described as representative for many other large mammals, in the same manner as the need for the acceptance and support of local people can. As demonstrated in the two flagship projects in Poland and Germany the human dimension plays the strongest role in establishing a sustainable coexistence of people and bison in anthropogenic landscapes.

For the long-term conservation of the global population of European bison it is essential to increase the population further. Its broad ecological valence would allow the opportunistic European bison to inhabit large areas within its former area. However, against the background of conflict avoidance and acceptance, particularly in the context of agriculture, forestry or traffic, suitable habitats are reduced dramatically and few areas remain for potential reintroductions. Reintroduction attempts need to be well prepared in order to be accepted by the local community. Pilot projects as described above are of outstanding importance in developing sound and adaptive management concepts to serve as models and 'icebreakers' in facilitating further reintroduction initiatives elsewhere. Additionally these projects assist in learning anew the routine coexistence with a large herbivore in the same ecosystem.

Conclusions: What Can We Learn?

The history of various species shows that it is quite easy to decimate a species and very difficult to get it back—a truism but so often forgotten. Our analyses, case studies, and the examples of threatened or extinct mammals given in Table 2.1 all indicate that there are usually several detrimental factors acting in parallel or synergy, which contribute jointly to the decline or extermination of the species. Because of habitat loss and degradation, many species of both herbivores and carnivores have to compete with humans for space and food. The competition of larger species with people may take the form of direct conflicts, as in the case of the Asian elephant or the tiger, which have lost most of their habitat and must therefore obtain food within a limited area shared with a dense human population. Some species are considered a threat to crops or domestic livestock and the magnitude of their persecution, often disproportional to the actual damage they cause, may bring the species to the brink of extinction (Tasmanian tiger, prairie dogs) and may also influence ecosystem equilibrium and thus indirectly harm other species dependent on those being persecuted (black-footed ferret). Extinctions of species may disrupt essential ecological processes such as seed dispersal and thus lead to cascading losses, ecosystem instability, and a higher general rate of extinction (Sodhi et al. 2009). The loss of entire functional groups, for example, frugivores, is likely to have severe consequences for ecosystem function (e.g. Grelle 2005). Trophic cascade effects have also been demonstrated in a wide variety of systems, where removal or reintroduction of a single apex predator species causes a trickle-down effect throughout the entire ecosystem (Ripple et al. 2014). Introduction of alien species or diseases caused significant problems for the survival of half of the species described in our case studies (mountain pygmy-possum, black-footed ferret, lesser bilby, Tasmanian tiger). Political instability, both war and civil unrest, frequently resulting in extensive poaching, forms another major threat which is often underestimated (European bison). All the species from the case studies have low reproductive rates so their potential for recovery is low, and those which are also habitat specialists (black-footed ferret) or species with highly restricted ranges (mountain pygmy-possum) have even less chance of surviving (Sodhi et al. 2009). A species that was close to extinction usually has a severely limited gene pool and so it may never be completely safe in the future (the European bison, black-footed ferret).

If we want to preserve biodiversity in our world, a compromise has to be reached between the needs of people and the needs of nature. Global biodiversity has exceptional value, for which substitutes cannot be found. Biodiversity gives us opportunities for education, relaxation and pleasure, and also for practical use. Sustainable use of biodiversity as a renewable natural resource may help in its conservation, as this means that species are not over-used and both people and the environment benefit from it. Individual conservation programmes are necessary for each threatened species based on a clear understanding of species' needs and threats. Protection of a species has to also include protection of its feeding base. Coexistence of humans and large animals within the same area may be sometimes difficult but is possible. Among solutions proposed to reduce dramatic conflicts with problematic large species are:

- Maintaining larger intact habitats at the cost of smaller fragmented areas.
- Development of agricultural lands outside of areas where they may attract large herbivores (for example, not on the way to a watering place).
- Creation or enlargement of protected areas for conflict-arousing species to separate them from people while providing them with food and space.
- Securing ecological corridors among the areas to create metapopulations and allow gene exchange.

In places where conflicts are rather seldom, a system of compensating for damages done by the species may be a good solution. In our times of fast development, human acceptance of the close proximity of sometimes problematic wildlife is often as important as suitable habitat; therefore, education, especially for local communities, is vital to change their negative attitude and convince them of benefits connected with the species' presence. An interesting solution, used in many African countries, is a system of returning benefits from wildlife resources (such as tourist park entrance revenue) to rural communities within community-based natural resource management schemes (Zhang and Wang 2003). This is intended to motivate people in rural areas to discourage poaching inside protected areas and to protect wildlife resources outside them. It is also recommended to involve local communities in programmes to keep the problematic species away from their crops or homes, such as maintaining barriers (MacKenzie 2012), as well as in the management of the given population, for example, in capture and translocation of aggressive individuals

(Sukumar 2000). It is essential to involve local people in protection and management of threatened species because nobody will take better care of them and nobody could be a worse enemy for them than the people living in the same area.

Acknowledgments We would like to express our gratitude to Cathy Jenks, Joanne Inchbald and two anonymous referees for their critical help and valuable suggestions which greatly improved the final version of the manuscript.

References

Barlow ACD, Smith JLD, Ahmad IU et al (2011) Female tiger *Panthera tigris* home range size in the Bangladesh Sundarbans: the value of this mangrove ecosystem for the species' conservation. Oryx 45:125–128

Belant J, Gober P, Biggins D (2008) *Mustela nigripes*. The IUCN Red List of Threatened Species. Version 2014.3. http://www.iucnredlist.org. Accessed 17 Dec 2014

Benecke N (2005) The Holocene distribution of European bison–the archaeozoological record. Munibe (Antropologia-Arkeologia) 57:421–428

Biggins DE, Schroeder MH (1988) Historical and present status of the Black-Footed Ferret. In: Proceedings of the eighth Great Plains Wildlife Damage Control Workshop, Rapid City, SD, 28–30 April 1987

Biggins DE, Livieri TM, Breck SW (2011) Interface between black-footed ferret research and operational conservation. J Mammal 92:699–704

Biswas S, Sankar K (2002) Prey abundance and food habits of tigers (*Panthera tigris tigris*) in Pench National Park, Madhya Pradesh, India. J Zool 256:411–420

Bodmer RE, Eisenberg JF, Redford KH (1997) Hunting and the likelihood of extinction of Amazonian mammals. Conserv Biol 11:460–466

Broom R (1896) On a small fossil marsupial with large grooved premolars. Proc Linnean Soc NSW 10:563–567

Broome LS (2001) Intersite differences in population demography of Mountain Pygmy-possums *Burramys parvus* Broom (1986–1998): implications for metapopulation conservation and ski resorts in Kosciuszko National Park, Australia. Biol Conserv 102:309–323

Broome LS (2008) Mountain Pygmy-possum, *Burramys parvus*. In: Van Dyck S, Strahan R (eds) The mammals of Australia, 3rd edn. Reed New Holland, Sydney, pp 210–212

Buczacki S (2002) Fauna Britanica. Hamlyn, London

Burbidge AA, Johnson KA, Fuller PJ et al (1988) Aboriginal knowledge of the mammals of the central deserts of Australia. Aust Wildl Res 15:9–39

Cardillo M, Mace GM, Jones KE et al (2005) Multiple causes of high extinction risk in large mammal species. Science 309(5738):1239–1241

Caughley J (1986) Distribution and abundance of the mountain pygmy-possum, *Burramys parvus* Broom, in Kosciusko National Park. Aust Wildl Res 13(4):507–516

Choudhury A (2004) Human–elephant conflicts in northeast India. Hum Dimens Wildl 9:261–270

Choudhury A (2009) Highest elevations reached by Asian elephants *Elephas maximus* Linn—a review. J Bombay Nat Hist Soc 106:92–94

Choudhury A, Lahiri Choudhury DK, Desai A et al (2008) *Elephas maximus*. The IUCN Red List of Threatened Species. Version 2014.3. http://www.iucnredlist.org. Accessed 16 Dec 2014

Chundawat RS, Habib B, Karanth U et al (2011) *Panthera tigris*. The IUCN Red List of Threatened Species. Version 2014.3. http://www.iucnredlist.org. Accessed 18 Feb 2015

Clark TW (1986) Some guidelines for management of the black-footed ferret. Great Basin Natl Memoirs 8:160–168

Collins LR (1973) Monotremes and marsupials: a reference for zoological institutions. Smithson Institution Press, Washington, DC

Daleszczyk K, Bunevich AN (2009) Population viability analysis of European bison populations in Polish and Belarusian parts of Białowieża Forest with and without gene exchange. Biol Conserv 142:3068–3075

Daleszczyk K, Krasińska M, Krasiński ZA et al (2007) Habitat structure, climatic factors, and habitat use by European bison (*Bison bonasus*) in Polish and Belarusian parts of the Białowieża Forest, Poland. Can J Zool 85:261–272

Decker SE, Bath AJ, Simms A et al (2010) The return of the king or bringing snails to the garden? The human dimensions of a proposed restoration of European Bison (*Bison bonasus*) in Germany. Restor Ecol 18:41–51

Dinerstein E, Loucks C, Wikramanayake E et al (2007) The fate of wild tigers. BioScience 57:508–514

FAO (2010) FAO forest resources assessment 2010. FAO forestry paper 163, Food and Agriculture Organisation of the United Nations, Rome

Fernando P, Wickramanayake E, Weerakoon D et al (2005) Perceptions and patterns in human-elephant conflict in old and new settlements in Sri Lanka: insights for mitigation and management. Biodivers Conserv 14:2465–2481

Forrest SC, Biggins DE, Richardson L et al (1988) Population attributes for the black-footed ferret (*Mustela nigripes*) at Meeteetse, Wyoming, 1981–1985. J Mammal 69:261–273

Fritz SA, Purvis A (2010) Selectivity in mammalian extinction risk and threat types: a new measure of phylogenetic signal strength in binary traits. Conserv Biol 24:1042–1051

Gębczyńska Z, Gębczyński M, Martynowicz E (1991) Food eaten by free-living European bison in Białowieża forest. Acta Theriol 36:307–313

Geiser F, Broome LS (1991) Hibernation in the Mountain Pygmy-possum *Burramys parvus* (Marsupialia). J Zool (Lond) 223:593–602

Gibbs KE, Mackey RL, Currie DJ (2009) Human land use, agriculture, pesticides and losses of imperilled species. Divers Distrib 15:242–253

Goodrich JM (2010) Human-tiger conflict: a review and call for comprehensive plans. Integr Zool 5:300–312

Goodrich JM, Seryodkin I, Miquelle DG et al (2011) Conflicts between Amur (Siberian) tigers and humans in the Russian Far East. Biol Conserv 144:584–592

Grelle CEV (2005) Predicting extinction of mammals in the Brazilian Amazon. Oryx 39: 347–350

Gurung B, Smith JLD, McDougal C et al (2008) Factors associated with human-killing tigers in Chitwan National Park, Nepal. Biol Conserv 141:3069–3078

Hedges S (2006) Conservation. In: Fowler ME, Mikota SK (eds) Biology, medicine and surgery of elephants. Blackwell, Oxford, pp 475–490

Hedges S, Tyson MJ, Sitompul AF (2005) Distribution, status and conservation needs of Asian elephants (*Elephas maximus*) in Lampung Province, Sumatra, Indonesia. Biol Conserv 124: 35–48

Heinze DA, Broome L, Mansergh I (2004) A review of the ecology and conservation of the mountain pygmy possum *Burramys parvus*. In: Goldingay R, Jackson S (eds) The biology of the Australian possums and gliding possums. Surrey Beatty, Chipping Norton, pp 254–267

Heptner VG, Nasimovich AA, Bannikov AG (1966) Die Säugetiere der Sowietunion. I Paarhufer und Unpaarhufer. G Fisher, Jena

Hetherington DA, Lord TC, Jacobi RM (2005) New evidence for the occurrence of Eurasian lynx (*Lynx lynx*) in medieval Britain. J Quat Sci 21:3–8

Hill CM, Osborn FV, Plumptre AJ (2002) Human-wildlife conflict: identifying the problem and possible solutions. Albertine Rift technical report series volume 1, Wildlife Conservation Society, New York

Hillman CN, Clark TW (1980) Mustela nigripes. Mamm Species 126:1–3

Hilton-Taylor C, Pollock CM, Chanson JS et al (2009) State of the world's species. In: Vié J-C, Hilton-Taylor C, Stuart SN (eds) Wildlife in a changing world—an analysis of the 2008 IUCN Red List of Threatened Species. IUCN, Gland, pp 15–41

Hofmann RR (1989) Evolutionary steps of ecophysiological adaptation and diversification of ruminants: a comparative view of their digestive system. Oecologia 78:443–457

Hogan CM, Caley KJ, McGinley M (2010) Causes of extinction. In: Cleveland CJ (ed) Encyclopedia of earth environmental information coalition. National Council for Science and the Environment, Washington, DC. http://www.eoearth.org/article/Causes_of_extinction?topic=58074. Accessed 6 Oct 2011

Inskip C, Zimmermann A (2009) Human-felid conflicts: a review of patterns and priorities. Oryx 43:18–34

IUCN Red List version 2014.3. http://www.iucnredlist.org. Accessed 16 Dec 2014

Jachowski DJ, Lockhart JM (2009) Reintroducing the black-footed ferret *Mustela nigripes* to the Great Plains of North America. Small Carniv Conserv 41:58–64

Jerozolimski A, Peres CA (2003) Bringing home the biggest bacon: a cross-site analysis of the structure of hunter-kill profiles in Neotropical forests. Biol Conserv 111:415–425

Johnson KA (1989) Thylacomyidae. In: Walton DW, Richardson BJ (eds) Fauna of Australia. AGPS, Canberra, pp 1–21

Johnson KA (2008) Lesser Bilby, *Macrotis leucura*. In: Van Dyck S, Strahan R (eds) The mammals of Australia, 3rd edn. Reed New Holland, Sydney, pp 194–195

Jones ME, Stoddart DM (1998) Reconstruction of the predatory behaviour of the extinct marsupial thylacine (*Thylacinus cynocephalus*). J Zool (Lond) 246:239–246

Kerley GIH, Kowalczyk R, Cromsigt JPGM (2012) Conservation implications of the refugee species concept and the European bison: king of the forest or refugee in a marginal habitat? Ecography 35:519–529

Knowles CJ, Proctor JD, Forrest SC (2002) Black-tailed prairie dog abundance and distribution in the great plains based on historic and contemporary information. Great Plains Res 12:219–254

Kowalczyk R, Ławreszuk D, Wójcik JM (2010) Layman report on LIFE06NAT/PL/000105 project "European bison conservation in the Białowieża Forest—BISON LAND". Mammal Research Institute Polish Academy of Sciences, Białowieża

Kowalczyk R, Taberlet P, Coissac E et al (2011) Influence of management practices on large herbivore diet—case of European bison in Białowieża Primeval Forest (Poland). Forest Ecol Manag 261:821–828

Krasińska M, Krasiński ZA (2002) Body mass and measurements of the European bison during postnatal development. Acta Theriol 47:85–106

Krasińska M, Krasiński ZA (2007) The European bison: a nature monograph. Mammal Research Institute Polish Academy of Sciences, Białowieża

Kuemmerle T, Perzanowski K, Akçakaya HR et al (2011) Cost-effectiveness of strategies to establish a European bison metapopulation in the Carpathians. J Appl Ecol 48(2):317–329

Larionov KO, Dzhapova RR, Rozenfeld SL, Abaturov BD (2008) Feeding of Saiga (*Saiga tatarica*) on pastures of Kalmykian Chernye Zemli under plant succession and steppefication. Zool Zhurnal 87:1259–1269

Lawreszuk D (2012) Social aspects of European bison conservation—experiences of the Land of Bison—LIFE project from Bialowieza Forest. In: Book of abstracts of the international workshop on the management of captive and free-ranging European bison, Bad Berleburg, 12–14 March 2012

Leimgruber P, Gagnon JB, Wemmer C et al (2003) Fragmentation of Asia's remaining wildlands: implications for Asian elephant conservation. Anim Conserv 6:347–359

Lepers E, Lambin EF, Janetos AC et al (2005) A synthesis of information on rapid land-cover change for the period 1981–2000. BioScience 55:115–124

Lindner U, Bunzel-Drüke M, Reisinger R et al (2010) "Die Rückkehr des Königs"—Freisetzung von Europäischen Wisenten (*Bison bonasus* Linnaeus, 1758) im Rothaargebirge. Natur und Landschaft 85(12):532–537

Luo S-J, Kim J-H, Johnson WE et al (2004) Phylogeography and genetic ancestry of tigers (*Panthera tigris*). PLoS Biol 2, e442

Mace GM, Masundire H, Baillie J et al (2005) Chapter 4: Biodiversity. In: Hassan R, Scholes R, Ash N (eds) Ecosystems and human well-being: current state and trends. Island Press, Washington, DC, pp 77–122

MacKenzie CA (2012) Trenches like fences make good neighbours: Revenue sharing around Kibale National Park, Uganda. J Natl Conserv 20:92–100

Mansergh IM, Scotts DJ (1990) Aspects of the life history and breeding biology of the Mountain Pygmy-possum, (*Burramys parvus*) (Marsupialia: Burramyidae) in alpine Victoria. Aust Mamm 13:179–191

Matjuschkin EN, Zhivotchenko VI, Smimov EN (1980) The Amur tiger in the U.S.S.R. Unpublished report, IUCN, Gland

Maxwell S, Burbidge AA, Morris K (1996) The 1996 action plan for Australian Marsupials and Monotremes. Australasian Marsupial and Monotreme Specialist Group, IUCN Species Survival Commission, Gland

Mazák V (1981) Panthera tigris. Mamm Species 152:1–8

McCarthy MA, Broome LS (2000) A method for validating stochastic models of population viability: a case study of the mountain pygmy-possum (*Burramys parvus*). J Anim Ecol 69:599–607

McDougal C (1991) Chuchchi: the life of a tigress. In: Seidensticker J, Lumpkin S (eds) Great cats. Merchurst, London, p 104

Menkhorst P, Broome L, Driessen M (2008) *Burramys parvus*. The IUCN Red List of Threatened Species. Version 2014.3. http://www.iucnredlist.org. Accessed 16 Dec 2014

Miller B, Reading RP, Forrest S (1996) Prairie night. Black-footed ferrets and the recovery of endangered species. Smithsonian Institution Press, Washington, DC

Miller BJ, Reading RP, Biggins DE et al (2007) Prairie dogs: an ecological review and current biopolitics. J Wildl Manage 71:2801–2810

Mitrovski P, Heinze DA, Broome L et al (2007) High levels of variation despite genetic fragmentation in populations of the endangered mountain pygmy-possum, *Burramys parvus*, in alpine Australia. Mol Ecol 16:75–87

Mitrovski P, Hoffmann AA, Heinze DA et al (2008) Rapid loss of genetic variation in an endangered possum. Biol Lett 4:134–138

Mooney N, Rounsevell DE (2008) Thylacine, *Thylacinus cynocephalus*. In: Van Dyck S, Strahan R (eds) The mammals of Australia, 3rd edn. Reed New Holland, Sydney, pp 167–168

Nowak RM (1991) Walker's mammals of the world. The Johns Hopkins University Press, Baltimore

Nowell K, Jackson P (eds) (1996) Wild cats. Status survey and conservation action plan. IUCN/SSC Cat Specialist Group, IUCN, Gland

NSW National Parks and Wildlife Service (2002) Approved recovery plan for the mountain Pygmy-possum, *Burramys parvus*. NSW National Parks and Wildlife Service, Hurstville

Paddle RN (2002) The last Tasmanian tiger: the history and extinction of the Thylacine. Cambridge University Press, Cambridge

Peres CA (2001) Synergistic effects of subsistence hunting and habitat fragmentation on Amazonian forest vertebrates. Conserv Biol 15:1490–1505

Prugh LR, Hodges KE, Sinclair ARE et al (2008) Effect of habitat area and isolation on fragmented animal populations. Proc Natl Acad Sci U S A 105(52):20770–20775

Pucek Z, Belousova IP, Krasińska M et al (2004) European bison. Status survey and conservation action plan. IUCN/SSC Bison Specialist Group, IUCN, Gland

Pullin AS (2002) Conservation biology. Cambridge University Press, Cambridge

Raczyński J (ed) (2014) European bison pedigree book 2013. Białowieża National Park, Białowieża

Ripple WJ, Estes JA, Beschta RL et al (2014) Status and ecological effects of the world's largest carnivores. Science 343:151–162

Sanderson E, Forrest J, Loucks C et al (2006) Setting priorities for the conservation and recovery of wild tigers: 2005–2015. The Technical Assessment, WCS, WWF, Smithsonian, and NFWF-STF, New York

Sankhala KS (1978) Tiger! The story of the Indian tiger. Collins, London
Seidensticker J (2010) Saving wild tigers: a case study in biodiversity loss and challenges to be met for recovery beyond 2010. Integr Zool 5:285–299
Shoshani J, Eisenberg JF (1982) Elephas maximus. Mamm Species 182:1–8
Smith JLD, McDougal C (1991) The contribution of variance in lifetime reproduction to effective populations size in tigers. Conserv Biol 5:484–490
Sodhi NS, Brook BW, Bradshaw CJA (2009) Causes and consequences of species extinctions. In: Levin SA (ed) The Princeton guide to ecology. Princeton University Press, Princeton, pp 514–520
Sommer RS, Benecke N, Lougas L et al (2011) Holocene survival of the wild horse in Europe: a matter of open landscape? J Quat Sci 26:805–812
Stork NE, Coddington JA, Colwell RK et al (2009) Vulnerability and resilience of tropical forest species to land-use change. Conserv Biol 23:1438–1447
Stromberg MR, Rayburn RL, Clark TW (1983) Black-footed ferret prey requirements: an energy balance estimate. J Wildl Manage 47:67–73
Sukumar R (2000) Asian elephant. In: Reading RP, Miller B (eds) Endangered animals. A reference guide to conflicting issues. Greenwood, Westport, pp 34–39
Sunquist ME (1981) The social organization of tigers (*Panthera tigris*) in Royal Chitwan National Park, Nepal. Smithson Contrib Zool 336:1–98
Tian Y, Wu J, Smith AT et al (2011) Population viability of the Siberian tiger in a changing landscape: going, going and gone? Ecol Model 222:3166–3180
Tillmann JE (2008) Der Wisent kehrt zurück—Die wechselvolle Geschichte des Wisents in Europa. Lebensraum—Zeitschrift für Naturschutz in der Kulturlandschaft 2/3:16–19
Tillmann J, Bunzel-Drüke M, Finck P et al (2012) Etablierung einer freilebenden Wisentherde im Rothaargebirge. Ein Entwicklungs- und Erprobungsvorhaben des Bundes. Naturschutz und Landschaftsplanung, Zeitschrift für angewandte Ökologie 44(9):267–272
Tillmann JE, Finck P, Riecken U (2013) Wisente im Rothaargebirge. Naturschutz und Biologische Vielfalt vol 133. Bundesamt für Naturschutz, Bonn-Bad Godesberg
Tokarska M, Pertoldi C, Kowalczyk R et al (2011) Genetic status of the European bison *Bison bonasus* after extinction in the wild and subsequent recovery. Mamm Rev 41:151–162
Vidya TNC, Sukumar R (2005) Social and reproductive behaviour in elephants. Curr Sci India 89(7):1200–1207
Vié J-C, Hilton-Taylor C, Stuart SN (eds) (2009) Wildlife in a changing world—an analysis of the 2008 IUCN Red List of Threatened Species. IUCN, Gland
Webb JK, Brook BW, Shine R (2002) What makes a species vulnerable to extinction? Comparative life-history traits of two sympatric snakes. Ecol Res 17:59–67
Wikramanayake E, Dinerstein E, Seidensticker J et al (2011) A landscape-based conservation strategy to double the wild tiger population. Conserv Lett 4:219–227
Wróblewski K (1927) Żubr Puszczy Białowieskiej. Wydawnictwo Polskie, Poznań
Wroe S, Clausen P, McHenry C et al (2007) Computer simulation of feeding behaviour in the thylacine and dingo as a novel test for convergence and niche overlap. Proc R Soc B 274: 2819–2828
Zhang L, Wang N (2003) An initial study on habitat conservation of Asian elephant (*Elephas maximus*), with a focus on human elephant conflict in Simao, China. Biol Conserv 11:453–459

Chapter 3
The Decline in the Lion Population in Africa and Possible Mitigation Measures

Martina Trinkel and Francesco Maria Angelici

Introduction

During the late Pleistocene lions *Panthera leo* spp. ranged all over Africa, Eurasia, and some parts of North America (Steele 2007; Barnett et al. 2009a, b). Today African lions only occur in Sub-Saharan Africa, where they have suffered a marked reduction of range and population sizes (Nowell and Jackson 1996). Over the last century the lion, classified as vulnerable in the IUCN Red List (Bauer et al. 2012), has lost about 82 % of its former distribution range (Ray et al. 2005). There are strong indications that lion populations are declining: current population estimates range from 23,000 to 38,000 free-ranging lions living in 68 lion areas (Fig. 3.1) (Chardonnet 2002; Bauer and Van Der Merwe 2004; Riggio et al. 2013). Approximately 24,000 lions are in strongholds, but about 6000 lions are living in populations with a high risk of local extinction, of which about 3000 are in West and Central Africa (Riggio et al. 2013).

Recently, Bauer et al. (2015) assessed the trend of 47 relatively well-monitored lions in Africa, and found an alarming population decline of about 38 % over 21 years (1993–2014). There were significant regional differences in lion population trends: in southern Africa, lion populations grew by 22 %. In contrast, lion populations decreased by 57 % in East Africa, and 66 % in West and Central Africa (Bauer et al. 2015).

In general, the largest lion populations occur in East and southern Africa with about 11,000 and 10,000 individuals, respectively (Bauer and Van Der Merwe 2004). Many of the large lion populations in southern Africa have been stable

M. Trinkel (✉)
School of Life Sciences, University of KwaZulu-Natal, Westville, Durban 4041, South Africa
e-mail: martina.trinkel@gmail.com

F.M. Angelici
Fondazione Italiana per la Zoologia dei Vertebrati (FIZV), Via Cleonia 30, Scala C, Rome 00152, Italy
e-mail: francescomariaangelici@gmail.com

F.M. Angelici (ed.), *Problematic Wildlife*, DOI 10.1007/978-3-319-22246-2_3

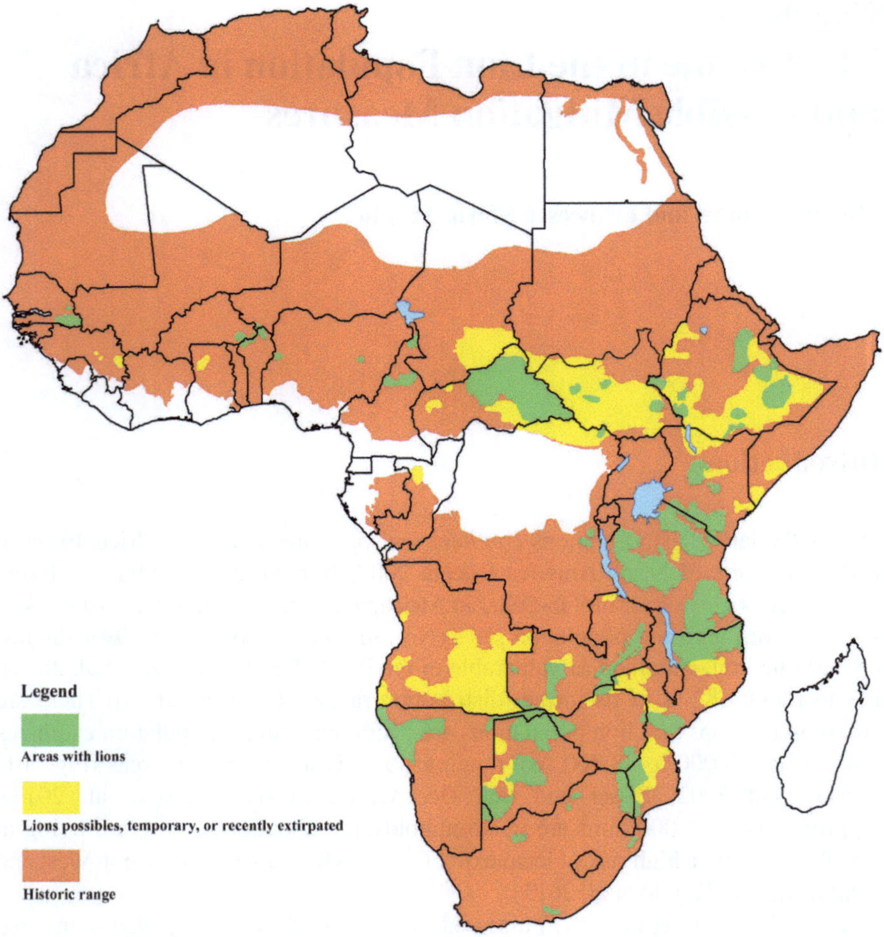

Fig. 3.1 Historic and present distribution of the lion *Panthera leo* in Africa

(or even increasing) over the last decades, where conservation of lions has benefited from the development of the ecotourism and trophy hunting industry (Chardonnet 2002). In West and Central Africa the most severe decline of the lions' range has occurred with estimates of only 850–950 individuals in 2001/2002 (Bauer and Van Der Merwe 2004). More recently, the West African lion population has declined drastically in terms of both numbers and range, resulting in a few small, geographically distant remaining populations (Henschel et al. 2014). In most non-protected areas in West and Central Africa, however, the lion has disappeared (Bauer and Van Der Merwe 2004).

All over Africa, the dramatic decline of the lion is mainly due to an increasing conflict between wildlife and humans over space (Nowell and Jackson 1996). Agriculture, human settlement, depletion of prey populations, and direct persecution have led to shrinking habitat for large carnivores (Nowell and Jackson 1996). Lions, therefore, more and more live in patchy and geographically isolated conservation areas (Nowell and Jackson 1996; Chardonnet 2002). Even national parks and protected areas cannot fully protect lions due to bushmeat hunting (Lindsey et al. 2013) and severe conflicts with humans on reserve borders (Woodroffe and Ginsberg 1998). When communities do not receive revenue from the species, it is difficult to find solutions for human–wildlife conflicts (Jorge et al. 2013). Therefore, it is vital for lion conservation to offer incentives for local communities and individual landowners, so that conserving lions becomes economically sustainable and profitable (Nelson et al. 2013).

In this review, we address problems that threaten the African lion population such as loss of habitat and connectivity, prey depletion, human–wildlife conflict, epidemic and diseases and the trades of lion parts. We summarize possible solutions to mitigate human–wildlife conflict, which were mainly investigates in parts of East, West, and Central Africa. Further, we address the issue of fencing conservation areas to separate humans and wildlife and the development of ecotourism and trophy hunting. In West Africa, where the species is Critically Endangered, we highlight which measures have to be taken quickly to prevent localized extinction.

Taxonomy

Analysis of mitochondrial DNA of lions from West and Central Africa showed that these populations are more closely related to Asiatic lions rather than to lions in East and southern Africa (Bertola et al. 2011). This may have resulted from an extinction of lions during the Pleistocene, followed by a re-colonization of North Africa, just after the late Pleistocene (Bertola et al. 2011). Re-colonization might have continued towards West and Central Africa (i.e., from Yankari Game Reserve, Eastern Nigeria to South Sudan) derived from the Middle East and Asia (Bertola et al. 2011). Besides lion populations in West and Central Africa, some other African lion populations, e.g., lions in Ethiopia, differ morphologically and genetically from lions in East and southern Africa (Bruche et al. 2013). For these reasons, taxonomy and sub-specific nomenclature of African lions should be reorganized, which would ultimately affect conservation strategies of the species both in situ and ex situ (Bertola et al. 2011; Henschel et al. 2014).

Threats

Habitat Loss, Isolation, and Prey Depletion

Large carnivores have suffered a marked decline due to increased human population pressure leading to habitat destruction (Patterson 2004), and patchy and geographically isolated conservation areas (Nowell and Jackson 1996). Human population growth and the expansion of agriculture has caused massive declines of almost all larger mammalian carnivores, resulting in range collapses down to a few per cent of their original distribution (Woodroffe et al. 2005).

In recent decades, lion distribution was drastically reduced and fragmented (Riggio et al. 2013). Nowadays, many areas where lions still occur are isolated without any connection to each other, in particular in West Africa and Ethiopia (Fig. 3.1). Many populations are small and may have insufficient genetic variability, which is essential for maintaining a healthy population in the long term (Björklund 2003). The most important protected areas for the conservation of large mammals, including the African lion, are concentrated in southern and East Africa (Wegmann et al. 2014). In West Africa, where lion numbers are low and ranges are isolated, conservation actions are urgently needed (Henschel et al. 2014). Based on recent surveys in West Africa, from Senegal to Western Nigeria (Tende et al. 2014), there are about 400 lions in the wild representing less than 250 mature individuals. These lions are concentrated in only four protected areas, forming two separate blocks (Fig. 3.1). To counteract genetic impoverishment of the critically endangered lion population in West Africa, translocations and reintroductions might be necessary (Trinkel et al. 2008). However, it is likely that there are more small lion populations in areas which have not yet been investigated and/or are currently monitored, i.e., in Guinea, Ivory Coast, and Ghana (Henschel et al. 2014; Angelici et al. submitted).

Case Study: Lions in the Mole National Park (Ghana)

The Mole National Park (MNP, size: 4600 km^2) in Ghana is one of the most important protected areas in West Africa, where lions have always been present (Angelici and Petrozzi 2010; Angelici et al. 2015). Recently, it was considered that lions in MNP and possibly in whole Ghana are extinct (Burton et al. 2011; Henschel et al. 2014). In 2004, a male lion was shot in MNP and this individual was thought to be the last lion of the area. In order to evaluate if there are remaining lions in the MNP and surrounding areas, the project "The Pride of Ghana" was initiated in 2005 (Angelici and Petrozzi 2010; Angelici et al. 2015). It is a cooperation between the Forestry Commission of the Ghanaian Ministry of Lands and Forestry, the Italian NGO "Ricerca e Cooperazione," and the University of Tuscia, Viterbo, Italy. The methods used in this project are camera traps, sound playbacks, and interviews with local people. Since 2005, lion tracks, prey presumably killed by lions, and roars were regularly registered by park staff and researchers (Angelici et al. 2015). Due to the large size of the MNP and the extremely scarce population, lions are difficult to count in this area (Myers 1975). Therefore, close collaboration with park staff and local people is of tremendous importance. Such collaborations with local people are often underrepresented or even ignored in field research resulting in the loss of important information (Black et al. 2013).

Wegmann et al. (2014) argued that for conservation of large mammals, i.e., ungulates and large carnivores, it is essential that protected areas throughout Africa are connected with each other. Even when connected, each protected area is vulnerable to habitat change which could limit the effectiveness of connectivity in a network of protected areas, and thus increase the risk of the extinction of species (Wegmann et al. 2014). Important parameters for connectivity are size and location of protected areas and, in general, habitats are better preserved inside rather than outside protected areas, where fast deterioration of habitat often occurs (Wegmann et al. 2014). Thus, it is essential to not only preserve protected areas, but also non-protected areas (Wegmann et al. 2014). To effectively conserve lions, it would be of high value to connect protected areas, whilst at the same time considering genetic differences (Bertola et al. 2011). In West and Central Africa, and in Ethiopia, where remaining lion populations are extremely small and isolated (Fig. 3.1), protected areas should be increased and connected through ecological corridors to maintain genetically viable populations. Failing this (or while it is being implemented), translocation schemes may be needed in the short term to conserve or increase genetic diversity (Hunter et al. 2007; Trinkel et al. 2008). A recent study of lions in Hwange National Park, Zimbabwe, demonstrated the importance of smaller-scale connectivity, showing that the genetic pattern of the Hwange lion population can be attributed to still existing high levels of habitat connectivity between protected areas (Morandin et al. 2014). In areas where direct connectivity cannot be employed, genetic connectivity could be achieved with meta-population management (Akçakaya et al. 2006; Gusset et al. 2008; Slotow and Hunter 2009; Lindsey et al. 2011).

Lions depend on medium and large prey species (Van Orsdorl et al. 1985; Hayward and Kerley 2009). Large mammals, however, were found to decline dramatically in most African countries (Craigie et al. 2010). Between 1970 and 2005, wildlife population abundance severely declined in West Africa (85 %) and East Africa (52 %); only southern Africa was able to maintain their wildlife populations (Craigie et al. 2010). In some areas, bushmeat trade, the illegal hunting and commercialization of bush meat, is a more pressing problem than the loss of habitat (Wilkie et al. 2011). Due to disappearing wildlife in unprotected areas (Newmark 2008), illegal bushmeat hunting has been increasing dramatically in protected savannah areas (Lindsey et al. 2013). The hunting and trade of bushmeat leads to edge effects around protected areas, disproportionate declines of particular species, and severe population declines in areas with inadequate anti-poaching (Lindsey et al. 2013), which all negatively affects lion population size (Packer et al. 2014 draft).

Human–Lion Conflict

When wildlife has little or no value outside protected areas, it rapidly disappears through habitat loss and direct persecution (Prins and Grootenhuis 2000). Anthropization, including expansion of agriculture, has caused massive declines of almost all larger mammalian carnivores (Woodroffe et al. 2005). In West and Central Africa, lion population density is low, with approximately 1–3 lions/100 km^2 (Bauer and Van Der Merwe 2004). Such low population densities are due to low prey densities (Hayward et al. 2007a; Bauer et al. 2008) and high human densities, which results in frequent and intense interactions between humans and lions (Bauer et al. 2010). Lions frequently kill livestock and are subsequently trapped, shot, or poisoned (Bauer and De Iongh 2005).

Even in conservation areas, large carnivores are not sufficiently protected. Poaching is a problem (e.g., Weladji and Tchamba 2003; Hilborn et al. 2006) and conflict with people at reserve borders negatively influences populations over wide areas (Weladji and Tchamba 2003). Increasing security through elevated anti-poaching can effectively reduce bushmeat hunting inside protected areas (Lindsey et al. 2013), which in turn reduces the decline of lions and wildlife populations in general. Edge-related mortality at reserve boundaries can lead to population declines or even extinction (Scheepers and Venzke 1995; Woodroffe and Ginsberg 1998; Balme et al. 2010). Problem animal control is a major source of mortality for lions outside of protected areas (Woodroffe and Ginsberg 1998), and is often carried out as a response to lions that prey on livestock (Kötting 2002; Woodroffe et al. 2006). For example, in the Kgalagadi Transfrontier Park, South Africa/Botswana, Van Vuuren et al. (2005) argued that, due to problem animal control, the lion population probably cannot be sustained in the long term.

In Namibia, the largest lion population occurs in the Etosha National Park, an arid habitat of immense size (approximately 23,000 km^2, about the same size as the Kruger National Park, South Africa) with an estimated population size of about 320 individuals (Stander 1991). Although lions are protected within the park, conflict is a problem where Etosha borders on surrounding commercial and communal farmland. More than 1000 lions were shot or destroyed in the last 20 years (Etosha Ecological Institute). These numbers, however, include only reported cases, and the actual number of killed lions is thought to be higher (Tana Burger, Personal Communication). Kötting (2002) investigated livestock losses due to problem animals, mainly lions and spotted hyenas, on commercial farms along the southern boundary of the Etosha National Park. Farmers on average lost livestock worth 2000€ per year to lions, and on average, one lion was shot on each farm per year. There existed a linear relationship between livestock losses and the number of destroyed lions, with one outlier: one farmer destroyed 400 lions in 55 years and this particular farmer is known to lure lions out of the park to subsequently destroy them (Kötting 2002). Despite such massive losses of lions through border conflicts, Trinkel (2013) found that the lion population size inside Etosha remained stable over a 20-year study. However, population size alone is not a good measure for the health of a population, especially not for lions, who have the ability to quickly recover from disturbances (Smuts 1978; Packer et al. 1990; Trinkel 2013). Recently, Trinkel et al. (submitted) found that the persecution of sub-adult males on farmland bordering Etosha was disproportionately high, and although it did not affect lion population size, it affected social dynamics of lions inside the national park. Due to a lack of competition from sub-adult males, Etosha males hold tenure over prides for nearly 9 years, which is four to five times longer than pride males in other ecosystems (Packer et al. 1988; Yamazaki 1996; Bauer et al. 2012; Trinkel et al. submitted). As young females usually first giving birth at 3–4 years of age and give birth every 2 years (Packer et al. 1988), there is consequently a high risk of them mating with direct relatives (Trinkel et al. 2010, submitted). Trinkel et al. (submitted) argue that if managed properly (Packer et al. 2009; Balme et al. 2012), wildlife tourism and sustainable utilization through trophy hunting in these areas may benefit species conservation (Selier et al. 2014) and could thus contribute to the genetic health and sustainability of the Etosha lion population.

Epidemics and Animal Diseases

Besides direct persecution, close contact between livestock and wildlife bears other risks: infectious diseases can have a dramatic impact on wildlife populations and is important in carnivore conservation (Scott 1988). Viruses have caused major declines in populations of large carnivores (Young 1994), such as canine distemper (Roelke-Parker et al. 1996), rabies (Sillero-Zubiri et al. 1996), and bovine tuberculosis (Keet et al. 1996). However, Ferreira and Funston (2010) evaluated the effect of bovine tuberculosis on estimates of lion density and survival in the Kruger National Park, South Africa, and found that the size of the lion population was stable, despite the high prevalence of bovine tuberculosis. Detailed analysis of the canine distemper virus epidemic in lions of the Serengeti and the Ngorongoro Crater, Tanzania, revealed that canine distemper is only a problem when lions are coinfected with high levels of Babesia (Munson et al. 2008).

Diseases can be transmitted through domestic animals, such as dogs or livestock, and can be lethal for populations of wild species such as lions (e.g., Roelke-Parker et al. 1996; Packer et al. 1999). Canine distemper—probably transmitted by dogs via wild canids and subsequently onto lions—killed over 35 % of lion in the Serengeti, East Africa, in 1994 (Roelke-Parker et al. 1996; Carpenter et al. 1998), but only if the lions were coinfected with Babesia (Munson et al. 2008). Bovine tuberculosis is a pathogen of growing concern in free-ranging wildlife in southern Africa (Trinkel et al. 2011) and was shown to be transmitted from livestock to wildlife populations (Michel et al. 2006). For example, in the Kruger National Park, South Africa, bovine tuberculosis has resulted in the mortality of buffalo *Syncerus caffer*, lion, and cheetah *Acinonyx jubatus* (Keet et al. 1996; De Lisle et al. 2002; Ferreira and Funston 2010). To prevent such fatal consequences, it is important that livestock and pets of humans living close to wildlife be vaccinated against diseases that can threaten wildlife populations (Harder et al. 1995). However, controlling canine distemper with a large-scale dog vaccination program around the Serengeti, Tanzania, did not prevent the spread of the disease onto lions (Viana et al. 2015). Despite the fact that direct vaccination of threatened wildlife has been heavily debated (Cleaveland et al. 2006), there has been considerable progress in the development of safe vaccines for large carnivores (Pardo et al. 1997; Wimsatt et al. 2003). Furthermore, mathematical models suggest that vaccinating only about 30–40 % of the individuals in a large carnivore population would be sufficient to effectively protect the population from serious diseases (Haydon et al. 2006; Vial et al. 2006).

Therefore, instead of a large-scale vaccination program for domestic animals, it might be more effective to directly vaccinate the lions themselves (Viana et al. 2015). In small reserves in South Africa, lions are frequently translocated to restore and/or maintain their health status and genetic diversity (Trinkel et al. 2008, 2011). To prevent negative consequences, lions are carefully screened for diseases prior to translocations (Trinkel et al. 2011).

Trade in Live Pets, Bones, Meat, and Skins

A problem that could have major implication on lion populations in the near future is an increasing trade of lion bones, skins, dried meat, and other parts of the lion's body for use in traditional medicine, mostly in Asia (Ellis 2005). As the tiger *Panthera tigris* is becoming rarer, the lion is replacing the tiger as an ingredient for traditional medicine and magic (Ellis 2005; Morell 2007; Gratwicke et al. 2008). Many captive lion breeders in South Africa obtain high incomes from the sale of lion carcasses to China, Vietnam, and other Asian countries (Lindsey et al. 2012a; Nowell 2012). Such trade, however, is dangerous, because it is difficult to control which lions were killed legally and which ones were poached. Poaching certainly is widely present in both the official and the illegal markets (Ellis 2005; Zhang et al. 2008). If the trade is not carefully controlled, lions could decline as dramatically as the tiger (Dinerstein et al. 2007). Therefore, any trade has to be seriously controlled and severe sanctions for poachers have to be installed (Gratwicke et al. 2008).

The trade of lions for pets seems to be declining (Hemley 1994). This is mainly due to legalities prohibiting the trade in wild animals for the pet trade (e.g., Nyhus et al. 2009). People who want to keep large carnivores at their home or in cages can mostly get lions from captive breeders or circuses (Hemley 1994; Bush et al. 2013). While this trade is of concern, it has less impact on the wild lion population than the trade in lion parts.

Possible Mitigations

Human–Lion Conflict: Possible Solutions

A reduction in livestock losses would ultimately reduce the number of lions and other predators that are killed benefitting both people and wildlife (Ogada et al. 2003; Woodroffe et al. 2006). Some conservationists, therefore, have investigated the effects of livestock husbandry for mitigating human–wildlife conflicts along reserve borders and outside protected areas (Ogada et al. 2003; Kissui 2008; Jorge et al. 2013). For example, Woodroffe et al. (2006) conducted a study in a non-protected area on communal land in Kenya, where attentive livestock husbandry is common: 97 % of livestock herds were accompanied by herders (Woodroffe et al. 2006).

The reason for continuing traditional husbandry in this area was mainly the high risk of livestock being stolen rather than being killed by predators. In such areas, small changes in livestock husbandry were shown to effectively reduce livestock losses (Ogada et al. 2003; Woodroffe et al. 2006). These changes included constructing denser boma walls and the presence of domestic dogs at bomas and with grazing herds (Ogada et al. 2003; Woodroffe et al. 2006). Lichtenfels et al. (2015) evaluated fortified bomas just east of Tarangire National Park on the Maasai Steppe, Tanzania, and found that carnivore attack rates at fortified bomas called Living Walls were significantly reduced. Living Walls are predator-proof enclosures that consist of fast-growing, thorny trees as fence posts with chain link fencing (Lichtenfels et al. 2015). A so-called Lion Guardians program has been successfully incorporated in the Massai society in Kenya, where lions killing livestock in pastoralist areas are traditionally poisoned and speared, which—besides retaliation—reinforces the role of warriors (Hazzah et al. 2014). Warriors in the Maasai society are well-respected people, and are responsible for protecting their community and livestock from carnivore attacks (Hazzah et al. 2014). Within the "Lion Guardians" program, warriors defend their community by "hunting" lions to radio-collar and monitor them, while receiving economic incentives and cultural recognition (Hazzah et al. 2014). As a result, no lions have been speared in areas where the Lion Guardians are working since the program started (Hazzah et al. 2014). Another study conducted in Kenya showed that a land use system based on temporary settlements and grazing areas allowed lions to co-occur with people and livestock at high density (Schuette et al. 2013). Dickman et al. (2011) reviewed the success of providing economic incentives for local people to tolerate the costs of living with wildlife. In particular, Dickman et al. (2011) investigated three major approaches: (a) the concept of compensation payments, which involves a payment to the livestock owner if the suspected incident of livestock depredation is attributed to a predator; (b) revenue sharing where local people directly benefit from wildlife as revenue generated from wildlife tourism are shared among communities; and (c) conservation payments, which are directly linked, e.g., to the maintenance of predators on private or community land. Dickman et al. (2011) found that the best way of "payments to encourage coexistence" might be a combination of compensation payments, revenue sharing, and conservation payments. Payments to encourage people to coexist with wildlife can also be a useful tool to decrease bushmeat hunting (Lindsey et al. 2013).

In West and Central Africa several methods for mitigating conflict between lions and humans were investigated, with encouraging results (Bauer et al. 2010). For example, in Pendjari National Park (Benin), enclosures of clay instead of thorny branches drastically reduced attacks of lions on livestock (Bauer et al. 2010). In Chad, protecting livestock herds with dogs and herdsmen, and creating settlements with houses surrounding livestock mitigated human–lion conflicts. The use of magic was found to be intensively practiced throughout West Africa, where rural people invest considerable amounts of money in the magical protection of their livestock (Bauer et al. 2010). Bauer et al. (2010) argue that although the effectiveness of such measures is not relevant, such practices should receive more attention when starting discussions with, and finding solution for, rural communities. Such mitigation measures could be used as a sensitisation instrument, e.g., by using Sourats on community radio (Bauer et al. 2010).

Improving livestock husbandry was shown to be a useful tool to mitigate human–wildlife conflict in many areas (Ogada et al. 2003; Woodroffe et al. 2006; Bauer et al. 2010; Lichtenfels et al. 2015; Hazzah et al. 2014). However, the effectiveness of the various techniques depends on the social and cultural susceptibility to such mitigation measures, and the applicability of each practice has to be evaluated from case to case. In many parts of Africa traditional husbandry has been abandoned, and it is therefore questionable if such mitigation measures will be useful in such areas (Breitenmoser et al. 2005).

Fences for Lion Conservation

Conservation fences are designed to separate biodiversity from factors threatening it (Hayward and Kerley 2009). For effective long-term conservation of African lions, Packer et al. (2013) recently recommended the erection of fences to separate protected areas from land used by humans. Packer et al. (2013) compared African lion population densities and population trends in fenced and unfenced reserves across 42 sites in 11 countries, and found that lion populations in fenced reserves were significantly closer to their carrying capacity than unfenced populations. Furthermore, lions in fenced reserves were mainly limited by density dependence (Packer et al. 2013). In contrast, lions in unfenced reserves were highly sensitive to human population densities in surrounding communities, and these lion populations were subjected to density-independent factors (Packer et al. 2013). Packer et al. (2013) predicted that nearly half the unfenced populations may decline to near extinction over the next 20–40 years.

Fences, however, can inhibit or prevent transmigration, i.e., immigration and emigration, and therefore, fenced reserves are vulnerable to problems associated with isolated populations (Lindsey et al. 2012b). Such populations are more susceptible to environmental, demographic, and genetic stochasticity (Caughley 1994; Lindsey et al. 2012b). Fences can lead to loss of dispersal and migration routes (De la Bat 1982; Creel et al. 2013), and prohibit the use of patches of primary productivity to wildlife populations (Williamson and Mbano 1988; Lindsey et al. 2012b). For example, the fence around the Etosha National Park reduced the number of blue wildebeest *Connochaetes taurinus* from more than 25,000 to 2500 animals (De la Bat 1982). The veterinary fences of Botswana reduced access of zebra *Equus burchelli*, blue wildebeest, and other wildlife to water and dry-season grazing areas and consequently caused a massive number of mortalities (Williamson and Mbano 1988; Lindsey et al. 2012b; Gadd 2012). For lion populations, depending on the size of the enclosed population, fencing often results in the need for genetic and demographic management of smaller populations via translocations (Trinkel et al. 2008; Miller et al. 2013).

Fencing is becoming an increasingly important conservation tool in southern Africa to contain animals inside protected areas and limit human–wildlife conflict at reserve borders, especially when reserves border on highly populated agricultural areas (Lindsey et al. 2012b). In South Africa, for example, all small wildlife reserves and even the 20,000 km^2 Kruger National Park are fenced by law (Anonymous 2003). This fencing has successfully reduced human–lion conflict (Hayward and Kerley 2009), and can also decrease the impact of bushmeat hunting onto wildlife populations (Lindsey et al. 2013) However, for small areas, the breakdown of natural processes that regulate lion population growth must also be considered (Miller et al. 2013). Namibia erected a game-proof fence around the 23,000 km^2 Etosha National Park, starting in 1961 and finishing in 1973 (Berry 1997). However, this fence has proved inadequate for large predators such lions and spotted hyenas *Crocuta crocuta* (Berry 1997). Warthogs *Phacochoerus africanus* and porcupines *Hystrix africaeaustralis* dig holes under the fence, which are then frequently used by lions and other predators to bypass the fence (T. Burger, Personal Communication). Besides these holes, the maintenance of the 850 km fence is a major problem (S. Kötting, Personal Communication). As a result, the conflict between humans and predators has not been limited (Trinkel et al. submitted), resulting in the killing of more than 1000 lions in 20 years (Etosha Ecological Institute).

So far, fencing has only been widely employed in a few countries in southern Africa, mainly because of aesthetic concerns, financial costs, and the impracticality of enclosing large-scale migratory ungulate populations (Packer et al. 2013). In West, Central, and most of East Africa, the use of fencing in wildlife management is rare (Lindsey et al. 2012b). For fencing to be an effective tool for limiting human–wildlife conflict, fences have to be well constructed (e.g., electrified) and maintained (Lindsey et al. 2012a, b). Fencing, however, has been considered on a case-by-case basis (Hayward and Kerley 2009).

Ecotourism and Trophy Hunting

Finding solutions for human wildlife conflicts is difficult, especially when communities do not receive revenue from the species (Jorge et al. 2013; Lindsey et al. 2013). Therefore, ecotourism and trophy hunting can be most beneficial to lion conservation where revenues and user rights over wildlife are devolved to local people (Nelson et al. 2013). The greatest threat to the sustainability of ecotourism and trophy hunting on communal land is the failure of governments, tour and hunting operators to devolve adequate benefits to local communities, which reduces incentives for rural people to conserve wildlife (Lewis and Alpert 1997; Mayaka et al. 2004; Child 2005; Lindsey et al. 2006) and promotes bushmeat hunting (Lindsey et al. 2013).

While West, Central, and East Africa have all witnessed a long-term decline in large mammal populations, southern Africa has generally experienced substantial recoveries in wildlife populations across large areas of private land and in some cases communal lands (Bond et al. 2004; Cragie et al. 2010; Nelson et al. 2013). In southern Africa, in particular South Africa, Zimbabwe, and Namibia, revenues from ecotourism and trophy hunting were largely responsible for the development of the game-farming industry (Bond et al. 2004).

Small Reserves in South Africa

In recent years there has been a rapid increase in the number of small (<1000 km^2), privately owned wildlife reserves in South Africa, many of which have been established for ecotourism and for biodiversity conservation (Hayward et al. 2007b). The development of the ecotourism industry led to an increase in the reintroduction of lions. More than 700 wild lions have been reintroduced into 45 small reserves in South Africa, adding considerable to two naturally occurring lion populations in the Kruger National Park and the Kgalagadi Transfrontier Park (Miller et al. 2013). However, many of these isolated lion populations have problems with inbreeding and overpopulation (Grubbich 2001; Vartan 2002; Trinkel et al. 2008, 2010; Miller and Funston 2014). Rapid population growth of lions is due to high levels of prey abundance (Vartan 2002) and a breakdown of natural processes to regulate lion population growth (Miller and Funston 2014). Due to a lack of guidelines on how to effectively manage these populations, these populations may be of limited value for the conservation of this species (Frankham 2009; Slotow and Hunter 2009). Therefore, Slotow and Hunter (2009) suggested to consider a meta-population management plan for lions in small reserves to insure high longer-term conservation potential. For wild dogs, meta-population management has already been successfully applied for the reintroduction and management of wild dogs in small reserves in South Africa (Akçakaya et al. 2006; Gusset et al. 2008). A viable wild dog population was established consisting of several sub-populations in a network of small fenced reserves (Davies-Mostert et al. 2009; Lindsey et al. 2012b). Cheetahs are also managed as a meta-population in South Africa (Lindsey et al. 2011). In 2010, a Lion Management Forum was formed to improve the conservation status of lions in small reserves by implementing a meta-population management plan for lions in South Africa (Miller et al. 2013).

Genetic Isolation and Inbreeding: Translocation as a Conservation Tool

A Case Study from the Hluhluwe iMfolozi Park

Small reserves may have major problems with inbreeding of lions (Vartan 2002). Through translocation, it was shown that it is possible to successfully restore the genetic health of a small, isolated lion population (Trinkel et al. 2008; Miller et al. 2013). The native Hluhluwe-iMfolozi Park (HiP) lion population descends from only three individuals, which repopulated the area in the 1960s (Trinkel et al. 2008).

In 1987 and 1999, the population consisted of about 140 and 80 individuals, respectively (Maddock et al. 1996; Balfour et al., unpublished data). The population declined since the early 1990 and showed little genetic variation and cub mortality was found to be high (Stein 1999). General poor condition of HiP lions and reduced immune competence were thought to be associated with inbreeding (Stein 1999).

To restore the genetic variation of the inbred HiP lion population, new (outbred) lions were translocated into the existing population (Trinkel et al. 2008): between 1999 and 2001, 16 new lions were translocated into HiP. Three different types of translocation in four separate releases were undertaken. First, a pride along with its resident coalition was translocated. Second, a pride of females was translocated. Third, new females were attempted to bond into preexisting native HiP prides. The translocations into the existing lion population in the Hluhluwe-iMfolozi Park were successful (Trinkel et al. 2008). They were designed so as to encourage the females to form four separate prides, but instead they split into six prides comprising of related and/or unrelated lionesses. Although it appeared as if a cohesive social bond had developed between all translocated female lions while they were still in the boma, many individuals separated after release. All translocated females established stable pride ranges. The three translocated males split into a pair (brothers) and a singleton, and associated with both native and new females. The pair was more successful than the solitary, gaining residence in a pride more easily and maintaining residence for longer. The native HiP population consisted of about 84 lions in 2000 but crashed to only 20 native individuals and their offspring by 2004, corresponding to 32 % of the total population. Offspring of translocated and native HiP lions totalled 29 individuals by the end of 2004 (47 %), and the translocated lions and their offspring totalled 13 individuals (21 %) (Trinkel et al. 2008). Thus, by the end of 2004, the native HiP population consisted of 20 lions, while translocated lion and offspring involving with at least one translocated animal comprised 42 animals. Descendants with at least one introduced lion replaced the entire purebred native lions by 2006 (Trinkel et al. 2008).

The translocations into the existing lion population in the Hluhluwe-iMfolozi Park were very successful. Such translocations may become an important adaptive management tool as lion populations become increasingly fragmented (Trinkel et al. 2008).

Commercial and Community Conservancies in Namibia

In Namibia, there has been an ongoing process of transforming former livestock farmland into wildlife and trophy hunting farms, where both humans and wildlife benefit (Göttert and Zeller 2008). User rights over wildlife were devolved to private landowners in the 1960s and 1970s (Bond et al. 2004). Since then, large areas of privately owned commercial cattle farms have been converted into wildlife farms. Due to such land conservation, wildlife has recovered tremendously and nowadays, 10–20 times more wild ungulates occur on private land than in protected areas

(Lindsey et al. 2013). Starting in the 1990s, Namibia—as the only country in Africa—devolved full user rights over wildlife to local communities. Such communities are able to establish "communal conservancies" where local residents can legally hunt wildlife for their own consumption or sell a quota to a hunting operator, keeping 100 % of the revenue (Jones and Weaver 2009; Nelson et al. 2013). Devolving user rights over wildlife to local communities has a large potential to reduce bushmeat poaching and increase wildlife populations (Lindsey et al. 2013). According to the Conservancy Association of Namibia, "A Conservancy is a legally protected area of a group of land-occupiers practicing cooperative management based on a sustainable utilization strategy, promotion of the conservation of natural resources and wildlife, and the desire to reinstate the original biodiversity with the basic goal of sharing resources amongst all members" (Shaw and Marker 2010).

To date, more than 70 community conservancies covering 160,000 km^2 have developed, resulting in rapid increases in wildlife populations (Lindsey et al. 2013). Through tourism and hunting, the Kunene lion population in north-western Namibia has been increasing continuously from about 15 individuals in 1999 to more than 130 individuals in 2010 (Stander 2010). Nowadays, the Kunene lion population is connected with the lion population of Etosha National Park, which holds the largest lion population in the country (Stander 2010; Trinkel et al. submitted). However, for many commercial wildlife farmers—similar to livestock farmers—large carnivores are competitors and are heavily persecuted (Lindsey 2005; Nelson et al. 2013). Besides providing space for wild ungulates, small wildlife reserves (with sizes usually between 200 and 600 km^2) at the interface between protected and non-protected areas have a high potential to provide additional space for lions: when landowners create collaboratively managed conservancies, land use tend to shift to high-value trophy hunting and ecotourism where lions and other predators are considered valuable (Lindsey et al. 2009; Nelson et al. 2013).

The Etosha Rand Lion Conservation Project

Wildlife–human conflict along borders between protected and non-protected areas is one of the main threats facing the African lion. The "Etosha Rand Lion Conservation Project" in Namibia focuses on providing a scientific basis for improved management of lions at the interface between protected and non-protected areas. Here, levels of human–lion conflict are particularly high and the resulting indiscriminate killing of lions represents one of today's main threats to the species. The "Etosha Rand Lion Conservation Project" aims to provide systematic data of lion population structure and distribution on a wildlife reserve in Namibia located at the protected/non-protected area interface. Over the last 20 years, more than 1000 lions have been killed by livestock farmers and wildlife reserve managers on lands bordering Etosha. In other small wildlife reserves in Namibia, culling—similar to small reserves in South Africa—is often carried out in response to overpopulation resulting from using electric fences to contain lions within reserve boundaries.

The "Etosha Rand Lion Conservation Project" is conducted on a wildlife reserve in Namibia, bordering on Etosha National Park. In contrast to other fully electrified small wildlife farms, the reserve is separated from the National Park by a semipermeable fence through which lions (but not prey species) can freely move and thus function as part of a much larger ecosystem. However, where the reserve borders onto livestock farms, an electrified fence has been erected. The "Etosha Rand Lion Conservation Project" aims to establish essential baseline data on population dynamics of lions on a reserve with semipermeable fencing and contribute to the development of a novel management approach for lions in non-protected reserves, and potentially reduce the extent of indiscriminate killing. The data will help provide the basis for developing new management approaches of lions outside protected areas, designed to increase the land area available to the species.

Trophy Hunting of Lions as a Conservation Tool

Trophy hunting can be a useful conservation tool with the potential to increase the range of African lions, but it can also be a threat, depending on how trophy hunting is regulated and managed (Whitman et al. 2004; Loveridge et al. 2007; Packer et al. 2011). Hunting of lions for sport is permitted in 23 African countries, with southern and eastern Africa having the largest hunting industries (Lindsey et al. 2006). Trophy hunting of lions plays a minor role in West and Central Africa (Lindsey et al. 2006). At present, African lions are permitted to be trophy hunted in an area of about 558,000 km^2 (Lindsey et al. 2013; Riggio et al. 2013). This represents 27–32 % of the range of the lion in countries where trophy hunting is generally allowed, and at least 16 % of the total lion distribution range in Africa (Lindsey et al. 2013; Riggio et al. 2013).

There are a number of characteristics making the hunting industry suitable to play an important role in conservation (Baker 1997; Bond 2004; Lindsey et al. 2006):

- Hunters have a lower impact on the environment than ecotourists in terms of disturbances, conversion of habitat, and the use of fossil fuel.
- The infrastructure required for ecotourism is more expensive than that required for hunting.
- The income per hunter is higher compared to the income per tourist. Thus, many more tourists than hunters are needed to generate the same revenue.
- Hunting operators can generate income in areas which are not suitable for ecotourism, such as remote areas, areas with low wildlife population densities, areas where—in addition to wildlife—livestock and local people are present, areas and/or countries which are politically unstable.
- Trophy hunting creates an economic incentive to generate and income from wildlife in areas that might otherwise be used for agriculture or livestock.

Problems with trophy hunting include unethical hunting practices and overharvesting (Loveridge et al. 2007; Packer et al. 2011). Unethical hunting practices include shooting from vehicles, shooting young individuals and "canned hunting," where captive-bred lions are hunted in small enclosures, mainly in South Africa (Damm 2005; Lindsey et al. 2006). Although executed only by a minor number of operators (Lindsey et al. 2006), such unethical hunting practices have contributed to a negative publicity of trophy hunting as a conservation tool (Vartan 2002). NGOs, mainly from Europe and North America, have undertaken several attempts to ban trophy hunting (Lindsey et al. 2013). Trade bans can be fatal for the retrieval and retention of land for wildlife and the tolerance towards lions (Lindsey et al. 2006).

Several countries have taken the decision to ban hunting completely. For example, in Kenya trophy hunting has been banned since 1977. In that time, wildlife numbers have declined by 60–70 % since the 1970s, mainly because of a lack of incentives to invest in wildlife as a form of land use (Norton-Griffiths 2007; Western et al. 2009; Croes et al. 2011; Lindsey et al. 2013; Nelson et al. 2013). Although ecotourism in Kenya generates large amounts of income, wildlife tourism is conducted in only 5 % of the land, which limits the distribution of income generated by wildlife (Norton-Griffiths 2007). Similar negative consequences on wildlife numbers following hunting bans were observed in Tanzania and Zambia, where sport hunting was forbidden from 1973 to 1978 and 2000 to 2003, respectively (Baker 1997; Leader-Williams and Hutton 2005; Lindsey 2005; Lewis and Jackson 1997). According to Lindsey et al. (2006), therefore, avoiding future hunting bans is vital for lion conservation.

Lions are one of the most economically valuable species in Africa's trophy hunting industry and are therefore most prone to over-harvesting (Lindsey et al. 2012b). In fact, recent studies demonstrated that over-harvesting may have contributed to lion declines in some areas in Tanzania (Packer et al. 2009, 2011). Furthermore, hunting activities in non-protected areas were found to extend into the unhunted lion population inside the protected area (Loveridge et al. 2007): excessive trophy hunting changed the demography of the lion population in Hwange National Park, Zimbabwe, although not reducing the viability of this lion population (Loveridge et al. 2007). Other areas of Zimbabwe have not faired so well, with lions in Gonarezhou National Park and Tuli Safari Area both suffering a collapse of numbers due to excessive trophy hunting (Groom et al. 2014). Groom et al. (2014) do recognize the importance of trophy hunting in the conservation of the lion; however, they call for stricter regulations to ensure the long-term survival of lions in hunted areas.

Whitman et al. (2004) suggested that implementing a 6-year age minimum would ensure safe harvest irrespective of population size. Some countries have made efforts to make lion hunting more sustainable in recent years and off-takes have declined significantly (Lindsey et al. 2013). For example, Tanzania and Mozambique's Niassa Reserve have successfully implemented a 6-year age minimum for hunted lions (Begg and Begg 2009; Packer et al. 2011). However, in many countries improved management of lion hunting to prevent negative impacts on hunted lion populations is still lacking (Lindsey et al. 2013). According to Loveridge et al. (2007) and Lindsey et al. (2013), main changes needed are as follows: (1) a reduction in quotas to realistic levels (no more than 10 % of adult males) based on robust population estimates in some countries, (2) an implementation of trophy monitoring and adequate quota management, (3) enforcing an age minimum with appropriate penalties for harvesting animals below a threshold age in all countries, and (4) a lion hunt should have a minimum length of 21 days. As lions in West and Central Africa are Critically Endangered, hunting, although playing a minor role in these areas, should be strictly prohibited.

Concluding Remarks

African lions have lost most of their historical range and continue to decline. Lion populations in West and Central Africa and in some parts of East Africa show an alarming downward trend which presses the need for urgent conservation actions.

In West and Central Africa, and in Ethiopia, where remaining lion populations are extremely small and isolated, conservation measures have to be taken quickly. Besides mitigating human–lion conflicts at reserve borders, the security of protected areas needs to be increased (e.g., by efficient prevention of poaching of lions and their prey) and protected areas should be connected through ecological corridors to maintain genetically viable populations. Failing this (or while it is being implemented), translocation schemes may be needed in the short term to conserve or increase genetic diversity. Hunting, although playing a minor role in West and Central Africa, should be strictly prohibited in protected areas and areas forming wildlife corridors. To counteract inbreeding of the Critically Endangered population in West and Central Africa, it might be essential to initiate a captive breeding program (ex situ) (Barnett et al. 2006) to increase the size of the population and its genetic diversity. In areas, where the species has disappeared and which meet the necessary ecological conditions (e.g., prey availability, protection from poaching), reintroduction of prey should be considered.

Conflict mitigation between humans and wildlife can be very effective and small changes in livestock husbandry can successfully reduce livestock losses and subsequently reduced the number of lions killed in many areas in East and West Africa. When developing such techniques, the tradition and culture of local communities have to be taken into consideration, and methods that proved to be successful in one area may not necessarily be applicable to other sites. Other approaches for effective long-term conservation of African lions favor the separation of land used by humans and conservation areas through the erection of fences. Fences have already been erected in many areas in southern Africa, where lion populations are still viable. In southern Africa and some parts of East Africa, where lions are still abundant, ecotourism and hunting have been successfully implemented to stabilize lion populations. To prevent negative impacts on hunted lion populations, however, improved management of lion hunting is necessary in many countries.

Since African lions occur in different macro-areas, i.e., in West and Central Africa, in the Ethiopian district, in East Africa and in southern Africa, these areas should be precisely re-defined geographically and conservation strategies for lions should be adjusted according to these geographical differences, based on a probable redefinition of the taxonomy of *Panthera leo*. This may result in different *P. leo* subspecies with the need to adjust conservation actions within an overall strategy for the conservation of the species as a whole (Walston et al. 2010). Such an overall conservation action plan should also include the allocation of funds by involved countries, international institutions, and private donors. For example, for the tiger *Panthera tigris* with its different populations and subspecies, a multiannual program was drawn up. The so-called Global Tiger Recovery Program was approved by all countries in which the tiger occurs and aims to achieve the objectives (e.g., fight against poaching or trade in body parts, regular monitoring of tiger populations, increasing the protection of tigers in both protected and non-protected areas) within 12 years (2010–2022) (World Bank 2011).

There are many factors threatening lions, which vary between sites. Multiple inventions will often be required and the suitability of adequate solutions will differ from area to area. Failure to address the problem will have fatal consequences for the survival of lions and wildlife in general.

Acknowledgments We wish to thank Matt W. Hayward, Susan M. Miller, and Craig Packer for comments and suggestions on an earlier draft which considerably improved this chapter. Lorenzo Rossi provided the figure.

References

Akçakaya HR, Mills MGL, Doncaster CP (2006) The role of metapopulations in conservation. In: Macdonald DW, Service K (eds) Key topics in conservation biology. Blackwell, Oxford, pp 64–84

Angelici FM, Petrozzi F (2010) Lions in the Mole National Park in Ghana, Northern Region. Cat News 53:28–31

Angelici FM., Mahama A, Rossi L (2015) The lion in Ghana: its historical and current status. Animal Biodiversity and Conservation 38.2: 151–162

Anonymous (2003) National principles, norms and standards for the sustainable use of large predators in South Africa. Government Gazette No. 25090, Appendix 2

Baker J (1997) Trophy hunting as a sustainable use of wildlife resources in southern and eastern Africa. J Sustain Tour 5:306–321

Balme GA, Slotow R, Hunter LTB (2010) Edge effects and the impact of non-protected areas in carnivore conservation: leopards in the Phinda–Mkhuze Complex, South Africa. Anim Conserv 13:315–323

Balme GA, Hunter L, Braczkowski AR (2012) Applicability of age-based hunting regulations for African leopards. PLoS One 7(4), e35209. doi:10.1371/journal.pone.0035209

Barnett R, Yamaguchi N, Barnes I, Cooper A (2009a) The origin, current diversity and future conservation of the modern lion (*Panthera leo*). Proc R Soc B 273:2119–2125

Barnett R, Shapiro B, Barnes I, Ho SYW, Burger J, Yamaguchi N, Higham TFG, Wheeler HT, Rosendahl W, Sher AV, Sotnikova M, Kuznetsova T, Baryshnikov GF, Martin LD, Harington CR, Burns JA, Cooper A (2009b) Phylogeography of lions (*Panthera leo* ssp.) reveals three distinct taxa and a late Pleistocene reduction in genetic diversity. Mol Ecol 18:1668–1677

Bauer H, De Iongh HH (2005) Lion (*Panthera leo*) home range and livestock conflicts in Waza National Park, Cameroon. Afr J Ecol 43:208–214

Bauer H, Van Der Merwe S (2004) Inventory of free-ranging lions *Panthera leo* in Africa. Oryx 38:26–31

Bauer H, Vanherle N, Di Silvestre I, De Iongh HH (2008) Lion–prey relations in West and Central Africa. Mamm Biol 73:70–73

Bauer H, De Iongh HH, Sogbohossou E (2010) Assessment and mitigation of human-lion conflict in West and Central Africa. Mammalia 74:363–367

Bauer H, Nowell K, Packer C (2012) *Panthera leo*. The IUCN Red List of Threatened Species. Version 2014.2. www.iucnredlist.org. Downloaded 14 Aug 2014

Bauer H, Packer C, Funston P, Henschel P, Nowell K (2015) Embargoed final draft: IUCN 2015 Red List—Lion (in press)

Begg C, Begg K (2009) Niassa carnivore project. Produced for SRN, Maputo

Berry HH (1997) Historical review of the Etosha Region and its subsequent administration as a National Park. Madoqua 20:3–12

Bertola LD, Van Hooft WF, Vrieling K, Uit De Weerd DR, York DS, Bauer H, Prins HHT, Funston PJ, Udo De Haesl HA, Leirs H, Van Haeringen WA, Sogbohossou E, Tumental PN, De Iongh HH (2011) Genetic diversity, evolutionary history and implications for conservation of the lion (*Panthera leo*) in West and Central Africa. J Biogeogr 38:1356–1367

Björklund M (2003) The risk of inbreeding due to habitat loss in the lion (*Panthera leo*). Conserv Genet 4:515–523

Black SA, Fellous A, Yamaguchi N, Roberts DL (2013) Examining the extinction of the Barbary lion and its implications for felid conservation. PLoS One 8(4), e60174. doi:10.1371/journal.pone.0060174

Bond I, Child B, De la Harpe D, Jones B, Barnes J, Anderson H (2004) Private land contribution to conservation in South Africa. In: Child B (ed) Parks in transition: biodiversity, rural development and the bottom line. Earthscan, London, pp 29–61

Breitenmoser U, Angst C, Landry J-M, Breitenmoser-Wursten C, Linnell JDC, Weber J-M (2005) Non-lethal techniques for reducing depredation. In: Woodroffe R, Thirgood S, Rabinowitz AR (eds) People and wildlife: conflict or coexistence? Cambridge University Press, Cambridge, pp 49–71

Bruche S, Gusset M, Lippold S, Barnett R, Eulenberger K, Junhold J, Driscoll CA, Hofreiter M (2013) A genetically distinct lion (*Panthera leo*) population from Ethiopia. Eur J Wildl Res 59:215–225

Burton AC, Buedi EB, Balangtaa C, Kpelle DG, Sam MK, Brashares JS (2011) The decline of lions in Ghana's Mole National Park. Afr J Ecol 49:122–126

Bush ER, Baker SE, Macdonald DW (2013) Global trade in exotic pets 2006-2012. Conserv Biol 28:663–676

Carpenter MA, Appel MJG, Roelke-Parker ME, Munsond L, Hofer H, East M, O'Brien SJ (1998) Genetic characterization of canine distemper virus in Serengeti carnivores. Vet Immunol Immunopathol 65:259–266

Caughley G (1994) Directions in conservation biology. J Anim Ecol 63:215–244

Chardonnet P (2002) Conservation of the African lion: contribution to a status survey. Foundation IGF, Paris

Child B (2005) Principles, practice and results of CBNRM in southern Africa. Sand County Foundation, Monona. http://sandcounty.net/assets/chapters/assets.pdf. Accessed 16 Jan 2015

Cleaveland S, Kaare M, Knobel D, Laurenson MK (2006) Canine vaccination-providing broader benefits for disease control. Vet Microbiol 117:43–50

Cragie ID, Baillie JEM, Balmford A, Carbone C, Collen B, Green RE, Hutton JM (2010) Large mammal population declines in Africa's protected areas. Biol Conserv 143:2221–2228

Creel S, Becker MS, Durant SM et al (2013) Conserving large populations of lions—the argument for fences has holes. Ecol Lett. doi:10.1111/ele.12145

Croes BM, Funston PJ, Rasmussen G, Buij R, Saleh A, De Iongh HH (2011) The impact of trophy hunting on lions (*Panthera leo*) and other large carnivores in the Benoué complex, northern Cameroon. Biol Conserv 144:3064–3072

Damm G (2005) Hunting in South Africa: facts, risks and opportunities. African Indaba 3:1–14

Davies-Mostert HT, Mills MGL, Macdonald DW (2009) A critical assessment of South Africa's managed metapopulation recovery strategy for African wild dogs and its value as a template for large carnivore conservation elsewhere. In: Hayward MW, Somers MJ (eds) The reintroduction of top-order predators. Blackwell, Oxford, pp 10–42

De la Bat BGJ (1982) Etosha 75 years. In: Southwest Africa annual. SWA, Windhoek, pp 11–22
De Lisle GW, Bengis RG, Schmitt SM, O'Brien DJ (2002) Tuberculosis in free-ranging wildlife: detection, diagnosis and management. Revue Scientifique et Technique de l'Office International des Epizooties 21:317–334
Dickman AJ, Macdonald EA, Macdonald DW (2011) A review of financial instruments to pay for predator conservation and encourage human-carnivore coexistence. Proc Natl Acad Sci U S A 108:13937–13944
Dinerstein E, Loucks C, Wikramanayake E, Ginsberg J, Sanderson E, Seidensticker J, Forrest J, Bryja G, Heydlauff A, Klenzndorf S, Leimgruber P, Mills J, O'Brien TG, Shrestha M, Simons R, Songer M (2007) The fate of wild tigers. BioScience 57:508–514
Ellis R (2005) Tiger bone & rhino horn: the destruction of wildlife for traditional Chinese medicine. Island Press, Washington DC
Frankham R (2009) Genetic considerations in reintroduction programs for large terrestrial predators. In: Hayward MW, Somers MJ (eds) The reintroduction of top-order predators. Blackwell, Oxford, pp 371–387
Gadd ME (2012) Barriers, the beef industry and unnatural selection: a review of the impacts of veterinary fencing on mammals in southern Africa. In: Hayward MW, Somers MJ (eds) Fencing for conservation. Springer, New York, pp 153–186
Göttert T, Zeller U (2008) Das Etosha Pufferzonenprojekt—ein Konzept zur Unterstützung der Bemühungen zur Anbindung des Etosha Nationalparks an das transnationale Netzwerk von Schutzgebieten im südlichen Afrika. Beiträge zur Jagd- und Wildforschung 33:283–292
Gratwicke B, Bennett EL, Broad S, Christie S, Dutton A, Gabriel G, Kirkpatric C, Nowell K (2008) The world can't have wild tigers and eat them, too. Conserv Biol 22:222–223
Groom JR, Funston PJ, Mandisozda R (2014) Surveys of lions *Panthera leo* in protected areas in Zimbabwe yield disturbing results: what is driving the population collapse? Oryx 48:385–393
Grubbich JD (2001) Genetic variation within and among fragmented populations of South African lions *Panthera leo*: implications for management. M.Sc. Thesis, University of Pretoria
Gusset M, Ryan SJ, Hofmeyr M, Van Dyk G, Davies-Mostert HT, Graf J, Owen C, Szykman M, Macdonald DW, Monfort S, Wildt D, Maddock A, Mills MGL, Slotow R, Somers M (2008) Efforts going to the dogs? Evaluating attempts to re-introduce endangered wild dogs in South Africa. J Appl Ecol 45:100–108
Harder TC, Kenter M, Appel MJC, Roelke-Parker ME, Barrett T, Osterhaus ADME (1995) Phylogenetic evidence of canine distemper virus in Serengeti's lions. Vaccine 13:521–523
Haydon DT et al (2006) Low-coverage vaccination strategies for the conservation of endangered species. Nature 443:692–695
Hayward MW, Kerley GIH (2009) Fencing for conservation: restriction of evolutionary potential or a riposte to threatening processes? Biol Conserv 142:1–13
Hayward MW, O'Brian J, Kerley GIH (2007a) Carrying capacity of large African predators: predictions and tests. Biol Conserv 139:219–229
Hayward MW, Adendorff J, O'Brien J, Sholto-Douglas A, Bissett C, Moolman LC, Bean P, Fogarty A, Howarth D, Slater R, Kerley GIH (2007b) Practical considerations for the reintroduction of large, terrestrial, mammalian predators based on reintroductions to South Africa's Eastern Cape Province. Open Conserv Biol J 1:1–11
Hazzah L, Dolrenry S, Naughton l, Edwards CTT, Mwebi O, Kearney F, Laurence F (2014) Efficacy of Two Lion Conservation Programs in Maasailand, Kenya. Conservation Biology 28: 851–86
Hemley G (ed) (1994) International wildlife trade: a CITES sourcebook. Island Press, Washington, DC
Henschel P, Coad L, Burton C, Chataigner B, Dunn A, Macdonald D, Saidu Y, Hunter LTB (2014) The lion in West Africa is critically endangered. PLoS One 9(1), e83500. doi:10.1371/ journal. pone.0083500
Hilborn R, Arcese P, Borner M, Hando J, Hopcraft G, Loibooki M, Mduma S, Sinclair ARE (2006) Effective enforcement in a conservation area. Science 314:1266
Hunter LTB, Skinner JD, Pretorius K, Carlisle LC, Rickelton M, Walker C, Slotow R (2007) Reintroducing lions *Panthera leo* to northern Kwazulu-Natal, South Africa: short-term biological and technical successes but equivocal long-term conservation. Oryx 41:196–204

Jones B, Weaver C (2009) CBNRM in Namibia: growth, trends, lessons and constraints. In: Suich H, Child B, Spenceley A (eds) Evolution and innovation in wildlife conservation in Southern Africa. Earthscan, London, pp 223–242

Jorge AA, Vanak AT, Thaker M, Begg C, Slotow R (2013) Costs and benefits of leopard to the sport hunting industry and local communities in Niassa National Reserve, Mozambique. Conserv Biol 27:832–843

Keet DF, Kriek NPJ, Penrith ML, Michel A, Huchzermeyer F (1996) Tuberculosis in buffaloes (*Syncerus caffer*) in the Kruger National Park: spread of the disease to other species. Onderstepoort J Vet Res 63:239–244

Kissui BM (2008) Livestock predation by lions, leopards, spotted hyenas, and their vulnerability to retaliatory killing in the Maasai steppe, Tanzania. Anim Conserv 11:422–432

Kötting B (2002) Problem animal incidents on Etosha National Park border farms: trends, causes and costs. Etosha Ecological Institute, Internal Report

Leader-Williams N, Hutton J (2005) Does extractive use provide opportunities to reduce conflicts between people and wildlife? In: Woodroffe R, Thirgood S, Rabinowitz A (eds) People and wildlife: conflict or coexistence. Cambridge University Press, Cambridge, pp 140–162

Lewis D, Alpert P (1997) Trophy hunting and wildlife conservation in Zambia. Conserv Biol 11:59–68

Lewis D, Jackson J (1997) Safari hunting and conservation on communal land in southern Africa. In: Woodroffe R, Thirgood S, Rabinowitz A (eds) People and wildlife: conflict or coexistence. Cambridge University Press, Cambridge, pp 239–251

Lichtenfeld LL, Trout C, Kisimir EL (2015) Evidence-based conservation: predatorproof bomas protect livestock and lions. Biodiversity and Conservation 24: 483–491

Lindsey P (2005) A review of the economic contribution of the trophy hunting industry in Africa. Safari Club International, Washington, DC

Lindsey P, Alexander R, Frank LG, Mathieson A, Romanach SS (2006) Potential of trophy hunting to create incentives for wildlife conservation in Africa where alternative wildlife-based land uses may not be viable. Anim Conserv 9:283–291

Lindsey PA, Masterson CL, Beck AL, Romanach S (2009) Ecological, social and financial issues related to fencing as a conservation tool in Africa. In: Hayward MW, Somers MJ (eds) The reintroduction of top-order predators. Blackwell, Oxford, pp 215–234

Lindsey PA, Tambling CJ, Brummer R, Davies-Mostert HT, Hayward MW, Marnewick KA, Parker DM (2011) Minimum prey and area requirements of cheetahs: implications for reintroductions and management of the species as a managed metapopulation. Oryx 45:587–599

Lindsey PA, Alexander R, Balme GA, Midlane N, Craig J (2012a) Possible relationships between the South African captive-bred lion hunting industry and the hunting and conservation of lions elsewhere in Africa. S Afr J Wildl Res 42:11–22

Lindsey PA, Balme GA, Booth VR, Midlane N (2012b) The significance of African lions for the financial viability of trophy hunting and the maintenance of wild land. PLoS One 7(1), e29332. doi:10.1371/journal.pone.0029332

Lindsey PA, Balme GA, Funston P, Henschel P, Hunter L, Madzikanda H, Midlane N, Nyirenda V (2013) The trophy hunting of African lions: scale, current management practices and factors undermining sustainability. PLoS One 8(9), e73808. doi:10.1371/journal.pone.0073808

Loveridge AJ, Searle AW, Murindagomo F, Macdonald DW (2007) The impact of sport-hunting on the population dynamics of an African lion population in a protected area. Biol Conserv 134:548–558

Maddock A, Anderson A, Carlisle F, Galli N, James A, Verster A, Whitfield W (1996) Changes in lion numbers in Hluhluwe-Umfolozi Park. Lammergeyer 44:6–18

Mayaka T, Hendriks T, Wesseler J, Prins H (2004) Improving the benefits of wildlife harvesting in northern Cameroon: a co management perspective. Ecol Econ 54:67–80

Michel AL, Bengis RG, Keet DF, Hofmeyr M, De Klerk LM, Cross PC, Jolles AE, Cooper D, Whyte IJ, Buss P, Godfroid J (2006) Wildlife tuberculosis in South African conservation areas: implications and challenges. Vet Microbiol 112:91–100

Miller SM, Funston PJ (2014) Rapid growth rates of lion (*Panthera leo*) populations in small, fenced reserves in South Africa: a management dilemma. S Afr J Wildl Res 44:43–55

Miller SM, Bissett C, Burger A, Courtenay B, Dickerson T et al (2013) Management of reintroduced lions in small, fenced reserves in South Africa: an assessment and guidelines. S Afr J Wildl Res 43:138–154
Morell V (2007) Can the wild tiger survive? Science 317:1312–1314. doi:10.1126/science.317.5843.1312
C. Morandin C, Loveridge AJ, Segelbacher G, Elliot N, Madzikanda H, Macdonald DW, Höglund J (2014) Gene flow and immigration: genetic diversity and population structure of lions (Panthera leo) in Hwange National Park, Zimbabwe. Conservation Genetics 15: 697-706.
Munson L, Terio KA, Kock R, Mlengeya T, Roelke ME, Dubovi E, Summers B, Sinclair ARE, Packer C (2008) Climate Extremes Promote Fatal Co-Infections during Canine Distemper Epidemics in African Lions. PLoS ONE 3(6): e2545. doi:10.1371/journal.pone.0002545
Myers N (1975) The Silent savannahs. Int Wildl 5:5–10
Nelson F, Lindsey P, Balme G (2013) Trophy hunting and lion conservation: a question of governance? Oryx 47:501–509
Norton-Griffiths M (2007) How many wildebeest do you need? World Econ 8:41–64
Nowell K (2012) Wildlife crime scorecard. Assessing compliance with and enforcement of CITES commitments for tigers, rhinos and elephants. WWF report 2012, Gland. http://plus.cdt.ch/files/docs/31e937b6c99b7047b702bff68ef64d93.pdf. Accessed 2 Feb 2015
Nowell K, Jackson P (1996) Wild cats: conservation status and species action plan. IUCN, Gland
Nyhus PJ, Ambrogi M, Dufraine C, Shoemaker A, Tilson R (2009) The status and evolution of laws and policies regulating privately owned tigers in the United States. J WildCat Conserv Legal Aid Soc 1:47–64
Ogada MO, Woodroffe R, Oguge NO, Frank LG (2003) Limiting depredation by African carnivores: the role of livestock husbandry. Conserv Biol 17:1521–1530
Packer C, Herbst L, Pusey AE, Bygott JD, Cairns SJ, Hanby JP, Borgerhoff-Mulder M (1988) Reproductive success in lions. In: Clutton-Brock TH (ed) Reproductive success. University of Chicago Press, Chicago, pp 363–383
Packer C, Scheel D, Pusey AE (1990) Why lions form groups: food is not enough. Am Nat 136:1–19
Packer C, Altizer S, Appel M, Brown E, Martenson J, O'Brien SJ, Roelke-Parker ME, Hofmann-Lehmann R, Lutz H (1999) Viruses of the Serengeti: patterns of infection and mortality in African lions. J Anim Ecol 68:1161–1178
Packer C, Ikanda D, Kissui B, Kushnir H (2005) Lion attacks on humans in Tanzania. Nature 436:927–928
Packer C, Kosmala M, Cooley HS, Brink H, Pintea L, Garshelis D, Purchase G, Strauss M, Swanson A, Balme G, Hunter L, Nowell K (2009) Sport hunting, predator control and conservation of large carnivores. PLoS One 4(6), e5941. doi:10.1371/journal.pone.0005941
Packer C, Brink H, Kissui BM, Maliti H, Kushnir H, Caro T (2011) Effects of trophy hunting on lion and leopard populations in Tanzania. Conserv Biol 25:142–153
Packer C, Loveridge A, Canney S et al (2013) Conserving large carnivores: dollars and fences. Ecol Lett 16:635–641
The IUCN Red List of Threatened Species. Version 2015-3. <www.iucnredlist.org>. Downloaded on 27 October 2015.
Pardo MC, Bauman JE, Mackowiak M (1997) Protection of dogs against canine distemper by vaccination with a canarypox virus recombinant expressing canine distemper virus fusion and hemagglutinin glycoproteins. Am J Vet Res 58:833–836
Bruce D. Patterson BD, Kasiki SM, Selempo E, Kays RW (2004) Livestock predation by lions (Panthera leo) and other carnivores on ranches neighboring Tsavo National Parks, Kenya. Biological Conservation 119: 507–516
Prins HHT, Grootenhuis JG (2000) Introduction: the value of priceless wildlife. In: Prins HHT, Grootenhuis JG, Dolan TT (eds) Wildlife conservation by sustainable use. Kluwer Academic, Norwell, pp 1–13
Ray JC, Hunter LTB, Zigouris J (2005) Setting conservation and research priorities for larger African carnivores. Wildlife Conservation Society, New York

Riggio J, Jacobson A, Dollar L, Bauer H, Becker M, Dickman A, Funston P, Groom R, Henschel P, de Iongh H, Lichtenfeld L, Pimm S (2013) The size of savannah Africa: a lion's (*Panthera leo*) view. Biodivers Conserv 22:17–35

Roelke-Parker ME, Munson L, Packer C, Kock R, Cleaveland S, Carpenter M, O'Brien SJ, Pospischil A, Hofmann-Lehmann R, Lutz H, Mwamengele GL, Mgasa MN, Machange GA, Summers BA, Appel MJ (1996) A canine distemper virus epidemic in Serengeti lions (*Panthera leo*). Nature 379:441–445

Sam M. Ferreira SM, Funston PJ (2010) Estimating lion population variables: prey and disease effects in Kruger National Park, South Africa. Wildlife Research 37: 194–206

Scheepers JL, Venzke KAE (1995) Attempts to reintroduce African wild dogs *Lycaon pictus* into Etosha National Park, Namibia. S Afr J Wildl Res 25:138–140

Schuette P, Creel S, Christianson D (2013) Coexistence of African lions, livestock, and people in a landscape with variable human land use and seasonal movements. Biol Conserv 157:148–154

Scott M (1988) The impact of infection and disease on animal populations: implications for conservation biology. Conserv Biol 2:40–56

Selier SAJ, Page BR, Vanak AT, Slotow R (2014) Sustainability of elephant hunting across international borders in southern Africa: a case study of the greater Mapungubwe Transfrontier Conservation Area. J Wildl Manage 78:122–132

Shaw DM, Marker L (eds) (2010) The Conservancy Association of Namibia: an overview of freehold conservancies. CANAM (Conservancy Association of Namibia), Windhoek

Sillero-Zubiri C, King AA, Macdonald DW (1996) Rabies and mortality in Ethiopian wolves (*Canis simensis*). J Wildl Dis 32:80–86

Skinner JD, Smithers RHN (1990) The mammals of the southern African subregion, New edition. University of Pretoria, Pretoria

Slotow R, Hunter LTB (2009) Reintroduction decisions taken at the incorrect social scale devalue their conservation contribution: the African lion in South Africa. In: Hayward MW, Somers MJ (eds) Reintroduction of top-order predators. Wiley–Blackwell, Oxford, pp 43–71

Smuts GL (1978) Effects of population reduction on the travels and reproduction of lions (*Panthera leo*) in Kruger National Park. Carnivore 1:61–72

Stander PE (1991) Demography of lions in the Etosha National Park, Namibia. Madoqua 18:1–9

Stander PE (2010) The impact of male-biased mortality on the population structure of desert-adapted lion in Namibia. Research Report 2010. http://www.desertlion.info/reports/dlion2010.pdf. Accessed 18 Jan 2015

Steele TE (2007) Late Pleistocene of Africa. In: Elias SA (ed) Encyclopedia of quaternary science. Elsevier, Oxford, pp 3139–3150

Stein B (1999) Genetic variation and depletion in a population of lions (*Panthera leo*) in Hluhluwe-iMfolozi Park. M. Agric. Thesis, University of Natal, Pietermaritzburg

Tende T, Bensch S, Ottosson U, Hansson B (2014) Dual phylogenetic origins of Nigerian lions (*Panthera leo*). Ecol Evol 4:2668–2674

Treves A, Naughton-Treves L (1999) Risk and opportunity for humans coexisting with large carnivores. J Hum Evol 36:275–282

Trinkel M (2013) Climate variability, human wildlife conflict and population dynamics of lions *Panthera leo*. Naturwissenschaften 100:345–353

Trinkel M, Fleischmann PH, Slotow R (submitted) Fences with holes: problem animal control threatens the long-term viability of a wild ranging lion population

Trinkel M, Ferguson N, Reid A, Reid C, Somers M, Turelli L, Graf J, Szykman M, Cooper D, Haverman P, Kastberger G, Packer C, Slotow R (2008) Translocating new lions into an inbred lion population in the Hluhluwe-iMfolozi Park, South Africa. Anim Conserv 11:138–143

Trinkel M, Funston P, Hofmeyr M, Hofmeyr D, Dell S, Packer C, Slotow R (2010) Inbreeding and density-dependent population growth in a small, isolated lion population. Anim Conserv 13:374–382

Trinkel M, Cooper D, Packer C, Slotow R (2011) Inbreeding depression increases susceptibility to bovine tuberculosis in lions: an experimental test using an inbred-outbred contrast through translocation. J Wildl Dis 47:494–500

Van Orsdol KG, Hanby JP, Bygott JD (1985) Ecological correlates of lion social organization (Panthers, leo). Journal of Zoology 206: 97–112.

Van Vuuren JH, Herrmann E, Funston PJ (2005) Lions in the Kgalagadi Transfrontier Park: modelling the effect of human-caused mortality. Int Trans Oper Res 12:145–147

Vartan S (2002) Overpopulation and inbreeding in small game reserves: the lion *Panthera leo* as a case study. M.Sc. Thesis, University of Cape Town, Cape Town

Viana M, Cleaveland S, Matthiopoulos J, Halliday J, Packer C, Craft ME, Hampson K, Czupryna A, Dobson AP, Dubovi EJ, Ernest E, Fyumagwa R, Hoare R, Hopcraft JGC, Horton DL, Kaare MT, Kanellos T, Lankester F, Mentzel C, Mlengeya T, Mzimbiri I, Takahashi E, Willett B, Haydon DT, Lembo T (2015)

Dynamics of a morbillivirus at the domestic–wildlife interface: Canine distemper virus in domestic dogs and lions.

Proceedings of the National Academy of Sciences of the United States of America 112: 1464–1469, doi:

10.1073/pnas.1411623112

Vial F, Cleaveland S, Rasmussen G, Haydon DT (2006) Development of vaccination strategies for the management of rabies in African wild dogs. Biol Conserv 131:180–192

Walston J, Robinson JG, Bennett EL, Breitenmoser U, Fonseca GAB et al (2010) Bringing the tiger back from the brink—the six percent solution. PLoS Biol 8(9), e1000485. doi:10.1371/journal.pbio.1000485

Wegmann M, Santini L, Leutner B, Safi K, Rocchini D, Bevanda M, Latifi H, Dech S, Rondinini C (2014) Role of African protected areas in maintaining connectivity for large mammals. Philos Trans R Soc B 369:20130193. http://dx.doi.org/10.1098/rstb.2013.0193. Accessed 18 Jan 2015 doi: 10.1098/rstb.2013.0193.%20Accessed%2018%20Jan%202015

Weladji RB, Tchamba MN (2003) Conflict between people and protected areas within the Bénoué Wildlife Conservation Area, North Cameroon. Oryx 37:72–79

Western D, Russell S, Cuthill I (2009) The status of wildlife in protected areas compared to non-protected areas of Kenya. PLoS One 4(7), e6140. doi:10.1371/journal.pone.0006140

Whitman K, Starfield AM, Quadling HS, Packer C (2004) Sustainable trophy hunting of African lions. Nature 428:175–178. doi:10.1038/nature02395

Williamson D, Mbano M (1988) Wildebeest mortality during 1983 at Lake Xau, Botswana. Afr J Ecol 26:341–344

David S. Wilkie DS, Bennett EL, Peres CA, Cunningham AA (2011) The empty forest revisited. ANNALS OF THE NEW YORK ACADEMY OF SCIENCES 1223: 120–128

Wimsatt J, Biggins D, Innes K, Taylor B, Garell D (2003) Evaluation of oral and subcutaneous delivery of an experimental canarypox recombinant canine distemper vaccine in the Siberian polecat (*Mustela eversmanni*). J Zoo Wildl Med 34:25–35

Woodroffe R, Ginsberg JR (1998) Edge effects and the extinction of populations inside protected areas. Science 280:2126–2128

Woodroffe R, Thirgood S, Rabinowitz A (2005) People and wildlife: conflict or coexistence? Cambridge University Press, Cambridge

Woodroffe R, Frank LG, Lindsey PA, ole Ranah SMK, Romanach S (2006) Livestock husbandry as a tool for carnivore conservation in Africa's community rangelands: a case–control study. Biodivers Conserv 16:1245–1260

World Bank (2011) Global tiger recovery program (2010-2022). World Bank, Washington, DC. http://documents.worldbank.org/curated/en/2011/11/16828660/global-tiger-recovery-program-2010-2022. Accessed 22 Jan 2015

Yamazaki K (1996) Social variation of lions in a male-depopulated area in Zambia. J Wildl Manage 60:490–497

Young TP (1994) Natural die-offs of large mammals: implications for conservation. Conserv Biol 8:410–418

Zhang L, Hua N, Sun S (2008) Wildlife trade, consumption and conservation awareness in southwest China. Biodivers Conserv 17:1493–1516

Chapter 4
Commodification of the Saker Falcon *Falco cherrug*: Conservation Problem or Opportunity?

Andrew Dixon

Conservation Status of the Saker Falcon

Distribution. The Saker Falcon (*Falco cherrug*) has a Palearctic breeding distribution (see map in Kovács et al. 2014), with its western limit in the lowlands of Central Europe, extending eastwards via fragmented or highly dispersed populations through Romania, Moldova, Ukraine, southwest Russia and Turkey to the open and mountainous landscapes of Central Asia. Here, breeding Saker Falcons can be found from the semi-deserts and deserts of Iran and Afghanistan, through the former Soviet Central Asian states to the southern steppes of Russia bordering Kazakhstan and Mongolia. A larger and generally contiguous population breeds throughout Mongolia, and across a large swathe of grassland and desert in China eastwards from the Qinghai-Tibetan plateau and Xinjiang to the eastern distribution limit of the species in Heilongjiang. Saker Falcons are 'partial migrants' and the proportion of birds making migratory movements differs between regions and across age classes, with birds tending to be more sedentary in the south and adults less likely to make long-distance migratory movements than juveniles (Baumgart 1978; Prommer et al. 2012). Migratory Saker Falcons typically move to more southerly wintering areas, either within the breeding distribution range of the species or beyond to North and sub-Saharan Africa, the Middle East and the Indian subcontinent (Orta et al. 2014).

Conservation status. The Saker Falcon has the dubious honour of being the only species in the genus *Falco* that is categorized as Globally Endangered on the IUCN Red Data List (Table 4.1). Following decline in the twentieth Century, the Saker Falcon population of the Pannonian Basin in Central Europe has been the focus of conservation efforts and has subsequently increased in recent decades (Chavko 2010;

A. Dixon (✉)
International Wildlife Consultants Ltd., Penllynin Farm, Llysonnen Road,
PO Box 33, Carmarthen SA33 5EH, UK
e-mail: falco@falcons.co.uk

F.M. Angelici (ed.), *Problematic Wildlife*, DOI 10.1007/978-3-319-22246-2_4

Table 4.1 IUCN Red List estimates of global population size (breeding pairs) of the Saker Falcon, median (minimum and maximum), and percentage population change over three generations, median (range)

Country (region)	Population size	% Change (range)
Mongolia	3464 (1765–5300)	–9.5 (–69.9 to +75.0)
Russia	1972 (1617–2345)	–64.7 (–73.2 to –53.9)
China	2851 (823–5262)	–38.8 (–84.6 to –48.6)
Kazakhstan	1046 (729–1370)	–75.4 (–83.2 to –66.6)
Ukraine	335 (288–383)	+128.5 (+87.9 to +175.6)
Hungary	194 (186–203)	+78.6 (+72.1 to +85.2)
Turkmenistan	125 (96–156)	0.0 (–32.2 to +47.6)
Other (Europe)	190 (119–287)	–14.1 (–49.8 to +90.4)
Other (Asia)	265 (71–725)	–60.4 (–90.3 to +58.0)
Total	10442 (5694–16031)	–47.6 (–75.2 to –2.3)

Data from the 2012 assessment by BirdLife International

Bagyura et al. 2012), whereas large-scale declines witnessed in the Balkans and Pontic-Caspian steppe have led to regional breeding extinction, e.g. Bulgaria and the Southern Federal District of Russia (Dixon et al. 2009; Ragyov et al. 2014) or diminished populations, e.g. across the steppic plains of Moldova, Ukraine and Crimea (Milobog et al. 2010). Accurate information on population status and trends is lacking for most Asian states, but there is evidence of significant declines in Kazakhstan, beginning during the Soviet-era in the west and then across most of the country following independence in 1991 (Levin 2011). Following dissolution of the USSR, similar population declines have taken place in the Russian Federation and probably also in other former Soviet states of Central Asia (Karyakin 2008). Nevertheless, large breeding populations exist in Mongolia, where numbers have probably increased, and across extensive areas of China, e.g. the Qinghai-Tibetan plateau (Dixon 2009).

The degree of spatial and temporal variation in the magnitude, rate and extent of population declines across the former Soviet states is largely unknown, as are the principal causal factors. For the observed population declines in Kazakhstan and Russia, speculation primarily focuses on the direct impact of trapping for the Arabian falconry trade and the indirect impact of electrocution at electricity distribution lines, together with wide-scale habitat change resulting from socio-economic changes in recent decades (Kenward et al. 2007; Kovács et al. 2014).

Saker Falcons and the Falconry Trade

Falconry is a deep-rooted and culturally significant practice in the Arabian Gulf, especially in Saudi Arabia, United Arab Emirates, Qatar, Kuwait and Bahrain. Falconry is recognized by UNESCO as part of humanity's 'Global Intangible Cultural Heritage' (Ceballos 2009; Wakefield 2012), and Saker Falcons are traditionally the main species used in Arabic falconry (Allen 1980; Upton 2002). Falconry is popular at all levels of society in the Gulf States and its influence extends to the wider Middle East, Central Asia and the Maghreb where hunting with falcons and trapping of falcons takes place. The demand for Saker Falcons from wealthy falconers in the Gulf States of Arabia has created a lucrative market (Seddon and Launay 2008). The demand for falcons in Arabic falconry is met by captive breeding and by harvesting from the wild, which can either be regulated and legal or unregulated and illegal (Dixon 2012). Trapping within end-user states in the Arabian Peninsula occurs but the main source of falcons comes through legal and illegal international trade (Barton 2000; Riddle and Remple 1994). The legal trade in both captive-bred and wild-harvested falcons is regulated by the Convention of International Trade in Endangered Species (CITES), whereas the extent of illegal trade is regulated by the efficacy of law enforcement.

Whilst traditional autumn trapping of 'passage' falcons for Arabian falconry has taken place for millennia the geographic scope, scale and nature of trapping has increased. However, we still have a very limited understanding of the drivers of the market resulting from the commodification of the Saker Falcon. On the supply side, the socio-economic upheaval associated with the dissolution of the Soviet Union is believed to have stimulated a surge in the supply of wild-caught Saker Falcons from Central Asian states that were formerly outside the geographic scope of the Arabian falconry market. In countries such as Kazakhstan and Kirghizstan, more people engaged in falcon trapping as a source of income and borders were opened to foreign trappers and traders (Levin 2011). The demand for wild-caught Saker Falcons in Arabian falconry was large enough to accommodate the increased supply from the newly opened source countries of the former Soviet Union. However, it is unlikely that such an illegal trade can be limited by market equilibrium because a lower price resulting from a surplus or increased supply will still be profitable when the marginal costs of trapping wild Saker Falcons are minimal and can further be absorbed by increasing supply chain efficiency. There is evidence to suggest that the illegal falcon trade has indeed evolved in such a way, becoming more organized and co-ordinated over time (Wyatt 2009). Furthermore, the trapping period has extended from the time of autumn passage when migrating birds were targeted, into the breeding season in the former Soviet states of Central Asia, and now includes the trapping of breeding birds and the removal of eggs and chicks from nests (Kenward et al. 2007).

In most range states, the trapping and trade of wild Saker Falcons is illegal under national laws (Kovács et al. 2014). Notable exceptions exist, including Saudi Arabia, which allows trapping and trade of wild Saker Falcons within the country and Mongolia, which issues permits for their harvest and international trade within CITES regulations. This legal, regulated trade involves trapping free-flying birds after the breeding season either during migration (as in Saudi Arabia; Shobrak 2014) or during post-breeding and post-fledging dispersal (as in Mongolia; Dixon et al. 2011). The development of the Mongolian Saker Falcon trade for the Arabian falconry market presents an interesting case study, providing insights into the conservation, ecological, economic and social aspects of this particular 'wildlife problem'.

Case Study I: The Mongolian Saker Falcon Trade

Transition from a National to an International 'Wildlife Problem'

The Mongolian Saker Falcon trade developed following the Democratic Revolution of 1990, when the harsh economic conditions of the early 1990s and the potential of international trade to the Gulf States provided the incentive for 'entrepreneurial' ornithologists to initiate a relatively small-scale private trade in wild-caught Saker Falcons. However, by the time the Mongolian government had become a signatory to CITES in 1996, this trade had been appropriated by the state and was controlled and regulated by government officials. Despite, or perhaps because of, governmental control together with an absence of transparency and accountability, the Saker Falcon trade became mired in controversy over issues of corruption and criminality. There were numerous media stories circulating about the trade and the issue became *cause célèbre* (Kohn 2006), where the protagonists in the debate freely exchanged a range of credible, specious and spurious accusations, making any reasonable assessment of the true position virtually impossible. This controversy resulted in the national and international conservation community looking askance at the Mongolian Saker Falcon trade (e.g. Boldbaatar 2009; Zahler et al. 2004), which no doubt contributed to its chequered history with CITES and other multilateral environmental agreements.

The CITES Review of Significant Trade

It was no mere coincidence that at its Asian regional meeting in Ulaanbaatar, Mongolia in August 2002, the CITES Secretariat proposed to organize a consultative meeting on the trade in falcons for falconry, a meeting that was subsequently held in Abu Dhabi, UAE in May 2004. At that time, the UAE had recently been subject to a CITES trade suspension, which had been withdrawn in 2002 following implementation of a series of measures, including a system of falcon registration. A heightened awareness of CITES trade issues, together with data on the scale of the falcon trade to the Gulf states and some preliminary survey data of wild Saker Falcon populations in Mongolia was the catalyst for a report from the CITES Scientific Authority of the UAE to the CITES Animal Committee in August 2003. This report resulted in the Saker Falcon immediately entering the CITES Significant Trade review process.[1] By its meeting in May 2005, the Animals Committee had provisionally categorized Saker Falcon range states as being of (1) least, (2) possible or (3) urgent concern with regard to engaging in trade that could be detrimental to the survival of the species and where monitoring of this trade may be inadequate.

[1] http://www.cites.org/eng/com/ac/19/summary_record.pdf

Mongolia was considered to be of 'urgent concern', having serious problems with regulating its trade in wild-caught Saker Falcons.[2]

Mongolia and eight other countries considered to be of 'urgent concern', were urged to suspend the issuance of export permits for Saker Falcons by September 2005 and, if they wished to resume the trade, conduct research on the distribution, abundance, population trends and threats facing the species in the country, and furthermore develop a science-based monitoring system and establish an adaptive management programme for the harvest and trade of Saker Falcons. In September 2005, the Mongolian government informed CITES that no further export permits would be issued until the problem of the Saker Falcon trade was 'resolved at the Animals Committee through the Secretariat'. Consequently, a notification was issued to all CITES Parties that Mongolia had suspended the issuance of export permits for Saker Falcons.[3] In the same year, the Mongolian government issued an order (#248) to regulate trapping and taking of wild Saker Falcons for research and scientific purposes, and to establish procedures for assessing the ecological and economic consequences of a harvest.

Meanwhile, apparently oblivious to their commitment to suspend the trade, the Mongolian government continued to issue CITES permits for the harvest and export of Saker Falcons. This dichotomous approach reflected, at least in part, a lack of capacity within the Mongolian government to adequately administer their obligations to the CITES convention and the absence of a formal structure of governance of the Saker Falcon trade. Alerted by press coverage in Mongolia, the CITES Secretariat made enquiries to the Mongolian CITES Management Authority and were informed that 167 and 407 Saker Falcons had been exported in 2006 and 2007, respectively, and that a quota of 300 specimens per year had been established based on the results of four surveys undertaken in the previous 6 years. Not satisfied that Mongolia was complying fully with the CITES recommendations, a trade suspension in relation to Saker Falcons was implemented in January 2009.[4]

A workshop meeting convened in Abu Dhabi in April 2009 at the request of CMS (see later), provided a forum where details of the Mongolian conservation management programme being developed by the Environment Agency—Abu Dhabi (EAD) could be transmitted to the CITES Secretariat. Consequently, the recommendation to suspend trade with Mongolia was withdrawn on condition that Mongolia maintained an export quota of 300 specimens in 2009 and 2010, before establishing a quota for 2011.[5] Subsequently, in July 2011, a report was presented to the CITES Animals Committee in Geneva detailing the conservation management activities being undertaken in Mongolia under an MoU between EAD and the Mongolian Ministry of Nature, Environment and Tourism (MNET), which outlined a programme for establishing a sustainable Saker Falcon harvest based on the use of artificial nest sites. CITES endorsed the positive management regime, agreed to an

[2] http://cites.org/sites/default/files/eng/com/ac/21/E21-10.1.1.pdf

[3] http://cites.org/sites/default/files/eng/notif/2006/E061.pdf

[4] http://www.cites.org/eng/notif/2009/E003.pdf

[5] http://www.cites.org/sites/default/files/eng/com/sc/58/E58-21-1.pdf

export quota of 300 Saker Falcons for 2011 and concluded the significant trade review, enabling Mongolia to set its own future harvest quotas.[6]

IUCN Red List, CMS Appendix I Listing and the SakerGAP

The document issued by the UAE in 2003 that instigated the CITES Significant Trade Review also triggered a review of the status of the Saker Falcon on the IUCN Red List of Threatened Species. Relying heavily on much of the survey data gathered and collected by the UAE, BirdLife International, the official listing authority for birds for the IUCN Red List, revised the status of the Saker Falcon from 'Least Concern' to 'Endangered' in 2004. Thus, whilst remaining on Appendix II of CITES, as a species that 'may become threatened with extinction unless trade is closely controlled', the Saker Falcon was now considered to be 'facing a very high risk of extinction in the wild' by the International Union for Conservation of Nature (IUCN).

In December 2008, there was a proposal to include the Saker Falcon on Appendix I of the Convention on the Conservation of Migratory Species of Wild Animals (CMS), which would have the effect of prohibiting a harvest of this species in Mongolia. The proposal was not adopted, but there was a resolution recommending that the species should be listed on Appendix I if the Saker Falcon was considered to be 'Threatened' in the IUCN Red List by the time of the next Conference of Parties in 2011. The resolution further recommended that Parties to CMS should support a workshop to consider the status and conservation needs of the species, to be held in the UAE in 2009. The role of the UAE, or more specifically Abu Dhabi, in the political machinations surrounding the Mongolian Saker Falcon trade was not because they were a major market for the trade (in fact very few Mongolian birds were destined for the emirate), but because the Environment Agency-Abu Dhabi (EAD) had been engaged in research for several years to investigate the potential of developing a conservation management strategy for the Saker Falcon in Mongolia (see Dixon et al. 2011).

As the Review of Significant Trade process came to a conclusion with CITES, the issue of the Saker Falcon was still very much alive with CMS. The IUCN Red List review initiated in response to the CMS resolution in Rome resulted in the Saker Falcon being downlisted from Endangered (EN) to Vulnerable (VU) in 2010. However, this revised status didn't last long, as BirdLife International undertook another review the following year, this time as part of the mandatory IUCN review process, and restored the former Endangered status as a precautionary measure. Whether classified as EN or VU, the Saker Falcon was still regarded as Globally Threatened, thus the proposal to list the species on Appendix I of CMS was made at the Tenth CMS Conference of Parties in Bergen in November 2011. However, in order to reflect the decision of CITES a few months earlier, the proposal was

[6] http://www.cites.org/eng/com/ac/25/sum/E25-SumRec.pdf

amended to exclude the Saker Falcon population in Mongolia. The proposal was accepted after some debate and was accompanied by another resolution establishing a Saker Falcon Task Force to produce a Global Action Plan for the species. In August 2014, the CMS Saker Falcon Global Action Plan (SakerGAP) was published, a document of over 200 pages that incorporates a management and monitoring system for the species (Kovács et al. 2014).

The Mongolian 'Commercial' Trade Suspension

Having commanded international attention for several years, in 2012 attention was again focused at a national level in Mongolia. Following parliamentary elections in the summer, in November 2012 the Saker Falcon was officially declared to be the national bird of Mongolia. The Saker Falcon was selected after coming top of a poll organized by the National University of Mongolia. Shortly after, amid turbulent debate about the exploitation of national resources, the Mongolian government used the designation of the Saker Falcon as the national bird to announce that it was implementing a 5-year moratorium on the 'commercial trade' in the species. However, what exactly constitutes 'commercial trade' is unclear: in 2013 and 2014 Mongolia continued to host international falcon trappers and provide CITES permits for the harvest and export of Saker Falcons. In 2013, it was announced in the Mongolian media that the Government would permit the export of 20 falcons to Qatar and Kuwait, whilst in 2014 licenced falcon trappers were again operating in Mongolia with unconfirmed reports of at least 30 birds exported to recipients in Dubai, Kuwait and Qatar (Table 4.2).

Conservation Management and the Potential for a Sustainable Harvest of Saker Falcons in Mongolia

The intense, and sometimes rancorous, debate about the Mongolian Saker Falcon trade in national and international forums was not particularly conducive to the establishment of conservation management projects for the species. From 1994, foreign researchers conducted expedition surveys (e.g. Ellis et al. 1997, 2010, 2011) with some experimentation in the use of artificial nest structures from 1997, initially using power poles as support structures (Ellis 2010), then later specially constructed 'tripods' in flat, open landscapes (Potapov et al. 2003). Over a decade, the raw data upon which so much of the debate about the Saker Falcon trade revolved was primarily based on rather piecemeal and unsystematic surveys (Gombobaatar et al. 2007). Whilst the existing survey data was clearly inadequate, it was unlikely that accurate population estimates would ever be achieved given the logistical constraints of surveying in Mongolia, but the preliminary work on artificial nests did offer the opportunity of developing a more extensive management programme for

Table 4.2 Number of wild-caught Saker Falcons reported as exported from Mongolia (2000–2013) based on data from the UNEP-WCMC CITES Trade Database

	2000	2001	2002	2003	2004	2005	2006	2007	2008	2009	2010	2011	2012	2013
Kuwait	50	102	121	171	180	131	41	141	185	151	105			
Saudi Arabia			87	85	77	151	100	60	30	25				
Qatar			15	82	49	5	26	40	51	81	88		39	
Syria		10	75	54	49	73								
UAE		75		10	30					35	44			
Non-Gulf			20											
Totals	50	187	318	402	385	360	167	241[a]	266	292	237	0[b]	39	0[b]

[a] Mongolian CITES Management Authority reported the export of 407 Saker Falcons to the CITES Secretariat in 2007
[b] Exports of Saker Falcons took place but no data available on quantity

the species. Thus, over a 5-year period from 2005 to 2009, a pilot study was established by EAD to examine the potential of using artificial nests to create readily monitored breeding 'populations' in nest-site limited habitats (Dixon et al. 2011; Rahman et al. 2014).

The programme set out to create a new, managed breeding population occupying artificial nests that are amenable to monitoring, which can provide the data required to determine a sustainable harvest quota. A monitored population occupying artificial nests can provide data on breeding productivity (incorporating annual and regional variation), adult survival and breeding dispersal (based on breeding turnover), natal recruitment and dispersal, and the age composition of the breeding population. These vital statistics can be used to accurately model a sustainable harvest quota based entirely on the managed and monitored population.

The Creation of a New, Managed and Monitored Breeding Population Occupying Artificial Nests

The 2005–2009 pilot study utilized artificial nesting boxes made from metal drums that were erected on metal poles at a height of 2.5 m (Fig. 4.1). Breeding density and productivity levels recorded during this pilot study suggested that at least 5000 artificial nests would be required to create a new Saker Falcon breeding population that could contribute to a viable and sustainable falconry harvest. In 2009, work began to identify areas of the central Mongolian steppe that could accommodate 5000 artificial nests. The decision was made to allocate 250 nests to each of 20 districts across five provinces, with the precise locations being determined after field surveys and a consultation meeting with district administrators (Fig. 4.2). Each district had one or two grids of nests spaced at 1.5 km intervals in areas of open steppe where few natural nest sites existed, which limited any breeding population of Saker Falcons. Nonetheless, the selected areas held prey resources, in the form of small mammals and birds, which probably supported a pre-existing non-breeding population of Saker Falcons as predicted for nest-site limited raptor populations (Newton 1979).

The erection of 5000 artificial nests was completed in 2010 and annual monitoring began the following year to record occupancy and breeding success of Saker Falcons. The number of Saker Falcons breeding in the artificial nests increased annually over the first 4 years until 2014 (Fig. 4.3). The observed incremental annual increase during the initial 'colonization phase' is predicted when the breeding population in the artificial nests is derived primarily via 'limited' recruitment from a local non-breeding population, rather than by 'unlimited' immigration via breeding dispersal from pre-existing populations elsewhere. Saker Falcons breeding at artificial nests in the Mongolian steppe produce an average of 3.2 (±0.3) fledglings per nesting attempt (Rahman et al. 2014). In 2014, the newly created, managed and monitored population breeding in artificial nests produced in the region of 2500 fledglings; this number may increase further if the 'colonization phase' continues. The existing programme has demonstrated that Saker Falcons can be practicably managed at a scale amenable to developing a sustainable harvest.

Fig. 4.1 Pair of Saker Falcons at an artificial nest in the Mongolian steppe, Sukhbaatar province

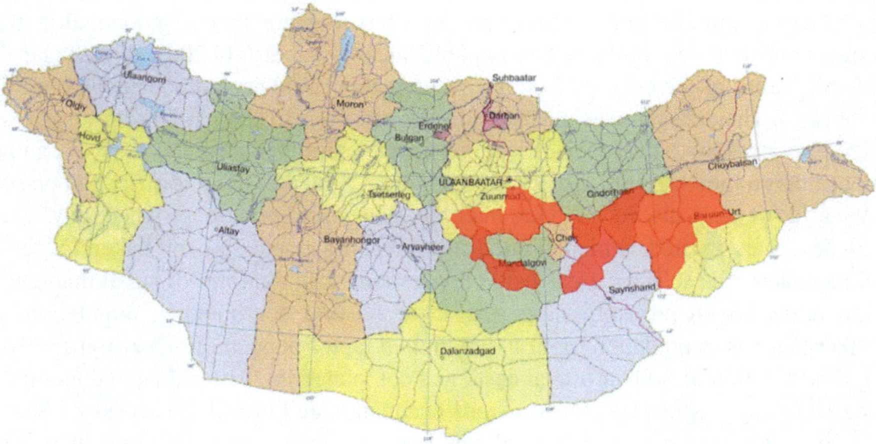

Fig. 4.2 Map of Mongolia showing districts where artificial nests were erected (*shaded red*; $N=250$ nests per district). Dornogovi province: Airag, Ikhkhet; Dundgovi province: Adaatsag, Saintsagaan/Mandalgovi, Gurvansaikhan; Khentiy province: Bayankhutag, Galshar, Bayanmunkh, Darkhan; Sukhbaatar province: Bayandelgar, Khalzan, Tuvshinshiree, Munkhkhaan, Uulbayan, Sukhbaatar and Töv province: Bayantsagaan, Bayan, Bayanjargalan, Bayanunjuul, Buren

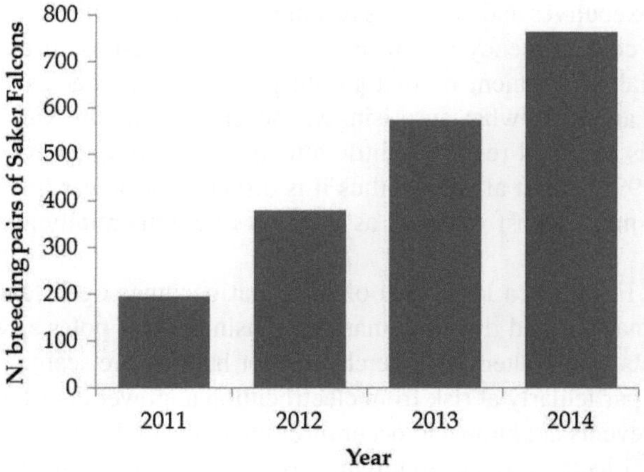

Fig. 4.3 Number of breeding pairs of Saker Falcons occupying artificial nests in Mongolia since their establishment in 2010

Saker Falcons and Electrocution at Power Lines

The expansion of electricity supply is synonymous with the socio-economic development of nations, and the last century has seen a dramatic and rapid increase in electricity infrastructure across the globe, including the erection of a vast network of electricity transmission and distribution lines. An electricity network typically comprises a series of high voltage transmission lines extending from a generation source, that then, via transformers, branches into lower voltage distribution lines that deliver power to human settlements and industrial centres. As more locations become connected to the power grid, the network of distribution lines grows, and it is these low- to medium-voltage distribution lines (≤15 kV) that pose the greatest electrocution risk to birds (Lehman et al. 2007). Birds that are large enough to span the distance between two phase conductor cables, or which are large enough to touch one phase conductor cable whilst standing on a grounded perch are electrocuted at power poles. Consequently, larger birds are more prone to electrocution than smaller birds, and the risk is higher for birds that frequently perch in elevated positions, such as raptors that use power poles as vantage points for hunting ground-dwelling prey (Bevanger 1998; Lehman et al. 1999; Janss 2000).

Medium-voltage electricity distribution lines can, where pole hardware is inappropriately configured, present a serious electrocution risk for birds of prey. The issue is considered to be serious enough for CMS to review the problem and produce guidance for mitigation (Prinsen et al. 2011a, b). Mitigation techniques are available to reduce the risk of electrocution for birds of prey at existing dangerous electricity distribution lines, which include deterrents that are designed to prevent birds perching in high-risk locations, insulation covers for live phases and reconfiguration of cable-carrying hardware. However, a lack of knowledge about the issue among executives and engineers responsible for power lines, together with the exigencies of cost efficiency has meant that in many circumstances mitigation, or even the initial deployment of raptor safe power lines, receives little attention. Furthermore, and somewhat surprisingly, the efficacy of the various mitigation techniques has received relatively little attention from researchers (but see Janss and Ferrer 1999; Guil et al. 2011), thus it is difficult for power line engineers and executives to make a cost v benefit assessment of commercially available mitigation products.

The Saker Falcon is a large bird of prey that occupies open landscapes and it often hunts small ground-dwelling mammals, using power poles as elevated perch sites in habitats where alternative perch sites for hunting are scarce. Consequently, the species is particularly at risk from electrocution at power distribution lines, and electrocution events are known to occur throughout the global breeding distribution of the species. Electrocution is highlighted as a mortality factor for Saker Falcons in Europe (Ragyov et al. 2012), whilst surveys in Russia, Kazakhstan, Mongolia and China indicate that many lines pose a significant risk and that mortality levels are high in Asia (Karyakin et al. 2008; Lasch et al. 2010; Dixon et al. 2013). The following case study reports on the scale of Saker Falcon mortality due to electrocution in Mongolia, which results in losses an order of magnitude greater than that arising from the Saker Falcon trade discussed earlier. I further discuss how the commodification of the Saker Falcon, often considered to be a 'wildlife problem', may in fact provide a mechanism whereby the species can generate the finances to pay for its own conservation.

Case Study II: Electrocution of the Saker Falcon in Mongolia

Factors Influencing Saker Falcon Electrocution Rates in Mongolia

Electrocution rates can be expressed in various ways, such as the number of birds killed per *N* poles or per *N* km of power line over a specified period of time. These rates can be determined by the integral structure of the power pole hardware, with some poles posing a higher risk of electrocution than others, whilst additional extraneous factors can also play a significant role such as surrounding landscape and habitat characteristics, and temporal or spatial variation in the number of birds exposed to the electrocution risk (Guil et al. 2011). Measuring these rates is further complicated by the fact that accurate quantification of electrocution events is dependent on search effort, efficacy and rates of carcass removal by scavengers (Ponce et al. 2010). Consequently, despite the geographical widespread nature of bird of prey electrocution and the large number of research studies undertaken, there are still relatively few studies that report quantitative measures for electrocution rates (Lehman et al. 2007).

Fig. 4.4 Adult male Saker Falcon electrocuted at an anchor pole with a jump wire passing over the crossarm, Mongolia March 2012

Pole design has been shown to be the main factor accounting for variation in electrocution rates between power poles, for example, grounded anchor poles with jump wires that pass over crossarms are especially dangerous for raptors (Fig. 4.4). Predictive models of avian electrocution risk at power lines typically incorporate parameters such as pole type and hardware configuration, as well as geographical location, surrounding habitat and topography, and prey availability (e.g. Tintó et al. 2010; Dwyer et al. 2013; Harness et al. 2013). However, the predictive ability of such models to identify priority high risk lines can be compromised if prey availability varies greatly in time and space. Prey abundance in the vicinity of power lines can influence electrocution rates by attracting birds of prey, where they may use the power poles as perches for hunting or loafing (Lammers and Collopy 2007). Diurnal small mammals prevalent in the Mongolian steppe, such as Brandt's Vole (*Lasiopodomys brandtii*), Mongolian Gerbil (*Meriones unguiculatus*) and Daurian Pika (*Ochotona dauurica*) exhibit large population fluctuations (Smith and Xie 2008), and their abundance can vary greatly over time and space.

Scale of Electrocution of Saker Falcons in Mongolia

How many Saker Falcons are electrocuted in Mongolia each year? This simple question does not have a simple answer. In the open steppe, the most important factor is the concomitance of abundant small mammal populations and dangerous power poles. Single-visit power line surveys targeted at 15 kV lines with poles known to be of a dangerous design, i.e. grounded steel-reinforced concrete poles

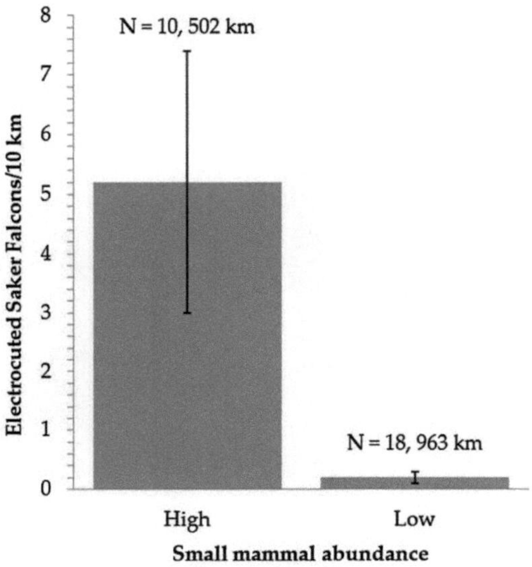

Fig. 4.5 The number of Saker Falcon carcasses found per 10 km during single-visit power line surveys across the Mongolian steppe in early autumn. Standard error bars are shown. N = the number of power lines surveyed, with their total distance (km). Local abundance of small mammals was assessed qualitatively as 'high' or 'low', based on the number of holes and animals seen during the survey visit

with metal crossarms, were conducted across Mongolia in late August 2014. During these surveys, 317 electrocuted carcasses of Saker Falcons were found below poles at power lines in areas of open steppe where small mammal densities were classified as high, yet just 25 were found at power lines where small mammal density was low (Fig. 4.5). Each carcass, based on its state of decay, was considered to have been electrocuted within the previous month (see Dixon et al. 2013).

Single-visit line surveys, whilst enabling extensive geographical coverage to allow comparison of power lines at a national scale, can only provide absolute minimum estimates of electrocution rates as it is not possible to take into consideration carcass removal by scavengers. However, high frequency survey visits can limit the influence of carcass removal on electrocution rate estimates. An additional study conducted over a 592 days with an average survey frequency of 1.1 days found 251 Saker Falcon carcasses at poles on a 26 km stretch of power line running through an area of the Mongolian steppe with a high density of small mammals. This survey found that in August, Saker Falcons were electrocuted at a mean rate of 8.5 (SE ± 5.2) birds per 10 km, which was within the range found for the single visit estimates (Fig. 4.6). Furthermore, there was marked variation in the number of Saker Falcons electrocuted during each calendar month of the year. Numbers increased from June to September, as fledglings dispersed from their nesting sites and aggregated in areas of high small mammal density, with a rapid decline from September to November as many birds, particularly juveniles, migrated to wintering areas in China. Electrocution rates remained low over winter but rose again in April when migrants, especially second-year birds, returned to the Mongolian steppe, with a subsequent gradual decline as the breeding season progressed.

The average length of 15 kV power lines in Mongolian is 52 km (N = 41 surveyed lines), and in a high density small mammal area an average line will kill an

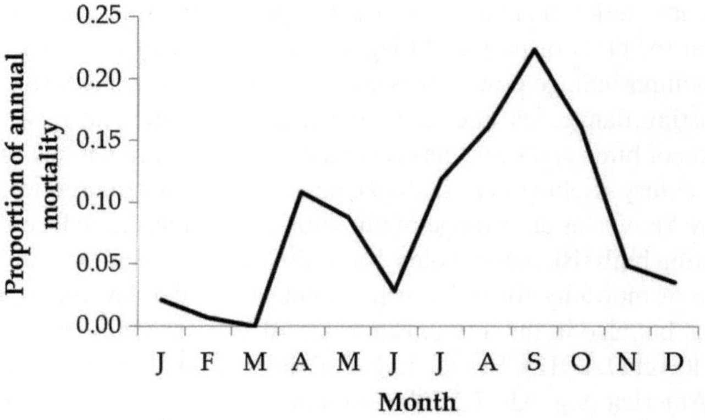

Fig. 4.6 Annual variation in monthly electrocution rates along 26 km of a power line in a region with a high small mammal density. Survey period was from 01 April 2013 to 15 August 2014, with 438 line surveys conducted at an average frequency of 1.1 day intervals. 251 electrocuted Saker Falcons were collected, equating to an average of 143 birds over a 12-month period

estimated 27 Saker Falcons per month in August (95 % CI=5–50 electrocutions/month), compared with just 1 at a corresponding power line in an area where small mammal density is low (95 % CI=0–3 electrocutions/month). August mortality represents ca. 16 % of annual electrocutions (Fig. 4.5), thus we can estimate that in areas of high small mammal density an average dangerous power line will kill 168 Saker Falcons (95 % CI=31–311 electrocutions/year) and in areas with low small mammal density an average line will kill an estimated 6 individuals (95 % CI=0–19 electrocutions/year). There are at least 65 'dangerous' 15 kV lines in Mongolia (Tserennyam 2013), with just over one-third occurring in open steppe areas with high small mammal densities (Fig. 4.5), thus an estimated 23 power lines are likely to coincide with high small mammal population areas and a further 42 are likely to traverse areas with lower small mammal densities. Consequently, the estimated number of electrocuted Saker Falcons across Mongolia over 1 year is 4116 individuals (90 % CI=713–7951 birds). To put this in context, the mean estimate of electrocution events in Mongolia exceeds by an order of magnitude the maximum number of Saker Falcons issued with CITES permits and exported annually from Mongolia for the Arabian falconry market.

Remediation and Reducing the Risk of Electrocution for Saker Falcons

While there is, at present, no evidence to demonstrate that the existing high level of electrocution is having an impact on the Saker Falcon population in Mongolia, the number of birds killed is alarming and the scale of the problem clearly has the potential to have a detrimental impact on the species. In order to reduce the risk of electrocution for birds of prey in Mongolia, it is necessary to (1) ensure that new low- and medium-voltage power lines comprise poles that are of a safe design, and that (2) existing dangerous lines are subject to remediation to make them safer. Electrocution of birds at power lines is not a new phenomenon, it has been recorded for over a century (Lehman et al. 1999), and the designs required to make power lines safe are known, as are a range of mitigation techniques to reduce electrocution risk on existing high risk power poles. Nevertheless, electrocution continues to be a major cause of mortality for birds of prey, not just in the developing countries of Central Asia, but also in India (e.g. Harness et al. 2013), Africa (e.g. Boshoff et al. 2011; Angelov et al. 2012), Europe (e.g. Guil et al. 2011; López-López et al. 2011) and North America (e.g. APLIC 2006; Kemper et al. 2013). Clearly, impediments exist across the globe that prevents the implementation of measures that could significantly diminish the risk of electrocution to birds posed by low- and medium-voltage power lines.

In Mongolia, it is apparent that there is a lack of awareness about the issue of bird of prey electrocution among key personnel involved in planning and constructing power lines; it is a relatively recent problem connected with the use of prefabricated reinforced concrete power poles and their associated hardware rather than the Soviet-style wooden poles favoured prior to the Democratic Revolution of 1990. The concrete poles and hardware are cheaper to purchase, more resilient to steppe fires, easier to erect and offer additional benefits associated with standardization of infrastructure and economies of scale. In a rapidly developing economy with an urgent need to create a modern power network to connect communities to the electricity grid and provide a secure power supply, it is perhaps not a surprise that the potential risk of electrocution to birds of prey has been overlooked in the process of infrastructure procurement in Mongolia. Furthermore, the procedure of undertaking environmental impact assessment is not well developed in the country at present (Dondov 2010), and there is an absence of statutory regulation pertinent to the issue of electrocution of birds at power lines.

A lack of awareness does not mean complete ignorance of the problem, as several of the state-operated power companies have instigated measures in order to reduce electrocution risk at many power lines, usually in the form of perch deflectors, such a spikes and brushes, or perch deterrents such as rotating mirrors. However, their installation is often haphazard, reflecting a lack of understanding of how these deflectors and deterrents are designed to work and in what circumstances. Consequently, it is not unusual to find large brush-spike perch deflectors designed to prevent birds perching in specific locations above chain insulars on high voltage transmission lines, instead being used on the crossarms of low- and medium-voltage lines. With this inappropriate placement, such deflectors not only potentially increase the risk of electrocution as they themselves are grounded but they can also pose an entanglement risk to perching birds. When positioned correctly on a crossarm, i.e. adjacent to insulators carrying the live phase wires, insulated single-spike deflectors can reduce electrocution rates (Dixon et al. 2013), but when inappropriately positioned they may increase electrocution risk by deflecting birds to perch closer to the phase wires (Amartuvshin and Gombobaatar 2012). Perch deterrents, such as rotating mirrors, are only effective as long the equipment remains functional, but often the extreme conditions experienced in Mongolia means that devices such as rotating mirrors frequently breakdown after relatively short periods.

At least some funding to the state-operated electricity companies is evidently available for the deployment of mitigation to reduce the risk of bird of prey electrocution. Certainly, electrocution events can potentially result in power outages that require line repairs, thus there is some economic cost associated with the problem. Nevertheless, cost is also an important factor governing the scale of any future remediation and mitigation programme across Mongolia. Government instability and a roller-coaster economic cycle based on the exploitation of mineral resources has characterized Mongolia in recent years (Rolle 2014), while a myriad pressing social and economic problems requires urgent action and resources. The electrocution of tens of thousands of birds of prey each year is just one issue among many that is competing for attention and funding from government officials; to date it is not a problem that has succeeded in making much headway in this competition for resources.

The Saker Falcon as a Commodity: Can It Pay for Its Own Conservation?

Saker Falcons have a monetary value; the Mongolian government has in recent years charged fees of ca. €10,000 per bird in order for trappers to catch and export birds for Arabian falconry. The cost to the end user is higher as they also have to cover the cost of employing trappers, and cover the logistical costs of trapping and transporting the falcons. This commodification of the species can be viewed as a problem, which is driving a market demand that has fuelled an unsustainable and largely illegal harvest across Central Asia, the Middle East, Maghreb and beyond, causing large-scale regional population declines. The response has typically been to instigate trade bans and demand greater compliance and enforcement of the associated national and international regulations that have been created to support such bans. However, this approach, in the case of the Saker Falcon and several other charismatic wildlife species that have become wildlife commodities, has largely failed to reduce their exploitation or market demand (Bennett 2011).

Arabian falconers, as the end users and ultimate cause of the commodification of Saker Falcons, have no incentive to see the species decline or become extinct in the wild. Quite the opposite, a thriving wild population has advantages in securing a long-term supply of falcons that have the phenotypically diverse characteristics desired by Arabian falconers. Commodification offers an opportunity, whereby Arabian falconers can contribute directly to conservation of the species by funding the procedures required in Mongolia to generate long-term, sustainable production via artificial nests and to limit mortality rates by remediation of dangerous power lines. Depending on the type of mitigation employed, electrocution can be significantly reduced at a cost of €20 to €200 per pole, thus the income generated by a single Saker Falcon could potentially pay for the mitigation of all poles on an average 52 km long 15 kV power line. Furthermore, this funding mechanism could have associated benefits of raising awareness of conservation issues in the countries of production and consumption.

The sustainable use of biological resources is enshrined within the Convention on Biological Diversity (CBD) and the concept of conservation through use for falconry species has been developed and articulated by representatives of bodies such as the IUCN's Sustainable Use and Livelihoods Specialist Group (Kenward 2004, 2009). The recently published Saker Falcon Global Action Plan produced by CMS (Kovács et al. 2014), perhaps represents early tentative steps towards a wider recognition among conservationists and policy makers that the exploitative use of wild-sourced falcons is not simply a conservation problem that requires prohibition of trade but rather that a regulated and sustainable trade potentially provides a mechanism for generating scarce conservation resources.

Acknowledgments My research on Saker Falcons has been undertaken for, and is funded by, the Environment Agency-Abu Dhabi. I thank HE Mohammed Al Bowardi for his interest and support. I wish to thank my colleagues and field assistants in Mongolia, particularly B. Galtbalt, N. Batbayar, B. Davaasuren, B. Bold, S. Inchinkherloo, A. Gunga, A. Saruul, A. Lhagvajav and G. Purev-Ochir. I also thank my colleagues M. L. Rahman and N. C. Fox at IWC Ltd. for discussions on the ecology, conservation and trade of Saker Falcons.

References

Allen M (1980) Falconry in Arabia. Stephen Greene Press, Lexington

Amartuvshin P, Gombobaatar S (2012) The assessment of high risk utility lines and conservation of globally threatened pole nesting steppe raptors in Mongolia. Ornis Mongolica 1:2–12

Angelov I, Hashim I, Oppel S (2012) Persistent electrocution mortality of Egyptian vultures Neophron percnopterus over 28 years in East Africa. Bird Conserv Int 23:1–6

APLIC—Avian Power Line Interaction Committee (2006) Suggested practices for avian protection on power lines: the state of the art in 2006. Edison Electric Institute, Washington, DC, Avian Power Line, Interaction Committee, and the California Energy Commission, Sacramento

Bagyura J, Szitta T, Haraszthy L et al (2012) Results of the Saker Falcon (Falco cherrug) conservation programme in Hungary between 1980–2010. Aquila 119:105–110

Barton NWH (2000) Trapping estimates for Saker and Peregrine Falcons used for falconry in the United Arab Emirates. J Raptor Res 34:53–55

Baumgart W (1978) Der Sakerfalke. Die Neue Brehm-Bücherei, Wittenberg Lutherstadt

Bennett EL (2011) Another inconvenient truth: the failure of enforcement systems to save charismatic species. Oryx 45:476–479

Bevanger K (1998) Biological and conservation aspects of bird mortality caused by electricity power lines: a review. Biol Conserv 86:67–76

BirdLife International (2014) Species factsheet: Falco cherrug. http://www.birdlife.org. Accessed 05/12/2014

Boldbaatar S (2009) Influence of commercial activity on number and condition of population Saker Falcon (Falco cherrug Gray, 1834) to Mongolia and problems of its protection. In: Andreeva TF, Vershinina TA, Karpov NV, Kuzmina LV (eds) Birds of prey and owls in zoos and breeding stations: 18th issue, Moscow, Russia

Boshoff AF, Minnie JC, Tambling CJ et al (2011) The impact of power line-related mortality on the Cape Vulture Gyps coprotheres in a part of its range, with an emphasis on electrocution. Bird Conserv Int 21:311–327

Ceballos J (2009) Falconry. Celebrating a living heritage. Motivate Publishing, Dubai

Chavko J (2010) Trend and conservation of Saker Falcon (Falco cherrug) in Western Slovakia between 1976 and 2010. Slovak Raptor J 4:1–19

Dixon A (2009) Saker Falcon breeding population estimates. Part 2: Asia. Falco 33:4–10

Dixon A (2012) Conservation of the Saker Falcon Falco cherrug and the use of hybrids for falconry. Aquila 119:9–19

Dixon A, Ragyov D, Ayas Z et al (2009) Population status of breeding Saker Falcons (Falco cherrug) in Turkey. Avian Biol Res 2:213–220

Dixon A, Batbayar N, Purev-Ochir G et al (2011) Developing a sustainable harvest of Saker Falcons (Falco cherrug) for falconry in Mongolia. In: Watson RT, Cade TJ, Fuller M, Hunt G, Potapov E (eds) Gyrfalcons and Ptarmigan in a changing World. Peregrine Fund, Boise. 10.4080/gpcw.2011.0315

Dixon A, Maming R, Gunga A et al (2013) The problem of raptor electrocution in Asia: case studies from Mongolia and China. Bird Conserv Int 23:520–529

Dondov Z (2010) Environmental impact assessment in Mongolia: critical analysis and recommendations for improvement. MSc thesis, Department of Environmental Sciences and Policy of Central European University, Budapest

Dwyer JF, Harness RE, Donohue K (2013) Predictive model of avian electrocution risk on overhead power lines. Conserv Biol 28:159–168

Ellis DH (2010) The Institute for Raptor Studies expeditions in Mongolia, 1994–2000. Erforschung biologische ressourcen der Mongolischen volkrepublik 11:189–212

Ellis DH, Ellis MH, Tsengeg P (1997) Remarkable Saker Falcon (Falco cherrug) breeding records for Mongolia. J Raptor Res 31:234–240

Ellis DH, Roundy TB, Ellis CH (2010) Raptor pit mortality in Mongolia and a call to identify and modify death traps wherever they occur. Ambio 39(4):349–351

Ellis DH, Kitowski I, Roundy TB (2011) Nomadism in large falcons: lessons from the Saker Falcon in Mongolia. In: Watson RT, Cade TJ, Fuller M, Hunt G, Potapov E (eds) Gyrfalcons and Ptarmigan in a changing world. Peregrine Fund, Boise

Gombobaatar S, Sumiya D, Potapov E et al (2007) Breeding biology of the Saker Falcon in Mongolia. Raptors Conserv 9:17–26

Guil F, Fernández-Olalla M, Moreno-Opo R et al (2011) Minimising mortality in endangered raptors due to power lines: the importance of spatial aggregation to optimize the application of mitigation measures. PLoS One 6(11), e28212. doi:10.1371/journal.pone.0028212

Harness RE, Juvvadi PR, Dwyer JF (2013) Avian electrocutions in Western Rajasthan, India. J Raptor Res 47:352–364

Janss GFE (2000) Avian mortality from power lines: a morphologic approach of a species-specific mortality. Biol Conserv 95:353–359

Janss GFE, Ferrer M (1999) Mitigation of raptor electrocution on steel power poles. Wildl Soc Bull 27:263–273

Karyakin IV (2008) Saker Falcon in Russia. Raptors Conserv 12:28–47

Karyakin IV, Levashkin AP, Glybina MA et al (2008) Use of GIS techniques in estimating the level of birds of prey electrocution on 6-10 kV power lines in the Kinel region of the Samara District. Raptors Conserv 14:50–58

Kemper CM, Court GS, Beck JA (2013) Estimating raptor electrocution mortality on distribution power lines in Alberta. Can J Wildl Manage. doi:10.1002/jwmg.586

Kenward RE (2004) Management tools for raptor conservation. In: Chancellor RD, Meyburg B-U (eds) Raptors worldwide. World Working Group on Birds of Prey and Owls, Berlin

Kenward RE (2009) Conservation values from falconry. In: Dickson B, Hutton J, Adams B (eds) Recreational hunting, conservation and rural livelihoods: science and practice, 1st edn. Blackwell, Hoboken

Kenward R, Katzner T, Wink M, Marcström V, Walls S, Karlbom M, Pfeffer R, Bragin E, Hodder K, Levin A (2007) Rapid sustainability modeling for raptors by radiotagging and DNA-fingerprinting. J Wildl Manage 71:238–245

Kohn M (2006) Dateline Mongolia. RDR Books, Muskegon

Kovács A, Williams NP, Galbraith CA (2014) Saker Falcon Falco cherrug Global Action Plan (SakerGAP), including a management and monitoring system, to conserve the species. Raptors MOU technical publication no. 2. CMS technical series no. 31. Coordinating Unit—CMS Raptors MOU, Abu Dhabi

Lammers WM, Collopy MW (2007) Effectiveness of avian predator perch deterrents on electric transmission lines. J Wildl Manage 71:2752–2758

Lasch U, Zerbe S, Lenk M (2010) Electrocution of raptors at power lines in Central Kazakhstan. Forest Ecol Landsc Nat Conserv 9:95–100

Lehman RN, Ansell AR, Garrett MG et al (1999) Suggested practices for raptor protection on power lines: the American story. In: Ferrer M, Janss GFE (eds) Birds and power lines: collision, electrocution and breeding. Quercus, Madrid

Lehman RN, Kennedy PL, Savidge JA (2007) The state of the art in raptor electrocution research: a global review. Biol Conserv 136:159–174

Levin AS (2011) Illegal trade and decrease in numbers of the Saker Falcon in Kazakhstan. Raptors Conserv 23:64–73

López-López P, Ferrer M, Madero A et al (2011) Solving man-induced large-scale conservation problems: the Spanish Imperial Eagle and power lines. PLoS One 6, e17196

Milobog YV, Vetrov VV, Strigunov VI et al (2010) The Saker (Falco cherrug Gray) in Ukraine and adjacent areas. Branta 13:135–159

Newton I (1979) Population ecology of raptors. T & AD Poyser, London

Orta J, Boesman P, Sharpe CJ (2014) Saker Falcon (Falco cherrug). In: del Hoyo J, Elliott A, Sargatal J, Christie DA, de Juana E (eds) Handbook of the birds of the world alive. Lynx Edicions, Barcelona

Ponce C, Alonso JC, Argandona G et al (2010) Carcass removal by scavengers and search accuracy affect bird mortality estimates at power lines. Anim Conserv 13:603–612

Potapov E, Sumya D, Shagdarsuren O et al (2003) Saker farming in wild habitats: progress to date. Falco 22:5–7

Prinsen HAM, Boere GC, Píres N et al (2011a) Review of the conflict between migratory birds and electricity power grids in the African-Eurasian region. CMS technical series no. XX, AEWA technical series no. XX, Bonn, Germany

Prinsen HAM, Smallie JJ, Boere GC et al (2011b) Guidelines on how to avoid or mitigate impact of electricity power grids on migratory birds in the African-Eurasian region. CMS technical series no. XX, AEWA technical series no. XX, Bonn, Germany

Prommer M, Bagyura J, Chavko J et al (2012) Migratory movements of Central and Eastern European Saker Falcons (Falco cherrug) from juvenile dispersal to adulthood. Aquila 119:111–134

Ragyov D, Koshev Y, Kmetova E et al (2012) Preparatory activities for Saker Falcon (Falco cherrug) reintroduction in Bulgaria: habitat management and electrocution risk assessment. Aquila 119:91–103

Ragyov D, Biserkov V, Gradev G et al (2014) Past and present status of the Saker Falcon, Falco cherrug (Aves: Falconidae) in Bulgaria. Acta Zool Bulg 66:299–308

Rahman ML, Purev-Ochir G, Etheridge M et al (2014) The potential use of artificial nests for the management and sustainable utilization of saker falcons (Falco cherrug). J Ornithol 155:649–656

Riddle KE, Remple JD (1994) Use of the Saker and other large falcons in Middle East falconry. In: Meyburg B-U, Chancellor RD (eds) Raptor conservation today. WWGBP, Berlin

Rolle K (2014) The Mongolian political administration: effectiveness and corruption. European Institute for Asian Studies, Briefing Paper 01, Feb 2014

Seddon PJ, Launay F (2008) Arab Falconry: changes, challenges and conservation opportunities of an ancient art. In: Lovelock B (ed) Tourism and the consumption of wildlife: hunting, shooting and sport fishing. Routledge, London

Shobrak MY (2014) Trapping of Saker Falcon Falco cherrug and Peregrine Falcon Falco peregrinus in Saudi Arabia: implications for biodiversity conservation. Saudi J Biol Sci. 10.1016/j.sjbs.2014.11.024

Smith AT, Xie Y (2008) A guide to the mammals of China. Princeton University Press, Princeton

Tintó A, Real J, Manosa S (2010) Predicting and correcting electrocution of birds in Mediterranean areas. J Wildl Manage 74:1852–1862

Tserennyam I (2013) Energy system of Mongolia. Gazryn Zurag, Sukhbaatar

Upton R (2002) Arab falconry. History of a way of life. Hancock House, Surrey

Wakefield S (2012) Falconry as heritage in the United Arab Emirates. World Archaeol 44:280–290

Wyatt T (2009) Exploring the organization of Russia Far East's Illegal wildlife trade: two case studies of the illegal fur and illegal falcon trades. Global Crime 10:144–154

Zahler P, Lhagvasuren B, Reading RP et al (2004) Illegal and unsustainable wildlife hunting and trade in Mongolia. Mongolian J Biol Sci 2(2):23–31

Chapter 5
Towards Extinction and Back: Decline and Recovery of Otter Populations in Italy

Alessandro Balestrieri, Luigi Remonti, and Claudio Prigioni

Introduction

Since the late 1950s, Eurasian otter (*Lutra lutra*) populations have crashed in large areas of the central-western part of their range because of a combination of factors such as water pollution, declining food resources, destruction of riparian vegetation and suitable breeding sites, hunting and road traffic mortality (Mason and Macdonald 1986; Macdonald and Mason 1994). POPs (Persistent Organic Pollutants) have been determined as a major factor (Ruiz-Olmo et al. 2000) and wherever their use has been forbidden or controlled by regulations, otter populations have gradually recovered, e.g. in the United Kingdom (Crawford 2003), Denmark (Elmeros et al. 2006), France (Janssens et al. 2006) and Spain (Cortés et al. 1998; Ruiz-Olmo and Delibes 1998).

The trend of the Italian otter population partially reflects this global pattern (Prigioni et al. 2007). In the first half of the nineteenth century, the otter was widespread along almost all main watercourses (Ghigi 1911; Cagnolaro et al. 1975). The first national survey carried out by questionnaires in 1968–1972 (Cagnolaro et al. 1975; Spagnesi and Cagnolaro 1981) confirmed positive sites for only a fraction of

A. Balestrieri (✉)
Department of Biosciences, University of Milan, via Celoria 26, Milan 20133, Italy
e-mail: alebls@libero.it

L. Remonti
Department of Earth and Environmental Sciences, University of Pavia, via Taramelli 24, Pavia 27100, Italy

Institut Agricole Régional, Regione La Rochère 1/A, Aosta 11100, Italy
e-mail: luigi.remonti@gmail.com

C. Prigioni
Department of Earth and Environmental Sciences, University of Pavia, via Taramelli 24, Pavia 27100, Italy
e-mail: claudio.prigioni@unipv.it

F.M. Angelici (ed.), *Problematic Wildlife*, DOI 10.1007/978-3-319-22246-2_5

the historical area of otter presence (56.4 % in northern Italy, 78 % in the central part of the peninsula and 70 % in the south; Prigioni et al. 2007).

Between 1968–1972 and 1984–1985, when the whole country was surveyed for otters by adopting the widely used "standard method" (Cassola 1986), otter range was shown to have suffered a more severe decline, particularly in the north of the country (Prigioni et al. 2007). In the 1980s, the Italian otter population consisted of five major isolated nuclei; Fumagalli and Prigioni (1993) predicted the extinction of the largest one, comprising more than 50 % of the whole population, within 43–53 years (Fig. 5.1).

At the turn of the twenty-first century, standard field surveys became more frequent and widespread, allowing a detailed and continuously updated picture of otter distribution to be obtained (Reggiani et al. 1997; Cripezzi et al. 2001; Loy et al. 2003; Fusillo et al. 2003; Prigioni et al. 2005, 2009; Balestrieri et al. 2008). The otter became extinct in northern Italy and further declined in the central part of the

Fig. 5.1 Otter range in the Italian peninsula in the 1980s

peninsula, excepting for the Molise region, while, unexpectedly, signs of a recovery were recorded in southern Italy. Here, a positive trend could be drawn for four regions: Campania, Basilicata, Apulia and Calabria (Prigioni et al. 2007). More recently, the otter is recolonizing north-eastern Italy from neighbouring countries.

Currently, because population size is lower than 1000 individuals (Prigioni et al. 2006a) and with a negative trend in northern and central Italy, the otter is listed in the Italian Red List of endangered species as "Endangered" (EN, according to IUCN criteria). River modification is still a major threat to otters in Italy, where morphological degradation and pollution of watercourses can affect otter occurrence and distribution (Panzacchi et al. 2010; Scorpio et al. 2014), and this is likely to drive the ongoing recolonization of empty river stretches.

We summarize the available knowledge on recolonizations using case studies covering the whole peninsula, which shed light on the factors affecting otter current distribution.

Case Studies

Otter Reintroduction on the River Ticino (N Italy)

In Lombardy (northern Italy), the Eurasian otter probably became extinct in the 1980s (Prigioni 1983, 1986a). Since the end of the 1970s, the reintroduction of the otter in Lombardy has been advocated many times (Prigioni et al. 1979) and a feasibility study was carried out in the valley of the River Ticino (Mason et al. 1985; Prigioni 1986b, 1995). This river had been identified as a potential reintroduction area in the Action Plan for the Conservation of European Otters (Macdonald and Mason 1991) and is currently protected by two Regional Parks, which cover a total area of 968.9 km^2.

A pair of otters was released in 1997 from a breeding centre located on the Piedmont side of the river (Montanari and Boffino 2000). Unfortunately, these individuals were not monitored to check their fate and assess the success of the reintroduction. A further pair possibly escaped before the end of the century (Prigioni, personal communication). Additionally, during floods in 1991 and 1993, respectively, a pair with a cub, and a pair with two sub-adults escaped from their breeding enclosure on the Lombardy side of the river ("La Fagiana", about 15 km downstream of the first centre). All these animals were, however, recaptured within a few months (Prigioni, unpublished report).

During summer 2008, an otter survey was carried out using the "standard method" with the aim of drawing a picture of the distribution of the species in the reintroduction area. Otter presence was recorded at only three out of ten sampling stations, approximately corresponding to a 2.6 km section of the River Ticino, next to the release site (Prigioni et al. 2009).

In 2010, the survey was extended to cover a 35 km-long stretch of the river centred on the release site. Otter spraints were found along a 7 km stretch of the river, mainly on canals and secondary arms. Finally, in late summer 2012, an adult otter was recorded about 30 km downstream the release site (Meriggi and Bellati, personal communication), while in winter 2013 two individuals were seen a further 10 km downstream (Cavalleroni, personal communication).

Currently, otter reinforcement in the River Ticino valley is still debated: the genetic composition of introduced animals has to be clarified, previous feasibility studies reviewed in the light of environmental changes that have occurred in the last 15 years (fish assemblage, pollutants, etc.), while a detailed survey of the river is needed to assess the current otter range and population size (Conroy, unpublished workshop report).

Otter Recolonization of NE Italy

Since legal protection was assured in the 1980s, otter populations in central and southern Austria have rapidly recovered, and, currently, in the Carinzia and Styria regions about 30 % of surveyed sites are positive for otters (Kranz and Polednik 2009). In contrast, on the Italian side of the eastern Alps, the otters went extinct in the Alto Adige region in the late 1950s and in the Friuli Venezia Giulia region in the 1960s (Lapini and Bonesi 2011).

As a consequence of otter range expansion in southern Austria, otter signs have been found since 2008 in South Tyrol, next to the Austrian border, (Kranz 2008), and nowadays the species occurs on the Italian stretch of the River Drava/Drau, a 707 km-long tributary of the River Danube, and one of its tributaries, the stream "Rio Sesto" (Righetti 2011).

In 1984 (Lapini 1986) and 2008 (Lapini and Bonesi 2011), otter spraints were found on the river Natisone, Friuli Venezia Giulia region. In the same region, an adult male otter was killed on the road in the catchment of the River Tagliamento in 2011, and a second otter in 2012, about 10 km away from the first record (Pavanello et al. 2015). Both animals were probably wandering individuals from the nearest otter population, which occurs in Slovenia, in the catchment of the River Drava, about 200 km away from the Italian border. Recently, an otter survey was carried out on the upstream and central catchment of the River Tagliamento and on some tributaries of the River Drava, searching for otter spraints at bridges, which are frequently used by otters as sprainting sites. Sites positive for otters were found for the latter area, in the catchment of the River Slizza-Gailitz, and on both the Italian and Slovenian sides of the Rateče swamp. The tracks of one adult otter with two cubs were found on a right-bank tributary of the River Slizza. Subsequently, camera-trapping confirmed the presence of a reproductive female (Pavanello et al. 2015).

The further expansion of the otter in NE Italy is probably hindered by the widespread alteration and urbanization of watercourses and their banks in the upper catchment of the River Tagliamento.

The Central-Italian Otter Population

In central Italy, otters were once common and widespread in almost all watercourses (Altobello 1921). The species' decline was first documented by Macdonald and Mason (1983), and later by Pellegrini and Febbo (1986). At the end of the 1980s, only a few individuals still occurred, on the rivers Volturno and Biferno.

In 2002–2004 and 2011–2013, two standard field surveys were carried out in Molise region to assess the current status and distribution of the otter. The first survey recorded the otter mostly along the same rivers. As these catchments are completely isolated from the main otter range in southern Italy, the authors concluded that this population was the result of the persistence of a remnant population and outlined its importance as a source of individuals for the recovery of the otter in central Italy (Loy et al. 2004). In 2006, a survey on the River Sangro found the first otter signs in the northern part of the region (De Castro and Loy 2007).

The latest survey produced a rather different picture, with otters widespread on the rivers Fortore (total length of the watercourse hosting otters: 140 vs. 28 km in 2004), Sangro (205 vs. 14 km) and Biferno (64 vs. 78 km), while a reduction in otter range was assessed for the River Volturno (44 vs. 121 km; Lerone et al. 2014a).

In contrast, otters are still absent in the catchments of the rivers Trigno, Saccione and Sinarca, although habitat conditions suggest that the first of these watercourses is suitable for otter colonization (Loy et al. 2009, 2010).

Surveying the Core Area of the Italian Otter Range: The Pollino Massif

In the 1980s, the core Italian otter range was reduced to a few southern river catchments, mostly in Basilicata region (Prigioni et al. 2007). Here, only one out of eight sampling stations was positive for otters in the catchment of the River Sinni and none of five in that of the River Mercure-Lao, the two major rivers of the Pollino Massif. On the rivers Rosa and Occido, two left-bank tributaries of the River Crati (Calabria region), otter occurrence was negligible, as only one spraint was found (Cassola 1986) on each.

During the following 15–20 years, the percentage of sites positive for otters has significantly increased for several watercourses, and otter populations progressively recovered, reinforcing their occurrences in previously occupied rivers and recolonizing new areas (Prigioni et al. 2007; Loy et al. 2009). In 2001–2002, an extensive field survey was carried out in the Pollino National Park (ca. 1930 km^2) and surrounding areas. Thirty-two river stretches (mean length: 673 m), spread over 17 rivers, were searched monthly for otter spraints (Prigioni et al. 2005). The monitored rivers showed the stable presence of otters: about 82 % of surveys proved positive for otters and 12 rivers (70 %) had positive otter presence in more than 70 % of surveys. Sprainting activity was much higher than that reported in 1987–1991 for other rivers of southern Italy: mean spraint density was 31.7 faecal samples/km, while reaching, for some intensively marked stretches, 378 faecal samples/km (Prigioni et al. 2005).

In the same area, in 2004, a non-invasive genetic survey was performed by typing 185 faecal samples collected along ten rivers at 12 microsatellite loci. Twenty-three different genotypes were identified, and a population of 34–37 individuals, corresponding to 0.18–0.20 otters/km of watercourse, was estimated by a rarefaction curve method (Prigioni et al. 2006b). More recently, these results have been confirmed by a similar non-invasive genetic survey conducted in the catchment of the River Sangro (central-Italian otter range) that assessed otter density at 0.16–0.17 otters/km of watercourse (Lerone et al. 2014a, b).

The Otter at the Southern Limit of Its Italian Range

Since the beginning of the twentieth century the Calabria region has been considered the southern limit of otter distribution in Italy (Cavazza 1911). In 1968–1972, the otter was still widespread in the northern and central parts of this region, "particularly in the watercourses of the Sila Massif" (Cagnolaro et al. 1975), where the species had been reported for 24 waterways—including the rivers Trionto, Mucone, Lese, Neto, Frappa, Savuto and Tacina, and five lakes.

In 1983–1985, the otter had disappeared from the central and southern parts of the region, while it was still reported for the lower course of the River Crati and some of its tributaries, with a total of only four positive sites and five spraints. Persecution (otter hunting was forbidden in 1977), the reduction of fish availability as a result of over-fishing and the alteration of riparian habitats were thought to be the main causes of otter reduction (Arcà 1986). Considering the few otter signs found and the sudden decline occurring within a few years, the authors indicated the threat of near extinction of the species in the region (Cassola 1986).

Contrary to these pessimistic expectations, in 2002–2003, the otter was recorded on the rivers Savuto and Neto, flowing on opposite sides of the massif, and on the River Crocchio, which flows southwards into the Ionian Sea (Fusillo et al. 2003).

In 2008, a survey for otters promoted by the Sila National Park and Calabria Region confirmed the stable occurrence of the species on the River Savuto, which flows on the western side of the Sila Massif (Balestrieri et al. 2008). More recently (summer 2014), otter occurrence has also been ascertained for the rivers Neto, Lese, Tacina and Simeri (Gariano and Balestrieri, personal communication). Overall, since the start of this century, otter signs have been recorded in 15 out of 46 sites surveyed (Fig. 5.2).

Recently (September 2014), a road-killed otter was found next to the mouth of the River Angitola, which flows in the Tyrrhenian Sea. The site is about 40 km south of the lower course of the River Savuto and may have been reached by the otter by sea.

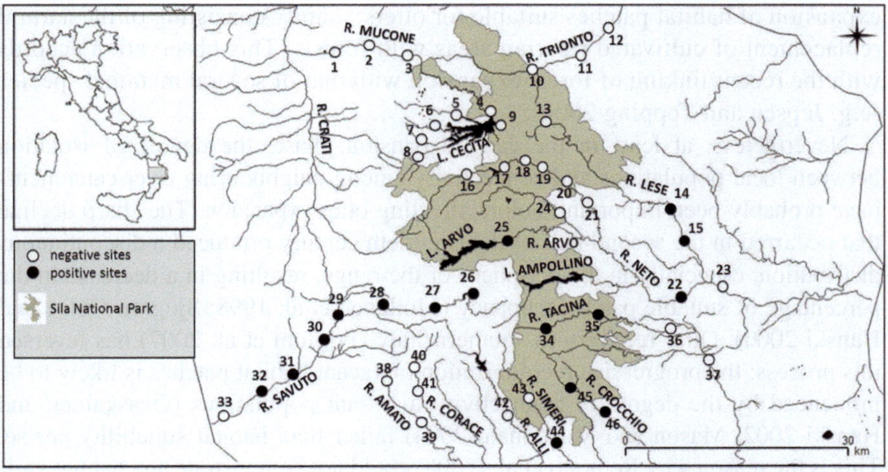

Fig. 5.2 Positive- (*black dots*) and negative- (*white dots*) for otters sites surveyed in the Sila Massif. The protected area of the Sila National Park is marked in grey

Factors Shaping the Current Otter Distribution in Italy

Ecological studies at local, regional or continental scales of the factors affecting the distribution of European otter populations have stressed the importance of both human and environmental variables (Barbosa et al. 2001, 2003; Ruiz-Olmo et al. 2001; Robitaille and Laurence 2002).

Among them, prey availability is a major factor affecting the use of river stretches by otters (Kruuk et al. 1990, 1993). In Mediterranean catchments, otter density has been positively related to fish biomass (Ruiz-Olmo et al. 2001) and otter sprainting activity has also been related to an index of fish biomass (Prenda and Granado-Lorencio 1996). The cover provided by trees and shrubs along riverbanks provides resting and breeding sites for otters, increases water quality and fish productivity, and improves connectivity among river basins (Mason 1995; Morrow and Fischenich 2000; Chanin 2003; Ruiz-Olmo et al. 2011; Carranza et al. 2012).

Accordingly, the suitability of the River Ticino for otters has been assessed based on available knowledge on the availability of fish, potential resting- and breeding sites, and water quality. The whole Italian course of the river (110 km) was split in 21, 5 km-long stretches and the overall habitat conditions of each stretch were compared to those of the one hosting reintroduced otters since 1997. Results suggest that the stretch of river flowing between the release site and about 15 km upstream of its confluence with the River Po may be suitable for otter expansion (Prigioni and Balestrieri 2011).

More recently, Carone et al. (2014) analysed the relationship between changes in land cover patterns and otter occurrence in southern Italy. By such a multi-temporal approach, they were able to observe, between 1985 and 2006, an expansion of habitat patches suitable for otters, mainly consisting of the natural replacement of cultivated riparian areas with forests. This observation accords with the recent linking of forest expansion with that of several mammal species (e.g. Jepsen and Topping 2004).

Nevertheless, at least in the early expansion phase, the degree of isolation between local populations and connectivity among neighbouring river catchments have probably been important factors affecting otter expansion. The sharp decline that occurred in the second half of the twentieth century produced a discontinuous distribution, especially in the periphery of the range, resulting in a decrease in the percentage of suitable patch occupancy (Moilanen et al. 1998; Sjogren-Gulve and Hanski 2000). Otter recovery in southern Italy (Prigioni et al. 2007) has reversed this process: the progressive recolonization of vacant habitat patches is likely to be influenced by the degree of connectivity to extant populations (Ovaskainen and Hanski 2002; Mason and Macdonald 2004) rather than habitat suitability per se. This is the reason why Remonti et al. (2008) could not demonstrate any habitat variable that clearly affected otter distribution in the Pollino massif. Here otter relative abundance varied along a geographical gradient, with the south-western, sparsely connected, peripheral strip of otter range showing unstable river occupancy by the species. It is probable that only a few roaming animals reach it occasionally, forming small populations that suffer a high risk of stochastic extirpation.

Discussion

Prior to hunting restrictions (1977), the decline in the otter range can be mostly ascribed to direct persecution, particularly in northern Italy, and habitat loss, which, again, was heavier in the intensively cultivated plain of the River Po (Prigioni 1997).

Moreover, the sharp reduction in otter numbers occurring in Italy throughout the 1970s and 1980s was synchronized with the species' decline over a wide area of central and western Europe (Foster-Turley et al. 1990; Macdonald and Mason 1994), suggesting that a common, continent-wide cause, namely persistent organic pollutants, played a major role among a wide range of threats acting synergistically (Ruiz-Olmo et al. 2000; Prigioni et al. 2007).

In Italy, the production and use of both polychlorinated biphenyl PCBs and DDT peaked in the 1970s (Maroni et al. 1991; Pacyna 1999). Available data for that decade confirm that in the rivers of northern Italy, POPs and heavy metals attained levels sometimes incompatible with the presence of otters (Viviani et al. 1974), while in southern Italy, where agriculture and industry played a minor role in the local economy, their levels were much lower (Prigioni 1995).

Otter range shrinkage in southern Italy was accordingly somewhat less drastic. In the Campania and Basilicata regions in particular, the availability of suitable habitats remained relatively high throughout the last decades of the twentieth century and, together with the establishing, in the early 1990s, of some protected areas, allowed otter recovery and expansion.

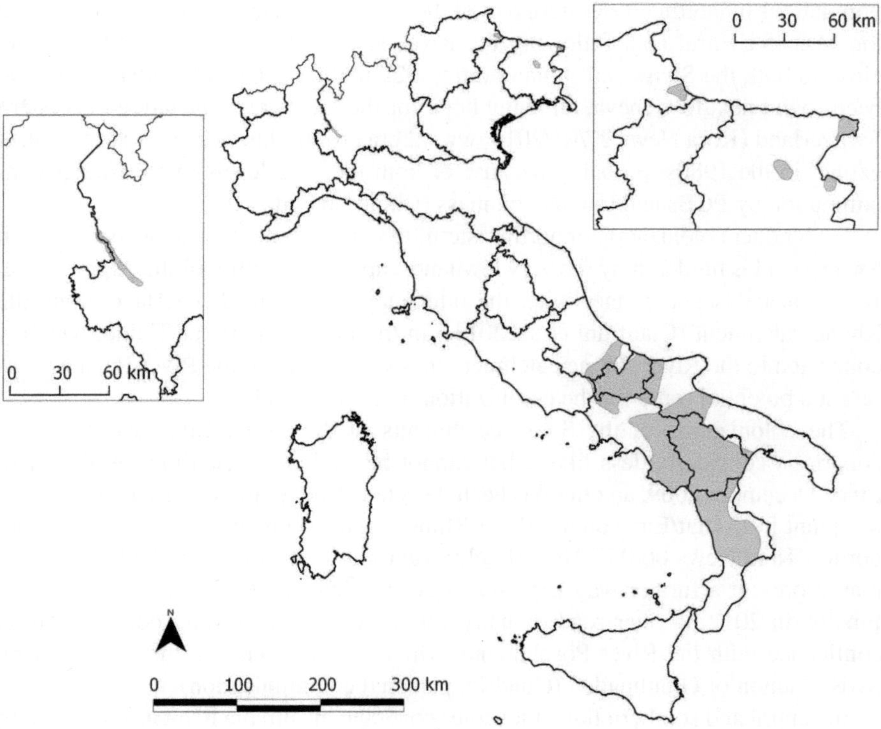

Fig. 5.3 Current otter range in the Italian peninsula with enlargements of the River Ticino Valley (*left*) and South Tyrol (Trentino Alto Adige region) and Friuli Venezia Giulia region (*right*)

Excluding reintroduction areas, the most recent estimate of otter range in Italy (occurrence data updated to 2007) is 29,000 km^2, disjointed into two areas: 5700 km^2 (11.6 %) in central Italy and 23,300 km^2 (88.4 %) in the south (Panzacchi et al. 2010). In northern Italy, the current otter range covers about 500 km^2, of which 150 km^2 is on the River Ticino and 350 km^2 in the north-eastern tip of the country (Fig. 5.3).

Until a few years ago, the return of the otter in northern Italy was considered a utopian ideal, the small population in the River Ticino valley being doomed to either represent a single example of the potential of habitat protection measures and environmental management for the restoration of past biodiversity or, at the worst, become extinct due to demographic stochasticity.

The westward expansion of otters in Austria (Conroy and Chanin 2000) has provided an unexpected opportunity for the natural recolonization of north-eastern Italy by this mustelid. The status of the Austrian and Slovenian otter populations together with the availability of a few suitable corridors may allow the stable colonization of South Tyrol and Friuli Venezia Giulia.

In north-western Italy, the potential for otter expansion from neighbouring countries currently seems much lower. In France, since the 1980s, the residual otter populations inhabiting the eastern part of the country have progressively recolonized the Massif Central and, following the River Rhone, the region of Haute-Savoie, close to both the Swiss and Italian border (Kuhn 2009). Recently, otter signs have been found next to Genèva, allowing hope for the future recolonization of western Switzerland (Kora News 27/05/2014, www.kora.ch). In this country, the otter went extinct in the 1980s, probably because of both the high levels of freshwater contamination by PCBs and low fish biomass (Cianfrani et al. 2013).

Direct otter colonization of northwestern Italy from Haute-Savoie or southwestern Switzerland is hindered by the very low landscape permeability of the Alpine ridge. Based on landscape connectivity, the otter may expand in Switzerland along the Rhone catchment (Cianfrani et al. 2013). In the future, the River Ticino, which is connected to the River Rhone catchment in Switzerland and the River Po, may represent a potential route for the colonization of northern Italy.

The colonization of the Swiss catchments of the rivers Rhine and Inn from Austria is considered less likely, but cannot be excluded (Cianfrani et al. 2013). Since December 2009, an otter has been detected close to the Reichenau hydroelectric plant in Domat/Ems, on the River Rhine, about 20 km away from the Austrian border (Kora News 06/05/2010). If colonization is going to occur, the River Rhine may represent a further way for otter expansion into the River Ticino catchment. Finally, in 2012 an otter road casualty was found on the River Adda, next to its confluence with the River Poschiavino, which has its source in the neighbouring Swiss Canton of Graubünden (Quadrio, personal communication).

In central and southern Italy, the larger coverage of suitable habitats compared to the heavily cultivated areas of the river Po plain, has further increased in the last decades as a consequence of the reduction of cultivated fields in riparian belts (Carone et al. 2014), favouring otter persistence and allowing its expansion since the end of the twentieth century.

Nonetheless, the recent colonization of less suitable river habitats and coastal stretches usually avoided by otters (Loy et al. 2010; Carone et al. 2014) may suggest that habitat quality is not as strong a barrier as previously believed. As an example, otters have been recorded twice, in 2011 and in 2014, on the River Lato, next to the gulf of Taranto (Ionic side of Apulia region), in a cultivated area not far from the industrial suburbs of the city (Marrese, personal communication).

Conclusion

Despite a general positive trend, otter populations of the Italian peninsula are completely isolated from other European populations.

Otter peripheral populations suffer a higher risk of extinction due to demographic stochasticity and population fragmentation still represents the main threat to the long-term survival of the otter in Italy (Ellstrand and Elam 1993). The degree of isolation between neighbouring populations and connectivity between river catchments have probably played a major role in driving the recolonization process and then shaping the current otter distribution (Remonti et al. 2008).

This critical situation led the Italian Ministry of Environment to promote the production of an Action Plan for the otter in Italy. The Action Plan has identified the priority areas for intervention and main goals to be accomplished in the short, medium and long term (reduction of mortality and disturbance, prevention of conflicts and impacts on human activities, enhancement of genetic flow among populations, habitat restoration and conservation) (Panzacchi et al. 2010).

The main goal for the immediate future is to encourage connection between the central- and south Italian populations, which may occur through the upper Volturno (Molise region) and Calore (Campania) catchments (Panzacchi et al. 2010). Habitat restoration on the banks of the River Crati (Calabria) may enhance the connectivity between the most southern population (Sila Massif) and the core area of otter range (Pollino Massif); although recent surveys have shown that in the first area the otter is more widespread than previously believed, the Sila population is probably currently the one suffering the highest risk of extinction and research is needed to draw a sound picture of its abundance and actual distribution.

Acknowledgements We thank Marco Pavanello and Luca Lapini for having shared their data about otter distribution in northeastern Italy with us. Pasquale Gariano mapped available data (partly still unpublished) about otter records in the Sila Massif. Lesley C. Wright (IUCN OSG) kindly revised the paper for English language use. Paul Chanin and Chris F. Mason improved the ms. with their comments and suggestions.

References

Altobello G (1921) Fauna Abruzzo e Molise. Mammiferi. Carnivori. Colitti, Campobasso

Arcà G (1986) La lontra in Calabria. In: Cassola F (ed) La Lontra in Italia. Serie Atti e Studi no. 5, WWF-Italia, Roma, pp 118–121

Balestrieri A, Remonti L, Smiroldo G, Prigioni C, Reggiani G (2008) Surveying otter *Lutra lutra* distribution at the southern limit of its Italian range. Hystrix It J Mamm 19:165–173

Barbosa AM, Real R, Marquez AL, Rendon MA (2001) Spatial, environmental and human influences on the distribution of otter (*Lutra lutra*) in the Spanish Provinces. Divers Distrib 7:137–144

Barbosa AM, Real R, Olivero J, Vargas JM (2003) Otter (*Lutra lutra*) distribution modelling at two resolution scales suited to conservation planning in the Iberian Peninsula. Biol Conserv 114:377–387

Cagnolaro L, Rosso D, Spagnesi M, Venturi B (1975) Inchiesta sulla distribuzione della Lontra (*Lutra lutra*) in Italia e nei Cantoni Ticino e Grigioni (Svizzera) 1971–1973. Ricerche di Biologia della Selvaggina 63:1–120

Carone MT, Guisan A, Cianfrani C, Simoniello T, Loy A, Carranza ML (2014) A multi-temporal approach to model endangered species distribution in Europe. The case of the Eurasian otter in Italy. Ecol Model 274:21–28

Carranza ML, D'Alessandro E, Saura S, Loy A (2012) Assessing habitat connectivity for semi-aquatic vertebrates. The case of the endangered otter in Italy. Landscape Ecol 27:281–290

Cassola F (1986) The otter in Italy. Status, distribution and conservation of an endangered species. WWF Italia, Serie Atti e Studi 5, p 135

Cavazza F (1911) Dei Mustelidi Italiani. Ann Mus Civ St Nat Genova 5:170–204

Chanin P (2003) Monitoring the otter *Lutra lutra*. Conserving Natura 2000 Rivers Monitoring Series No. 10. English Nature, Peterborough

Cianfrani C, Maiorano L, Loy A, Kranz A, Lehmann A, Maggini R, Guisan A (2013) There and back again? Combining habitat suitability modelling and connectivity analyses to assess a potential return of the otter to Switzerland. Anim Conserv 16:584–594

Conroy JWH, Chanin P (2000) The status of the Eurasian otter (*Lutra lutra*) in Europe—a review. J Int Otter Surv Fund 1:7–28

Cortés Y, Fernández-Salvador R, García FJ, Virgós E, Llorente M (1998) Changes in otter *Lutra lutra* distribution in Central Spain in the 1964–1995 period. Biol Conserv 86:179–183

Crawford A (2003) Fourth otter survey of England 2000–2002. Environmental Agency, Bristol, p 88

Cripezzi V, Dembech V, La Nave AM, Marrese M, Caldarella M (2001) La presenza della Lontra nel bacino del fiume Ofanto (Puglia, Basilicata e Campania). Convegno: La lontra (*Lutra lutra*) in Italia: distribuzione, censimenti e tutela, Montella, Avellino. Dryocopus IV(1)

De Castro G, Loy A (2007) Un nuovo censimento della lontra (*Lutra lutra*, Carnivora, Mammalia) nel fiume Sangro (Abruzzo): inizia la ricolonizzazione dell'Italia centrale? 68° Convegno Unione Zoologica Italiana, Lecce, 23–25 Sept 2007

Ellstrand NC, Elam DR (1993) Population genetic consequences of small population size: implications for plant conservation. Ann Rev Ecol Systematics 24:217–242

Elmeros M, Hammershøj M, Madsen AB, Søgaard B (2006) Recovery of the otter *Lutra lutra* in Denmark monitored by field surveys and collection of carcasses. Hystrix It J Mamm 17:17–28

Foster-Turley P, Macdonald SM, Mason CF (1990) An action plan for European Otters. IUCN/SSC Otter Specialist Group. Kelvin Press, Broadview

Fumagalli R, Prigioni C (1993) Evoluzione della distribuzione della Lontra (*Lutra lutra*) in Italia e possibilità di sopravvivenza dei nuclei residui. Supp. Ricerche di Biologia della Selvaggina XXI:189-199

Fusillo R, Marcelli M, Boitani L (2003) Progetto di ricerca sulla lontra nel Parco Nazionale del Cilento e Vallo di Diano. Relazione conclusiva. Parco Nazionale del Cilento e Vallo di Diano. Università di Roma "la Sapienza"

Ghigi A (1911) Ricerche faunistiche e sistematiche sui Mammiferi d'Italia che formano oggetto di caccia. Natura 2:289–337

Janssens X, Defourny P, De Kermabon J, Baret PV (2006) The recovery of the otter in the Cevennes (France): a GIS-based model. Hystrix It J Mamm 17:5–14

Jepsen JU, Topping CJ (2004) Modelling roe deer (*Capreolus capreolus*) in a gradient of forest fragmentation: behavioural plasticity and choice of cover. Can J Zool 82:1528–1541

Kranz A (2008) Sudtirol: Fischotter im Oktober 2008 nachgewiesen. In: Wild und Jagd in Europa. Anblick 11/2008

Kranz A, Polednik L (2009) Verbreitung und Erhaltungszustand 2009 in Karnten. Endbericht im Auftrag der Abteilung 20 des Amtes der Karntner Landesregierung

Kruuk H, Wansink D, Moorhouse A (1990) Feeding patches and diving success of otters, *Lutra lutra*, in Shetland. Oikos 57:68–72

Kruuk H, Carss DN, Conroy JWH, Durbin L (1993) Otter (*Lutra lutra* L.) numbers and fish productivity in rivers in north-east Scotland. Symp Zool Soc London 65:171–191

Kuhn R (2009) Plan National d'Actions pour la Loutre d'Europe (*Lutra lutra*), 2010–2015. Société Française pour l'Etude et la Protection des Mammifères/Ministère de l'Ecologie, de l'Energie, du Développement Durable et de la Mer, p 110

Lapini L (1986) La lontra in Friuli Venezia Giulia. In: La lontra in Italia—Censimento, distribuzione e problemi di conservazione di una specie minacciata, World Wildlife Fund, Rome

Lapini L, Bonesi L (2011) Evidence of the natural recovery of the Eurasian otter in NE Italy. 29th European Mustelid Colloquium, 3–4 Dec 2011, Southampton, UK

Lerone L, De Castro G, Imperi F, Marrese M, Carranza ML, Scorpio V, Loy A (2014) Population structure and range dynamics of an isolated otter population in Italy: lessons from the field. IUCN OSG XII International Otter Congress, 1–15 Aug 2014, Rio de Janeiro, Brazil

Lerone L, Mengoni C, Carpaneto GM, Randi E, Loy A (2014b) Procedures to genotype problematic non-invasive otter (*Lutra lutra*) samples. Acta Theriol 59:511–520

Loy A, Bucci L, Carranza ML, De Castro G, Di Martino P, Di Marzio P, Reggiani G (2003) Censimento della lontra euroasiatica (*Lutra lutra*) in Molise: primi risultati e considerazioni su un'area critica per la ricolonizzazione dell'Italia centrale. 64° Congresso Unione Zoologica Italiana, Varese, 21–25 Settembre 2003, p 115

Loy A, Bucci L, Carranza ML, De Castro G, Di Marzio P, Reggiani G (2004) Survey and habitat evaluation for a peripheral population of the Eurasian otter in Italy. IUCN Otter Spec Group Bull 21A:1–8

Loy A, Carranza ML, Cianfrani C, D'Alessandro E, Bonesi L, Di Marzio P, Minotti M, Reggiani G (2009) Otter *Lutra lutra* population expansion: assessing habitat suitability and connectivity in south-central Italy. Folia Zool 58:309–326

Loy A, Boitani L, Bonesi L, Canu A, Di Croce A, Fiorentino PL, Genovesi P, Mattei L, Panzacchi M, Prigioni C, Randi E, Reggiani G (2010) The Italian action plan for the endangered Eurasian otter *Lutra lutra*. Hystrix It J Mamm 21:19–33

Macdonald SM, Mason CF (1983) The otter *Lutra lutra* in Southern Italy. Biol Conserv 25:95–101

Macdonald SM, Mason CF (1991) Action Plan for European Otters. In: Foster-Turley P, Macdonald SM, Mason CF (eds) Otters. An Action Plan for their Conservation. IUCN/SSC Otter Specialist Group, Kelvin Press Inc., Broadview, IL, USA, p 213–217

Macdonald SM, Mason CF (1994) Status and conservation needs of the otter (*Lutra lutra*) in the western Palearctic. Nature Environment 67, Council of Europe, Strasbourg, p 53

Maroni M, Colombi A, Carrer P, Barbieri F, Foà V, Del Frate G, Bossi G (1991) Stima delle giacenze di PCB in Italia e loro soluzioni di smaltimento. Acqua-Aria 8:761–767

Mason CF (1995) Habitat quality, water quality and otter distribution. Hystrix 7:195–207

Mason CF, Macdonald SM (1986) Otters: ecology and conservation. Cambridge University Press, Cambridge

Mason CF, Macdonald SM (2004) Growth in otter (*Lutra lutra*) populations in the UK as shown by long-term monitoring. Ambio 33:148–152

Mason CF, Macdonald SM, Prigioni C, Meriggi A, Bogliani G (1985) Indagine sulla Lontra (*Lutra lutra*) nel Parco Lombardo della Valle del Ticino. Piano di settore fauna, stato di avanzamento delle ricerche, I° anno. Dipartimento di Biologia Animale dell'Università di Pavia, Pavia

Moilanen A, Smith AT, Hanski I (1998) Long-term dynamics in a metapopulation of the American Pika. Am Nat 152:530–542

Montanari M, Boffino G (2000) Il ritorno della Lontra. 1990–2000, il progetto di reintroduzione nel Parco Naturale della Valle del Ticino Piemontese. Parco Ticino Piemontese e Regione Piemonte

Morrow JV, Fischenich C (2000) Habitat Requirements for Freshwater Fishes. EMRRP Technical Note Collection (ERDC TN-EMRRP-SR-06), U.S. Army Engineer Research and Development Center, Vicksburg

Ovaskainen O, Hanski I (2002) Transient dynamics in metapopulation response to perturbation. Theor Pop Biol 61:285–295

Pacyna JM (1999) Environmental cycling of selected Persistent Organic Pollutants (POPs) in the Baltic Region. Technical Report of Environmental & Climate Research Programme (No. ENV4-CT96-0214)

Panzacchi M, Genovesi P, Loy A (2010) Piano d'Azione Nazionale per la Conservazione della Lontra (Lutra lutra). Ministero dell'Ambiente e della Tutela del Territorio e del Mare—Istituto Superiore per la Protezione e la Ricerca Ambientale, Roma

Pavanello M, Lapini L, Kranz A, Iordan F (2015) Rediscovering the Eurasian Otter (Lutra Lutra L.) in Friuli Venezia Giulia and Notes on its Possible Expansion in Northern Italy. IUCN Otter Spec. Group Bull. 32(1):12–20

Pellegrini M, Febbo D (1986) La lontra in Molise. In: Cassola F (ed) La lontra in Italia. Censimento, distribuzione e problemi di conservazione di una specie minacciata. WWF, Serie Atti e Studi 5, Roma, p 101–102

Prenda J, Granado-Lorencio C (1996) The relative influence of riparian habitat structure and fish availability on otter *Lutra lutra* L. sprainting activity in a small Mediterranean catchment. Biol Conserv 76:9–15

Prigioni C (1983) Confermata la presenza della lontra (*Lutra lutra*) nel Lago di Mezzola (Lombardia). Natura 74:125–126

Prigioni C (1986a) La lontra in Lombardia e nelle province di Asti, Alessandria, Parma e Piacenza. In: Cassola F (ed) La Lontra in Italia. Serie Atti e Studi no 5, WWF-Italia, Roma, p 25–28

Prigioni C (1986b) Studio di fattibilità della reintroduzione della Lontra *Lutra lutra* nel Parco del Ticino. Piano di settore fauna del Parco Lombardo della Valle del Ticino, Dipartimento di Biologia Animale dell'Università di Pavia, Pavia

Prigioni C (1995) Guidelines for the feasibility study of reintroduction of the otter *Lutra lutra* in Italy: the project of the Ticino Valley (north-western Italy). In: Prigioni C (ed) Proceedings II Italian Symposium on Carnivores, Hystrix 7:225–264

Prigioni C (1997) La lontra. Una vita silenziosa negli ambienti acquatici. Edagricole, Bologna

Prigioni C, Balestrieri A (2011) Piano di monitoraggio dei Vertebrati terrestri di interesse comunitario in Lombardia: lontra *Lutra lutra*. Dipartimento di Biologia Animale, Università di Pavia, Pavia

Prigioni C, Bogliani G, Meriggi A (1979) Proposta di reintroduzione e gestione di alcuni Mammiferi nel Parco Lombardo della Valle del Ticino. Atti VIII Simp. naz. Conserv. Natura, Bari, p 351–353

Prigioni C, Remonti L, Balestrieri A, Sgrosso S, Priore G, Misin CE, Viapiana M, Spada S, Anania R (2005) Distribution and sprainting activity of the otter (*Lutra lutra*) in the Pollino National Park (Southern Italy). Ethol Ecol Evol 17:171–180

Prigioni C, Balestrieri A, Remonti L, Sgrosso S, Priore G (2006a) How many otters are there in Italy? Hystrix It J Mamm 17:29–36

Prigioni C, Remonti L, Balestrieri A, Sgrosso S, Priore G, Mucci N, Randi E (2006b) Estimation of European otter (*Lutra lutra*) population size by fecal DNA typing in southern Italy. J Mammal 87:855–858

Prigioni C, Balestrieri A, Remonti L (2007) Decline and recovery in otter *Lutra lutra* populations in Italy. Mammal Rev 37:71–79

Prigioni C, Smiroldo G, Remonti L, Balestrieri A (2009) Distribution and diet of reintroduced otters (*Lutra lutra*) on the River Ticino (NW Italy). Hystrix It J Mamm 20:45–53

Reggiani G, Ciucci P, Boitani L, Rocca F (1997) Un approccio multicriteri alla conservazione della Lontra (*Lutra lutra*). Ricerche di Biologia della Selvaggina 27:751–757

Remonti L, Prigioni C, Balestrieri A, Sgrosso S, Priore G (2008) Distribution of a recolonising species may not reflect habitat suitability alone: the case of the Eurasian otter (*Lutra lutra*) in southern Italy. Wildlife Res 35:798–805

Righetti D (2011) Return of the otter in South Tyrol (NE Italy). In: Prigioni C, Loy A, Balestrieri A, Remonti L (eds) Abstracts 11th International Otter Colloquium. Hystrix It J Mamm Supp 2011:122

Robitaille JF, Laurence S (2002) Otter, *Lutra lutra*, occurrence in Europe and in France in relation to landscape characteristics. Anim Conserv 5:337–344

Ruiz-Olmo J, Delibes M (1998) La nutria en españa ante el horizonte del año 2000. SECEM Málaga 28:8

Ruiz-Olmo J, Lafontaine L, Prigioni C, López-Martín JM, Santos-Reis M (2000) Pollution and its effects on otter populations in south-western Europe. In: Conroy JWH, Yoxon P, Gutleb AC (eds) Proceedings of the first Otter Toxicology Conference. Journal of the International Otter Survival Fund 1:63–82

Ruiz-Olmo J, López-Martin JM, Palazón S (2001) The influence of fish abundance on the otter (*Lutra lutra*) populations in Iberian Mediterranean habitats. J Zool 254:325–336

Ruiz-Olmo J, Batet A, Mañas F, Martínez-Vidal R (2011) Factors affecting otter (*Lutra lutra*) abundance and breeding success in freshwaters habitats of the northeastern Iberian Peninsula. Eur J Wildl Res 57:827–842

Scorpio V, Loya A, Di Febbraro M, Rizzo A, Aucelli P (2014) Hydromorphology meets mammal ecology: river morphological quality, recent channel adjustments and otter resilience. River Res Applic. doi:10.1002/rra.2848

Sjogren-Gulve P, Hanski I (2000) Metapopulation viability analysis using occupancy models. Ecol Bulletins 48:53–71

Spagnesi M, Cagnolaro L (1981) Lontra *Lutra lutra* Linnaeus 1758. In: CNR Roma (ed) Distribuzione e Biologia di 22 specie di Mammiferi in Italia, Collana del Progetto finalizzato "Promozione della qualità dell'ambiente" AQ/1/142–164, p 95–101

Viviani R, Crisetig G, Cortesi P, Carpene L (1974) Résidus de polychlorobyphenyls (PCBs) dans les poissons et les oiseaux du delta du Po. Revista Internationalle de Océonografia della Méditerranea 35–36:79–90

Part III
When Wildlife Creates Problems for the Environment and Human Activities: General Features and Some Case Studies

This part examines the cases in which wildlife creates problems, either directly or indirectly, for man, for man's life and existence or for his economic activities.

There are an infinite number of cases related to this issue, bearing in mind that wildlife can become a 'problem' in stressed ecosystems which are altered quite easily (Peterson et al. 2010). There are countless examples of this, from pest species such as mice, rats, feral cats and so on to species that prey on cattle, such as wolves, coyotes and jackals. Moreover, generally speaking, when non-native or alien species are introduced into a new range, they often create problems with broad-ranging effects (Schmitz and Simberloff 1997).

The aforementioned are merely the most classic examples, yet there are certainly many other different cases, especially in areas where man is expanding his range of influence and stealing more and more land from nature to establish human settlements, especially where humans live in close contact with wildlife (Sillero-Zubiri et al. 2007).

There are three chapters in this part. The first chapter (Battisti and Amori 2016) takes a look at the general management of 'problematic wildlife' and the various kinds of problems that need to be addressed by the people or entities that have been searching for solutions, including the practical and legal tools currently available to take action that is in line with the international standards required by the international community and various conservation organisations, primarily the IUCN.

The second chapter covers a rather pressing issue that is becoming increasingly more widespread in Africa: the coexistence of humans and elephants (Le Bel et al. 2016). These large mammals, especially when they live near human settlements and agricultural crops, can be a serious problem that needs managing.

The third chapter (Rajaratnam et al. 2016) examines cases of livestock herd predation in the Kingdom of Bhutan by four large predators: the tiger (*Panthera tigris*), common leopard (*P. pardus*), snow leopard (*P. uncia*), and the dhole (*Cuon alpinus*). These predators strongly influence the economy of mountain communities. The challenge is to protect these species, which are threatened with extinction, and at the same time safeguard and guarantee future human activities.

References

Battisti C, Amori G (2016) Problem solving and decision making in project management of problematic wildlife: a review of some approaches and conceptual tools. In: Angelici FM (ed) Problematic wildlife: a cross-disciplinary approach. Springer, New York, p 109–122

Le Bel S, La Grange M, Czudek R (2016) Managing human elephant conflict in Zimbabwe: a boundary perspective rather than a problematic species issue. In: Angelici FM (ed) Problematic wildlife: a cross-disciplinary approach. Springer, New York, p 123–142

Peterson MN, Birckhead JL, Leong K, Peterson MJ, Peterson TR (2010) Rearticulating the myth of human–wildlife conflict. Conserv Lett 3:74–82

Rajaratnam R, Vernes K, Sangay T (2016) A review of livestock predation by large carnivores in the Himalayan Kingdom of Bhutan. In: Angelici FM (ed) Problematic wildlife: a cross-disciplinary approach. Springer, New York, p 143–171

Schmitz DC, Simberloff D (1997) Biological invasions: a growing threat. Issues Sci Technol 13:33–40

Sillero-Zubiri C, Sukumar R, Treves A (2007) Living with wildlife: the roots of conflict and the solutions. In: Macdonald D, Service K (eds) Key topics in conservation biology. Blackwell, Oxford, pp 253–270

Chapter 6
Problem Solving and Decision-Making in Project Management of Problematic Wildlife: A Review of Some Approaches and Conceptual Tools

Corrado Battisti and Giovanni Amori

Introduction

Wildlife managers and practitioners are faced with many problems. Their actions include the selection of targets (e.g. species, communities, ecosystems, ecological processes) and priorities and the making of diagnosis, predictions and decisions aimed to achieve specific goals. Therefore, they should use conceptual tools belonging to the interdisciplinary arenas of problem solving and decision-making.

Problem solving consists of using generic or ad hoc methods, in an orderly manner, to find solutions to problems (Wang and Chiew 2010). Decision-making theory is a framework (largely used in economic and engineering disciplines) within which people responsible for management attempt to achieve explicitly stated objectives while acknowledging the levels of uncertainty involved with the decision process (Clemen 1996). Decision-making includes quantitative and qualitative tools (Possingham et al. 2001; Pullin and Knight 2003).

As with other sectors of wildlife management, the practices of problem solving and decision-making in managing problematic wildlife are difficult due a multifaceted complexity that characterize the systems where practitioners work. In particular, both practitioners and problematic wildlife are embedded in three inter-acting complex systems: (1) an ecological system (problematic species and communities relating with other species, communities and processes), (2) a social system (the anthropic world, as the main driving force directly or indirectly causing the presence

C. Battisti (✉)
"Torre Flavia" LTER (Long Term Ecological Research) Station, Protected Areas Service, Città metropolitana di Roma, via Tiburtina, 691, Rome 00159, Italy
e-mail: c.battisti@cittametropolitanaroma.gov.it

G. Amori
CNR-ISE, Institute for Ecosystem Studies, viale dell'Università, 32, Rome 00185, Italy
e-mail: giovanni.amori@uniroma1.it

F.M. Angelici (ed.), *Problematic Wildlife*, DOI 10.1007/978-3-319-22246-2_6

of problematic wildlife, including political forces) and (3) an organizational system (that characterize the multidisciplinary project team managing problematic wildlife and that include all the managers, practitioners and people involved in the project and the related human dimensions). Social and organizational complexity includes many aspects of human dimensions. It has been highlighted that the lack in awareness and consideration of human dimensions in conservation biology may induce failures (Jacobson and McDuff 1998).

In complex systems, diagnosis and predictions may be affected from a high uncertainty that subsequently affects the decision process (see Regan et al. 2002; Hey et al. 2003; Ascough II et al. 2008): a lack in reliable knowledge on components, relationships, processes and feedbacks may contribute to the ineffectiveness of many strategies, projects and actions. Therefore, many disciplinary contexts (e.g. among economists and engineers) have developed, in the last few decades, many approaches and conceptual tools that help decrease the level of uncertainty, thereby facilitating the development of strategies. These approaches also may be useful in wildlife management.

In this contribution, we review a selection of approaches, criteria and conceptual tools developed from researchers and organizations (also not strictly related to the conservation world) referring to problem solving and decision-making. We hope that this review will be useful to wildlife managers and practitioners and, particularly, to all people working in projects focused on problematic wildlife.

Frequently, conceptual tools use models. Here, we point out that models are only an approximation of reality because they are a "four-S" (synthetic, symbolic, simplistic and selective) representation of the real world because the impossibility of representing the complete complex systems.

We have included all the reviewed approaches having in mind a "project-cycle" scenario. Starting from the project cycle defined by IUCN (World Conservation Union) to monitor the effectiveness of management in natural protected areas (Hockings et al. 2006a, b), we have included the different conceptual tools and approaches in different stages of this framework (Fig. 6.1).

We started with the organizational systems because managers first encounter this intrinsic complexity, e.g. when they build their project teams, select practitioners and evaluate resources. Then, we reviewed other approaches and conceptual tools useful to analyse and control the external socio-ecological complexity.

The Project Cycle Framework as a Basic Framework

Recently, the IUCN has developed a project cycle framework aimed to a management effectiveness evaluation, here defined as the assessment of how well protected areas are being managed (Hockings et al. 2006a, b). This framework refers to the design issues relating to individual sites and protected area systems, the adequacy and appropriateness of management, and delivery of protected area objectives

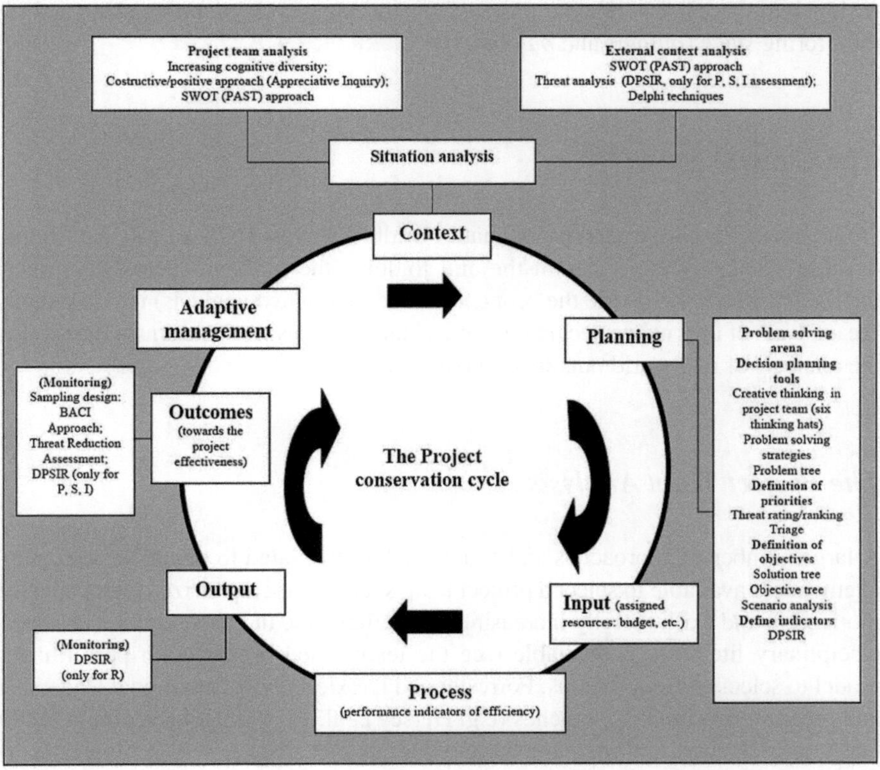

Fig. 6.1 The framework based on the IUCN project conservation cycle (Hockings et al. 2006a, b, modified) including the selected approaches and conceptual tools

including conservation of biodiversity values. The framework also provides broad criteria for assessment, using a range of evaluation "tools" that can be used to conduct evaluations at different scales. It is based on the idea that the management follows a process with six distinct stages (hereafter, in *italic*). It begins with reviewing the *context*, so establishing a vision for site management (within the context of existing status of site and their biodiversity and of threats and pressures), monitoring progress through a *planning* process and an allocation of resources (*inputs*), and as a result of management actions (*process*), eventually produces goods and services (e.g. the results of the management actions or *outputs*). Finally, these outputs should allow obtaining positive impacts on socio-ecological targets defined as *outcomes* (Hockings et al. 2006a, b).

Although this framework has been designed for protected areas, we think that it is logical that it may apply to any site where a problem in wildlife occurs. Therefore, we apply this framework for areas where problematic wildlife are present and where projects aimed to solve specific problems were defined. Particularly, we divide this review into three parts, related to the: (1) *context* stage; (2) *planning* stage and (3) monitoring stage (*outputs* and *outcomes*).

The Context Stage

Upon having a site where problematic wildlife represents a threat, we should develop a team of experts with the aim to solve the problem (hereafter, project team). Therefore, we divide the "context" stage (or context analysis) into two parts: the analysis of internal project team context and the analysis of external context (i.e. the analysis of real world outside to project team).

The Project Team Analysis

A large number of approaches and conceptual tools related to organizational management are available to select a project team such that it emphasizes their capacity, motivation and skill, thereby increasing their efficiency. In this sense, a wide interdisciplinary literature is available (see the team-based organization performance model to select working groups: Forrester and Drexler 1999). Situational leadership and other motivational approaches (e.g. Hersey et al. 1979) may be useful to facilitate team selection and increase the motivation and efficiency of a project team. Here below, we reported only a limited selection.

Increasing the Cognitive Diversity

Increasing the cognitive diversity in a project team is strategic because counter-intuitively, diverse groups solve problems more effectively than homogeneous ones (i.e. including only experts of a single discipline). In this sense Page (2007) highlighted that people could be considered as individual collections of "mental" problem-solving tools that can be applied to many different domains. Moreover, diverse values and preferences in a group can hinder decision-making. Therefore, project teams should encourage and use cognitive diversity (Hong and Page 2004). More concisely, a project team that works in the area of problematic wildlife should include individuals with different skills and cognitive diversity and perception. For example, in a project focused on the coypu (*Myocastor coypus*) in a remnant wetland, other than zoologists and animal ecologists, landscape ecologists and planners, engineers, fishermen, sociologists and stakeholders also should be included. They could aid to solve problems and make decisions on related topics as how long ago the wetland has been reclaimed, how to manage the hydrological regime where this exotic rodent lives and the human dimensions related to it.

Favouring a Constructive and Positive Approach

When selecting a project team, it is important to distinguish among people that have different approaches (e.g. negative, positive, constructive or destructive) to solve problem and how they relate with colleagues. The "Appreciative Inquiry" approach attempts to use ways of asking questions and envisioning the future to foster positive relationships and build on the present potential of a given person, organization or situation (Cooperrider et al. 2008). The aim is to build project teams around what works, rather than trying to fix what doesn't. Following this approach, positive sentiments that increase creativity, openness to new ideas and people, and cognitive flexibility should be favoured. This approach may promote the strong connections and relationships between people, particularly between groups in conflict, required for collective inquiry and change (see the "Positive principle" in Cooperrider et al. 2008; for creative problem solving: see Seely et al. 2003). In the arena of problematic wildlife, new "heretic" ideas should be criticized (constructively and positively) inside a dialectical process.

Searching the Conditional Factors Internal and External to the Project Team

An useful and largely used conceptual tool is the SWOT, an acronym for Strengths, Weaknesses, Opportunities, Threats. This is a framework proposed as an analytical tool which should be used to categorize significant environmental factors (internal and external to the organization) that may affect the process of management (Hill and Westbrook 1997). The SWOT analysis has been adopted for very different purposes and can actually be used in every decision process if a final status or objective has been defined. In the SWOT descriptive approach, the strengths and weaknesses, as internal conditioning factors, and threats and opportunities, as external conditioning factors should be defined. For example, if in a strategy the goal is "decrease the population density of *Procambarus clarkii* from two streams", the project team may check, as internal conditioning factors, the following strengths: high budget and high number of experts with different skills, and the following weaknesses: high level of hierarchy (intrinsic to the organization) and consequent demotivation and frustration of practitioners; among the external conditioning factors: the following opportunities: a large number of stakeholders (e.g. fishermen) interested in the project and possibility to increase awareness of the general public on the problematic wildlife; among the threats: antagonist individual behaviour (e.g. people illegally releasing *Procambarus*).

Many rules have been suggested to matching strengths and opportunities to determine some advantages and to convert threats into opportunities and weaknesses into strengths (Orr 2011; Battisti et al. 2013), for example, conducted a study in a remnant wetland with many threats, including the presence of exotic species. Orr (2011) suggested also a PAST analysis that allows separating strengths, weaknesses, opportunities and threats in Political, Economic, Social and Technological sectors.

The Context Analysis

When the project team was structured, it should conduct a thorough analysis of the context where the problematic wildlife occurs. This corresponds to the first stage of the IUCN project cycle, where a vision was established and where targets and threats were identified. Often, when the project team starts with the analysis of the local site situation, data are not available or difficult to obtain. Nevertheless is in this phase of scanty knowledge that the more important decisions should be made (Christensen and Kreiner 1991). Therefore, when problems and objectives are not easy understandable, data are not available and the uncertainty is high, the project teams may adopt rapid expert-based approaches to assign scores of evaluation, selecting and ranking the different problems, objectives and overall priorities. There is much literature on the expert-based approaches (or Delphi approaches introduced by Linstone and Turoff 1975) also applied to conservation biology and wildlife management. In particular, the Delphi technique is "a method for structuring a group communication process so that the process is effective in allowing a group of individuals, as a whole, to deal with a complex problem" (Linstone and Turoff 1975). Although there are many limitations in these expert-based techniques (e.g. the "leader/boss effect"), they have been largely used in many crises where analytical data are difficult to obtain and uncertainty is high.

In the context analysis it is important to understand what's causing the problem and to do this some approaches were developed to search solutions (e.g. root cause analysis; see also section "Other problem-solving strategies").

More in particular, problem tree analysis (or situation analysis) helps to find solutions by mapping out the anatomy of cause and effect around an issue in a similar way to a Mind map, but with more structure. In this sense, this logic approach allows to break the problem down into manageable and definable topics. This enables a clearer prioritization of factors and helps focus objectives. An objectives tree is a visual representation of objectives. It is the positive opposite of the problem tree, and helps to give us a clear idea of all objectives, which are more important, which need to be achieved first and the relationship between them all.

The Threat Analysis

Knowing human-induced events, their type and regime allows defining appropriate strategies and management measures aimed at mitigating or eliminating their impact on ecosystem components and ecological processes. For this reason, over the last decade a specific approach in conservation biology, named "threat analysis", has been developed (Salafsky et al. 2003, 2008). Such approach is characterized by its own terminology, theoretical framework and principles. It aims at standardizing concepts, methodologies and operational procedures to facilitate the definition of effective actions against human-induced threats. Threat analysis has been applied to evaluate the causes of threats to single species in a holistic dimension facing extinction (species-based approach) and constitutes part of the methodological aspects of Red Lists.

In this sense, problematic wildlife represents a threat for native species, communities and processes. Therefore, problematic wildlife has been included in the recent IUCN taxonomy (IUCN 2005a, b) with a proper nomenclature and systematics. The disciplinary arena of threat analysis is very useful to define logically the role of a threat in the complex socio-ecological world. A so-named "direct" threat (e.g. a coypu population) has a specific location in a cause–effect relationship (causal chain threat-target). This direct threat is caused by an indirect threat that is caused by a driving force. The logic of threat analysis also provides a quantification of the regime of a threat in terms of scope, frequency, intensity, duration, reversibility, etc. Adopting this approach in a site where a large suite of threats occur (e.g. Battisti et al. 2008), it is possible to develop a threat ranking, so facilitating the process of priority selection (e.g. managers develop a plan focused on the prioritized threats; see below).

The Planning Stage

Starting with Problem Solving

Use of Different Type of Thinking

de Bono (1987) asserted that increased competence in thinking can become easy by the conscious use of six types of thinking when appropriate to the occasion. He used the metaphor of the "six thinking hats", each one linked to a rule for thinking (neutral and objective, emotional, negative, positive, creative, cold and controlled, organizational). This approach may be used when complex situations appear on a wildlife scene. For example, a project team might analyse the problem represented by an exotic plant species on a dune (e.g. *Carpobrotus edulis*) by approaching with these six types of thinking. Neutrally thinking, it may be obtained through qualitative and quantitative information on parameters of state, pressure and impact (cover, density, competition on other sensitive targets, etc.); emotionally thinking, the experts in the project team may express their emotive-based perceptions of the problem; negatively thinking, the experts should express their pessimistic evaluation on the presence and impact of the *Carpobrotus* emphasizing the implication on biodiversity and socio-ecological systems; positively thinking, the project team may have the opportunity of highlighting the possible (and also in an anti-conformist way) positive implications of the presence of this exotic species on the site (an ecological vicariant? A possible aesthetic role of flowers on dunes? An opportunity to increase awareness of exotic species on general people frequenting the beaches?); creatively thinking, it should stimulate the experts on the project team to define innovative ideas in dealing with this species (e.g. eradication of *Carpobrotus* promoted during an educational stage for children; implication on the beach people, economic use of the plant, etc.); cold and controlled thinking, it should organize all the positive, negative, creative inputs in a technical-scientific scenario, deleting not realizable or naïve ideas; finally, organizational thinking, it should conclude the process, organizing all the inputs in an effective arrangement.

Other Problem-Solving Strategies

There are many techniques to solve problems used in a large number of disciplines (from psychology, to computer sciences, to engineering, marketing and economical areas) usually called problem-solving strategies. Some of them are simple qualitative techniques that may be used quickly by wildlife managers and practitioners. They may be classified into the following categories (Wang and Chiew 2010): (1) abstraction: i.e. solving the problem in a model of the system before applying it to the real system; (2) analogy: i.e. using a solution that solves an analogous problem; (3) brainstorming (a way to come up with the solutions): i.e. among groups of people, suggesting a large number of solutions or ideas and combining and developing them until an optimum solution is found; (4) divide and conquer (a way to implement solutions): i.e. breaking down a large, complex problem into smaller, solvable problems; (5) hypothesis testing: i.e. assuming a possible explanation to the problem and trying to prove (or, in some contexts, disprove) the assumption; (6) lateral thinking: i.e. approaching solutions indirectly and creatively (see also de Bono 1987); (7) means–ends analysis: i.e. choosing an action at each step to move closer to the goal; (8) method of focal objects: i.e. synthesizing seemingly non-matching characteristics of different objects into something new; (9) morphological analysis: i.e. assessing the output and interactions of an entire system; (10) proof: try to prove that the problem cannot be solved. The point where the proof fails will be the starting point for solving it; (11) reduction: transforming the problem into another problem for which solutions exist; (12) research: using existing ideas or adapting existing solutions to similar problems; (13) root cause analysis: identifying the cause of a problem (yet cited for the context analysis); (14) trial-and-error: testing possible solutions until the right one is found. Many of them, or the logics inside them, were translated in the conservation biology and wildlife management arenas and may be useful to solve problems in projects (Shea et al. 1998; Possingham et al. 2001), also in a problematic wildlife scenario.

From the Problems to Solutions: Definition of the Priorities

Prioritization: Triage and Threat Analysis (Ranking)

Prioritization is a process largely used in conservation, especially in conservation planning (Moilanen and Wilson 2009). In wildlife management, this stage is adopted using many approaches, often using expert-based methods and a score evaluation, when data are lacking. However, the process of prioritizing actions is known as triage (Bottrill et al. 2008). Triage allows for an efficient allocation of conservation resources, when resources are scarce. An interesting further approach is known as threat analysis, as highlighted before. This approach to prioritize allows an expert-based evaluation using scores, thus allowing a threat rating and ranking (Salafsky et al. 2008). Having a ranking among a set of threats in a site facilitates the prioritizing among different problems. Consequently, the project team may select a limited number of threats that may represent the object of the strategy.

From the Problems to Solutions: Definition of the Objectives

Once concluded, the context analysis (or situation analysis) defines a general mission and a vision aimed to solve a main problem (e.g. control an invasive species in a site) focused on specific targets and threats. Finally, when the project team has selected a set of priorities, it may better define the end and main objectives of their project (Sutherland 2000). For example, once defining which targets are threatened by an invasive species, and once defining how to investigate the local or general status of an invasive species and the relative threatened targets (sampling design, methods, protocols), managers should decide which actions to develop (and where and when), predicting also possible future scenarios, etc. In a synthesis, they should develop a set of main objectives, relating each one to a single problem.

In management and, in our case, in problematic wildlife contexts, we widely use the approach that uses the decision-tree concept (based on "tree problems" and "tree objectives"; see cap.) and facilitating the definition of "tree solutions" (Magee 1964). A problem tree provides an overview of all the known causes and effects to an identified problem. This is important as it establishes the context in which a project is to occur. Understanding the context helps reveal the complexity of the systems which is essential in planning a successful change project.

A problem tree involves writing causes in a negative form (e.g. increasing rate of pressure of an exotic fish impacting a native crustacean community). Reversing the problem tree, by replacing negative statements with positive ones, creates a solution tree (e.g. reduce the increasing rate of pressure, etc.). A solution tree identifies means–ends relationships as opposed to cause–effects. This provides an overview of the range of projects or interventions that need to occur to solve the core problem. In other words, a problem tree analysis helps to plan a project, providing a guide to the complexity of a problem by identifying the multiple causes and finally, identifies particular lines of intervention inside the project.

The Scenario Analysis

A key challenge in project management is to examine the range of plausible futures emerging in socio-ecosystems under conditions of uncertainty and complexity. Scenario analysis provides a powerful tool for integrating knowledge, scanning the future in an organized way and internalizing human choice into sustainable science (Swart et al. 2007). Also in the wildlife management arena, decisions about how, when, and where to act are typically based on our expectations for the future. When the world is highly unpredictable and we are working from a limited range of expectations, our expectations will frequently be proved wrong. Scenario planning offers a framework for developing conservation policies when faced with uncertainty, and relating to the future consequences of a decision. Ideally, scenarios should be constructed by a cognitively diversified project team for a single purpose. The participation of a diverse group of people (increase of cognitive diversity, see above) in a systemic process of collecting, discussing and analysing scenarios builds shared understanding (Peterson et al. 2003).

Towards the Monitoring

There are many approaches to monitoring and evaluating conservation and wildlife management (review in Stem et al. 2005). Here, we would provide some suggestions and approaches that may be useful in problematic wildlife management.

Define Indicators Not Only Among Impacted Targets

Noss (1990) defined a useful framework to define indicators at different hierarchical ecological levels. These indicators coincide with components of biodiversity and ecological processes. Enlarging this arrangement, Salafsky and Margoluis (1999) stated that some stages of project management may be more effective in select indicators, even components not contributing to biodiversity, such as among human-induced threats (i.e. among abiotic processes and factors). Indeed, they stated that: (1) it may be often difficult to obtain data from biodiversity targets; (2) responses of biodiversity indicators may be delayed and not linear, thus being difficult to obtain clear indications from it. Therefore, when a project team works to solve a problem (e.g. reduce the intensity of a threat), it may be more easy to obtain robust information directly from the metrics of threat (e.g. extension, duration, frequency, intensity of the threat) and on their change following the project (see the Threat Reduction Assessment, below). In the arena of problematic wildlife, for example we may obtain reliable information directly measuring the expansion of the range of the exotic tree *Ailanthus altissima* (using range size as a metric of threat). This metric is very easy to detect when comparing to metrics on the biodiversity targets impacted from this exotic species (often very difficult to calculate).

Decide Indicators in Many Steps of the Project Management: The DPSIR (Driving Forces, Pressures, State, Impacts, Responses) Approach

This framework has been adopted by the European Environment Agency. Generally speaking, DPSIR is a causal framework for describing the interactions between society and the environment. The components of this model are: Driving forces, Pressures, States (of targets), Impacts, and Responses. Once defined these components will be possible to define useful set of related indicators of Driving Forces, Pressures, etc.. As a first step, data and information on all the different elements in the DPSIR chain are collected. Then, possible connections between these different aspects are postulated. Using the DPSIR modeling framework, it is possible to gauge the effectiveness of responses put into place, because, as stated before, for each component we may select a number of indicators (indicators of Driving forces, of Pressures, etc.). This approach can encourage and support decision-making, by pointing to clear steps in the causal chain where the chain can be broken by policy action. The DPSIR represents a systems analysis view: social and economic developments exert pressures on the environment and, as a consequence, the state of the environment changes. This leads to impacts on ecosystems that may elicit a societal response that feeds back on the driving forces, on the pressures or on the state or impacts directly, through specific projects (EEA 2007).

For example, wild boar (*Sus scrofa*) is a mammal frequently released for hunting. This species could become problematic in some contexts, impacting different socio-economic targets (stakeholders, plants, reptiles, small mammals). A DPSIR approach applied to this species may consider: (1) number of hunting organizations, as indicator of Driving forces; (2) density of wild boar, as indicator of Pressure; (3) density of sensitive small mammal species or reptiles (or the cover of plant species) that suffer the impact, as indicators of State; (4) change in density (or cover) of the previous targets, as indicators of Impact and, finally, (5) number of actions (laws, projects, etc.) contemplated to control the species, as indicators of Response.

Using a Monitoring Design: The BACI (Before-After-Control-Impact) Approach

The purpose of impact assessment is to evaluate if a stressor (e.g. in this case, a problematic target) has changed the environment, which components are adversely affected, and to estimate the magnitude of the effects. To make this strategic, we must define a robust sampling design including sampling in a treatment area (the area where a stressor act or Impact area) and a similar area where the stressor is absent (or Control area). When information is available prior to the potential impact, we have the possibility to sample an a-priori (Before) and *a-posteriori* context (After). Therefore, this design is often referred to as a *Before–After-Control-Impact* (BACI) design (Smith 2013). With this monitoring design, it is possible to control the environmental factors not directly connected to the stressor. BACI can include experimental, quasi-experimental, non-experimental and qualitative design. For example, to assess the impact of the American mink (*Neovison vison*) on a small rodent community along a river of central Europe, practitioners could use this approach. If field sampling data are available before the introduction (*Before* phase), they may be compared to comparable data (using the same reliable sampling design) obtained after that mink individuals were introduced (*After* phase). Moreover, it may be possible to compare these data sets to others belonging to comparable control study areas (i.e. areas where minks not occur), before and after that this exotic species has been introduced. This design allows to detect patterns in abundance or diversity metrics correlating them with a set of environmental conditions and constraints (the presence/absence of species, context and temporal factors, etc.), so obtaining reliable information on the relationships among the problematic target and a set of variables.

The Qualitative TRA (Threat Reduction Assessment)

As stated before, current biologically based indicators aimed to measure conservation outcome have a number of practical limitations that preclude their use by typical project teams. As a result, most project teams do not measure project outcome and thus find it difficult to determine if their interventions are working. To address this problem, Salafsky and Margoluis (1999) have developed an qualitative assessment approach called Threat Reduction Assessment (TRA) that measures project outcomes using indicators selected among threats (not among impacted targets). We found that although the TRA approach has the theoretical disadvantages of being a proxy measurement of biodiversity and is subject to bias, it has the theoretical advantages of being sensitive to changes over short time periods and throughout a project site, and of allowing comparison among projects in different settings. Furthermore, it is practical and cost-effective because (it is based on data collected through simple techniques), it is directly related to project interventions and project team can readily interpret it.

Conclusions

Many recent conceptual tools and approaches are now available for practitioners and managers that work in problematic wildlife scenarios. They belong to the wide interdisciplinary sectors of problem solving and decision-making and their use and improvement should make actions more effective in conservation biology and wildlife management. In this work, we have not mentioned additional tools available in the adaptive management arena (Lancia et al. 1996) and many other useful concepts (e.g. threshold models: Suding and Hobbs 2009) that together with those which have been mentioned above may better improve the practices of management on problematic wildlife.

Acknowledgment We would acknowledge Caroline Steen and Bruce Leopold have carefully reviewed the first draft of the manuscript.

References

Ascough JC II, Maier HR, Ravalico JK et al (2008) Future research challenges for incorporation of uncertainty in environmental and ecological decision-making. Ecol Model 219:383–399

Battisti C, Luiselli L, Pantano D et al (2008) On threats analysis approach applied to a Mediterranean remnant wetland: is the assessment of human-induced threats related into different level of expertise of respondents? Biodiv Conserv 16:1529–1542

Battisti C, Franco D, Luiselli L (2013) Searching the conditioning factors explaining the (in)effectiveness of protected areas management: a case study using a SWOT approach. Environ Pract 15:401–407

Bottrill MC, Joseph LN, Carwardine J et al (2008) Is conservation triage just smart decision making? Trends Ecol Evol 23:649–654

Christensen S, Kreiner K (1991) Prosjektledelse under usikkerhet, Universitetsforlaget A/S, Oslo (Title in English: "Project Management under Uncertainty")

Clemen RT (1996) Making hard decision: an introduction to decision analysis. Duxbury Press, Pacific Grove

IUCN (World Conservation Union) (2005a) Threats authority file. Version 2.1. IUCN Species Survival Commission. Cambridge. http://iucn.org/webfiles/doc/ssc/RedList/AuthorityF/threats.rtf

IUCN (World Conservation Union) (2005b) Conservation actions authority file. Version 1.0.IUCN Species Survival Commission. http://iucn.org/webfiles/doc/ssc/RedList/AuthorityF/consaction.rtf

Cooperrider DL, Whitney D, Stavrros JM (2008) Appreciative inquiry handbook, 2nd edn. Crown Custom, Brunswick

de Bono E (1987) Six thinking hats. Penguin, London

EEA (2007) The DPSIR framework used by the EEA. http://root-Devel.ew.eea.europa.eu/ia2dec/knowledge_base/Frameworks/doc101182

Forrester R, Drexler AB (1999) A model for team-based organization performance. Acad Manag Perspect 13:36–49

Hersey P, Blanchard KH, Natemeyer WE (1979) Situational leadership, perception, and the impact of power. Group Organ Manag 4:418–428

Hey J, Waples RS, Arnold ML et al (2003) Understanding and confronting species uncertainty in biology and conservation. Trends Ecol Evol 18:597–603

Hill T, Westbrook R (1997) SWOT analysis: it's time for a product recall. Long Range Plann 30:46–52

Hockings M, Stolton S, Leverington F et al (2006a) Evaluating effectiveness: a framework for assessing management effectiveness of protected areas, 2nd edn. IUCN, Cambridge

Hockings M, Stolton S, Dudley N et al (2006b) Evaluating effectiveness: a framework for assessing the management of protected areas, 2nd edn. IUCN, Cambridge

Hong L, Page SE (2004) Groups of diverse problem solvers can outperform groups of high-ability problem solvers. Proc Natl Acad Sci 101:16385–16389

Jacobson SK, McDuff MD (1998) Training idiot savants: the lack of human dimensions in conservation biology. Conserv Biol 12:263–267

Lancia RA, Braun CE, Collopy MW et al (1996) ARM! For the future: adaptive resource management in the wildlife profession. Wildl Soc Bull 24:436–442

Linstone HA, Turoff M (1975) Delphi method: techniques and applications. Addison-Wesley, Boston

Magee JF (1964) Decision trees for decision-making. Harv Bus Rev 42:126–138
Moilanen A, Wilson KA (2009) Spatial conservation prioritization: quantitative methods and computational tools. Oxford University Press, Oxford
Noss RF (1990) Indicators for monitoring biodiversity: a hierarchical approach. Conserv Biol 4:355–364
Orr SK (2011) The private sector on public land: policy implications of a SWOT analysis of Banff National Park. J Nat Res Policy Res 3:341–354
Page SE (2007) The difference—how the power of diversity creates better groups, firms, schools, and societies. Princeton University Press, Princeton
Peterson GD, Cumming GS, Carpenter SR (2003) Scenario planning: a tool for conservation in an uncertain world. Conserv Biol 17:358–366
Possingham H, Andelman SJ, Trombulak SBR et al (2001) Making smart conservation decisions. In: Soule ME, Orians GH (eds) Conservation biology: research priorities for the next decade. Island Press, Washington, DC, pp 225–244
Pullin AS, Knight TM (2003) Support for decision making in conservation practice: an evidence-based approach. J Nat Conserv 11:83–90
Regan HM, Colyvan M, Burgman MA (2002) A taxonomy and treatment of uncertainty for ecology and conservation biology. Ecol Appl 12:618–628
Salafsky N, Margoluis R (1999) Threat reduction assessment: a practical and cost-effective approach to evaluating conservation and development projects. Conserv Biol 13:830–841
Salafsky N, Salzer D, Ervin J et al (2003) Conventions for defining, naming, measuring, combining, and mapping threats in conservation. An initial proposal for a standard system. Conservation Measures Partnership, Washington, DC. www.fosonline.org/images/Documents/Conventions_for_Threats_in_Conservation.pdf
Salafsky N, Salzer N, Stattersfield AJ et al (2008) A standard lexicon for biodiversity conservation: unified classifications of threats and actions. Conserv Biol 22:897–911
Seely MK, Zeidler J, Henschel JR et al (2003) Creative problem solving in support of biodiversity conservation. J Arid Environ 54:155–164
Shea KS, Amarasekare P, Kareiva P et al (1998) Management of populations in conservation, harvesting and control. Trends Ecol Evol 13:371–375
Smith EP (2013) BACI design. Ecological statistics. Encyclopedia of environmetrics. http://onlinelibrary.wiley.com/doi/10.1002/9780470057339.vab001.pub2/abstract?deniedAccessCustomisedMessage=&userIsAuthenticated=false
Stem C, Margoluis R, Salafsky N et al (2005) Monitoring and evaluation in conservation: a review of trends and approaches. Conserv Biol 19:295–309
Suding KN, Hobbs J (2009) Threshold models in restoration and conservation: a developing framework. Trends Ecol Evol 24:271–279
Sutherland WJ (2000) The conservation handbook. Blackwell Science, Massachusetts
Swart RJ, Raskin P, Robinson J (2007) The problem of the future: sustainability science and scenario analysis. Glob Environ Change 14:137–146
Wang Y, Chiew V (2010) On the cognitive process of human problem solving. Cogn Syst Res 11:81–92

Chapter 7
Managing Human–Elephant Conflict in Zimbabwe: A Boundary Perspective Rather Than a Problematic Species Issue

Sébastien Le Bel, Mike La Grange, and René Czudek

Background of Human–Wildlife Conflict in Zimbabwe

Zimbabwe is a landlocked country in southern Africa with a total surface area of 391,000 km^2; much of the country is semi-arid. The estimated human population was about 12 million in 2010 with 70 % of the Zimbabwean population living in the rural areas (UN 2011). Poverty continues to be a great concern especially where rain fed crop production, the most common method of meeting livelihood demands, is constrained by a low and erratic rainfall. Livestock farming is also an equally risky enterprise due to endemic diseases, unpredictable droughts and limited access to good quality grazing and markets (Mitchell 2001). As a result of living in such fragile agro-ecological conditions characterized by uncertainty and limited livelihood options, most households still depend largely on wild food resources, including bushmeat, and assistance from wealthier households, government and non-governmental organizations (Cunliffe 2010). With 15 % of its land protected as National Parks, Forest Reserves and Botanical Gardens (Fig. 7.1), two successful

S. Le Bel (✉)
Cirad Department ES UPR BSEF, TA C-22/D101, Campus International de Baillarguet, Montpellier Cedex 5 34398, France
e-mail: sebastien.le_bel@cirad.fr

M. La Grange
AWMC (Pvt) Ltd, Suite 1-5 Murandy Square West Newlands, Harare, Zimbabwe
e-mail: mhiripiribomber@gmail.com

R. Czudek
Food and Agriculture Organization of the United Nations, Sub-regional Office for Southern Africa, Block 1 Tendeseka Office Park, Corner Samora Machel Ave/Renfrew Rd., Eastlea, P.O. Box 3730, Harare, Zimbabwe
e-mail: Rene.Czudek@fao.org

F.M. Angelici (ed.), *Problematic Wildlife*, DOI 10.1007/978-3-319-22246-2_7

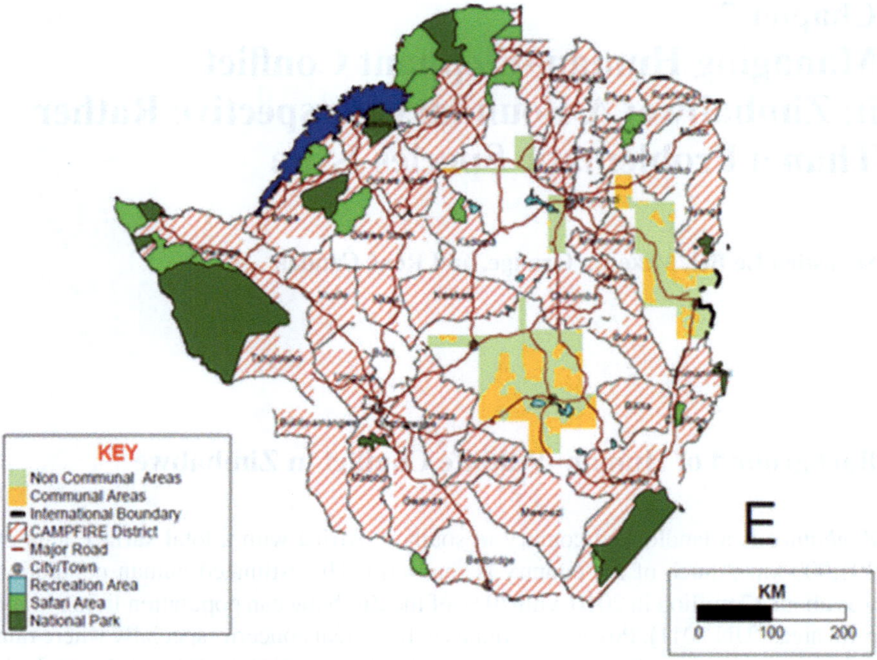

Fig. 7.1 Map of Zimbabwe showing key protected areas and CAMPFIRE Districts (Taylor 2006)

ventures were launched with the aim of reconciling conservation challenges with rural development issues.

Wildlife harvesting which started in the late 1960s was pioneered by game cropping ventures on commercial ranches and was reinforced in 1975 by the Parks and Wildlife Act conferring 'privileges on owners or occupiers of alienated land as custodians of wildlife' (Bond et al. 2004). The result was the establishment of a flourishing wildlife business, employing community members, with the creation of large cooperatively managed wildlife areas called conservancies (PriceWaterhouse 1994).

CAMPFIRE: The second pillar of this wildlife success story was the involvement of local communities through the Communal Areas Management Program for Indigenous Resources (CAMPFIRE) in an attempt to share the benefits of the wildlife revenue with the rural people, but also to reduce the acute cost of Human–Wildlife Conflict (HWC) on the boundaries of National Parks and safari areas. Conceptually, the philosophy behind CAMPFIRE was that local communities of dedicated districts (Fig. 7.1) would sustainably manage wildlife and other resources when: (1) rights and responsibilities to protect and use wildlife were devolved to them as managers (Martin 1986); (2) benefits of managing the natural resources exceed the costs (Murphree 1990); (3) communities entered into business partnerships with the private sector (Katerere 2002); (4) benefits of wildlife conservation were captured by the local communities as resource managers and communities were small enough to be cohesive (Murphree 1991). Rural people will protect and sustainably use wildlife and related natural resources when the benefits of doing so are perceived to exceed the costs (Murphree 1991; Duffy 2000; Baldus 2009), and this will in turn increase tolerance towards wildlife species, especially elephants (Taylor 1993). As Bond (2001) points out, economic benefits are considered to provide the most important motivational factor influencing local people's interest and participation in community-based wildlife management.

During the first 15 years of CAMPFIRE (1989–2006), over USD30 million was generated for the participating communities, most of this income came from sport hunting (80–90 %) and mostly (60–65 %) from elephant sport hunting (Taylor 2009; Frost and Bond 2008; Taylor and Cumming 1993; Bond 1994; Maveneke 1996).

Despite the early promise of this seemingly idyllic situation, over the last 15 years two issues have exacerbated the occurrence and magnitude of HWC, and have resulted in a decreased tolerance of local communities towards wildlife in general.

Wildlife-Based Land Reform Policy (*WBLRP*): Since the year 2000, the Land Reform Program has resulted in communities settling on former commercial farms as well as sections of National Parks and opening up new, but scattered, cultivated lands adjacent to Protected Areas. Following some early studies that concluded that land reform and wildlife management could be reconciled (Wolmer et al. 2003), the government drafted the WBLRP of 2004 which recognized that wildlife is a viable land use option especially in agriculturally marginal areas. Its principal objectives were to ensure more equitable access by the majority of Zimbabweans to land, wildlife resources and business opportunities that stem from the use of these resources, and to develop and implement appropriate institutional arrangements for wildlife-based land reform.

However, the communities turned to an emphasis on traditional crop and livestock production, to the exclusion of wildlife, and hence fuelled poaching (Lindsey et al. 2011b), habitat degradation and woodland loss in newly settled areas (Wolmer et al. 2003). This often took place in areas of virgin bush that were unsuitable for agricultural use, and resulted in a greater incidence of illegal off-take of wildlife in the form of poaching. This rejection of wildlife management was also linked to the collapse of law enforcement and a general economic break down within Zimbabwe. As an example, from August 2001 to July 2009, the Savé Valley Conservancy recorded 10,520 illegal hunting incidents and at least 6454 wild animals were killed with an estimated future financial loss predicted to exceed USD1.1 million per year (Lindsey et al. 2011a, b).

CAMPFIRE's failure since 2006: CAMPFIRE felt the impact of the political and economic crises prevalent throughout the country, and the quality of governance and community benefits became drastically reduced (Balint and Mashinya 2008). There is currently a general awareness that the programme is failing to resolve HWC, particularly Human–Elephant Conflict (HEC) when operational costs of managing wildlife far exceed the benefits to local communities (Gadzirayi 2007; Le Bel et al. 2011), and especially when promised benefits are not received (Fischer et al. 2005; Rihoy et al. 2010). Household incomes from CAMPFIRE participation were often insufficient to offset crop damage and loss, which was aggravated by the disproportionate allocation of safari revenues in favour of the Rural District Councils (Logan and Moseley 2002). The recent decline in distribution of revenues to sub-district levels was largely due to a deterioration of the national economy since the 2000s. This was associated with exchange rate distortions and the lack of commitment of rural district councils to disburse scarce financial resources (Taylor 2009; Rihoy et al. 2010).

Human–Elephant Conflict

Local overabundance: Paradoxically, at the same time as the poaching upsurge, Zimbabwe had to cope with an overabundance of its elephant population that encroached onto farming land, thus placing elephants as the primary conflict species.

Southern Africa holds the largest population of elephants within the continent; this currently numbers 300,000 animals, three quarters of which live within the boundaries of two countries, Botswana and Zimbabwe. In Zimbabwe, the latest aerial counts place the national population at 83,000 (LaGrange pers. com), this being similar to the 2006 estimation of 84,416 elephants (Blanc et al. 2007). The range area for elephant in Zimbabwe is 77,000 km^2 (29 % of the county's land mass) and 58 % of this (45,000 km^2) is within protected areas. It is estimated that 10–15 % of the country's elephant population is now living outside of protected areas and therefore in direct contact with subsistence farmers (Dunham and Mackie 2002; Foggin 2003) with an estimated density of 0.43 elephants/km^2 in CAMPFIRE hunting areas (Taylor 2009).

With many new settlers having moved into areas contiguous with reserves or National Parks (following the Fast Track Land Reform programme started in 2000) to practice subsistence farming, Zimbabwe now has all the ingredients to fuel HWC hot spots (Le Bel et al. 2011; FAO 2009). This overabundance and encroachment onto particularly new farming land has resulted in increased HEC (Nelson et al. 2003; WWF 2005). Elephant bulls (Karidozo and Osborn 2005) have predominated in problem animal control (PAC) cases concerning crop raiding (cotton, maize, sorghum, etc.) with frequent losses of both human and elephant lives (Nelson et al. 2003). Between 2002 and 2006 more than 5000 cases of HEC were recorded in Zimbabwe which resulted in the killing of 774 elephants during subsequent PAC operations (Campfire 2007).

In planning elephant ranges for the future, cognizance should be given to the spatial relationships between elephant and human populations (Hoare and du Toit 1999). Furthermore, attention should also be given to the rural people who live alongside and in competition with these animals, allowing them to benefit more from the wildlife rather than rousing animosity (Schmidtz 2002). In response to the increase in HEC, the wildlife authority (PWMA) decided, as a shortcut, to allocate a nationwide management quota of 1000 elephant per year from 2008; this was in addition to the 500 elephants already sold as trophy hunts each year from 1997. To-date, this management option has not been properly implemented; only one trial of managing local overabundance of elephants through game meat supply was conducted in SVC in 2009–2010 (Le Bel et al. 2013a, b). The cost (USD550 per carcass) and logistic constraints needed for the organization and implementation of such a control measure limit the use of this option as a viable HEC control at the national level (Le Bel et al. 2013a, b).

Spatial and social drivers: HWC exists when the needs and behaviour of wildlife impact negatively on the goals of human beings (Cumming and Jones 2005). It tends to manifest itself in scenarios, where human strategies affect the free movement of wild animals, and thus it becomes inevitable in all communities, where human and wildlife coexist.

HEC, caused by the so-called problem elephants, is considered as a major threat and is a challenge to elephant conservation programmes throughout Africa (Lee and Graham 2006; Hoare 2001; Taylor and Martin 1987; WWF 2005; FAO 2009). Problems have been reported from most of the 37 countries on the African continent where elephants range (Hoare 2000). In southern Africa, the SADC[1] Technical Committee on Wildlife declared HWC and its elephant component to be one of the main problems for Africa's rural populations with regard to personal security and economic loss (Le Bel 2011). In central Africa, the COMIFAC[2] (Kamga Kamdem 2012) urged member states to develop national HEC mitigating strategies similar to the one recently developed in Gabon (MekuiBiyogo 2010).

If we consider the various and well-documented HEC hot spots in Botswana (Gupta 2013; Hanks 2006), Ghana (Danquah et al. 2006), Kenya (Gichohi et al. 2013; Kamweya et al. 2012; Graham et al. 2009; Sitati et al. 2005), Mozambique (Osborn and Anstey 2002; Le Bel 2011), Namibia (Hanks 2006), Tanzania (Malima et al. 2005), Zambia (Nyirenda et al. 2012; Hanks 2006) and Zimbabwe (Le Bel 2011; Hanks 2006), we can expect the occurrence, and therefore the management, of hundreds of crop raids at district level each year.

Within a project implemented through FAO Technical Cooperation Programme in Zimbabwe, which aimed to improve both food security and the management of natural resources, a study was conducted in November 2010 with the objective of understanding the dynamics of HWC at the edge of protected areas, National Parks and Conservancies (Le Bel et al. 2011). In this study, a spatially and socially explicit approach was chosen to map HWC hot spots in order to understand the characteristics of the conflict. Questionnaires were delivered to 613 households randomly selected in such a way that some were closer to the boundaries of protected areas while others were further away. The objectives of the analysis were to understand the changing temporal and spatial nature of HWC and to determine and explain its geographic distribution in relation to key household characteristics such as longevity and location of settlement. Preliminary results showed that HWC hot spots in Chiredzi and Hwange districts were adjacent to protected areas: namely Hwange and Gonarezhou National Parks (Fig. 7.2). In the third site, Mbire district, no clusters existed and HWC seemed to be spread all over the district with an apparent link to the finding that elephants are not permanent residents but transit through this area to surrounding wildlife areas. As most HWC incidents occur in agricultural fields, it was found that conflict was correlated to distance from conservation areas and was a function of vegetation density. HWC hot spots highlight the existence of

[1] SADC: Southern African Development Community.

[2] COMIFAC: Commission des Forêts d'Afrique Centrale.

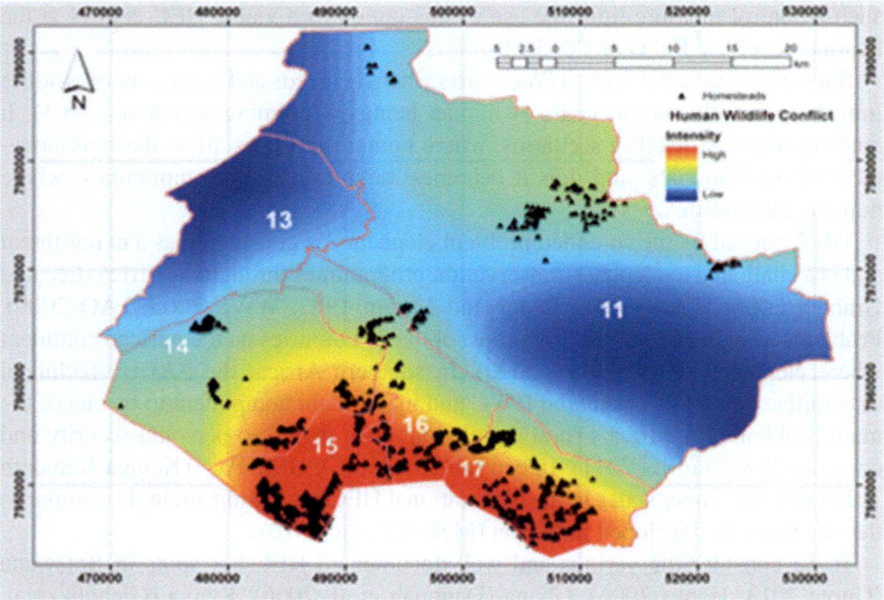

Fig. 7.2 Distribution of the intensity of HWC in the study Wards of the Hwange study site based on fitting a kernel function on the occurrences of HWC (Le Bel et al. 2011)

boundaries of dispute. These interfaces are characterized by temporal and social patterns. Most HEC raids occur at night, when the boundaries are more difficult to maintain. Other observations have revealed that recent settlers are more likely to report incidents of HEC because their settlements are within old memory fences of wildlife areas.

This observation raised the hypothesis that the deep issues of HWC and especially HEC are strongly correlated to boundaries. In other words, the establishment and respect of boundaries between problematic animals and human population could be a new approach to managing HEC; this would be implemented through changing the methods of controlling problematic animals rather to controlling problematic interfaces. This finding could explain why the encroachment of new settlers in wildlife areas, driven by unplanned land reform, is increasing the magnitude of HEC; the settlers are constantly establishing new disputed and unstable boundaries that are unrecognized by wildlife and make the management of HWC almost impossible.

HEC Mitigation

Lessons from 15 years of HEC mitigation have indicated areas of progress with a better understanding of the behaviour of problem elephants and an assessment of the various mitigation methods available. Taken from various manuals (Parker et al. 2007;

WWF 2005), mitigation measures were recently compiled as a set of handy solutions in an HWC tool box developed by FAO (Le Bel et al. 2010a, b). Today, more than 30 solutions are available for HEC mitigation. For ease of utilization, they were classified according to the five types of conflict local communities are facing: human threat, crop raiding, infrastructure damage, water competition and livestock threat. Accordingly, the expected outcomes may also be grouped into four categories: awareness measures, preventing access, chasing away and removing problem animals. For chasing away problem elephants, all nine solutions suggest the use of chilli pepper as an efficient deterrent.

Chilli pepper: The deterrent effect of chilli pepper (fruits of the *Capsicum* genus) on elephants has been studied since the 1990s and is based on the presence of *capsaicin*, a compound that causes the sensation of heat by stimulating noci-receptors of the trigeminal system (watering eyes, burning sensation in the trunk mucosa, trigeminal pain).

Low-tech and sustainable defences using chilli pepper-based olfactory repellents have produced some promising results in deterring elephants from entering crop fields or human habitations (Hoare 2012). The most popular methods are to use chilli pepper grease on traditional fences (Sitati and Walpole 2006) or to burn elephant dung mixed with chilli pepper powder. Mitigation packages of chilli pepper-based measures suitable for small farmers were developed and disseminated during training courses organized by NGOs (Osborn and Parker 2002; Parker and Anstey 2002). Although chilli pepper has been tested with success on crop-raiding elephants (Osborn 2002; Osborn and Parker 2002; Osborn and Rasmussen 1995), its use on a larger scale has been limited, probably due to the lack of widespread knowledge of it in Africa (Sitati and Walpole 2006) and because some unreliable effects were reported in Asia (Hedges and Gunaryasi 2009). The local production of strong chilli pepper oil extract with a high deterrent effect remains a challenge, as does the production of reliable chilli pepper dispensers for use by communities on crop-raiding elephants.

If chilli pepper is to be seriously considered as a local deterrent, this cannot be based solely on chilli pepper by-products grown by the affected farmers, but will require the formal establishment of nurseries or greenhouses dedicated to HEC mitigation purposes. Its production remains a labour-intensive cash crop (FINTRAC 2009) with an expected yield of 400–1200 kg of fresh African bird's eye chilli pepper per hectare (KHDP 2008). With sun drying reducing the average moisture from 75 % (fully ripened chilli pepper) to 10 % (dry chilli pepper) (Wiriya et al. 2009), a maximum of about 100 g of dry chilli pepper is expected per m^2, with an average planting density of 3.5 plants per m^2. In practice, an HEC hot spot needs to regularly harvest a minimum size plot of 25 m^2.

The extraction of capsaicin and the production of locally made chilli oil extract can be done with minimal equipment. After grinding dry chilli pepper as finely as possible with a pestle and mortar, the powder is placed in a sealed bottle and then soaked with unleaded petrol for 48 h. The chilli pepper residue is filtered out using a perforated tin with cotton wool and further washed by adding more fuel to it until the dark redness of the drained liquid is reduced. The liquid is then placed in a large open container in the shade to evaporate slowly to half of its volume. Subsequently, the solution is diluted with locally made vegetable oil at a ratio of ½ a volume of vegetable oil to one volume of fuel. The final solution is then bottled and stored in the shade. The product rates at about 300,000 Scoville Heat Units[3] which is extremely irritating.

Chilli pepper dispenser: Pilot projects to introduce chilli pepper have been slow to develop; they have focused on passive measures with no reliable chilli pepper dispensers for use on crop-raiding elephants. The use of a 'paint-ball' type projectile was suggested, but its utilization in African rural areas encountered problems (Nelson et al. 2003). As a response to this challenge, a first chilli pepper gas dispenser was recently developed in Zimbabwe; it was made with two pieces of PVC and matched the financial and technical capabilities of local communities and individuals (Fig. 7.3). The dispenser propels a standard table tennis ball (henceforth known as a Ping-Pong ball) filled with chilli pepper oil extract. Its description and use are detailed in a paper reporting on the preliminary results of a field test conducted in 2007 in Hwange National Park (Le Bel et al. 2010a, b).

The main result confirmed the combined deterrent effect of the noise produced at firing (100 dB), the hit (>16 J) and the release of hot chilli pepper. From these first attempts, it was agreed to conduct further field trials to improve the system and separate out the discrete effects of projectile impact, the bang produced on firing and the chilli pepper itself. Real-life field testing against crop-raiding elephants,

Fig. 7.3 The initial chilli pepper gas dispenser being demonstrated in Limpopo National Park, Mozambique © S Le Bel

[3] The number of Scoville Heat Units indicates the amount of capsaicin present.

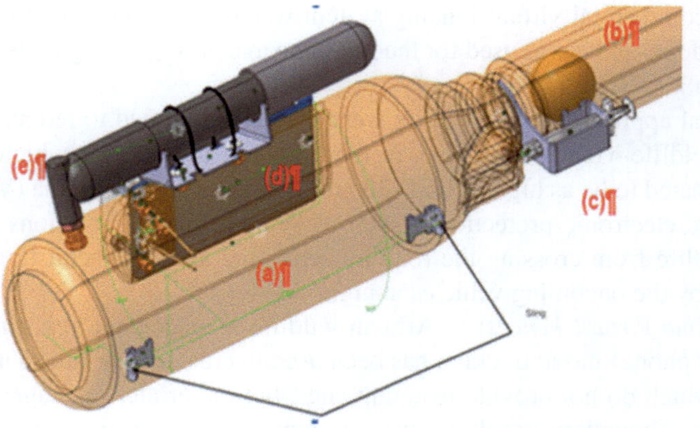

Fig. 7.4 EL@OUT™ chilli dispenser: (**a**) the combustion chamber, (**b**) a telescopic barrel for quick reloading, (**c**) an injection system regulating the volume of standard petroleum fuel, (**d**) an ignition system providing a powerful spark to ignite the fuel-air mix, (**e**) a pump to vent exhaust gases

conducted in Mozambique, Zambia and Zimbabwe indicated that the release of chilli pepper provided the strongest deterrent effect (Le Bel et al. 2014).

These tests also highlighted the need to improve the prototype. The production of an advanced prototype, named EL@OUT, using an industrial moulding process incorporated a set of modifications simplifying the loading process for subsequent shots and allowing flawless function in the darkness of night when most crop raiding occurs (Fig. 7.4)

Memory Fence Dynamics

Women and elephants never forget an injury (Munro 2013)

With a high brain-to-body mass ratio (second only to humans), a long-life, a complex social structure and a large home range, elephants have the reputation of a long-term memory hence the saying that 'an elephant never forgets' (Dale 2008). Their remarkable faculty for learning in either captivity or in the wild highlights elephants' capacity to acquire knowledge and to share it amongst themselves (Goldsworthy 2010). This finding echoes methods of controlling domestic animals without ground-based fencing by placing a non-visual boundary around individual animals or on landscapes (Anderson 2007).

Virtual fencing: Virtual fencing is based on the animals position with respect to the fence line; when approaching the perimeter, the animal is exposed to stimuli which makes them move away (Butler et al. 2004). In this way, the fencing capitalizes on low stress handling principles, in which the animal's behaviour is to move away from a stimulus that has penetrated its fight-flight zone (Anderson 2001).

The first commercial virtual fencing system was patented in 1973 for controlling domestic dogs, and it was used for the first time to control grazing livestock in 1987 (Anderson 2007).

Practical applications for wildlife were mainly developed to reduce the magnitude of wildlife–vehicle collisions in the Northern hemisphere which costs have been estimated to be as high as USD 200 million annually in Canada (Vanlaar et al. 2012). The electronic protection system of Wildlife Safety Solutions (2015) prevents wildlife from crossing the road by activating an audio-acoustic alert devise triggered by the oncoming vehicles at night.

Real-Time Virtual Fences: In African wildlife management, real-time (satellite or cellular phone)-based tracking has been used to create Real-Time Virtual Fences (RTVF) which do not provide irritating cues, but rather alert managers when animals cross a boundary, resulting in a subsequent management action (Jachowski et al. 2014). Commercial products exist, such as the one developed by VECTRONIC Aerospace (2014) that has an option in their GPS collars for alerting wildlife managers [via e-mail and text messages (SMS)] when tracked animals are moving close to a no-go area (e.g. settlements, paddocks, crops). Species that have tight social structures, such as elephants, are ideal for virtual fence designs because a single satellite transmitting collar on the matriarch represents the larger family group's movements (Jachowski et al. 2012).

Another way of establishing RTVF was tested directly with the use of mobile phones. In the vicinity of Laikipia in Kenya (Graham et al. 2011), farmers or rangers are alerted when elephants approach the boundary of the protected areas for crop raiding (BBC 2005; Turrettini 2008); in India, a similar set-up informs people about elephant movement and operates red-flashing LED lights to deter elephants (Saju 2012).

Memorizing human boundaries: The expected outcome of a virtual fencing approach is the training of individual animals or group of animals to understand and respect boundaries without the need for repeated management action.

Over the years of early problem animal control and veterinary fence protection in the late 1980s (Taylor and Martin 1987), it was noticed that elephants could be trained to respect fences via a policy of shooting and thus removing habitual fence-breakers (Nelson et al. 2003). This lethal practice has not proved to be an effective long-term deterrent (Hoare 2012), as it is no longer acceptable to most Wildlife Authorities. However, it formed a basis from which to promote an array of non-lethal memories, which, if repeatedly applied at a boundary, could train crop raiders to stay away from farming ventures. The ability to associate and remember conflict areas means that elephants could be taught to respect boundaries and stay clear of crops; this hypothesis is supported by recent observations during capture and tracking operations.

A first set of observations supporting this memory fence approach came from capture operations. For some time now, routine animal capture has revealed the difficulty, if not the impossibility, of driving elephants through fence lines even where the fences have been removed. This was apparent only when the animals had become habituated to the fence. To illustrate this, a Zimbabwean capture company

(AWMC) was contracted to remove 19 elephant bulls from a small private game farm into the adjacent Hwange National Park in the West of Zimbabwe. Despite the removal of a 500 m section of the fence some weeks prior to the date of capture and the use of a helicopter, the elephants failed to be driven across the fence line, and therefore needed to be captured individually and moved out one by one.

Another example to illustrate this soft memory dynamic came from the monitoring of elephants in the Sikumi Forest Reserve adjacent to Hwange National Park. Sikumi Reserve's common boundary with settled Communal Land was demarcated by a well-maintained cable and wire veterinary fence supported on steel posts. The fence had been in existence for some 30 years and was enforced by regular patrols and the shooting of some elephants and particularly buffalo by Zimbabwean Park and Wildlife Management Authority, where they attempted to break through. In the years 2000–2003, as a result of political change, the fence was systematically entirely removed and was no longer minded and policed. Cropping had previously been discouraged near the fence in order to provide a buffer zone; however, this was subsequently ignored and settlements became established right up to the now none existent fence and became attractive to crop-raiding elephants. A CNRS project was launched in 2010 to examine the feasibility of bee fences as a deterrent for elephant crop raiding, and eight elephant bulls from within the area of the previously existing fence were fitted with satellite radio collars (Guerbois and Fritz 2011). Studying their movement patterns over the entire 2010–2011 cropping season showed conclusively that they still respected the existence of the now removed fence. The elephants often crowded against it but only ventured for a short distance of less than 1 km and for only for a couple of hours, indicating the significance of the 'soft boundary effect' in the elephants' memory, despite the fence having been entirely removed 10 years previously (Fig. 7.5).

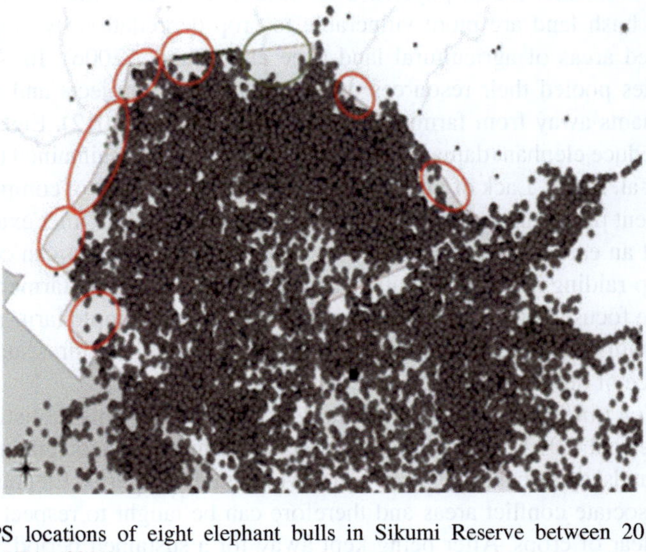

Fig. 7.5 GPS locations of eight elephant bulls in Sikumi Reserve between 2010 and 2011 (Guerbois and Fritz 2011)

Integrating the Chilli Dispenser in a Boundary Strategy

Based on these findings, our hypothesis proposes that the management of HEC may be more effectively based on the establishment and maintenance of a memory fence dynamic. Practical applications are proposed to modify the area of conflict rather killing problem animals as a quick fix solution. The proposed strategy is aimed at establishing and reinforcing the existence of boundaries to exclude problematic animals from human settlements. This can be related to the 'soft fence hypothesis' where human disturbances, or landscape features can act as soft barriers.

The implications of this knowledge for HWC is considerable, providing the potential to train wildlife to respect boundaries as they move around, but more importantly to establish exclusion areas. Mitigation should identify animal movement patterns already in existence and then separate agricultural areas from wildlife corridors, in particular those used by elephants. The proposed protocol would establish, maintain and defend boundaries from elephants by the use of traditional approaches reinforced by innovative tools such as the EL@OUT chilli pepper dispenser.

The mitigation approach developed in the Bio-Hub/FAO HWC toolkit offers a range of simple and low-cost solutions affordable at community level, which will help elephant and other species to develop the memory fence dynamic. The ultimate goal would be for HEC mitigation to be achieved by local communities themselves through the use of passive measures such as repellents rather than involving national agents to remove problem animals.

Man-made boundaries, particularly strong fences or concentrated disturbances will achieve this result with time, as animals learn to avoid them. Depending on the intensity of disturbance, response may be immediate, but generally it will take time and effort to achieve long-term results with the boundary eventually becoming a memory fence rather than a physical one. Small, scattered settlements surrounded by natural bush land are more vulnerable to crop depredation by elephants than consolidated areas of agricultural land (Lee and Graham 2006). In Kenya, local communities pooled their resources in ongoing fencing projects and managed to keep elephants away from farming areas (Kamweya et al. 2012). Electric fencing seems to reduce elephant damage, but it has to be very well maintained (O'Connell-Rodwell et al. 2000). Lack of maintenance, vandalism and theft of components have been frequent problems at the community level (Hoare 2012). Other examples have shown that an early warning system with a guarding component can contribute to reduce crop-raiding incidents (Sitati et al. 2005). For small-scale farmers, resources can only be focused on cheap fencing with the support of simple farm-based deterrents (Graham et al. 2009, 2010); however, this strategy still requires support for the maintenance of deterrent devices.

Improving traditional fences as a boundary perspective: This 'holistic' approach (La Grange et al. 2012) is based on the observation that elephants remember boundaries and landscapes that affect their general movement patterns. Elephants remember and associate conflict areas and therefore can be taught to respect boundaries and stay clear of crops. After being kept away for a sustained period, a boundary could become firmly established in their memory and remain in place even after the removal of the physical boundary. Additional deterrent action has to be administered at the interface, where the elephants are not wanted. We suggest that permanent traditional boundary fences, combined with the use of a passive chilli pepper repellent on twine or cloth (as an olfactory reminder), and the use of the EL@OUT chilli pepper gas dispenser to target persistent offenders, can be successful in reinforcing boundary dynamics.

In practice, we suggest a step-by-step approach combining different mitigation measures described in the FAO HWC toolkit (Czudek 2014). Currently, on most crop/wildlife interfaces there are simply no strategies in place to separate farming and wildlife, which leads to increased tensions and direct encounters between man and elephants. The following picture of a devastated plantain plantation in Gabon (Fig. 7.6) illustrates a typical situation, where risky human activity creates a situation that invites crop raiding.

The first step is to establish a buffer zone between the domestic and the wild areas, which will alert the intruder that he is entering a new territory. The principle of a proper zoning of cropping areas that are separated from elephant movement corridors is a long-term method for helping to reduce HWC (WWF 2005). An effective barrier, such as a traditional brush fence (Fig. 7.7), will set boundaries where the elephants are not wanted.

To alert the presence of intruders, an early warning system can be made from a single wire or rope arrangement supporting empty tin cans containing pebbles. The activation of this alarm system will also alert the elephant that they are penetrating an unfamiliar area (Fig. 7.8).

Fig. 7.6 Plantain field raided by forest elephant (*Loxodonta cyclotis*) in Gabon© S Le Bel

Fig. 7.7 Traditional fence filled thorn bush (*Acacia tortilis*) in Chiredzi, Zimbabwe © S Le Bel

Fig. 7.8 Early warning system made of a single wire with empty cans containing pebbles. Holes made in the can prevent it from becoming a breeding-ground for mosquitos© S Le Bel

To reinforce the boundary during the learning curve, the EL@OUT chilli pepper dispenser, in both its manual and ambush versions, can issue disciplinary action to individuals that persist in breaking through the fence, helping them to respect the boundary (Fig. 7.9).

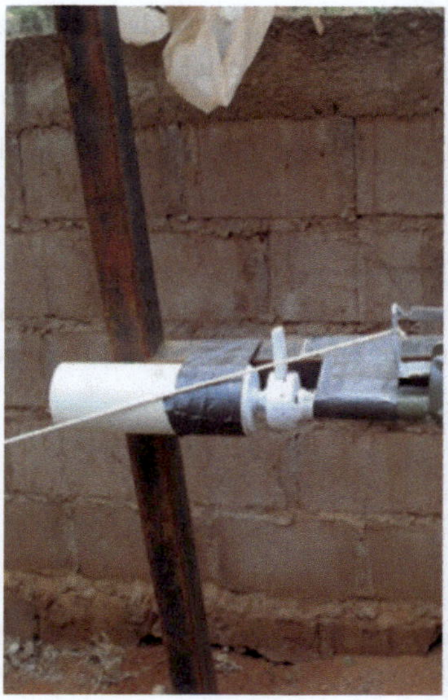

Fig. 7.9 The two types of EL@OUT in action: while the manual version is peppering an undesirable individual, the automatic version will be trigged by the elephant itself when entering in the no-go area © S Le Bel © PPP

The duration of the active deterrent period will be site dependent according to the level of attractiveness of the area to be protected (e.g. crops, orchards, granaries, settlements, water points). It will also depend on the magnitude of the passive application of the chilli pepper extract as an olfactory reminder to activate the memory of the elephant that he is approaching a no-go area. This process of teaching problem elephants to avoid farming areas will work well if the concerned communities or individuals commit to adopting the practice.

Conclusion

Many challenges face today's wildlife managers in combating the relatively new development of HEC. Information relating to the nature of HEC, as well as the reasons behind the occurrence of hot spots, is now critical if this problem is to be effectively dealt with.

As HEC mitigation seeks to increase human tolerance towards wildlife species and decrease negative interactions with them (FAO 2009), the improvement of community tolerance towards wildlife must start by enabling them to protect themselves and to adopt less risky attitudes when confronting dangerous animals.

As no 'stand-alone' solution to HEC exists (Sand and Wikenros 2006), new mitigation measures should be built on existing traditional approaches, thus enabling targeted intervention on specific problem elephants at the crop interface.

The strategy of virtual fencing using active chilli pepper dispensers will help crop-raiding elephants to avoid human settlements through a discipline learning curve. Animal behaviour is never 100 % predictable, therefore virtual fencing should not be used if absolute animal control is required (Anderson 2007).

All the tools and strategies developed should not overshadow the promotion and improvement of wildlife-based revenue ventures, which are essential for ensuring long-term human–wildlife coexistence.

References

Anderson DM (2001) Virtual fencing—a prescription range animal management tool for the 21st century. In: Institute TMLUR (ed) Tracking animals with GPS, Aberdeen, 12–13 March 2001, pp 85–94

Anderson DM (2007) Virtual fencing—past, present and future. Rangeland J 29:65–78

Baldus RD (2009) A practical summary of experiences after three decades of community-based wildlife conservation in Africa "What are the Lessons Learnt?". FAO/CIC, Budapest

Balint PJ, Mashinya J (2008) CAMPFIRE during Zimbabwe's National crisis: local impacts and broader implications for community-based wildlife management. Soc Nat Res 21(9):783–796

BBC (2005) Texting to save Kenyan elephants. http://news.bbc.co.uk/cbbcnews/hi/newsid_4340000/newsid_4347300/4347363.stm.

Blanc JJ, Barnes RFW, Craig GC, Dublin HT, Thouless CR, Douglas-Hamilton I, Hart JA (2007) African elephant status report 2007: an update from the African elephant database. Occasional Paper Series of the IUCN Species Survival Commission No. 33. IUCN/SSC. African Ele. Spe. Group. IUCN, Gland, 276p

Bond I (1994) The importance of sport-hunted African elephants to CAMPFIRE in Zimbabwe. Traffic Bulletin 14(3):117–119

Bond I (2001) CAMPFIRE and incentives for institutional change. In: Hulme D, Murphree M (eds) African wildlife and livelihoods. James Currey, Oxford, pp 227–243

Bond I, Child B, Ddl H, Jones B, Barnees J, Anderson H (2004) Private land contribution to conservation in South Africa. In: Child B (ed) Parks in transition: biodiversity, rural development, and the bottom line. Earthscan, London, pp 29–61

Butler Z, Corkey P, Peterson R, Rus D (2004) Virtual fences for controlling cows. In: ICRA '04. 2004 IEEE International Conference, 26 Apr–1 May 2004. Robotics and Automation, pp 4429–4436. http://dx.doi.org/10.1109/ROBOT.2004.1302415

Campfire (2007) PAC statistics 2002–2006. Campfire Association, Harare

Cumming D, Jones B (2005) Elephants in southern Africa: management issues and options. WWF/SARPO Occasional Paper

Cunliffe R (2010) Community resource use assessments for selected wards within Chiredzi District. Thematic Bulletin Series 1, no. 4 Wild Meat Component, PARSEL Project. Harare, 72p

Czudek R (2014) 1st World Wildlife Day sees launch of human-wildlife conflict management centre in Zimbabwe. http://www.fao.org/archive/from-the-field/detail/en/c/215424/. Accessed 3 Mar 2014

Dale RHI (2008) The spatial memory of African elephants (Loxodonta Africana): durability, interference & response biases. In: Innis NK (ed) Reflections on adaptive behavior: essays in honor of JER Staddon. MIT, Cambridge, pp 143–169. http://digitalcommons.butler.edu/facsch_papers/339

Danquah E, Oppong SK, Sam MK (2006) Aspects du Comportement des Éléphants qui Ravagent les Cultures dans l'Aire de Conservation de Kakum, au Ghana. Nat Faune 21(2):17–21
Duffy R (ed) (2000) Killing for conservation: wildlife policy in Zimbabwe. Indiana University Press/Weaver International African Institute, Bloomington
Dunham K, Mackie C (2002) National summary of aerial census results for elephant in Zimbabwe: 2001. DNPWLM/WWF-SARPO, Harare
FAO (2009) Human-wildlife conflict in Africa: causes, consequences and management strategies. Forestry paper, vol 157. FAO, Rome
FINTRAC (2009) Kenya horticultural development program October 2003–March 2009. USAID-KHDP, Washington, DC
Fischer C, Muchapondwa E, Sterner T (2005) Shall we gather 'round the campfire'?: Zimbabwe's approach to conserving indigenous wildlife. Res Future 158:12–15
Foggin CM (2003) The elephant population problem in Zimbabwe: can there be any alternative to culling?. In: B Colenbrander, de Gooijer J R Paling, S Stout, T Allen (ed) Managing African elephant populations: act or let die? Proceedings of an Expert Consultation on Control of Wild Elephant Populations, Utrecht University, The Netherlands. Beekbergen 7–8 Nov 2003. Faculty of Veterinary Medicine, pp 17–21
Frost PGH, Bond I (2008) The CAMPFIRE programme in Zimbabwe: payments for wildlife services. Ecol Econ 65:776–787
Gadzirayi EMCT (2007) Impact of community-based approaches to wildlife management: case study of the CAMPFIRE programme in Zimbabwe. Int J Sustain Dev World Ecol 14(4):336–344
Gichohi PM, Muthoni DM, Karienye DK (2013) Mapping of human-elephant conflict hotspots for conservation planning and management in Aberdare National Park in Kenya. Int J Innov Res Stud 2(5):14
Goldsworthy A (2010) Effective leadership: learn from the animals. Dog Ear, Indianapolis
Graham MD, Gichohi N, Kamau F, Aike G, Craig B, Douglas-Hamilton I, Adams WM (2009) The use of electrified fences to reduce human elephant conflict: a case study of the Ol Pejeta Conservancy, Laikipia District, Kenya. Working Paper 1. Laikipia Elephant Project, Nanyuki
Graham MD, Notter B, Adams WM, Lee PC, Ochieng TN (2010) Patterns of crop-raiding by elephants, Loxodonta Africana, in Laikipia, Kenya, and the management of human–elephant conflict. System Biodivers 8(4):435–445
Graham MD, Adams WM, Kahiro GN (2011) Mobile phone communication in effective human-elephant conflict management in Laikipia, Kenya. Oryx 46(1):137–144
Guerbois C, Fritz H (2011) Collared Bulls Report—March 2011—Patch selection on the edge. CNRS, HERD Program, Khatschana Lodge Dete Zimbabwe
Gupta AC (2013) Elephants, safety nets and agrarian culture: understanding human-wildlife conflict and rural livelihoods around Chobe National Park, Botswana. J Polit Ecol 20:238–254
Hanks J (2006) Mitigation of human-elephant conflict in the Kavango-Zambezi Transfrontier Conservation Area through community based problem animal control with particular reference to the use of chilli peppers (Final report)
Hedges S, Gunaryasi D (2009) Reducing human-elephant conflicts: do chillies help deter elephants from entering crop field? Oryx 44(1):139–146
Hoare R (2000) Humans and elephants in conflicts: the outlook for coexistence. Oryx 34(1):34–38
Hoare R (2001) A decision support system for managing human-elephant conflict situations in Africa. The African Elephant Specialist Group (AfESG), Nairobi
Hoare R (2012) Lessons from 15 years of human elephant conflict mitigation: management considerations involving biological, physical and governance issues in Africa. Pachyderm 2012:60–74
Hoare RE, Du_Toit JT (1999) Coexistence between people and elephants in African savannas. Conserv Biol 13(3):633–639
Jachowski DS, Slotow R, Millspaugh JJ (2012) Physiological stress and refuge behavior by African elephants. PLoS One 7(2), e31818. doi:10.1371/journal.pone.0031818

Jachowski DS, Slotow R, Millspaugh JJ (2014) Good virtual fences make good neighbors: opportunities for conservation. Animal Conserv 17:187–196

Kamga Kamdem SL (2012) Stratégie d'atténuation des conflits hommes-éléphants en Afrique Centrale 2010 –2014. Comifac & Rapac, Libreville, Gabon, Novembre 2012

Kamweya AM, Ngene SM, Muya SM (2012) Occurrence and level of elephant damage to farms adjacent to mount Kenya forests: implications for conservation. J Biol Agric Healthc 2(5):41–55

Karidozo M, Osborn FV (2005) Can bees deter elephants from raiding crops? An experiment in the communal lands of Zimbabwe. Pachyderm 39:26–32

Katerere Y (2002) Community-public-private partnerships in CBNRM: the real challenges. In: Benjaminsen TA (ed) Contested resources: challenges to the governance of natural resources in southern Africa. PLAAS, University of Zimbabwe, Cape Town, pp 20–38

KHDP (2008) Production guidelines for African bird's eye chili. Horticulture Development Centre, USAID, Kenya

La Grange M, Le Bel S, Czudek R, Stansfield F (2012) Managing human wildlife conflict in Zimbabwe: a boundary perspective rather than a problematic species issue. In: IVth International Wildlife Management Congress 2012, Durban, 9–12 July 2012

Le Bel S (2011) International consultant on Human-Wildlife Conflict mitigation and HWC database development. Projects TCP/MOZ/3301 and TCP/ZIM/3301. FAO, Maputo

Le Bel S, Taylor R, Lagrange M, Ndoro O, Barra M, Madzikanda H (2010a) An easy-to-use capsicum delivery system for crop-raiding elephants in Zimbabwe: preliminary results of a field test in Hwange National Park. Pachyderm 47:80–89

Le Bel S, Mapuvire G, Czudek R (2010b) Human-wildlife conflict toolkit: comprehensive solutions for farmers and communities. Unasylva 236(61):12–13

Le Bel S, Murwira A, Mukamuri B, Czudek R, Taylor R, LaGrange M (2011) Human wildlife conflicts in southern Africa: riding the whirl wind in Mozambique and in Zimbabwe. In: Pujo JL (ed) The importance of biological interactions in the study of biodiversity. InTech, Rijeka, pp 283–322

Le Bel S, Mombeshora S, Goredema L, Mapuvire G, Jiri A (2013a) Involving communities in wildlife ranching in Zimbabwe, a grass-roots initiative. Journal of Wildlife Management. Proceedings of the IVth International Wildlife Management Congress. Cooperative wildlife management across borders: learning in the face of change, 9–12 July 2012, Durban, pp 53–63

Le Bel S, Stansfield F, LaGrange M, Taylor R (2013b) Managing local overabundance of elephants through game meat supply, the case of Savé Valley Conservancy in Zimbabwe. S Afr J Wildl Res 43(2):103–119

Le Bel S, LaGrange M, Drouet-Hoguet N (2014) Repelling elephants with a chili pepper gas dispenser: field tests and practical use in Mozambique, Zambia and Zimbabwe from 2009 to 2013. Pachyderm

Lee PC, Graham MD (2006) African elephants Loxodonta africana and human–elephant interactions: implications for conservation. Int Zoo Yearb 40:9–19

Lindsey P, Romanach S, Matema S, Matema C, Mupamhadzi I, Muvengwi J (2011a) Dynamics and underlying causes of illegal bushmeat trade in Zimbabwe. Oryx 45(1):84–95

Lindsey P, Romanach S, Tambling C, Chartier K, Groom R (2011b) Ecological and financial impacts of illegal bushmeat trade in Zimbabwe. Oryx 45(1):96–111

Logan BI, Moseley WG (2002) The political ecology of poverty alleviation in Zimbabwe's communal areas management programme for indigenous resources (CAMPFIRE). Geoforum 33(1):1–14. doi:10.1016/s0016-7185(01)00027-6

Malima C, Hoare RE, Blanc JJ (2005) Systematic recording of human-elephant conflict: a case study in south-eastern Tanzania. Pachyderm 38:29–38

Martin RB (1986) Communal areas management programme for indigenous resources (CAMPFIRE). Branch of Terrestrial Ecology, Department of national Parks and Wildlife Management, Harare

Maveneke TN (1996) The sustainable use of elephants and its importance to CAMPFIRE. In: The elephant workshop, Etosha, Namibia, 26–29 Feb 1996, p 5

MekuiBiyogo AP (2010) Stratégie Nationale et Plan d'Actions de Gestion des Conflits Homme-Faune au Gabon. Ministère des Eaux et Forêts, République Gabonnaise, Libreville, Gabon, Septembre 2010

Mitchell TW (2001) The land crisis in Zimbabwe: getting beyond the myopic focus upon Black & White. Indiana Int Comp Law Rev 11(3):587–603

Munro HH (2013) Reginald on besetting sins: the woman who told the truth. In: Munro HH (ed) The Saki Megapack: 140 classic novels and short stories. Wildside, Rockville, pp 378–380

Murphree MW (1990) Decentralising the proprietorship of wild resources in Zimbabwe's communal lands. CASS, University of Zimbabwe, Harare

Murphree MW (1991) Communities as institutions of natural resources management. CASS, University of Zimbabwe, Harare

Nelson A, Bidwell P, Sillero-Zubiri C (2003) A review of human-elephant conflict management strategies: people and wildlife initiative. Wildlife Conservation Research Unit, Oxford University, Oxford

Nyirenda VR, Myburgh WJ, Reilly BK (2012) Predicting environmental factors influencing crop raiding by African elephants (Loxodonta africana) in the Luangwa Valley, eastern Zambia. Afr J Environ Sci Technol 6(10):391–400

O'Connell-Rodwell CE, Rodwell T, Rice M, Hart LA (2000) Living with the modern conservation paradigm: can agricultural communities co-exist with elephants? A five-year case study in East Caprivi, Namibia. Biol Conserv 93:381–391

Osborn FV (2002) Capsicium oleoresin as an elephant repellent: field trials in the communal lands of Zimbabwe. J Wildl Manage 66(3):674–677

Osborn FV, Anstey S (2002) Elephant/human conflict and community development around the Niassa Reserve, Mozambique. Mid Zambezi Elephant Project, Harare

Osborn FV, Parker GE (2002) Living with elephants II: a manual for implementing an integrated programme to reduce crop loss to elephants and to improve livelihood security of small-scale farmers. Mid Zambezi Elephant Project, Harare. www.elephantpepper.org/downloads/manual/%202.2.pdf

Osborn FV, Rasmussen LEL (1995) Evidence for the effectiveness of an oleoresin capsicum aerosol as a repellent against wild elephants in Zimbabwe. Pachyderm 20:55–64

Parker GE, Anstey SG (2002) Human-elephant conflict and community development in Niassa Province, Mozambique. WWF, Mid-Zambezi Elephant Project, Harare

Parker GE, Osborn FV, Hoare RE, Niskasen LS (eds) (2007) Human-elephant conflict mitigation: a training course for community-based approaches in Africa. Trainer's manual. Elephant Pepper Development Trust & IUCN/SSC AfESG, Livingstone

PriceWaterhouse (1994) The lowveld conservancies: new opportunities for productive and sustainable land-use. Savé Valley, Bubiana and Chiredzi River Conservancies, Chiredzi

Rihoy L, Chirozva C, Anstey S (2010) "People are not happy": crisis, adaptation and resilience in Zimbabwe's CAMPFIRE programme. In: Nelson F (ed) Community rights. Conservation and contested land. Earthscan, London, pp 174–201

Saju MT (2012) How SMS saves humans from elephants in the tea meadows of Valparai. The Time of India. http://timesofindia.indiatimes.com/city/coimbatore/How-SMS-saves-humans-from--elephants-in-the-tea-meadows-of-Valparai/articleshow/16921832.cms. Accessed 12 Feb 2015

Sand H, Wikenros C (2006) Controlling elephants with electric fences in Sri Lanka. Proc Biol Sci 273(1592):1087

Schmidtz D (2002) When preservationism doesn't preserve. In: Schmidtz D, Willott E (eds) Environmental ethics, 1st edn. Oxford University Press, Oxford, pp 320–329

Sitati NW, Walpole MJ (2006) Assessing farm-based measures for mitigating human-elephant conflict in Transmara District, Kenya. Oryx 40(3):279–286

Sitati NW, Walpole MJ, Leader-Williams N (2005) Factors affecting susceptibility of farms to crop raiding by African elephants: using a predictive model to mitigate conflict. J Appl Ecol 42:1175–1182

Taylor RD (1993) Elephant management in Nyaminyami District, Zimbabwe: turning a liability into an asset. Pachyderm 17:19–29

Taylor RD (2006) Case studies on successful southern African NRM initiatives and their impacts on poverty and governance. Case study: Campfire (communal areas management programme for indigenous resources), Zimbabwe. IUCN-TPARI (Trans-boundary Protected Areas Research Initiative)

Taylor RD (2009) Community based natural resource management in Zimbabwe: the experience of CAMPFIRE. Biodivers Conserv 18(10):2563–2583. doi:10.1007/s10531-009-9612-8

Taylor RD, Cumming DHM (1993) Elephant management in southern Africa. WWF/Zimbabwe DNPWLM, Harare

Taylor RD, Martin RB (1987) Effects of veterinary fences on wildlife conservation in Zimbabwe. Environ Manage 11(3):327–334

Turrettini E (2008) Text messages save elephants' lives, villagers' crops. Textually.org. http://www.textually.org/textually/archives/2008/10/021426.htm. Accessed 12 Feb 2015

UN (2011) Total population, both sexes combine. Last update in UN data: 25 Aug 2011. United Nations Population Division, New York

VECTRONIC Aerospace (2014) Virtual fence: a location specific GPS mode for GPS PLUS collars. http://www.vectronic-aerospace.com/files/Flyer_VIRTUAL_FENCE.pdf. Accessed 11 Mar 2015

Vanlaar WGM, Gunson KE, Brown SW, Robertson RD (2012) Wildlife-vehicle collisions in Canada: a review of the literature and a compendium of existing data sources. Traffic Injury Research Foundation, Ottawa

Wildlife Safety Solutions (2015) Virtual fencing. http://www.wildlifesafetysolutions.com.au/. Accessed 11 Mar 2015

Wiriya P, Paiboon T, Somchart S (2009) Effect of drying air temperature and chemical pretreatments on quality of dried chilli. Int Food Res J 16:441–454

Wolmer W, Chaumba J, Scoones I (2003) Wildlife management and land reform in Southeastern Zimbabwe: a compatible pairing or a contradiction in terms? Sustainable livelihoods in southern Africa: Institutions, Governance and Policy Processes. Institute of Development Studies, Brighton

WWF (2005) Human wildlife conflict manual. WWF-SARPO, Harare

Chapter 8
A Review of Livestock Predation by Large Carnivores in the Himalayan Kingdom of Bhutan

Rajanathan Rajaratnam, Karl Vernes, and Tiger Sangay

Introduction

Regional Context

Covering an area of 38,394 km^2 (MoAF 2011), Bhutan is a predominantly Buddhist country located between India and China in the Eastern Himalayas (Fig. 8.1). It is mountainous with 95 % of its land area located more than 600 m above sea level (asl) (MoA 2009; MoAF 2011). Altitudes range from 80 m asl (Tempa et al. 2011) to 7500 m asl (Sherub 2004) while annual precipitation ranges from 5500 mm in the south (MoA 2002) to a relatively lower 500 mm in the alpine areas (Sangay and Wangchuk 2005). Bhutan encompasses two bio-geographical realms: the Oriental and the Sino-Japanese (Holt et al. 2013), resulting in rich biodiversity. This biodiversity is distributed across subtropical forests, warm and cool broad-leaved forests, to temperate conifer forests and alpine meadows. More than 5600 species of angiosperms and gymnosperms have been recorded in Bhutan including 369 species of orchids and 46 species of rhododendrons. Bhutan has at least 105 endemic

R. Rajaratnam (✉)
Department of Geography and Planning, University of New England, Armidale, NSW 2351, Australia
e-mail: rrajarat@une.edu.au

K. Vernes
Ecosystem Management, University of New England, Armidale, NSW 2351, Australia
e-mail: kvernes@une.edu.au

T. Sangay
Ecosystem Management, University of New England, Armidale, NSW 2351, Australia

Ugyen Wangchuk Institute for Conservation and Environment, Bumthang, Bhutan
e-mail: tagsangay@gmail.com

F.M. Angelici (ed.), *Problematic Wildlife*, DOI 10.1007/978-3-319-22246-2_8

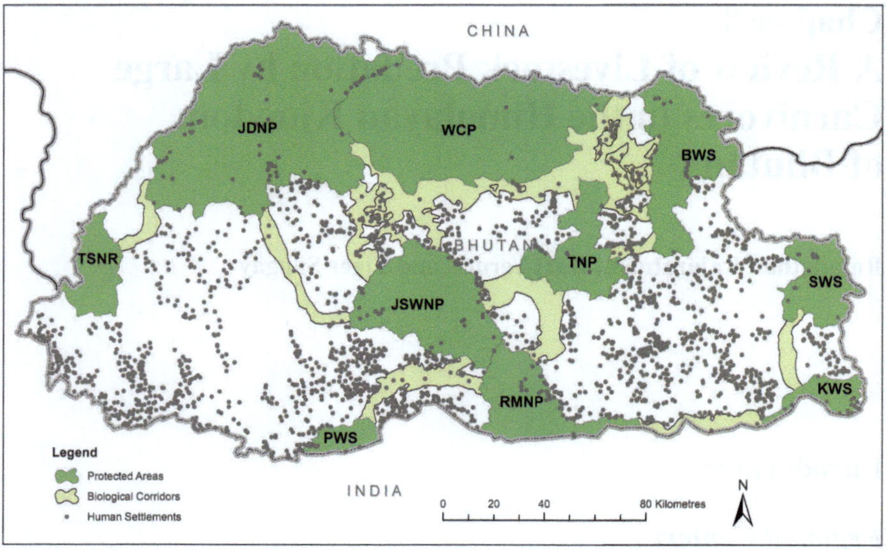

Fig. 8.1 The protected area network of Bhutan, showing National Parks and Biological Corridors. The distribution of human settlements is also shown, demonstrating the occurrence of people within the protected area network. (Key to Protected Areas: *BWS* Bumdeling Wildlife Sanctuary, *JDNP* Jigme Dorji National Park, *JSWNP* Jigme Singye Wangchuck National Park, *JWS* Jomotshangkha Wildlife Sanctuary, *PWS* Phibso Wildlife Sanctuary, *RMNP* Royal Manas National Park, *SWS* Sakteng Wildlife Sanctuary, *PNP* Phrumsengla National Park, *JKSNR* Jigme Khesar Strict Nature Reserve, *WCP* Wangchuck Centennial National Park)

plant species (MoA 2009) and a rich bird (686 species; Norbu 2012) and mammal (160 species; MoA 2009) diversity. Fittingly, Bhutan is recognised as a global biodiversity 'hot spot' (Myers et al. 2000) and one of the Global 200 priority ecoregions for conservation (Olson and Dinerstein 2002), further exemplified by the fact that 27 of Bhutan's 160 mammal species are globally threatened (MoA 2009) including IUCN red-listed species such as the red panda (*Ailurus fulgens*), takin (*Budorcas taxicolor*), bharal (*Psuedois nayaur*), Himalayan serow (*Capricornis thar*), Himalayan goral (*Naemorhedus goral*), Asian elephant (*Elephas maximus*), and golden langur (*Trachpithecus geei*). Contextually, Bhutan has a diverse carnivore community (39 recorded species; Wangchuk et al. 2004), which includes the endangered tiger (*Panthera tigris*), snow leopard (*Panthera uncia*) and dhole (*Cuon alpinus*), the vulnerable Himalayan black bear (*Ursus thibetanus*) and clouded leopard (*Neofelis nebulosa*), and the near threatened common leopard (*Panthera pardus*).

Gross National Happiness and Biodiversity Conservation

Bhutan's population stands at approximately 673,000 people of whom 69 % are rural based (Tshering 2009) with a subsistence agriculture livelihood based upon crops, livestock, and forest products (Katel and Schmidt-Vogt 2011). Bhutan is unique with regard to its economic development policy in that the concept of Gross National Happiness (GNH) takes priority over the more established Gross Domestic Product (GDP) as a measure of prosperity. GNH was conceptualised in the 1970s by the fourth King of Bhutan to promote the greater importance of the people's happiness over the growth in GDP (Katel et al. 2014). GNH is now preserved in Bhutan's constitution and based upon four pillars: the promotion of sustainable development, preservation and promotion of cultural values, conservation of the natural environment, and the establishment of good governance.

Because environmental conservation is central to GNH and firmly backed by the Buddhist philosophical respect for all life, there is strong political support for biodiversity conservation. A constitutional mandate ensures that at least 60 % of the country remains as wildlife habitat under natural forest cover, through a system of protected areas (RGoB 2008) secure from population pressure (Katel and Schmidt-Vogt 2011). Furthermore, an integral domain of the GNH philosophy is ecological diversity and resilience with wildlife being a key indicator (Pennock and Ura 2011). Currently, protected areas cover 42.7 % of Bhutan's territory, with an additional 9 % declared as biological corridors connecting the protected areas (Fig. 8.1; NCD 2004; Sangay and Wangchuk 2005; DoFPS 2009). Protected areas capture large tracts of all representative forest types and elevations throughout Bhutan, making it a world leader in protected area planning and management. Additionally, more than 95 % of Bhutan remains vegetated, of which approximately 70 % constitutes natural forest cover (MoAF 2011). This landscape, now known as the Bhutan Biological Conservation Complex (B2C2), covers almost 6000 km^2 and was bequeathed as a 'Gift to the Earth' from the people of Bhutan in 1999 (CEPF 2005).

People, Predators, and Protected Areas

Bhutan is distinctive with regard to its rural people and their spatial relationship with the conservation landscape. Rural settlements are very much a component of the B2C2 (Fig. 8.1) unlike the orthodox approach in protected area management where human settlements and anthropogenic activities are often excluded from a conservation area. However, protected areas in developing countries are crucial for the provision of ecosystem services to local people through livestock grazing and collection of forest products. Mitigating human–wildlife conflict and controlling resource extraction are management challenges wherever people live inside protected areas (Karanth and Nepal 2012), and this applies to Bhutan. For example, more than 320,000 cattle (RGoB 2000) and 23,000 horses (Sangay and Vernes 2008) are grazed in the countryside in conjunction with localised timber and fuel wood extraction, and the collection of fodder and other minor forest products to support an agrarian economy. Cattle numbers have now increased by 3.5 % while yak populations have increased by 55.9 % (RGoB 2009). Livestock grazing in the B2C2 complex has inadvertently led to human–wildlife conflict through livestock predation by carnivores. This has attracted increased attention (e.g. Wang et al. 2006; Wang and MacDonald 2006; Sangay and Vernes 2008, 2014; Katel et al. 2014), being seen as a constraint on rural development and a serious threat to rural livelihoods (RGOB 2004; Wang and Macdonald 2006; Wang et al. 2006).

Livestock predation by mammalian carnivores is a global issue and a frequent source of conflict with humans (Mazzolli et al. 2002) because many large carnivore species specialise on ungulate prey (Treves and Karanth 2003) and, therefore, opportunistically prey on domesticated ungulates (Karanth et al. 1999) in a shared landscape. It becomes prominent and often controversial when the predators are globally threatened or endangered, and have legal protection. Global livestock predation is evidenced through sheep killing by wolves (*Canis lupus*) and lynx (*Lynx lynx*) in Europe and North America (Meriggi and Lovari 1996; Ciucci and Boitani 1998; Stahl et al. 2002; Musiani et al. 2003; Herfindal et al. 2005; Odden et al. 2002, 2008); lions (*Panthera leo*), leopards (*Panthera pardus*), spotted hyenas (*Crocuta crocuta*), and African wild dog (*Lycaon pictus*) killing cattle, sheep, pigs, and goats in Africa (Rasmussen 1999; Ogada et al. 2003; Kolowski and Holekamp 2006; Holmern et al. 2007; Sogbohossou et al. 2011); jaguars (*Panthera onca*), pumas (*Puma concolor*), and coyotes (*Canis latrans*) taking cattle in South and Central America (Mazzolli et al. 2002; Polisar et al. 2003; Soto-Shoender and Giuliano 2011); golden jackals (*Canis aureus*) preying on cattle in the Middle East (Yom-Tov et al. 1995); dingoes (*Canis lupus dingo*) and wild dogs (*Canis lupus familiaris*) attacking cattle and sheep in Australia (Fleming et al. 2014); and tigers killing livestock in Asia (Johnson et al. 2006; Nugraha and Sugardjito 2009; Li et al. 2009; Pettigrew et al. 2012). In the Himalayan region of South Asia, tiger, snow leopard, leopard, Tibetan grey wolf (*Canis lupus chanco*), Eurasian lynx (*Lynx lynx isabellina*), Himalayan black bear, and dhole prey on a variety of livestock most notably cattle, yak, horses, sheep, goats, and domestic dogs (Oli et al. 1994; Mishra 1997; Bagchi and Mishra 2006; Wang et al. 2006; Wang and MacDonald 2006; Mishra et al. 2006; Namgail et al. 2007; Sangay and Vernes 2008, 2014; Shah et al. 2009; Li et al. 2013; Katel et al. 2014; Shehzad et al. 2014).

In this chapter, we provide an overview of livestock predation in Bhutan and how it relates in the global context, with an emphasis on four internationally prominent carnivore species: tiger, snow leopard, common leopard, and dhole. The tiger, leopard, and dhole are sympatric throughout much of their geographical range in South and South East Asia (Corlett 2011; Ramesh et al. 2012; Steinmetz et al. 2013; Selvan et al. 2013), while the leopard and snow leopard share 10,000 km^2 of mountainous range in Asia (Lovari et al. 2013). Uniquely, all four species are sympatric between 3000 and 4000 m asl in Bhutan (Wangchuk et al. 2004) and totally protected under Schedule 1 of Bhutan's Forest and Nature Conservation Act of 1995 (Wang and Macdonald 2006). We present the contributing factors to this livestock predation and the preponderance of particular livestock species to predation, along with information on the socio-economic impacts and cultural perception. Management strategies and their effectiveness to alleviate and/or offset the economic consequences of livestock predation are also presented and discussed.

Large Predators of Livestock in Bhutan

Tiger

The tiger is the largest terrestrial predator in Asia (Corlett 2011), distributed across 13 range states (see Chundawat et al. 2011 for a full list) in Southeast Asia through to Indo China, South Asia, and Far East Asia. It occurs in a wide variety of habitats from lowland tropical forests, to mixed evergreen and deciduous forests, grasslands, mangroves, and high altitude conifer forests (see O'Brien et al. 2003; Johnson et al. 2006; Barlow et al. 2010; Steinmetz et al. 2013; Sangay et al. 2014). Tigers are solitary and nocturnal, primarily preying on ungulate species like deer, wild cattle, and wild pigs (Corlett 2011; Selvan et al. 2013; Hayward et al. 2012). They also prey on domestic livestock (see Table 3 in Inskip and Zimmermann 2009) particularly where resource use overlap between wildlife and humans has increased (see Karanth and Chellam 2009; Inskip and Zimmermann 2009; Pettigrew et al. 2012).

Tigers are endangered (Chundawat et al. 2011), having had their historical range reduced by 93 % (Wang and Macdonald 2009) because of poaching and illegal trade in body parts, decline in natural prey, habitat loss, and conflict with humans through livestock predation (Karki et al. 2013). It is a conservation flagship (Karanth and Chellam 2009: Karanth and Nepal 2012) and an umbrella species for the protection of biodiversity in Asia (Dinerstein et al. 2007), with Bhutan possibly being an important stronghold (Wang and Macdonald 2009).

Tigers occur at varied altitudes and habitats in Bhutan (Plate 8.1), from 100 m asl in the southern subtropical forests to northern alpine zones at 4500 m asl (Tharchen 2013). They avoid human settlements and their distribution overlaps closely with sambar deer (*Rusa unicolor*), a principal prey species (Wangchuk et al. 2004). Other prey include wild pig (*Sus scrofa*) and gaur (*Bos gaurus*), opportunistic smaller prey animals (Wangchuk et al. 2004), and livestock (Wang and Macdonald 2009; Sangay and Vernes 2008; Katel et al. 2014). Currently there are an estimated 115–150 tigers in Bhutan (Sangay and Wangchuk 2005).

Plate 8.1 Tiger (*Panthera tigris*) photographed in Ha district, Bhutan (see Fig. 8.2). Photo credit: Tiger Sangay

Snow Leopard

The snow leopard is restricted to Central Asia and patchily distributed across a 6000-km wide arc along the high elevation mountainous borders of 12 range countries (see Jackson et al. 2008 for a full list), between 900 and 4500 m asl (McCarthy et al. 2003). It occupies a variety of habitats from cold, arid, and semi-arid shrubland, grassland, and barren rocky areas (Jackson 1996) to open coniferous forest in China (McCarthy et al. 2003). It is crepuscular (Wangchuk et al. 2004) and principal prey species are bharal and ibex (*Capra sibirica*), with considerable obligate predation on high altitude domestic livestock (Mishra et al. 2003) grazed in snow leopard habitat also favoured by natural prey like bharal.

Like the tiger, the snow leopard is endangered (Jackson et al. 2008) and also a conservation flagship and umbrella species to protect alpine biodiversity (Wikramanayake et al. 2006). Major threats to this species include depleting prey populations, poaching and illegal trade, conflict with local people through livestock predation, and the lack of conservation capacity, policy, and awareness (McCarthy et al. 2003).

Plate 8.2 A group of three snow leopards (*Uncia uncia*) photographed by camera trap at Wangchuck Centennial National Park (see Fig. 8.1). Photo credit: Tenzin/Wangchuck Centennial National Park

The snow leopard in Bhutan (Plate 8.2) occurs in the sub-alpine and alpine zones with steep broken rocky areas between 3000 and 5400 m asl (Wangchuk et al. 2004). It is mainly solitary but persistent small groups have been recorded (see Sangay et al. 2014). Prey of the snow leopard in Bhutan are bharal, Himalayan marmots (*Marmota himalayana*), pika (*Ochotona* spp.), Himalayan snowcock (*Tetraogallus himalayensis*), partridges (Wangchuk et al. 2004), and livestock (Sangay and Vernes 2008). There is no reliable estimate of snow leopard population size in Bhutan although McCarthy et al. (2003) place the population at 100–200 individuals.

Common Leopard

The common leopard is arguably the most widespread felid in the world, geographically distributed from sub-Saharan Africa through to the Arabian Peninsula including Turkey, Southwest Asia and the Caucasus, the Himalayas, India, China, the Russian Far East, mainland South East Asia, and the islands of Sri Lanka and Java (Sunquist and Sunquist 2002). Because it is sympatric with larger felid species (lions *Panthera leo*; tigers; cheetahs *Acinonyx jubatus*), it is highly adaptable, being also able to climb trees after arboreal prey like primates (Corlett 2011). The leopard occurs in a variety of habitats from rainforests, coastal scrub, and semi-deserts to alpine mountains and the edges of human settlements (Henschel et al. 2008; Balme et al. 2010). Leopards exhibit either nocturnal or diurnal activity patterns in response to varying abundance and activity of prey species (Ramesh et al. 2012; Selvan et al. 2013) and spatiotemporal partitioning with other overlapping large carnivore species (Steinmetz et al. 2013). They tend to avoid areas frequented by tigers possibly to avoid intraguild predation by the more socially dominant tiger (Carter et al. 2015) because tigers are known to attack and kill leopards (Corlett 2011).

The common leopard is a solitary generalist predator, consuming a variety of prey ranging from small arthropods, fish, birds, rodents, wild pig, and small deer to large ungulates like the eland (*Taurotragus oryx*) and sambar deer (Balme et al. 2007; Shehzad et al. 2014). Kleptoparasitism is avoided by caching carcasses in trees, caves, and dense vegetation (Sunquist and Sunquist 2002). Leopards also closely associate with human settlements and livestock grazing areas, preying extensively on domestic dogs and livestock (Edgaonkar and Chellam 2002; Sangay and Vernes 2008; Shehzad et al. 2014) although contrary evidence now suggests avoidance of areas with human activities (see Carter et al. 2015).

The common leopard is listed as near threatened but may be upgraded to 'vulnerable' due to declining numbers in the wild (Henschel et al. 2008). It faces threats from poaching, prey depletion, habitat loss and fragmentation, trade in body parts, and trophy hunting (Karanth and Chellam 2009). Leopards in Bhutan (Plate 8.3) survive in varied habitats up to 3300 m asl including forests, open country, scrub,

Plate 8.3 A melanistic form of the common leopard (*Panthera pardus*) photographed by camera trap at Trongsa district, Bhutan (see Fig. 8.2). Photo credit: DoFPS/UNE

and areas near human settlements, preying on reptiles, birds, primates, antelopes, larger deer, and domestic livestock (Wangchuk et al. 2004; Wang and Macdonald 2009; Sangay and Vernes 2008; Katel et al. 2014). Leopards avoid areas frequented by tigers thereby tending to occupy the fringes of tiger territory (Wangchuk et al. 2004). There are no current estimates of the total population size of leopards in Bhutan.

Dhole

The dhole is a native canid in Asia with a wide distribution across 17 countries in North, Central, South, and South East Asia (see Durbin et al. 2008 for a full list). It inhabits a diverse vegetation types from tropical and subtropical forests to degraded secondary forests, dry and moist deciduous forests, dry thorn forests, grassland-scrub-forest mosaics, and alpine steppes above 3000 m asl (Durbin et al. 2008). Dholes hunt in packs with pack sizes ranging from 3–12 to a maximum of 40 (see Corlett 2011). They are diurnal but largely crepuscular in their foraging, primarily preying on medium-sized ungulates such as barking deer (*Muntiacus muntjac*) and wild pig but able to take larger ungulates like sambar deer and gaur (see Hayward et al. 2014 for a detailed prey review). Dholes are also known to prey on livestock such as sheep and cattle (Wang and Macdonald 2009; Selvan et al. 2013; Katel et al. 2014; Lyngdoh et al. 2014). The dhole is endangered and facing threats from prey depletion, poisoning, and disease (Durbin et al. 2008), as well as hunting and retaliatory killing (Lyngdoh et al. 2014).

The dhole is found throughout Bhutan (Plate 8.4) in all forest types from the subtropical zone to the limit of the temperate zone at 4000 m asl, and at the edge of human settlements (Wangchuk et al. 2004). They mainly prey on wild pig (Wangchuk 2004) and are known predators of livestock (Wang and Macdonald 2009; Katel et al. 2014). Their historical predation of livestock led to dholes being poisoned and exterminated by farmers in a massive poisoning campaign in the early 1980s (Wangchuk et al. 2004), which resulted in an increase in wild pig populations leading to serious crop damage (Wang 2004). Consequently, dholes were reintroduced in the 1990s and populations are being re-established (Wang and Macdonald 2009) but with no overall population estimate currently available.

Livestock Predation in Bhutan

Conflicts between humans and predators primarily arise when people and predators compete for shared, limited resources in the landscape (Graham et al. 2005). Livestock predation by carnivores occurs when their large home ranges overlap with human-dominated landscapes through habitat loss and fragmentation and/or the reduction of wild prey through anthropogenic exploitation (Mishra 1997).

Plate 8.4 Dhole (*Cuon alpinus*) photographed by camera trap at Trongsa district, Bhutan (see Fig. 8.2). Photo credit: DoFPS/UNE

Livestock predation is also present in areas where prey species naturally occur at low densities (Bhattarai and Fischer 2014). Occasionally, conflict between predators and humans can arise through a change in land use, e.g. regrowth of forests in the USA allowing the recolonisation of extirpated predators in specific regions, i.e. puma recovering in areas where farms have given way to woodland (Karanth and Chellam 2009), or the reintroduction of extirpated carnivores through successful recovery programmes (Breitenmoser 1998). The recovery of threatened carnivores in Europe such as the lynx, wolf, and brown bear (*Ursus arctos*) through antihunting sentiments, animal welfare concerns, improved legislative protection, and better socio-economic management (Karanth and Chellam 2009; Chapron et al. 2014) can also lead to increased conflict between humans and predators.

Livestock predation is an inherent issue in Bhutan (Plate 8.5 and 8.6), but with forcing mechanisms like habitat fragmentation or loss of prey not being the major drivers. Although the road network is expanding (Ito 2011) and habitat quality deteriorating from livestock grazing and forest produce collection (Wang and Macdonald 2006), there is no substantial fragmentation of carnivore habitat associated with major forest conversion as seen elsewhere (see O'Brien et al. 2003 and Linkie et al. 2003). As a Buddhist country with a strong belief in the sanctity of life, hunting of ungulates for human consumption is negligible and there is evidence of a diverse

Plate 8.5 A cattle killed by a large mammalian carnivore, being examined by the Tiger Survey Team. Photo credit: Sangay Dorji/Tiger Survey Team, DoFPS, Bhutan

Plate 8.6 The result of tiger predation upon a yak, being examined by the Tiger Survey Team. Photo credit: Sangay Dorji/Tiger Survey Team, DoFPS, Bhutan

ungulate population albeit at relatively low densities (Wang 2010), in contrast to Laos where active hunting of ungulates has seen increased predation on livestock by tigers (Johnson et al. 2006).

The propensity for livestock predation in Bhutan is, undoubtedly, due to a high reliance on rural livestock for economic subsistence in a rugged landscape unsuitable for a full dependence on cropping agriculture. Approximately 69 % of Bhutan's population are farmers practising a combination of small-scale cropping and animal husbandry (Katel and Schmidt-Vogt 2011) with livestock integral towards their socio-economy (Katel et al. 2014). While human–wildlife conflict tends to occur near protected areas in remote regions (Pettigrew et al. 2012), this is further magnified in Bhutan by the widespread grazing of livestock herds within the protected B2C2 complex (Wang and Macdonald 2006), increasing the likelihood of contact between predators and livestock amidst adequate cover for stalking. Furthermore, most domesticated livestock breeds may have lost their anti-predator behaviour, making them especially vulnerable to carnivores specialised for ungulate predation (Polisar et al. 2003). It is probable that where high densities of predisposed livestock are readily accessible, predators have the opportunity for predation as noted by Wang and Macdonald (2006). There is also a matching stratification between livestock and predators in Bhutan in terms of altitude. Semi-nomadic livestock herders seasonally move yak and/or pack horses along an altitudinal gradient between lower elevation subtropical (<2000 m asl; Tharchen 2013) winter pastures to higher elevation, summer alpine grasslands (>4000 m asl; Tharchen 2013; Moktan et al. 2006), making livestock susceptible to predation at high altitudes by snow leopards from 3000 to 5000 m asl, and by tigers to at least 4000 m asl (Wangchuk et al. 2004; Sangay et al. 2014). Similarly, seasonal semi-nomadic cattle herding between the temperate zone (2000–4000 m asl; Tharchen 2013) and the subtropical zone (Moktan et al. 2006) is further exposed to predation by the altitudinal overlap between tigers, snow leopards, common leopards, and dhole between 3000 and 4000 m asl (Wangchuk et al. 2004). Livestock grazing at lower elevations below 3000 m asl are exposed to predation by tigers, common leopards, and dhole.

Despite comparable densities of wild pigs, muntjac, and sambar deer relative to other areas in the Indian subcontinent, Wang (2010) reports the ungulate biomass in Bhutan to be the lowest for Asia, resulting in livestock supplementing the diet of large predators. Wang and Macdonald (2006) and Wang (2010) further suggest lowered densities of natural ungulate prey through deteriorated habitat quality from overgrazing livestock, making livestock more prone to predation. This has been especially noted in the northern regions of Bhutan (Sangay and Vernes 2008) and regionally in trans-Himalayan India, where widespread overstocking led to a reduction in native herbivore numbers (Mishra et al. 2001), through competition for resources (Katel et al. 2014) resulting in increased carnivore predation of livestock (Bagchi and Mishra 2006).

Poor animal husbandry practices involving herd management and human negligence also contribute significantly to livestock predation in Bhutan. Wang and Macdonald (2006) noted heightened predation of unguarded cattle by tigers and leopards in remote pastures and forest habitats with less human activity, compared to livestock grazed closer to villages. Katel et al. (2014) also noted a high level of cattle predation by dholes in forests than in villages or farms, directly linking this to the absence of herders. Low vigilance over livestock has also been linked to seasonality. Cropping agriculture in Bhutan traditionally peaks in summer and autumn (June to September) with inadequate human resources to guard livestock freely grazing in forests and subject to increased predation (Sangay and Vernes 2008; Katel et al. 2014). Inadequate herding practices have been strongly linked to livestock predation throughout the Himalayan region particularly with regard to livestock regularly untended (Ikeda 2004; Namgail et al. 2007) or untended seasonally during cropping activities (Shehzad et al. 2014). Similarly, untended flocks of sheep in Europe are subject to increased predation by lynx (Ciucci and Boitani 1998; Espuno et al. 2004). Lack of vigilance when herding livestock has also contributed to increased predatory attacks in Africa (Kolowski and Holekamp 2006), while limited vigilance on grazing livestock in Central America during the wet season has led to increased predation (Soto-Shoender and Giuliano 2011).

Another factor contributing to livestock predation in Bhutan is poor management of livestock at night. Wang and Macdonald (2006) observed livestock not penned at night or penned in poor constructions, potentially subject to break-ins by predators. Inadequate corralling of livestock in Central America (Soto-Shoender and Giuliano 2011) has contributed to predation by jaguars, pumas, and coyotes, while poor livestock penning in Pakistan has led to increased attacks by leopards (Shehzad et al. 2014). Poorly designed pens (low height, lack of ceiling) in Nepal have resulted in successful snow leopard attacks that often involve surplus killing of livestock (Namgail et al. 2007). Several authors have speculated that the quality and type of materials used in traditional African 'bomas', i.e. corals, including their designs, are linked to predation by lions, leopards, cheetahs, and hyenas (e.g. Ogada et al. 2003; Patterson et al. 2004: Kolowski and Holekamp 2006; Sogbohossou et al. 2011).

Wang and Macdonald (2006) and Wang et al. (2006) noted legislation such as the Nature Conservation Act of 1995 which fully protected the tiger, snow leopard and common leopard, possibly increasing predator populations leading to increased livestock predation. Wang and Macdonald (2006) observed livestock predation by leopards in Jigme Singye Wangchuck National Park (Fig. 8.1) increasing significantly between 1993 (when the park was established) and 2001. Since 2002, the Royal Government of Bhutan has received overwhelming complaints of livestock predation by large carnivores across the country (Sangay and Wangchuk 2005), 7 years after the enactment of the Nature Conservation Act of 1995. Increases in human wildlife conflict have been noted elsewhere after the establishment of protected areas with dedicated conservation policies e.g. India (Saberwal et al. 1994; Mishra 1997; Maikhuri et al. 2000; Karanth and Nepal 2012), Nepal (Oli et al. 1994; Studsrød and Wegge 1995; Karanth and Nepal 2012; Bhattarai and Fischer 2014), Pakistan (Shehzad et al. 2014), Mongolia (McCarthy 2000), and Benin (Sogbohossou et al. 2011). In many cases, this conflict escalates when people feel that their needs are less considered to those of wildlife (Pettigrew et al. 2012).

Lastly, it must also be noted that livestock predation is not purely driven by poor animal husbandry. Mishra et al. (2001) noted ecological factors like individual predator behaviour, natural prey densities, predator–prey interactions, and predator densities, all contributing to livestock predation.

Predation Hotspots, Prey Preference, and Prey Vulnerability in Bhutan

Wildlife distribution is often linked to patchy 'hotspot' areas exhibiting higher animal density, greater diversity, or more concentrated use (Nur et al. 2011). Predation hotspots are associated with carnivores selecting areas with high prey abundance to maximise encounter rates (Wolf et al. 2015), which in some areas also has abundant livestock that are attacked through increased encounter (Yom-Tov et al. 1995). An example of this is lynx in the French Jura and Norway preying on grazing sheep in predator hotspots with higher roe deer (*Capreolus capreolus*) abundance (Stahl et al. 2001; Herfindal et al. 2005; Odden et al. 2008). Similarly, snow leopards in the Himalayas preyed heavily on grazing livestock in areas harbouring ungulate prey like the bharal, ibex (*Capra ibex*), and Ladakh urial (*Ovis orientalis vignei*) (Oli et al. 1994; Mishra et al. 2001; Bagchi and Mishra 2006; Namgail et al. 2007).

There are 20 administrative districts in Bhutan called 'dzongkhags' which are further subdivided into 205 'gewogs' or sub-districts (Fig. 8.2). Sangay and Vernes (2008) noted widespread livestock predation in 18 (90 %) of the districts but more extensive in Wangduephodrang, Lhuentse, Yangtse, and Trashigang (Fig. 8.3a), all of which collectively accounted for 47 % of the total livestock lost to predation. Leopards were the major predators of livestock (Fig. 8.3a), accounting for 70 % of all livestock kills (Sangay and Vernes 2008), possibly due to their widespread

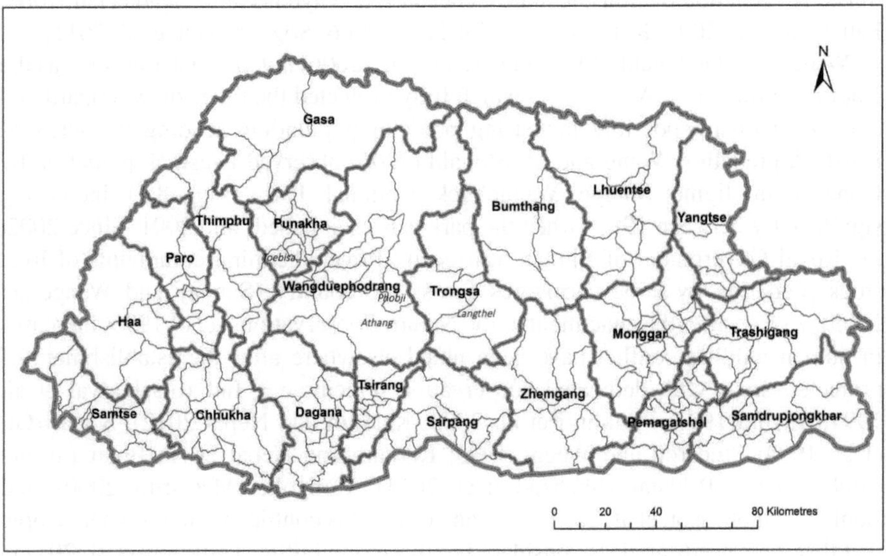

Fig. 8.2 The distribution of the 20 administrative districts ('dzongkhags') that comprise Bhutan. Each district is further subdivided into sub-districts ('geogs'). Sub-districts mentioned in the text are named on the map

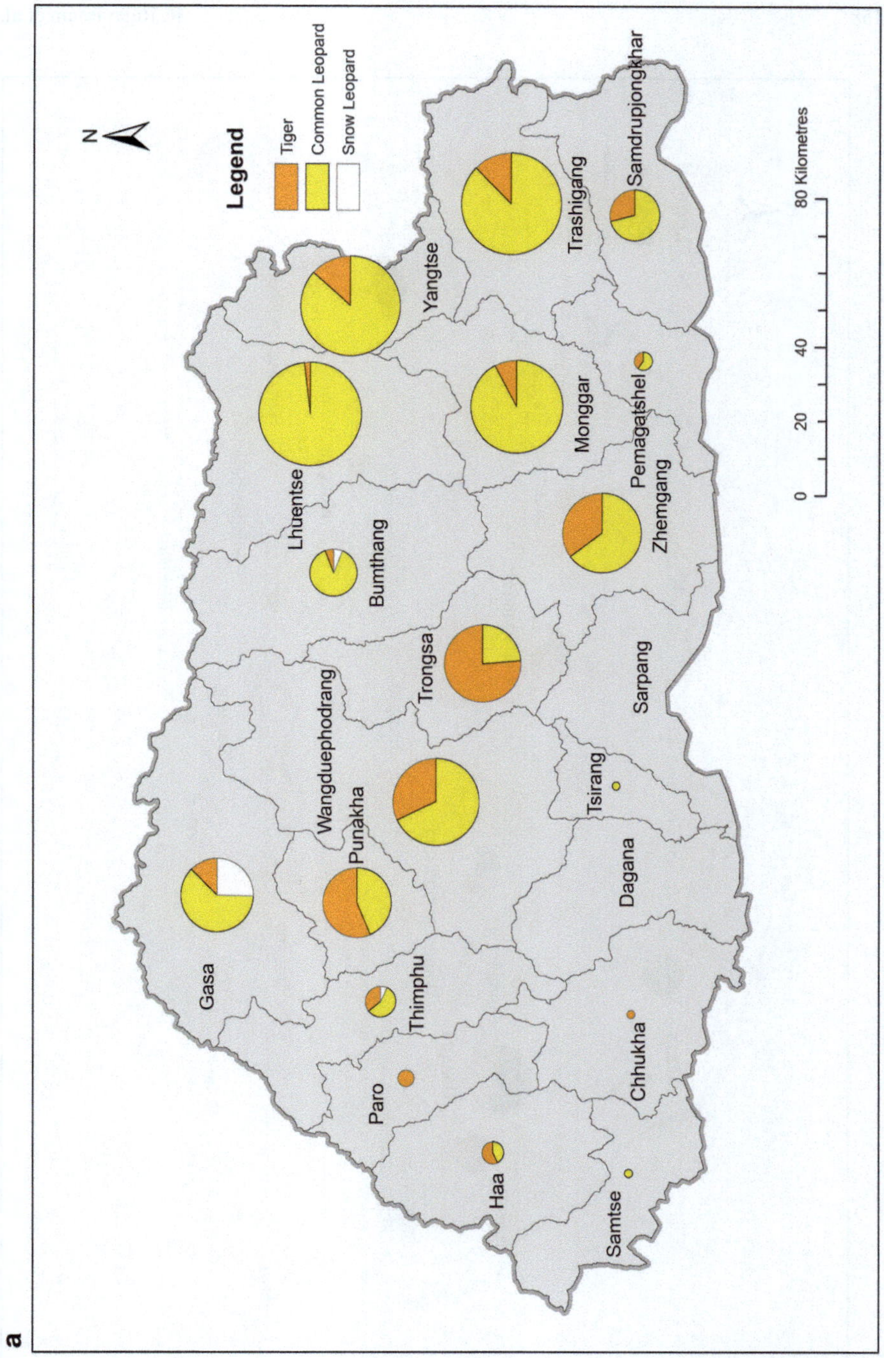

Fig. 8.3 Distribution of (**a**) livestock kills by tigers, leopards, and snow leopards between 2003 and 2005, (**b**) livestock types held, and (**c**) predation hotspots (livestock kills/number of livestock held) according to the 20 administrative districts in Bhutan. For each map, pie graphs show the relative distribution of each variable while the size of the circle shows the relative abundance. Data from Sangay and Vernes (2008)

b

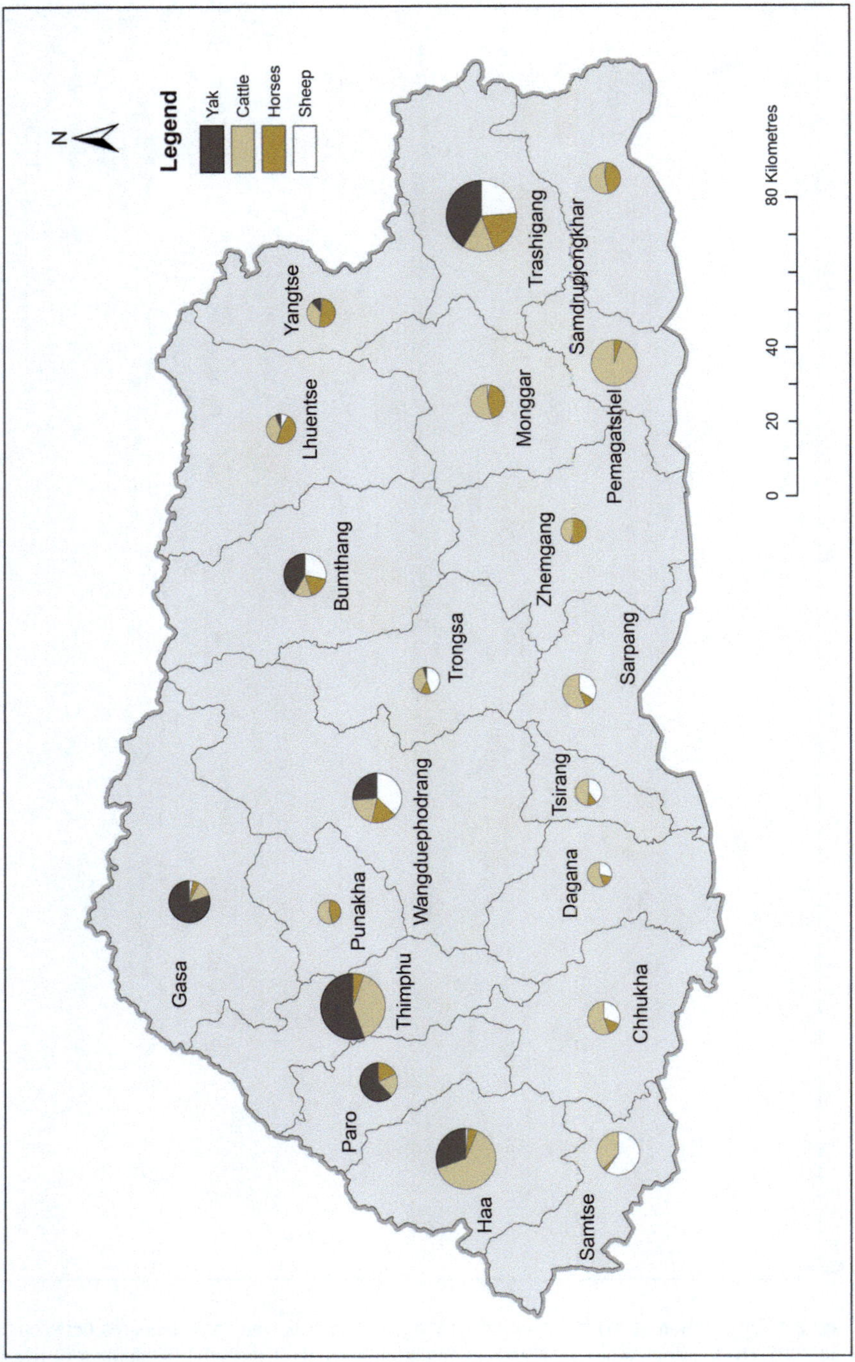

Fig. 8.3 (continued)

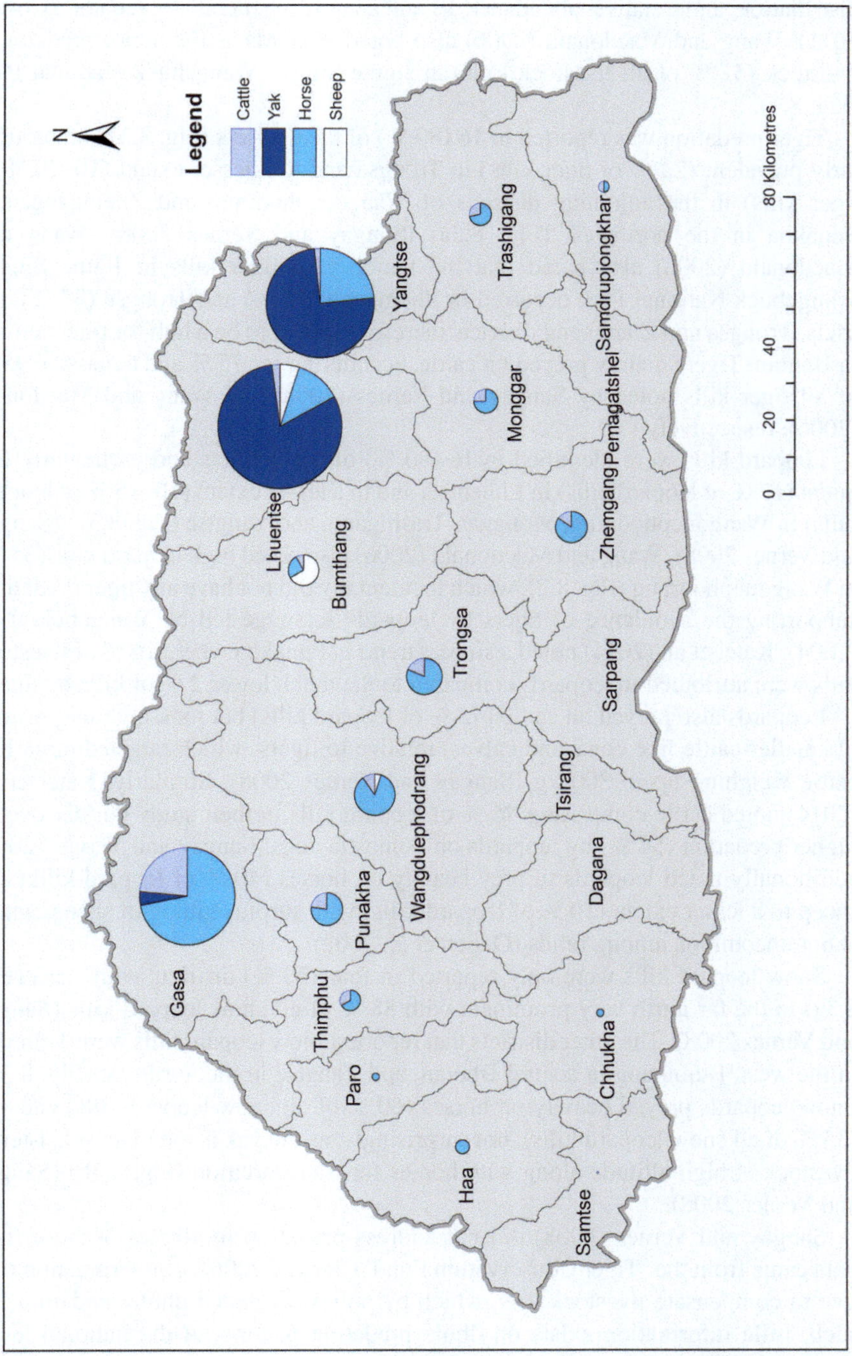

Fig. 8.3 (continued)

distribution and relative abundance as an adaptive generalist predator (Corlett 2011). Wang and Macdonald (2006) also noted leopards as the major predator of livestock (52 % of all livestock kills) in Jigme Singye Wangchuck National Park (Fig. 8.1).

Tiger predation was reported in 16 (80 %) of the districts (Fig. 8.3) and particularly prevalent (22 % of tiger kills) in Trongsa and to a lesser extent (10–20 % of tiger kills) in the adjoining districts of Wangduephodrang and Zhemgang, and Punakha in the northwest (Fig. 8.3a) (Sangay and Vernes 2008). Wang and Macdonald (2006) also noted that the majority of tiger kills in Jigme Singye Wangchuck National Park occurred in Zhemgang (42 %) and Trongsa (37 %) districts. Trongsa and Zhemgang districts therefore appear to be a hub for tiger activity in Bhutan. Tigers mainly preyed on cattle, accounting for 75 % and a massive 95 % of all tiger kills noted by Sangay and Vernes (2008) and Wang and Macdonald (2006), respectively.

Leopard kills were identified in 16 (80 %) of the districts and particularly frequent (17 % of leopard kills) in Lhuentse, and to a lesser extent (10–15 % of leopard kills) in Wangduephodrang, Monggar, Trashigang, and Yangtse (Fig. 8.3a) (Sangay and Vernes 2008). Wang and Macdonald (2006) also noted high leopard kills (35 %) in Wangduephodrang (Fig. 8.2), which incidentally did not have any tiger predation, supporting the avoidance of tigers by leopards as suggested by Wangchuk et al. (2004). Katel et al. (2014) noted a similar trend in Punakha where 16 % of livestock kills were attributed to leopards compared to the much lower 2 % of kills by tigers.

Leopards also preyed on cattle (45 % of leopard kills) but took high proportions of smaller cattle like cows and calves, relative to tigers, which targeted large bull cattle weighing up to 300 kg (Sangay and Vernes 2008). Similarly, Katel et al. (2014) noted cattle comprising 46 % of leopard kills in their study but discovered higher predation (54 %) by leopards on domestic dogs. Sangay and Vernes (2008) additionally noted leopards to prey heavily on horses (40 % of leopard kills) and sheep to a lesser extent (10 % of leopard kills) with surplus killing of sheep, which is not uncommon among felids (Odden et al. 2002).

Snow leopard kills were only reported in four (20 %) districts with Gasa (Fig. 8.3a) in the far north very prominent with 88 % of all snow leopard kills (Sangay and Vernes 2008). The other districts that reported snow leopard kills were Thimphu in the west, Bumthang in central Bhutan, and Yangtse in the northeast (Fig. 8.3a). Snow leopards preyed heavily on horses (60 % of all snow leopard kills) and yak (35 % of all snow leopard kills), not surprisingly, as the yak is the main subsistence livestock at high altitude along with horses for transportation (Fig. 8.3b) (Sangay and Vernes 2008).

Sangay and Vernes (2008) did not address predation by dholes because their data came from the 'Tiger Conservation Fund' (Sangay 2006), a government initiative to compensate livestock loss, which by policy, excluded dhole predation. As such, little information exists on dhole predation patterns at the national level. However, limited information exists on dhole predation in Central Bhutan from Wang and Macdonald's (2006) study within the districts of Trongsa, Punakha, Wangduephodrang, and Zhemgang and more recently from Katel et al.'s (2014) study in Punakha (Fig. 8.2). Wang and Macdonald (2006) noted that dholes mainly killed sheep especially in the sub-districts of Phobji (40 % of livestock predation) within Wangduephodrang, and Langthel (30 % of livestock predation) within Trongsa (Fig. 8.2). On the contrary, Katel et al. (2014) noted exceptionally high levels of dhole predation (82 % of all livestock kills) in the Toebesa sub-district of Punakha (Fig. 8.2), with cattle comprising the sole prey. Being pack hunters, dholes are able to chase down faster large ungulates like adult male sambar deer (Selvan et al. 2013; Hayward et al. 2014) making ponderous cattle prey especially vulnerable. It could also be indicative that dhole populations in Toebesa, which were once poisoned and killed, have successfully re-established and are relatively abundant following their reintroduction to control high wild pig numbers causing substantial crop damage (Katel et al. 2014).

As the most dominant livestock in the Bhutanese landscape (Sangay and Vernes 2008) with loss of anti-predator behaviour (Polisar et al. 2003), it is not surprising that domestic cattle are heavily preyed by tigers, leopards, and dhole (Wang and Macdonald 2006; Sangay and Vernes 2008; Katel et al. 2014). Katel et al. (2014) also note more unguarded cattle killed in forests than in human settlements. However, Sangay and Vernes (2008) stipulate that horses which are less abundant than cattle (320,000 cattle vs. 23,000 horses; RGoB 2000; Sangay and Vernes 2008) comprised almost one-third of all reported livestock kills between 2003 and 2005, with predation strongly correlated with horse density. Approximately 42 % of Bhutanese farm households depend on horses for transportation to the nearest motorable road especially in remote districts like Lhuentse, Trashigang, and Yangtse (Fig. 8.3b), which reported much higher horse predation relative to abundance. Furthermore, horses are less economically viewed than cattle (Sangay and Vernes 2008), often untethered at night and untended in fields when not used for portage, further increasing their vulnerability to leopards occurring closer to human settlements (Wangchuk et al. 2004) and snow leopards at higher altitudes. Yaks were more predisposed to predation by snow leopards in the summer when herded to high altitude meadows (Sangay and Vernes 2008) where they range freely with natural prey like bharal. Snow leopard predation on high altitude livestock is especially prevalent in the Himalayan region given the spatial overlap between livestock and ungulate prey like bharal and ibex (Bagchi and Mishra 2006; Namgail et al. 2007).

Predation patterns in Bhutan are subject to the regional intensity, abundance, and type of livestock grazed, as well as the distribution of predators. Based on the ratio of reported kills to relative livestock abundance, predation prevalence, and the proportion of livestock lost per year to predation, Sangay and Vernes (2008) identified several districts as 'livestock predation hotspots' in northern Bhutan notably Gasa, Lhuentse, and Yangtse (Fig. 8.3c). Among all livestock classes, horses were the dominant livestock lost to predation in most districts (Fig. 8.3c). Cattle comprised a sizeable proportion of kills in many districts, and yak predation was rampant in Lhuentse and Yangtse. Sheep predation was notably confined to Bumthang and Lhuentse (Fig. 8.3c). See Sangay and Vernes (2008) for more comprehensive details on predation rates, type of preyed livestock, seasonality of predation, and regional differences in predation.

Socio-economic Consequences ofLivestock Predation

Almost every rural Bhutanese family owns a parcel of land for subsistence agriculture and for keeping livestock for grazing in pasture or open forest. The repercussion from livestock predation is not just restricted to livestock loss but rather measured by the impact on household socio-economy. Farming and animal husbandry provides employment and livelihood to 69 % of rural Bhutanese, with significant contribution (33 %) to the Gross Domestic Product (GDP) (Sangay and Vernes 2008). Livestock contributes to 7 % of the GDP and accounts for 22 % of rural income (MoA 2009). Livestock predation can therefore take its toll, imposing heavy financial burden on farmers (Namgail et al. 2007). The loss of a ploughing ox or buffalo can have dire economic consequences to a farmer (Bhattarai and Fischer 2014) notably in Bhutan where approximately 23 % of the predominantly rural population subsist below the poverty line (BTN 1100, or USD25, per person per month; NSBoB 2009).

Average annual household cash income from livestock in Bhutan ranges from USD192 to 250 (Wang and Macdonald 2006; Sangay and Vernes 2014; Katel et al. 2014). At the site-specific level, Wang and Macdonald (2006) reported a 17 % loss in the annual household cash income to livestock predation, mainly by leopards and tigers, in Jigme Singye Wangchuck National Park (Fig. 8.1). Conversely, Katel et al. (2014) noted a 11 % annual household cash income loss, primarily from dhole predation, in the Toebesa sub-district in Punakha (Fig. 8.2). The difference was attributed to higher livestock numbers killed in Jigme Singye Wangchuck National Park and a lower livestock dependency in Toebesa (Katel et al. 2014). At the national level, Sangay and Vernes (2014) observed a much higher 64–72 % annual household income loss from livestock predation, noting a pronounced multiplying effect from surplus killing. For example, a leopard that killed 21 sheep in a single attack caused a USD966 household income loss equivalent to 3.5 years of annual household income. A tiger attack that killed eight yaks caused a considerably higher loss of USD2048 or 7.4 years of annual income.

Economic losses are more pronounced when collectively totalled on an annual basis across households that lost livestock to predators. Quite often, the value of the livestock killed is the benchmark to gauge the economic impact of predation (Baker et al. 2008). Wang and Macdonald (2006) reported an annual monetary loss of USD12,252 due to livestock predation in Jigme Singye Wangchuck National Park. From 2003 to 2005, livestock predation loss in Bhutan was valued at a substantial USD389,879 or an average of USD129,959 per year, based on market value (Sangay and Vernes 2014). Fluctuating losses have been reported elsewhere in neighbouring India. For example, the yearly average market value of livestock lost to predation in the Nanda Devi Biosphere Reserve, India, was only USD3539 per year (Maikhuri et al. 2000). Namgail et al. (2007) and Lyngdoh et al. (2014) reported much higher yearly economic loss of USD12,120 and USD67,247 for livestock preyed in India's Gya-Miru Wildlife Sanctuary and villages around three protected areas (Pakke Tiger Reserve, Itanagar Wildlife Sanctuary, Talle Valley Wildlife Sanctuary), respectively. More conservative estimates of livestock loss are available for Africa. Patterson et al. (2004) reported annual livestock predation loss to the value of USD8749 in villages adjacent to Tsavo National Park, Kenya, while Holmern et al. (2007) estimated annual livestock predation loss at a higher USD12,846 in villages outside Serengeti National Park, Tanzania. Much higher estimates have been noted in China. For example, in the Sanjiangyuan region in Qinghai Province, Li et al. (2013) estimated a yearly livestock loss of USD92,957 and USD398,320 to predation by snow leopards and wolves, respectively. When combined with crop damage, the overall economic loss of human wildlife conflict in China can cost up to a massive USD8.75 million per year (see Table 1 in Pettigrew et al. 2012). Economic loss from livestock predation is very substantial in the USA with national sheep losses to coyotes valued up to USD83 million (Baker et al. 2008). Although Bhutan is a developing country, livestock predation is nonetheless significant in terms of rural socio-economic repercussions.

The Role of Culture and Religion, in Tolerance of Large Predators

Religion and culture should not be downplayed in livestock predation as these can influence the willingness of people to share a common landscape with predators. This is evident in Bhutan as a direct significance from the Buddhist ethos of valuing the sanctity of life, which has provided a safe refuge for biodiversity (Wang et al. 2006) and where religious ethics are resonant with environmental protection (Wang and Macdonald 2006). The tiger in particular is firmly enshrouded in Buddhist religion and Bhutanese culture and has a unique national identity as a symbol of majestic power as one of the four 'protector' animals of Bhutan. Along with the leopard, tigers feature conspicuously in local culture and religion (Wang and Macdonald 2009). As such, Bhutanese suffering from wildlife predation do not blame the predators but rather accept it as 'karma' or destiny (Seeland 2000). Katel et al. (2014) further noted that Bhutanese farmers preferred to chase away dholes killing their livestock rather than kill them, with some invoking the Buddhist ban on killing as a reason for their tolerance despite prejudice to dholes because of their predisposition to disembowelling and feeding on prey still alive (Wangchuk et al. 2004). In the predominantly Hindu region of Western Nepal, there is a positive attitude to tigers despite its role in livestock predation, largely due to the belief that tigers are the vehicle of 'Durga' the goddess of might (Bhattarai and Fischer 2014). Bagchi and Mishra (2006) noted that despite resentment to the presence of large carnivores in Nepali pastures, people did not actively persecute them because of cultural and religious reasons. In Tibet, the snow leopard is seen positively in local culture, being considered the guardians of sacred mountains with fewer snow storms and no livestock diseases where snow leopards dwell (Li et al. 2013).

Despite this perceived cultural tolerance, there is growing resentment to livestock predators in Bhutan, now facing human–wildlife conflicts that were absent two decades ago (Wang et al. 2006). Human wildlife conflict causes significant economic losses (Pettigrew et al. 2012) which lead to serious retaliatory killing of livestock predators (Ogada et al. 2003; Holmern et al. 2007). Wang et al. (2006) noted increased hostility by farmers to livestock predators in Jigme Singye Wangchuck National Park with 68 % of interviewed farmers expressing strong desire to exterminate problematic wildlife. Farmers went on a mass poisoning of dholes in Bhutan and virtually exterminated dhole populations because the species was perceived as a livestock predator (Wang and Macdonald 2006). Despite perceived tolerance to dholes by Bhutanese farmers, Katel et al. (2014) also noted a degree of resentment and calls for retaliatory measures. Such retaliation is also occurring regionally and globally. Snow leopards, wolves, and dholes are killed in retaliation for livestock predation in India and Tibet (Mishra et al. 2003, 2006; Li et al. 2013; Lyngdoh et al. 2014). Similarly, lions, leopards, cheetahs, and spotted hyenas are killed in retaliation for livestock predation in Africa (Ogada et al. 2003) while recreational hunting of lynx in Europe is encouraged to offset sheep predation (Herfindal et al. 2005).

Management Strategies to Reduce Livestock Predation

Various control measures have been proposed to mitigate livestock predation. Predator control is one of the most widespread forms of wildlife management (Berger 2006) comprising both lethal and non-lethal measures (Baker et al. 2008). Proactive culling removes problematic animals prior to the period of greatest risk while reactive culling removes individuals that have been identified as culprits behind the conflict (Baker et al. 2008). Such lethal measures would be totally unpragmatic in Bhutan given the strong Buddhist ethos for the sanctity of life.

Mitigating human–wildlife conflict in Bhutan can only be solved via an integrative approach (Li et al. 2013) through cost-efficient incentive programmes (Shehzad et al. 2014) and improved animal husbandry (Wang and Macdonald 2006; Sangay and Vernes 2008; Sogbohossou et al. 2011). A common scheme is through monetary compensation for livestock loss, the rationale being that compensation payments will improve tolerance and prevent retaliatory kills (Naughton-Treves et al. 2003). A compensation scheme was successfully initiated by the Bhutanese government from 2003 to 2005 with assistance from the World Wide Fund for Nature (WWF) Bhutan (Sangay and Vernes 2014). Called the 'Tiger Conservation Fund' (TCF), the scheme paid out a total of USD138,454 to compensate agro-pastoralists facing livestock predation, through a rigorous verification system. However, the TCF relied on external funding which was not financially sustainable in the long term, resulting in the discontinuation of the scheme in 2006 (Sangay and Vernes 2008). Compensation schemes are often difficult to implement and suffer from funding constraints, determining fair payments, long delays, low compensation rates relative to market value of livestock, and bureaucracy (Nyhus et al. 2003; Namgail et al. 2007). Negative impacts on conservation efforts can arise when livestock owners begin to lose faith in failed compensation schemes. However, financially sustainable compensation schemes work well under strict restrictions. For example, farmers may be required to actively tend livestock, use guard dogs, enclose livestock at night, and respect grazing zones within protected areas, as exemplified in Switzerland where compensation is not provided unless livestock owners show evidence of effective animal husbandry practices (Nyhus et al. 2003).

A community-based insurance scheme currently being piloted across ten sub-districts in Bhutan may be effective in circumventing common complaints associated with traditional compensation schemes like time delays between reporting and payment, excessive bureaucracy, and great distances to report livestock attacks (Sangay and Vernes 2014). Funded by the government, seed money (USD4800) has been invested and the generated interest used to pay compensation to members who pay a nominal membership fee of USD3.00 and an additional premium of USD1.60–2.40 per livestock owned. However, the seed money may not be sufficient in sustainably paying insurance claims, requiring a top up to make this scheme functional (S. Dorji, National Tiger Coordinator, personal communication). Community-based insurance schemes to offset economic loss from livestock predation are also being successfully implemented in China (Pettigrew et al. 2012; Li et al. 2013).

In Bhutan, improved animal husbandry through better herding practices and effective corralling (Wang and Macdonald 2006; Sangay and Vernes 2008; Katel et al. 2014) are paramount as a preventative measure against livestock predation. Vigilance and tending of herds is very important and can be very effective. For example, Katel et al. (2014) noted that the people of Eyamoo village in Toebesa did not lose a single livestock although cattle were grazed in the forest during the day because cattle were always guarded and stall fed at night. Katel et al. (2014) also advocates promoting better cattle breeds in Bhutan like the jersey and jersey-cross, which can be stall fed and therefore safe within the confines of a village. Additionally, migratory cattle herds have reduced in the last decade, and the expanding road network now connect many villages resulting in less reliance on horses for transport (S. Dorji, National Tiger Coordinator, personal communication).

Guard dogs can be viable in livestock protection due to their inexpensiveness and long-term deployment (Baker et al. 2008), and indeed, most Bhutanese herders keep dogs for this very reason. However, dogs are not effective in guarding free ranging livestock (Baker et al. 2008) and in the case of Bhutan and elsewhere, subject to predation themselves (Musiani et al. 2003; Katel et al. 2014). Dogs do not chase predators, but rather, alert people to the presence of predators (Ogada et al. 2003). Therefore the best use of dogs as a deterrent against livestock predators in Bhutan is to strictly confine them along with corralled/tethered livestock in a village or farmhouse. This also prevents the transmission of potential disease and dog predation on native species, which is a viable threat to some threatened species such as the vulnerable red panda (Dorji et al. 2012) and calves of the Bhutan takin (*Budorcas taxicolor whitei*; Tiger Sangay, personal observation).

A feasible approach to promote the tolerance of predators is to use the strong cultural and religious status of predators like the tiger and leopard in 'religious' education to the general populace by the Buddhist clergy. This is currently being practiced in Tibet in collaboration with four monasteries, whereby Buddhist priests emphasise the special value of snow leopards through religious sermons in combination with cultural folklore (Li et al. 2013). A similar approach in using the Buddhist philosophy on compassion and tolerance to all forms of life has been advocated for Arunachal Pradesh in India (Mishra et al. 2006).

Farmers should be educated on known predation hotspots and encouraged to not graze their livestock in these areas (Wang and Macdonald 2006; Sangay and Vernes 2008). Integrated Conservation Development Programmes (Dorji et al. 2012) that improve the livelihood of rural people through community-based ecotourism and handicraft making for tourist sale should be actively encouraged and expanded. Such programmes can alleviate rural poverty and reduce the dependence on livestock and/or allow farmers to afford sturdier material in the construction of enclosures. Wang (2010) advocates the banning of livestock grazing in designated zones within the protected areas to promote the recolonisation of wild ungulate prey, thereby reducing the frequency and intensity of predator attacks on livestock.

Finally, more ecological research is needed in Bhutan following on from the work done by Wang and Macdonald (2006) and Wang (2010). Additionally, Sangay and Vernes (2008) point the need for more research on livestock numbers, grazing pressure, and animal husbandry practices to fully understand and address the scope of livestock predation in Bhutan.

Conclusion

Bhutan is at the cross roads between nature conservation and improving the socioeconomy of its rural citizens. However, persistent livestock predation remains a clear and present threat to rural livelihood. Unless this is alleviated, the very fabric that binds the coexistence of people and nature in a benevolent kingdom is under serious jeopardy. Nature conservation within the confines of GNH would lose its credibility with rural people if they perceive that conservation of large predators is being practiced at their socioeconomic cost.

Acknowledgements We thank the Wildlife Conservation Division (WCD) under the Ministry of Agriculture (Bhutan), the Ugyen Wangchuk Institute for Conservation and Environment (UWICE), and the University of New England (UNE) for supporting our work in Bhutan. We also thank the Tiger Survey Team (Department of Forest and Park Services, Bhutan) for providing images of livestock kills gathered during their 2014–2015 field work and appreciate the constructive comments received from Dr. Sandro Lovari and an anonymous referee on an earlier draft of the manuscript.

References

Bagchi S, Mishra C (2006) Living with large carnivores: predation on livestock by the snow leopard (*Uncia uncia*). J Zool 268:217–224

Baker PJ, Boitani L, Harris S et al (2008) Terrestrial carnivores and human food production: impact and management. Mammal Rev 38:123–166

Balme GA, Hunter LTB, Slotow R (2007) Feeding habitat selection by hunting leopards in a woodland savanna: prey catchability versus abundance. Anim Behav 74:589–598

Balme GA, Slotow R, Hunter LTB (2010) Edge effects and impact of non-protected areas in carnivore conservation: leopards in the Phinda-Mkhuze Complex, South Africa. Anim Conserv 2010:1–9

Barlow ACD, Greenwood CJ, Ahmad IU et al (2010) Use of an action-selection framework of human–carnivore conflict in the Bangladesh Sundarbans. Conserv Biol 24:1338–1347

Berger KM (2006) Carnivore–livestock conflicts: effects of subsidized predator control and economic correlates on the sheep industry. Conserv Biol 20:751–761

Bhattarai BR, Fischer K (2014) Human-tiger *Panthera tigris* conflict and its perception in Bardia National Park, Nepal. Oryx 48(4):522–528

Breitenmoser U (1998) Large predators of the Alps: the fall and rise of man's competitors. Biol Conserv 83:279–289

Carter N, Jasny M, Gurung B et al (2015) Impacts of people and tigers on leopard spatiotemporal activity patterns in a global biodiversity hotspot. Global Ecol Conserv 3:149–162

CEPF (2005) Ecosystem profile: eastern Himalayas region. Critical Ecosystem Partnership Fund. WWF-US Asia Program, Washington, DC

Chapron C, Kaczensky P, Linnell JDC et al (2014) Recovery of large carnivores in Europe's modern human-dominated landscapes. Science 346(6216):1517–1519

Chundawat RS, Habib B, Karanth U et al (2011) *Panthera tigris*. The IUCN Red List of Threatened Species. Version 2014.3. http://www.iucnredlist.org. Accessed 20 Apr 2015

Ciucci P, Boitani L (1998) Wolf and dog depredation on livestock in central Italy. Wildl Soc Bull 26:504–514

Corlett RT (2011) Vertebrate carnivores and predation in the Oriental (Indomalayan) region. Raffles Bull Zool 59(2):325–360

Dinerstein E, Loucks C, Wikramanayake E (2007) The fate of wild tigers. Bioscience 57:508–514

DoFPS (2009) Circular notifying revised areas coverage in percentage of protected areas and biological corridors. Department of Forests and Park Services, Ministry of Agriculture and Forests, Thimphu

Dorji S, Rajaratnam R, Vernes K (2012) The vulnerable red panda *Ailurus fulgens* in Bhutan: distribution, conservation status and management recommendations. Oryx 46(4):536–543

Durbin LS, Hedges S, Duckworth JW et al (2008) *Cuon alpinus*. The IUCN Red List of Threatened Species. Version 2014.3. http://www.iucnredlist.org. Accessed 20 Apr 2015

Edgaonkar A, Chellam R (2002) Food habit of the leopard, *Panthera pardus*, in the Sanjay Gandhi National Park, Maharashtra, India. Mammalia 66:353–360

Espuno N, Lequette B, Poulle ML et al (2004) Heterogeneous response to preventive sheep husbandry during wolf recolonization of the French Alps. Wildl Soc Bull 32:1195–1208

Fleming PJS, Allen BL, Allen LR et al (2014) Management of wild canids in Australia: free-ranging dogs and red foxes. In: Glen AS, Dickman CR (eds) Carnivores of Australia: past, present and future. CSIRO Publishing, Collingwood, pp 105–109

Graham K, Beckerman AP, Thirgood S (2005) Human-predator-prey conflicts: ecological correlates, prey losses and patterns of management. Biol Conserv 122:159–171

Hayward MW, Jędrzejewski W, Jędrzejewska B (2012) Prey preferences of the tiger *Panthera tigris*. J Zool 286:221–231

Hayward MW, Lyngdoh S, Habib B (2014) Diet and prey preferences of dholes (*Cuon alpinus*): dietary competition with Asia's apex predator guild. J Zool 294(4):255–266

Henschel P, Hunter L, Breitenmoser U et al (2008) *Panthera pardus*. The IUCN Red List of Threatened Species. Version 2014.3. Accessed 20 Apr 2015

Herfindal I, Linnell JDC, Moa PF et al (2005) Does recreational hunting of lynx reduce depredation losses of domestic sheep? J Wildl Manage 69:1034–1042

Holmern T, Nyahongo J, Roskaft E (2007) Livestock loss caused by predators outside the Serengeti National Park, Tanzania. Biol Conserv 135:518–526

Holt BG, Jean-Philippe L, Borregaard MK et al (2013) An update of Wallace's zoogeographic regions of the World. Science 339:74–77

Ikeda N (2004) Economic impacts of livestock depredation by snow leopard *Uncia uncia* in the Kanchenjunga Conservation Area, Nepal Himalaya. Environ Conserv 31:322–330

Inskip C, Zimmermann A (2009) Human–felid conflict: a review of patterns and priorities worldwide. Oryx 43:18–34

Ito T (2011) Road expansion and its influence on trail sustainability in Bhutan. Forests 2:1031–1048

Jackson RM (1996) Home range, movements and habitat use of snow leopard (*Uncia uncia*) in Nepal. Dissertation, University of London

Jackson R, Mallon D, McCarthy T et al (2008) *Panthera uncia*. The IUCN Red List of Threatened Species. Version 2014.3. Accessed 20 Apr 2015

Johnson A, Vongkhamheng C, Hedemark M et al (2006) Effects of human-carnivore conflict on tiger (*Panthera tigris*) and prey populations in Lao PDR. Anim Conserv 9:421–430

Karanth KU, Chellam R (2009) Carnivore conservation at the crossroads. Oryx 43:1–2

Karanth KK, Nepal SK (2012) Local residents perception of benefits and losses from protected areas in India and Nepal. Environ Manage 49:372–386

Karanth KU, Sunquist ME, Chinnappa KM (1999) Long-term monitoring of tigers: lessons from Nagarahole. In: Seidensticker J, Christie S, Jackson P (eds) Riding the tiger: tiger conservation in human-dominated landscapes. Cambridge University Press, Cambridge, pp 114–122

Karki JB, Pandav B, Jnawali, SR et al (2013) Estimating the abundance of Nepal's largest population of tigers *Panthera tigris*. Oryx: 1–7

Katel ON, Schmidt-Vogt D (2011) Forest resource use by local residents in Jigme Singye Wangchuk National Park, Bhutan: practices and perceptions of constraint. Mt Res Dev 31:325–333

Katel ON, Pradan S, Schmidt-Vogt D (2014) A survey of livestock losses caused by Asiatic wild dogs, leopards and tigers, and the impact of predation on the livelihood of farmers in Bhutan. Wildl Res 41:300–310

Kolowski JM, Holekamp KE (2006) Spatial, temporal, and physical characteristics of livestock depredations by large carnivores along a Kenyan reserve border. Biol Conserv 128:529–541

Li B, Zhang E, Liu Z (2009) Livestock depredation by Amur tigers in Hunchun Nature Reserve, Jilin, China. Acta Theriol Sin 29:231–238

Li J, Yin H, Wang D et al (2013) Human-snow leopard conflicts in the Sanjiangyuan region of the Tibetan Plateau. Biol Conserv 166:118–123

Linkie M, Martyr DJ, Holden J (2003) Habitat destruction and poaching threaten the Sumatran tiger in Kerinci Seblat National Park, Sumatra. Oryx 37:41–48

Lovari S, Ventimiglia M, Minder I (2013) Food habits of two leopard species, competition, climate change and upper treeline: a way to the decrease of an endangered species. Ethol Ecol Evol 25(4):305–318

Lyngdoh S, Gopi GV, Selvan KM et al (2014) Effect of interactions among ethnic communities, livestock and wild dogs (*Cuon alpinus*) in Arunachal Pradesh, India. Eur J Wildl Res 60:771–780

Maikhuri RK, Nautiyal S, Rao KS et al (2000) Analysis and resolution of protected area-people conflicts in Nanda Devi Biosphere Reserve, India. Environ Conserv 27(1):43–53

Mazzolli M, Graipel ME, Dunstone N (2002) Mountain lion depredation in southern Brazil. Biol Conserv 105:43–51

McCarthy TM (2000) Ecology and conservation of snow leopards, Gobi brown bears and wild Bactrian camels in Mongolia. Dissertation,University of Massachusetts

McCarthy TM, Allen PM, Fox J et al (2003) Snow leopard survival strategy. ISLT, Seattle
Meriggi A, Lovari S (1996) A review of wolf predation in southern Europe: does the wolf prefer wild prey to livestock. J Appl Ecol 33:1561–1571
Mishra C (1997) Livestock depredation by large carnivores in the Indian Trans-Himalaya: conflict perceptions and conservation prospects. Environ Conserv 24:338–343
Mishra C, Prins HHT, Van Wieran SE (2001) Overstocking in the trans-Himalayan rangelands of India. Environ Conserv 28(3):279–283
Mishra C, Allen P, McCarthy T et al (2003) The role of incentive programs in conserving the snow leopard. Conserv Biol 117:1512–1520
Mishra C, Madhusudan MD, Datta A (2006) Mammals of the high altitudes of western Arunachal Pradesh, eastern Himalaya: an assessment of threats and conservation needs. Oryx 40:29–35
MoA (2002) Biodiversity action plan for Bhutan 2002. Ministry of Agriculture, Royal Government of Bhutan, Thimphu
MoA (2009) Biodiversity action plan 2009. Ministry of Agriculture, Royal Government of Bhutan, Thimphu
MoAF (2011) Land cover atlas of Bhutan. National Soil Services Centre and Planning and Policy Division, Ministry of Agriculture and Forests, Thimphu
Moktan MR, Norbu L, Nirola H et al (2006) Migratory cattle grazing: an ecosystem approach to livelihood. Bhutan J Renew Nat Res 2(1):18–28
Musiani M, Mamo C, Boitani L et al (2003) Wolf depredation trends and the use of fladry barriers to protect livestock in western North America. Conserv Biol 17:1538–1547
Myers N, Mittermeier RA, Mittermeier C et al (2000) Biodiversity hotspots for conservation priorities. Nature 403:853–858
Namgail T, Fox JL, Bhatnagar YV (2007) Carnivore-caused livestock mortality in Trans-Himalaya. Environ Manage 39:490–496
Naughton-Treves L, Grossberg R, Treves A (2003) Paying for tolerance: rural citizens' attitudes toward wolf depredation and compensation. Conserv Biol 17:1500–1511
NCD (2004) Bhutan's biological corridor complex (living in harmony with the nature). Nature Conservation Division, Department of Forestry Services, Ministry of Agriculture and WWF Bhutan, Thimphu
Norbu P (2012) The 686th bird species. http://www.kuenselonline.com. Accessed 10 Feb 2015
NSBoB (2009) Statistical yearbook of Bhutan. National Statistics Bureau, Royal Government of Bhutan, Thimphu
Nugraha RT, Sugardjito J (2009) Assessment and management options of human-tiger conflicts in Kerinci Seblat National Park, Sumatra, Indonesia. Mammal Study 34(3):141–154
Nur N, Jahncke J, Herzog MP et al (2011) Where the wild things are: predicting hotspots of seabird aggregations in the California Current System. Ecol Appl 21:2241–2257
Nyhus P, Fischer H, Madden F et al (2003) Taking the bite out of wildlife damage: the challenges of wildlife compensation schemes. Conserv Prac 4:37–43
O'Brien T, Kinnaird M, Wibisono H (2003) Crouching tigers, hidden prey: Sumatran tiger and prey populations in a tropical forest landscape. Anim Conserv 6:131–139
Odden J, Linnell JDC, Moa PF et al (2002) Lynx predation on domestic sheep in Norway. J Wildl Manage 66:98–105
Odden J, Herfindal I, Linnel JDC et al (2008) Vulnerability of domestic sheep to lynx predation in relation to roe deer density. J Wildl Manage 72(1):276–282
Ogada MO, Woodroffe R, Oguge NO et al (2003) Limiting depredation by African carnivores: the role of livestock husbandry. Conserv Biol 17:152–1530
Oli MK, Taylor IR, Rogers ME (1994) Snow leopard *Panthera uncia* predation of livestock: an assessment of local perceptions in the Annapurna conservation area, Nepal. Biol Conserv 68:63–68
Olson DM, Dinerstein E (2002) The Global 200: priority ecoregions for global conservation. Ann MO Bot Gard 89(2):199–224
Patterson BD, Kasiki SM, Selempo E et al (2004) Livestock predation by lions (*Panthera leo*) and other carnivores on ranches neighboring Tsavo National Park, Kenya. Biol Conserv 119:507–516

Pennock M, Ura K (2011) Gross national happiness as a framework for health impact assessment. Environ Impact Assess Rev 31:61–65

Pettigrew M, Xie Y, Kang A et al (2012) Human-carnivore conflict in China: a review of current approaches with recommendations for improved management. Int Zool 7:210–226

Polisar J, Maxit I, Scognamillo D et al (2003) Jaguars, pumas, their prey base, and cattle ranching: ecological interpretations of a management problem. Biol Conserv 109:297–310

Ramesh T, Kalle R, Sankar K et al (2012) Spatio-temporal partitioning among large carnivores in relation to major prey species in Western Ghats. J Zool 287:269–275

Rasmussen GSA (1999) Livestock predation by the painted hunting dog *Lycaon pictus* in a cattle ranching region of Zimbabwe: a case study. Biol Conserv 88:133–139

RGoB (2000) Renewable natural resources statistics of Bhutan. Ministry of Agriculture, Royal Government of Bhutan, Thimphu

RGOB (2004) Bhutan biological conservation complex. Nature Conservation Division, Department of Forestry Services, Ministry of Agriculture and Forests, Royal Government of Bhutan, Thimphu

RGoB (2008) The constitution of the Kingdom of Bhutan. Royal Government of Bhutan, Thimphu

RGoB (2009) Bhutan biodiversity action plan 2009. National Biodiversity Center, Royal Government of Bhutan, Thimphu

Saberwal VK, Gibbs JP, Chellam R et al (1994) Lion-human conflict in the Gir Forest, India. Conserv Biol 8(2):5501–5507

Sangay T (2006) Tiger (*Panthera tigris*) and its conservation efforts in the Kingdom of Bhutan. Dissertation, University of New England, Armidale, NSW, Australia

Sangay T, Vernes K (2008) Human-wildlife conflict in the Kingdom of Bhutan: Patterns of livestock predation by large mammalian carnivores. Biol Conserv 141:1272–1282

Sangay T, Vernes K (2014) The economic cost of wild mammalian carnivores to farmers in the Himalayan Kingdom of Bhutan. Proc Bhutan Ecol Soc 1:98–111

Sangay T, Wangchuk T (2005) Tiger action plan for the Kingdom of Bhutan 2006–2015. Nature Conservation Division, Department of Forests and Park Services, Thimphu

Sangay T, Rajaratnam R, Vernes K (2014) Wildlife camera trapping in the Himalayan kingdom of Bhutan with recommendations for the future. In: Meek P, Fleming P (eds) Camera trapping: wildlife management and research. CSIRO Publishing, Collingwood, pp 87–98

Seeland K (2000) National Park policy and wildlife problems in Nepal and Bhutan. Popul Environ 22:43–62

Selvan KM, Gopi GV, Lyngdoh S et al (2013) Prey selection and food habits of three sympatric large carnivores in a tropical lowland forest of the Eastern Himalayan Biodiversity Hotspot. Mammal Biol 78:296–303

Shah GM, Jan U, Bhat BA et al (2009) Food habits of the leopard *Panthera pardus* in Dachigam National Park, Kashmir, India. J Threat Species 1(3):184–185

Shehzad W, Nawaz MA, Pompanon F et al (2014) Forest without prey: livestock sustain a leopard *Panthera pardus* population in Pakistan. Oryx: 1–6

Sherub (2004) Bird report on Royal Manas National Park. Nature Conservation Division, Department of Forests, Thimphu

Sogbohossou EA, de Longh HH, Sinsin B et al (2011) Human-carnivore conflict around Pendjari Biosphere Reserve, northern Benin. Oryx 45:569–578

Soto-Shoender JR, Giuliano WM (2011) Predation on livestock by large carnivores in the tropical lowlands of Guatemala. Oryx 45:561–568

Stahl P, Vandel JM, Herrenschmidt V et al (2001) Predation on livestock by an expanding reintroduced lynx population: long-term trend and spatial variability. J Appl Ecol 38:674–687

Stahl P, Vandel JM, Ruette S et al (2002) Factors affecting lynx predation on sheep in the French Jura. J Appl Ecol 39:204–216

Steinmetz R, Seuaturien N, Chutipong W (2013) Tigers, leopards, and dholes in a half-empty forest: assessing species interactions in a guild of threatened carnivores. Biol Conserv 163: 68–78

Studsrød JE, Wegge P (1995) Park-people relationships: the case of damage caused by park animals around the Royal Bardia National Park, Nepal. Envron Conserv 22(2):133–142
Sunquist ME, Sunquist F (2002) Wild cats of the world. University of Chicago Press, Chicago
Tempa T, Norbu N, Dhendup P et al (2011) Results from a camera trapping exercise for estimating tiger population size in the lower foothills of Royal Manas National Park. Ugyen Wangchuck Institute for Conservation and Environment and Royal Manas National Park, Royal Government of Bhutan, Lamai Gompa
Tharchen L (2013) Protected areas and biodiversity of Bhutan. CDC Printers, Kolkata
Treves A, Karanth KU (2003) Human-carnivore conflict and perspectives on carnivore management worldwide. Conserv Biol 17:149–1499
Tshering K (2009) Agriculture and usage of natural resources in Bhutan. J Fac Agric Shinshu Univ 45(1):33–42
Wang SW (2004) The impacts of wildlife damage and conservation policies on farmer attitudes in Jigme Singye Wangchuck National Park, Bhutan. Dissertation, Cornell University
Wang SW (2010) Estimating population densities and biomass of ungulates in the temperate ecosystem of Bhutan. Oryx 44:376–382
Wang SW, Macdonald DW (2006) Livestock predation by carnivores in Jigme Singye Wangchuck National Park, Bhutan. Biol Conserv 129:558–565
Wang SW, Macdonald DW (2009) The use of camera traps for estimating tiger and leopard populations in the high altitude mountains of Bhutan. Biol Conserv 142:606–613
Wang SW, Lassoie JP, Curtis P (2006) Farmer attitudes towards conservation in Jigme Singye Wangchuck National Park, Bhutan. Environ Conserv 33:1–9
Wangchuk T (2004) Predator–prey dynamics: the role of predators in the control of problem species. J Bhutan Studies 10:68–89
Wangchuk T, Thinley K, Tshering K et al (2004) A field guide to the mammals of Bhutan. Department of Forestry, Ministry of Agriculture, Royal Government of Bhutan, Thimphu
Wikramanayake E, Moktan V, Aziz T et al (2006) The WWF snow leopard action strategy for the Himalayan Region. World Wildlife Fund, Bhutan
Wolf PJ, Taylor CA, Heske EJ et al (2015) Habitat selection by American mink during summer is related to hotspots of crayfish prey. Wildl Biol 21:9–17
Yom-Tov Y, Ashkenazi S, Viner O (1995) Cattle predation by the golden jackal *Canis aureus* in the Golan heights, Israel. Biol Conserv 73:19–22

Part IV
Managing Problematic Species: Case Studies from Protected Areas and Areas Subject to Other Kinds of Management (Rural, Forest, Hunting and Urban Areas). Introductions, Reintroductions and Restocking

As the title indicates, this part focuses on the management of species that become problematic and, as a result, a true challenge to be faced (Conover 2002; Woodroffe et al. 2005). We know how these situations can be created both in areas that are not protected in any particular way, such as urban, agricultural and industrial areas, and in protected areas that are ecologically important, such as national and regional parks or various types of nature reserves. A typical example of the kind of wildlife interaction that generates a great deal of problems often relates to the introduction, reintroduction or restocking of wildlife species (Angelici 2016).

There are four chapters in this part. The first chapter is a review examining the operations undertaken in the USA by commercial wildlife control operators (WCO) when they create situations of conflict or interaction between humans and wildlife in urban or urbanised environments (Vantassel and Groepper 2016). However, one of the things underlined in the chapter is that the actions planned and carried out by organisations which have not been formally trained to carry out these types of tasks, such as the initiatives of private individuals or amateurs, can often be destructive and lead to ineffective or even negative effects.

The second chapter (Capizzi et al. 2016) addresses an issue that is very important in island ecology, i.e. rat control and eradication (*Rattus* sp.), primarily the black rat (*R. rattus*) which preys on both the eggs and nestlings of seabirds (the most frequently preyed upon species being *Calonectris diomedea* and *Puffinus yelkouan*) nesting on islands, which is devastating and often leads to their total disappearance. The chapter examines the case of some Mediterranean islands where the success of such operations has been lauded. This chapter demonstrates how the eradication of the black rat has not only led to the recovery of the successful reproduction of nesting seabirds but also contributed to the general recovery of the ecology of the island.

The third chapter (Meriggi et al. 2016) analyses wild boar management (*Sus scrofa*) in protected areas. It is known that this species heavily impacts the environment, i.e. the vegetation, fauna and human activities (Barrios-Garcia and Ballari 2012).

This chapter specifically looks at Elba Island, in the Tyrrhenian Sea in Italy, where the species was introduced for hunting purposes. Naturally, this chapter also highlights the appropriate strategies and approaches to contain, to a large extent, the boar population to make it acceptable and ecologically sustainable.

The last chapter (Menchetti et al. 2016) discusses the case of a species of parrot, the ring-necked parakeet (*Psittacula krameri*), which is currently expanding its range into many areas of the world where it was not originally present. This has been facilitated by individual parrots which have escaped from captivity (this parakeet is a very popular cage bird). These birds are easily naturalized into their surroundings and in recent decades have also become favourites, probably due to climate change (Sutherland 2004). The chapter examines the environmental effects of this species in recently colonised areas.

References

Angelici FM (2016) Problematic wildlife at the beginning of the twenty-first century: introduction. In: Angelici FM (ed) Problematic wildlife: a cross-disciplinary approach. Springer, New York, p 3–18

Barrios-Garcia MN, Ballari SA (2012) Impact of wild boar (*Sus scrofa*) in its introduced and native range: a review. Biol Invasions 14:2283–2300

Capizzi D, Baccetti N, Sposimo P (2016) Fifteen years of rat eradication on Italian islands. In: Angelici FM (ed) Problematic wildlife: a cross-disciplinary approach. Springer, New York, p 205–227

Conover MR (2002) Resolving human-wildlife conflicts: the science of wildlife damage management. CRC, Boca Raton

Menchetti M, Mori E, Angelici FM (2016) Effects of the recent world invasion by ring-necked parakeets *Psittacula krameri*. In: Angelici FM (ed) Problematic wildlife: a cross-disciplinary approach. Springer, New York, p 253–266

Meriggi A, Lombardini M, Milanesi P, Brangi A, Lamberti P, Giannini F (2016) Management of wild boar in protected areas: the case of Elba island. In: Angelici FM (ed) Problematic wildlife: a cross-disciplinary approach. Springer, New York, p 229–251

Sutherland WJ (2004) Climate change and coastal birds: research questions and policy responses. Ibis 146(suppl 1):120–124

Vantassel SM, Groepper SR (2016) A survey of wildlife damage management techniques used by wildlife control operators in urbanized environments in the United States. In: Angelici FM (ed) Problematic wildlife: a cross-disciplinary approach. Springer, New York, p 175–204

Woodroffe R, Thirgood S, Rabinowitz A (2005) People and wildlife, conflict or co-existence? Cambridge University Press, Cambridge

Chapter 9
A Survey of Wildlife Damage Management Techniques Used by Wildlife Control Operators in Urbanized Environments in the USA

Stephen M. Vantassel and Scott R. Groepper

Introduction

Wildlife damage management (WDM) is a field of human endeavor that employs various methods and tools to reduce or prevent the effects of activities by vertebrate animal species that impede human interests. Though sharing many similarities with traditional pest control, WDM does not address damage by invertebrate animals such as insects, or fish (contra Reidinger and Miller 2013), except in an indirect manner such as reducing white-tailed deer (*Odocoilleus virginianus*) populations to reduce ticks (*Ixodes* spp.). The goal of WDM is to resolve human–wildlife conflicts, or prevent potential conflicts, and reduce them to a level tolerable to those affected by the damage (Peterson et al. 2010).

The focus of this chapter will be on WDM in urbanized areas, namely those areas that generally consist of population densities of at least 1000 people per square mile with surrounding blocks with densities of at least 500 people per square mile (Nowak et al. 2005) and are serviced by private wildlife control operators (WCOs).

While human conflicts with wildlife have occurred as far back as ancient Israel (1 Sam 17:34ff), in the USA the specialized role of WCO has developed only in the last 40 years (Bluett et al. 2003). In the 1980s, the increase in the number and impact of human–wildlife conflicts coupled with budgetary cuts caused states to cede control of problematic wildlife species to the private sector (Messmer et al. 1999; Miller 2007). The entry of WCOs into the wildlife management profession has hastened the development of new methods and tools to control human–wildlife conflicts.

S.M. Vantassel (✉)
Wildlife Control Consultant, LLC, Lewistown, MT, USA
e-mail: stephenvantassel@hotmail.com

S.R. Groepper
Wildlife Biologist, Mitchell, SD, USA
e-mail: scott.groepper@yahoo.com

F.M. Angelici (ed.), *Problematic Wildlife*, DOI 10.1007/978-3-319-22246-2_9

This chapter will survey WDM by WCOs because researchers generally have overlooked WCOs in favor of studying the role of USDA-APHIS-Wildlife Services or sport hunting (Bruggers et al. 2002; Conover 2002). The authors' insights into the industry that follow flow from personal knowledge and interaction in the industry and with WCOs.

Wildlife Damage

What do we mean by damage by wildlife? The question of when wildlife activity rises to the level of causing damage remains a controversial one and depends on factors including individual perceptions. Some adopt an objective standard in that wildlife damage only occurs when wildlife negatively impacts property, production, or human health and safety in a measurable way. Thus a coyote (*Canis latrans*) howling in the distance and frightening nearby residents would be a nuisance, but would not be considered damage. However, if the same coyote or pack began to prey on pets, then the activity would constitute damage.

Others take a perspectival approach, arguing that damage exists whenever humans perceive the animal's behavior or presence in a negative way (Hadidian et al. 2000). For example, one might consider the presence of a snake, even a nonvenomous snake such as the common garter snake (*Thamnophis sirtalis*), as wildlife damage due to the emotional toll or fear that it provokes in many people. While the issue of when humans have the right to control wildlife is an important one, this chapter will focus on the management of objective, or measurable damage by wildlife, and avoid discussion of damage that can be considered perspectival or emotional in nature.

The types of damage caused by wildlife are as diverse as they are far reaching. While the true extent of wildlife damage in the USA is unknown because most damage is unreported, surveys clearly show that the economic cost is significant. Pimentel et al. (2005) estimated that the cost of damage caused by nonnative rats in the USA was 19 billion dollars per year. Wildlife can consume and contaminate food and fiber, damage landscapes and pristine habitats (Diamond 1992), and damage structures. In addition, wildlife can threaten human health and safety by attacks, vehicle strikes, and transmission of diseases. Though the extent and cost of wildlife damage in urbanized areas also is unknown, data do suggest it too is a significant subset of wildlife damage in general (Seuffert 2007; Responsive Management 2012; Pest Management Professional 2014).

The reason wildlife activity can affect human interests in so many ways stems from two key facts. First, wildlife encompasses a variety of species (i.e., birds, reptiles, and mammals) with different life histories and levels of adaptability. For example, a garden can be damaged by rodents (i.e., voles, *Microtus* spp.), mesocarnivores (i.e., raccoons, *Procyon lotor*), and birds. A gardener may be diligent in controlling damage by birds but discover that techniques effective on birds will have little to no effect on mesocarnivores and rodents. To complicate matters further, actions to manage one type of damage may actually create conditions that favor the viability of another species whose activities may also negatively impact human interests (Cf. Messmer 2009). The second problem with wildlife is that they do not have the capability to recognize borders or other forms of human-constructed boundaries. They do not understand that urban dwellers think "wildlife belongs in the woods" or if a textbook says that corn and apples are not part of their traditional food choices.

Damage by Wildlife

To help readers appreciate the extent wildlife activity has on human interests, we have provided a brief survey of wildlife damage in urbanized landscapes. The examples we present only scratch the surface and are in no way comprehensive (Conover 2002).

Damage to Structures and Landscapes

Various avian species can cause damage in the urban and suburban environments. European starlings (*Sturnus vulgaris*), house sparrows (*Passer domesticus*), pigeons (*Columba livia*), crows (*Corvus* spp.), and others commonly cause damage by deposition of feces (Fig. 9.1).

Bird nests in buildings may lead to fire or cause other problems (VerCauteren et al. 2010; Fig. 9.1b). Woodpeckers (Family: Picidae) may damage wood siding on houses or other buildings by boring holes (Belant et al. 1997; Fig. 9.2).

Resident Canada geese (*Branta canadensis*) deposit large amounts of feces on lawns at parks and can cause significant damage to golf course turf by grazing (Smith et al. 1999). Migratory geese often avoid urban areas, but resident geese will continually use the same locations, unless prompted to move, concentrating damage.

Deer can cause significant damage to urban and suburban environments. Deer may damage ornamental plants through browsing and antler-rubbing (Maas-Hebner et al. 2005). Wildlife, particularly burrowing animals, can undermine the integrity of earthen structures, such as dams and levees, and structural failure may occur if not addressed (Bayoumi and Meguid 2011). Wild turkeys in urban and suburban environments often become acclimated to human activities and, in the absence of hunting, may lose fear of humans (Groepper et al. 2013). These birds may appear tame and, given their proximity to humans, may be more prone to inflict damage to golf courses, gardens, and lawns by scratching turf or mulch in search of insects, peck at cars, and chase or frighten people (Sanford et al. 2004). Damage to automobile mirrors, paint, and other parts has been reported when male turkeys exhibited aggressive behavior towards their reflection (Miller et al. 2000).

Fig. 9.1 (**a**) Feces of Canada geese on a roof can clog drains and lead to roof collapse from the weight of accumulated rain water. Photo by Carla Wagner. (**b**) Starling (*Sturnus vulgaris*) nest has plugged this dryer vent which could have caused a fire. Photo by Kevin Cornwell

Fig. 9.2 Damage to this shed was caused by a flicker or other woodpecking bird. Photo by Stephen M. Vantassel

Wildlife Strikes

As populations of humans and large vertebrates expand and spatial overlap increases, the probability of wildlife collisions with vehicles also increases. State Farm Insurance, a predominant car insurance provider in the USA, estimated that 1.2 million deer–vehicle collisions occurred between July 2012 and June 2013, causing an average of \$3414 in damage per vehicle (Insurance Information Institute 2014; Fig. 9.3a). Several studies (Etter et al. 2000; Mastro et al. 2008) have found deer–vehicle collisions increase with increased deer densities. Likewise, reduction in deer populations led to a reduction in deer–vehicle collisions (DeNicola and Williams 2008). Wildlife strikes also include bird strikes to aircraft but this area of WDM is still primarily handled by USDA-APHIS-Wildlife Services (Cleary and Dolbeer 2005; Cleary et al. 2006).

Diseases

Human face certain risks from wildlife-borne diseases at the urban/wildland interface typical in many suburban areas. Most emerging infectious diseases in humans result from exposure to zoonotic pathogens, and wildlife plays a role by providing the zoonotic pool (Daszak et al. 2000).

Fig. 9.3 (**a**) Deer struck by a vehicle. Photo by Stephen M. Vantassel. (**b**) Raccoon feces on a roof. Photo by Jesse Tenley

Lyme disease is one of the most prevalent vector-borne diseases in North America (Bacon et al. 2008). The disease is caused by bacteria which are transmitted to humans through tick bites (Barbour and Fish 1993). The life cycle of Lyme disease depends on interactions of ticks, various small mammals, canids, such as coyotes, and deer. The emergence of Lyme disease has been attributed to high populations of deer (Wilson et al. 1988), but others have implicated increases in small mammal abundance (Ostfeld et al. 2001), perhaps caused by decreases in predator abundance (Levi et al. 2012). The mosaic landscape of streams, forests, parks, and other green spaces in urban areas provides high-quality habitat for deer (Grund et al. 2002) and concentrated resources influence host migration into urban landscapes, increasing contact rates between humans and wildlife (Bradley and Altizer 2006). Deer often coexist in close proximity with humans in suburban areas. Reduction of natural predators and limited hunting has led to increases in deer populations (Cote et al. 2004).

Because raccoons often use attics and backyards for latrines, WCOs have been involved in the removal of feces to protect homeowners from raccoon roundworm (*Baylisascaris procyonis*; Fig. 9.3b) which can cause blindness or severe neurological injuries, including death, to victims (Vantassel 2011). Standards governing cleaning and removal of raccoon toilettes have not been established, but interest in professional standards is growing.

Rabies is an acute, progressive, incurable viral encephalitis. Rabies infection has the highest case fatality ratio in humans (Hemachudha et al. 2002). Variants of rabies virus are maintained in the wild by terrestrial carnivore species including raccoons, skunks (*Mephitis mephitis*), and bats (Guerra et al. 2003). Incubation periods for naturally infected animals may be 6 months or longer and incubation periods for humans may be 6 years. Every mammal is susceptible and the domestic dog is the main reservoir worldwide (Rupprecht et al. 2002). Dogs cause the majority of the annual human deaths (Fekadu 1993), but cats are effective vectors, although they do not seem to serve as reservoirs (Rupprecht et al. 2002). Skunks and raccoons coexist in the same geographical areas, but raccoons occur in higher densities in close proximity to humans (Hoffmann and Gottschang 1977), while skunks prefer interfaces between agricultural and nonagricultural land (Pool and Hacker 1982). Both have the potential to come into contact with humans or companion animals in urban or suburban areas as evidenced in the Mid-Atlantic rabies epidemic of the 1980–1990s (Wyatt et al. 1999). Humans generally only encounter bats that are sick or incapacitated and infection is common among such animals (Rupprecht et al. 2002). But since many of the relatively few human rabies cases in the USA have come from encounters with bats, the Centers for Disease Control (CDC) has tightened protocols for handling bats inside the living areas of homes (Blanton et al. 2010).

Wildlife Control Methods Used by WCOs

WDM requires a variety of methods and tools to reduce wildlife conflicts to tolerable levels because typically no single technique can address all problems with wildlife damage at any given locale.

Methods for WDM fall into the broad categories listed below organized by their respective lethality and amount of contact with the animal (Hygnstrom et al. 1994). Just as in pest control, an integrated approach that addresses the environmental conditions, population, and actions of problem species often provides the best results (VerCauteren et al. 2010).

Habitat Modification

All animals require water, food, and shelter. Areas that provide these needs are called habitat for our purposes (Cf. Hall et al. 1997). Habitat modification addresses all three of these life requisites; efforts to reduce any one of these will ultimately reduce the carrying capacity of the land for that species. Most habitat modifications utilized by WCOs and homeowners are subtle, but they play an important role in long-term WDM. Even small modifications to the quality of the habitat can enhance the effectiveness of other techniques. Habitat modification can provide long-term solutions to difficult wildlife conflicts. Typical modifications include modifying bird feeders to reduce spilled seed from reaching the ground (Pennisi and Vantassel 2012), thus reducing the role of food in attracting other wildlife, cutting back bushes and trees to reduce cover and access to structures, and removing brush, woodpiles, and debris to eliminate potential harborage for prey or predators.

While habitat modification is an important wildlife control method, its use in urban environments to manage wildlife damage often is hindered by two problems. First, habitat modification is most effective when applied over large areas. Small plots, characteristic of urbanized landscapes, means animals can easily move through a modified yard to reach a yard that is not modified. For example, a landowner can experience damage from fox squirrels (*Sciurus niger*) entering the attic of her home even though they were attracted to the area because of the adjacent neighbor's feeding of squirrels. Second, modifications to limit the damage of one species can create favorable conditions for another. For example, letting grass grow >6 in. will discourage foraging by Canada geese (Smith et al. 1999), but will encourage the presence of voles (*Microtus* spp., Merwin et al. 1999). While WCOs often advise clients to modify habitat, suggestions typically focus on removal of favorable conditions to animals rather than implementing tactics that would seek greater harmony with wildlife while reducing damage to tolerable levels. For example, most will advise the removal of a bird feeder rather than provide instructions on how to modify the feeder to reduce attracting unwanted animals. One exception to this is the increase, albeit small, in the use of beaver pipes to control water levels caused by beaver dams.

Exclusion

Exclusion involves the use of physical barriers to prevent unwanted wildlife from accessing areas and causing damage. Though exclusion provides immediate, long-term, and high levels of protection, high cost of materials and installation limit its adoption by customers (Vantassel et al. 2012). WCOs also must ensure that the exclusion method will not trap wildlife inside of a structure or otherwise cause physical harm. Failure to ensure all wildlife is removed from an enclosed area can result in excessive animal suffering as well as significant property damage as animals struggle to exit.

Fig. 9.4 Electric shock track used to keep birds off of the building's ledge. Photo by Carla Wagner

The greatest advances in exclusion products have been among those created for the control of birds. For example, while expensive to install, nets are considered the gold standard to prevent birds from accessing sensitive locations. Additionally, a wide variety of ledge exclusion products have been developed to prevent birds from perching, including spikes, coils, and wires. Supply catalogs, such as one published by Bird Barrier® details the various products available. In recent years, electric shock strips have gained popularity due to improvements in ease of installation (Fig. 9.4).

Another common exclusion method, fences, can prevent terrestrial animals from gaining access to landscapes (e.g. fields, gardens, airports) and structures (e.g. decks, porches, buildings). Fences may not need to enclose the entire site to achieve significant protection (Hildreth et al. 2012). Two types of fences are typically used: electric and nonelectric (McKillop and Sibly 1988). Electric fences are both physical and psychological barriers to wildlife. They affect the movement of animals through a painful but harmless shock. Nonelectric fences are physical barriers that block the movement of wildlife. While requiring lower maintenance than electric fences, they often cost significantly more to install (VerCauteren et al. 2006).

Covers, caps, and screens prevent wildlife from using potential access points in structure, such as chimneys, attics, vents, doors, windows, and window wells, where they can damage insulation, wires, and contaminate areas with urine and feces (Figs. 9.5a, b). In some instances, these access points can also inadvertently prevent wildlife from leaving the structure as well skunks (*Mephitis mephitis*) fall in a window well and become trapped (Fig. 9.6). Crevice sealers include materials such as caulk, foam, mortar, fabric (such as, Xcluder™ by Global Material Technologies,

Fig. 9.5 (**a**) Framing, insulation, and wire damaged by a gray squirrel (*Sciurus carolinensis*). Photo by Kevin Cornwell. (**b**) Bat droppings staining the outside of this house. Photo by Stephen M. Vantassel. (**c**) Bat droppings covered this pink insulation enough to turn it black. Photo by Jason Reger

Buffalo Grove, IL, USA), and other sealants to prevent animals from entering structures. In recent years, the WCO industry has been endeavoring to make repairs aesthetically pleasing as well as ensuring wildlife cannot reenter the structure (Budenski 2014).

Frightening Devices

Frightening devices deter wildlife use from a location through nonchemical means. Frightening devices fall into four categories: visual, audio, audiovisual, and biological (Vantassel et al. 2012). Frightening devices have two drawbacks. First, wildlife

Fig. 9.6 Striped skunk trapped in a window well. Photo by Stephen M. Vantassel

often habituate to frightening devices unless the device includes movement or are actively manipulated by a WCO. Second, it is illegal to haze migratory bird species during their nesting or molt periods (U.S. Migratory Bird Treaty Act; 16 U.S.C. 703–712). Otherwise, birds may be actively harassed provided such actions do not physically harm the birds and such activities are not prohibited by additional legislation, such as the Bald and Golden Eagle Protection Act (16 U.S.C. 668–668c), the Endangered Species Act (16 U.S.C. 1531–1544), and state legislation for the species causing issues.

Visual Frightening Visual repellents range dramatically in price, sophistication, and effectiveness, but are most effective on birds. Visual repellents include effigies (e.g. plastic owls), scary-eye balloons, Mylar® tape, and lasers. Stationary visual frightening devices, such as the ubiquitous plastic owl, are the least effective, as birds tend to habituate to these devices quickly. Animated or moving devices are typically more effective than stationary devices. While gruesome, use of turkey vulture (*Cathartes aura*) carcasses suspended near roosts have been shown to be effective at frightening their conspecifics (Ball 2009). Geese (Blackwell et al. 2002) and crows can be dispersed from a night-time roost by pointing a spotlight, laser pointer, or laser pistol (such as the Avian Dissuader® by Feather-Light Technologies, Louisville, KY, USA) at them. Strobe lights (such as Evictor Strobe Lights by Pest Tools®, Philadelphia, PA, USA) are marketed to frighten wildlife such as squirrels. It is claimed that the lights irritate the animals and cause them to leave the structure. Presently, claims of the efficacy of this type of product have not been tested under scientific and peer-review conditions.

Audio Frightening Audio frightening devices use sound to deter wildlife use of sensitive areas. Devices include propane cannons, distress calls, and Long Range Acoustic Devices (LRAD; LRAD® Corp., San Diego, CA, USA; Fig. 9.7). However, in residential areas volume can be a concern for neighbors, and run afoul of local

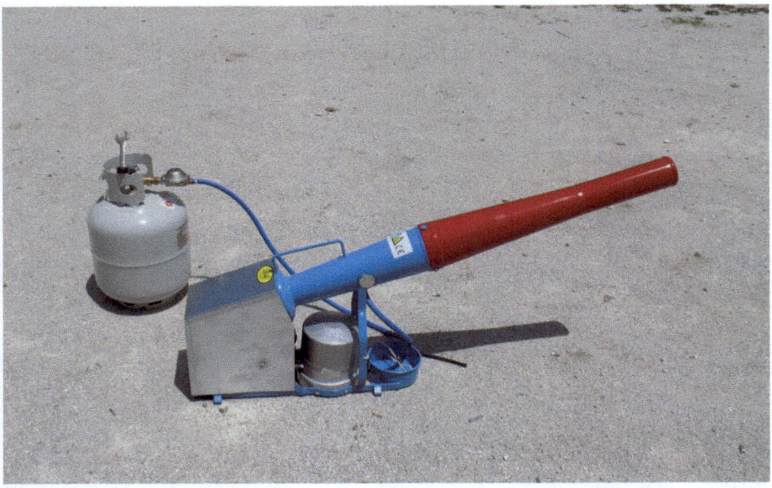

Fig. 9.7 Propane cannon. Photo by Stephen M. Vantassel

noise ordinances. Habituation is a problem with these devices, as wildlife tends to ignore repetitive sounds (Bomford and O'Brien 1990). Propane cannons emit a large boom that is suitable only for rural settings. Recorded distress calls that are played at irregular intervals have more versatility and can target specific species, such as crows (Gorenzel and Salmon 1993). Research has shown that use of white-tailed deer distress sounds frighten deer (Gilsdorf et al. 2004; Hygnstrom et al. 2013). LRADs emit a concentrated cone of high decibel sound that can maintain a high decibel level at long distances. LRADs have been used successfully at airports to frighten birds. Despite intense marketing, no peer-reviewed evidence is available to support the claim that ultrasonic devices are effective in repelling wildlife (Bomford and O'Brien 1990). Adoption of these devices by WCOs has been limited due to the problem noise would cause with neighbors.

Audiovisual Devices Audiovisual devices combine sound and sight to frighten wildlife. Fireworks-based noisemakers (a.k.a. "pyrotechnic devices") are the most common items in this category. Pyrotechnic devices differ based on sound, distance, and firing mechanism (Margo Supplies, LTD, margosupplies.com). Users of pyrotechnics must follow federal and local laws and abide by safety recommendations to avoid damage to self, bystanders, and property (OSHA 2007).

Quieter devices are available such as the Scarey Man® (R. Royal, Midnight, MS, USA; Andelt et al. 1997) and movable effigies from Custom Robotic Wildlife (Mosinee WI, USA). Remote control devices, such as toy helicopters and boats, have been used successfully to haze Canada geese at parks, golf courses, and other suburban areas.

Biological Frightening Guard animals, such as dogs (Andelt and Hopper 2000) and llamas (Meadows and Knowlton 2000), are used to protect livestock, especially sheep, from predators. Livestock guard dogs may be aggressive towards humans or other domestic animals, so the livestock and the guard animal may need to be confined

Fig. 9.8 Border collie stalking birds. Photo by Carla Wagner

within a fenced area. Dogs can protect orchards, Christmas tree plantations, or vineyards from deer. Dogs contained within an invisible-fence system may reduce deer damage to home garden and landscape plantings.

While not as common as other methods, raptors have been used to haze gulls near landfills (Baxter and Allan 2006). The level of success with the raptors is mixed and often requires a significant investment of time from the handler. Border collies have been shown to be effective in evicting Canada geese from areas (Castelli and Sleggs 2000; Fig. 9.8). Whenever using predatory animals, such as raptors, to haze wildlife, be sure to have any required depredation permits in case the predator harms a protected species. Some WCOs have opted to specialize in providing goose control using dogs.

Repellents

Repellents are chemicals that deter animal activity through pain, fear, touch, or aversive conditioning. The active ingredients in repellents are considered pesticides and are therefore under the jurisdiction of the U.S. Environmental Protection Agency (1970) (7 U.S. Code Chapter 6 Subchapter II-Environmental Pesticide

Control-136; http://www.epa.gov/agriculture/lfra.html#Reregistration%20of%20 Pesticides). In most states WCOs must have a pesticide applicator license to use these products. Since obtaining pesticide applicator licenses is often restricted by experience requirements and increased costs (i.e., training and insurance), many WCOs have opted not to use pesticides. WCOs are frequently hired after clients have tried ineffective home remedies, such as mothballs, to manage a wildlife problem (Fitzwater 1990).

While many repellent products are available, there are relatively few active ingredients, and even fewer that have been scientifically shown to be effective. The effectiveness of repellents often is highly variable, depending on the motivation of problem animals (e.g., nesting and feeding), alternative resources, previous experience, and active ingredients of the repellent. Deer repellents have been researched but often fail to achieve significant effectiveness (Conover 1984). Capsaicin is an active ingredient in some repellents that causes pain to animals that eat it. Other repellents exploit a perceived fear response, such as using coyote urine, or that of other predators, to cause herbivores to avoid an area. The sticky material used in polybutene-based bird repellents is a tactile repellent that birds avoid stepping on. Anthraquinone is an aversive conditioning repellent that causes nausea when consumed. Geese that consume grass sprayed with anthraquinone (Flight Control™) are sickened and quickly learn to avoid eating treated grass in the future. Methyl anthranilate is a chemical derived from grapes which is unpalatable and aversive (odor) to birds such as Canada geese (Cummings et al. 1991). Efficacy of the product on turf, however, is questioned (Belant et al. 1996).

Development of new active ingredients is expensive and there is limited availability of new products (Vantassel et al. 2013). Obtaining an experimental use permit costs >$77,000 even after a 75 % fee waiver (http://www2.epa.gov/pria-fees/r090-pria-fee-category). The pesticide law is at times irrational. For example, if a WCO uses coyote urine to attract a coyote, the urine is not considered a pesticide. But if the WCO uses the same urine to evict an attic-dwelling raccoon, the urine is a pesticide, and its use likely requires a pesticide license.

Toxicants

Toxicants are chemical compounds used to kill problem animals such as house mice (*Mus musculus*), Norway rats (*Rattus norvegicus*), and pigeons. Like repellents, toxicants are regulated by the U.S. EPA. Commercial applicators of toxicants (and fumigants) are required to have a pesticide applicators license.

Only a few dozen active ingredients are available for the control of vertebrates (Mallis 2010). Of those, the vast majority are for the management of commensal rodents. It is illegal to use to toxicants on most wildlife species in the USA. In addition, fumigants have restrictions against their use near structures due to safety concerns, thus they cannot be used for control of burrowing animals near structures. In light of regulatory issues and safety concerns of users and clients, it is unlikely that WCOs will employ toxicants to any great extent.

Shooting

The use of firearms such a pistols, shotguns, rifles, and high-velocity projectiles from air rifles to manage wildlife damage has been long accepted in rural areas. Safety concerns, however, have limited this technique in urbanized landscapes. Shooting safely requires skill, but in urban environments the skill requirements are higher due to the risk from a bullet that may accidentally strike a person or property. It is no surprise that many communities prohibit the discharge of any projectile, whether propelled by gunpowder or compressed air, within their borders.

Control of white-tailed deer populations has conventionally focused on lethal removal (i.e., professional sharpshooting or public hunting opportunities; Boulanger et al. 2012). In an increasing number of communities, however, lethal management strategies are rejected based on legal, safety, or ethical concerns (Decker and Connelly 1989; Wright 1993; McCullough et al. 1997). Social and demographic changes have altered contemporary wildlife management in the USA and have led to the declining public tolerance of lethal methods of wildlife management (Curnow 2001). Interestingly, issues with wildlife damage (i.e., deer) have caused communities to grant exemptions to these restrictions (Williams et al. 2013; DeNicola and Williams 2008). Schwartz et al. (1997) recommended rifle shots to the head as the most efficient and humane method of euthanasia for urban deer. The National Wildlife Control Operators Association (NWCOA; NWCOA.com) has developed a training program and certification for shooting in sensitive environments. We believe that if the needs of clients for rapid removal of problem wildlife are met with improved training standards the regulatory bodies will ease restrictions on the use of this important tool.

Trapping

Traps are devices that capture animals without the WCO present. Since traps are time multipliers, they are the most frequently used tool by WCOs. Traps come in two forms: live and kill. Live traps capture animals without killing them. Live traps include cage traps, box traps, multiple-capture traps, foothold traps, nets, cable-restraints, and a variety of bird traps (Vantassel 2012). Kill traps are designed to capture and cause the quick death of the animal. Kill traps include Conibear-style traps, mouse snap traps, and snares.

Cage and Box Traps Cage and box traps are the main stock-in-trade of most WCOs. Cage traps are made of wire mesh while box traps are made of solid material (Fig. 9.9a). The traps are available in many sizes and styles and may include one or more entrances (Wildlife Control Supplies, LLC 2014). Cage and box traps are easy to set and can serve as a carrier. An animal enters a cage or box trap and steps on a treadle, which causes the door(s) at the end(s) of the trap to close. Typically, little site preparation is needed and these traps pose little threat for children and pets. When properly set, traps are humane, but animals may harm themselves due to the stress of being restrained, or while trying to escape. For example, plastic-walled

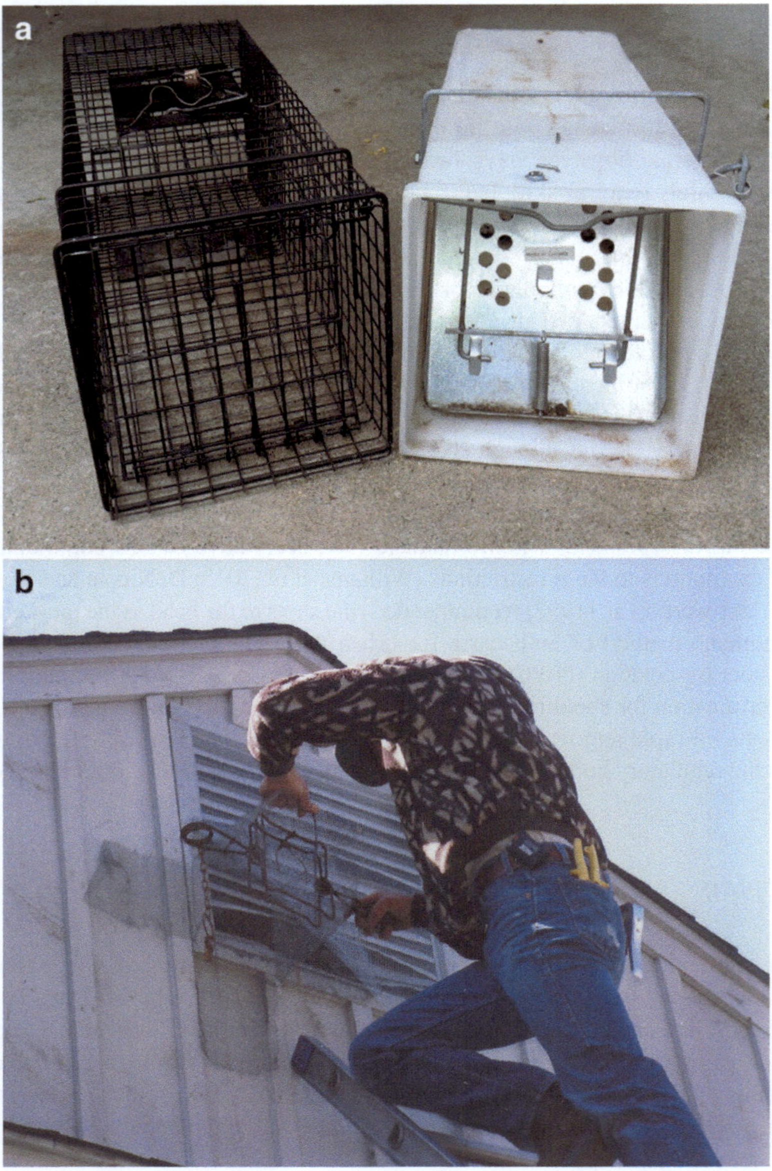

Fig. 9.9 (**a**) Cage trap (*left*) and box trap (*right*). Photo by Stephen M. Vantassel. (**b**) Rich Daniotti sets a 220 Conibear-style trap outside an attic vent to capture a raccoon. Photo by Stephen M. Vantassel

traps are almost 7 °C hotter than a comparable cage trap covered with a cloth along 50 % of its length (Vantassel, unpublished data). This temperature difference could benefit or harm the trapped animal depending on environmental conditions. Most cage and box traps are bulky and difficult to conceal because of their size and shape, making them vulnerable to theft and vandalism. They are also more expensive than other devices, such as footholds.

Cage and box traps are not universally effective in capturing animals. Some species, such as coyotes, may avoid them. Even individuals of species that generally are easy to catch in a cage or box trap, such as raccoons and gray squirrels (*Sciurus carolinensis*), can become "trap-shy," if they have been previously captured.

WCOs, however, have encouraged the development of new styles, add-ons, and uses of cage traps. Today, WCOs have access to professional grade cage traps which are constructed with heaver gauge wire, stronger doors/locking mechanisms, and narrow weave mesh (Tomahawk Live Trap, LLC 2014). The add-ons for cage traps are quite varied, but most concentrate on ways to direct target animals into the trap, ensure target animals are captured, mounting traps on structures, and safety/animal welfare (Tomahawk Live Trap, LLC 2014).

Multiple-Capture Traps Multiple-capture traps are designed to catch more than one animal without having to be reset. Most multiple-capture traps are designed for mice (e.g., Ketch-All, Kness Mfg. Co., Inc., Albia, IA, USA). Multiple-catch traps have also been placed in runs or in front of dens to capture muskrats (*Ondatra zibethicus*). WCOs have used multiple-catch traps to capture flying squirrels (*Glaucomys* spp.), juvenile gray and fox squirrels, and birds, such as pigeons. Unfortunately, some states have not properly distinguished WCOs from recreational trappers. Thus, regulations designed for recreational harvest (such as bans on multi-catch traps to prevent overharvest of furbearing species) may prevent WCOs from resolving client problems dealing with urban wildlife that are, also classified as game animals even when it is the most efficient method possible (see Massachusetts Fish and Wildlife Trapping Regulations (2014) http://www.eregulations.com/massachusetts/14maab/trapping/ Accessed 29 Dec 2014).

Foothold Traps Foothold traps are live traps designed to capture and restrain animals by the foot. Leg hold, while a common term, actually is inaccurate, as animals should not be captured by the leg due to risk of breaking bones. When used properly, foothold traps generally are not capable of capturing many species higher on the leg than the foot. Footholds are very versatile traps capable of being used effectively in land sets, water sets (streams, lakes), or beneath ice. Unfortunately, their use is illegal in some states despite their being the most efficient tools for catching coyotes (Way et al. 2002) and useful for capturing raccoons, beavers (Castor canadensis), muskrats (*Ondatra zibethicus*), nutria (*Myocastor coypus*), and other species.

Several designs of foothold traps are available, including the coil-spring trap, which is the most commonly used foothold trap, and the long-spring trap. In some cases, modifying a standard coil-spring trap by padding or laminating (i.e., increasing jaw thickness) the jaws will reduce the chance of injuring the captured animal, while also possibly increasing the effectiveness of the trap (Linhart and Dasch 1992; Huot and Bergman 2007). Commercial models are available with these modifications (Victor Soft Catch®, Woodstream Corp., Lititiz, PA, USA). Legal restrictions on the use of foothold traps including jaw spread and where and how they may be set are common in many states. The jaw spread refers to the distance between the two jaws when the trap is set (not counting the thickness of each jaw's gripping surface). The technical elements of ideal foothold use has progressed remarkably (Noonan 2014).

Enclosed-foot Traps Enclosed-foot traps, sometimes called dog-proof traps (mistakenly) or species-specific traps (Hubert et al. 1999), are another type of foothold traps. Foothold traps rely on the animal pressing the pan to fire the trap. Enclosed-foot traps place the trigger (tension pan) in a tube, which reduces nontarget capture, because the width of the tube limits the animal's capability of reaching inside to trigger the trap. The earliest version was the Egg trap® (EGG Trap Co., Ackley, IA, USA). While effective, WCOs resisted use of the trap, at least in part, because the trap had to be dismantled to release a captured animal (Cf. Austin et al. 2004). Fortunately, manufacturers have developed several variations of this style of trap that are significantly more user-friendly, while still being effective. Traps include the Lil Grizz® (Qutad Performance Products, Co. Bonnots Mill, MO USA), Duffer's Raccoon Trap (Duffer's, Bern, KS, USA), Coon Dagger (Sudden Valley Trap Supply, Warrenton, MO, USA), and others. We are not aware of peer-reviewed research on these newer variations of enclosed foot traps.

Snares and Cable-Restraint Traps Snares are a type of capture device that use multi-stranded wire, or wire rope, designed and set, to close around the neck, torso, foot, or leg of an animal (Group 2009). Cable-restraints, also known as live-capture snares, are just like snares except they hold animals without killing them. Sometimes the difference between a lethal snare and a cable-restraint lies only in the way the device is set, but equipment selection can play a significant role in lethality (Pruss et al. 2002). Snares and cable-restraints are either passive (gravity operated) or active (spring operated). Brands of cable-restraints include Belisle Footsnare® and Collarum® (Wildlife Control Supplies, LLC 2014). Many states have restrictions on the use of cable-restraints for wildlife control because they have not fully appreciated the advances in cable-restraint technology (Vantassel et al. 2010). We are hopeful that as more regulators understand the distinction between snares and cable-restraints, regulations will be modified.

Kill Traps Kill traps are spring-loaded traps designed to strike the animal (typically in the chest or neck) in a manner that causes death. The familiar mouse snap trap is a form of kill trap. WCOs have used kill traps to control voles, weasels, and tree squirrels, but typically when WCOs speak of kill traps they are referring to the Conibear-style body-gripping trap (named after the inventor, Frank Conibear). Several sizes of Conibear trap are available. Typically, they are square, but some models are round (e.g., Bill Crum, Hillsdale, IN, USA).

"Magnum," "zero tolerance," or "Canadian-bend" body-gripping traps are also available (e.g., Butera Manufacturing Ind. Wickliffe, OH, USA). Magnum-style traps offer the potential for a quicker kill and kill more consistently than standard traps (Proulx and Barrett 1993) because the trap jaws close very tightly in comparison to standard body-gripping traps. They may increase the chance of a proper strike with squirrels, raccoons, or other small, flexible animals that might pull back if there is a slight gap between the jaws. Magnum versions are very strong, so WCOs must use care when using magnum traps.

Although the USA does not have standards requiring minimum levels of lethality for kill traps, the WCS Tube Trap (Wildlife Control Supplies, LLC 2014) has met the Spring Traps Approval (England) Order 2012 Standard of the United Kingdom (http://www.legislation.gov.uk/uksi/2012/13/schedule/made). The WCS Tube Trap is very popular among WCOs for the control of gray and fox squirrels.

As with all lethal techniques, WCOs must use caution to ensure that only the intended nuisance animals are caught. WCOs have adopted techniques from the trapping industry (Indiana Department of Natural Resources 2014) as well as developed their own techniques to reduce and/or eliminate risk to nontargets. For example, WCOs have developed ways to place cage and box traps in front of the animal's entrance hole while simultaneously preventing animals outside the building from being trapped. This set is known as a positive set because it will catch only the animals exiting the structure. WCOs also set Conibear-style traps over holes at elevated positions such as attic and roof vents (Vantassel 1999; Fig. 9.9b).

Mole Traps Mole traps are a special type of kill trap designed to capture ground moles (e.g., *Scalopus aquaticus*). There are a wide variety of types developed over the years (Marsh 1995). Despite the assertions of many landowners, the mole traps are exceedingly effective provided the user follows simple but essential steps (Ferraro et al. 2010).

Trapping Sets

A trap set encompasses the use of the trap and its placement. Trapping sets can be described as either lethal or nonlethal. Footholds, cable-restraints, snares, and, in some cases, cage traps can be situated in ways that allow the captured animal to die or live. Footholds, for example, can be attached to slide wires that only allow the trap to move towards deeper water increasing the chances that the animal will drown. This technique is often utilized for species such as beavers, which may offer an appropriate alternative to live capture on land using foothold traps under some conditions. Cable-restraints can be set in areas with physical structures (e.g., shrubs, fences) that cause the animal to entangle and strangle.

Sets are categorized as blind, positive, or baited. Blind sets rely on the tendency of animals to follow trails to enter and trigger the trap (Fig. 9.10a). No attractant is used. Cable-restraints set along trails are classic examples of blind sets. Blind sets can be used with foothold, cage, Conibear-style, snares, cable-restraints, and box traps.

Positive sets place a trap directly over a den hole so that only animals emerging from the den will be caught. Cage traps with double doors can be positioned directly in front of a den hole and barricaded to direct the animal into the trap. Positive sets have an advantage in that you do not need bait. In addition, they substantially reduce the risk of capturing nontargets.

Baited sets are traps accompanied by lures and/or bait to entice the animal to approach or enter them. The principle is well understood by anyone who has placed cheese or peanut butter on the pan of a snap trap to attract a house mouse.

Fig. 9.10 (**a**) Two-door cage trap set on a trail. Photo by Stephen M. Vantassel. (**b**) A positive set to capture the gray squirrel living inside this structure. Photo by Jesse Tenley

Lures and Baits

WCOs have dramatically advanced lures and baits both in development and use. This advancement was partly driven by the need for WCOs to attract wildlife into cage or box traps (Ron Jones, ACP Wildlife Control Services, Monroeville, NJ, personal communication) while avoiding the capture of domestic species such as house cats (*Felis catus*). Today, WCOs have access to a variety of paste baits which can be suspended from the rear of a cage trap. The development of trailing lures allows WCOs to entice an animal to a trap when the trap cannot be set in an ideal location due to safety or theft concerns.

Robert Erickson, On Target ADC (Cortland, IL, USA), was an early innovator of baits and lures specifically for WDM. One particular advance was his creating baits and lures for woodchucks. The invention of the Collarum™ trap required the development of a bait that caused the coyote to bite rather than roll or urinate/defecate. Tim Julien of Proline™ (exclusively distributed by Wildlife Control Supplies, LLC 2014) developed baits meet that need. We expect that WCOs will continue testing new baits and lures to help capture wildlife, particularly as some states expand the species which WCOs may control.

Other Control Methods

Direct Capture Direct capture plays an important role in WDM as it allows a WCO to remove the offending animal immediately. Typically, direct capture occurs when the animal is less mobile due to illness, environmental constraint (i.e., trapped inside a structure), or molt. WCOs have adopted equipment used by animal control officers (i.e., dog catchers) such as catch poles, cat tongs, and throw nets. During the molt, WCOs drive flightless geese like cattle into temporary catch pens made of panels with sides made of nets (Smith et al. 1999; Fig. 9.11). Boats or radio-controlled boats can be used to drive the geese off water bodies to land where a group of people then guide them to the catch pen.

The most interesting advances in direct capture equipment have occurred with nets projected by compressed air. The Super Talon Ultra Net Launcher (Advanced Weapons Technology, Inc. La Quinta, CA, USA) uses a carbon-dioxide cartridge to propel a net about 30 ft from a hand-held device. The device is suitable for capture of birds or mammals up to the size of a Canada goose. The WCS Net Blaster™ (Wildlife Control Supplies, LLC 2014) uses compressed air to launch a 40×60 ft net. It can capture groups of animals up to the size of Canada geese and possibly white-tailed deer. The WCS Net Blaster does not require federal permits and is safe to use around areas with electromagnetic radiation, such as power lines, because it does not use explosives.

One-way Doors One-way doors allow an animal to leave but not reenter a building (Tomahawk Live Trap, LLC 2014). They come in a variety of designs, sizes, and materials. For example, plastic check valves or nets are used for control of bats (BatCone™, LLC, Brewster, NY, USA; Fig. 9.12), and wire frames are specially

Fig. 9.11 Round up of Canada geese. Photo by Stephen M. Vantassel

Fig. 9.12 BatCone™ installed to exclude bats. Photo by Jim Dreisaker

designed to work with certain species and/or fit certain cage traps. Some WCOs build their own one-way doors. The advantage of one-way doors is that they do not require any handling of animals. They are relatively new for wildlife control, but use will likely increase in use as WCOs discover the financial advantages of the technique and as they learn how to use the technique while preventing excluded animals from chewing back into the structure.

Chemical Immobilization Chemical agents can be used to immobilize animals so they are unable to escape. Included are products such as alpha-chlorolose (O'Hare et al. 2007), ketamine (Administration, U. S. D. E. (2014) Drug Scheduling. Title 21 Code of Federal Regulations (C.F.R.) §§ 1308.11 through 1308.15 http://www.deadiversion.usdoj.gov/21cfr/cfr/2108cfrt.htm Accessed 23 October 2015), and others that typically are available only to USDA-APHIS-WS, veterinary, state government, and academic personnel. Unfortunately, due to the difficulty WCOs have in obtaining restricted drugs and chemicals capable of incapacitating wildlife, they are not commonly used in the industry.

Biological Control

Biological control involves the introduction of a disease or predator to manage a target population. Due to regulatory issues, difficulty of targeting nuisance species, lack of experience, and difficulty with including the method in a business model, WCOs have not employed this type of control method. For instance, clients could easily install perches for raptors (Hall et al. 1981) or obtain house cats to manage rodents on their own (Van Sambeek et al. 1995).

Fertility Control

Fertility control, essentially, is birth control for wildlife. The oldest form of birth control is egg addling and oiling eggs, techniques used to manage resident Canada geese in undesired nesting areas. Permits from the USFWS are required (https://epermits.fws.gov/eRCGR/geSI.aspx). The USDA-APHIS-WS (http://www.aphis.usda.gov/wildlife_damage/index.shtml) administers the program.

Research into the development of chemical contraceptives has occurred, but the challenges to a safe and cost-effective method are daunting (Fagerstone et al. 2002). Most contraceptive methods require specialized permits available only to researchers and veterinarians. However, Innolytics, LLC (San Clemente, CA, USA) manufactures OvoControl®-P, an oral bait, with the active ingredient nicarbazin, to control fertility of feral pigeons. The version for Canada geese (OvoControl-G) was taken off the market due to insufficient sales (Vantassel personal communication). The delivery method of some contraceptives is difficult, as with white-tailed deer. The deer must be chemically immobilized, implanted, and released. This technique is expensive, time consuming, and requires special permits to obtain the drugs. In addition, administering contraceptives to only a portion of the target population may not give the desired population reduction effect.

Challenges Facing WCOs

The phrase human–wildlife conflict management is used to describe techniques and strategies that are applied to manage situations that involve any negative interactions between humans and wildlife (Messmer 2000). It was a combination of cultural, environmental, and legal factors that gave rise to the WCO profession. Stricter environmental policies (EPA 1972) coupled with the increased protections on threatened wildlife (Endangered Species Act 1973) encouraged the restoration of depleted species. Many wildlife populations have increased with these protective measures as well as limitations placed on management approaches (Messmer et al. 1997; Fall and Jackson 1998). Regulations have become more restrictive through federal, state, local, and international laws that govern techniques, materials, and timing of management actions for problem wildlife (Fall and Jackson 1998). There has been growing public opposition to use and management of wildlife, resulting in legislation that regulates and restricts control methods (Fall and Jackson 2002).

Despite the growth in the WCO industry, it continues to face threats and challenges from numerous fronts. Various animal rights and protection groups have condemned some tools preferred for the efficient management of species that cause damage (Vantassel 2009), thereby making WDM even more difficult. For example, in the mid-1990s animal protectionist groups successfully used ballot initiatives in Massachusetts and Washington to ban body-gripping traps, including those used to trap moles (subfamily Scalopinae) (Minnis 1998; Massachusetts Fish and Wildlife Trapping Regulations (2014) http://www.eregulations.com/massachusetts/14maab/trapping/ and Washington Department of Fish & Wildlife (2014) Living with Wildlife Moles http://wdfw.wa.gov/living/moles.html Accessed 29 Dec 2014). Interestingly, toxicants for moles were not banned. Furthermore, poorly written legislation can result in prohibiting methods that are more humane than animal rights and animal protection groups claim (Vantassel 2008). We contend that efficiency, effectiveness, and environmental sensitivity should be valued just as much as humaneness.

Conversely, WCOs confront a lack of regulation that permits them to access equipment that would help them perform wildlife control while ensuring humanness. For example, skunks pose special concerns due to their ability to spray. While trapping is often used to remove skunks, how should WCOs remove skunks that enter structures where clients need them removed quickly and without odor, such as a clothing store? Acetone and chloroform are highly effective ways to dispatch skunks in an odor-free manner (Erickson 2005; Noonan 2013), but the American Veterinary Medical Association (AVMA, Euthanasia Panel 2013) considers both methods to be unacceptable. The authors appreciate the AVMA's desire to list techniques that are safe, effective, and humane. But we wonder why the AVMA rejected acetone without any experimental review (Vantassel personal communication). In addition, the authors are unaware of any attempts by the AVMA to use its lobby power to help WCOs gain access to acceptable methods.

Likewise, state wildlife control agencies are woefully slow in updating regulations to account for developments in technology. For example, poorly worded regulation or legislation in regards to snares prevents several states from permitting Collarum™ traps from being used by WCOs (Vantassel et al. 2008). In addition, many states have not updated their trap check rules to allow for the development of remote electronic trap check devices to tell WCOs of the trap's status (National Pest Management Association 2013).

Conclusion

WDM in urban environments has evolved dramatically over the past 40 years. Not only has the field become a stand-alone specialty, but tools and methods have improved in their humanness and target specificity (Wildlife Control Technology 1994–2015). We fully expect the industry to continue to innovate and improve the efficiency and humanness of its methods. In particular, WCOs have developed techniques that greatly reduce the capture of non-offending animals while improving the likelihood of capturing the offending ones. Challenges remain, however, regarding the regulatory oversight of the wildlife control industry. Too many states fail to provide any meaningful standards or training requirements to license WCOs. Some states treat WCOs as year-round fur trappers and do not fully appreciate the challenges facing urban wildlife control. The question is whether the public and government officials will allow the industry the legal space to continue to improve or will it adopt an animal rights perspective and regulate it to ineffectiveness, with the resultant damage to wildlife (Cf. DeStefano and Deblinger 2005).

References

Administration, US Drug Enforcement Administration. (2015) Title 21 Code of Federal Regulations Part 1308--Schedules of Controlled Substances Section 1308.11-1308.15. http://deadivers.usdoj.gov/21cfr/cfr/210cfrt.htm Accessed 23 October 2015.

Agency, Environmental Protection (1972) The challenge of the environment: a primer on EPA's statutory authority. U.S. Environmental Protection Agency, Washington D.C.

Andelt WF, Hopper SN (2000) Livestock guard dogs reduce predation on domestic sheep in Colorado. J Range Manage 53:259–267

Andelt WF, Woolley TP, Hopper SN (1997) Effectiveness of barriers, pyrotechnics, flashing lights, and scary man for deterring heron predation on fish. Wildl Soc Bull 25:686–694

Association of Fish and Wildlife Agencies Furbearer Conservation Technical Work Group (2009) Modern snares for capturing mammals: definitions, mechanical attributes and use considerations, 10th edn. Association of Fish and Wildlife Agencies Furbearer Conservation Technical Work Group, Washington, DC

Austin J, Chamberlain MJ, Leopold BD et al (2004) An evaluation of EGG and wire cage traps for capturing raccoons. Wildl Soc Bull 32:351–356

Bacon RM, Kugeler KJ, Mead PS (2008) Surveillance for Lyme disease: United States 1992–2006. Morbidity and Mortality Weekly Report. Centers for Disease Control, Atlanta

Ball SA (2009) Suspending vulture effigies from roosts to reduce bird strikes. Human-Wildl Confl 3:257–259

Barbour AG, Fish D (1993) The biological and social phenomenon of Lyme disease. Science 260:1610–1616

Baxter AT, Allan JR (2006) Use of raptors to reduce scavenging bird numbers at landfill sites. Wildl Soc Bull 34:1162–1168

Bayoumi A, Meguid MA (2011) Wildlife and safety of earthen structures: a review. J Fail Anal Prev 11:295–319

Belant JL, Seamans TW, Tyson LA, Ickes SK (1996) Repellency of methyl anthranilate to pre-exposed and naive Canada geese. J Wildl Manag 60(4):923–928

Belant JL, Seamans TW, Dolbeer RA et al (1997) Evaluation of methyl anthranilate as a woodpecker repellant. Int J Pest Manag 43:59–62

Blackwell BF, Bernhardt GE, Dolbeer RA (2002) Lasers as nonlethal avian repellents. J Wildl Manag 66:250–258

Blanton JD, Palmer D, Rupprecht CE (2010) Rabies surveillance in the United States during 2009. J Am Vet Med Assoc 237(6):646–657

Bluett RD, Hubert GF Jr, Miller CA (2003) Regulatory oversight and activities of wildlife control operators in Illinois. Wildl Soc Bull 31(1):104–116

Bomford M, O'Brien PH (1990) Sonic deterrents in animal damage control: a review of device tests and effectiveness. Wildl Soc Bull 18:411–422

Boulanger JR, Curtis PD, Cooch ED et al (2012) Sterilization as an alternative deer control technique: a review. Human-Wildl Interact 6:273–282

Bradley CA, Altizer S (2006) Urbanization and the ecology of wildlife diseases. Trends Ecol Evol 22(2):95–102. doi:10.1016/j.tree.2006.11.001

Bruggers RL, Owens R, Hoffman T (2002) Wildlife damage management research needs: perceptions of scientists, wildlife managers, and stakeholders of the USDA/wildlife services program. Int Biodeterior Biodegradation 49:213–223

Budenski J (2014) Hands-on exclusion learning session. Talk presented at the Wildlife Expo, New Orleans, 9–10 Jan 2014

Castelli PM, Sleggs SE (2000) Efficacy of border collies to control nuisance Canada geese. Wildl Soc Bull 28:385–392

Cleary EC, Dolbeer RA (2005) Wildlife hazard management at airports: a manual for airport personnel, 2nd edn. USDA-APHIS-Wildlife Services, Washington, DC

Cleary EC, Dolbeer RA, Wright SE (2006) Wildlife strikes to civil aircraft in the United States: 1990–2005. Federal Aviation Administration, Washington, DC

Conover MR (1984) Effectiveness of repellents in reducing deer damage in nurseries. Wildl Soc Bull 12:399–404

Conover MR (2002) Resolving human-wildlife conflicts: the science of wildlife damage management. Lewis Publishers, New York

Cote SD, Rooney TP, Tremblay JP et al (2004) Ecological impacts of deer overabundance. Ann Rev Ecol Evol Syst 35:113–147

Cummings JL, Mason JR, Otis DL, Heisterberg JF (1991) Evaluation of dimethyl and methyl anthranilate as a Canada goose repellent on grass. Wildl Soc Bull 19(2):184–190

Curnow RD (2001) What are the research needs and skills of the future? In: MC Brittingham, J Kays, R McPeake (eds) Proceedings of the 9th Wildlife Damage Control Conference, pp 18–22

Daszak P, Cunningham AA, Hyatt AD (2000) Emerging infectious diseases of wildlife: threats to biodiversity and human health. Science 287:443–449

Decker DJ, Connelly NA (1989) Deer in suburbia-pleasures and pests. Conserv 43:46–49

DeNicola AJ, Williams SC (2008) Sharpshooting suburban white-tailed deer reduces deer-vehicle collisions. Human-Wildl Confl 2:28–33
Destefano S, Deblinger RD (2005) Wildlife as valuable natural resources vs. intolerable pests: a suburban wildlife management model. Urban Ecosyst 8(2):179–190
Diamond J (1992) Must we shoot deer to save nature? Nat Hist 8:2–8
Endangered Species Act of 1973 (16 U.S.C. 1531–1544, 87 Stat. 884), as amended—Public Law 93–205, approved 28 December 1973, repealed the Endangered Species Conservation Act of 5 December 1969 (P.L. 91–135, 83 Stat. 275). The 1969 Act had amended the Endangered Species Preservation Act of 15 October 1966 (P.L. 89–669, 80 Stat. 926). http://www.fws.gov/laws/lawsdigest/esact.html. Accessed 24 Dec 2014
Environmental Protection Agency (1970). http://www2.epa.gov/aboutepa/epa-history. Accessed 24 Dec 2014
Erickson R (2005) Innovative skunk control. RJE Publications, DeKalb
Etter DE, VanDeelen TR, Ludwig DR et al (2000) Management of white-tailed deer in Chicago, IL, forest preserves. Vertebr Pest Conf 19:190–196
AVMA Euthanasia Panel (2013) AVMA guidelines on euthanasia of animals: 2013 Edition
Fagerstone KA, Coffey MA, Curtis PD et al (2002) Wildlife fertility control. Wildl Soc Tech Rev 02–2:29pp
Fall MW, Jackson WB (1998) A new era of vertebrate pest control? An introduction. Int Biodeterior Biodegradation 42:85–91
Fall MW, Jackson WB (2002) The tools and techniques of wildlife damage management-changing needs: an introduction. Int Biodeterior Biodegradation 49:87–91
Fekadu M (1993) Canine rabies. J Vet Res 60:421–427
Ferraro DM, Vantassel SM, Hygnstrom SE (2010) Mole control. University of Nebraska-Lincoln Extension NebGuide, Lincoln, pp 1–4
Fitzwater WD (1990) Mythology of vertebrate pest control. Vertebr Pest Conf 14:12–15
Gilsdorf JM, Hygnstrom SE, VerCauteren KC et al (2004) Evaluation of a deer-activated bio-acoustic frightening device for reducing deer damage in cornfields. Wildl Soc Bull 32:515–523
Gorenzel WP, Salmon TP (1993) Tape-recorded calls disperse American crows from urban roosts. Wildl Soc Bull 21:334–338
Groepper SR, Hygnstrom SE, Houck B et al (2013) Real and perceived damage by wild turkeys: a literature review. J Integr Pest Manage 4. doi:10.1603/IPM12013
Grund MD, McAninch JB, Wiggers EP (2002) Seasonal movements and habitat use of female white-tailed deer associated with an urban park. J Wildl Manag 66:123–130
Guerra MA, Curns AT, Rupprecht CE et al (2003) Skunk and raccoon rabies in the eastern United States: temporal and spatial analysis. Emerg Infect Dis 9:1143–1150
Hadidian J, Childs M, Perry N et al (2000) Resolving conflicts with Canada geese: an animal welfare perspective. In: MC Brittingham, J Kays, R McPeake (eds) The 9th Wildlife Damage Management Conference Proceedings, pp 293–299
Hall TR, Howard WE, Marsh RE (1981) Raptor use of artificial perches. Wildl Soc Bull 9:296–298
Hall LS, Krausman PR, Morrison ML (1997) The habitat concept and a plea for standard terminology. Wildl Soc Bull 25(1):173–182
Hemachudha T, Laothamatas J, Rupprecht CE (2002) Human rabies: a disease of complex neuropathogenetic mechanisms and a diagnostic challenge. Lancet 2:101–109
Hildreth AM, Hygnstrom SE, Blankenship EE et al (2012) Use of partially fenced fields to reduce deer damage to corn. Wildl Soc Bull 36:199–203
Hoffmann CO, Gottschang JL (1977) Numbers, distribution, and movements of a raccoon population in a suburban residential community. J Mammal 58:623–636
Hubert GF Jr, Wollenberg GK, Hungerford LL et al (1999) Evaluation of injuries to Virginia opossums captured in the EGG™ trap. Wildl Soc Bull 27:301–305
Huot AA, Bergman DL (2007) Suitable and effective coyote control tools for the Urban/Suburban Setting. In: R Timm (ed) Proceedings of the 12th Wildlife Damage Management Conference, pp 312–322

Hygnstrom SE, Timm RM, Larson GE (eds) (1994) Prevention and control of wildlife damage. University of Nebraska-Lincoln, Lincoln

Hygnstrom SE, Hildreth AM, VerCauteren KC (2013) Deer-activated bio-acoustic frightening device deters white-tailed deer. In: JB Armstrong, GR Gallagher (eds) 15th Wildlife Damage Management Conference, p 117

Indiana Department of Natural Resources (2014) Avoiding otter while trapping beaver and raccoon. http://www.in.gov/dnr/fishwild/3353.htm. Accessed 29 Dec 2014

Insurance Information Institute (2014) Auto crashes. Insurance Information Institute. http://www.iii.org/issue-update/auto-crashes. Accessed 25 Dec 2014

Levi T, Kilpatrick AM, Mangel M et al (2012) Deer, predators and the emergence of Lyme disease. Proc Natl Acad Sci 109:10942–10947

Linhart SB, Dasch GJ (1992) Improved performance of padded jaw traps for capturing coyotes. Wildl Soc Bull 20:63–66

Maas-Hebner KG, Emmingham WH, Larson DJ et al (2005) Establishment and growth of native hardwood and conifer seedlings underplanted in thinned Douglas-fir stands. For Ecol Manage 208:331–345

Mallis A (2010) Handbook of pest control, 10th edn. Mallis Handbook, Cleveland

Management R (2012) Public attitudes toward and expectations regarding management of wildlife problems in the Northeast United States. Responsive Management, Harrisonburg

Marsh RE (1995) Mole traps: a collector's manual. Rex Marsh, Davis

Mastro L, Conover MR, Frey SN (2008) Deer-vehicle collision prevention techniques. Human-Wildl Confl 2:80–92

McCullough DR, Jennings KW, Gates KB et al (1997) Overabundant deer populations in California. Wildl Soc Bull 25:478–483

McKillop IG, Sibly RM (1988) Animal behaviour at electric fences and the implications for management. Mammal Rev 18(2):91–103

Meadows LE, Knowlton FF (2000) Efficacy of guard llamas to reduce canine predation on domestic sheep. Wildl Soc Bull 28:614–622

Merwin IA, Ray JA, Curtis PD (1999) Orchard groundcover management systems affect meadow vole populations and damage to apple trees horticultural. Science 34:271–274

Messmer TA (2000) The emergence of human-wildlife conflict management: turning challenges into opportunities. Int Biodeterior Biodegradation 45:97–102

Messmer TA (2009) Human-wildlife conflicts: emerging challenges and opportunities. Human-Wildl Interact 3:10–17

Messmer TA, George S, Cornicelli L (1997) Legal considerations regarding lethal and non-lethal approaches to managing urban deer. Wildl Soc Bull 25:424–429

Messmer TA, Brunson MW, Reiter D et al (1999) United States public attitudes regarding predators and their attitudes regarding predators and their management to enhance avian recruitment. Wildl Soc Bull 27:75–85

Miller JE (2007) Evolution of the field of wildlife damage management in the United States and future challenges. Human-Wildl Interact 1:13–20

Miller JE, Tefft BC, Eriksen RE et al (2000) Turkey damage survey: a wildlife success story becoming another wildlife damage problem. Proc Wildl Damage Manag Conf 9:24–32

Minnis DL (1998) Wildlife policy-making by the electorate: an overview of citizen-sponsored ballot measures on hunting and trapping. Wildl Soc Bull 26(1):75–83

National Pest Management Association (2013) NPMA's wildlife committee promoting increased acceptance of remote trap technology. Natl Pest Manag Assoc. http://www.multibriefs.com/briefs/npma/121713.html. Accessed 20 Feb 2015

Noonan B (2013) Odorless skunk removal. Bob Noonan, Canaan

Noonan B (2014) Reducing foothold damage: keeping capture humane and ethical, vol 6. Trapper's Post, Canaan, pp 23–30

Nowak DJ, Walton JT, Dwyer JF et al (2005) The increasing influence of urban environments on US forest management. J For 103(8):377–382

O'Hare JR, Eisemann JD, Fagerstone KA, et al (2007) Use of alpha-chlorolose by USDA Wildlife Services to Immobilize Birds. In: DL Nolte, WM Arjo, DH Stalman (eds) Proceedings of the 12th Wildlife Damage Management Conference, pp 103–113

OSHA (2007) Protect yourself: pest control pyrotechnics. U.S. Department of Labor, Washington, DC

Ostfeld RS, Schauber EM, Canham CD et al (2001) Effects of acorn production and mouse abundance on abundance and *Borrelia burgdorferi* infection and prevalence of nymphal *Ixodes scapularis* ticks. Vector Borne Zoonotic Dis 1:55–63

Pennisi L, Vantassel SM (2012) Selective bird feeding: deterring nuisance wildlife from bird feeders. EC1783. University of Nebraska-Lincoln Extension, Lincoln, pp 1–12

Pest Management Professional (2014) Bird management survey. Pest Manag Prof 82:BG12, BG14, BG16

Peterson MN, Birckhead JL, Leong K et al (2010) Rearticulating the myth of human-wildlife conflict. Conserv Lett 3:74–82

Pimentel D, Zuniga R, Morrison D (2005) Update on the environmental and economic costs associated with alien-invasive species in the United States. Ecol Econ 52(3):273–288

Pool GE, Hacker CS (1982) Geographic and seasonal distribution of rabies in skunks, foxes, and bats in Texas. J Wildl Dis 24:434–449

Proulx G, Barrett MW (1993) Field testing the C120 magnum trap for mink. Wildl Soc Bull 21(4):421–426

Pruss SD, Cool NL, Hudson RJ et al (2002) Evaluation of a modified neck snare to live-capture coyotes. Wildl Soc Bull 30:508–516

Reidinger RF Jr, Miller JE (2013) Wildlife damage management: prevention, problem solving and conflict resolution. The Johns Hopkins University Press, Baltimore

Rupprecht CE, Hanlon CA, Hemachudha T (2002) Rabies re-examined. Lancet 2:327–343

Sanford RK, Armstrong T, Jurczak A et al (2004) Wild turkey management plan. New York State Department of Environmental Conservation, Albany

Schwartz JA, Warren RJ, Henderson DW et al (1997) Captive and field tests of a method of immobilization and euthanasia for urban deer. Wildl Soc Bull 25:532–541

Seuffert M (2007) Meet your new neighbors: NPMA survey results. Wildl Manage Prof (Supplement to Pest Management Professional, October):10–12, 18, 20–21

Smith AE, Craven SR, Curtis PD (1999) Managing Canada geese in urban environments. Jack Berryman Institute Publication 16, Cornell University Cooperative Extension, Ithaca

Tomahawk Live Trap, LLC (2014). Hazelhurst. http://livetrap.com

Van Sambeek F, McMurray BL, Page RK (1995) Incidence of *Pasteurella multocida* in poultry house cats used for rodent control programs. Avian Dis 39:145–146

Vantassel SM (1999) Wildlife removal handbook: a guide to the capture and control of wild urban animals, 2 (rev.)th edn. Vantassel, Springfield

Vantassel S (2008) Ethics of wildlife control in humanized landscapes: a response. Vertebr Pest Conf 23:294–300

Vantassel SM (2009) Dominion over wildlife? An environmental-theology of human-wildlife relations. Wipf and Stock, Eugene

Vantassel SM (2011) A draft proposal for an industry protocol for the cleanup of raccoon latrines. In: JB Armstrong, GR Gallagher (eds) Proceedings of the 14th Wildlife Damage Management Conference, pp 1–11

Vantassel SM (2012) Wildlife management professionals need to redefine the terms: lethal control, nonlethal control, and live trap. Human-Wildl Interact 6:335–338

Vantassel SM, Powell KDJ, Hiller TL (2008) Survey of changes to cable-trap regulations in the United States during 1980–2007. Internet Center for Wildlife Damage Management and the University of Nebraska-Lincoln, Lincoln, p 25

Vantassel SM, Hiller TL, Powell KDJ et al (2010) Using advancements in cable-trapping to overcome barriers to furbearer management in the United States. J Wildl Manag 74(5):934–939

Vantassel SM, Curtis PD, Hygnstrom SE (2012) National wildlife control training program: core principles and information, vol 1. University of Nebraska-Lincoln Cornell University, Ithaca

Vantassel SM, Hygnstrom SE, Hiller TL (2013) Efficacy of two raccoon eviction fluids. In: JB Armstrong, GR Gallagher (eds) Proceedings of the 15th Wildlife Damage Management Conference, pp 108–112

VerCauteren KC, Lavelle MJ, Hygnstrom SE (2006) Fences and deer-damage management: a review of designs and efficacy. Wildl Soc Bull 34(1):191–200

VerCauteren KC, Dolbeer RA, Gese EM (2010) Identification and management of wildlife damage (Chapter 34). In: Silvy NJ (ed) The wildlife techniques manual. Johns Hopkins University Press, Baltimore

Washington Department of Fish & Wildlife (2014) Living with wildlife: moles. http://wdfw.wa.gov/living/moles.html. Accessed 29 Dec 2014

Way JG, Ortega IM, Auger PJ et al (2002) Box-trapping eastern coyotes in southeastern Massachusetts. Wildl Soc Bull 30(3):695–702

Wildlife Control Technology (1994–2015). WCT Group, Inc, Sharon Center. http://www.wctech-magazine.com

Wildlife Control Supplies, LLC (2014). East Granby. http://wildlifecontrolsupplies.com

Massachusetts Fish and Wildlife Trapping Regulations (2014). http://www.eregulations.com/massachusetts/14maab/trapping/. Accessed 29 Dec 2014

Williams SC, DeNicola AJ, Almendinger T et al (2013) Evaluation of organized hunting as a management technique for overabundant white-tailed deer in suburban landscapes. Wildl Soc Bull 37:137–145

Wilson ML, Telford SR, Piesman J et al (1988) Reduced abundance of immature *Ixodes dammini* (Acari: Ixodidae) following elimination of deer. J Med Entomol 25:224–228

Wright RG (1993) Wildlife management in parks and suburbs: alternatives to sport hunting. Renew Res J 11:18–22

Wyatt JD, Barker WH, Bennett NM et al (1999) Human rabies postexposure prophylaxis during a raccoon rabies epizootic in New York, 1993 and 1994. Emerg Infect Dis 5(3):415–423

Chapter 10
Fifteen Years of Rat Eradication on Italian Islands

Dario Capizzi, Nicola Baccetti, and Paolo Sposimo

Background

Island ecosystems are especially prone to the negative consequences of alien species introductions (e.g. Manne et al. 1999; Baillie et al. 2004). Rats are known to be among the worst invaders of island ecosystems, being responsible for a variety of impacts, ranging from predation upon seabirds at all life stages including eggs, nestlings and adults (Jones et al. 2008), and of a large range of other vertebrate (i.e. mammals and reptiles, e.g. see Harris 2009; Whitaker 1978), invertebrate (St. Clair 2011; Towns et al. 2009) and plant taxa (Palmer and Pons 2001), and also affecting also ecosystem functions (Towns et al. 2006). Global evidence, including from Mediterranean islands (Traveset et al. 2009), where the observed population decline of burrowing seabirds (i.e. Cory's shearwater *Calonectris diomedea*, Yelkouan shearwater *Puffinus yelkouan*, Balearic shearwater *Puffinus mauretanicus*, storm petrel *Hydrobates pelagicus*) was mainly attributed to alien predators, especially rats (e.g. Thibault 1995; Penloup et al. 1997; Martin et al. 2000; Igual et al. 2006; Baccetti et al. 2009; Ruffino et al. 2009).

D. Capizzi (✉)
Latium Region—Regional Park Agency, via del Pescaccio 96, Rome 00166, Italy
e-mail: dcapizzi@regione.lazio.it

N. Baccetti
Istituto Superiore per la Protezione e la Ricerca Ambientale (ISPRA),
via Ca' Fornacetta 9, Ozzano Emilia (Bologna) 40064, Italy
e-mail: nicola.baccetti@isprambiente.it

P. Sposimo
NEMO Srl, Piazza D'Azeglio 11, Florence 50121, Italy
e-mail: sposimo@nemoambiente.com

F.M. Angelici (ed.), *Problematic Wildlife*, DOI 10.1007/978-3-319-22246-2_10

Due to their high reproductive potential and opportunism in exploiting a wide range of food sources (Macdonald and Fenn 1994; Varnham 2010), as well as to the lack of predators, competitors, parasites and diseases, rats have spread with great success on island ecosystems all over the world. The black rat (*Rattus rattus*) spread throughout the western Mediterranean basin about 2000 years ago (Ruffino et al. 2009), although recent studies highlighted that current population may have originated from a single recent colonization event (Colangelo et al. 2015). However, the black rat is perhaps the most widespread mammal on Mediterranean islands (Amori 1993; Sarà 1998). In Italy, for example, it is present on about 80 % of the islands, being absent only on small or very isolated ones (Baccetti et al. 2009).

When feasible, eradication is judged as the ideal tool to manage invasive species (e.g. Myers et al. 2000; Veitch et al. 2011). Bomford and O'Brien (1995) defined eradication as "the complete and permanent removal of all wild populations from a defined area by a time-limited campaign" (p. 249). They also listed six criteria to be met for a successful eradication campaign, three of them being essential: (1) rate of removal exceeds rate of increase; (2) immigration prevented and (3) all reproductive animals must be at risk.

With regard to the second criterion, natural isolation makes islands ideal places for implementing eradication programmes, and in last decades considerable efforts have been made in order to eradicate rats (mostly brown, black and pacific rat) from islands all over the world (e.g. Howald et al. 2007; Genovesi and Carnevali 2011; Capizzi et al. 2014). In order to eliminate or mitigate the detrimental impacts of the black rat on native ecosystems in Italy, several restoration projects were implemented over the last 15 years on Italian islands. Although multiple rodent species may be present on Mediterranean islands (e.g. house mouse *Mus musculus* and brown rat *Rattus norvegicus*) and an eradication programme have sometimes been carried out against these species (e.g. see for *R. norvegicus* V. Di Dio, unpubl. data, for *M. musculus* Baccetti and Sposimo, unpubl. data), the focus of this chapter is on the black rat, as it is believed to be the most detrimental to seabirds more widespread (Amori 1993; Sarà 1998). Over the last 15 years, a large amount of data were collected focusing on aspects of rat ecology, impacts of rat predation on native species (i.e. seabirds), the monetary costs of implementing rat eradication or control and the impact of these projects on non-target species. In this chapter, we summarize the most applicable results, highlighting future research and management priorities in an attempt to fill the knowledge gap about several aspects of rat presence, impact and management on Mediterranean islands. Despite the wide distribution and long presence of black rats, detailed information on their ecology and their impacts on island ecosystems have been, until recently, largely unavailable for Mediterranean islands. Furthermore, the outcome of rat eradication projects—both positive and negative—can provide valuable lessons for future projects.

Field Techniques for Rat Eradication

Since 2000, rat eradication has been carried out on 14 islands (Fig. 10.1), using two different techniques, bait stations or broadcast (hand or aerial), either exclusively or in combination with each other (Table 10.1). The first technique consisted in of securing extruded bait blocks inside bait stations (Thomas and Taylor 2002), i.e. plastic boxes fixed to a shrub, tree or other substratum, with the dual purpose of protecting bait from environmental conditions and from consumption by non-target animals (e.g. mouflons, wild goats, rabbits, seagulls, ravens). The second technique was the aerial distribution of pelleted bait by helicopter (Broome et al. 2014). Pelleted bait rapidly degraded after exposure to moisture and rain. Hand broadcast of pellets as secondary method was used only on limited sectors of islands to reinforce application of the other methods.

The first rat eradication projects (Perfetti et al. 2001) were carried out using bait stations, at an approximate density of about 10 bait stations/ha (distance apart of about 30 m), which has limited utility to small islands. Bait formulations included durable extruded bait blocks containing Brodifacoum or Bromadiolone as active ingredients (concentration: 50 ppm; trade names: Solo® Blox, Notrac® Blox, Bell Laboratories).

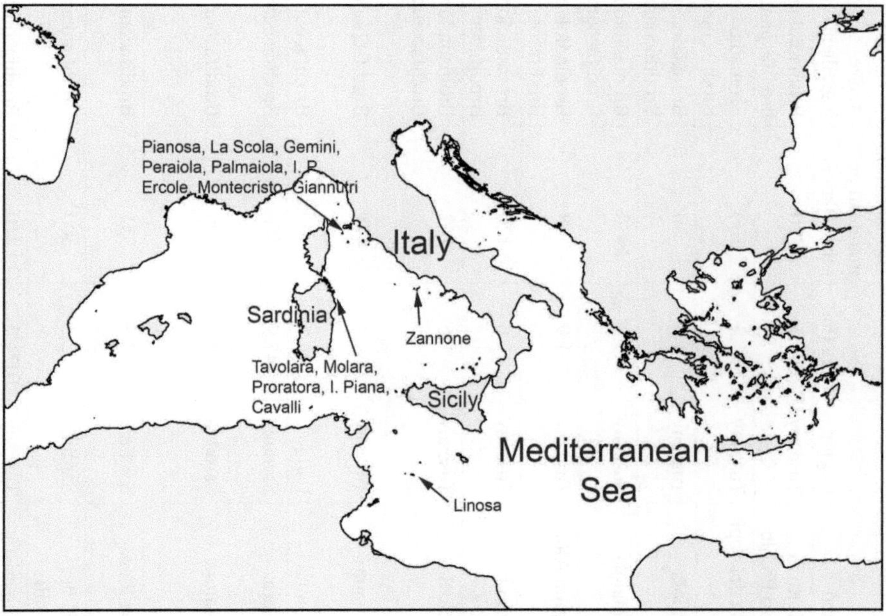

Fig. 10.1 Map showing the geographic position of the islands mentioned in the text

Table 10.1 Italian islands where rat eradication was carried out from 1999 to 2012

Year	Island	Region	Area (ha)	Distance from mainland or other islands	Active ingredient	Bait distribution	Responsible (funding)	Outcome
1999	Isolotto di Porto Ercole	Tuscany	6.5	320	Bromadiolone, brodifacoum	Bait station	National Park of Tuscan Archipelago (LIFE)	Successful, reinvaded
1999	Isola dei Topi	Tuscany	1.3	300	Bromadiolone, Brodifacoum	Bait station	National Park of Tuscan Archipelago (LIFE)	Successful, reinvaded
1999	Peraiola	Tuscany	1	30	Bromadiolone, Brodifacoum	Bait station	National Park of Tuscan Archipelago (LIFE)	Successful
1999	Palmaiola	Tuscany	7.2	2950	Bromadiolone, Brodifacoum	Bait station	National Park of Tuscan Archipelago (LIFE)	Successful
1999	Gemini Alta	Tuscany	1.9	48	Bromadiolone, Brodifacoum	Bait station	National Park of Tuscan Archipelago (LIFE)	Successful, reinvaded
1999	Gemini Bassa	Tuscany	1.6	120	Bromadiolone, Brodifacoum	Bait station	National Park of Tuscan Archipelago (LIFE)	Successful, reinvaded
2001	La Scola	Tuscany	1.6	242	Bromadiolone, Brodifacoum	Bait station	National Park of Tuscan Archipelago (LIFE)	Successful, new incursions (3) promptly eradicated
2006	Giannutri	Tuscany	239.4	11,471	Brodifacoum	Bait station	National Park of Tuscan Archipelago (LIFE)	Successful
2007	Zannone	Latium	104.7	5700	Brodifacoum	Bait station	Circeo National Park	Successful
2008	Molara	Sardinia	347.9	1400	Brodifacoum	Aerial	Marine Protected Area of Tavolara—Punta Coda Cavallo	Successful, reinvaded in 2010
2008	Proratora	Sardinia	4.5	200	Brodifacoum	Bait station	Marine Protected Area of Tavolara—Punta Coda Cavallo	Successful, immediately reinvaded, eradicated 2010, reinvaded in 2010
2010	Isola Piana	Sardinia	13.6	551	Brodifacoum	Bait station	Marine Protected Area of Tavolara—Punta Coda Cavallo	Successful
2010	Isola dei Cavalli	Sardinia	2.2	300	Brodifacoum	Bait station	Marine Protected Area of Tavolara—Punta Coda Cavallo	Successful, new incursions (2) promptly eradicated
2012	Montecristo	Tuscany	1071.7	29,410	Brodifacoum	Aerial	National Park of Tuscan Archipelago (LIFE)	To be confirmed

The active ingredient, the bait administration method and the final outcome are, respectively, indicated in the last three columns

By the mid-2000s (2006–2007) projects aimed at eradicating rats from much larger islands were a focus on conservation efforts (i.e. Giannutri and Zannone, respectively, 239 and 104 ha: see Sposimo et al. 2008; Francescato et al. 2010). In these projects, a more precise calibration of bait station density and bait application rates were facilitated by specific studies on rat relative abundance in different habitat types. Extruded blocks containing Brodifacoum (concentration: 50 ppm; trade name: Solo® Blox) was used exclusively by securing in bait stations to prevent the removal of the bait. Bait station placement was four per hectare (distance apart of 50 m), and each station monitored for bait uptake during each visit. Overall, bait was applied on four different occasions, using a pulsed baiting technique (Dubock 1984; Buckle 1994). On the largest island (Giannutri, size ha 239; about 950 bait stations), we set out bait at rate of ~2.4 kg per ha, followed after 15 days by a second administration of about 2.4 kg per ha, with a third and a fourth administration of 1.2 kg per ha each. The total amount of bait administrated was 7.2 kg. Not unexpectedly, not all bait was exclusively consumed by rodents, but also by invertebrates and snails. Residual bait in the stations was replaced at each pulse event.

On the larger, more rugged islands of Molara (2008: Sposimo et al. 2012) and Montecristo (2012: Sposimo 2014) with foot inaccessible areas, such as steep areas or cliffs, bait was delivered by broadcast buckets suspended under a helicopter, and bait stations were used sporadically.

Two applications by aerial broadcast of pellets containing 50 ppm Brodifacoum were spread from buckets suspended under helicopters guided by a GPS-based control system on two islands, at intervals of 20 days (Molara) or 1 month (Montecristo, where the second distribution was carried out on only 10 % of the island surface). The use of helicopters increased efficiency of the eradication, and reduced the financial costs of the eradications facilitating the support for these projects. For example, the use of bait stations on Montecristo would be impractical, unsafe and improbable because of the steep terrain (up to 650 m a.s.l.) and size (over 1000 ha) of the island.

What to Do When Eradication Is Not Feasible?

Eradication of rats from islands is an ideal option to protect breeding seabird species; however, it is not always feasible (Bomford and O'Brien 1995; Myers et al. 2000). Eradication feasibility is often limited by its high costs (due to the relevant amount of materials and labour), and either the risk of reinvasion or the hazard from the rodenticide to native species. When eradication is not feasible, an alternative strategy of controlling rodents should be put into practice in order to mitigate the impact of rats on breeding seabird species (Corbi et al. 2005; Igual et al. 2006; Pascal et al. 2008; Baccetti et al. 2009).

Local control is initially easier and less expensive to implement on annual basis with respect to eradication, but it has to be implemented every year to give results with increasing costs over time. Therefore, in the long term it is not cost-effective (e.g. Capizzi et al. 2006; Pascal et al. 2008). However, results at several Italian islands

(Zannone, Ponza, Pianosa, Molara, Tavolara, Linosa) showed that the seabirds' reproductive success was actually improved by undertaking rat control in the surroundings of the colonies (Corbi et al. 2005; Igual et al. 2006; Baccetti et al. 2009).

A hypothetical example of an evaluation of the two alternatives is given in Fig. 10.2. Monetary costs (in €) of eradication and control, in relation, respectively, to island size and the number of seabird colonies to be protected, are shown in

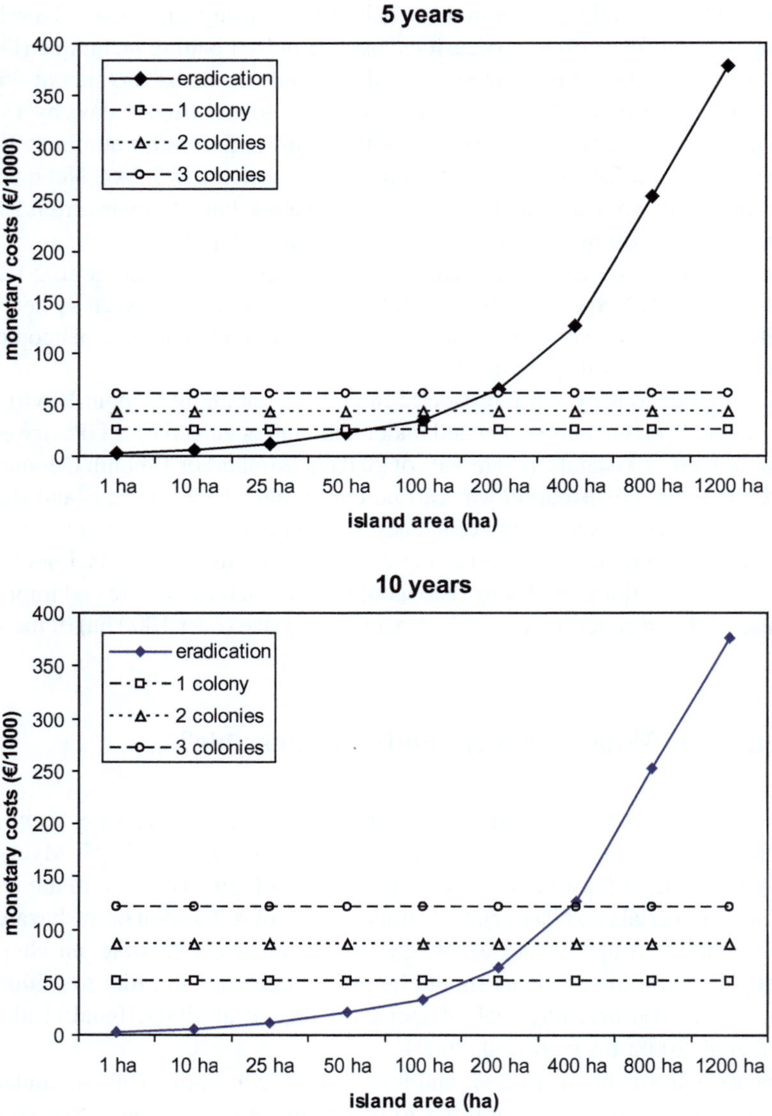

Fig. 10.2 Comparison of monetary costs to be sustained for eradicating rats from islands of different size and those necessary for controlling rats around seabird breeding colonies. We assessed three different scenarios, depending on whether the island is reinvaded by rats after 5, 10 or 20 years

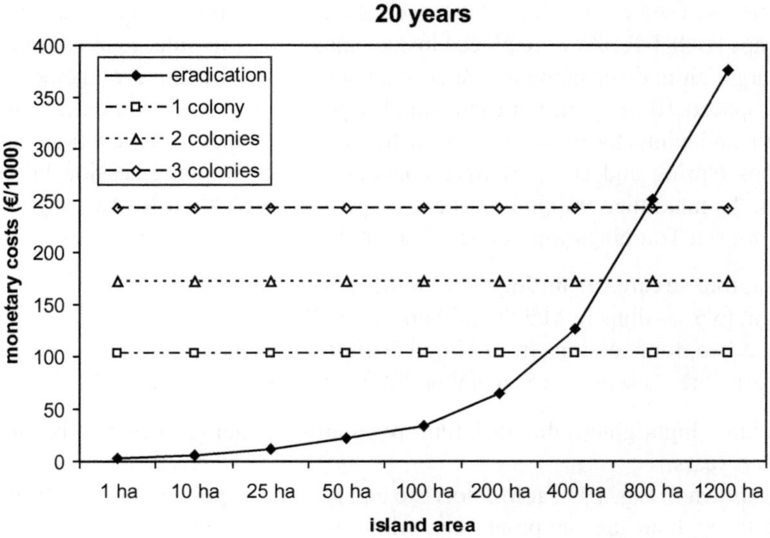

Fig. 10.2 (continued)

Fig. 10.2a, b and c, where the respective costs of a rat eradication are compared with the estimated costs of control for 5, 10 and 20 years, respectively. As shown in the graphics, the equilibrium point is achieved at different areas depending on the number of colonies and from the years before reinvasion. For example, for islands at high risk of reinvasion (i.e. assuming benefits lasting for only 5 years) the equilibrium point between eradication and local control is achieved at 50, 100 and 200 ha in the presence of one, two or three bird colonies, respectively. Conversely, when reinvasion is not likely and/or may be effectively prevented (benefits lasting 20 years or more), eradication is always the most appropriate and cost-effective option for islands from 400 ha (one colony) to 800 ha (three colonies).

However, in this example monetary costs for eradication are estimated assuming the use of bait stations, which is often impractical or even unfeasible on large islands, where aerial baiting is a more cost-effective method. Furthermore, the analysis does not take into account benefits to other components impacted by rats, such as other vertebrates, invertebrates, plants and ecosystem functions, as well as benefits to local human population derived from rat removal (Bell 2011; Oppel et al. 2011).

Studies on Rat Ecology

Seasonal Abundance

Conducting studies on the ecology of the black rat is required to fill the lack of knowledge with regard to this species, as well as informing about more effective control strategies (Ringler et al. 2014).

Patterns of seasonal rat abundance were studied on four islands by snap-trapping. Rat traps (Bell T-Rex®) were placed inside bait stations, in order to avoid injuries to non-target animals or humans. We adopted a transect-like trap arrangement of ten traps, spaced 10 m apart. On each island, four trapping sessions (spring, summer, autumn and winter) were carried out, with the exception of Molara, where only two sessions (spring and autumn) were conducted. Each trapping session lasted five nights. To maximize trapping success, traps were pre-baited for two nights, with traps not set. Trappings were undertaken on the various islands in different years:

Linosa, four sessions from July 2013 to May 2014
Molara, two sessions in March and September 2008
Montecristo, four sessions from March 2010 to February 2011
Tavolara, three sessions in September 2009, April 2010, December 2013

Results highlighted the different population patterns between islands and between years.

As shown in Fig. 10.3, at Montecristo and Linosa the population remained rather stable throughout the sampling year, while at Tavolara and Molara different seasonal patterns were recorded. However, a new trapping session in 2014 at Tavolara showed a very different scenario, with a much lower overall capture rate, stable throughout the year. This suggests the presence of inter-islands and inter-annual differences in the patterns of rat abundance.

Reproductive Period

Rats sampled (above) were evaluated for age and reproductive status and compared with seasons and islands, also recording contrasting patterns (Fig. 10.4). At Tavolara, the higher proportion of juveniles was in early winter, suggesting that reproduction occurred mainly during autumn. At Linosa, the highest proportion of juveniles occurred in autumn, with peak of reproduction in late summer, but juveniles were present throughout the year. In Molara, we recorded juveniles in both spring and autumn, indicating reproduction occurred in both winter and summer. Finally, our data suggest that in Montecristo reproduction occurred mainly in spring and was interrupted in the period from November to March, in Tavolara between late summer and autumn. It is very difficult to explain these differences between islands, and it is possible that they might be better understood by collecting long term both population and climate data. However, these findings indicate that population data are always needed before undertaking an island eradication project.

Rat Density

On Ponza Island (Capizzi et al. in prep.), we estimated the rat density in the surroundings of a shearwater colony (Cory's shearwater *Calonectris diomedea*) in October 2009 and January 2010. The two seasons were chosen according to the

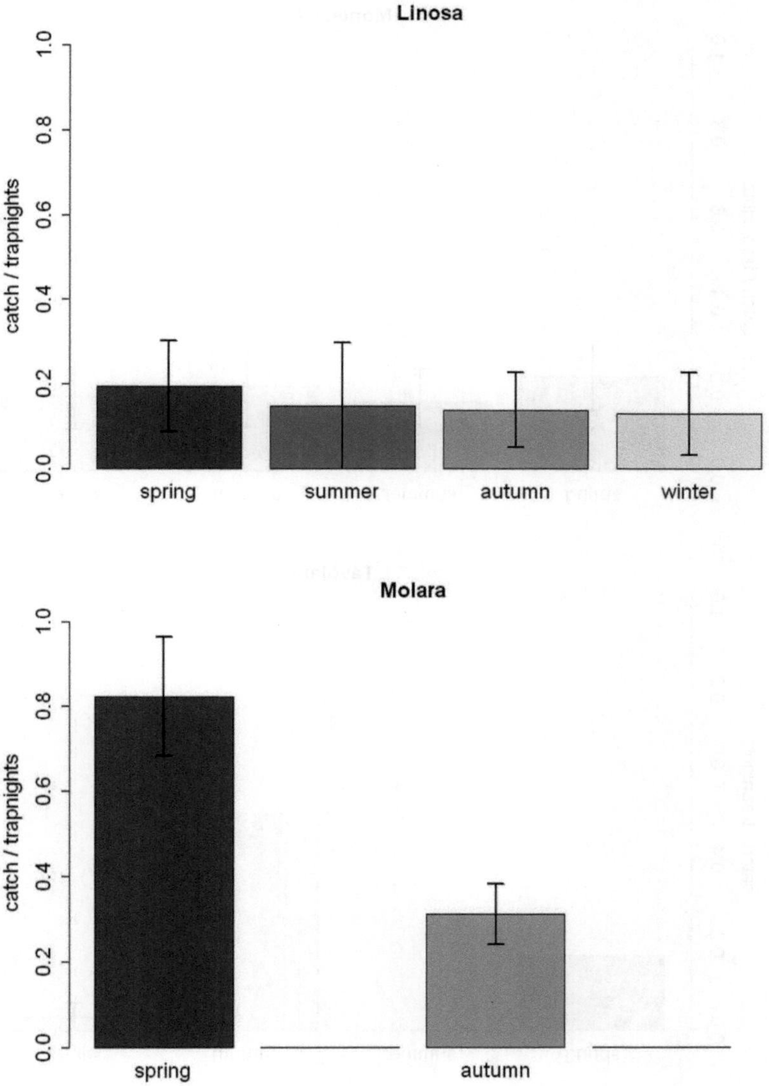

Fig. 10.3 Trapping success recorded during live-trapping sessions carried out on four Italian islands before rat eradication

usual timing of eradication programmes, which are typically performed in autumn or early winter. Overall, 48 Sherman traps arranged to form a 4×12 rectangular grid, with traps spaced 15 m apart. To overcome neophobia, at the beginning of each capture session two nights of pre-baiting (traps baited but with shutter locked) followed by five and four night trapping, respectively. Rats were marked by cutting a few tufts of hair on the right thigh of individuals captured. For the data analysis, Noremark software and the Join hypergeometric Estimator (JHE) for closed population were used, then assuming that the assumptions for a closed population were

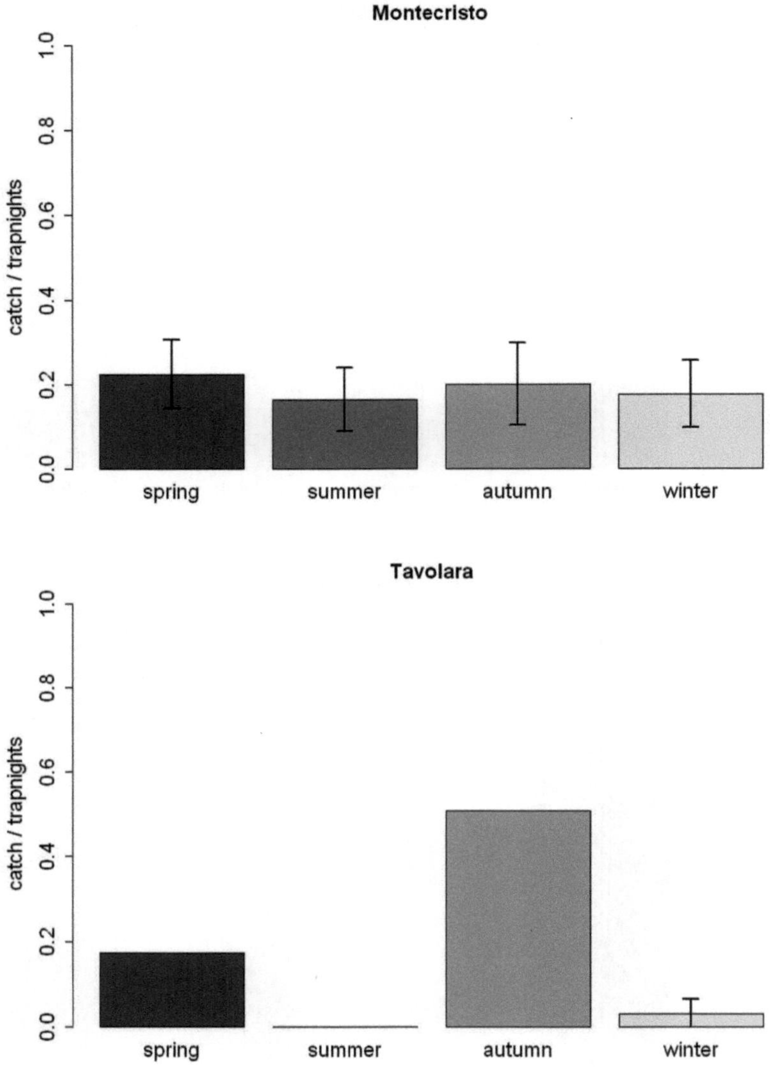

Fig. 10.3 (continued)

met. Density was estimated using an area 48 traps placed by the method of the MCP (Minimum Convex Polygon) with a buffer of around 15 m per side. For statistical tests, SPSS (version 12.0) was used.

In autumn, using the estimator JHE, the estimated population for the study area was 59 individuals with a minimum of 46 and a maximum of 78, for a confidence interval of 95 %. The estimated density for the study area (MCP traps + buffer = 34,465.5 m^2) is 17.1 individuals/ha with a minimum density of 13.3 rats per hectare and a maximum density of 22.6. The corresponding Minimum Number

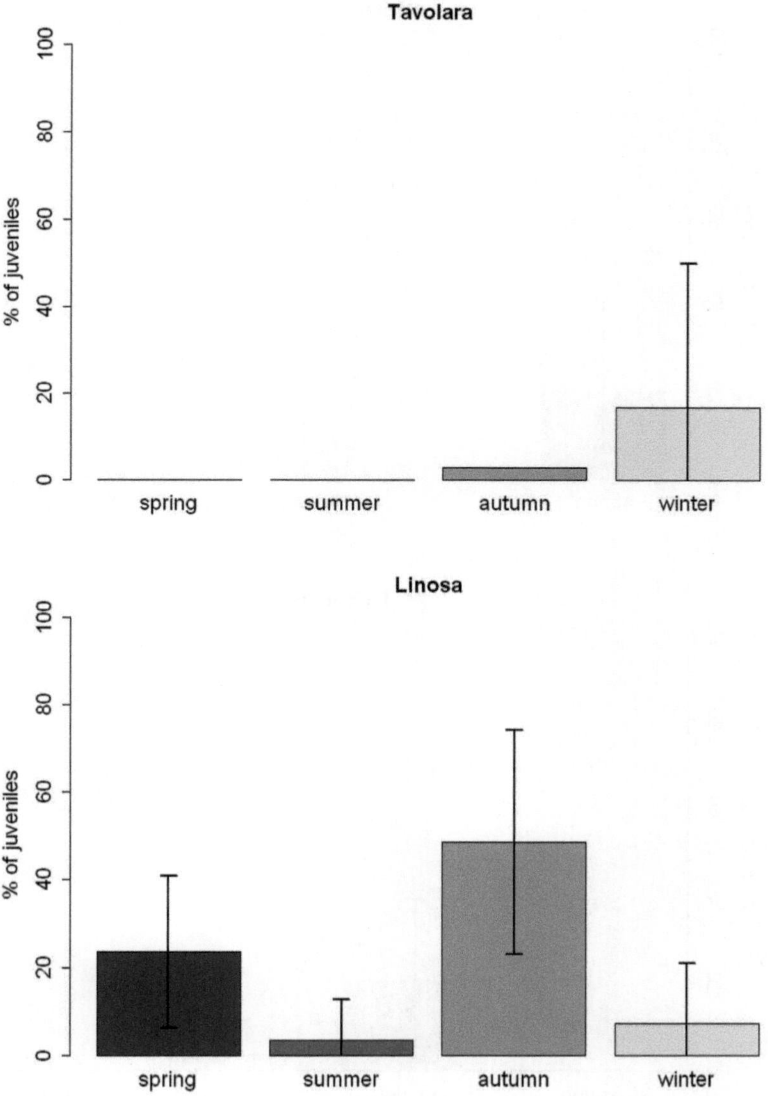

Fig. 10.4 Percentage of juveniles recorded during live-trapping sessions carried out on four Italian islands before rat eradication. For each island, the estimated reproductive period is also indicated

Alive (i.e. the number of rats actually captured) was of 39 rats within the study area, corresponding to a density of 11.3 ind/ha. Sex ratio was of 0.62, i.e. strongly skewed towards females.

In winter, the estimated population by JHE was 34 individuals with a minimum of 15 and a maximum of 180, for a confidence interval of 95 %. The estimated density for the study area (MCP traps + buffer = 34,465.5 m^2) is 9.7 ind/ha with a minimum density of 4.4 ind/ha and a maximum density of 52.2 ind/ha. The corresponding

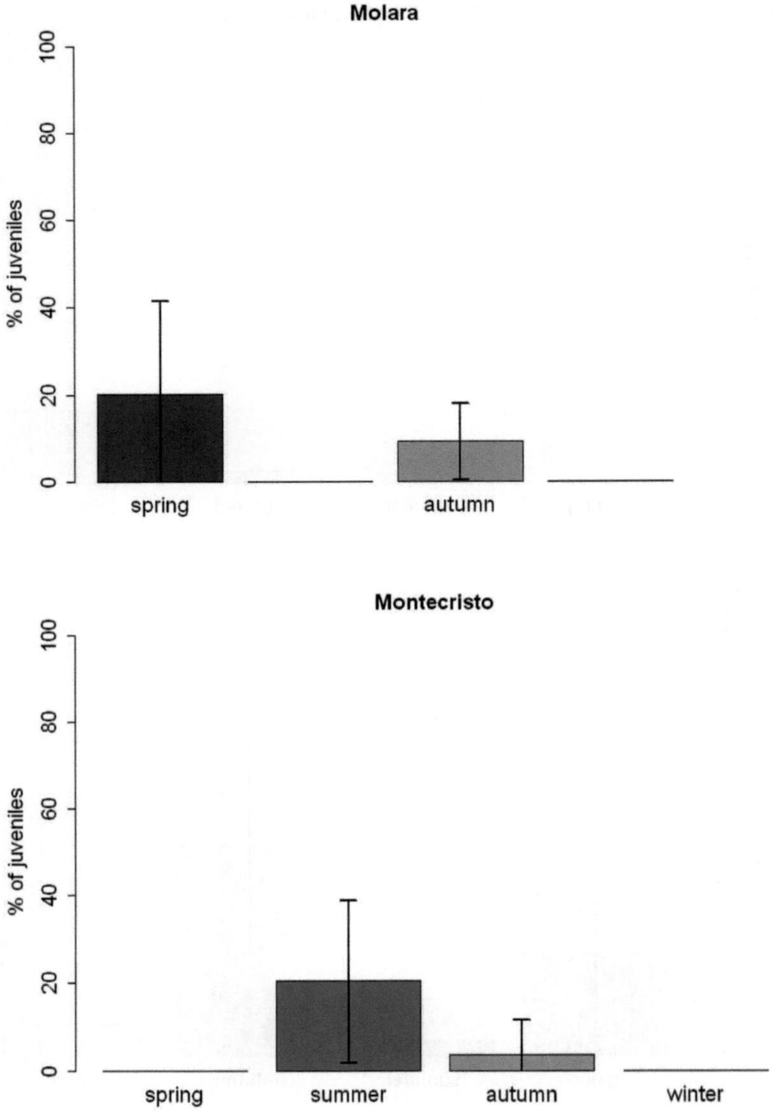

Fig. 10.4 (continued)

Minimum Number Alive (i.e. the number of rats actually captured) was of nine rats within the study area, corresponding to a density of 2.6 ind/ha. Sex ratio of the nine captured individuals was of 0.8.

Home Range and Movements

Rat movements by radio tracking were utilized to evaluate home range (Capizzi et al. in prep.). In autumn 2009, radio collars (manufactured by Sirtrack Ltd.) were fitted to nine adult females and six adult males. In winter, four new individuals were collared, in addition to the eight survivors from the autumn session already fitted with radio collar.

Rats had an average home range of 1685.36 m^2, with a 95 % confidence interval ranging from 1171.14 to 2200.19 m^2. The average home range was also calculated separately for the two study seasons, but the difference between the two values was not statistically significant (Student's t test: $t=0.971$; $p=0.34$). Similarly, no significant difference emerged from the comparison of the average home range of males and females (Kruskal–Wallis test: $\chi^2=1.125$, $p=0.29$), although males home range (1882.0 ± 412.1 st. error) were on average larger than those of females (1447.9 ± 274.1 st. error) (Fig. 10.5). It is likely that the lack of statistical significance was due to the low number of sampled rats, especially males ($n=6$). However, this pattern was partly in agreement with what was found in other studies on black rat movements from New Zealand (Hooker and Innes 1995; Ringler et al. 2014), which have shown that male home range are much larger than those of females.

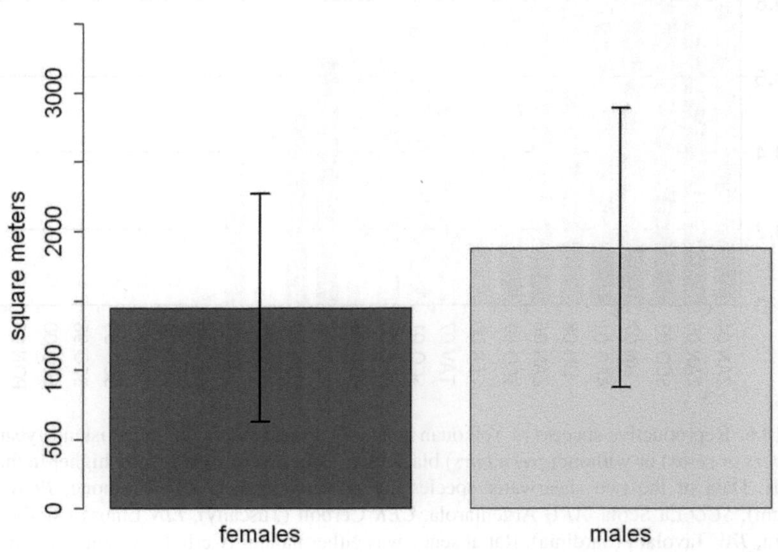

Fig. 10.5 Mean and standard deviation of home-range size of black rats at Ponza (nine females and six males) (method: MCP) (from Capizzi et al. in prep.). Differences between sexes were not statistically significant

Benefits from Rat Eradication and Control to Seabirds

The detrimental impact of invasive rats on nesting success of shearwaters has been highlighted by several studies (see Introduction). Detailed surveys on Italian islands corroborated the evidences, showing a strong difference in reproductive success between islands with or without rats (Fig. 10.6). Pooling together data of both shearwater species, it was determined that pairs breeding on islands without rats attained a much higher reproductive success (0.78 ± 0.17, $n = 15$) than those breeding on islands with rats (0.14 ± 0.25, $n = 11$), and the difference was statistically significant (one-way ANOVA, $F_{1,24} = 60.66$, $P < 0.00001$). The benefits derived from rat removal (either by eradicating or locally controlling them) to seabirds were confirmed by monitoring programmes, showing that controlling or eradicating rats significantly improved shearwater reproductive success, as well as enabling an increase in colony size on islands, where eradication was carried out. At La Scola, following rat eradication, Cory's shearwater reproductive success increased from zero (i.e. total reproduction failure) to about 0.8, and the size of the colony from 70–100 nesting pairs in 1999 to 150–250 in 2010. At Zannone, local control (2004–2006) and eradication (2007) allowed an increase of Cory's shearwater reproductive success from zero (2003) to over 0.8. At Montecristo, the Yelkouan shearwater colony showed maximum productivity values as well as signs of population increase

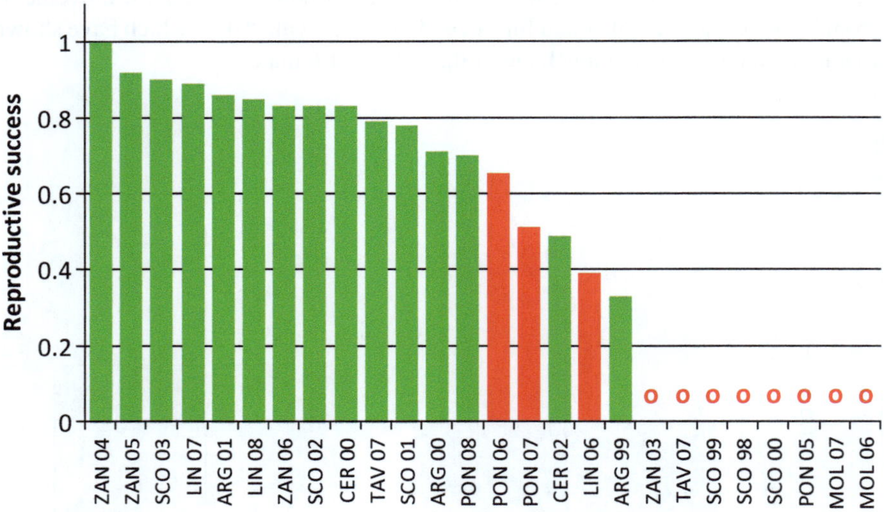

Fig. 10.6 Reproductive success of Yelkouan and Cory's shearwaters on Italian islands/years with (*red bars* or *zeros*) or without (*green bars*) black rats. Values were significantly higher in the latter islands. Data of the two shearwater species are pooled together. *ZAN* Zannone, *PON* Ponza (Latium), *SCO* La Scola, *ARG* Argentarola, *CER* Cerboli (Tuscany), *LIN* Linosa (Sicily), *MOL* Molara, *TAV* Tavolara (Sardinia). Rat absence was either natural (Cerboli, Argentarola) or due to local control or eradication (Zannone after 2003, Linosa after 2006, La Scola after 2000, Ponza after 2007, part of Tavolara in 2007)

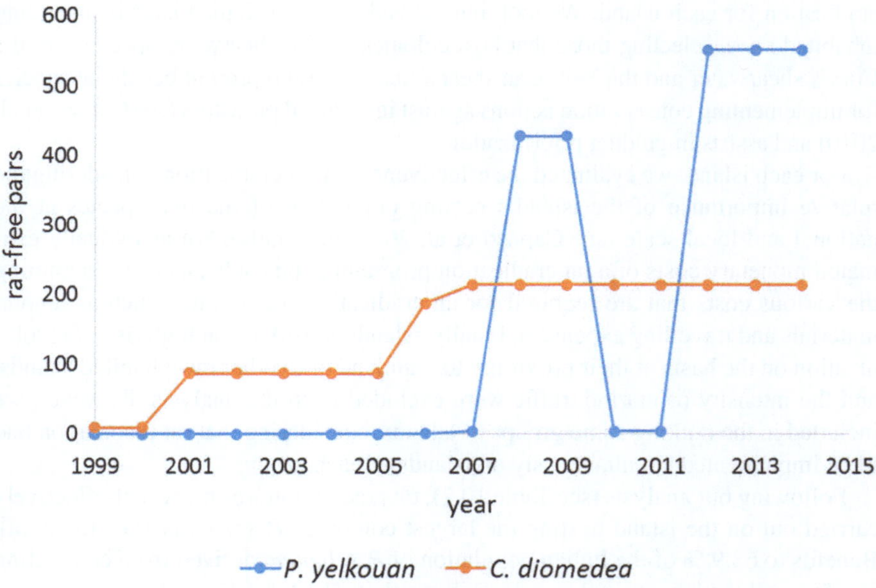

Fig. 10.7 Number of pairs of both Yelkouan and Cory's shearwater released from rat predation during the period 1999–2014 following the various rat eradications. The number of rat-free pairs of the former species increased in 2008 after eradication at Molara, but decreased in 2010, following rat reinvasion

and the occupation of previously unused burrows from the first season after rats were eradicated (Gotti et al. 2014).

The number of shearwater pairs released from rat predation pressure following rat eradication is shown in Fig. 10.7. Early eradication projects have been performed at islands hosting colonies of Cory's shearwater only (La Scola, Giannutri, Zannone), but after 2007 these projects included islands with important colonies of Yelkouan shearwater (i.e. Molara and Montecristo). The number of shearwater pairs on rat-free islands will increase greatly for both species when planned rat eradications are implemented on Tavolara and Linosa (both in 2015, area of 602 and 545 ha, respectively), the most important colonies, respectively, of Cory's and Yelkouan shearwater.

Establishing Priorities

An important issue was the identification of priority islands for rat eradication—i.e. which ones should be implemented. As the implementation of a management strategy can be quite demanding in terms of monetary, safety and image costs, restoration programmes on various islands are inevitably in competition for the same limited monetary budget (Dawson et al. 2015). We addressed this problem by comparing effectiveness (i.e. importance of shearwater populations) and estimated monetary costs for rat eradication for each island. We took into consideration all Italian islands, including inhabited ones, selecting those that host colonies of two shearwater species, i.e. the Cory's shearwater and the Yelkouan shearwater, as these represent beneficial species for implementing conservation actions against introduced predators (see Capizzi et al. 2010) and assists in guiding prioritization.

For each island, we evaluated the effectiveness of rat eradication considering the relative importance of the island's nesting population of the two species at the national and local scale (see Capizzi et al. 2010 for details). We analytically estimated monetary costs of a rat eradication programme on each island (i.e. summing the various costs that are required for an eradication programme, such as labour, materials and travelling expenses). Finally, islands considered at high risk of recolonization on the basis of their proximity to mainland or to other rat-inhabited islands, and the intensity of marine traffic were excluded from the analysis. However, we included in the ranking some groups of islands, considering that rat eradication had to be implemented simultaneously on islands of each group.

Following our analysis (see Table 10.2), rat eradication was most cost-effectively carried out on the island hosting the largest colony of *P. yelkouan* (i.e. Tavolara). Benefits to 63.9 % of the Italian population of *P. yelkouan* derived from eradicating rats from all the islands in the ranking, but only to 7.1 % of *C. diomedea*.

Table 10.2 Islands' ranking according to cost-effectiveness of rat eradication, excluding islands at high risk of recolonization, but recovering four groups of islands, where rat eradication has to be performed simultaneously (from Capizzi et al. 2010)

Islands (or groups)		Area (ha)	Actions implemented or planned
1	Tavolara	602.0	Eradication planned (2015?)
2	Palmarola	125.1	Feasibility study available
3	Barrettini	10.3	
4	Montecristo	1071.7	Eradication (2012)
5	Giannutri	239.5	Eradication (2006)
6	Zannone	104.7	Eradication (2007)
7	Soffi Group	53.4 (four islands)	
8	Santo Stefano Ponziane	31.0	Feasibility study in progress
9	Molara	347.8	Eradication (2008), reinvaded (2009)
10	Mortorio	55.7	
11	La Vacca	9.1	
12	Santa Maria Group	556.1 (14 islands)	
13	Pianosa+La Scola	1028.4 (two islands)	Eradication in La Scola (2001), eradication planned in Pianosa (2016)
14	Rossa di Teulada	10.5	
15	Spargi	421.9	
16	Serpentara	31.3	
17	Cavoli	42.1	
18	Corcelli Group	16.7 (three islands)	

The analysis has not been just a theoretical exercise. Comparing the ranking with Table 10.1, it is noted that rat eradication had already been carried out on many of the islands (Montecristo, Giannutri, Zannone, La Scola, Molara). Furthermore, rat eradication on Tavolara is planned for autumn 2015, and planning for a project aimed at removing rats from Palmarola and S. Stefano is currently under way. Finally, knowledge gained about the effectiveness of quarantine measures (e.g. Dilks and Towns 2002; Russell et al. 2008) has led to programming rat eradication on those islands that host important Mediterranean colonies (e.g. Linosa, the main Italian colony of Cory's shearwater: Baccetti et al. 2009) that are also subject to human pressure.

Rat Reinvasion

Rat reinvasion following an eradication programme is a distinct threat (Russell and Clout 2005), risking the great financial, time and field effort investments clearing the island of rats in the first place. Rat reinvasion has occurred on six islands, five of them being very small and closer than 500 m to mainland or other rat-infested islands, and in two cases it occurred more than once: La Scola (three times in a 15-year period since 2001, i.e. in 2005, 2009 and 2011) and Cavalli (at least twice in the period 2010–2014). The only reinvaded island more than 500 m distant from mainland was Molara. Genetic analyses highlighted the difference between eradicated population and the new invaders, thus supporting the evidence of a successful eradication, and indicated that reinvasion may be caused by a rat exchange between the Molara Island and Sardinia mainland populations (Ragionieri et al. 2013). The considerable distance from the mainland (more than 1000 m) led us to exclude the possibility that rats may have reached the island by swimming (Russell et al. 2008), thus hypothesizing that reinvasion was probably driven by humans (Ragionieri et al. 2013; Sposimo et al. 2012). At Barrettini (Sardinia, distant 700 m from the nearest island), we recorded a black rat reinvasion at least 12 years after from its natural extinction (Baccetti, pers. obs.). It is likely that reinvasion occurred by swimming, since the islet benefits from a high level of protection under the national park zonation and landing of boats is not allowed (Cecere and Nissardi, pers. obs.). This also suggests that islands within 700 m of a rat population are at risk of reinvasion. These reinvasions highlight the importance of strengthening biosecurity measures to both protect the investment in conducting eradications and secure the conservation benefits accruing to seabird species.

Impact on Non-target Species

An important concern in all rat eradication projects is the impact of rodenticides on non-target species (Fisher et al. 2011; Masuda et al. 2015). Here, we present evidences (before and after eradications, see Table 10.3) outlining impacts on several

Table 10.3 Summary table showing the observed impacts on non-target species following rat eradication on the various islands

Species	Island(s)	Observed impact
Mouflon (*Ovis aries*)	Zannone	Population stable (about 45 individuals before and after)
Goats (*Capra hircus*)	Montecristo	Population decrease after eradication (2012), then fully recovered after 2 years (Gotti et al. 2014)
Asp viper (*Vipera aspis*)	Montecristo	No direct impact observed
Western whip snake (*Hierophis viridiflavus*)	Giannutri, Molara, Proratora	No impact observed
Italian wall/Wall lizards (*Podarcis* spp.)	All islands	No impact observed, increase in La Scola and Zannone
Turkish gecko (*Hemidactylus turcicus*)	Palmaiola, Giannutri, Montecristo	No impact observed
Common wall gecko (*Tarentola mauritanica*)	Giannutri, Montecristo	No impact observed
European leaf-toed gecko (*Euleptes europaea*)	Isola dei Topi, Palmaiola, Gemini Alta, Gemini Bassa, Giannutri, Molara, Proratora, Isola Piana, Montecristo	No impact observed
Ocellated bronze skink (*Chalcides ocellatus*)	Molara, Proratora, Isola Piana	No impact observed
Fitzinger's algyroides (*Algyroides fitzingeri*)	Molara	No impact observed
Tyrrhenian painted frog (*Discoglossus sardus*)	Montecristo	No impact observed
Yellow-legged gull (*Larus michahellis*)	All islands	No impact observed in all island but Montecristo, where a decrease in population has been recorded
Common raven (*Corvus corax*)	La Scola, Molara, Montecristo	1–2 pairs possibly extinct at Montecristo, no impact observed elsewhere
Peregrine falcon (*Falco peregrinus*)	La Scola, Isola dei Topi, Palmaiola, Giannutri, Zannone, Molara, Montecristo	No impact observed
Barn owl (*Tyto alba*)	Giannutri, Molara	Extinction of 1–3 pairs in Molara, maybe one pair in Giannutri (uncertain presence before eradication)

Only vertebrate species were considered (mammals, reptiles, amphibians and birds)

species of mammals, reptiles and birds from 12 islands, showing no significant impact at the population level, with the exception of *Larus michahellis* and wild goats *Capra hircus* at Montecristo (population decrease after the eradication, now recovered to the levels present prior to the intervention). Furthermore, we guess that Barn owl *Tyto alba* became possibly extinct at both Giannutri (one pair, but the presence before eradication was uncertain) and Molara (1–3 pairs). However, it has to be noted that Barn owl may not be able to survive on islands without rodent prey; therefore, their extinction may be due either to effects of anticoagulants or to island abandonment due to lack of suitable prey. This also raises the question of whether barn owls existed on the islands prior to their colonization by rats, but this goes far beyond the aims of this chapter. No impact has been recorded on reptiles (geckos, lizards and western whip snakes) or amphibians, as well as on ravens (with the exception of one or two pairs which were possibly impacted by primary or secondary poisoning at Montecristo) and diurnal raptors (*Falco tinnunculus* and *F. peregrinus*). A detailed report of the fauna present on the 12 islands and the evidences regarding the impact on their populations are shown in Table 10.3.

Conclusions and Management Perspectives for Italian Islands

Rat eradication proved to be a valuable tool for restoring island ecosystems and provided actual benefits to rat-impacted native species. Although the cost of eradication is often perceived to be high because it incurs a one-off cost, by comparing available management options (i.e. control or eradication) our analysis suggested that it may be cheaper than the cost of long-term control or the cost of doing nothing. Projects carried out on Italian islands are an example of an effective and lasting solution to an environmental problem and demonstrate a cost-effective conservation management action. The conservation status of species such as Cory's shearwater, Yelkouan shearwater and Storm petrel is closely linked to the outcome of these restoration projects. As it hosts important nesting seabird colonies, Italy has a major responsibility with regard to their conservation. However, problems may come from a new Italian regulation on the use of rodenticides, allowing aerial distribution, but imposing the use of low persistence active ingredients. According to this regulation, the only allowed active ingredients are first-generation anticoagulants (e.g. chlorophacinone and warfarin), ineffective for eradicating rats from islands by aerial baiting (see Parkes et al. 2011) and hardly available on the market. This restriction would significantly reduce our ability to achieve meaningful conservation outcomes on islands with invasive rodent populations.

The opposition of animal rights movements may also hamper the implementation of such projects. However, although present, such opposition is not as strong as in the case of Italian projects aimed at managing other invasive species (e.g. Grey squirrel *Sciurus carolinensis*, see Bertolino and Genovesi 2003).

The challenge for the future is twofold. First, there is the need to improve the effectiveness of biosecurity (quarantine) measures, thus achieving protection from the risk of reinvasion (e.g. Dilks and Towns 2002; Russell and Clout 2007). This will allow the potential to eradicate rats even in islands connected to mainland or other islands by regular boat service, thus extending the benefits to other important colonies (e.g. Linosa, currently the most important European colony of Cory's Shearwater). Secondly, implementing the appropriate biosecurity measures may also allow eradication of rats from islands with small human settlements, considering the benefits in terms of the welfare of the resident population (Oppel et al. 2011; Hilton and Cuthbert 2010; Bell 2011). This may be a strategic advantage, as it may strengthen public support for these strategic projects. Furthermore, if residents appreciate the increased natural biodiversity after removal of rats, they may encourage other island communities to support the same measures—this seems to be happening on the Scilly Isles (UK) at the moment (Bell 2011).

Acknowledgements Thanks are due to all friends and colleagues who have worked during these 15 years in these restoration projects. In particular to Ferdinando Corbi, Samantha Francescato, Silvio D'Alessio, Michele Giunti, Bruno Massa, Sergio Nissardi, Danilo Pisu, Max Putzu, Giovanna Spano, Viviana Stanzione, Toni Torre, Marco Zenatello and Carla Zucca. The directors and staff members of many protected areas facilitated and stimulated our work on many levels, and we are indebted especially to Francesca Giannini, Augusto Navone, Franca Zanichelli and Sergio Zerunian. Finally, we wish to thank two anonymous referees for their helpful comments on an earlier draft of the manuscript.

References

Amori G (1993) Italian insectivores and rodents: extinction and current status. Suppl Ric Biol Selv 21:101–121

Baccetti N, Capizzi D, Corbi F, Massa B, Nissardi S, Spano G, Sposimo P (2009) Breeding shearwater on Italian islands: population size, island selection and co-existence with their main alien predator. Riv Ital Orn 78:83–99

Baillie JEM, Hilton-Taylor C, Stuart SN (2004) 2004 IUCN red list of threatened species: a global species assessment. IUCN, Gland

Bell E (2011) Isles of Scilly Seabird Recovery Project: improving rodent control on uninhabited islands, assessment of the feasibility of rat removal across the Isles of Scilly archipelago and feasibility of rat removal from St Agnes and Gough. http://www.ios-aonb.info/wp-content/uploads/2012/08/IOS-Report-Summary-Report-21-2-11.pdf

Bertolino S, Genovesi P (2003) Spread and attempted eradication of the grey squirrel (*Sciurus carolinensis*) in Italy, and consequences for the red squirrel (*Sciurus vulgaris*) in Eurasia. Biol Conserv 109:351–358

Bomford M, O'Brien P (1995) Eradication or control for vertebrate pests? Wildl Soc Bull 23: 249–255

Broome KG, Cox A, Golding C, Cromarty P, Bell P, McClelland P (2014) Rat eradication using aerial baiting: current agreed best practice used in New Zealand (version 3.0). New Zealand Department of Conservation internal document, Wellington, New Zealand, 24pp

Buckle AP (1994) Rodent control methods: chemical. In: Buckle AP, Smith RH (eds) Rodent pests and their control. Cab International, University Press, Cambridge, pp 117–160

Capizzi D, Baccetti N, Corbi F, Giannini F, Giunti M, Perfetti A, Sposimo P, Zerunian S (2006) Eradication versus local control of *Rattus rattus* on Tyrrhenian islands: ecological aspects, field techniques and economics. In: Prigioni C, Nieder L, Colli L (eds) Proceedings of 10th rodents & spatium. Hystrix Ital J Mammal (n.s.), Suppl, p 33

Capizzi D, Baccetti N, Sposimo P (2010) Prioritizing rat eradication on islands by cost and effectiveness to protect nesting Seabirds. Biol Conserv 14:1716–1727

Capizzi D, Bertolino S, Mortelliti A (2014) Rating the rat: global patterns and research priorities in impacts and management of rodent pests. Mammal Rev 44:148–162

Clair JJH (2011) The impacts of invasive rodents on island invertebrates. Biol Conserv 144:68–81

Colangelo P, Abiadh A, Aloise G, Amori G, Capizzi D, Vasa E, Annesi F, Castiglia R (2015) Mitochondrial phylogeography of the black rat supports a single invasion of the western Mediterranean basin. Biol Invasions 17(6):1859–1868. doi:10.1007/s10530-015-0842-2

Corbi F, Francescato S, Pinos F, Baccetti N, Capizzi D, Forcina G, Sposimo P, Zerunian S (2005) Intervento di controllo dei ratti nell'isola di Zannone (Parco Nazionale del Circeo) a tutela di una colonia di Berta maggiore *Calonectris diomedea*. In: Zerunian S (ed) Habitat, flora e fauna del Parco Nazionale del Circeo. Uff. gestione Beni ex ASFD di Sabaudia. Parco Nazionale del Circeo, pp 245–252

Dawson J, Oppel S, Cuthbert RJ, Holmes N, Bird JP, Butchart SHM, Spatz DR, Tershy B (2015) Prioritizing islands for the eradication of invasive vertebrates in the UK overseas territories. Conserv Biol 29:143–153

Dilks P, Towns D (2002) Developing tools to detect and respond to rodent invasions of islands: workshop report and recommendations. DOC Science Internal Series 59, Wellington, New Zealand, 19pp

Dubock AC (1984) Pulsed baiting. A new technique for high potency slow acting rodenticides. In: Dubock AC (ed) Proceedings of conference on organisation and practice of vert. Pest Control., Elvetham Hall, Hampshire, pp 105–142

Fisher P, Griffiths R, Speedy C, Broome K (2011) Environmental monitoring for brodifacoum residues after aerial application of baits for rodent eradication. In: Veitch CR, Clout MN, Towns DR (eds) Island invasives: eradication and management. IUCN, Gland, pp 300–304

Francescato S, Capizzi D, Cecchetti M, Forcina G, Mastrobuoni G, Noal A, Sposimo P, Zerunian S (2010) L'intervento di eradicazione del Ratto nero dall'Isola di Zannone. In: Zerunian S (ed) L'eradicazione del Ratto nero dall'Isola di Zannone. Ufficio Territoriale per la biodiversità di Fogliano, Parco Nazionale del Circeo, pp 37–110

Genovesi P, Carnevali L (2011) Invasive alien species on European islands: eradications and priorities for future work. In: Veitch CR, Clout MN, Towns DR (eds) Island invasives: eradication and management. IUCN, Gland, pp 56–62

Gotti C, Cozzo M, De Faveri A, Zenatello M, Baccetti N., Lazzaro L, Ferretti G, Foggi B (2014) Il monitoraggio della fauna e della flora a Montecristo. In: Quaderni del Parco, documenti tecnici 2, PROGETTO LIFE+ MONTECRISTO 2010, Eradicazione di componenti florofaunistiche aliene invasive e tutela di specie e habitat nell'Arcipelago Toscano, Parco Nazionale Arcipelago Toscano, Portoferraio, pp 54–69

Harris DB (2009) Review of negative effects of introduced rodents on small mammals on islands. Biol Invasions 11:1611–1630

Hilton GM, Cuthbert RJ (2010) The catastrophic impact of invasive mammalian predators on birds of the UK Overseas Territories: a review and synthesis. Ibis 152:443–458

Hooker S, Innes J (1995) Ranging behaviour of forest-dwelling ship rats, *Rattus rattus*, and effects of poisoning with brodifacoum. New Zeal J Zool 22:291–304

Howald G, Donlan CJ, Galván JP, Russell JC, Parkes J, Samaniego A, Wang Y, Veitch D, Genovesi P, Pascal M, Saunders A, Tershy B (2007) Invasive rodent eradication on islands. Conserv Biol 21:1258–1268

Igual JM, Forero MG, Gomez T, Orueta JF, Oro D (2006) Rat control and breeding performance in Cory's shearwater (*Calonectris diomedea*): effects of poisoning effort and habitat features. Anim Conserv 9:59–65

Jones HP, Tershy BR, Zavaleta ES, Croll DA, Keitt BS, Finkelstein ME, Howald GR (2008) Severity of the effects of invasive rats on seabirds: a global review. Conserv Biol 22:16–26

Macdonald DW, Fenn MGP (1994) The natural history of rodents. In: Buckle AP, Smith RH (eds) Rodent pests and their control. Cab International, University Press, Cambridge, pp 1–22

Manne LL, Brooks TM, Pimm SL (1999) Relative risk of extinction of passerine birds on continents and islands. Nature 399:258–261

Martin JL, Thibault JC, Bretagnolle V (2000) Black rats, island characteristics, and colonial nesting birds in the Mediterranean: consequences of an ancient introduction. Conserv Biol 14:1452–1466

Masuda BM, Fisher P, Beaven B (2015) Residue profiles of brodifacoum in coastal marine species following an island rodent eradication. Ecotoxicol Environ Saf 113:1–8

Myers JH, Simberloff D, Kuris AM, Carey JR (2000) Eradication revisited: dealing with exotic species. Trends Ecol Evol 15:316–320

Oppel S, Beaven BM, Bolton M, Vickery J, Bodey TW (2011) Eradication of invasive mammals on islands inhabited by humans and domestic animals. Conserv Biol 25:232–240

Palmer M, Pons GX (2001) Predicting rat presence on small islands. Ecography 24:121–126

Parkes J, Fisher P, Forrester G (2011) Diagnosing the cause of failure to eradicate introduced rodents on islands: brodifacoum versus diphacinone and method of bait delivery. Conserv Evid 8:100–106

Pascal M, Lorvelec O, Bretagnolle V, Culioli JM (2008) Improving the breeding success of a colonial seabird: a cost-benefit comparison of the eradication and control of its rat predator. Endang Species Res 4:267–227

Penloup A, Martin JL, Gory G, Brunstein D, Bretagnolle V (1997) Distribution and breeding success of pallid swift, *Apus pallidus*, on Mediterranean islands: nest predation by the roof rat, Rattus rattus, and nest quality. Oikos 80:78–88

Perfetti A, Sposimo P, Baccetti N (2001) Il controllo dei ratti per la conservazione degli uccelli marini nidificanti nelle isole italiane e mediterranee. Avocetta 25:126

Ragionieri L, Cutuli G, Sposimo P, Spano G, Navone A, Capizzi D, Baccetti N, Fratini S (2013) Establishing the eradication unit of Molara Island: a case of study from Sardinia, Italy. Biol Invasions 15:2731–2742

Ringler D, Russell J, Jaeger A, Pinet P, Bastien M, Le Corre M (2014) Invasive rat space use on tropical islands: implications for bait broadcast. Basic Appl Ecol 15:179–186

Ruffino L, Bourgeois K, Vidal E, Duhem C, Paracuellos M, Escribano F, Sposimo P, Baccetti N, Pascal M, Oro D (2009) Invasive rats and seabirds after 2,000 years of an unwanted coexistence on Mediterranean islands. Biol Invasions 11:1631–1651

Russell JC, Clout MN (2005) Rodent incursions on New Zealand islands. In: Parkes J, Statham M, Edwards G (eds) Proceedings of the 13th Australasian vertebrate pest conference. Landcare Research, Lincoln, pp 324–330

Russell JC, Clout MN (2007) Early detection and eradication of invading rats. In: Witmer et al. (eds) Managing vertebrate invasive species. USDA/APHIS/WS, National Wildlife Research Center, Fort Collins, pp 268–272

Russell JC, Beaven BM, MacKay JWB, Towns DR, Clout MN (2008) Testing island biosecurity systems for invasive rats. Wildlife Res 35:215–221

Sarà M (1998) I mammiferi delle isole del Mediterraneo. L'Epos, Palermo, 168pp

Sposimo P (2014) L'eradicazione del Ratto nero a Montecristo. In: Zanichelli et al. (eds) I Quaderni del Parco, documenti tecnici volume 2 PROGETTO LIFE+ MONTECRISTO 2010, Eradicazione di componenti florofaunistiche aliene invasive e tutela di specie e habitat nell'Arcipelago Toscano. Parco Nazionale Arcipelago Toscano, Portoferraio, pp 20–25

Sposimo P, Capizzi D, Giannini F, Giunti M, Baccetti N (2008) L'eradicazione del ratto nero (*Rattus rattus*) nell'isola di Giannutri. I Quaderni del Parco, Docum. Tecn., 1 Progetto LIFE Natura, Isole di Toscana: nuove azioni per uccelli marini e habitat. Parco Naz. Arcipelago Toscano, pp 33–38

Sposimo P, Spano G, Navone A, Fratini S, Ragionieri L, Putzu M, Capizzi D, Baccetti N, Lastrucci B (2012) Rat eradication at Yelkouan Shearwater Puffinus yelkouan colonies on NE Sardinian islets: success followed by unexplained re-appearance. In: Yésou P, Baccetti N, Sultana J (eds) Ecology and Conservation of Mediterranean Seabirds and other bird species under the Barcelona Convention. Proceedings of the 13th Medmaravis Pan-Mediterranean symposium. Medmaravis, Alghero, pp 58–64

Thibault JC (1995) Effect of predation by the Black rat *Rattus rattus* on the breeding success of Cory's Shearwater *Calonectris diomedea* in Corsica. Mar Ornithol 23:1–10

Thomas BW, Taylor RH (2002) A history of ground-based rodent eradication techniques developed in New Zealand, 1959–1993. In: Veitch CR, Clout MN (eds) Turning the tide: the eradication of invasive species. IUCN, Gland, pp 301–310

Towns D, Atkinson IAE, Daugherty CH (2006) Have the harmful effects of rats been exaggerated? Biol Invasions 4:863–891

Towns D, Wardle DA, Mulder CPH, Yeates GW, Fitzgerald BM, Parrish GR, Bellingham PJ, Bonner KI (2009) Predation of seabirds by invasive rats: multiple indirect consequences for invertebrate communities. Oikos 118:420–430

Traveset A, Nogales M, Alcover JA, Delgado JD, Lopez-Darias M, Godoy D, Igual JM, Bover P (2009) A review on the effects of alien rodents in the Balearic (Western Mediterranean Sea) and Canary Islands (Eastern Atlantic Ocean). Biol Invasions 11:1653–1670

Varnham K (2010) Invasive rats on tropical islands: their history, ecology, impacts and eradication. RSPB research report no. 41. Royal Society for the Protection of Birds, Sandy, Bedfordshire, 54pp

Veitch CR, Clout MN, Towns DR (eds) (2011) Island invasives: eradication and management. In: Proceedings of the international conference on Island Invasives. IUCN, Gland, 542pp

Whitaker AH (1978) The effects of rodents on reptiles and amphibians. In: Dingwall PR, Atkinson IAE, Hay C (eds) The ecology and control of rodents in New Zealand nature reserves. Department of Lands and Survey Information Series No. 4, Wellington, New Zealand, pp 75–86

Chapter 11
Management of Wild Boar in Protected Areas: The Case of Elba Island

Alberto Meriggi, Marco Lombardini, Pietro Milanesi, Anna Brangi, Paolo Lamberti, and Francesca Giannini

Introduction

From the 1960s onwards, the wild boar (*Sus scrofa* Linnaeus, 1758) has become a species of social and economic interest in Europe. In recent decades in fact, European populations have increased greatly (Apollonio et al. 2010), leading to an increase in problems, such as damage to croplands, collisions with vehicles, an impact on plant communities and ecosystems and transmission of diseases to livestock and humans (e.g. Gortázar et al. 2007; Bueno et al. 2009; Lagos et al. 2012; Puerta-Piñero et al. 2012; Li et al. 2013).

In Italy the population of *Sus scrofa* has dramatically increased in the last 40 years, growing from an estimate of 1900 individuals in 1977 to about 667,000 in 2004 (Meriggi et al. 2011). Moreover, the species has expanded its range, occupying all the hilly and mountainous areas of the Italian peninsula, and, more recently, several zones of the Alps and intensively cultivated plains (Merli and Meriggi 2006; Monaco et al. 2006; Carnevali et al. 2011). The wild boar is present even on many Italian islands, including Elba (Angelici et al. 2009).

This demographic explosion is due to various reasons: the extensive recovery of natural woodlands, the adaptability of the species to a wide range of environmental

A. Meriggi (✉) • M. Lombardini • P. Milanesi • P. Lamberti
Department of Earth and Environmental Sciences, University of Pavia,
Via Ferrata 1, Pavia 27100, Italy
e-mail: meriggi@unipv.it; zarc00@yahoo.it; pietro.milanesi@unipv.it;
paolo.lamberti@shgroupmilan.com

A. Brangi
Province of Pavia, Via Taramelli 2, Pavia 27100, Italy
e-mail: anna.brangi@provincia.pv.it

F. Giannini
Tuscan Archipelago National Park, Loc. Enfola, Portoferraio, Livorno, Italy
e-mail: giannini@islepark.it

F.M. Angelici (ed.), *Problematic Wildlife*, DOI 10.1007/978-3-319-22246-2_11

conditions, its great reproductive capacity, repeated releases of hand-reared animals for hunting and escapes of individuals from farms (Merli and Meriggi 2006; Apollonio et al. 2010; Ficetola et al. 2014; Massei et al. 2015).

The growth of the population has produced an increase of damage to croplands; at present, compensation payments for wild boar damage in Italy amount to about nine millions of Euros per year (Riga et al. 2011). The problem of damage is particularly important in protected areas (Parks, Natural Reserves, Wildlife Refuges). In these areas hunting is forbidden; moreover, hunting activity near boundaries induces the boar to concentrate within protected areas, with a consequent increase in damage to agricultural activities and natural habitats (Monaco et al. 2010), often unsustainable for the financial resources available. In Italy, the wild boar is the most important game ungulate; 200,000 wild boars were harvested in 2012, mainly in Northern Apennines and central Italy, but this number probably represents an underestimate of the true number of heads shot per year (Apollonio et al. 2010; Massei et al. 2015). All these factors demonstrate the need for effective management strategies that reconcile the objective of conservation in protected areas and the presence of the wild boar as a big game species in surrounding hunting districts.

In Italian protected areas, the response to damage is prevention, compensation and population control by capture, culling and shooting (Monaco et al. 2010). Shooting is used as a control method for wild boar in many parts of the world (Massei et al. 2011). It has been demonstrated to be very effective, particularly in high-density areas and with a great hunting effort, but its success can be limited by compensatory responses exhibited by boar, such as increased immigration and reproduction (Hanson et al. 2009). Therefore, effective numerical control of wild boar in protected areas could be economically unsustainable, because the reduction of density is only temporary, and control shooting may become so intense as to become the main activity of a protected area. Prevention with electrified fences can be very effective (e.g. Santilli and Mazzoni Della Stella 2006; Reidy et al. 2008), but, considering the high cost for buying and maintaining the materials, it is normally only recommended for fields of high economic value or for high-risk situations (Santilli and Mazzoni Della Stella 2006).

To improve cost-effective strategies for damage prevention, it is important to identify which factors affect the risk of damage caused by wild boar. Species distribution models allow identification of relationships between species occurrence (e.g. presence, abundance, damage) and environmental factors (e.g. land use, topography, landscape features, climate) (Elith and Leathwick 2009; Jiménez-Valverde et al. 2011).

Wild boar damage is affected by many factors, including both security and food availability. Safety factors include the distance to the edge of the nearest forests, roads, rivers, and the human presence (Calenge et al. 2004; Cocca et al. 2007; Honda and Sugita 2007; Thurfjell et al. 2009). Food factors include types, abundance, maturation time of crops and availability of agricultural lands (Herrero et al. 2006; Schley et al. 2008; Li et al. 2013).

Our aim is to define a management model for wild boar populations in protected areas able to mitigate social and economic problems. Moreover, we would like to (1) verify the effectiveness of shooting as a prevention method, (2) define the minimum control effort that could lead to a relevant population decline and (3) identify the areas most at risk of damage with a predictive model.

Study Area

The Elba Island

The study was conducted on Elba Island (in the Tuscan Archipelago, Central Italy), with an area of 223.2 km^2 and a maximum altitude of 1016 m a.s.l. (Mount Capanne) (Fig. 11.1). The climate is Mediterranean, with a mean yearly temperature of 16.5 °C (minimum 10 °C in January, maximum 24.5 °C in July) and a mean yearly precipitation of 595 mm (minimum 13 mm in July, maximum 86 mm in November) (Chiatante et al. 2013). The island is covered by Mediterranean maquis with strawberry tree (*Arbutus unedo*), heather (*Erica arborea*), *Cistus* spp., rosemary (*Rosmarinus officinalis*) and lavender (*Lavandula stoechas*) (37.2 %), holm-oak (*Quercus ilex*) forests (26.2 %), urban areas (11.2 %), agricultural areas (10.2 %), pine plantations (7.0 %) and meadows (4.1 %). The surface of the island used for farming is partitioned among cereal crops, vineyards, orchards and olive groves; animal husbandry is based primarily on goats (Fifth National Census of Agriculture—Istituto Nazionale di Statistica 2000).

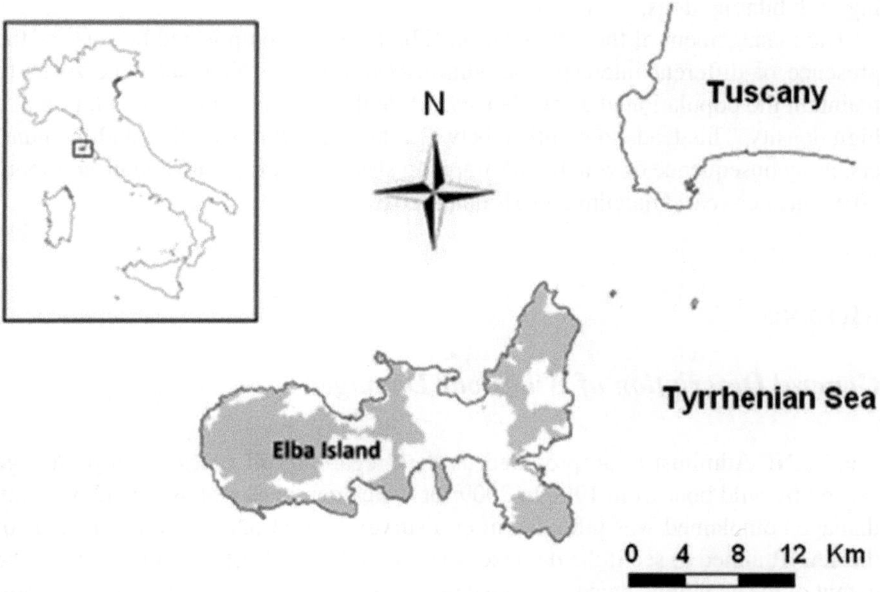

Fig. 11.1 Location of Elba Island (Central Italy, Tyrrhenian Sea). The Tuscan Archipelago National Park is in *grey*

Over half of the island (127.4 km^2) was designated as Tuscan Archipelago National Park (thereafter: TANP) in 1996, the remaining part (95.8 km^2) is a hunting district (ATC LI 10) (Fig. 11.1). In the National Park hunting is forbidden, although control of wild boar and mouflon (*Ovis orientalis musimon*) is carried out. Outside the National Park, hunting is mainly for wild boar, from October to January.

Hunting is performed by drives with hunting dogs. The hunt takes place in areas of approximately 50–120 ha; 25–30 hunters, armed and located in fixed positions, wait for the arrival of boars that are moved from their resting places by teams of dogs conducted by 6–8 persons.

History of Wild Boar on Elba Island

The presence of the wild boar on Elba Island has been reported for the first time at the beginning of the twentieth century (Damiani 1923). Then, some individuals were released for hunting in 1963 (Angelici et al. 2009). From the 1960s onwards, the presence of the species has caused a lot of problems: damage to the traditional agricultural system, the destruction of dry walls, negative effects on plant and animal biodiversity and collisions with vehicles (Giannini and Montauti 2010). Traffic accidents are particularly critical in summer, when Elba Island is visited by about two millions of tourists (Giannini and Montauti 2010).

Since 1997, to limit the negative impact of the wild boar, the administration of the TANP has developed a control programme, mostly based on trapping and hunting with hunting dogs.

The management of the wild boar on Elba Island is complicated because of the presence of different interests: the administration of the National Park wants to maintain the population at a low density, while the hunters want to maintain it at a high density. This leads to conflicts between the Park administration and the hunters, as a consequence of which Park trapping structures are often vandalized (about 30 events per year; Giannini and Montauti 2010).

Methods

General Description of Wild Boar Damages

The TANP Administration provided us data regarding all cases of crop damage caused by wild boar from 1999 to 2009 for which compensation was paid out. Any damage complained was refunded after a survey carried out by the technicians of the TANP, aimed to see if the damage was caused by wild boars and to estimate the extent of the economic damage. Each case was georeferred, and the amount of compensation and the type of crop/structure damaged (vineyards, orchards, vegetables, meadows and dry walls) were recorded.

To evaluate the trend of damage events and compensation payments during the study period, we performed regression analyses with curve estimation, putting damage variables (the number of events per year and compensation payments per year) as dependent variables and time as independent variable. A Spearman rank correlation test was performed to evaluate the relationship between (1) compensation payments and the number of damage events recorded and (2) the damage area and the number of damage events recorded. Finally, we looked for a seasonality of the distribution of damage events using a χ^2 goodness-of-fit test.

Relation between the Number of Wild Boars Removed and Damage

We collected hunting data (outside the National Park) and control data (inside the National Park) from 1998 to 2009. We analysed the trend of bags using regression analyses with curve estimation, putting the number of wild boars shot per year as dependent variable and time as independent variable.

Pearson's correlations were used to evaluate (1) relations between damage events and the number of wild boars removed in the same year and (2) relations between damage events and the number of wild boars removed in the previous year. We performed correlation analyses considering the number of wild boars shot only in the National Park and those from the whole Elba Island.

Wild Boar Population Estimate

Hunting and control statistics were used to estimate the wild boar population in every year by removal methods: we used Leslie–Davis regression model (Leslie and Davis 1939) and Ricker model (Ricker 1954). Both methods require that three assumptions are satisfied: (1) the population is closed; (2) the probability of each individual being shot is constant throughout the experiment and (3) all individuals have the same probability of being shot in the time *t*. Tacking account that the area under study is an island and the hunting method adopted, we considered these assumptions realistic.

According to Leslie–Davis model, the population size at the time *t* (before starting to shoot boars) could be estimated with a linear regression between catch per unit effort and accumulated catch. In our case, we considered the accumulated harvest (x) and the harvest per unit effort (y). Ricker model presupposes a linear relationship between the natural logarithm of catch per unit effort (y) and the accumulated effort (x) (Krebs 1999).

Population Viability Analysis

Hunting data were useful to assess the demographic parameters of the wild boar on Elba Island (Meriggi et al. 2010) (Table 11.1). We used these parameters to run PVAs, which have been demonstrated to be an accurate tool for predicting population viability (Brook et al. 2000). The PVA is a quantitative approach useful to assess the effects of demographic, environmental and genetic stochasticity and different management practices on wildlife populations. It is largely used in conservation biology to predict extinction risks for threatened species and to compare alternative options for their management (e.g. Galimberti et al. 2001; Chilvers 2012; Carroll et al. 2013). In our case, we run PVAs with the main aim of determining the harvest effort necessary to cause population extinction in the TANP over a 30-year period.

We carried out PVAs starting with a population of 2500 wild boars (12.6 per km^2) and estimating a carrying capacity of 3000 animals (15.1 per km^2). We calculated the initial population size considering the number of boars harvested in 2009 (1500 heads) and a harvesting rate of 60 %; we estimated the carrying capacity considering the density range recorded for wild boar populations in Mediterranean protected areas in Italy (13–25 per km^2, Massei et al. 1996; Massei and Genov 2000). We run PVAs simulating five different scenarios: (1) maintaining the present

Table 11.1 Demographic input values used for the population viability analyses (taken from Meriggi et al. 2010)

Parameters	Values
Extinction	Only one sex remaining
Lethal equivalent	3.14
Proportion of lethal genetic load	0.5
Mating system	Polygynous
Age of the first reproduction (females)	1 year
Age of the first reproduction (males)	2 years
Maximum reproduction age	>3 years
Sex ratio at birth	1:1
Maximum litter size	12
Mean litter size (SD)	5.0 (1.7)
Reproductive success % (SD)	26.3 (9.8)
Mortality year 1 to year 2 (age 1)	1.5 (57.2)
Mortality year 2 to year 3 (age 2)	34.1 (19.7)
Mortality year 3 to year 4 (age 3)	58.8 (24.3)
Mortality after year 4 (age 4)	22.8 (39.5)
Starting population	2500
Percentage of males and females of age 1	M 21.2, F 17.4
Percentage of males and females of age 2	M 16.2, F 15.2
Percentage of males and females of age 3	M 10.9, F 9.3
Percentage of males and females of age 4	M 2.3, F 7.5
Carrying capacity (SD)	3000 (807.0)

harvesting rate; (2) increasing the harvesting rate by 10 % in the first 5 years equally distributed on all age classes and on both sexes; (3) increasing the harvesting rate by 30 % in the first 5 years equally distributed on all age classes and on both sexes; (4) increasing the harvesting rate by 40 % in the first 5 years only on reproductive females and (5) increasing the harvesting rate by 40 % in the first 10 years only on reproductive females. PVAs were carried out with the software VORTEX (version 9.98) (Lacy 2000).

Factors Influencing Damage Distribution

In order to define the areas with the highest risk of wild boar damage, we applied a 1 km^2-spaced grid to the study area that identified 278 individual cells. In all cells, using ArcMap v. 9.3 GIS software (ESRI, Redlands, USA), we measured the proportion of 17 habitat variables concerning vegetation cover, 11 variables of elevation (corresponding to 11 different classes of altitude), 7 variables of slope (corresponding to 7 different classes of slope) and 9 variables of exposure (Table 11.2). Data on habitat types were taken from a 1:25,000 digital vegetation map (Foggi et al. 1996), while elevation, slope and exposure data were derived from a Digital Terrain Model (DTM; cell size 20 m) produced by the Italian Military Geographic Institute.

Table 11.2 Ecogeographical variables used in the analyses

Land use variables	Exposure variables	Elevation variables (m a.s.l.)	Slope variables
Cork-oak woods	No exposure	0–100	0–10° of slope
Chestnut woods	North	101–200	11–20° of slope
Holm-oak woods	North-east	201–300	21–30° of slope
European hop-hornbeam woods	East	301–400	31–40° of slope
Pine plantations	South-east	401–500	41–50° of slope
Lentisk-dominated maquis	South	501–600	51–60° of slope
Heather-dominated maquis	South-west	601–700	61–70° of slope
Shrublands	West	701–800	
Pastures	North-west	801–900	
Rocky areas		901–1000	
Coastal areas		1001–1100	
Wetlands			
Arable lands			
Grasslands			
Vineyards			
Orchards			
Urban areas			

We predicted the areas with the highest risk of damage following a model-averaging approach, combining the results obtained with two different methods: (1) RSPFs, following a use vs. availability approach obtained by Binary Logistic Regression Analysis (thereafter: BLRA) (Pearce and Ferrier 2000; Boyce et al. 2002; Keating and Cherry 2004) and (2) Maximum Entropy algorithm (MaxEnt version 3.3.3) (Phillips et al. 2006; Elith et al. 2011), which requires only presence data.

Resource Selection Probability Functions

We carried out BLRA comparing the features of damaged cells and of an equal number of cells randomly placed over the island. We defined the cells as "damaged" by the species if at least one damage event fell inside. We built models with all possible combinations of habitat variables. Inference from models was made according to the Information-Theoretic Approach (Anderson et al. 2000, 2001; Richards et al. 2011), using corrected Akaike Information Criterion (AICc) because of the small ratio of sample/parameters (Akaike 1973; Anderson et al. 2000, 2001; Johnson and Omland 2004). Models were ranked and scaled by the differences with minimum AICc (ΔAICc) and Akaike weights (ωi) for each i-model (Anderson et al. 2000; Burnham and Anderson 2002). Models with $\Delta \text{AICc} \leq 2$ were considered the best ones and used to develop model averaging (Burnham and Anderson 2002; Massolo and Meriggi 2007). The relative importance of predictor variables (ω) was measured, as resulted from the best models, by the sum of Akaike weights of the models in which each variable appeared (Burnham and Anderson 2002; Merli and Meriggi 2006).

Model validation was achieved through a k-fold crossvalidation process. We partitioned our data into two random subsets (Fielding and Bell 1997; Boyce et al. 2002; Chiatante et al. 2013); the former was used to calibrate the model and the latter was used to evaluate the result. We replicated this process two times, using each subset in turn for validation purpose, and we measured the model accuracy using the Boyce index (Boyce et al. 2002).

Maxent

Maxent is a general-purpose machine learning method (Phillips et al. 2006), widely used in species distribution modelling in a large range of taxons and across many disciplines (i.e. biological invasions, evolution, conservation; Ward 2007; Monterroso et al. 2009; Isaac et al. 2013; Ficetola et al. 2014), which follows a presence-only modelling approach.

Presence-only data consist of records describing known occurrences (presence) of species without information about known absences. The collection of real absence data is one of the major problems to solve; while collecting reliable data on animal presence is straightforward in most case studies, it is more difficult to verify the true absence of a species in a given habitat, especially when the target species occurs at low densities, is cryptic or elusive, thus resulting in a low detection probability (Santos et al. 2006; Isaac et al. 2013). For all these reasons, the absence of observations at a given location cannot be reliably interpreted as a true absence, thus it is necessary to rely only on presence data.

Maxent approach assesses the likelihood of presence in a given cell on the basis of the environmental features of that cell, establishing flexible relationships between the dependent and independent variables (Elith et al. 2011). It finds the most uniform species distribution with the constraint that the expected value for each variable should match the average value of a set of sample points taken from the target-species distribution (Phillips et al. 2006). The probability distribution is exponential, ranging from 0 to 1. This is achieved by dividing the sum of weighted probability values by a scaling constant.

The area under the Receiver Operating Characteristics curve (AUC) was used to examine model performance. AUC provides a measure of discrimination ability, varying from 0.5 for a model with discrimination ability no better than random to 1.0 for a model with perfect discriminatory ability.

A rough guide for classifying the accuracy of a diagnostic test is the traditional academic point system (Swets 1988): 0.91–1.00 = excellent; 0.81–0.90 = good; 0.71–0.80 = fair; 0.61–0.70 = poor; 0.51–0.60 = fail.

Damage Risk Map

For both methods, we computed a damage risk (DR) map, which indicates how the combination of the ecogeographical variables of a single cell determines the probability of risk of damage. All maps were classified in five risk groups: null (0.00–0.20), low (0.21–0.40), medium (0.41–0.60), high (0.61–0.80) and maximum (0.81–1.00).

Finally, to reduce the model-based uncertainty in range predictions, we used an ensemble forecasting approach, which represents one of the most efficient consensus methods (Araújo and New 2007), computing the average probability of damage predicted by the two methods.

Results

General Description of Wild Boar Damage

From 1999 to 2009 we recorded a total of 256 damage events in the TANP (mean per year ± SD: 23.3 ± 13.0 events). Damage was concentrated mainly on vineyards (46 % of total events) and dry walls (27 %), but was also reported on orchards (11 %), meadows (9 %) and vegetables (7 %).

There was a significant negative trend of damage during the study period; both the number of events and compensation payments decreased following a logarithmic model (number of events: $y = 43.2 - 12.5 \ln(x)$, $r^2 = 0.462$, $F = 9.57$, $p = 0.013$; compensation payments: $y = 19.5 - 6.4 \ln(x)$, $r^2 = 0.447$, $F = 9.09$, $p = 0.015$).

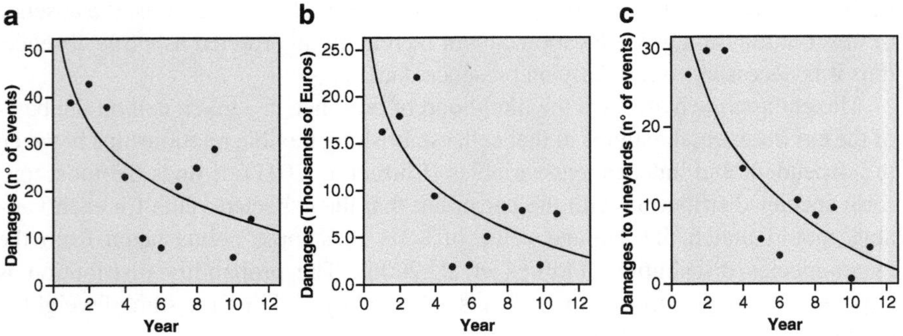

Fig. 11.2 (**a**) Trend of wild boar damage (1999–2009), (**b**) trend of refunds paid for wild boar damage (1999–2009) and (**c**) trend of wild boar damage to vineyards (1999–2009)

Both the amount of compensation payments and the damage area were strongly correlated with the number of damage events (rho = 0.89, $p = 0.001$ and rho = 0.92, $p = 0.001$, respectively).

We observed a significant logarithmic reduction in the number of damage events to vineyards ($y = 32.9 - 12.2\ln(x)$, $r^2 = 0.692$, $F = 23.50$, $p = 0.001$), although for the other types of fields and for dry walls there were not significant trends ($p > 0.05$) (Fig. 11.2).

There were strong monthly differences in the distribution of damage events ($\chi^2 = 400.03$; df = 11; $p < 0.001$), with a peak recorded in August and September and a minimum in March, November and December. Damage to vineyards almost exclusively occurred in August and September, damage to orchards was concentrated from July to September, damage to vegetables occurred mainly in summer, damage to meadows was recorded throughout the year, while damage to dry walls almost exclusively occurred in spring and summer. Few events were recorded in autumn and winter (Fig. 11.3).

Relation between the Number of Wild Boars Removed and Damage

The number of wild boars shot increased significantly from 1998 to 2009, following a linear model. This growth was observed cumulating data ($y = 584.4 + 53.9x$, $r^2 = 0.409$, $F = 8.61$, $p = 0.015$) and considering only the animals killed inside the TANP ($y = 292.2 + 43.1x$, $r^2 = 0.282$, $F = 5.32$, $p = 0.044$) and in the hunting district ($y = 240.4 + 13.6x$, $r^2 = 0.329$, $F = 6.39$; $p = 0.030$) (Fig. 11.4).

No significant correlations were found between the amount of damage and the number of wild boars removed (Table 11.3).

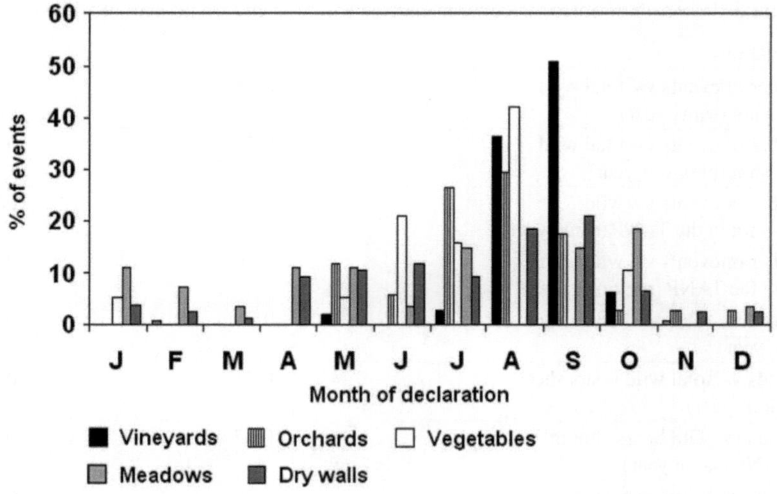

Fig. 11.3 Monthly distribution of wild boar damage

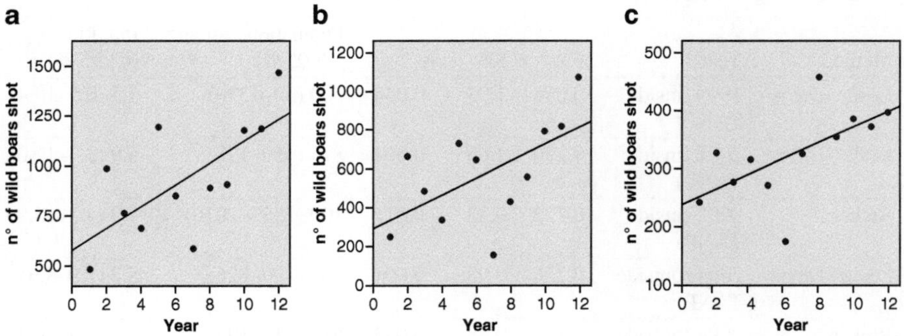

Fig. 11.4 Trend in the number of wild boar killings on (**a**) the whole of Elba Island (1998–2009), (**b**) Tuscan Archipelago National Park (1998–2009) and (**c**) the hunting district of Elba Island (1998–2009)

Wild Boar Population Estimate

Considering harvest data, we obtained significant regressions only for 2004 and 2006 with both Leslie–Davis and Ricker method; considering control data, we obtained significant regressions for 1999 and 2002 with Leslie–Davis method and only for 2002 with Ricker method (Table 11.4). We did not find significant differences between the estimates obtained with the two methods.

Table 11.3 Pearson's correlations between damage variables and the number of wild boars shot

Correlations	r	p
Number of events vs. total wild boars shot (same year)	0.23	0.50
Number of events vs. total wild boars shot (previous year)	–0.52	0.10
Number of events vs. wild boars shot in the TANP (same year)	–0.17	0.61
Number of events vs. wild boars shot in the TANP (previous year)	–0.46	0.16
Refunds vs. total wild boars shot (same year)	0.21	0.52
Refunds vs. total wild boars shot (previous year)	–0.41	0.20
Refunds vs. wild boars shot in the TANP (same year)	–0.17	0.62
Refunds vs. wild boars shot in the TANP (previous year)	–0.29	0.38

Table 11.4 Significant regressions for wild boar population estimates

Method	Year	r^2	SE	p	Population estimate (CI 95 %)	Ind./km^2 (CI 95 %)
Leslie–Davis	1999 (inside TANP)	0.058	1.93	0.044	700 (70–1330)	5.5 (0.5–10.4)
Leslie–Davis	2002 (inside TANP)	0.296	1.66	0.004	875 (469–1281)	6.9 (3.7–10.0)
Ricker	2002 (inside TANP)	0.222	0.33	0.015	1008 (553–2030)	7.9 (4.3–15.9)
Leslie–Davis	2004 (outside TANP)	0.256	0.98	0.001	500 (180–826)	5.2 (1.9–8.6)
Ricker	2004 (outside TANP)	0.127	0.58	0.033	615 (318–1923)	6.4 (3.3–20.1)
Leslie–Davis	2006 (outside TANP)	0.399	0.99	<0.001	476 (232–718)	5.0 (2.4–7.5)
Ricker	2006 (outside TANP)	0.350	0.51	<0.001	474 (175–1289)	4.9 (1.8–13.4)

Population Viability Analysis

Maintaining the current harvesting rate, the wild boar population decreased by about 30 % within the first 5 years. In the following years, the population still decreased, but more slowly, and stabilized after about 15 years. The survival probability decreased linearly over the simulation period, but it always maintained a high value (more than 0.6) (Fig. 11.5).

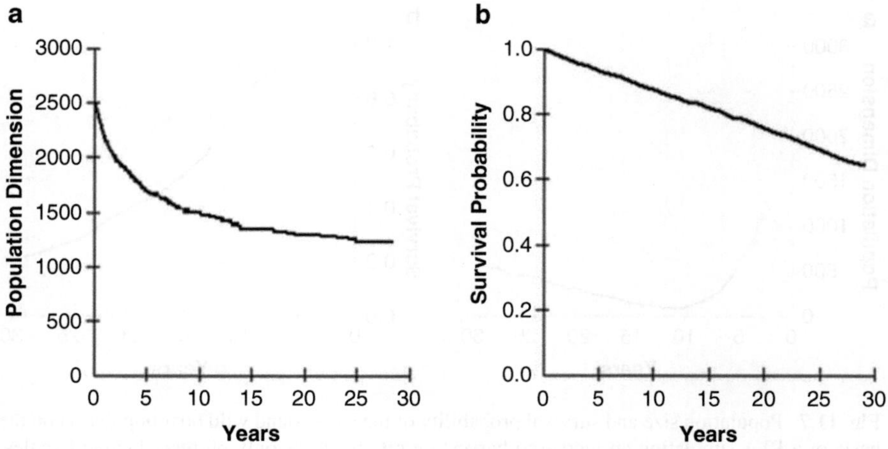

Fig. 11.5 Population dimension and survival probability of the Elba Island wild boar population on the basis of a PVA simulating the current harvesting rate

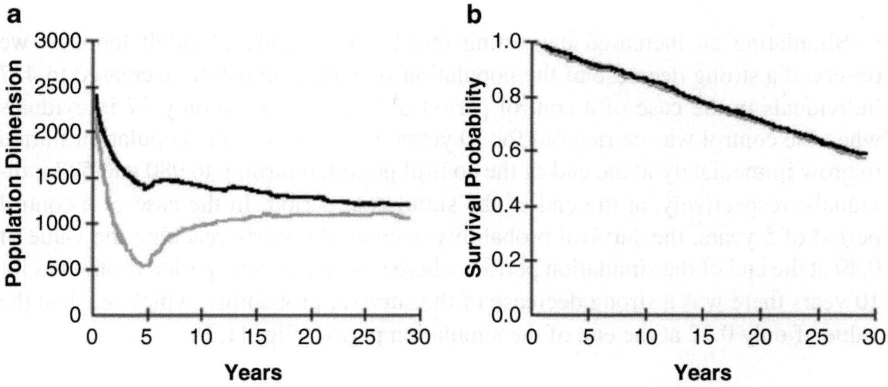

Fig. 11.6 Population size and survival probability of the Elba Island wild boar population on the basis of a PVA simulating an increased harvesting rate equally distributed on all age classes and on both sexes. The *black line* indicates an increased rate of 10 %, the *grey line* indicates an increased rate of 30 %

Simulating an increased harvesting rate of 10 and 30 % on all the age classes and on both sexes, we observed a strong decrease in the population size only at the 30 % level. Nevertheless, the effect of this stronger control action had little long-term impact: the population started to grow up immediately at the end of the control period, returning to values of about 1000–1200 individuals in a few years. The survival probability was very similar comparing the two different scenarios: it decreased linearly during the simulation period, but it maintained a high value, of about 0.6 (Fig. 11.6).

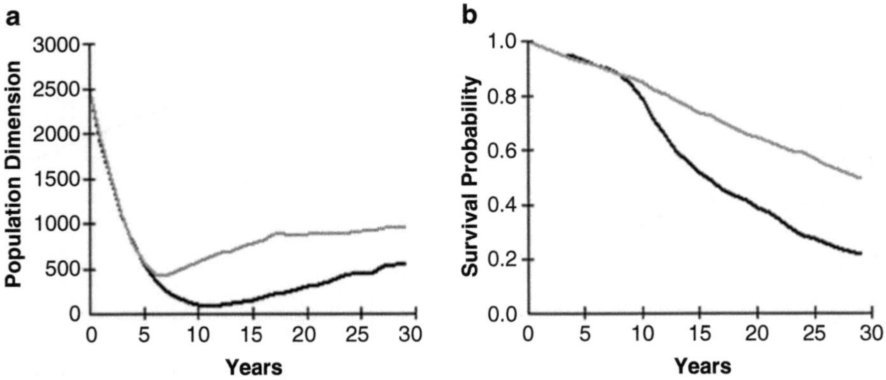

Fig. 11.7 Population size and survival probability of the Elba Island wild boar population on the basis of a PVA simulating an increased harvesting rate by 40 % only on reproductive females. The *black line* indicates an increased rate for 10 years, the *grey line* indicates an increased rate for 5 years

Simulating an increased harvesting rate by 40 % only on adult females, we observed a strong decrease of the population size; the population decreased to 437 individuals in the case of a control period of 5 years, and to only 97 individuals when the control was carried out for 10 years. In both cases, the population started to grow immediately at the end of the control period, returning to 980 and 573 individuals, respectively, at the end of the simulation period. In the case of a control period of 5 years, the survival probability decreased linearly reaching the value of 0.49 at the end of the simulation period, whereas with a control period protracted for 10 years there was a strong decrease of the survival probability, which reached the value of only 0.22 at the end of the simulation period (Fig. 11.7).

Factors Influencing Damage Distribution

Resource Selection Probability Functions

In the study area we found a total of 56 cells with presence of wild boar damage. By multimodel inference we selected three models, which included a total of seven variables (Table 11.5). The most important variable positively affecting the probability of wild boar damage were the areas between 10° and 20° of slope, while for the other variables the effect was uncertain (Table 11.6). Crossvalidation results indicated an excellent predictive power of the model: calculating the Boyce Index we measured a Spearman rank coefficient of 0.99 ($p<0.001$).

Table 11.5 Results of multimodel inference obtained by information-theoretic approach on binary logistic regression analysis of wild boar damage presence cells ($n=56$) vs. control ones ($n=56$) (only models with $\Delta AICc \leq 2$ are shown)

Habitat variables	AICc	ΔAICc	ωi
41–50° slope, grasslands, heather-dominated maquis, lentisk-dominated maquis, shrublands	129.26	0.00	0.414
11–20° slope, heather-dominated maquis, lentisk-dominated maquis	129.43	0.17	0.379
31–40° slope, 41–50° slope, grasslands, shrublands, heather-dominated maquis, lentisk-dominated maquis	130.62	1.36	0.207

Table 11.6 Relative importance of habitat variables for wild boar damages

Habitat variabless	β	β SE	Lower CL (95 %)	Upper CL (95 %)	ω
Heather-dominated maquis	10.70	9.44	–7.81	29.21	1.00
Lentisk-dominated maquis	–11.61	13.92	–38.89	15.67	1.00
Grasslands	6.30	6.17	–5.79	18.39	0.621
Shrublands	–27.26	28.64	–83.40	28.89	0.621
Slope 41–50°	179.47	173.38	–160.36	519.31	0.621
Slope 11–20°	1.23	0.62	0.01	2.46	0.379
Slope 31–40°	–3.48	1.86	–7.12	0.16	0.207

β coefficients of the variables, β *SE* standard error, *CL* confidence limit, ω relative importance of habitat variables

Maxent

The areas between 100 and 300 m a.s.l., European hop-hornbeam woods, shrublands, rocky areas, arable lands, grasslands and lentisk-dominated maquis were the most important variables in the prediction of the areas most at risk of damage. All these variables positively affected the probability of wild boar damages (Table 11.7).

The evaluation of the model performance with AUC indicated a fair predictive power of the model, with an area under the ROC curve of 0.78 ($p<0.001$).

Damage Risk Map

Combining the two models more than 45 % of Elba Island was classified either in the high or maximum classes of risk (respectively, 29.1 % and 18.8 %). Of the remaining area, 29.1 % was classified in the medium risk, 19.4 % in the low risk and only 3.6 % in the null risk class. The areas with the highest risk of damage (medium, high or maximum) were distributed throughout the island (Fig. 11.8) but particularly in the south, in the central-eastern part.

Table 11.7 Relative contribution of the most important environmental variables to the Maxent model

Habitat variables	Coefficients of the model
101–200 m a.s.l	1.96
201–300 m a.s.l.	1.78
European hop-hornbeam woods	1.66
Shrublands	1.49
Rocky areas	1.17
Arable lands	1.00
Grasslands	0.96
Lentisk-dominated maquis	0.82

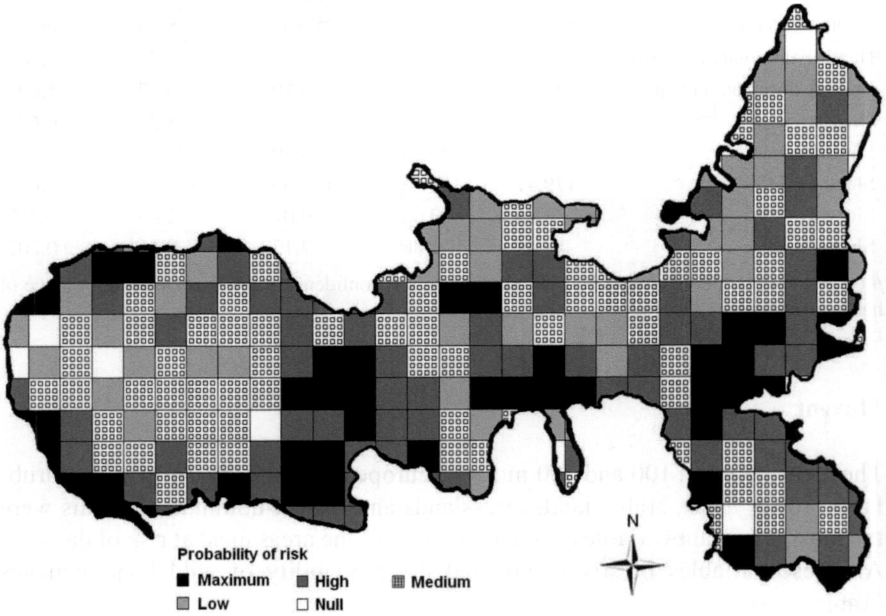

Fig. 11.8 Damage risk map for wild boar damages on Elba Island obtained following an ensemble forecasting approach

Discussion

Removal methods failed to provide a precise estimate of the population present on the Elba Island because of the high variability of the catch per unit effort; moreover, even when we obtained significant regressions, confidence intervals were very wide. The lack of an accurate measure of the population size could represent a great problem for wild boar management. Probably only with a heavy hunting and control pressure it is possible to use removal methods to obtain precise and accurate population estimates, otherwise it is necessary to adopt other census methods. But methods such as spotlight counts, aerial surveys and vantage point counts are difficult to carry out in Mediterranean environments, because the habitat is characterized by tall and dense vegetation. Thus, it is necessary to use different methods, like tracking plots and dung counts. These methods have been successfully used to monitor wild boar abundance (Engeman et al. 2001, 2013), a parameter that is often correlated with density (Hone 1995).

Hunting data are often used to assess density and relative abundance of wild boar populations (e.g. Boitani et al. 1994; Fernández-Llario et al. 2003; Bosch et al. 2012; Sarasa and Sarasa 2013). The application of this technique requires several assumptions; the existence of a linear relationship between catch and population abundance, the use of standardized methods of removal, the knowledge of all removals (Mitchell and Balogh 2007; Engeman et al. 2013). Unfortunately, most of these assumptions are difficult to meet or unknown, potentially biasing results.

To reduce bias, hunt indices require large samples (Siren et al. 2004). In our case, we collected systematically data regarding hunting bags over the 11 years of study, and the TANP administration did the same regarding claims for compensation. For these reasons, we are confident that these data could be used for the interpretation of our results.

Our results show a negative trend of wild boar damage on Elba Island from 1999 to 2009. Vineyards were the most frequently damaged crops. Damage monitoring in Europe mainly regards maize, other cereals and grasslands, while vineyards are less recorded, but comparisons with other countries are difficult because of the differences between their agricultural mosaics (Herrero et al. 2006; Schley et al. 2008; Novosel et al. 2012). On Elba Island, damage to vineyards almost exclusively occurred from August to October, in correspondence of the ripening period of grapes.

The number of boars shot increased during the study period, but the lack of correlation between wild boars removed and damage does not allow us to assume a positive effect of shooting in determining the reduction of damage events, damage area and compensation payments. In some cases, hunting has been demonstrated to be an effective solution in reducing damage to crops (Mazzoni Della Stella et al. 1995; Geisser and Reyer 2004). However, combination of shooting with prevention methods could increase the efficiency in damage reduction (McCann and Garcelon 2008; Massei et al. 2011).

Supplementary feeding (e.g. with maize) has been demonstrated to be a very effective tool to reduce the level of damage to vineyards in southeastern France (Calenge et al. 2004), which is very similar to Elba Island regarding the habitat composition. It is essential to distinguish between the "dissuasive" and the "attractive" food supply. In several areas, hunters spread maize throughout the year to attract boars to their hunting territories. This additional food may result in an increase in population size and leads to a long-term increase in the number of damage events (Geisser and Reyer 2004; Schley et al. 2008). On Elba Island, boars are not artificially fed. We strongly warn against this practice, and we stress that the dissuasive spreading of maize is different both in its aim and in its results. When used as a deterrent, the maize is spread only over a very short period, as long as the grapes are ripe. In fact, the dissuasive spreading of maize is a replacement food source rather than an additional food source (Calenge et al. 2004).

PVAs showed that only a heavy control effort concentrated on reproductive females is effective in markedly decreasing the population size and the survival probability of wild boar. Increasing the harvesting effort on reproductive females has been demonstrated to be the most efficient way to stop wild boar population growth (Bieber and Ruf 2005; Servanty et al. 2011). In our case, the effect is stronger simulating a control effort protracted for a longer period; we observed a reduction of the population size of 96 % after 10 years, and a reduction of the survival probability of 78 % after 30 years. This great result, however, is very difficult to reach in practice because of the great engagement and economic effort needed.

Our model identified landscape features that increase the risk of crop damage: the risk was highest in low altitude areas (100–300 m a.s.l.) with a high presence of arable lands, grasslands, European hop-hornbeam woods and shrublands. The high relative importance of arable lands and grasslands could be related to wild boar nutrition. Agricultural plants represent the main food resource for wild boar in several areas (Schley and Roper 2003; Herrero et al. 2006); wild boar damage to grasslands is caused by direct consumption and by rooting activity linked to the search for invertebrates, roots and bulbs (Baubet et al. 2004; Schley et al. 2008). Acorns and maize are higher in carbohydrates and fats, but have lower crude protein content than grasslands and earthworms (Massei et al. 1996; Baubet et al. 2004; Schley et al. 2008), so wild boars could have to counterbalance their diet with animal foods and graminoids. The positive relationship with shrublands and hop-hornbeam woods likely occurs because of the high suitability of this habitat for wild boars (Meriggi and Sacchi 2001; Schley et al. 2008; Honda and Kawauchi 2011). In fact, they can find alternative shelter areas in the garrigue and in the Mediterranean maquis, which are rife in the study area and, like woodlands, can provide a good cover.

Damage modelling produced a map of the risk with a satisfactory predictive power; the risk maps could be used to plan preventive measures with the aim of improving the effectiveness of management both in the short- and long-term perspectives. First, these maps can help to identify the areas more suitable for population monitoring schemes, which will allow a better knowledge of population dynamics, and can provide essential information for a more effective management of populations. Collecting standardized data and regularly updating the databases is necessary for an adaptive management of populations (Ficetola et al. 2014). Second, the maps could allow to identify crops most at risk that can be protected through fences (mechanical or electrified), feeding plots in forested areas or numerical shooting control.

Our ensemble forecasting approach will help to overcome reliance on a single modelling method. With recent advances in sophisticated species distribution modelling approaches (e.g. BIOMOD, Thuiller et al. 2009), wildlife ecologists will be able to construct robust ensemble models of up to a dozen different modelling approaches.

Conclusions

The goal of the wild boar management of the TANP is a drastic reduction of damage and of conflicts between the Park administration and the local population. This study, both with correlation analyses between killings and damages and with PVAs, highlighted the importance of a heavy control targeted on specific age or size classes as a powerful instrument in the reduction of damage. However, a sufficiently heavy numerical control is difficult to sustain in the long term, so control activity should be combined with damage prevention by stable or electrified fences and supplementary feeding, especially to protect vineyards, which represent the main crop damaged and, consequently, the main source of conflicts.

Models of damage risk can be an effective tool to address prevention effort with more effectiveness and lower costs. In particular, the ensemble forecasting approach, with its great robustness given by the use of different models which are based on different principles, seems to be an efficient method in predicting the areas most at risk of damage.

More research is needed to improve management strategies of wild boar on Elba Island; firstly, it is necessary to provide accurate density or abundance estimates, with drive counts, tracking plots or dung counts. Secondly, it is necessary to study the diet of the wild boar on Elba Island; only the knowledge of the foraging behaviour of the wild boar will allow understanding clearly the impact of the species on croplands and natural ecosystems. Besides the study of the feeding habits, it should be important monitoring the yearly availability of natural food resources (i.e. acorns and chestnuts), which often influences wild boar population dynamics and crop damage (Bieber and Ruf 2005; Servanty et al. 2009; Cutini et al. 2013).

Acknowledgements The research was funded by the Tuscan Archipelago National Park. We thank the Hunting District n. 10 of Livorno province, that provided the hunting data from 1999 to 2009, and two anonymous referees that reviewed an early draft of the manuscript.

References

Akaike H (1973) Information theory as an extension of the maximum likelihood principle. In: Petrov BN, Csaki F (eds) Second international symposium on information theory. Akademiai Kiado, Budapest, pp 267–281

Anderson DR, Burnham KP, Thompson WL (2000) Null hypothesis testing: problems, prevalence and an alternative. J Wildl Manag 64:912–923

Anderson DR, Link WA, Johnson DH, Burnham KP (2001) Suggestion for presenting the results of data analyses. J Wildl Manag 65:373–378

Angelici FM, Laurenti A, Nappi A (2009) A checklist of the mammals of small Italian islands. Hystrix 20:3–27

Apollonio M, Andersen R, Putman R (2010) European ungulates and their management in the 21st century. Cambridge University Press, Cambridge

Araújo MB, New M (2007) Ensemble forecasting of species distributions. Trends Ecol Evol 22:42–47

Baubet E, Bonenfant C, Brandt S (2004) Diet of the wild boar in the French Alps. Galemys 16:101–113

Bieber C, Ruf T (2005) Population dynamics in wild boar *Sus scrofa*: ecology, elasticity of growth rate and implications for the management of pulsed resource consumers. J Appl Ecol 42:1203–1213

Boitani L, Mattei L, Nonis D, Corsi F (1994) Spatial and activity patterns of wild boars in Tuscany. J Mammal 75:600–612

Bosch J, Peris S, Fonseca C, Martinez M, De la Torre A, Iglesias I, Muñoz MJ (2012) Distribution, abundance and density of the wild boar on the Iberian Peninsula, based on the CORINE program and hunting statistics. Folia Zool 61:138–151

Boyce MS, Vernier PR, Nielsen SE, Schmiegelow FKA (2002) Evaluating resource selection functions. Ecol Model 157:281–300

Brook BW, O'Grady JJ, Chapman AP, Burgman MA, Akçakaya HR, Frankham R (2000) Predictive accuracy of population viability analysis in conservation biology. Nature 404:385–387

Bueno CG, Alados CL, Gómez García D, Barrio IC, García-González R (2009) Understanding the main factors in the extent and distribution of wild boar rooting on alpine grasslands. J Zool 279:195–202

Burnham KP, Anderson DR (2002) Model selection and multimodel inference. A practical information—theoretic approach, 2nd edn. Springer, New York

Calenge C, Maillard D, Fournier P, Fouque C (2004) Efficiency of spreading maize in the garrigues to reduce wild boar (*Sus scrofa*) damage to Mediterranean vineyards. Eur J Wildl Res 50:112–120

Carnevali L, Di Luzio P, Riga F (2011) Ungulati in Italia, Distribuzione e consistenza. In: Riga F, Genghini M, Cascone C, Di Luzio P (eds) Impatto degli ungulati sulle colture agricole e forestali: proposta per linee guida nazionali. Manuali e Linee Guida ISPRA 68/2011, pp 3–19

Carroll C, Fredrickson RJ, Lacy RC (2013) Developing metapopulation connectivity criteria from genetic and habitat data to recover the endangered Mexican wolf. Conserv Biol 28:76–86

Chiatante G, Meriggi A, Giustini D, Baldaccini NE (2013) Density and habitat requirements of red-legged partridge on Elba Island (Tuscan Archipelago, Italy). Ital J Zool 80:402–411

Chilvers BL (2012) Population viability analysis of New Zealand sea lions, Auckland Islands, New Zealand's sub-Antarctics: assessing relative impacts and uncertainty. Polar Biol 35:1607–1615

Cocca G, Sturaro E, Dal Compare L, Ramanzin M (2007) Wild boar *Sus scrofa* damages to mountain grassland: a case study in the Belluno province, Eastern Italian Alps. Ital J Anim Sci 6:845–847

Cutini A, Chianucci F, Chirichella R, Donaggio E, Mattioli L, Apollonio M (2013) Mast seeding in deciduous forests of the northern Apennines (Italy) and its influence on wild boar population dynamics. Ann For Sci 70:493–502

Damiani G (1923) La Fauna. In: L'Elba illustrata, pp 103–129

Elith J, Leathwick JR (2009) Species distribution models: ecological explanation and prediction across space and time. Annu Rev Ecol Evol Syst 40:677–697

Elith J, Phillips SJ, Hastie T, Dudík M, En Chee Y, Yates CJ (2011) A statistical explanation of MaxEnt for ecologists. Divers Distrib 17:43–57

Engeman RM, Constantin B, Nelson M, Woolard J, Bourassa J (2001) Monitoring changes in feral swine population and spatial distribution. Environ Conserv 28:235–240

Engeman RM, Massei G, Sage M, Gentle MN (2013) Monitoring wild pig populations: a review of methods. Environ Sci Pollut Res 20:8077–8091

Fernández-Llario P, Mateos-Quesada PM, Silvério A, Santos P (2003) Habitat effects and shooting techniques on two wild boar (*Sus scrofa*) populations in Spain and Portugal. Z Jagdwiss 49:120–129

Ficetola GF, Bonardi A, Mairota P, Leronni V, Padoa-Schioppa E (2014) Predicting wild boar damages to croplands in a mosaic of agricultural and natural areas. Curr Zool 60:170–179

Fielding AH, Bell JF (1997) A review of methods for the assessment of prediction errors in conservation presence/absence models. Environ Conserv 24:38–49

Foggi B, Cartei L, Pignotti L, Signorini MA, Viciani D, Dell'Olmo L, Menicagli E (1996) Il paesaggio vegetale dell'Isola d'Elba (Arcipelago Toscano): studio fitosociologico e cartografico. Fitosociologia 43:3–94

Galimberti F, Sanvito S, Boitani L, Fabiani A (2001) Viability of the southern elephant seal population of the Falkland Islands. Anim Conserv 4:81–88

Geisser H, Reyer HU (2004) Efficacy of hunting, feeding, and fencing to reduce crop damage by wild boars. J Wildl Manag 68:939–946

Giannini F, Montauti G (2010) Controllo del Cinghiale e conflitti sociali: il caso del Parco Nazionale Arcipelago Toscano. In: Monaco A, Carnevali L, Toso S (eds) Linee guida per la gestione del cinghiale (*Sus scrofa*) nelle aree protette, pp 86–87

Gortázar C, Ferroglio E, Höfle U, Frölich K, Vicente J (2007) Diseases shared between wildlife and livestock: a European perspective. Eur J Wildl Res 53:241–256

Hanson LB, Mitchell MS, Grand JB, Jolley DB, Sparklin BD, Ditchkoff SS (2009) Effect of experimental manipulation on survival and recruitment of feral hogs. Wildl Res 36:185–191

Herrero J, García-Serrano A, Couto S, Ortuño VM, García-González R (2006) Diet of wild boar *Sus scrofa* L. and crop damage in an intensive agroecosystem. Eur J Wildl Res 52:245–250

Honda T, Kawauchi N (2011) Methods for constructing a wild boar relative-density map to resolve human-wild boar conflicts. Mamm Stud 36:79–85

Honda T, Sugita M (2007) Environmental factors affecting damage by wild boars (*Sus scrofa*) to rice fields in Yamanashi Prefecture, central Japan. Mamm Stud 32:173–176

Hone J (1995) Spatial and temporal aspects of vertebrate pest damage with emphasis on feral hogs. J Appl Ecol 32:311–319

Isaac B, White J, Ieradiaconou D, Cooke R (2013) Response of a cryptic apex predator to a complete urban to a forest gradient. Wildl Res 40:427–436

Jiménez-Valverde A, Peterson AT, Soberón J, Overton JM, Aragón P, Lobo JM (2011) Use of niche models in invasive species risk assessments. Biol Invasions 13:2785–2797

Johnson JB, Omland KS (2004) Model selection in ecology and evolution. Trends Ecol Evol 19:101–108

Keating KA, Cherry S (2004) Use and interpretation of logistic regression in habitat-selection studies. J Wildl Manag 68:774–789

Krebs CJ (1999) Ecological methodology, 2nd edn. Addison-Wesley, Toronto

Lacy RC (2000) Structure of the VORTEX simulation model for population viability analysis. Ecol Bull 48:191–203

Lagos L, Picos J, Valero E (2012) Temporal pattern of wild ungulate-related traffic accidents in northwest Spain. Eur J Wildl Res 58:661–668

Leslie PH, Davis DHS (1939) An attempt to determine the absolute number of rats on a given area. J Anim Ecol 8:94–113

Li L, Shi J, Wang J, Gao Y, Wang L, Wang J, Ying X (2013) Factors influencing wild boar damage in Taohongling National Nature Reserve in China: a model approach. Eur J Wildl Res 59:179–184

Massei G, Genov PV (2000) Il Cinghiale. Calderini Edagricole, Bologna, 189pp

Massei G, Genov PV, Staines BW (1996) Diet, food availability and reproduction of wild boar in a Mediterranean coastal area. Acta Theriol 41:307–320

Massei G, Roy S, Bunting R (2011) Too many hogs? A review of methods to mitigate impact by wild boar and feral hogs. Hum Wildl Interact 5:79–99

Massei G, Kindberg J, Licoppe A, Gačić D, Šprem N, Kamler J, Baubet E, Hohmann U, Monaco A, Ozoliņš J, Cellina S, Podgorski T, Fonseca C, Markov N, Pokorny B, Rosellp C, Nahlikq A (2015) Wild boar populations up, numbers of hunters down? A review of trends and implications for Europe. Pest Manag Sci 71:492–500. doi:10.1002/ps.3965

Massolo A, Meriggi A (2007) I modelli di valutazione ambientale per le potenzialità faunistiche: aspetti metodologici e stato in Italia. In: Lucifero M, Genghini M (eds) Valorizzazione agroforestale e faunistica dei territori di collina e montagna. INFS, MIPAF, S.t.e.r.n.a. Ed. Grafiche 3B, Toscanella di Dozza (BO), pp 43–77

Mazzoni Della Stella R, Calovi F, Burrini L (1995) The wild boar management in a province of central Italy. J Mt Ecol 3:213–216
McCann BE, Garcelon DK (2008) Eradication of feral hogs from Pinnacles National Monument. J Wildl Manag 72:1287–1295
Meriggi A, Sacchi O (2001) Habitat requirements of wild boars in the northern Apennines (N Italy): a multi-level approach. Ital J Zool 68:47–55
Meriggi A, Milanesi P, Brangi A, Lamberti P (2010) Indagine sui danneggiamenti da cinghiale (Sus scrofa) nel Parco Nazionale dell'Arcipelago Toscano e sull'efficacia dei metodi di controllo della popolazione. Università di Pavia, 100 pp
Meriggi A, Brangi A, Schenone L, Signorelli D, Milanesi P (2011) Changes of wolf (*Canis lupus*) diet in relation to the increase of wild ungulate abundance. Ethol Ecol Evol 23:1–17
Merli E, Meriggi A (2006) Using harvest data to predict habitat-population relationship of the wild boar *Sus scrofa* in Northern Italy. Acta Theriol 51:383–394
Mitchell B, Balogh S (2007) Monitoring techniques for vertebrate pests—feral pigs. NSW Department of Agriculture, Orange
Monaco A, Carnevali L, Riga F, Toso S (2006) Il cinghiale sull'arco alpino: status e gestione delle popolazioni. Centro di Ecologia Alpina 38:5–23
Monaco A, Carnevali L, Toso S (2010) Linee guida per la gestione del cinghiale (Sus scrofa) nelle aree protette. Quaderni di Conservazione della Natura 34, Ministero dell'Ambiente—ISPRA
Monterroso P, Brito JC, Ferreras P, Alves PC (2009) Spatial ecology of the European wildcat in a Mediterranean ecosystem: dealing with small radio-tracking datasets in species conservation. J Zool 279:27–35
Novosel H, Piria M, Safner R, Kutnjak H, Sprem N (2012) The game damages on agricultural crops in Croatia. J Cent Eur Agric 13:631–642
Pearce J, Ferrier S (2000) Evaluating the predictive performance of habitat models developed using logistic regression. Ecol Model 133:225–245
Phillips SJ, Anderson RP, Schapire RE (2006) Maximum entropy modeling of species geographic distributions. Ecol Model 190:231–259
Puerta-Piñero C, Pino J, Gómez JM (2012) Direct and indirect landscape effects on *Quercus ilex* regeneration in heterogeneous environments. Oecologia 170:1009–1020
Reidy MM, Campbell TA, Hewitt DG (2008) Evaluation of electric fencing to inhibit feral pig movements. J Wildl Manag 72:1012–1018
Richards SA, Whittingham MJ, Stephens PA (2011) Model selection and model averaging in behavioural ecology: the utility of the IT-AIC framework. Behav Ecol Sociobiol 65:77–89
Ricker WE (1954) Stock and recruitment. J Fish Res Board Can 11:559–623
Riga F, Genghini M, Cascone C, Di Luzio P (2011) Impatto degli ungulati sulle colture agricole e forestali: proposta per linee guida nazionali. Manuali e Linee Guida ISPRA 68/2011
Santilli F, Mazzoni Della Stella R (2006) Electrical fencing of large farmland area to reduce crop damages by wild boars *Sus scrofa*. Agr Med 136:79–84
Santos X, Brito JC, Sillero N, Pleguezuelos JM, Llorente GA, Fahd S, Parellada X (2006) Inferring habitat-suitability areas with ecological modelling techniques and GIS: a contribution to assess the conservation status of *Vipera latastei*. Biol Conserv 130:416–425
Sarasa M, Sarasa J-A (2013) Intensive monitoring suggests population oscillations and migration in wild boar *Sus scrofa* in the Pyrenees. Anim Biodivers Conserv 36:79–88
Schley L, Roper T (2003) Diet of wild boar *Sus scrofa* in Western Europe, with particular reference to consumption of agricultural crops. Mammal Rev 33:43–56
Schley L, Dufrene M, Krier A, Frantz AC (2008) Patterns of crop damage by wild boar (*Sus scrofa*) in Luxembourg over a 10-year period. Eur J Wildl Res 54:589–599
Servanty S, Gaillard J-M, Toïgo C, Brandt S, Baubet E (2009) Pulsed resources and climate-induced variation in the reproductive traits of wild boar under high hunting pressure. J Anim Ecol 78:1278–1290

Servanty S, Gaillard J-M, Ronchi F, Focardi S, Baubet E, Gimenez O (2011) Influence of harvesting pressure on demographic tactics: implications for wildlife management. J Appl Ecol 48:835–843

Siren A, Hambäck P, Machoa J (2004) Including spatial heterogeneity and animal dispersal when evaluating hunting: a model analysis and empirical assessment in an Amazonian community. Conserv Biol 18:1315–1329

Swets KA (1988) Measuring the accuracy of diagnostic systems. Science 240:1285–1293

Thuiller W, Lafourcade B, Engler R, Araujo MB (2009) BIOMOD—a platform for ensemble forecasting of species distributions. Ecography 32:369–373

Thurfjell H, Ball JP, Åhlén PA, Kornacher P, Dettki H, Sjöberg K (2009) Habitat use and spatial patterns of wild boar *Sus scrofa* (L.): agricultural fields and edges. Eur J Wildl Res 55:517–523

Ward DF (2007) Modelling the potential geographic distribution of invasive ant species in New Zealand. Biol Invasions 9:723–735

Chapter 12
Effects of the Recent World Invasion by Ring-Necked Parakeets *Psittacula krameri*

Mattia Menchetti, Emiliano Mori, and Francesco Maria Angelici

Introduction

One of the main components of human-induced global change is represented by the translocation of species from their native ranges to alien environments, where they may exert significant damages (IUCN 2000; Mack et al. 2000; Genovesi and Shine 2004; Wonham 2006). Throughout the history of life on Earth, geographic isolation facilitated the diversification of animal and plant taxa (Sanmartin et al. 2001; Dirzo and Raven 2003). One of the main drivers of evolutionary patterns is given by slow processes of natural dispersal and colonization of new geographic areas. Since the beginning of the Holocene, humans began a voluntary or involuntary transport of species (Vitousek et al. 1997; Hulme 2009; Ellis 2011). The number of these species has increased dramatically over the last 200 years, together with the expansion of the rate of human migration and trade (Mack et al. 2000; Hulme 2009). Three main introduction pathways are currently identified: (1) accidental escapes from captivity, (2) accidental introductions through trade and tourist routes and (3) intentional introductions for hunting, ornamental purposes or individuals released by 'animal rights' groups, and biological control (Pimentel 2002; Hulme 2009).

M. Menchetti (✉)
Department of Biology, University of Florence,
Via Madonna del Piano 6, Sesto Fiorentino (Florence) 50019, Italy
e-mail: mattiamen@gmail.com

E. Mori
Di.S.A.F.A., Entomology and Zoology, University of Turin,
Largo Paolo Braccini, 2, Grugliasco (Turin) 10095, Italy
e-mail: moriemiliano@tiscali.it

F.M. Angelici
Fondazione Italiana per la Zoologia dei Vertebrati (FIZV), Via Cleonia 30,
Scala C, Rome 00152, Italy
e-mail: francescomariaangelici@gmail.com

F.M. Angelici (ed.), *Problematic Wildlife*, DOI 10.1007/978-3-319-22246-2_12

Many species kept as pets, as attractions in urban parks, and for fur farms have escaped from captivity, occasionally establishing self-sustainable populations (Reino and Silva 1996; Duncan et al. 2003). Birds are remarkably recurrent among all introduced animals, with more than 1400 attempts to introduce 400 species recorded worldwide (Lever 1987; Lockwood 1999; Duncan et al. 2003). Among those, parrots are one of the most popular groups because of their colourful plumage and for collecting purposes (Cassey et al. 2004; Menchetti and Mori 2014). It has been estimated that about two-thirds of all parrot species are commonly transported outside their natural range (Cassey et al. 2004; Menchetti and Mori 2014), corresponding to approximately four million parrots per year taken from the wild to sustain the pet market, with an annual industry of 1.4 billion dollars (Drews 2001; Mori et al. 2013a). Moreover, many species are decreasing even inside their natural range because of continuous capture and nest robbing (Drews 2001; Cassey et al. 2004). More than 16 % of living Psittaciformes species (60 out of 335 species) has currently established exotic breeding populations (Menchetti and Mori 2014). Identifying those species that have a higher predisposition to establish themselves in a territory and become invasive (Daehler and Strong 1993; Duncan et al. 2003; Blackburn et al. 2009) is a challenging task. Parrots have all of the features that increase bird establishment success, e.g. wide ecological tolerance, highly synanthropic behaviour, high number of individuals traded (Duncan et al. 2003; Shwartz et al. 2008; Menchetti and Mori 2014). Thus, they often have a high probability of developing self-maintaining populations, even starting from a few released individuals (Cassey et al. 2004).

The Ring-Necked Parakeet *Psittacula krameri*

The ring-necked parakeet *P. krameri* (Fig. 12.1) is a widely distributed gregarious parrot, naturally distributed in tropical Sub-Saharan Africa and in the Indian subcontinent (Fig. 12.2), with four subspecies recognized on morphological and geographical bases (Cramp 1985; Del Hoyo et al. 1997; Juniper and Parr 1998: Fig. 12.2):

- *P. k. krameri*: African ring-necked parakeet. This subspecies is generally pale green, with black chin and continuing black across lower cheek. Collar on hind neck is pinkish. The tail is darker than the rest of the body, with central feathers blue and yellow-green tipped. The upper mandible is red with a black tip; the lower is blackish-red. Black neck and cheeks are not present in the female as well as the collar. Tail feathers are longer in males than in females. It is naturally distributed from Western Africa (Senegal, Guinea, Southern Mauritania) to Uganda and Southern Sudan
- *P. k. parvirostris*: Abyssinian ring-necked parakeet. Head and cheek are less yellowish, and the bill smaller than in the nominate subspecies. The upper mandible is bright red, although less blackish towards the tip with respect to *P. k. krameri*. It is distributed in Southern Sudan, Eritrea, Ethiopia and Djibouti

Fig. 12.1 A male ring-necked parakeet in an urban park in Rome (Italy)

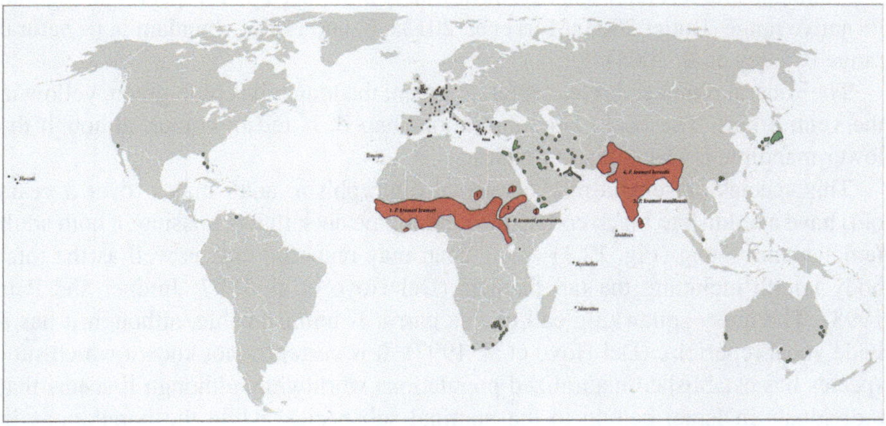

Fig. 12.2 Global distribution of the ring-necked parakeet (Del Hoyo et al. 1997; Juniper and Parr 1998; Latsoudis 2007; DAISIE 2008). In *red*, native range: subspecies are reported. In *green*, introduced range. Distribution of the alien population has been implemented with occurrences from the Alien Parrot Observatory (http://www.inaturalist.org/projects/alien-parrots-observatory)

- *P. k. borealis*: Boreal ring-necked parakeet or Neumann's ring-necked parakeet. This subspecies is larger than *P. k. krameri*, with wholly red upper mandible and black markings on lower mandible. It is naturally present in Pakistan. Northern India, Bangladesh, Nepal and Myanmar
- *P. k. manillensis*: Indian ring-necked parakeet. It is the largest subspecies, slightly paler and yellower than *P. k. borealis*. Its distribution range includes Southern India and Sri Lanka

Fig. 12.3 An adult couple of *P. krameri* in an urban habitat. Male on the *left*, female on the *right*

Psittacula krameri represents the most widely introduced parrot species throughout its native range (Butler 2003; Mori et al. 2013a, b) and is also abundant in its natural range (Cassey et al. 2004).

The general plumage is clear emerald green; the long tail is blue-green, yellow in the ventral part. The beak, large and hook-shaped, is red in colour, although the lower mandible is black in adult males.

This species presents a marked sexual dimorphism: adult males (over 3 years old) have a reddish to black collar surrounding the neck that is missing in both adult females and young (Fig. 12.3). Wing span may reach 40 cm, as well as the total body length including the tail feathers (Del Hoyo et al. 1997; Juniper and Parr 1998). The noisy squawking call of this parrot is unmistakable, although it has a wide vocal repertoire (Del Hoyo et al. 1997). It is currently not known which subspecies has established naturalized populations worldwide, although it seems that individuals in Israel belong to the nominal subspecies, while the populations in other countries belong to *P. k. manillensis* (Del Hoyo et al. 1997; Juniper and Parr 1998; Scalera 2001; BirdLife International 2012). Furthermore, it seems that ring-necked parakeets in the UK probably stem from both *P. k. borealis* and *P. k. manillensis*, based on morphological measurements (Pithon and Dytham 2001).

The ring-necked parakeet may be found in a variety of habitat types, including woodlands, urban parks and cultivated areas surrounded by trees, from 0 to 1600–2000 m a.s.l. It is considered one of the few parrot species successfully adapted to living in anthropic habitats, having withstood the onslaught of urbanization and deforestation. According to this consideration, it may be considered as a habitat-generalist, although it depends on trunk cavities for reproduction (Del Hoyo et al. 1997; Khan et al. 2004). Breeding period is related to the latitude, from November to June in Asia, from August to November in Africa (Del Hoyo et al. 1997; Scalera 2001). In Europe, breeding period is set up in late winter-early spring (Butler et al. 2013). It is a resident, herbivorous parrot, which feeds mainly on buds, fruits, vegetables, blossoms and seeds. In their native range, ring-necked parakeets may also fly several miles in large flocks to forage in crops and orchards causing damages (Khan et al. 2004; Ahmad et al. 2011). As a matter of fact, the ring-necked parakeet is one of the worst pest species of the Indian subcontinent (Dhindsa and Saini 1994); corn (losses up to 81 %: Reddy 1998a), sorghum (losses up to 74 %: Reddy 1998b) and sunflowers constitute the staple of its diet in summer, fruits in winter (Rao and Shivanarayan 1981; Luft 1994).

Ring-necked parakeets are among the most popular pets with the pet trade luckily keeping the native populations lower, thus limiting damages (Scalera 2001). In its native range, many individuals are still caught from natural environments despite widely practiced breeding in captivity. In Northern India, according to Scalera (2001), up to 5000 ring-necked parakeets are caught per day and sold to wholesale at less than a dollar each. The species is not included in any CITES Appendices, with the only exception being the Ghana population, listed in Appendix III (BirdLife International 2012), although the species in this country cannot be considered as rare (Borrow and Demey 2010).

As a consequence to this wide trade, since the 1970s, escaped individuals have colonized a number of cities in many European (Portugal, Spain, France, Switzerland, Italy, Austria, Germany, Belgium, the Netherlands and Greece), Asian (Turkey, Israel, Syria, Iraq, Iran, Saudi Arabia, Kuwait, Bahrein, Qatar, Oman, Yemen, Maldives, Singapore, Macau, Hong Kong, China and Japan), African (Algeria, Egypt, Kenya, Mauritius, Seychelles and South Africa), North American (USA: California, Florida, Hawaii), South American (Venezuela) and Oceanian (Australia) countries (Del Hoyo et al. 1997; Nebot 1999; Leven and Corlett 2004; Fellous et al. 2005; BirdLife International 2012; Fig. 12.1). No breeding success has yet been recorded in Romania (Liviu Parau, *in litteris*), Poland (Piotr Trijanowski, *in litteris*) and Bulgaria (Boris Nikolov, *in litteris*), although some individuals have been observed in these countries. Physiological tests suggested that the ring-necked parakeet has peculiar seasonal thermoregulatory responses that make it tolerant to low environmental temperatures, allowing it to better cope with a wide range of climatic conditions, and thus explaining its success as an invader species (Thabethe et al. 2013). Such a population increase brought the IUCN to classify this species as least concern (LC) (BirdLife International 2015), and its populations, as well as the extension of natural geographical range, appear to be rising in their native range, both in Africa and in Asia, probably associated with the increase in intensive agricultural crops (Juniper and Parr 1998).

To summarize, the ring-necked parakeet is currently considered the most effective parrot species to colonize new territories where it has been historically absent (Del Hoyo et al. 1997; Menchetti and Mori 2014).

Impacts of the Ring-Necked Parakeet

All introduced species may experience a 'lag period' (i.e. up to some decades) after their release (e.g. Zocchi and Panella 1978; Angelici 1986), and then increase exponentially in both range and population size (Keikl 2001). The ring-necked parakeets took higher latitudes almost by surprise quickly establishing in a high number of areas outside their native range. Such increased amounts of breeding colonies have begun to exert serious problems to native environments, wildlife and humans (Menchetti and Mori 2014, for a review).

Impacts on Native Fauna and Flora

Although the ring-necked parakeet is considered a non-territorial species (Del Hoyo et al. 1997, during the breeding period, it displays aggressive behaviours against other birds which fly in the surrounding of its nests. Noisy and physical intimidations, as well as direct killings, have been observed against raptors (e.g. *Falco tinnunculus*, *Athene noctua*) and corvids (UK: Cramp 1985; France: Dubois 2007; Italy and Algeria: Menchetti and Mori 2014). Flocks of ring-necked parakeets may also mob larger birds (e.g. seagulls, herons: Dubois 2007). Feeding mostly on unripe fruits and seeds, the ring-necked parakeet may take food before other species consume it, thus enhancing competition with native birds (Fletcher and Askew 2007; Lin Neo 2012). Furthermore, the presence of ring-necked parakeet significantly reduced feeding rates of native birds and increased their vigilance (Clergeau and Vergnes 2011; Peck et al. 2014). Trunk cavities are used by a variety of species for nesting and roosting purposes (Newton 1994; Cornelius et al. 2008; Hernández-Brito et al. 2014a), thus eliciting competition between introduced parrots and native hole-nester birds. Parakeets seem not to compete with native woodpeckers (Strubbe et al. 2009; Newson et al. 2011; Orchan et al. 2012), although harassments may be rarely recorded (Keikl 2001). Woodpeckers are primary cavity nesters; it means that they excavate new holes at each reproduction event. By contrast, ring-necked parakeets rely on existing cavities: so they are classified as secondary cavity nesters, as well as tits, nuthatches and doves. Moreover, woodpeckers tend to nest at lower altitudes on trunks with respect to parakeets (Strubbe et al. 2009; Newson et al. 2011; Orchan et al. 2012). Although ring-necked parakeets are advantaged in competition, being early breeders (Cramp 1985), antagonistic behaviours by ring-necked parakeets are displayed against secondary hole-nesters which select cavities larger than those used by parakeets (Fletcher and Askew 2007; Orchan et al. 2012; Hernandez-Brito et al. 2014). In such cases, displacements and/or nest destructions by parakeets were observed in Europe (Cramp 1985; Strubbe and Matthysen 2007; Czajka et al. 2011; Hernandez-Brito et al. 2014), Venezuela (Nebot 1999) and Israel (Shwartz et al. 2008). Displacement behaviour by introduced ring-necked parakeets is particularly noteworthy if exerted against native, threatened parrots in oceanic islands, such as the endemic *Psittacula eques* in Mauritius (Jones 1980), *Coracopsis nigra* in Mahé Island (Fanchette 2012), *Calyptorhynchus latirostris* in Oceania (Chapman 2005) and *Cyanoramphus cooki* in Norfolk Island (Lever 2005). Despite differences in body size necessarily reflecting different cavity preferences (Martin et al. 2004; Fletcher and Askew 2007; Orchan et al. 2012), different diameters of hole-nests do not exclude possible competition with native species (e.g. European starlings *Sturnus vulgaris*: Braun et al. 2009). An experimental study showed that ring-necked parakeets may also compete with the nuthatch *Sitta europaea* (Strubbe et al. 2009). This native species may choose large cavities, idoneous for breeding parakeets, and then apply mud to reduce their size (Cramp 1985). Newson et al. (2011) claimed that this competition phenomenon disappears when the degree of urbanization is taken into account. Furthermore, despite being secondary cavity nesters, ring-necked parakeets can enlarge smaller tree holes with their strong beaks to better satisfy their needs for nesting purposes. In Israel, cavity enlargement by ring-necked parakeets improves the breeding success of an alien invasive species, the common myna *Acridotheres tristis* (Orchan et al. 2012). Trunk cavities may also provide parakeets with the occasion to outcompete other native species. Gebhardt (1996) claimed that ring-necked parakeets may displace dormice and bats (*Myotis* sp.) from tree-holes and nest-boxes, without providing further details; Haarsma and Van der Graaf (2013) reported a number of anecdotal observations of dead or weakened bats close to trees used by ring-necked parakeets. In Central Italy, an individual of Leisler's bat *Nyctalus leisleri* roosting inside a cavity had been killed by a pair of ring-necked parakeets who tried to build their nest inside the cavity (Menchetti et al. 2014). A similar behaviour has been recorded against honeybees *Apis mellifera* in Germany (Menchetti and Mori 2014). Such behaviours are only rarely observed because they usually happen in high cavities and possibly at the immediate start of the breeding period. As a result, spatial segregation between alien parrots and native species may take place. Populations of the threatened giant noctule bat *Nyctalus lasiopterus* decreased in an urban park in Sevilla in parallel to the increase in the ring-necked parakeet population (Hernández-Brito et al. 2014a). Active displacement by the parakeets is suggested to occur, triggering conservation problems for this rare bat species (Hernández-Brito et al. 2014a). Ring-necked parakeets may harass and push away European squirrels, only rarely touching them. In France, a group of 3–4 ring-necked parakeets was observed once attacking and killing a European red squirrel *Sciurus vulgaris* (Japiot 2005; Menchetti and Mori 2014). Squirrels are listed among the main predators of the chicks of ring-necked parakeets in Europe (Mori et al. 2013b), so these attacks may be related to a nest defence behaviour by the birds (Philippe Clergeau, *in litteris*). As well, direct killings of black rats *Rattus rattus* have been reported by breeding parakeets around nests in Spain (Hernández-Brito et al. 2014b). Experimental evidence is still required to support the hypothesis that ring-necked parakeets reduce the impact of some invasive predators on native fauna. For instance, Shwartz et al. (2008) observed grey squirrels *Sciurus carolinensis* preying on nests of ring-necked parakeets in the UK. Figure 12.4 summarizes the species impacted by introduced ring-necked parakeets (birds, mammals and insects).

The effect on native flora by introduced ring-necked parakeets has been poorly studied. The role of parrots in seed dispersion is still debated (Wunderle 1997; Norconk et al. 1998; Loope et al. 2001), although the involvement of some parrot species in spreading weeds or alien plants in new environments has been hypothesized (see Menchetti and Mori 2014 for a review). Although Tella et al. (in press) suggested that *P. krameri* may act as seed disperser on long distances for many alien

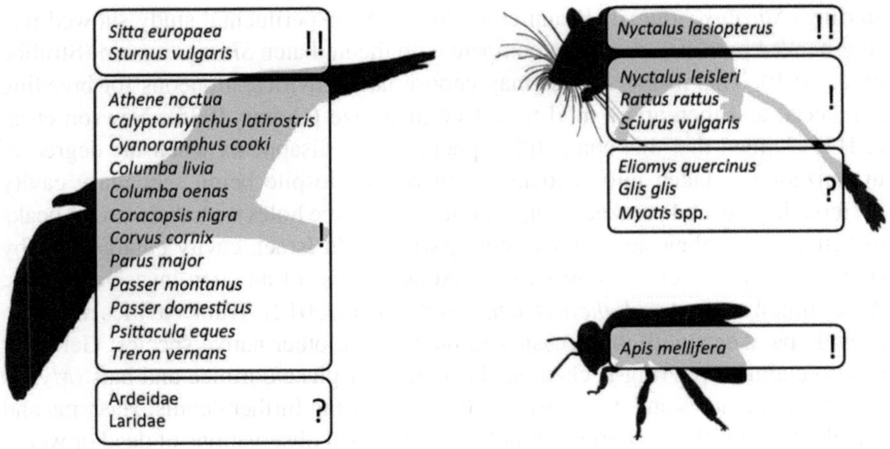

Fig. 12.4 Summary of species impacted by ring-necked parakeet within the introduced range: !!, experimental studies; !, anecdotal observations; ?, suggested impacts

plants, quantitative studies on its seed-dispersal ability in the introduced range are lacking and would be beneficial.

By contrast, droppings by ring-necked parakeet under their roosts may have altered the herbaceous vegetation composition in Southern England, but a direct cause–effect relationship cannot be assessed as no data are available about floral composition before the invasion of parakeets (Fletcher and Askew 2007). Furthermore, defoliation of ornamental trees, resulting in health problems for the plant, may occur (Van Kleunen et al. 2010).

Impacts on Human Activities

In the native range, damages by ring-necked parakeets on crops/orchards, greenhouses as well as on human facilities have been observed. As for the introduced ranges, records are locally significant and increasing in the grey literature (e.g. agriculture bulletins, unpublished technical reports), but damages are rarely quantified (Butler 2003; Bauer and Woog 2008). Orchard damage by ring-necked parakeets has been recorded since the 1950s in the UK (Yealland 1958), and the cost of vineyard damages in Surrey is estimated at about £5000 per year (Fletcher and Askew 2007). In Germany, 10–15 % of apples were damaged in 0.8 ha large orchards, mainly high in the tree (Van Kleunen et al. 2010). On the contrary, crop damages are cited for many European countries, but without any quantification (Spanò and Truffi 1986; Andreotti et al. 2001; Associazione Faunisti Veneti 2004; Vidal Rodriguez 2004; Dubois 2007). Orchards and cultivated fields (corn, vine, *Hordeum* spp., *Pisum sativum*, *Pistacia vera*: Borgo et al. 2005; Latsoudis 2007; Tayleur 2010) are often selected as feeding sites, even when other plants are present within the same study area (Fletcher and Askew 2007). Barns may also be invaded and grain bags torn by the ring-necked parakeets (Andreotti et al. 2001). While most ring-necked parakeet colonies located in the surroundings of airports never cause problems (e.g. Montemaggiori 1998), three birdstrikes involved this parakeet species in Heathrow Airport (England) from 2004 to 2005, with an average cost of ca. £20,000 each (Fletcher and Askew 2007).

Health Impacts

Parakeets are reservoirs of a plethora of bacterial and viral diseases (Fletcher and Askew 2007; Menchetti and Mori 2014). Thus, free-ranging alien populations may threaten the fitness of native wild species, as well as aviculture and human health. Diseases are transmitted by direct contact with infected birds (handling of plumage and tissues), contaminated equipment, droppings and aerosols of the secretions through nostrils, mouth and eyes. Ring-necked parakeets are known reservoirs of chlamydiosis and other diseases (Menchetti and Mori 2014). Psittacine beak and feather disease was detected in two wild ring-necked parakeets in England through PCR and histopathological examination of affected skin, and this disease could pose a threat to wild birds and captive psittacines (Sa et al. 2014). In 1997–1998, the influenza virus H9N2 was isolated from the respiratory organs of two ring-necked parakeets imported from Pakistan to Japan (Mase et al. 2001). Diseases carried by the ring-necked parakeet may harm the indigenous Vasa parrot *Coracopsis nigra* in Mahé Island (Seychelles: Fanchette 2012). Roost sites of ring-necked parakeet can cause noise pollution (Van Kleunen et al. 2010).

Positive or Neutral Impacts/Interactions

Enlargement of cavities by ring-necked parakeets seems to increase the breeding chances of native species, e.g. *Columba oenas* (Czajka et al. 2011). The general attitude of the public towards ring-necked parakeets is positive. Most people find that parakeets bring colour to urban environments or enjoy observing them flying in the sky (Menchetti and Mori 2014).

Neutral to positive interactions were observed with some coexisting species, e.g. the Alexandrine parakeet *Psittacula eupatria* (Khalegizadeh 2004; Van Kleunen et al. 2010; Weiserbs 2010; Angelici and Fiorillo in press). The few feral individuals of *P. eupatria* in Rome (Italy) are living almost constantly with the most numerous *P. krameri*, forming heterospecific flocks (Angelici and Fiorillo in press). A similar form of apparent mutualism has been observed between *P. krameri* and the Barraband's parakeet *Polytelis swainsonii* in Southern Tuscany (Central Italy). Interactions with other introduced parrot species, e.g. the monk parakeet, were never recorded: even where in syntopy, *P. krameri* and *M. monachus* seem to be mutually exclusive and never observed in heterospecific flocks or in the same micro-areas (Alessandro Fiorillo, *in verbis*).

Eradication and Numerical Control

Eradication and numerical control of alien/invasive species are always expensive procedures. Furthermore, lethal control against feral/escaped pets is unpopular and may result in public protests (Menchetti and Mori 2014), mainly if these operations regard charismatic species. In detail, common people enjoy the sight of bright parrots flying in courtyards or urban parks (Lever 1987; Polkanov and Greene 2000; Van Kleunen et al. 2010). Thus, eradication programmes against alien Psittaciformes have been started and interrupted several times in North America and the Oceanic islands, such that populations rapidly grew (Lever 1987; Van Bael and Pruett-Jones 1996). In the native range, the use of ultrasonic sound players seems to be an effective way to reduce crop damages and economic losses (Khan et al. 2013). Application of chemosterilants to reduce fitness has been considered, but no data on the effectiveness of this technique is available yet (Lambert et al. 2009). Strategies to prevent new escapes and establishments of self-maintaining populations should also include public education and breeding controls in aviaries (cf. Chapman 2005). From a feasibility point of view, eradication of small populations is still possible, but requires logistically and economically intense efforts. Strict controls are necessary to prevent future invasions and monitoring population increases within adaptive management programmes is more highly recommended (Genovesi and Shine 2004).

Conclusions

According to scientific and grey literature, the ring-necked parakeet appears to be a species characterized by a high ecological success, currently undergoing a range expansion both in its native range and in many areas of the world (Juniper and Parr 1998), where it has been voluntarily introduced or escaped from captivity (BirdLife International 2012). This is possible due to the plasticity of this parrot species, within increase its natural range favoured by anthropization, habitat alteration, an increase of intensive agriculture (Juniper and Parr 1998) and perhaps by climate change, tending towards a gradual increase in mean global temperatures (cf. Huntley et al. 2006).

Acknowledgements Authors would like to thank Philippe Clergeau, Boris Nikolov, Liviu Parau and Piotr Trijanowski for the precious information provided. Dylan Beal revised the first draft of this manuscript and improved English. Alessandro Fiorillo has kindly provided new information and some of his original photographs (Figs. 12.1, and 12.3). Sabrina Kumschick and Diederik Strubbe kindly provided useful comments and improvements on the manuscript.

References

Ahmad S, Khan HA, Javed M (2011) Roost composition and damage assessment of ring-necked parakeet (*Psittacula krameri*) on maize and sunflower in agro-ecosystem of Central Punjab, Pakistan. Int J Agric Biol 13:731–736

Andreotti A, Baccetti N, Perfetti A, Besa M, Genovesi P, Guberti V (eds) (2001) Mammiferi ed Uccelli esotici in Italia: analisi del fenomeno, impatto sulla biodiversità e linee guida gestionali. Ministero dell'Ambiente e Istituto Nazionale per la Fauna Selvatica 'A. Ghigi', Ed. Quaderni di Conservazione della Natura 2. Ministero dell'Ambiente, Istituto Nazionale Fauna Selvatica, Rome

Angelici FM (1986) Uccelli esotici in libertà nella città di Roma. Gli Uccelli d'Italia 11:46–50

Angelici FM, Fiorillo A (in press) Repeated sightings of Alexandrine parakeet (*Psittacula eupatria*) in Rome (Central Italy) and its likely acclimatization. Riv Ital Ornitol (in press)

Associazione Faunisti Veneti (2004) Rapporto ornitologico per la regione Veneto. Anno 2003. Boll Mus Civ Stor Nat Venezia 55:171–200

Bauer HG, Woog F (2008) Nichtheimische Vogelarten (Neozoen) in Deutschland, Teil I: Auftreten, Bestände und Status. Vogelwarte 46:157–194

BirdLife International (2012) *Psittacula krameri*. The IUCN red list of threatened species. Version 2014.3. www.iucnredlist.org. Accessed 16 Feb 2015

BirdLife International (2015) *Psittacula krameri*. IUCN red list for birds. http://www.birdlife.org. Accessed 7 Jan 2015

Blackburn T, Lockwood J, Cassey P (2009) Avian invasions: the ecology and evolution of exotic birds. Oxford University Press, Oxford

Borgo E, Galli L, Galuppo C, Maranini N, Spanò S (eds) (2005) Atlante ornitologico della città di Genova (1996–2000). Boll Mus Istit Biol Univ Genova, Genova

Borrow N, Demey R (eds) (2010) Birds of Ghana. Cristopher Helm, London

Braun M, Czajka C, Wink M (2009) Gibt es eine Brutplatzkonkurrenz zwischen Star und Halsbandsittich? Vogelwarte 47:361–362

Butler CJ (2003) Population biology of the introduced ring-necked parakeet *Psittacula krameri* in UK. University of Oxford Dissertation, Edward Gray Institute of Ornithology, Department of Zoology, Oxford

Butler CJ, Cresswell W, Gosler A, Perrins C (2013) The breeding biology of ring-necked parakeets *Psittacula krameri* in England during a period of rapid population expansion. Bird Study 60:527–532

Cassey P, Blackburn TM, Russel GJ, Jones KE, Lockwood JL (2004) Influences on the transport and establishment of exotic bird species: an analysis of the parrots (Psittaciformes) of the world. Glob Chang Biol 10:417–426

Chapman T (2005) The status and impact of the rainbow lorikeet (*Trichoglossus haematodus moluccanus*) in South-Western Australia. Miscellaneous Publication, Department of Agriculture, Government of Western Australia

Clergeau P, Vergnes A (2011) Bird feeders may sustain feral Ring-necked parakeets *Psittacula krameri* in temperate Europe. Wildl Biol 17:248–252

Cornelius C, Cockle K, Politi N, Berkunsky I, Sandoval L, Ojeda V, Rivera L, Hunter M, Martin K (2008) Cavity-nesting birds in neotropical forests: cavities as a potentially limiting resource. Ornitol Neotrop 19:253–268

Cramp S (ed) (1985) Handbook of the birds of Europe, the Middle East and North Africa. The birds of the Western Palearctic. Volume IV. Terns to woodpeckers. Oxford University Press, New York

Czajka C, Braun MP, Wink M (2011) Resource use by non-native ring-necked parakeets (*Psittacula krameri*) and native starlings (*Sturnus vulgaris*) in central Europe. Open Ornithol J 4:17–22

Daehler CC, Strong DR Jr (1993) Prediction and biological invasions. Trends Ecol Evol 8:380–381

Del Hoyo J, Elliott A, Sargatal J (eds) (1997) Handbook of the birds of the world. Sangrouse to cuckoos, vol 4. BirdLife International, Cambridge

Dhindsa MS, Saini HK (1994) Agricultural ornithology: an Indian perspective. J Biosci 19:391–402

Dirzo R, Raven PH (2003) Global state of biodiversity and loss. Annu Rev Environ Resour 28:137–167

Drews C (2001) Wild animals and other pets kept in Costa Rican household: Incidence, species and numbers. Soc Anim 9:107–126

Dubois PJ (2007) Les oiseux allochtones en France: status et interactions avec les espèces indigènes. Ornithos 14:329–364

Duncan RP, Blackburn TM, Sol D (2003) The ecology of bird introductions. Annu Rev Ecol Evol Syst 34:71–98

Ellis EC (2011) Anthropogenic transformation of the terrestrial biosphere. Philos Trans R Soc A 369:1010–1035

DAISIE European Invasive Alien Species Gateway (2008) *Psittacula krameri*. http://www.europe-aliens.org. Accessed 15 Feb 2015

Fanchette R (2012) Invasive alien species: threats to biodiversity and human well-being. Seychelles strategy. In: IAS workshop IUCN, 23–27th January 2012, Mayotte

Fellous A, Moulai R, Jacob JP (2005) Introduction et nidification de la Perruche à collier *Psittacula krameri* en Algérie. Aves 42:272–277

Fletcher M, Askew N (2007) Review of the status, ecology and likely future spread of parakeets in England. CSL, York, England. http://archive.defra.gov.uk/wildlife-pets/wild life/management/non-native/documents/csl-parakeet-deskstudy.pdf. Accessed 15 July 2013

Gebhardt H (1996) Ecological and economics consequences of introductions of exotic wildlife (birds and mammals) in Germany. Wildl Biol 2:205–211

Genovesi P, Shine C (2004) European strategy on invasive alien species, final version. Convention on the conservation of European wildlife and natural habitats. Council of Europe, Strasbourg

Haarsma AJ, Van der Graaf C (2013) Halsbandparkieten een bedreiging voor Rosse vleermuizen. Levende Nat 114:10–13

Hernández-Brito D, Carrete M, Popa-Lisseanu AG, Ibañez C, Tella JL (2014a) Crowding in the city: losing and winning competitors of an invasive bird. PLoS One 9, e100593

Hernández-Brito D, Luna A, Carrete M, Tella JL (2014b) Alien rose-ringed parakeets (*Psittacula krameri*) attack black rats (*Rattus rattus*) sometimes resulting in death. Hystrix Ital J Mammal 25:121–123. doi:10.4404/hystrix-25.2-10992

Hulme PE (2009) Trade, transport and trouble: managing invasive species pathways in an era of globalization. J Appl Ecol 46:10–18

Huntley B, Collingham YC, Green RE, Hilton GM, Rahbek C, Willis SG (2006) Potential impacts of climatic change upon geographical distributions of birds. Ibis 148(Suppl s1):8–28

IUCN (2000) IUCN guidelines for the prevention of biodiversity loss caused by alien invasive species. In: Approved by the 51st meeting of the IUCN Council, Gland, Switzerland

Japiot X (eds) (2005) Psittacides en villes d'Europe. Mairie de Paris, Direction des Parcs, Jardin and Espaces Verts, Service de l'Ecologie Urbaine, Section Etudes et Prospectives Environnementales, Pole Biodiversité, Paris

Jones CG (1980) Parrot on the way to extinction. Oryx 15:350–354

Juniper T, Parr M (1998) Parrots: a guide to parrots of the world. Christopher Helm, London

Keikl GO (2001) Halsbandparkieten *Psittacula krameri* in Amsterdam, 1976–2000. Limosa 74:29–31

Khalegizadeh A (2004) On the diet and population of the Alexandrine parakeet, *Psittacula eupatria*, in the urban environment of Tehran, Iran. Zool Middle East 32:27–32

Khan HA, Beg MA, Khan AA (2004) Breeding habitats of the Ring-necked Parakeet (*Psittacula krameri*) in the cultivations of central Punjab. Pak J Zool 36:133–138

Khan HA, Javed M, Tahir A, Kanwal M (2013) Limiting the ring-necked parakeet (*Psittacula krameri*) damage on guava (*Psidium guajava*) and mango (*Mangifera indica*) with an ultrasonic sound player in a farmland in Faisalabad, Pakistan. Afr J Agric Res 8:6608–6614

Lambert MS, Massei G, Bell J, Berry L, Haigh C, Cowan DP (2009) Reproductive success of rose-ringed parakeets *Psittacula krameri* in a captive UK population. Pest Manag Sci 65:1215–1218
Latsoudis P (2007) Preliminary distribution study of the Ring-necked Parakeet *Psittacula krameri* (Scopoli 1769) in Greece. http://www.ornithologiki.gr/page_cn.php?aID=929. Accessed 15 July 2013
Leven MR, Corlett RT (2004) Invasive birds in Hong Kong, China. Ornithol Sci 3:43–55
Lever C (ed) (1987) Naturalized birds of the world. Longman, Harlow
Lever C (2005) Naturalized birds of the world. Poyser, London
Lin Neo M (2012) A review of three alien parrots in Singapore. Nat Singapore 5:241–248
Lockwood JL (1999) Using taxonomy to predict success among introduced avifauna: relative importance of transport and establishment. Conserv Biol 13:560–567
Loope LL, Howarth FG, Kraus F, Pratt TK (2001) Newly emergent and future threats to pacific birds and ecosystems. Stud Av Biol 22:291–304
Luft S (1994) Die Papageien Sri Lankas. Gefiederte Welt 11:383–385
Mack RN, Simberloff D, Londsdale M, Evans H, Clout M, Bazzaz FA (2000) Biotic invasions: causes, epidemiology, global consequences, and control. Ecol Appl 10:689–710
Martin K, Aitken KEH, Wiebe KL (2004) Nest sites and nest webs for cavity-nesting communities in interior British Columbia, Canada: nest characteristics and niche partitioning. Condor 106:5–19
Mase M, Imada T, Sanada Y, Itoh E, Sanada N, Tsukamoto K, Kawaoka Y, Yamaguchi S (2001) Imported parakeets harbor H9N2 influenza A viruses that are genetically closely related to those transmitted to humans in Hong Kong. J Virol 75:3490–3494
Menchetti M, Mori E (2014) Worldwide impact of alien parrots (Aves Psittaciformes) on native biodiversity and environment: a review. Ethol Ecol Evol 26:172–194
Menchetti M, Scalera R, Mori E (2014) First record of a possibly overlooked impact by alien parrots on a bat (*Nyctalus leisleri*). Hystrix Ital J Mammal 25:61–62
Montemaggiori A (1998) The importance of bird monitoring at airports: the case of Fiumicino, Rome. IBSC24/WP17 Stara Lesna
Mori E, Di Febbraro M, Foresta M, Melis P, Romanazzi E, Notari A, Boggiano F (2013a) Assessment of the current distribution of free-living parrots and parakeets (Aves: Psittaciformes) in Italy: a synthesis of published data and new records. Ital J Zool 80:158–167
Mori E, Ancillotto L, Menchetti M, Romeo C, Ferrari N (2013b) Italian red squirrels and introduced parakeets: victims or perpetrators? Hystrix Ital J Mammal 24:195–196
Nebot JC (1999) First report on the ring-necked parakeet (*Psittacula krameri*) in Venezuela and preliminary observations on its behavior. Ornitol Neotrop 10:115–117
Newson SE, Johnston A, Parrott D, Leech DI (2011) Evaluating the population-level impact of an invasive species, ring-necked parakeet *Psittacula krameri*, on native avifauna. Ibis 153:509–516
Newton I (1994) The role of nest sites in limiting the numbers of hole-nesting birds: a review. Biol Conserv 70:265–276
Norconk MA, Grafton BW, Conklin-Brittain NL (1998) Seed dispersal by neotropical seed predators. Am J Primatol 45:103–126
Orchan Y, Chiron F, Shwartz A, Kark S (2012) The complex interaction network among multiple invasive bird species in a cavity-nesting community. Biol Invasions 15:429–445
Peck HL, Pringle HE, Marshall HH, Owens IP, Lord AM (2014) Experimental evidence of impacts of an invasive parakeet on foraging behavior of native birds. Behav Ecol 25:582–590
Pimentel D (ed) (2002) Encyclopedia of pest management. Dekker, New York
Pithon JA, Dytham C (2001) Determination of the origin of British feral Rose-ringed Parakeets. Br Birds 94:74–79
Polkanov A, Greene T (2000) Pest Psittacenes: the rainbow lorikeet in New Zealand. Eclectus 9:2–5
Rao GS, Shivanarayan N (1981) Note on the food of nestlings of ring-necked parakeets in Hyderabad. Pavo 19:97–99

Reddy VR (1998a) Bird damage to maize crop on the student's research farm at Rejendranagar, Hyderabad, Andra Pradesh. Pavo 36:77–78

Reddy VR (1998b) Studies on damages to sorghum by the ring-necked parakeet, *Psittacula krameri*, at Rejendranagar, Hyderabad, Andra Pradesh. Pavo 36:79–80

Reino LM, Silva T (1996) Distribution and expansion of the common waxbill (*Estrilda astrild*) in Portugal. In: Holmes JS, Simons JR (eds) The introduction and naturalization of birds. Stationery Office Publications Centre, London, pp 103–106

Sa RC, Cunningham AA, Dagleish MP, Wheelhouse N, Pocknell A, Borel N, Peck HL, Lawson B (2014) Psittacine beak and feather disease in a free-living ring-necked parakeet (*Psittacula krameri*) in Great Britain. Eur J Wildl Res 60:395–398

Sanmartin I, Enghoff H, Ronquist F (2001) Patterns of animal dispersal, vicariance and diversification in the Holarctic. Biol J Linn Soc 73:345–390

Scalera R (2001) Il parrocchetto dal collare *Psittacula krameri.* In: Scalera R (ed) Invasioni biologiche. Le introduzioni di vertebrati in Italia: un problema tra conservazione e globalizzazione. Corpo Forestale dello Stato e Ministero delle Politiche Agricole e Forestali, Rome, pp 195–199

Shwartz A, Strubbe D, Butler CJ, Matthysen E, Kark S (2008) The effect of enemy-release and climate conditions on invasive birds: a regional test using the ring-necked parakeet (*Psittacula krameri*) as a case study. Divers Distrib 15:310–318

Spanò S, Truffi G (1986) Il parrocchetto dal collare, *Psittacula krameri*, allo stato libero in Europa, con particolare riferimento alle presenze in Italia e primi dati sul pappagallo monaco, *Myiopsitta monachus*. Riv Ital Ornithol 56:231–239

Strubbe D, Matthysen E (2007) Invasive ring-necked parakeets *Psittacula krameri* in Belgium: habitat selection and impact on native birds. Ecography 30:578–588

Strubbe D, Matthysen E, Graham CH (2009) Assessing the potential impact of invasive ring-necked parakeets *Psittacula krameri* on native nuthatches *Sitta europaea* in Belgium. J Appl Ecol 47:549–557

Tayleur JR (2010) A comparison of the establishment, expansion and potential impacts of two introduced parakeets in the United Kingdom. BOU proceedings, the impacts of non-native species. http://www.bou.org.uk/bouproc-net/non-natives/tayleur.pdf. Accessed 15 July 2013

Tella JL, Baños-Villalba A, Hernandez-Brito D, Rojas A, Pacifico E, Diaz-Luque J, Carrete M, Blanco G, Hiraldo F (2015) Parrots as overlooked seed dispersers. Front Ecol Environ (in press)

Thabethe V, Thompson LJ, Hart LA, Brown M, Downs CT (2013) Seasonal effects on the thermoregulation of invasive ring-necked parakeets (*Psittacula krameri*). J Therm Biol 48:553–559

Van Bael S, Pruett-Jones S (1996) Exponential population growth of monk parakeets in the United States. Wils Bull 108:584–588

Van Kleunen A, Van den Bremer L, Lensink R, Wiersma P (2010) De Halsbandparkiet, Monniksparkiet en Grote Alexanderparkiet in Nederland: risicoanalyse en beheer. SOVON-onderzoeksrapport 2010/10 Dit rapport is samengesteld in opdracht van Team Invasieve Exoten van het Ministerie van Landbouw, Natuur en Voedselkwaliteit

Vidal Rodriguez C (2004) Situación das aves exóticas asilvestradas en Galicia. Chioglossa 2:79–87

Vitousek PM, D'Antonio CM, Loope LL, Rejmanek M, Westbrooks R (1997) Introduced species: a significant component of human-caused global change. N Z J Ecol 21:1–16

Weiserbs A (2010) Espèces invasives: le cas des Psittacidès en Belgique. Incidences, èvaluation des risques et ènventail de mesures. Aves 47:21–35

Wonham M (2006) Species invasions. In: Groom MJ, Meffe GK, Carroll CR (eds) Principles of conservation biology. Sinauer Associates, Sunderland, pp 209–227

Wunderle JM (1997) The role of animal seed dispersal in accelerating native forest regeneration on degraded tropical lands. For Ecol Manag 99:223–235

Yealland JJ (eds) (1958) Cage birds in colour. H.F. & G. Witherby, London

Zocchi A, Panella M (1978) Osservazioni ornitologiche nella Villa Doria Pamphili a Roma. Riv Ital Ornithol 48:253–255

Part V
Genetic Contributions to the Management of Problematic Species

An interesting topic of the utmost importance concerns the genetic aspects of problematic species. In fact, if, for example, the genetics of invasive species is analysed, their origin can be reconstructed by examining their mitochondrial DNA (e.g., Dejean et al. 2012) or by using other techniques normally used in the study of wildlife biology (Deyoung and Honeycutt 2005).

Moreover, using the genetic characterisation of populations, the appropriate conservation measures can be taken (e.g., restocking, reintroduction) to protect species in danger of extinction (Piggott and Taylor 2003) and, more generally, to comprehensively understand the processes that lead to the numerical explosion of a population (Frankham 2005) or, on the contrary, to the decline of certain species which impact or interact with humans in other ways (O'Brien 1994).

In this part, there are three chapters, one of which is general in nature and two refer to more specific cases. The first chapter (Kekkonen 2016) specifically highlights how utterly essential genetic studies are to the monitoring of the various possible problematic species, from invasive species to species in rapid decline. Some specific case studies are also presented.

The second chapter (Gaubert 2016) deals with the dissemination and invasion of mongooses (*Herpestes* spp.) and common genet (*Genetta genetta*) in the Mediterranean basin. All these species have been introduced in historical time and are generally regarded as harmful and in need of containment or eradication. The chapter looks at the history of the introduction of these species and examines in detail whether, how and to what extent these small carnivores are truly problematic, proposing possible solutions to attempt to resolve these situations.

The third chapter (Gentile et al. 2016) examines a very special, emblematic case. In 2009 a new species of iguana was described from the Galapagos Archipelago, i.e., the Galapagos pink land iguana (*Conolophus marthae*), an endemic species that is already in danger of extinction (Critically Endangered according to IUCN categories) because of its very small population, its restricted area and many other looming problems. This chapter discusses the various efforts being taken to protect this species, including the intention of making this iguana a flagship species.

References

Dejean T, Valentini A, Miquel C, Taberlet P, Bellemain E, Miaud C (2012) Improved detection of an alien invasive species through environmental DNA barcoding: the example of the American bullfrog *Lithobates catesbeianus*. J Appl Ecol 49:953–959

Deyoung RW, Honeycutt RL (2005) The molecular toolbox: genetic techniques in wildlife ecology and management. J Wildl Manage 69:1362–1384

Frankham R (2005) Resolving the genetic paradox in invasive species. Heredity 94:385. doi:10.1038/sj.hdy.6800634. Published online 15 December 2004. doi:6800634A/sj.bjp.0704832

Gaubert P (2016) Fate of the mongooses and the genet (Carnivora) in Mediterranean Europe: none native, all invasive? In: Angelici FM (ed) Problematic wildlife—a cross-disciplinary approach. Springer, New York, p 295–314

Gentile G, Marquez C, Snell H, Tapia W, Izurieta A (2016) Conservation of a new flagship species: the Galápagos pink land iguana (Conolophus marthae Gentile and Snell, 2009). In: Angelici FM (ed) Problematic wildlife—a cross-disciplinary approach. Springer, New York, p 315–336

Kekkonen J (2016) Temporal genetic monitoring of declining and invasive wildlife populations—current state and future directions. In: Angelici FM (ed) Problematic wildlife—a cross-disciplinary approach. Springer, New York, p 269–294

O'Brien SJ (1994) Genetic and phylogenetic analyses of endangered species. Annu Rev Genet 28:467–489

Piggott MP, Taylor AC (2003) Remote collection of animal DNA and its applications in conservation management and understanding the population biology of rare and cryptic species. Wildl Res 30:1–13

Chapter 13
Temporal Genetic Monitoring of Declining and Invasive Wildlife Populations: Current State and Future Directions

Jaana Kekkonen

Introduction to Genetic Monitoring and Its Usefulness as a Wildlife Management Tool

As human actions are the biggest factors changing the environment also for other organisms, we need to monitor any possible deteriorating changes and use all means possible to detect signs which may require management actions. Populations should be monitored for risks of becoming too small, fragmented, or their viability becoming compromised. This is because in small populations genetic diversity is lost rapidly due to random genetic drift, environmental stochasticity, and inbreeding (e.g., Frankham and Ralls 1998). The loss of genetic diversity may reduce population fitness (reviewed by Reed and Frankham 2003; Briskie and MacIntosh 2004), which can in turn impede population survival due to lowered adaptation to changing conditions (Frankham 2005; Sarre and Georges 2009). Monitoring can identify fitness risks that precede severe losses in abundance (e.g., Antao et al. 2011). On the other hand, humans have also introduced, intentionally or by accident, numerous animal, pathogen, and plant species that have become invasive in the ecosystems, thus creating whole new challenges to managing wildlife populations.

The means to monitoring species are plentiful, but among the most important advancements in recent decades is the development of genetic monitoring methods (Schwartz et al. 2007). When using genetic tools, insights into demographic and evolutionary processes in natural populations are gained. These insights are difficult or even impossible to obtain using traditional methods. Thus, genetic methods should be integrated with demographic data to improve monitoring programs to help ensure population viability in the long run (Schwartz et al. 2007). In this chapter, genetic monitoring is defined as repeated sampling effort over time. Long-term

J. Kekkonen (✉)
University of Helsinki, Viikinkaari 1, Helsinki 00014, Finland
e-mail: jaana.kekkonen@helsinki.fi

F.M. Angelici (ed.), *Problematic Wildlife*, DOI 10.1007/978-3-319-22246-2_13

monitoring, in particular, is recognized as a necessary component when assessing population trends in changing environmental conditions. As the anthropogenic changes including climate change, selective harvesting, and landscape alterations greatly affect most organisms, how animals and plants adapt to these altered environments via contemporary evolution is of strong interest (Hansen et al. 2012).

In temporal genetic monitoring, population genetic parameters are followed by sampling the same population at different points in time, as opposed to looking at populations in a single snapshot (Schwartz et al. 2007; Ramakrishnan et al. 2005). One time surveys or assessments are also highly useful in biological studies, but this chapter is focused specifically on measuring the change and understanding the mechanisms behind it. Even just two points in time can potentially give invaluable information on how the population has changed, how these changes relate to altering conditions, and what might be the implications to the management and future of the population (Leonard 2008). For example, in declining species, it can be reliably determined whether a low level of contemporary genetic diversity is a result of population declines (e.g., Hector's dolphin *Cephalorhynchus hectori*, Pichler and Baker 2000) or if the level was already low in historical samples (e.g., Morro Bay kangaroo rat *Dipodomys heermanni morroensis*, Matocq and Villablanca 2001). With invasive species, on the other hand, combining knowledge of changes in allele frequencies with invasion patterns can help to predict the future trajectories and plan preventive measures (Wares et al. 2005). Monitoring genetic changes over time is especially important because some of the processes can happen with a time lag. For example, potential adverse effects of population declines may manifest only after a delay. The genetic diversity falls below a threshold value, below which the low diversity affects negatively on population viability and in the end, survival (Frankham 2005). Or on the other hand, in species invasions the actual expansions can explode after a slow start because a small founding population overcomes first obstacles of establishing itself to a new area but after some time the number of individuals is enough for exponential population growth rate (Sax et al. 2007). Even though genetic tools have become cost-effective, reliable, and provide information relevant to ecological and evolutionary time frames, many monitoring programs have not taken advantage of complementing traditional methods by using molecular tools. This should be addressed in wildlife management strategies.

The time frames used in population monitoring depend on the questions asked. Nowadays, however, the ample amount of genetic tools available enable the use of many types of organic samples (Beja-Pereira et al. 2009). Thus, all sorts of samples collected for purposes other than genetic monitoring can be used if they have been well documented (see Jackson et al. 2012 for guidelines) and the same populations are currently accessible. Many authors have highlighted the possibilities of museum collections to be used as reference points in population genetic studies. Indeed the vast collections of natural history museums, other institutes, or even private collections can provide remarkable opportunities to study genetic change through time (Austin and Melville 2006; Wandeler et al. 2007; Leonard 2008). In many cases these collections go well back in time before anthropogenic changes of the industrialized period, thus giving appropriate points of comparison (e.g., Groombridge et al. 2000; Austin and Melville 2006). Thus, for wildlife managers there would be ample opportunities for establishing historical baselines. Samples collected in the recent past, by, for example, researchers in Universities or game managers, can also provide valuable information on the rate and nature of changes to genetic population parameters.

In this chapter we present the vast possibilities and implications genetic monitoring provides for the conservation of threatened species and the management of problematic species. The genetic markers and techniques are briefly introduced and the applications of temporal genetic monitoring to managing wild populations are presented through interesting study cases from different animal taxa. The house sparrow *Passer domesticus* is presented as an example of a species which is not yet rare but nevertheless has significantly declined in number. Lastly, the future of genetic monitoring is discussed with an emphasis on modern molecular techniques. The aim of this chapter is to show the great potential in genetic and genomic monitoring to wildlife managers and to offer useful insights into the fields of conservation and invasion biology.

Methods and Markers Commonly Used in Genetic Monitoring

Recent advancements in genetics have brought genetic monitoring within the reach of many management programs in terms of both feasibility and affordability. Moreover, with modern genetic tools many types of organic samples can be used as a source of DNA. This is especially important in genetic monitoring where sampling is most often done noninvasively. Suitable samples include blood and tissue puncture samples (requires capturing of animals) or noninvasively collected hair, fecal, scale, feathers, urine, saliva from chewed material, scent marks, eggshells, and sloughed skin samples (Beja-Pereira et al. 2009). All of these sample types have their own special practices on how they are handled and how the DNA is extracted and preserved, but the techniques are widely available and used in biological research (Avise 2004; Beja-Pereira et al. 2009). Also Jackson et al. (2012) provide detailed guidelines on how to collect and maintain archives for genetic monitoring. By following this advice there should be good opportunities for monitoring programs to run their genetic analysis on many kinds of samples over many timescales. Moreover, when using museum samples, particular problems due to the various preservatives used in conserving the specimens may arise, but most of these issues can nowadays be overcome by careful laboratory work (Wandeler et al. 2007). All in all, the use of various types of samples in genetic monitoring should not be a preventing factor these days (however, see section Potential Caveats in Genetic Monitoring for issues that should be taken into account).

The molecular markers that seem to be most often used for the types of analyses discussed in this chapter are microsatellites, but allozymes, mitochondrial DNA, and amplified fragment length polymorphisms (AFLPs) are also often used (Avise 2004; Selkoe and Toonen 2006) (see Table 13.1 about the suitability of marker types for different analyses). Microsatellites are sections of DNA that consist of very

Table 13.1 Genetic marker types used in temporal genetic monitoring and their suitability for different analyses (marked by X)

Marker type	Genetic diversity	Population structure and gene flow	Mixture	Effective population size	Abundance	Distribution changes	Hybridization	Adaptive variation	Published examples
Microsatellite	X	X	X	X	X		X		Tracy and Jamieson (2011) (diversity, structure)
									Jolly et al. (2011) (hybridization)
									Paulus et al. (2014) (detection, distribution)
Allozyme	X	X	X	X		X	X		Berg et al. (2002) (diversity, structure)
									Palmé et al. (2013) (structure, N_e)
mtDNA	X	X	X			X	X		Ivkosic et al. (2014) (diversity, structure)
									Johnson et al. (2008) (diversity)
									Ou et al. (2014) (structure, distribution)

AFLP	X	X	X	X	X		X		Honnay et al. (2009) (diversity, structure)
SNP	X	X	X	X	X		X	X	Nielsen et al. (2009) (selection, adaptive)
									Therkildsen et al. (2013) (structure, adaptive)
Genomic sequence	X	X	X	X	X	X	X	X	Hansen et al. (2012) (for ideas)

Some examples of published studies given for each marker type short repetitive motifs, with the number of repeats varying between alleles. Microsatellites are mostly located in noncoding areas of the genome and are thus assumed to be selectively neutral. Selective neutrality makes microsatellites suitable for various population genetic analyses, from genetic diversity to population structure to hybridization (Table 13.1). Allozymes are variants of an enzyme and are produced by different alleles at a single locus. Their expression can be used to measure population structure, diversity, and divergence. Mitochondrial DNA (mtDNA) in eukaryotes is found in the cell organelle, separate from the nuclear DNA. When mtDNA sequences are compared they produce information valuable to, e.g., evolutionary relationships and relatedness of populations. A special feature of the mtDNA is that it is inherited only through the matrilineage. This makes it useful for the detection of geographic variation because of traceability. Moreover, it exists in the cell in more copies than nuclear DNA and thus can be extracted from degraded material more easily. The AFLP technique is based on the selective polymerase chain reaction (PCR) amplification of restriction fragments from a total digest of genomic DNA. Using this method, sets of restriction fragments may be visualized without knowledge of the nucleotide sequence. The benefit of AFLP is that large amounts of fragments can be produced in one go and the method is quite sensitive in detecting population structure (Avise 2004).

The development of laboratory techniques like next generation sequencing is now enabling genome-wide approaches for genetic monitoring (Allendorf et al. 2010). Moreover, the costs for running large numbers of samples and markers are going down as the methods are getting more affordable. Genome-wide coverage is especially important when studying the adaptation of populations to environmental changes over time. Even though some studies published have thus far found ways to measure adaptive responses to changing conditions (e.g., Coltman et al. 2003; Nielsen et al. 2009; Therkildsen et al. 2013), examples are quite scarce and the straightforward implementations in management issues are highly underused (reviewed by Hansen et al. 2012). However, this should change in the near future (see section "Future Directions of Genetic Monitoring"). To help wildlife managers in choosing and implementing genetic monitoring methods, as part of "Genetic Monitoring Working Group," an interesting web site "Genetic monitoring for managers" has been established (Stetz et al. 2011).

Potential Caveats in Genetic Monitoring

Like with any biological research, genetic monitoring methods are not without their caveats. Certain aspects need to be taken into account when dealing with long-term monitoring. First of all, sampling design needs to be such that it addresses the monitoring needs: how many and which populations are sampled, what types of samples are collected (noninvasive or invasive), and how many samples is the target amount. The number of samples needed depends on the questions asked but in general some tens of samples per time point is required. Larger coverage often guarantees better reliability but the quality of samples also counts. Simulations by Tallmon et al. (2010) suggest that genetic samples of 10–60 % of the total population may be required to detect biologically realistic changes in abundance and N_e for populations of 100–500 individuals. However, this number will vary with organism life history, the extent of population change, and the detection period (Tallmon et al. 2010). Moreover, Hoban et al. (2014) used simulations to study the detection of genetic erosion following demographic decline with different markers, genetic metrics, and sampling protocols. Interestingly, the type and severity of demographic decline strongly affected power, while the number and arrangement of temporal samples (interval between samples) had little effect. Sampling 50 individuals at as few as two time points with 20 microsatellites performed well. This sampling and genotyping effort should be affordable often. Power increased substantially with more samples or markers, and it was observed that the power of 2500 SNPs was nearly equivalent to 250 microsatellites. Higher resolution also means that lesser samples are needed. The time frame for temporal sampling depends also on the organisms' life history, i.e., generation time and generation overlap (Jackson et al. 2012). Optimal sampling may not always be possible despite planning and in that case, in particular, these issues need be addressed when interpreting the results.

DNA degradation can be problematic in both historical and contemporary samples. Museum samples may have various preservatives used in the conservation of specimens and these can cause DNA degradation and PCR inhibition (Hall et al. 1997). Thus, using historical collections may require sample selection as well as additional time and processing steps. Moreover, with historical samples the contamination risk is bigger and thus the isolation of laboratory areas is important. There are a growing number of laboratory techniques which help to overcome potential problems related to historical samples (Wandeler et al. 2007). However, degradation problems relating to contemporary samples are not often taken into consideration. Thus, with sampling design, the careful planning of the storage and transport conditions of samples is of great importance. If collecting noninvasive samples (e.g., feces, hair), correct storage on the field is even more crucial because DNA degradation will occur easier than for e.g. blood and tissue samples. Moreover, long-term storage in the laboratory is important; there is a direct relationship between the cost of specimen preservation and quality for genetic archives (Jackson et al. 2012). Thus, it is worthwhile to consider how sample preservation is done (e.g., dried or in preservative, temperature, sealed and protected from contaminants, UV light). Beja-Pereira et al. (2009) provides detailed information concerning these issues.

Marker choice needs to be such that it best applies to the biological problems in question (Larsson et al. 2009). The number of markers needed depends on the analyses but financial issues often limit the possibility to run any number of markers. However, marker quality and quantity should be such that they give enough power for the statistical analyses of the genetic monitoring data. In the genetic analysis phase, there is always potential for genotyping errors, large allele dropouts, and null alleles, and these need to be controlled by replicates and negative controls (Sefc et al. 2003; Bonin et al. 2004). When a reliable data set is achieved, statistical analyses carefully selected for the biological questions are employed. The data and analyses need to have enough power to reliably test the hypothesis (Ryman et al. 2006; Larsson et al. 2009).

Many of the potential caveats in genetic monitoring can be taken into account with careful planning. When the grounds are properly set, implementation of the results to the actual management of declining or invasive populations will deliver maximal benefits.

Measures Followed in Populations

In monitoring, either one population or groups of populations can be sampled over time, depending on which questions are of interest. In some cases there may be only one population threatened or introduced and this is then followed. However, if there are connected populations, then the whole group of populations is of interest, since the pieces naturally affect each other. The genetic estimates evaluated by genetic monitoring are genetic variation, degree of admixture of populations, effective population size, population structure, and connectivity, and how these estimates change in time in relation to changes in the environment.

Genetic diversity is measured in terms of allelic richness, and expected and observed heterozygosity. Both empirical and simulation studies demonstrate (e.g., Luikart et al. 1999; Spencer et al. 2000; Ramakrishnan et al. 2005) that allelic richness is more sensitive to the decline in population size than expected heterozygosity. This is because it is sensitive to the loss of rare alleles while expected heterozygosity is less affected. As such, these measures reflect somewhat different processes. Heterozygosity can be used as a predictor of a population's capacity to respond to selection immediately after a bottleneck. Allelic diversity, on the other hand, predicts a population's ability to respond to long-term selection over many generations (Allendorf and Leary 1986). Both estimates should be reported and used in genetic monitoring programs and are equally relevant in conservation and invasions (Kekkonen and Brommer 2014). Furthermore, monitoring changes in allele frequencies through time can be an even more sensitive indicator of population decline than loss of allelic richness is (Spencer et al. 2000).

Estimating population size through traditional census methods has proven difficult with many species that are very elusive, rare, difficult to observe, difficult to count, or even too numerous. Moreover, the actual number of organisms in a population is not equivalent to the breeding capacity of the population. Wildlife managers should also monitor the number of breeding individuals that produce offspring that survive to a reproductive age. This relates to the so-called effective population size (N_e), which is the corresponding population size of an idealized (Wright–Fisher) population that would function in the same way with respect to genetic drift and inbreeding as the focal population under investigation. The effective population size is often much smaller than the census population size and provides a more accurate description of the breeding capacity of the population (Frankham 2005). Comparing the effective population size to the actual census size can also provide valuable information. Most of the techniques for calculating effective population sizes require two samplings apart in time (Leberg 2005; Charlesworth 2009) and monitoring N_e thus require at least three samplings. Especially in threatened species, monitoring the changes to the effective population size and early interference allows for rapid management actions to avoid any irreversible loss of genetic variation and increased risk of extinction due to genetic and demographic factors (e.g., simulation studies by Antao et al. 2011). However, it must be stressed that N_e estimates are based on a number of assumptions such as random mating, no mutation, migration, and selection—almost all of which are violated in natural populations (Clarke 2000). Skewed allele frequencies, overlapping generations, and iteroparity would upwardly bias N_e when applied to temporal samples taken over short time periods. Gene flow from a genetically not very dissimilar source population also increases estimates of N_e. Thus, adequate sampling needs to be performed and quantification of life history parameters and gene flow need to be incorporated into the N_e estimation (Serbezov et al. 2012).

The level of population structuring and the gene flow between populations or subpopulations is of great interest when considering human-altered environments such as fragmented habitat (Frankham 2005). If populations become smaller, the connectivity to adjacent populations becomes more important for population survival, because gene flow can retain sufficient levels of genetic diversity and thus adaptability (Ryman et al. 1995; Koskinen et al. 2002). Measuring the level of population structure at different points in time and determining how the level of gene flow has changed provides invaluable information on whether the changes seen in the environment may have affected the viability of the populations. Furthermore, if the structure existed historically and there are underlying threats that are not observable with traditional monitoring methods, genetic monitoring may identify them before it is too late. Moreover, genetic methods can also be used to predict how planned changes to the landscape could affect the gene flow patterns over time (Hansen et al. 2012).

For wildlife managers the estimation of population admixture, i.e., whether the individuals within a population are from a single ancestry or possibly from several different populations, is also interesting. Managers can also monitor whether commercial stock populations are mixing with natural populations and potentially override the natural genetic material over time (e.g., Beacham et al. 2004). In the case of invasive species, the interest is in whether the founder population is a mixture of several source populations and how it affects the invasiveness of the species. Moreover, hybridization between species can also become problematic for both the conservation and invasion perspectives (e.g., Hitt et al. 2003; Schwartz et al. 2004).

Temporal Genetic Monitoring of Declining Species

Genetic diversity is recognized as one of the three levels of biodiversity that should be protected (Convention on Biological Diversity; Laikre et al. 2010). Thus, the implications of ongoing changes in the level of genetic diversity through time are apparent for conservation management. Moreover, when studying the impacts of environmental changes at the level of neighboring populations, the potential losses in genetic diversity can also lead to decreased gene flow and increased structure. However, since spatial genetic patterns across landscapes can arise from different processes and during different timescales (Reding et al. 2010), a contemporarily observed structure could be the result of historically reduced gene flow due to natural barriers, as well as a consequence of recent change. Thus, the use of historic samples collected prior to fragmentation allows for the origin of genetic structure to be established and appropriate management steps to be determined. Though the potential is not yet so well utilized in conservation management strategies, some examples from the wild highlight the usefulness of these methods in detecting warning signals in the deteriorating population viability. Research in natural settings is highly important as simulations or laboratory studies cannot accurately predict the outcomes in wild populations.

For example, Athrey et al. 2011 used historical and contemporary samples (69 and 65 individuals) of the endangered golden-cheeked warbler *Dendroica chrysoparia* for genetic monitoring. They documented a steep decline in the genetic diversity of this endangered species over a 115-year period (1890–2005), along with an increase in genetic differentiation. There was no difference between the 1900 and 1915 estimates of allelic richness and heterozygosity. However, both estimates were significantly higher than in the contemporary 2005 samples from the same sites, with average declines of 20 and 13 % (all declines in every location statistically significant), respectively. The measured levels of pairwise differentiation had increased significantly when comparing the historic and contemporary levels of the same population pairs. These reported changes reflect the habitat fragmentation that has been going on in the environment. Because this species lives in a very specific habitat and thus occurs only in confined areas in Texas and nowhere else, the genetic changes found by Athrey et al. (2011) should be considered in the management plans. In a similar study, Tracy and Jamieson in 2011 compared the historic and current levels of genetic diversity and structure of an endangered passerine, the New Zealand mohua or yellowhead *Mohoua ochrocephala*. They used 56 historical samples (collected in 1872–1939) and 157 contemporary (2006–2007) with 11 microsatellites. A significant amount of allelic richness (22.6 %, $t = -2.405$, $p = 0.019$) was lost over the 100 years. Interestingly, almost no structure was found between the historical populations, but during the study period statistically significant genetic structure appeared as genetic diversity was lost. The historical sample size is much smaller than the contemporary one but the analyses were performed so that these uneven sample sizes were carefully taken into account. Thus, the emerged structure of the yellowheads was most likely due to habitat fragmentation caused by burning, clear-felling of forests, and predation by the introduced mammals. This knowledge helps management planning to justify the need for conservation actions and focus the funds on the most promising conservation measures in order to retain the viability of the island population system.

In case conditions improve, a declining population trend can be reversed. However, reversing a genetic diversity loss may not be as straightforward a process and thus long-term genetic monitoring is called for in these situations. An extreme example of a population bottleneck and recovery is the Mauritius kestrel *Falco punctatus*. The species at first suffered from a small population of less than 50 individuals due to the adverse effects of pesticides in 1940–60, then subsequently went through a very narrow bottleneck of a single breeding pair in 1974 (Groombridge et al. 2000). An intensive conservation management program saved the kestrels and led to the recovery of the population to a couple of hundred birds. The program included the captive rearing of eggs removed from nests, captive breeding, supplementary feeding of wild breeders, predator control, nest site enhancement, and so forth. The loss of genetic variation that resulted from the bottleneck was evaluated by typing 12 microsatellite loci in museum skins up to 170 years old and in modern kestrels. Across all loci, allelic diversity fell by 55 % and heterozygosity by 57 %. However, the ancestral variation was remarkably high (in comparison to continental kestrel species) and thus despite the heavy bottleneck, the remaining diversity likely enabled the population to recover. The unexpected resilience of the population could not have been due to historic small population size or newly added genetic variation. Thus, it seems, at the moment, that productivity was only weakly affected by the bottleneck. It is encouraging that a population can recover from a bottleneck this severe but temporal genetic monitoring should definitively be continued since some of the effects of the lowered diversity may lag behind. In particular, the capability for adaptive responses to environmental changes is not known.

Along the same lines, Ortego et al. (2011) used 28 microsatellites to investigate changes in genetic diversity over 14 cohorts in a small and relatively isolated population of mountain goats *Oreamnos americanus* during a period of increasing abundance. Despite population growth, the offspring heterozygosity actually decreased and inbreeding coefficients increased, which is contrary to expectations ($F_{1,120}=8.16$, $p=0.005$). However, immigrants, which naturally had migrated to the population, introduced three new alleles into the population and matings between the residents and immigrants produced more heterozygous offspring than local crosses ($F_{1,120}=4.75$, $p=0.03$). This suggested that immigration can indeed increase the population's genetic variability. Nevertheless, there was an overall decline in genetic diversity that probably resulted from a combination of genetic drift due to small effective population size, inbreeding, and insufficient immigration. A similar case was the European bison *Bison bonasus*, which was reintroduced to Poland in the 1950s. Tokarska et al. (2009) compared the genetic diversity of the bottlenecked population before and after reintroduction by monitoring five temporal groups of animals born between 1955 and 2005 (total number of individuals 178). The variability of 12 microsatellite markers showed low overall allelic richness per locus and low overall expected heterozygosity. The inbreeding coefficient was significantly different from zero in individuals born between 1955 and 1965, which was around the time of the founding event. Moreover, relatively small effective population sizes ranging from 7.9 to 28.4 were observed in the temporal groups. The low N_e values confirm that, despite a rapid growth in the bison population following the founder event, N_e increased only slowly. This means that in spite of the absence of any signs of inbreeding depression, the low adaptive potential of the European bison may substantially hinder its survival in changing environments. Again, these studies highlight the importance of genetic monitoring also in the cases where populations are recovering from small sizes, because the genetic effects may not recover at the same pace as population census size would allow to assume.

The changes in genetic diversity and population structure are not the only concerns for conservation managers. Hybridization with other, often introduced, species can eradicate native species. Genetic monitoring is at times the only way to detect the hybridization process before the native species has disappeared. However, there is also concern that human-induced disturbance can lead to hybridization among naturally occurring species pairs (Jolly et al. 2011). This can happen if, for example, the reproductive isolation that was previously maintained by direct environment-dependent sexual selection is lost. This means that in different habitats different traits are selected in breeding, maintaining species boundaries. Habitat alterations have been implicated as driving hybridization between two shad species *Alosa alosa* and *Alosa fallax*, i.e., overharvesting, water pollution, and prevention of access to breeding sites likely forced the species into the same breeding areas. Jolly et al. (2011) used eight microsatellite loci and 268 samples collected between 1989 and 2008 in the UK to genetically identify hybrids and study temporal changes in their allelic frequencies. The results showed that genetic diversity had remained largely the same within both species, but there were declines in the inbreeding coefficients. Moreover, hybrids showed no evidence of lower fecundity than the purebred individuals. Together the results suggested that hybridization between the shad species has been ongoing for at least two decades. The authors were wondering why the species had remained genetically separate despite the hybrid fertility but this remained unclear. Indeed, for evolutionary biologists and managers, the extent and consequences of hybridization cannot yet be easily predicted. In this case, further genetic monitoring could provide more answers and help create management plans so that these endangered species would remain separate. In the case of the shads, the use of selective markers could really prove to be useful in making management decisions.

As seen from these case studies, without taking the genetic measurements of populations at several points in time, the long-term evolutionary patterns would not have been revealed. Combining the genetic data into ecological information will give good inference for management planning. However, careful planning of such monitoring programs and taking the potential caveats into account are very important so that the results are accurate and can be successfully implemented into actual management (Fig. 13.1). Furthermore, in many cases the management strategies would benefit further from knowledge of the adaptive genetic variation, which could be the next step in planning the conservation of a threatened species.

A Common Species in Decline: The Case of the House Sparrow

Like the examples above, conservation concerns are often raised when considering species that are becoming rare and, thereby, potentially threatened (Soulé 1987). As such, genetic monitoring methods should also be applied to species

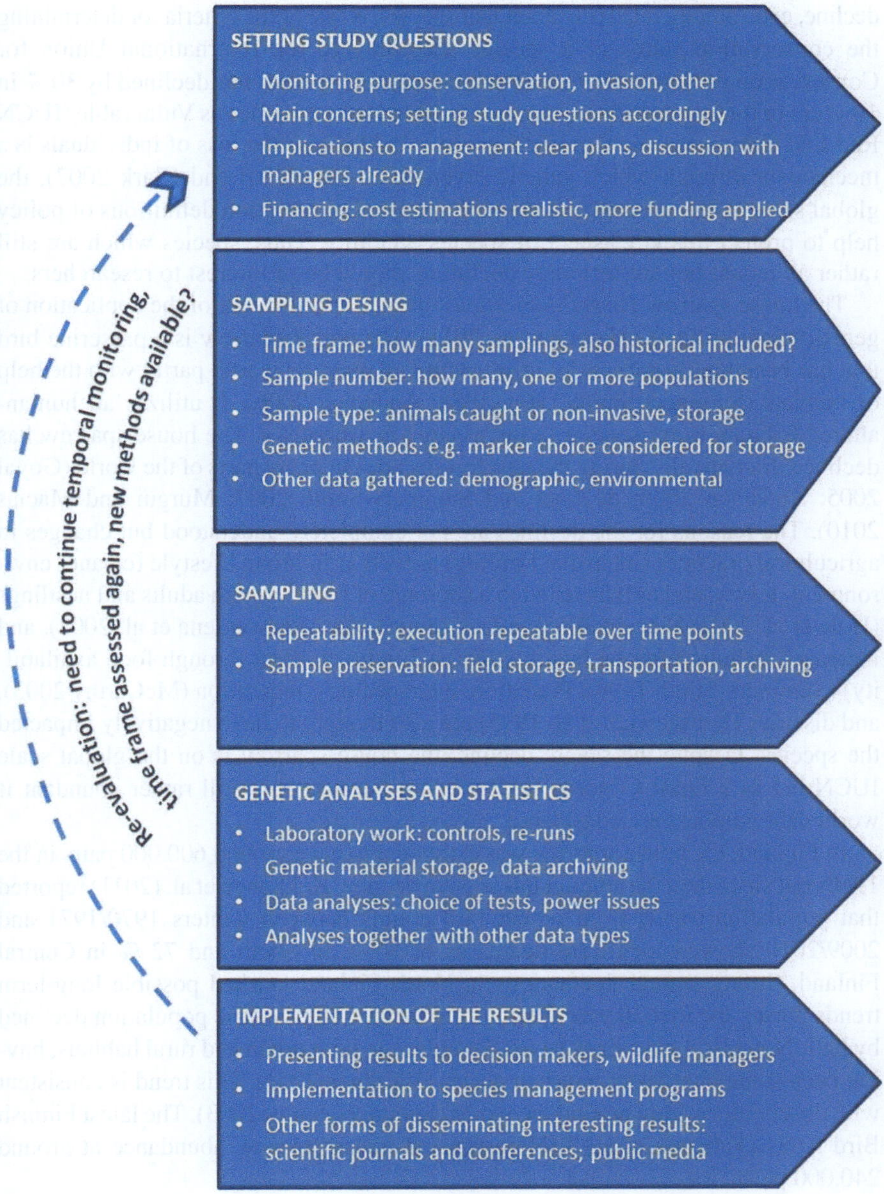

Fig. 13.1 Tips for designing a genetic management program

which are still fairly common but declining in number. This is because the very essence of monitoring is not to just rescue what is left of diversity but to detect the earlier signals of genetic deterioration and to initiate management actions to ensure population viability and adaptability to changing conditions. Population decline, even among relatively abundant species, is one of the criteria for determining the conservation status of a species according to the International Union for Conservation of Nature (IUCN). For example, if a species has declined by 30 % in 10 years or three generations, its conservation status qualifies as Vulnerable (IUCN Red List Categories and Criteria Version 3.1). Because the loss of individuals is a mechanism through which genetic diversity is lost (Hartl and Clark 2007), the global scale IUCN criteria and country-specific conservation definitions of policy help to protect this key aspect of species' viability. Thus, species which are still rather abundant but are seriously declining should be of interest to researchers.

The house sparrow *Passer domesticus* offers a specific case of the application of genetic monitoring (Kekkonen et al. 2011). The house sparrow is a passerine bird that has been highly successful in spreading all over the world, partly with the help of humans (Summers-Smith 1963, 1988; Anderson 2006). It utilizes all human-altered habitats from isolated farms to large metropolises. The house sparrow has declined dramatically during the past few decades in many parts of the world (Goyal 2005; Anderson 2006; de Laet and Summers-Smith 2007; Murgui and Macias 2010). The reasons for the declines are not completely understood but changes in agricultural practices (intensive farming) as well as in urban lifestyle (cleaner environment, less wasteland) have led to a shortage of food for both adults and nestlings (Hole et al. 2002; Vincent 2005), a lack of nest sites (Siriwardena et al. 2002), and increased pollution (direct health effect and indirect effect through food availability) (Summers-Smith 1999). Predation, interspecific competition (McCarthy 2003), and diseases (Kruszewicz et al. 1995) are also thought to have negatively impacted the species. Despite the severe decline, the house sparrow is on the global scale IUCN list as a Least Concern (LC) species, because it is still rather abundant if worldwide numbers are considered.

In Finland, the house sparrow was estimated to have around 600,000 pairs in the 1960s but since then the amount has been in decline. Kekkonen et al. (2011) reported that population trends from winter bird counts between winters 1970/1971 and 2009/2010 showed long-term decreases of 86 % in South and 72 % in Central Finland. Strong annual fluctuations in North Finland masked possible long-term trends during the first 30 years, but during the last 10 years the population declined by half. Notably, the declines have taken place in both urban and rural habitats, having been somewhat faster in urban places (Väisänen 2003). This trend is consistent with observations from around the world (Summers-Smith 2003). The latest Finnish Bird Atlas (Valkama et al. 2011) estimated house sparrow abundance at around 240,000 pairs.

Kekkonen et al. (2011) sampled house sparrows in the same 12 locations in 2009 as where museum samples were collected in the 1980s to compare the genetic diversity within and across the populations before and after the population decline (Fig. 13.2). The two data sets separated by approximately 25 years included almost 1000 individuals typed at 13 microsatellite loci. The results suggested that genetic diversity was being lost but none of the populations had become inbred. However, the genetic structure of the more or less panmictic house sparrow population

Fig. 13.2 House sparrow being sampled for temporal genetic monitoring (© Hannu Lehikoinen)

increased threefold (overall F_{ST} from 0.0051 to 0.0146, 95 % confidence intervals nonoverlapping). Further, an AMOVA test showed that the populations were genetically more different at the two sampling time points than what the geographic groups were different from each other.

Given that the declines took place during only approximately 12.5 generations (generation time ~ 2 years, Jensen et al. 2008), it is surprising that evidence of a change in the population genetics was detectable. Although this species is considered to disperse only short distances (Altwegg et al. 2000; Siriwardena et al. 2002; Tufto et al. 2005), it seems that the level of dispersal over a homogeneous landscape together with high numbers had been enough to maintain a high level of gene flow prior to the 1980s. This had prevented population differentiation but the large declines have now reversed this situation. Little over ten generations is, in many cases, a short time for populations to reach a new equilibrium in genetic diversity and differentiation (Varvio et al. 1986). In addition, the house sparrow has remained a relatively abundant species in Finland and its number has not fallen as low as classical examples of bottlenecked avian populations (for example ones presented in this chapter). The fact that the population declines have left detectable and significant footprints in the genetic structure of this species is therefore highly interesting from a conservation perspective. Although the current situation seems fairly good, the observed change in genetic differentiation should be taken as an indication that important population genetic processes have been affected and a loss of genetic diversity is likely to be ongoing. Crow and Aiko (1984) as well as Varvio et al. (1986) demonstrated with simulation studies that in recently fragmented populations genetic differentiation reaches a new equilibrium state much faster than within-population diversity. Thus, the decline in genetic diversity may be in progress and a new genetic monitoring program would reveal it.

Despite the fact that in Finland the house sparrow is listed as a Least Concern species and it is still relatively abundant, it remains clear that the house sparrow has experienced massive declines during the last four decades. Results from this study highlight the fact that the final impact on genetic diversity may yet to be seen. Therefore, this particular case study highlights the importance of considering population genetic consequences in declining but still relatively common species (Fig. 13.2).

Genetic Monitoring of Invasive Species

Another highly relevant application for genetic monitoring is the controlling of invasive species, which are recognized to be the second largest threat to biodiversity after habitat loss (Lowe et al. 2000) and have thus become a major risk to ecosystems (Duncan et al. 2003). However, the control and management of invasive species is often hampered by a lack of understanding of the history and origins of the populations, number of introduction events, connectivity, and the genetic potential to adapt (Mack et al. 2000). In some cases, once the invasion has progressed far enough, controlling it has become nearly impossible (Sax et al. 2007). If information on the species and its invasion dynamics would be collected early on, controlling the invasion could be more effective.

As an interesting example, Paulus et al. (2014) applied genetic monitoring to reconstruct the colonization history of the quagga mussel *Dreissena rostriformis* in German river systems. This mussel has become one of the most successful invasive species in Western European freshwaters. Paulus et al. used cryogenically stored zebra mussel *D. polymorpha* samples which had been collected for water pollution monitoring but considered that quagga mussels might have been collected as well due to their similarity. Microsatellites were run to reconstruct quagga range expansion from 2004 to 2011. The discovery of quagga mussels in 2004 in the Rhine River presented the first record of this species in Western Europe. Inferring invasion histories is a basic requirement for predicting their future dispersal and developing suitable management strategies. However, emerging biological invasions often remain undiscovered until the invaders become apparent due to their numbers or their deleterious effect on the environment or other species (Lockwood et al. 2007; Estoup and Guillemaud 2010). Thus, this study highlights the innovative use of biological material collected for potentially other purposes but found to be useful for the systematic monitoring of an invasion process. Another interesting development is the recently discovered potential of using environmental DNA (eDNA) for detection of species invasions (Ficetola et al. 2008). eDNA means collecting genetic material from the environment without sampling whole adult organisms and it is now being used for example for the invasive quagga and zebra mussel detection in the Great Lakes area (Michigan State University, eDNA team). Sources of eDNA can be, e.g., scraped-off cells, excrement, eggs, larva, juveniles, and free-floating DNA released from any cell lysis. The use of eDNA is being developed in particular for the detection of invasive species from water samples (Dejean et al. 2012; Thomsen et al. 2012). In temporal genetic monitoring it could be used for validating the presence or absence of an organism in an area and creating distribution maps to determine how widespread the invasion is.

In invasions, adaptation capability to new conditions is often a key for success, and therefore knowledge of the evolutionary changes of an invasive population can provide useful information for understanding the genetic patterns associated with each stage of the invasion process. This way management, including monitoring and control, can be targeted to the invasion dynamics of the invasive species in question (Sakai et al. 2001). As seen from conservation studies, retaining genetic diversity is highly important for adaptation to changing conditions. Many populations expanding into novel environments not only thrive but also exhibit rapid evolutionary changes in crucial traits such as dispersal ability, reproductive output, phenotypic plasticity, and size (Whitney and Gabler 2008). This empirical evidence is supported by simulations, demonstrating that evolution may move at a faster rate when an organism's environment changes (Kashtan et al. 2007). Evidence of rapid evolution in novel environments supports the idea that genetic diversity is important to the success of introduced populations because adaptations are more likely to be derived from existing genetic variation rather than mutation (Barrett and Schluter 2008). Some studies have demonstrated rapid evolution despite the presence of low genetic diversity (Dlugosch and Parker 2008), but more often the invasion success has been explained by high propagule pressure, multiple introductions, or genetic admixture, all of which are processes that maintain high genetic diversity and therefore enable adaptation (Allendorf and Lundquist 2003; Stepien et al. 2005; Roman and Darling 2007). However, due to the complexity of evolutionary processes, it would be imperative to genetically monitor the populations in order to recognize the underlying evolutionary patterns and manage them accordingly.

In one invasive species case study where temporal monitoring was implemented, the level of genetic diversity was found to have increased over time. Berg et al. (2002) used allozymes in spiny water flea *Bythotrephes longimanus* to examine changes in the genetic structure over time of recently founded North American populations. Interestingly, the founder effects that could be detected in the samples from 1989 disappeared by 1996, so that the genetic structure of North American populations in 1996 was similar to that of native European populations. Moreover, the Lake Ladoga population in Russia was significantly more closely related to North American populations than to other European populations (Mann–Whitney test $p = 0.01$). This was consistent with nongenetic evidence indicating that Lake Ladoga would be the source of North American populations. The likely route of the spiny water flea is in the transport ship ballast water from the Russian lake to the Great Lakes of North America. This study shows that founder effects can erode in an invasive species over a short period of time and therefore there are probably no limitations to success due to low initial genetic diversity for further range expansions in North America.

Another interesting study case comes from Europe where the western corn rootworm, *Diabrotica virgifera virgifera LeConte*, has established itself via the USA. This species is a major pest for maize *Zea mays* in North America and has spread to Europe through five separate invasion events (Ciosi et al. 2010). Damage and control costs are huge on both continents. Microsatellite and mitochondrial

DNA studies were done in Croatia and Serbia from 1996 to 2011 and in the USA in 2011 (Ivkosic et al. 2014). As expected, the seven USA populations had greatest allelic diversity. In Croatia and Serbia, the highest number of alleles was found in two sites closest to international airports and the rootworm was identified to have expanded from Serbia to Croatia. Pairwise population comparisons indicated that genetic differentiation slightly decreased over time for all Croatian populations and for the Serbian population. In this study, although the genetic structure indicated a single population in Croatia, the presence of genetic bottlenecks among populations sampled in 2011 illustrated the negative effect crop rotation is having in reducing the overall genetic diversity of the rootworm. As a future management plan, the authors suggested that routine genetic monitoring on both continents would be done so that the tracking of possible resistant and susceptible alleles and haplotypes and their changes throughout populations worldwide would be possible.

Besides understanding the invasion histories, following the species also on the invasion fronts is important. Rollins et al. (2009) used common starling *Sturnus vulgaris* populations in Australia to demonstrate that genetic techniques can provide information on invasion pathways and connectivity to aid management, even when applied to highly vagile species over very large geographical scales. The common starling was originally introduced to Australia to consume insect pests from crops but it was soon considered an invasive pest itself. Samples were taken from individuals captured by control agencies for destruction between 2003 and 2007, from 17 localities across the Australian range. In addition, historical samples were included from a museum collection representing ten previous invasions between 1914 and 1998. Based on 11 microsatellites there were four populations found. One population consisted of all sampling sites from the expansion front in Western Australia and, even though there has been genetic exchange over both contemporary and historical timescales, gene flow was low between this population and all three more easterly populations (Rollins et al. 2009). This suggests that the localized control of starlings on the expansion front may be an achievable goal. However, even with low levels of gene flow, successful control of starlings on the expansion front will be demanding, and thus the genetic monitoring of this population will be important to managers. Thus, long-term genetic monitoring should be used to assess patterns of dispersal, changes in population size and the effectiveness of ongoing control efforts. Because not everything can be controlled, e.g., dispersal patterns and routes can help to determine what is most efficiently controlled to restrict range expansion and population growth.

These examples flag the important role that temporal genetic monitoring has within invasive species management and control programs. Yet the examples where these methods have been actually implemented in management programs are still rather scarce, the sample sizes are sometimes rather modest, and the diversity measured is mostly that of neutral markers. Thus, implementing novel techniques (next generation sequencing, whole genome scans, and gene expression analysis) revealing adaptation to novel environments could create valuable tools to help curtail the huge problems that human-induced invasion processes cause in nature. Moreover, museum samples will prove invaluable for building points of comparison to help with the restriction of invasion processes (Fig. 13.1).

Future Directions of Genetic Monitoring

Conservation of threatened populations and detection of invasive species are just two areas where temporal genetic monitoring can be of use. The methods described in this chapter can also be implemented in game and fishing management, and have been to some extent. Wildlife managers can benefit a great deal in getting more accurate estimates of, for example, the effective population size, sex distribution, or genetic diversities of populations over time. Such information could reduce the risk of overharvesting, thereby keeping game populations productive and healthy in terms of stochastic events, pathogen resistance, and other threats. Moreover, in many countries, hunters or game managers already collect samples every year and with modern methods these could be used retrospectively to establish baselines for comparisons. Temporal genetic monitoring methods, like any other scientific techniques, are not without their caveats and managers need to pay attention when choosing the particular methods, so that the marker selection, sampling, lab methods, and statistical analyses all fit in with the species and problems in question. However, when taking these aspects carefully into consideration, temporal monitoring can provide highly beneficial tools for wildlife managers (Schwartz et al. 2007; Wandeler et al. 2007; Jackson et al. 2012; Stetz et al. 2011).

Advancements in molecular techniques will add to the realization of this potential as they enable the examination of thousands of genetic markers with relative ease. This will make it possible to answer many important questions that have been intractable until now (Allendorf et al. 2010). Importantly, the transition from genetics to genomics leads to insights into the dynamics of selectively important variation, its interaction with changes in the environment, and into the mechanisms behind this interaction. The key developments are the genome-wide estimates which will provide both new levels of power and accuracy to neutral marker estimates, but are also needed for detecting functional level variation. Thus, adaptive genes can be found and their connection to fitness and demographic vital rates can be estimated (Allendorf et al. 2010; Ouborg et al. 2010). Techniques like next generation sequencing, whole genome scans, and gene-expression pattern analysis have made it possible to sequence an organism's full genome more easily and analyze functional DNA. In particular, the applicability of these new approaches to non-model organisms opens tremendous possibilities from the perspective of genetic monitoring programs. A key point is that these methods can also be applied to populations in the wild (Van Straalen and Roelofs 2006). A crucial component of all genomic techniques is bioinformatics. Bioinformatics is the application of advanced computing techniques to manage the massive amounts of biological information from data acquisition to analysis to data storage. Temporal monitoring with genomics will require time and expertise in bioinformatics and this should be understood from early on. However, these tools are already readily available (Mount 2001).

Advancements have happened both in developing sequence variation and on the transcriptomics level, and both pathways present highly interesting and also nonexclusive possibilities for genetic monitoring. High resolution sequence variation can be used in population genomics in two ways. First, by having a considerably larger coverage over the genome it will provide a representative view on the genetic variation within individuals and populations. This, in turn, enables much more accurate inferences about population history, demography, and structure. Second, sequence variation can be used for finding 'footprints of selection,' i.e., detecting deviating distribution patterns for markers which can be indicative of past or ongoing selection at loci linked to the markers (Luikart et al. 1999). Comparison of spatial and temporal patterns of these markers with neutral markers makes it possible to separate genetic drift and selection effects. This enables the study of the effects of changing conditions on the different regions of the genomes. Moreover, by detecting areas in the genome that are under selection, candidate genes which are under selection can also be found. Although identifying selective effects in natural populations is a challenge, the methods have developed so that this has become possible (Allendorf et al. 2010).

One important type of marker for sequence variation that will be increasingly used also in genetic monitoring is single nucleotide polymorphisms (SNPs). Recent improvements in the speed, cost, and accuracy of next generation sequencing and associated bioinformatics tools are revolutionizing the discovery of these markers. SNPs are versatile and sensitive markers, evenly spread throughout the genome, and have the potential to substantially expand the ability to analyze both noncoding (neutral) and coding (genes under selection) regions in populations, also providing a broader genome coverage than, for example, microsatellites (Morin et al. 2004; Ouborg et al. 2010). This use of genome-wide SNPs will improve estimates of genetic and demographic parameters including genetic diversities, past population growth rates, current abundances, population structure, and hybridization.

In addition to DNA sequence variation approaches, the application of transcriptomics presents great potential for monitoring adaptive change on the functional level. It enables the study of variation in the level of gene activity as a function of changing conditions. Transcriptomes comprise the analysis of gene expression and they can differ between individuals under different environmental conditions. Gene expression during organisms' development gives rise to different phenotypes but it also affects organisms' metabolism constantly when conditions change. The most common technologies used to investigate gene expression changes are DNA microarrays, quantitative real-time PCR (qRT-PCR) (Lettieri 2006), and RNA sequencing (Montgomery et al. 2010). The uptake of transcriptomic approaches is a promising step towards understanding the mechanisms underlying adaptation and maladaptation. These approaches will help answer questions like how many genes, and of what function, are differentially expressed under which conditions, and are these always the same genes across populations (Ouborg et al. 2010). These methods are not as readily available or cost-effective as the methods based on, for example, SNPs and thus may be out of reach for some monitoring and management programs. Nonetheless, these methods should be explored as they are rapidly evolving and provide information unattainable through other methods.

In addition to considering one species at a time, a future perspective on genetic monitoring could be the introduction of metagenomics. In metagenomics, multiple taxa can be combined in a single sequencing analysis using genomic techniques like bar coding, and from the sample (e.g., soil or water sample or combined insect sampling) all species can be identified by sequencing against a reference database (Hebert et al. 2003). Following changes in species compositions over time with this technique could be beneficial in genetic monitoring programs for difficult species compositions, but the method is currently very complex.

Concluding Remarks

Looking at the examples of temporal genetic monitoring and the future possibilities arising from advancing technologies, the potential in these methods for wildlife management is tremendous, but they are not used nearly as often as they could be. Implementing genetic methods in management programs is of course not trivial but with modern molecular technology, reduced costs, and greater acceptance of the methods this is highly feasible and most analyses can also be performed by commercial labs. Despite all the benefits, genetic monitoring should be coupled with ecological, demographic, and environmental research when possible to gain as comprehensive a picture of the managed species as possible, in order to take the proper management actions. As the temporal genetic monitoring techniques are readily available, I encourage scientists, managers, and decision makers to take the next step and take advantage of the many resources out there and explore the methods for themselves more to determine if and how their objectives can be served by implementing these rapidly evolving techniques!

Acknowledgements I would like to thank the two anonymous referees who really put the time and effort in commenting this chapter which improved it remarkably.

References

Allendorf FW, Leary RF (1986) Heterozygosity and fitness in natural populations of animals. In: Soule M (ed) Conservation biology: the science of scarcity and diversity. Sinauer, Sunderland, pp 57–76

Allendorf FW, Lundquist L (2003) Introduction: population biology, evolution and control of invasive species. Conserv Biol 17:24–30

Allendorf FW, Hohenlohe PA, Luikart G (2010) Genomics and the future of conservation genetics. Nat Rev Genet 11:697–709

Altwegg R, Ringsby TH, Saether BE (2000) Phenotypic correlates and consequences of dispersal in a metapopulation of house sparrows Passer domesticus. J Anim Ecol 69:762–770

Anderson TR (2006) The ubiquitous house sparrow: from genes to populations. Oxford University Press, Oxford

Antao T, Pérez-Figueroa A, Luikart G (2011) Early detection of population declines: high power of genetic monitoring using effective population size estimators. Evol Appl 4:144–154

Athrey G, Lindsay DL, Lance RF, Leberg PL (2011) Crumbling diversity: comparison of historical archived and contemporary natural populations indicate reduced genetic diversity and increasing genetic differentiation in the golden-cheeked warbler. Conserv Genet 12:1345–1355

Austin JJ, Melville J (2006) Incorporating historical museum specimens into molecular systematic and conservation genetics research. Mol Ecol Notes 6:1089–1092

Avise JC (2004) Molecular markers, natural history, and evolution, 2nd edn. Sinauer, Sunderland

Barrett RDH, Schluter D (2008) Adaptation from standing genetic variation. Trends Ecol Evol 23:38–44

Beacham TD, Lapointe M, Candy JR, Miller KM, Withler RE (2004) DNA in action: rapid application of DNA variation to sockeye salmon fisheries management. Conserv Genet 5:411–416

Beja-Pereira A, Oliviera R, Alves PC, Schwartz MK, Luikart G (2009) Advancing ecological understandings through technological transformations in noninvasive genetics. Mol Ecol Res 9:1279–1301

Berg DJ, Garton DW, MacIsaac HJ, Panov VE, Telesh IV (2002) Changes in genetic structure of North American Bythotrephes populations following invasion from Lake Ladoga, Russia. Freshw Biol 47:275–282

Bonin A, Bellemain E, Bronken Eidesen P, Pompanon F, Brochmann C, Taberlet P (2004) How to track and assess genotyping errors in population genetics studies. Mol Ecol 13:3261–3273

Briskie JV, MacIntosh MM (2004) Hatching failure increases with severity of population bottlenecks in birds. Proc Natl Acad Sci U S A 101:558–561

Charlesworth B (2009) Fundamental concepts in genetics: effective population size and patterns of molecular evolution and variation. Nat Rev Genet 10:195–205

Ciosi M, Miller NJ, Toepfer S, Estoup A, Guillemaud T (2010) Stratified dispersal and increasing genetic variation during the invasion of Central Europe by the western corn rootworm, Diabrotica virgifera virgifera. Evol Appl 4:54–70

Clarke GM (2000) Inferring demography from genetics: a case study of the endangered golden sun moth, Synemon plan. In: Young AG, Clarke GM (eds) Genetics, demography and viability of fragmented populations. Cambridge University Press, Cambridge, pp 213–226

Coltman DW, O'Donoghue P, Jorgenson JT, Hogg JT, Strobeck C, Festa-Bianchet M (2003) Undesirable evolutionary consequences of trophy hunting. Nature 426:655–658

Crow JF, Aiko K (1984) Group selection for a polygenic behavioral trait: estimating the degree of population subdivision. Proc Natl Acad Sci U S A 19:6073–6077

De Laet J, Summers-Smith JD (2007) The status of the urban house sparrow Passer domesticus in north-western Europe: a review. J Ornithol 148(Suppl 2):275–278

Dejean T, Valentini A, Miquel C, Taberlet P, Bellemain E et al (2012) Improved detection of an alien invasive species through environmental DNA barcoding: the example of the American bullfrog Lithobates catesbeianus. J Appl Ecol 49:953–959

Dlugosch KM, Parker IM (2008) Founding events in species invasions: genetic variation, adaptive evolution, and the role of multiple introductions. Mol Ecol 17:431–449

Duncan RP, Blackburn TM, Sol D (2003) The ecology of bird introductions. Annu Rev Ecol Evol Syst 34:71–98

Estoup A, Guillemaud T (2010) Reconstructing routes of invasion using genetic data: why, how and so what? Mol Ecol 19:4113–4130

Ficetola GF, Miaud C, Pompanon F, Taberlet P (2008) Species detection using environmental DNA from water samples. Biol Lett 4:423–425

Frankham R (2005) Genetics and extinction. Biol Conserv 126:131–140

Frankham R, Ralls K (1998) Inbreeding leads to extinction. Nature 392:441–442

Genetic monitoring for managers. http://www.fws.gov/alaska/gem/mainPage_1.htm

Goyal M (2005) M.Sc. Dissertation, Preliminary survey of house sparrow (Passer domesticus) in three different areas of Haridwar, Uttaranchal. Wildlife Institute of India, Dehradun. Gurukul Kangri University, Haridwar

Groombridge JJ, Jones CG, Bruford MW, Nichols RA (2000) 'Ghost' alleles of the Mauritius kestrel. Nature 403:616

Hall LM, Willcox MS, Jones DS (1997) Association of enzyme inhibition with methods of museum skin preparation. Biotechniques 22:928–934
Hansen MM, Olivieri I, Waller DM, Nielsen EE, The GeM Working Group (2012) Monitoring adaptive genetic responses to environmental change. Mol Ecol 21:1311–1329
Hartl DL, Clark AG (2007) Principles of population genetics, 4th edn. Sinauer, Sunderland
Hebert PDN, Ratnasingham S, deWaard JR (2003) Barcoding animal life: cytochrome c oxidase subunit 1 divergences among closely related species. Proc R Soc B 270:96–99
Hitt NP, Frissell CA, Muhlfeld CC, Allendorf FW (2003) Spread of hybridization between native westslope cutthroat trout, Oncorhynchus clarkii lewisi, and nonnative rainbow trout, Oncorhynchus mykiss. Can J Fish Aquat Sci 60(12):1440–1451
Hoban S, Arntzen JA, Bruford MW, Godoy JA, Hoelzel RA, Segelbacher G, Vilà C, Bertorelle G (2014) Comparative evaluation of potential indicators and temporal sampling protocols for monitoring genetic erosion. Evol Appl 7:984–998
Hole DG, Whittingham MJ, Bradbury RB, Anderson GQA, Lee PLM, Wilson JD (2002) Widespread local house sparrow extinctions. Nature 418:139–149
Honnay O, Jacquemyn H, Van Looy K, Vandepitte K, Breyne P (2009) Temporal and spatial genetic variation in a metapopulation of the annual Erysimum cheiranthoides on stony river banks. J Evol 97:131–141
IUCN (2010) IUCN red list of threatened species. Version 2010.4. http://www.iucnredlist.org
Ivkosic SA, Gorman J, Lemic D, Mikac KM (2014) Genetic monitoring of western corn rootworm Coleoptera: Chrysomelidae) populations on a microgeographic scale. Environ Entomol 43:804–818
Jackson JA, Laikre L, Baker CS, Kendall KC (2012) Guidelines for collecting and maintaining archives for genetic monitoring. Conserv Genet Resour 4:527–536
Jensen H, Steinsland I, Ringsby TH, Saether B-E (2008) Evolutionary dynamics of a sexual ornament in the house sparrow (Passer domesticus): the role of indirect selection within and between sexes. Evolution 62:1275–1293
Johnson JA, Gilbert M, Virania MZ, Asima M, Mindell DP (2008) Temporal genetic analysis of the critically endangered oriental white-backed vulture in Pakistan. Biol Conserv 141:2403–2409
Jolly MT, Maitland PS, Genner MJ (2011) Genetic monitoring of two decades of hybridization between allis shad (Alosa alosa) and twaite shad (Alosa fallax). Conserv Genet 12:1087–1100
Kashtan N, Noor E, Alon U (2007) Varying environments can speed up evolution. Proc Natl Acad Sci U S A 104:13711–13716
Kekkonen J, Brommer JE (2014) Reducing the loss of genetic diversity associated with assisted colonization-like introductions of animals. Curr Zool. http://www.currentzoology.org/site_media/onlinefirst/downloadable_file/2014/11/19/Kekkonen.pdf
Kekkonen J, Hanski IK, Jensen H, Väisänen RA, Brommer JE (2011) Increased genetic differentiation in house sparrows after a strong population decline: from panmixia towards structure in a common bird. Biol Conserv 144:2931–2940
Koskinen MT, Sundell P, Piironen J, Primmer CR (2002) Genetic assessment of spatiotemporal evolutionary relationships and stocking effects in grayling (Thymallus thymallus, Salmonidae). Ecol Lett 5:193–205
Kruszewicz AG, Kruszewicz AH, Pawiak R, Mazurkiewicz M (1995) Bacteria in House Sparrow (Passer domesticus) and Tree Sparrow (Passer montanus) nestlings: occurrence and influence on growth and mortality. In: Pinowski J, Kavanagh B, Pinowska B (eds) Nestling mortality of granivorous birds due to microorganisms and toxic substances: Synthesis. PWN – Polish Scientific Press, Warsaw, pp 267–282
Laikre L, Allendorf FW, Aroner LC, Baker CS, Gregovich DP, Hansen MM, Jackson JA, Kendall KC, McKelvey K, Neel MC, Ryman N, Schwartz MK, Shortbull R, Stetz JB, Tallmon DA, Taylor BL, Vojta CD, Waller DM, Waples RS (2010) Neglect of genetic diversity in implementation of the conservation on biological diversity. Conserv Biol 24:86–88

Larsson LC, Charlier J, Laikre L, Ryman N (2009) Statistical power for detecting genetic divergence—organelle versus nuclear markers. Conserv Genet 10:1255–1264
Leberg P (2005) Genetic approaches for estimating the effective size of populations. J Wildl Manage 69:1385–1399
Leonard JA (2008) Ancient DNA applications for wildlife conservation. Mol Ecol 17:4186–4196
Lettieri T (2006) Recent applications of DNA microarray technology to toxicology and ecotoxicology. Environ Health Perspect 114:4–9
Lockwood JL, Hoopes MF, Marchetti MP (2007) Invasion ecology. Blackwell, Malden
Lowe S, Browne M, Boudjelas S, De Poorter M (2000) 100 of the World's Worst Invasive Alien Species A selection from the Global Invasive Species Database. Published by The Invasive Species Specialist Group (ISSG) a specialist group of the Species Survival Commission (SSC) of the World Conservation Union (IUCN), 12pp. First published as special lift-out in Aliens 12, December 2000. Updated and reprinted version: November 2004
Luikart G, Cornuet J-M, Allendorf FW (1999) Temporal changes in allele frequencies provide estimates of population bottleneck size. Conserv Biol 13:523–530
Mack RN, Simberloff D, Lonsdale WM, Evans H, Clout M, Bazzaz FA (2000) Biotic invasions: causes, epidemiology, global consequences, and control. Ecol Appl 10:689–710
Matocq MD, Villablanca FX (2001) Low genetic diversity in an endangered species: recent or historic pattern? Biol Conserv 98:61–68
McCarthy M (2003) Wood Pigeons' grain diet linked to the demise of House Sparrows. The Independent, 3rd September
Montgomery SB, Sammeth M, Gutierrez-Arcelus M, Lach RP, Ingle C, Nisbett J, Guigo R, Dermitzaki ET (2010) Transcriptome genetics using second generation sequencing in a Caucasian population. Nature 464:773–777
Morin PA, Luikart G, Wayne RK, Grp WSNP (2004) SNPs in ecology, evolution and conservation. Trends Ecol Evol 19:208–216
Mount DW (2001) Bioinformatics—sequence and genome analysis. Cold Spring Harbor Laboratory Press, New York
Murgui E, Macias A (2010) Changes in the House Sparrow Passer domesticus in Valencia (Spain) from 1998 to 2008. Bird Study 57:281–288
Nielsen EE, Hemmer-Hansen J, Poulsen N, Loeschcke V, Moen T, Johansen T, Mittelholzer C, Taranger G-L, Ogden R, CarvalhoGR (2009) Genomic signatures of local directional selection in a high gene flow marine organism; the Atlantic cod (Gadus morhua). BMC Evolutionary Biology 9:276
Ortego J, Yannic G, Shafer AB, Mainguy J, Festa-Bianchet M, Coltman D, Cöte SD (2011) Temporal dynamics of genetic variability in a mountain goat (Oreamnos americanus) population. Mol Ecol 20:1601–1611
Ou W, Takekawa S, Yamada T, Terada C, Uno H, Nagata J, Masuda R, Kaji K, Saitoh T (2014) Temporal change in the spatial genetic structure of a sika deer population with an expanding distribution range over a 15-year period. Popul Ecol 56:311–325
Ouborg NJ, Pertoldi C, Loeschcke V, Bijlsma R, Hedrick PW (2010) Conservation genetics in transition to conservation genomics. Trends Genet 26:177–187
Palmé A, Laikre L, Ryman N (2013) Monitoring reveals two genetically distinct brown trout populations remaining in stable sympatry over 20 years in tiny mountain lakes. Conserv Genet 14:795–808
Paulus M, Teubner D, Hochkirch A, Veith M (2014) Journey into the past: using cryogenically stored samples to reconstruct the invasion history of the quagga mussel (Dreissena rostriformis) in German river systems. Biol Invasions 16:2591–2597
Pichler FB, Baker CS (2000) Loss of genetic diversity in the endemic Hector's dolphin due to fisheries-related mortality. Proc R Soc B 267(1438):97–102
Ramakrishnan U, Hadley EA, Mountain JL (2005) Detecting past population bottlenecks using temporal genetic data. Mol Ecol 14:2915–2922
Reding DM, Freed LA, Cann RL, Fleischer RC (2010) Spatial and temporal patterns of genetic diversity in an endangered Hawaiian honeycreeper, the Hawaii Akepa (Loxops coccineus coccineus). Conserv Genet 11:225–240

Reed DH, Frankham R (2003) The correlation between population fitness and genetic diversity. Conserv Biol 17:230–237
Rollins LA, Woolnough AP, Wilton AN, Sinclair R, Sherwin WB (2009) Invasive species can't cover their tracks: using microsatellites to assist management of starling (Sturnus vulgaris) populations in Western Australia. Mol Ecol 18:1560–1573
Roman J, Darling JA (2007) Paradox lost: genetic diversity and the success of aquatic invasions. Trends Ecol Evol 22:454–464
Ryman N, Utter F, Hindar K (1995) Introgression, supportive breeding, and genetic conservation. In: Ballou JD et al (eds) Population management for survival and recovery. Columbia University Press, New York, pp 341–365
Ryman N, Palm S, André C, Carvalho G, Dahlgren T, Jorde PE, Laikre L, Larsson L, Palmè A, Ruzzante D (2006) Power for detecting genetic divergence: differences between statistical methods and marker loci. Mol Ecol 15:255–261
Sakai AK, Allendorf FW, Holt JS, Lodge DM, Molofsky J, With KA, Baughmann S, Cabin RJ, Cohen JE, Ellstrand NC, McCauley DE, O'Neil P, Parker IM, Thompson JN, Weller SG (2001) The population biology of invasive species. Annu Rev Ecol Syst 32:305–332
Sarre SD, Georges A (2009) Genetics in conservation and wildlife management: a revolution since Caughley. Wildl Res 36:70–80
Sax DF, Stachowicz JJ, Brown JH, Dawson MN, Bruno JF, Gaines SD, Grosberg RG, Hastings A, Holt RD, Mayfield MM, O'Connor MI, Rice WM (2007) Ecological and evolutionary insights from species invasions. Trends Ecol Evol 22:466–471
Schwartz MK, Pilgrim KL, McKelvey KS, Lindquist EL, Claar JJ, Loch S, Ruggiero LF (2004) Hybridization between Canada lynx and bobcats: genetic results and management implications. Conserv Genet 5:349–355
Schwartz MK, Luikart G, Waples RS (2007) Genetic monitoring as a promising tool for conservation and management. Trends Ecol Evol 22:25–33
Sefc KM, Payne RB, Sorenson MD (2003) Microsatellite amplification from museum feather samples: effects of fragment size and template concentrations on genotyping errors. Auk 120:982–989
Selkoe KA, Toonen RJ (2006) Microsatellites for ecologists: a practical guide to using and evaluating microsatellite markers. Ecol Lett 9:615–629
Serbezov D, Jorde PE, Bernatchez L, Olsen EM, Vøllestad LA (2012) Short-term genetic changes: evaluating effective population size estimates in a comprehensively described brown trout (Salmo trutta) population. Genetics 191:579–592
Siriwardena G, Robinson R, Crick H (2002) Status and population trends of the House Sparrow Passer domesticus in Great Britain. In: Crick H, Robinson R, Appleton G, Clark N, Rickard A (eds) Investigation into the causes of the decline of Starlings and House Sparrows in Great Britain. BTO/DEFRA, Bristol. BTO research report number 290
Soulé ME (1987) Viable populations for conservation. Cambridge University Press, Cambridge
Spencer CC, Neigel JE, Leberg PL (2000) Experimental evaluation of the usefulness of microsatellite DNA for detecting demographic bottlenecks. Mol Ecol 9:1517–1528
Stepien CA, Brown JE, Neilson ME, Tumeo MA (2005) Genetic diversity of invasive species in the Great Lakes versus their Eurasian source populations: insights for risk analysis. Risk Anal 25:1043–1060
Stetz JB, Kendall KC, Vojta CD, Genetic Monitoring (GeM) Working Group (2011) Genetic monitoring for managers: a new online resource. J Fish Wildl Man 2:216–19
Summers-Smith JD (1963) The house sparrow. Collins, London
Summers-Smith JD (1988) The sparrows. T. & A.D. Poyser, Staffordshire
Summers-Smith JD (1999) Current status of the house sparrow in Great Britain. Br Wildl 10:381–386
Summers-Smith D (2003) Decline of the House sparrow: a review. Br Birds. www.ndoc.org.uk
Tallmon D, Gregovich D, Waples RS, Baker CS, Jackson J, Taylor BL, Archer E, Martien KK, Allendorf FW, Schwartz MK (2010) When are genetic methods useful for estimating contemporary abundance and detecting population trends? Mol Ecol Res 10:684–692

Therkildsen NO, Hemmer-Hansen J, Hedeholm RB, Wisz MS, Pampoulie C, Meldrup D, Bonanomi S, Retzel A, Olsen SM, Nielsen EE (2013) Spatiotemporal SNP analysis reveals pronounced biocomplexity at the Northern range margin of Atlantic cod Gadus morhua. Evol Appl 6:690–705

Thomsen PF, Kielgast J, Iversen LL, Wiuf C, Rasmussen M, Gilbert MTP, Orlando L, Willerslev E (2012) Monitoring endangered freshwater biodiversity using environmental DNA. Mol Ecol 21:2565–2573

Tokarska M, Kawalko A, Wójcik JM, Pertoldi C (2009) Genetic variability in the European bison (Bison bonasus) population from Białowiez˙a forest over 50 years. Biol J Linn Soc 97: 801–809

Tracy LN, Jamieson IG (2011) Historic DNA reveals contemporary population structure results from anthropogenic effects, not pre-fragmentation patterns. Conserv Genet 12:517–526

Tufto J, Ringsby TH, Dhondt AA, Adriaensen F, Matthysen E (2005) A parametric model for estimation of dispersal patterns applied to five passerine spatially structured populations. Am Nat 165:13–26

Väisänen R (2003) Linnut 38(3):26–27

Valkama J, Vepsäläinen V, Lehikoinen A (2011) Suomen III Lintuatlas. Luonnontieteellinen keskusmuseo ja ympäristöministeriö. http://atlas3.lintuatlas.fi (viitattu [15.12.2014]) ISBN 978-952-10-6918-5

Van Straalen NM, Roelofs D (2006) An introduction to ecological genomics. Oxford University Press, Oxford

Varvio SL, Chakraborty R, Nei M (1986) Genetic variation in subdivided populations and conservation genetics. Heredity 57:189–198

Vincent K (2005) PhD Thesis, Investigating the causes of the decline of the urban House Sparrow Passer domesticus population in Britain. De Montfort University, Great Britain

Wandeler P, Hoeck PEA, Keller LF (2007) Back to the future: museum specimens in population genetics. Trends Ecol Evol 22:634–642

Wares JP, Hughes AR, Grosberg RK (2005) Mechanisms that drive evolutionary change: insights from species introductions and invasions. In: Species invasions: insights into ecology, evolution and biogeography. Sinauer, Sunderland

Whitney KD, Gabler CA (2008) Rapid evolution in introduced species, 'invasive traits' and recipient communities: challenges for predicting invasive potential. Divers Distrib 14:569–580

Chapter 14
Fate of the Mongooses and the Genet (Carnivora) in Mediterranean Europe: None Native, All Invasive?

Philippe Gaubert

Introduction: The Mediterranean Basin Context

Superimposed forces including climatic fluctuations, natural processes of colonization and human-mediated introductions have deeply impacted the biodiversity of the Mediterranean Basin (MB) (Blondel et al. 2010). Palaeontological and genetic studies have suggested that sweepstake migrations of nonflying vertebrates from North Africa to southwestern Europe (and vice versa) had occurred across the Strait of Gibraltar during sea-level fluctuations associated with the Pleistocene last glaciations (e.g., Pinho et al. 2007; Geraads 2010). At that time, the sea-level depression reaching 140 m below present made large vegetated islands emerged across the Strait of Gibraltar, rending possible natural crossings by nonflying vertebrates such as mammals (Masseti 2009).

The MB, connected by cultural exchanges since prehistoric times, provides an outstanding framework to study species introductions, notably in mammals (Dobson 1998). However, this task is rendered complicated by the fact that early human-mediated translocations of mammals from both sides of the western Mediterranean were sometimes contemporaneous with natural sweepstake dispersals between the two continents (see Zeder 2008). Consequently, it remains difficult to trace back the events and processes at the origin of the extant mammalian fauna of the MB, without adopting an integrative approach that combines a wide spectrum of evidence.

Distinguishing between natural and human-mediated dispersals of mammals into Europe is a critical issue, since the MB is one of the IUCN biodiversity hotspots that most suffers from the pressure of introduced species (Cuttelod et al. 2008), whereas paradoxically, several mammals supposed to have originated from historical

P. Gaubert (✉)
Institut des Sciences de l'Evolution de Montpellier (ISEM)—UM2-CNRS-IRD-EPHE-CIRAD, Université de Montpellier, Place Eugène Bataillon—CC 64, Montpellier Cedex 05 34095, France
e-mail: philippe.gaubert@umontpellier.fr

F.M. Angelici (ed.), *Problematic Wildlife*, DOI 10.1007/978-3-319-22246-2_14

introductions are considered as a component of the Mediterranean bio-cultural heritage (Gippoliti and Amori 2006). Human-mediated introductions since the end of the Würmian glaciations (14–12 kyr ago) have deeply impacted current patterns of biodiversity in the MB (Vigne et al. 2009). These led to dramatic levels of endemic extinction, at the same time counterbalanced by the establishment of various allochthonous taxa (Masseti 2009).

The intensity of introductions significantly increased from the first millennium BC, following massive human migrations from eastern to western borders of the Mediterranean Sea that opened several dozens of potential routes to the human-mediated dispersal of species across the MB (Ciolek 2011). Historical introductions had motives mostly related to agricultural practices (domestication, pest control) but also to more "esthetic" and political considerations, including entertainment, cultural exchanges and pet trade (Hughes 2003; Morales et al. 1995). More recent introductions (twentieth to twenty-first centuries) of mammals still originated from such motives (e.g., Delibes 1977). Thus, the long history of introductions in the MB has resulted in serial faunal turnovers involving local extinctions of endemic fauna and serial establishments of introduced species, with new "invaders" regularly entering the native fauna (Cuttelod et al. 2008). Assessing the impact of such introductions on the Mediterranean fauna is politically and economically crucial, but has proven a difficult task that may deserve a "case-by-case" (i.e., taxonomically and/or geographically) approach. By focusing on a group of Afro-Asian small Carnivores present in southern Europe, we intend to provide an exhaustive reassessment of their status that shall clarify their ecological impact in the MB.

Study Model and Objectives: Status of the Mongooses and the Genet Present in Europe

Carnivores are among the most successful mammalian invaders, with species such as the domestic cat (*Felis silvestris catus*) and dog (*Canis lupus familiaris*), the American mink (*Neovison vison*), and the small Indian mongoose (*Herpestes auropunctatus*; see below) each established in more than 30 countries or islands around the world (Clout and Russell 2007). Middle-sized representatives ("mesocarnivores") such as those above-mentioned have been pinpointed for their deleterious impacts on the native communities of Carnivores and their preys (Bonesi and Palazon 2007), notably in the context of endemic fauna (Medina et al. 2011). In the MB, mustelids (martens, weasels, badgers) seem to have been the earliest Carnivores transported on islands (Masseti 1995). This pattern is congruent with the earliest molecular estimate of transportation of weasels ca. 10 kya (Lebarbenchon et al. 2010). However, given their natural, circum-Mediterranean distribution at the Pleistocene period, it is unclear whether all the mustelids present on Mediterranean islands were introduced or natural dispersers (Masseti 1995).

A contrario, the establishment of another lineage of small Carnivores including mongooses (Herpestidae) and genets (Viverridae) in Europe has traditionally been considered as more recent. Four species of herpestids and viverrids are or have recently been recorded from Mediterranean Europe and none of them has been considered native: the Indian grey mongoose *Herpestes edwardsii*, the small Indian mongoose *H. auropunctatus*, the Egyptian mongoose *H. ichneumon*, and the common genet *Genetta genetta* (Long 2003). Those species are medium-sized predators naturally distributed in the tropical and subtropical zones of the Old World. They were supposedly introduced in Europe at various historical times, from the Middle Age to the twentieth century. Because such small Carnivores actively predate on species that can have a deleterious role in agriculture (e.g., rodents) or be directly harmful to humans (e.g., snakes), and also because they are commensal and can be kept as pets, they were good candidates to be spread through Mediterranean's trading and political networks.[1]

Although the introductions of the Indian grey and small Indian mongooses in Europe are quite well documented, the introduction history of the Egyptian mongoose and the common genet has remained highly speculative. Importantly, it is unclear whether those four Carnivores have or had deleterious impacts on the native European fauna, and how their niches/ranges in the MB are characterized. Despite such lack of empirical data, local control operations—notably targeting mongooses in Portugal and Croatia—have been attempted with various levels of "success" (Hays and Conant 2007; Barun et al. 2011; Beja et al. 2009). In order to clarify their history of introduction and their status within the European fauna, I will (1) briefly review the natural history of the Afro-Asian herpestids and viverrids present in the MB and their relationships with humans in their native ranges, (2) review their history of introduction, and for the lesser known species the speculations that have surrounded the factors promoting their possible introduction in Europe, (3) detail the recent enlightenments—and sometimes contradictions—brought by genetic analyses (mostly phylogeography) as to the dispersal histories of those small Carnivores, and (4) summarize their range dynamics and ecological interactions with the European fauna. In view of the reviewed evidence, I will then conclude on the expected "invasiveness" of those species in Europe and will eventually open a prospective on the strategies that could be adopted to improve our understanding of small Carnivores' establishments in the MB.

[1] Another viverrid, the African civet *Civettictis civetta* was known in Europe from the fifteenth century. It was used as a political gift between southern (Mediterranean) and northern states because of its great value related to the musk produced by its perineal glands (Dannenfeldt 1985; Morales Muñiz 2000). Although African civets were kept alive at some royal European menageries (Dannenfeldt 1985), there has never been any evidence of escaped individuals that would have established in Europe.

Natural History of the Mongooses and the Genet in Europe and Their Relationships with Humans in Their Native Ranges

The Indian grey mongoose occurs in the Indian subcontinent and at the eastern fringe of the Middle East (Fig. 14.1). It seems preferentially commensal with humans as it is often recorded near human settlements in central India where it frequently scavenges on carrion. The species is most common in disturbed areas, in dry secondary forests and thorn forests. It is generally diurnal, goes solitary or by mating pair and mainly feeds on small mammals, insects and reptiles (Santiapillai et al. 2000; Choudhury et al. 2013). Litter size is 2–4 and there are 2–3 litters a year (Gilchrist et al. 2009).

The small Indian mongoose ranges from the eastern fringe of the Middle East to the Indian subcontinent and southern China (Veron et al. 2007; Gilchrist et al. 2009) (Fig. 14.2). The species is found in a variety of open habitats and tolerates a large degree of habitat conversion (notably in its introduced range). It seems quite resistant to persecution and is still recorded from intensely hunted and cultivated areas (Wozencraft et al. 2008). It goes solitary or by pair and feeds during both day and night on a wide range of items including arthropods, small mammals, birds, reptiles, frogs and crustaceans. Mean litter size is 2 (range = 1–5) and there are 2–3 litters a year (Gilchrist et al. 2009).

Because they are natural predators of snakes, mongooses can be very popular animals in Asia, in contrast with them being viewed as pests in most parts of their introduced ranges. It is, in fact, mainly for this reason (snake killing, but also rodent-killing) that Asian mongooses have been introduced in various parts of the world (Hays and Conant 2007). Archaeological evidence from Harappan sites (western

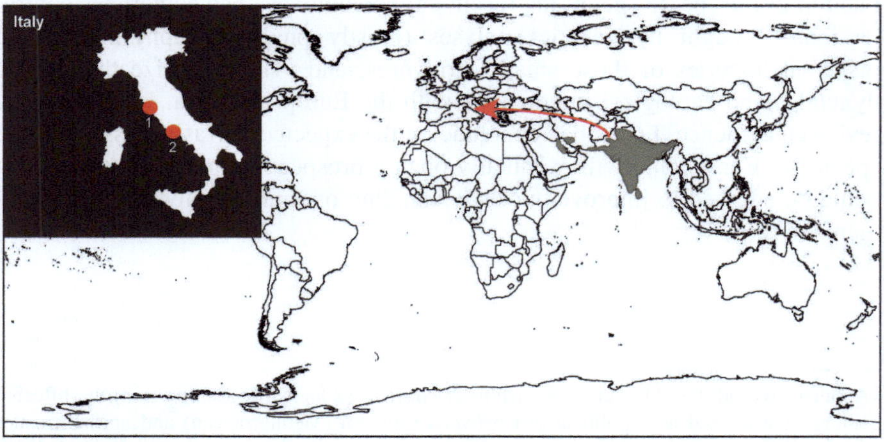

Fig. 14.1 Distribution of the Indian grey mongoose *Herpestes edwardsii*. *Grey areas* represent the native range of the species. *Red arrow* points to the location where the species was introduced in Europe. The *upper left inset* shows the two locations where the species was introduced (*red circles*) in Italy (Angelici 2003): *1* Capalbio; *2* Circeo NP

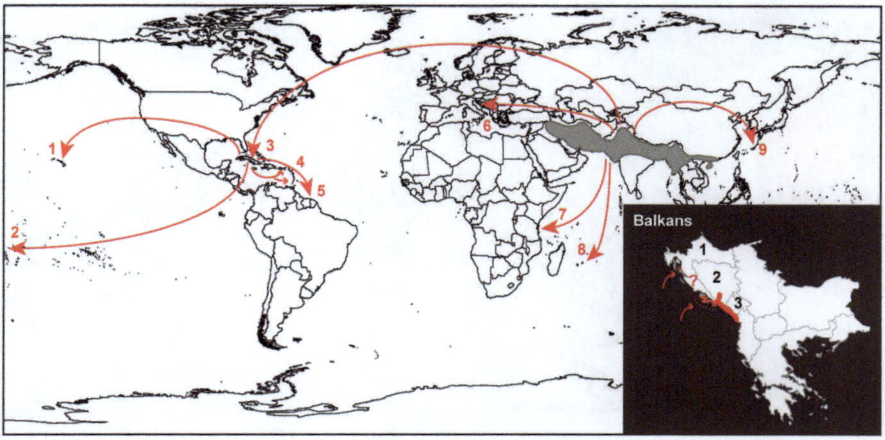

Fig. 14.2 Distribution of the small Indian mongoose *Herpestes auropunctatus. Grey areas* represent the native range of the species. *Red arrows* point to the main paths and locations where the species was introduced (Hays and Conant 2007; Thulin et al. 2006): *1*, Hawaii Isl.; *2*, Fiji Isl.; *3*, Jamaica; *4*, Puerto Rico and Lesser Antilles Isl.; *5*, Guianas; *6*, Balkans; *7*, Mafia Isl.; *8*, Mauritius Isl.; *9*, Okinawa—Amami-Oshima Isl. The *lower right inset* shows the potential range (in *red*) of the species in the Balkans (Ćirović et al. 2011): *1*, Croatia; 2, Bosnia-Herzegovina; *3*, Montenegro. *Small red arrows* indicate introductions on several Adriatic islands

Indian subcontinent; fourth to first millennia BC) supports the idea that mongooses frequented human habitations, possibly as semidomesticated animals (Lodrick 1982). The "Brahmin and the Mongoose" is a famous folktale from India that is another token of the good consideration that the mongoose benefits there. It describes the impulsive killing of a loyal mongoose that had protected a baby from snakes, and thus is a warning against hasty actions (Emeneau 1940). In India, the mongoose is also associated to opulence and generally represents the god of wealth in the Buddhist iconography (Lodrick 1982).

Small Indian mongooses were possibly introduced during the second or first millennium BC in eastern Arabia and Bahrain (Uerpmann 1995) and were found buried in the temple of Saar (Dobney and Jaques 1994) where they were probably linked to religious rituals. As of today, the small Indian and the Indian grey mongooses are frequently captured and sold as pets, notably in India and Nepal. In central India, people consider the two species of mongooses to be sacred (Wozencraft et al. 2008; Choudhury et al. 2013). Another use is made by the Jogi tribes in Pakistan, whom capture the small Indian mongoose for stage fights with cobras (Gilchrist et al. 2009).

The Egyptian mongoose is widely distributed in northern and sub-Saharan Africa and the coastal Near East, avoiding deserts, high rainfall forest areas and the southern African steppe (Fig. 14.3). It primarily occurs in habitats having dense understory vegetation, but is also frequent in cultivated zones. The species is generally solitary, although pairs and families of 4–6 individuals can be observed. It is mostly diurnal and has an opportunistic, omnivorous diet including small mammals, birds,

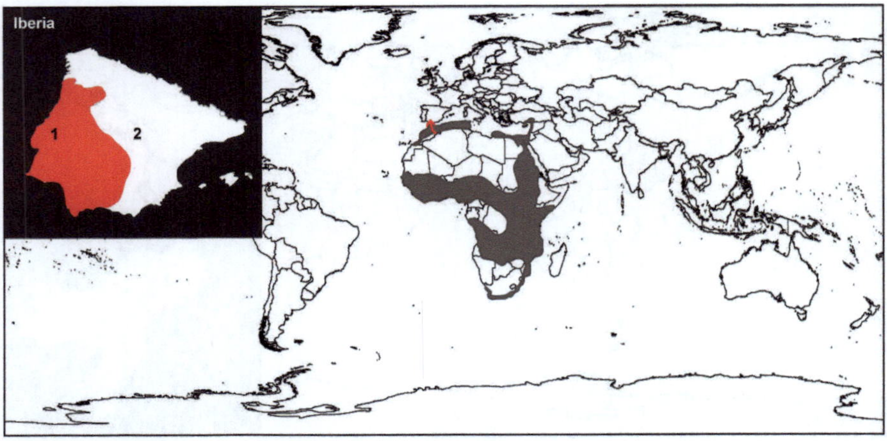

Fig. 14.3 Distribution of the Egyptian mongoose *Herpestes ichneumon*. *Grey areas* represent the native range of the species. *Red arrow points* to the location from which the species dispersed in Europe. The *upper left inset* shows the potential range of the species (in *red*) in Iberia (Balmori and Carbonell 2012): *1*, Portugal; *2*, Spain

arthropods, amphibians, reptiles, fish, gastropods, carrion, fungi, fruit and other plant material. Mean litter size is estimated to be 3.3 (range = 1–4), and there is usually a single litter per year (Palomares 2013).

Ichneumon is derived from the ancient Greek for "tracker," possibly originating from the mongoose's supposed ability to track crocodile dens and feed on their eggs. In addition to this ancient belief related to the sacred crocodiles, its capacity of preying on snakes made the Egyptian mongoose played an important part in the bestiary of the Pharaonic Egypt. Representations of the species can be found on the walls of tombs and temples in Thebes and Saqqara as early as the period of the Old Kingdom (2800–2150 BC). The Egyptian mongoose was associated to several deities including Atum, Re and Horus. It was also sacred to Mafdet, the goddess providing protection from snakebite, and mummified Egyptian mongooses were discovered inside bronze statues of the lion-headed goddess Uto. A legend relates the defeat of the thunder snake Apophis by the mongoose as the surrogate of the god Letopolis, reflecting again the representation of the species as a beneficial snake-killer. In Ancient Egypt and later in the Arabic culture (as late as nineth century AD), the Egyptian mongoose was believed to alter its size between day and night, becoming very small at night (as a shrew or a mouse) and being able to kill snakes by suddenly increasing its size if captured (Stuart 1988). Depictions of mongooses hold by their tail or on a leash suggest that the species was tamed in Ancient Egypt, possibly as a household pet, biocontrol agent, or hunting animal. Egyptian mongooses were kept in temples as votive offerings until the Greco-Roman times. However, they seem not to have been domesticated or bred in captivity (Osborn and Osbornova 1998).

The status of the species is said to have moved from beneficial to pest once the domestic fowl, on which it can prey, was introduced in Egypt (Osborn and Osbornova

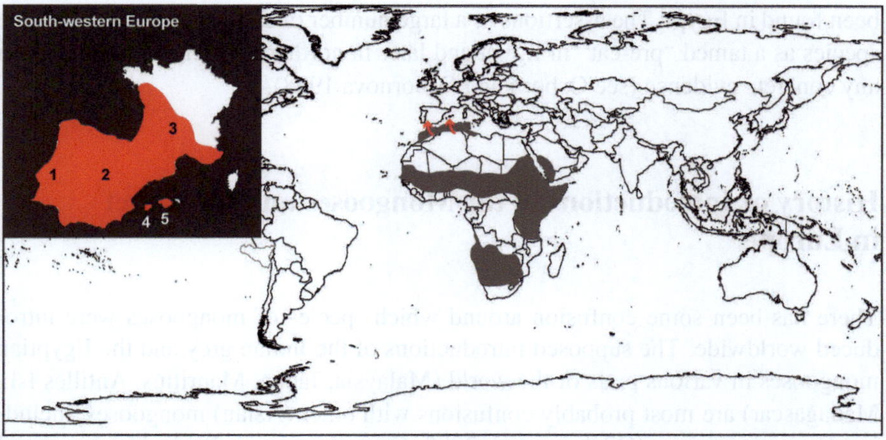

Fig. 14.4 Distribution of the common genet *Genetta genetta*. *Grey areas* represent the native range of the species. *Red arrows* point to the locations where the species was introduced in Europe. The *upper left inset* shows the potential range of the species (in *red*) in southwestern Europe (Gaubert et al. 2008; Delibes and Gaubert 2013): *1*, Portugal; *2*, Spain, *3*, France; *4*, Ibiza Isl.; *5*, Mallorca + Cabrera Isl

1998). More likely, the arising of the domestic cat as the preferred household pet and biocontrol agent against rodents all around the MB together with the spread of monotheist religions should have brought forward the discredit on the Egyptian mongoose. There is, to date, no concrete evidence for the domestication or taming of *H. ichneumon* by post-Roman North African cultures, although the discovery of a tibia from Punic Sardinia ca. fifth to fourth century BC shows that episodic, historical translocations of the species might have occurred as early as the Carthaginian period (Campanella and Wilkens 2004).

The common genet has a wide distribution in northern and sub-Saharan Africa and is also present in southern Peninsular Arabia (Fig. 14.4). It avoids deserts, rainforests, dense woodlands and woodland-moist savannah mosaics, and is mostly found in open savannahs, oak forests and bushy areas with woody or rocky shelters. The species has a predisposition to live in the vicinity of human settlements. It is solitary and nocturnal, and is a generalist feeder consuming small mammals, arthropods, birds, eggs, reptiles, amphibians, fish, fruits, mushrooms and garbage. Mean litter size is 2–2.6 (range = 1–4), and there seems to be a single litter per year (Delibes and Gaubert 2013).

Remains of common genets associated to hunting sites were found in the Late Pleistocene of northern Africa (Ouchaou and Amani 2002). Although possible associations between common genets and humans have been the subject of numerous speculations, the archaeological and historical evidence is scarce. The species was pictured in swamp scenes climbing papyrus stems—often in association with the Egyptian mongoose—on the walls of tombs and temples of Ancient Egypt (ca. 3000–2200 BC). The common genet was seldom represented in later periods (i.e., the XVIIIth dynasty: 1600–1300 BC). No mummies of the species have ever been found in Egypt. The assertions of a large number of authors as to the use of the species as a tamed "pre-cat" in Egypt and later in northern Africa are not based on any concrete evidence (see Osborn and Osbornova 1998).

History of Introductions of the Mongooses and the Genet in Europe

There has been some confusion around which species of mongooses were introduced worldwide. The supposed introductions of the Indian grey and the Egyptian mongooses in various parts of the world (Malaysia, Japan, Mauritius, Antilles Isl., Madagascar) are most probably confusions with other (Asian) mongooses, including *H. auropunctatus* (Gilchrist et al. 2009; Choudhury et al. 2013). The history of introduction of the small Indian mongoose in Italy is relatively well documented. Probably in 1952, the owner of a hotel in San Felice Circeo (Central Italy) released a few mongooses—acquired from the Giardino Zoologico di Roma—in her park with the aim of removing adders. Those animals escaped during the second half of the 1950s, and a population established into nearby areas of the Circeo National Park. At the end of the 1970s, mongooses reached their maximum range, which covered ca. 15 km^2 from the whole Circeo promontory to most of the protected area south of Molella Bay (Carpaneto 1990; Angelici 2003). The collection of a specimen of *H. edwardsii* from Capalbio (Tuscany), ca. 200 km northward of Circeo, shows that at least two events of introduction occurred in Italy at ca. 10 years of interval, probably from the same captive stock. The reasons for this second introduction remain unknown. However, a letter from 1966 written by the director of the Capalbio hunting estate quoted a negative advice provided by the Laboratorio di Zoologia Applicata alla Caccia (future Istituto Superiore per la Protezione e la Ricerca Ambientale) about the introduction of mongooses to control populations of adders, reinforcing the idea that using these animals to control venomous snakes was rather widespread at that time. The Indian grey mongoose is now considered extinct in Italy (Angelici 2003), a unique fate among the herpestids and viverrids introduced in Europe (see below).

The small Indian mongoose is considered as one of the world's 100 worst invasive species. It was introduced to many islands in the Pacific and Indian Oceans and the Caribbean Sea, mostly in the late nineteenth and early twentieth centuries, to control rats and poisonous snakes in sugar cane fields (Lowe et al. 2000). The introduction of *H. auropunctatus* in Europe is well documented. Seven males and four females purchased in India were released in 1910 on the island of Mljet (current Croatia), formerly known as the "island of snakes." At that time, the Austro-Hungarian authorities had decided to introduce wild mongooses in order to exterminate the horned viper *Vipera ammodytes* from the island. Given the rapid decrease in the number of snakes and the growing numbers of mongooses on Mljet Isl., the species was soon introduced on nearby islands (Korcula, Peljesac, Brac) between 1921 and 1927. Afterwards, several attempts of introduction—with various levels of success—were planned onto a series of Adriatic islands but also in the mainland of former Yugoslavia (currently Bosnia-Herzegovina and Macedonia) until the 1970s (Tvrtkovic and Krystufek 1990; Krystufek and Tvrtkovic 1992). Interestingly, the historical record kept track of a transportation of ca. 100 mongooses from Mljet Isl. to Venezuela around 1926, thus suggesting that part of the small Indian mongooses introduced in the Caribbean Isl. originated from Europe (Tresic Pavicic 1936).

Conversely, the scenario of introduction of the Egyptian mongoose in Europe is highly speculative. The absence of paleontological records in southwestern Iberia—the European range of the species—and the existence of archaeological remains dated from the Arab occupation led some authors to postulate an introduction of the Egyptian mongoose associated to migrant Berber farmers between the eighth and thirteenth centuries AD (Riquelme-Cantal et al. 2008; Detry et al. 2011). However, this scenario is somewhat contradicted by the discovery of an Egyptian mongoose remain in a Carthaginian site from the fifth to fourth centuries BC in Sardinia (Campanella and Wilkens 2004), evidencing the historical transportation of the species by an earlier civilization. Besides, it has long been said that there is no evidence for the domestication or taming of *H. ichneumon* by North African people during historical times (Geoffroy Saint-Hilaire 1813). As a matter of fact, traces of manipulation or taming as could be the case with tooth abrasion, presence of associated artifacts or intentional disposal practices, have never been observed on any of the archaeological remains found in Europe.

The scenario of introduction of the common genet in southwestern Europe (the species is present from southern Iberia to southwestern France and in several Balearic islands) is highly speculative and calls to earlier periods than the Arab conquest. The Greek historian Herodotus (fifth century BC) mentioned a "weasel from Tartessos" (southwestern Iberia) similar to the one found in Libya, which has been identified as a common genet (Amigues 1999). This led authors to suggest an early introduction of the species through the political network between the Greek colonies and the Kingdom of Tartessos. According to Posidonius (first century BC), this "weasel" was used in southern Iberia as a bio-control agent against rabbit proliferation (Amigues 1999). The common genet was also mentioned in a faunistic list from the Gallic site of Ambrussum (southern France) as associated with domestic furniture from the third century BC (Columeau 1979). However, the remains of the animal have been lost.

Despite such possibilities for anterior introductions, the common genet has traditionally been associated to the Arab conquerors of Europe. A legend relates that after the defeat of Moor armies near Poitiers, France (732 AD), the King's Majordomo Charles Martel found in the loot of the defeated armies such a great quantity of furs—but also living animals—belonging to the common genet that he decided to create the "Ordre de la Genette" (Favyn 1620). Although this order of chivalry is a total myth, this narrative long stood as the main evidence supporting the introduction of the species through Arab invasions. Such hypothesis was further supported by the fact that there is no fossil record of the species in Europe and the only known archaeological remain dates back to the Almohads—an Arab dynasty—in Portugal, at the thirteenth century AD (Morales 1994).

Dispersal Histories of the Mongooses and the Genet in Europe: What Does Genetics Say?

The material representing the extinct Indian grey mongooses in Italy is very scarce. Only four specimens are known to be preserved in collections: three flat skins from the early 1960s are kept at the ISPRA museum (Ozzano dell'Emilia, Italy) and one mounted specimen from the 1970s is exhibited at the museum visitors' centre of Sabaudia at Circeo NP (Angelici 2003). Those represent the two distinct sites where the species was introduced, including Circeo NP and Capalbio (Tuscany), ca. 200 km north of the former. To date, a single genetic analysis based on mitochondrial DNA (mtDNA) and including the four remaining specimens of the mongooses introduced in Italy has been conducted. It clearly confirmed the Indian grey mongoose as the species having been introduced in Italy during the twentieth century and traced the introduced pool's origin to Pakistan or India, which is the core distribution of the species in its native range. The mtDNA diversity of Italian mongooses was null, thus suggesting a very limited number of founders (Gaubert and Zenatello 2009). It is likely that the low number of preserved Italian individuals will significantly limit the contribution of future genetic studies on the assessment of the species' introduction in Europe.

The introduction history of the small Indian mongoose has been assessed in more details. Genetics—and notably, rapidly evolving markers such as microsatellites—has shown its utility in tracing the introductions of the small Indian mongoose worldwide and allowed to discover new paths of transportations (Thulin et al. 2006; Watari et al. 2011). It has also provided insights into the demographic characteristics of introduced populations and potential admixture with *H. edwardsii* in the species' native range that may be used to better delineate the dynamics of the introduced populations (Thulin et al. 2006). Despite this, no detailed genetic study has so far been conducted on the small Indian mongooses introduced in Europe. To date, only a study on the systematic status of the small Indian mongoose and the Javan mongoose *H. javanicus* confirmed that the former was the species introduced in Croatia (Veron et al. 2007). Thus, further genetic investigations will have to be undertaken to characterize in detail the introduction patterns of the small Indian mongoose in the Balkans.

The dispersal history of the Egyptian mongoose has been assessed by a mitochondrial analysis based on ca. 90 samples from Africa, the Middle East and Europe (Gaubert et al. 2011). The results of this study radically contradicted the established idea that the Egyptian mongoose was introduced in Europe. Instead, Gaubert et al. (2011) proposed a natural crossing of the Mediterranean Sea by *H. ichneumon* via the Strait of Gibraltar during the Middle Pleistocene, long before the earliest (Paleolithic) human exchanges between North Africa and Europe. The strong genetic differentiation between European and North African haplogroups, the significant level of genetic diversity found in Europe, and the important phenotypic differences between European and North African mongooses all pointed to a scenario of long-term in situ evolution of European populations. These molecular results supported the hypothesis that natural dispersal across the Strait of Gibraltar was possible for nonflying vertebrates during the Pleistocene cyclical lowering of sea levels. The swimming abilities of the Egyptian mongoose make plausible a sweepstake migration using a partially emerged shoal such as the archipelago of Cape Spartel (where the mythic city of *Atlantis* was possibly located) that is now 56 to 200 m below sea level (Collina-Girard 2001). The long-term stability of mongooses' effective population size in Europe was supported by various genetic indices and the remarkable correspondence between the limits of the proposed ice age refugium in southwestern Iberia (Hewitt 1996) and the distribution of suitable ecological conditions for the species (specifically, low rainfall and warm temperatures; Borralho et al. 1996). Niche modelling approaches have since supplied independent evidence for the long-term stability (climatic niche conservatism) of the Egyptian mongoose in southwestern Iberia (Papeş et al. 2015).

The introduction scenario of the common genet in Europe has been assessed by the genetic analysis of ca. 180 individuals from the native and introduced species' ranges, using mtDNA (Gaubert et al. 2009; Gaubert et al. 2011) and more recently, microsatellite markers (Gaubert et al. 2015). The combined evidence supported multiple introductions from North Africa into Europe, including the Balearic Isl. (with three distinct introduction events on Ibiza, Mallorca and later Cabrera), southwestern Iberia (corresponding to the Tartessian Kingdom's zone of influence), and possibly northeastern Spain and southwestern France (secondary introduction from Iberia for the latter). Those studies suggest that the common genet was intentionally introduced in southern Iberia at a time (<300 bc) antedating the Arab invasion, possibly via Phoenicians' commercial routes. Subsequent introduction in France, long-term genetic drift, and admixture between the Iberian and French pools likely shaped the species' genetic variation currently observed in continental Europe. The mtDNA-based demographic scenario of multiple, historical introductions of common genets in Europe followed by sudden population expansion is characteristic of populations at disequilibrium (Gaubert et al. 2009; Gaubert et al. 2011). Such scenario was supported by niche modelling analysis through the detection of a climatic niche shift in the northern European range of the species (Papeş et al. 2015). Altogether, these results suggest that an exceptional combination of factors including multiple introductions, local admixture, and ecological adaptation promoted the successful spread of the common genet in continental Europe.

Range Dynamics of the Mongooses and the Genet in Europe and Assessment of Their Ecological Interactions with the European Fauna

The number of Indian grey mongooses in Italy abruptly decreased from the early 1980s, and the species was considered extinct by 1984. A survey conducted in summer 1984 failed to recover any evidence for the presence of the mongoose (Biondi 1985). Although poorly documented, it is probable that the extinction process of

H. edwardsii in Italy might have taken place quickly after the species reached its maximum range between 1978 and 1980 (Carpaneto 1990), possibly due to harsh winters. Mongooses were seen wandering in villages close to the Circeo promontory, apparently searching for food and shelter. The species showed a tame, diurnal behavior, consuming tourists' leftovers and accepting direct feeding from humans. Its sudden extinction in Italy fits with crashes observed in populations with very restricted ranges within the 25 years following their time of introduction (Duncan and Forsyth 2006). Because the mitochondrial diversity among Italian individuals was null, it is reasonable to conclude that a combination of deleterious factors including low genetic diversity, restricted range, and non-adaptation to western Palearctic winter conditions was likely responsible for the extinction of the species in Italy (Gaubert and Zenatello 2009). Documentation of interspecific competition with native Carnivores is scarce, although a dominance of the Indian grey mongoose over polecats *Mustela putorius* was suggested during the years of mongooses' maximal expansion. During the brief establishment of the Indian grey mongoose in Circeo NP, no impact on the density of black rats *Rattus rattus* was observed (Carpaneto 1990).

The small Indian mongoose successfully established and spread in Europe, with the notable exception of the island of Brac where it went extinct for unknown reasons. The two introduction sites on the continent (Peljesac Peninsula, Bosnia-Herzegovina and Mostar, Macedonia) are supposed to be the sources of the populations having spread ca. 150 km southwards into Montenegro. At present, the European range of the species includes the thick Mediterranean vegetation of the Adriatic coast, from Skrda Isl. and the Neretva River in the north to Albania in the south (Barun et al. 2010; Ćirović et al. 2011). Given the favorable (higher) mean annual temperatures in southern Europe, the further spread of the species' range should be expected towards southern Albania and Greece (Ćirović et al. 2011). On European islands, the small Indian mongoose can show drastic annual fluctuations of population densities. Because the species' range in Europe is characterized by temperatures well below its previously known isothermal limit (10 °C), episodic cold winters could be the cause of such large density fluctuations (Tvrtkovic and Krystufek 1990).

In comparison with *H. edwardsii*, the small Indian mongoose shows a series of characteristics that may promote invasive success: (1) efficient physiological mechanisms for dealing with hot and moderately cold environments; (2) aggressive behavior against direct competitors/predators such as domestic cats; and (3) wide range of deleterious pathogens, including rabies (Gaubert and Zenatello 2009). The success of the small Indian mongoose as a biocontrol agent is questionable because the species is a generalist predator preying on other species than rodents and snakes (Hinton and Dunn 1967). Reductions or extinctions of populations of birds, reptiles, and amphibians caused by *H. auropunctatus* have been reported on islands worldwide, although there is controversy over whether the small Indian mongoose has genuinely been the main culprit (Lewis et al. 2011; Hays and Conant 2007). A secondary aspect of the deleterious impact of the species resides in its role of main reservoir for viruses (e.g., rabies) and parasites (e.g., Weil's disease) impacting wildlife and humans in several parts of the Caribbean (Hatcher et al. 2012; Everard and Everard 1992). This latter point remains undocumented in Europe. In Croatia, the species is accused of having a deleterious impact on wild fowl, poultry and several cultivars, and so is subject to extermination campaigns led by hunting federations (Tvrtkovic and Krystufek 1990). It has also been speculated that the species could have "catastrophic consequences" on the Balkan continental herpetofauna (Ćirović et al. 2011). However, a study conducted on Korcula Isl. evidenced a low consumption of reptiles and amphibians by the small Indian mongoose, and in comparison a high consumption of small mammals, birds, arthropods and plants (Cavallini and Serafini 1995). The minor representation of the herpetofauna in the mongoose's diet is actually a general trend throughout its introduced range (Table 1 in Hays and Conant 2007). On the other hand, it was observed that reptiles and amphibians were generally rare or absent from the islands occupied by the species whereas they were common on the mongoose-free island of Brac (Barun et al. 2010). Unfortunately, predation of the native herpetofauna by other invasive species such as the black rat and the domestic cat has not been evaluated.

The Egyptian mongoose must have occurred in the papyrus swamps of the Nile valley at the time of Ancient Egypt. The reasons for its extinction are unknown, but were probably linked to its artificial maintenance as a semi-domestic animal or to the progressive disappearance of such habitat (Osborn and Osbornova 1998). In Europe, suitability models predicted the expansion of the species in southern and central Spain in areas with a high rabbit abundance, thus foreseeing the existence of large regions of potential conflict with hunting interests (Recio and Virgos 2010). Climatic niche modelling outputs were less conservative and predicted most of the Iberian Peninsula as potentially suitable for the Egyptian mongoose (Papeş et al. 2015). Because there is no significant change in the composition of the Carnivore community at the northern fringe of the European range of the species (Wilson & Mittermeier, 2009), interspecific competition cannot be considered a limiting factor. In addition, release from biotic/historical constraints, including habitat disruption and climate warming, could lead to local or temporary range expansion of the Egyptian mongoose, as reflected by its recent spread into northern Portugal related to rural depopulation (Barros 2009). Whether this current trend can be assimilated to a colonization front and whether the recent records of the species in northwestern Spain (Balmori and Carbonell 2012) reflect a genuine increase of northern dispersals will have to be evaluated.

In northwestern Portugal, mongooses prey mostly upon mammals (especially lagomorphs) but also on reptiles and arthropods, with males preferentially consuming mammals (Rosalino et al. 2009). In case of competition with other Carnivores, the Egyptian mongoose may modify its realized niche by having more diurnal activities (Santos et al. 2007) and can shift its microhabitat use (e.g., by preferentially using thicker scrubland) to prevent deadly encounters with dominant species such as the Iberian lynx *Lynx pardinus* (Viota et al. 2012). So far, the role of *H. ichneumon* in carrying zoonotic diseases seems very limited. In Europe, rabies spillover infection from red foxes (*Vulpes vulpes*) was not detected (Müller et al. 2015). In Portugal, a high prevalence of parvovirus DNA was detected in mongooses (58 %), potentially carrying a risk to susceptible populations at the wildlife–domestic interface and to threatened species of sympatric Carnivores (Duarte et al. 2013).

The common genet was probably present in the papyrus swamps of the Nile Valley further north from its current range, as suggested by remains found in the South Galala Plateau cave, Egypt, and illustrations on papyrus and in stone reliefs from the Nile Valley (Osborn and Osbornova 1998). The reasons for its extinction are unknown, but were probably similar to those of the Egyptian mongoose (i.e. artificial maintenance by humans or progressive disappearance of the habitat). In Europe, the species has recently crossed its traditional range barriers of the Rhône (southeastern France) and Loire (northwestern France) rivers (Gaubert et al. 2008; Léger and Ruette 2010). Climatic niche modelling predicted a large portion of Europe as suitable for the species, Italy being the best candidate for a near future colonization via the Liguria–southern Piedmont corridor (Papeş et al. 2015).

The common genet is an opportunistic Carnivore that may expand its trophic niche on the Mediterranean islands (Ibiza and Cabrera) where it is the sole mesopredator (Virgós et al. 1999). Conversely, marked trophic differentiation occurs on another island (Mallorca) where the species coexists with the pine marten *Martes martes* (Clevenger 1995). In continental Europe, niche overlap among the common genet, the Egyptian mongoose, and other Carnivores is generally high, but subtle and dynamic (i.e., seasonal) adjustments in foraging behavior and in the use of microhabitats and main prey items seem to balance the coexistence of such small Carnivores' communities (Zabala et al. 2009; Lopez-Martin 2006; Carvalho and Gomes 2004; Melero et al. 2008; Santos-Reis et al. 2005; Monterroso et al. 2014; Zapata et al. 2007). The common genet eats significantly more fruits than the Egyptian mongoose and shows little overlap (in terms of fruit diversity) with the other Mediterranean Carnivores (Rosalino and Santos-Reis 2009). Interestingly, the invasive American mink seems to have a deleterious impact on the abundance of the common genet in northeastern Spain because of high niche overlap (Melero et al. 2012).

Similarly to the Egyptian mongoose, the common genet avoids suitable habitats where densities of Iberian lynxes are high, suggesting a "mesopredator release" when larger Carnivores competing for food and interspecies-killing disappear (Palomares and Caro 1999). The role of *G. genetta* in carrying zoonotic diseases is unproven. Asian viverrids such as the masked palm civet *Paguma larvata* were identified as the source of SARS cases with mild symptom in 2004 in China (Shi and Hu 2008), but so far, no similar coronaviruses were detected in the common genet. On the other hand, in Portugal and southwestern France, the species suffers from a high prevalence of a host-adapted canine parvovirus (Santos et al. 2009).

Conclusion on the "Invasiveness" of the Mongooses and the Genet in Europe

The species of herpestids and viverrids present in Europe fall into three categories: (1) introduced and spreading (*G. genetta*, *H. auropunctatus*), (2) introduced and extinct (*H. edwardsii*), and (3) natural disperser and spreading (*H. ichneumon*).

Usually, species introduced within the last century are considered deleterious ("invasive") *by nature*, whereas species having naturally dispersed or introduced during historical times (i.e., before 1500 AD) have been considered as "naturalized." Thus, in our case, only the introduction of the small Indian mongoose in the Balkans has been envisaged in an invasive framework.

From the above-mentioned amount of evidence, there is weak support for a deleterious impact of herpestids and viverrids on the European fauna (except possibly on the herpetofauna of small Mediterranean islands in the case of the small Indian mongoose), notably in comparison with genuine invasive species such as the black rat and the domestic cat. In fact, the small Indian mongoose is only 11th on the list of alien species affecting native species in Europe, far behind the American mink, the domestic cat, the domestic goat, the European hedgehog *Erinaceus europaeus*, and rats (Genovesi et al. 2012). Coexistence among native European Carnivores seems to occur through a dynamic adjustment of their niches, and there is no body of evidence to refute the fact that the mongooses and genet have fitted this framework without disrupting the equilibrium of Carnivores' communities.

My conclusions should have some impact on the way mongooses and genets are considered and managed in European countries. Indeed, the episodic, local control operations of those Carnivores—notably of mongooses in Portugal and Croatia—have been shown to be expensive, inefficient, and/or potentially deleterious for the rest of the Carnivores' communities, while favoring the pullulating of the species on which they prey (e.g., rabbits) (Hays and Conant 2007; Barun et al. 2011; Beja et al. 2009). Eradication successes of mongooses seem somehow limited to small islands up to 1.15 km^2 (Barun et al. 2011), whereas extirpation from larger islands or areas might require enormous means not affordable by most governments (see Abe et al. 2006; Fukasawa et al. 2013 for an example on another species of mongoose in Japan).

Prospective

Predator control is a contentious issue that is becoming under the scrutiny of the general public, with sectors of the society expressing ethical and biological arguments against the killing of predators (Barun et al. 2011). Instead, we suggest that the attention of governments should be focused on restoration programs including (1) large Carnivores that are natural regulators of mesocarnivore populations (Palomares and Caro 1999), (2) small Carnivores' communities, and (3) their natural habitats, which all may be a safe buffer to the deleterious impacts potentially related to introduced small Carnivores (Letnic et al. 2009; McDonald et al. 2007).

We urge ecologists to conduct long-term surveys on the population dynamics and trophic overlap of the small Indian mongoose with sympatric Carnivores and invasive species in Europe, in order to provide scientifically based guidelines on the attitude to adopt for the management of the species (notably on Adriatic islands). Future studies should also be directed on the beneficial aspects of herpestids and viverrids on European ecosystems, including their role as seed dispersers and as regulators of potential pest species such as native and invasive rodents and insects.

The potential colonization fronts of the mongooses and genet in Europe provide a tremendous framework for studying the dynamics of mesopredators at disequilibrium with their environment. Yet, there is a crucial need for comparative studies in areas such as northwestern Iberia (*H. ichneumon*), northwestern and southeastern France (*G. genetta*), and the Balkans (*H. auropunctatus*) to better understand the processes behind the spread of herpestids and viverrids in Europe.

Eventually, a global perspective on the natural history of those small Carnivores in their native ranges (including reproduction strategies, interspecific competition with other Carnivores, and zoonotic prevalence) would help understanding their successful establishment—or not (*H. edwardsii*)—in Europe, but is still lacking.

References

Abe S, Yamada F, Handa Y, Takatsuki Y, Abe Y, Yamashita R, Fukuda M (2006) Reproductive responses of the mongoose (*Herpestes javanicus*), to control operations on Amami-oshima Island, Japan. In: Koike F, Clout MN, Kawamichi M, De Poorter M, Iwatsuki K (eds) Assessment and control of biological invasion risks. Shoukadoh Book Sellers, Kyoto, pp 157–164

Amigues S (1999) Les belettes de Tartessos. Anthropozool 29:55–64

Angelici FM (2003) *Herpestes edwardsii*. In: Boitani L, Lovari S, Vigna Taglianti A (eds) Fauna d'Italia. Mammalia III. Carnivora—Artiodactyla. Calderini, Bologna, Italy, pp 201–205

Balmori A, Carbonell R (2012) Expansion and distribution of the Egyptian mongoose (*Herpestes ichneumon*) in the Iberian Peninsula. Galemys 24:83–85

Barros T (2009) Estatuto e distribuição do Sacarrabos (*Herpestes ichneumon*) em Portugal. Universidade de Aveiro, Portugal

Barun A, Simberloff D, Budinski I (2010) Impact of the small Indian mongoose on native amphibians and reptiles of the Adriatic islands, Croatia. Anim Conserv 13(6):549–555. doi:10.1111/j.1469-1795.2010.00374.x

Barun A, Hanson CC, Campbell KJ, Simberloff D (2011) A review of small Indian mongoose management and eradications on islands. In: Veitch CR, Clout MN, Towns DR (eds) Island invasives: eradication and management. IUCN, Gland, pp 17–25

Beja P, Gordinho L, Reino L, Loureiro F, Santos-Reis M, Borralho R (2009) Predator abundance in relation to small game management in southern Portugal: conservation implications. Eur J Wildl Res 55(3):227–238. doi:10.1007/s10344-008-0236-1

Biondi M (1985) Aspetti faunistici del parco Nazionale del Circeo. Quaderni del Parco, 6. Ministero Agricoltura e Foreste, Parco Nazionale del Circeo, Sabaudia, Italy

Blondel J, Aronson J, Bodiou JY, Boeuf G (2010) The Mediterranean region: biological diversity in space and time, 2nd edn. Oxford University Press, Oxford

Bonesi L, Palazon S (2007) The American mink in Europe: status, impacts, and control. Biol Conserv 134(4):470–483

Borralho R, Rego F, Palomares F, Hora A (1996) The distribution of the Egyptian Mongoose *Herpestes ichneumon* (L.) in Portugal. Mammal Rev 26(1):1–8

Campanella L, Wilkens B (2004) Una mangusta egiziana (*Herpestes ichneumon*) dall'abitato fenicio di Sant'Antioco. Riv St Fen 32(1):25–48

Carpaneto GM (1990) The Indian Grey Mongoose (*Herpestes edwardsii*) in the Circeo National Park: a case of incidental introduction. Mustelid Viverrid Conserv 2:10

Carvalho JC, Gomes P (2004) Feeding resource partitioning among four sympatric carnivores in the Peneda-Gerês National Park (Portugal). J Zool 263(3):275–283. doi:10.1017/S0952836904005266

Cavallini P, Serafini P (1995) Winter diet of the small Indian mongoose, *Herpestes auropunctatus*, on an Adriatic island. J Mammal 76(2):569–574

Choudhury A, Wozencraft C, Muddapa D, Yonzon P, Jennings A, Veron G (2013) Herpestes edwardsii. In: The IUCN red list of threatened species, v. 2014.3. www.iucnredlist.org. Downloaded on 18 Feb 2015. In: Veitch CR, Clout MN, Towns DR (eds) (2011) Island invasives: eradication and management. IUCN, Gland, pp 17–25

Ciolek TM (ed) (2011) Georeferenced historical transport/travel/communication routes and nodes. www.ciolek.com—Asia Pacific Research Online. www.ciolek.com/OWTRAD/DATA/oddda.html. Accessed 6 Apr 2011

Ćirović D, Raković M, Milenković M, Paunović M (2011) Small Indian mongoose *Herpestes auropunctatus* (Herpestidae, Carnivora): an invasive species in Montenegro. Biol Invasions 13(2):393–399. doi:10.1007/s10530-010-9831-7

Clevenger AP (1995) Seasonality and relationships of food resource use of *Martes martes*, *Genetta genetta* and *Felis catus* in the Balearic Islands. Rev Ecol Terre Vie 50:109–131

Clout MN, Russell JC (2007) The invasion ecology of mammals: a global perspective. Wild Res 35(3):180–184. doi:10.1071/WR07091

Collina-Girard J (2001) L'Atlantide devant le détroit de Gibraltar ? Mythe et géologie. CR Acad Sci Ser IIA Earth Planet Sci 333(4):233–240

Columeau P (1979) Sondage au sommet de la colline d'Ambrussum. Documents d'Archéologie Méridionale 2:51–52

Cuttelod A, García N, Abdul Malak D, Temple H, Katariya V (2008) The Mediterranean: a biodiversity hotspot under threat. In: Vié J-C, Hilton-Taylor C, Stuart SN (eds) The 2008 review of the IUCN red list of threatened species. IUCN, Gland

Dannenfeldt KH (1985) Europe discovers civet cats and civet. J Hist Biol 18(3):403–431

Delibes M (1977) Sobre las ginetas de la isla de Ibiza (*Genetta genetta isabelae* n. ssp.). Doñana Acta Vertebrata 4:139–160

Delibes M, Gaubert P (2013) *Genetta genetta* Common Genet (Small-spotted Genet). In: Kingdon JS, Hoffmann M (eds) The mammals of Africa, vol 5, Carnivores, Pangolins, Equids and Rhinoceroses. Bloomsburry, London, pp 224–229

Detry C, Bicho N, Fernandes H, Fernandes C (2011) The Emirate of Córdoba (756–929 AD) and the introduction of the Egyptian mongoose (*Herpestes ichneumon*) in Iberia: the remains from Muge, Portugal. J Archeol Sci 38(12):3518–3523. doi:10.1016/j.jas.2011.08.014

Dobney KM, Jaques D (1994) Preliminary report on the animal bones from Saar. Arab Archaeol Epigr 5(2):106–120. doi:10.1111/j.1600-0471.1994.tb00060.x

Dobson M (1998) Mammal distributions in the western Mediterranean: the role of human intervention. Mammal Rev 28(2):77–88

Duarte MD, Henriques AM, Barros SC, Fagulha T, Mendonça P, Carvalho P, Monteiro M, Fevereiro M, Basto MP, Rosalino LM, Barros T, Bandeira V, Fonseca C, Cunha MV (2013) Snapshot of viral Infections in wild carnivores reveals ubiquity of parvovirus and susceptibility of Egyptian mongoose to feline panleukopenia virus. PLoS One 8(3), e59399. doi:10.1371/journal.pone.0059399

Duncan RP, Forsyth DM (2006) Modelling population persistence on islands: mammal introductions in the New Zealand archipelago. Proc R Soc B Biol Sci 273(1604):2969–2975

Emeneau MB (1940) A classical Indian folk-tale as a reported modern event: the brahman and the mongoose. Proc Am Philos Soc 83(3):503–513. doi:10.2307/985117

Everard COR, Everard JD (1992) Mongoose rabies in the Caribbean. Ann N Y Acad Sci 653(1):356–366. doi:10.1111/j.1749-6632.1992.tb19662.x

Favyn A (1620) Le Théâtre d'honneur et de chevalerie, ou l'Histoire des ordres militaires des roys et princes... de l'institution des armes et blasons... duels, joustes et tournois et de tout ce qui concerne le faict du chevalier de l'ordre. Robert Fouet, Paris

Fukasawa K, Hashimoto T, Tatara M, Abe S (2013) Reconstruction and prediction of invasive mongoose population dynamics from history of introduction and management: a Bayesian state-space modelling approach. J Appl Ecol 50(2):469–478. doi:10.1111/1365-2664.12058

Gaubert P, Zenatello M (2009) Ancient DNA perspective on the failed introduction of mongooses in Italy during the XXth century. J Zool 279:262–269. doi:10.1111/j.1469-7998.2009.00614.x

Gaubert P, Jiguet F, Bayle P, Angelici FM (2008) Has the common genet (*Genetta genetta*) spread into south-eastern France and Italy? Ital J Zool 75(1):43–57

Gaubert P, Godoy JA, del Cerro I, Palomares F (2009) Early phases of a successful invasion: mitochondrial phylogeography of the common genet (*Genetta genetta*) within the Mediterranean Basin. Biol Invasions 11:523–546

Gaubert P, Machordom A, Morales A, López-Bao JV, Veron G, Amin M, Barros T, Basuony M, Djagoun CAMS, Do Linh San E, Fonseca C, Geffen E, Ozkurt SO, Cruaud C, Couloux A, Palomares F (2011) Comparative phylogeography of two African carnivorans presumably introduced into Europe: disentangling natural versus human-mediated dispersal across the Strait of Gibraltar. J Biogeogr 38:341–358

Gaubert P, Del Cerro I, Centeno-Cuadros A, Palomares F, Fournier P, Fonseca C, Paillat J-P, Godoy JA (2015) Tracing historical introductions in the Mediterranean Basin: the success story of the common genet (*Genetta genetta*) in Europe. Biol Invasions 17:1897–1913

Genovesi P, Carnevali L, Alonzi A, Scalera R (2012) Alien mammals in Europe: updated numbers and trends, and assessment of the effects on biodiversity. Integr Zool 7(3):247–253. doi:10.1111/j.1749-4877.2012.00309.x

Geoffroy Saint-Hilaire E (1813) Description des mammifères qui se trouvent en Egypte. Imprimerie Impériale, Paris

Geraads D (2010) Biogeographic relationships of Pliocene and Pleistocene North-western African mammals. Quatern Int 212(2):159–168

Gilchrist JS, Jennings AP, Veron G, Cavallini P (2009) Herpestidae (Mongooses). In: Wilson DE, Ruff S (eds) Handbook of mammals of the world, vol 1, Carnivores. Lynx Edicions, Barcelona, pp 262–329

Gippoliti S, Amori G (2006) Ancient introductions of mammals in the Mediterranean Basin and their implications for conservation. Mammal Rev 36(1):37–48

Hatcher MJ, Dick JTA, Dunn AM (2012) Disease emergence and invasions. Funct Ecol 26(6):1275–1287. doi:10.1111/j.1365-2435.2012.02031.x

Hays WST, Conant S (2007) Biology and impacts of Pacific island invasive species. 1. A worldwide review of effects of the small Indian mongoose, *Herpestes javanicus* (Carnivora: Herpestidae). Pac Sci 61(1):3–16

Hewitt GM (1996) Some genetic consequences of ice ages, and their role in divergence and speciation. Biol J Linn Soc 58:247–276

Hinton HE, Dunn AMS (1967) Mongooses: their natural history and behaviour. Oliver & Boyd, Edinburgh

Hughes JD (2003) Europe as consumer of exotic biodiversity: Greek and Roman times. Landsc Res 28(1):21–31

Krystufek B, Tvrtkovic N (1992) New information on the introduction into Europe of the Small Indian mongoose, *Herpestes auropunctatus*. Small Carniv Conserv 7:16

Lebarbenchon C, Poitevin F, Arnal V, Montgelard C (2010) Phylogeography of the weasel (*Mustela nivalis*) in the western-Palaearctic region: combined effects of glacial events and human movements. Heredity 105:449–462

Léger F, Ruette S (2010) La répartition de la genette en France. Faune Sauvage 287:16–22

Letnic M, Koch F, Gordon C, Crowther MS, Dickman CR (2009) Keystone effects of an alien top-predator stem extinctions of native mammals. Proc R Soc B Biol Sci 276:3249–3256. doi:10.1098/rspb.2009.0574

Lewis D, van Veen R, Wilson B (2011) Conservation implications of small Indian mongoose (*Herpestes auropunctatus*) predation in a hotspot within a hotspot: the Hellshire Hills, Jamaica. Biol Invasions 13(1):25–33. doi:10.1007/s10530-010-9781-0

Lodrick DO (1982) Man and mongoose in Indian culture. Anthropos 77(1/2):191–214

Long JL (2003) Introduced mammals of the world: their history, distribution and influence. CSIRO, Collingwood

Lopez-Martin JM (2006) Comparison of feeding behaviour between stone marten and common genet: living in coexistence. In: Santos-Reis M, Birks JDS, O'Doherty EC, Proulx G (eds) Martes in carnivore communities. Alpha Wildlife, Sherwood Park, pp 137–155

Lowe S, Browne M, Boudjelas S, De Poorter M (2000) 100 of the world's worst invasive alien species. A selection from the Global Invasive Species Database. The Invasive Species Specialist Group/IUCN, Gland

Masseti M (1995) Quaternary biogeography of the Mustelidae family on the Mediterranean islands. Hystrix 7(1–2):17–34

Masseti M (2009) Mammals of the Mediterranean islands: homogenisation and the loss of biodiversity. Mammalia 73(3):169–202

McDonald RA, O'Hara K, Morrish DJ (2007) Decline of invasive alien mink (*Mustela vison*) is concurrent with recovery of native otters (*Lutra lutra*). Divers Distrib 13(1):92–98. doi:10.1111/j.1366-9516.2006.00303.x

Medina FM, Bonnaud E, Vidal E, Tershy BR, Zavaleta ES, Josh Donlan C, Keitt BS, Le Corre M, Horwath SV, Nogales M (2011) A global review of the impacts of invasive cats on island endangered vertebrates. Glob Change Biol 17(11):3503–3510. doi: 10.1111/j.1365-2486.2011.02464.x

Melero Y, Palazón S, Bonesi L, Gosàlbez J (2008) Feeding habits of three sympatric mammals in NE Spain: the American mink, the spotted genet, and the Eurasian otter. Acta Theriol 53(3):263–273. doi:10.1007/BF03193123

Melero Y, Plaza M, Santulli G, Saavedra D, Gosàlbez J, Ruiz-Olmo J, Palazón S (2012) Evaluating the effect of American mink, an alien invasive species, on the abundance of a native community: is coexistence possible? Biodivers Conserv 21(7):1795–1809. doi:10.1007/s10531-012-0277-3

Monterroso P, Alves P, Ferreras P (2014) Plasticity in circadian activity patterns of mesocarnivores in Southwestern Europe: implications for species coexistence. Behav Ecol Sociobiol 68(9):1403–1417. doi:10.1007/s00265-014-1748-1

Morales A (1994) Earliest genets in Europe. Nature 370(6490):512–513

Morales Muñiz DC (2000) La fauna exótica en la Península Ibérica: apuntes para el estudio del coleccionismo animal en el Medievo hispánico. Espacio Tiempo Forma Ser III Ha Medieval 13:233–270

Morales A, Riquelme JA, Liesau C (1995) Dromedaries in antiquity: Iberia and beyond. Anglais 69(263):368–375

Müller T, Freuling CM, Wysocki P, Roumiantzeff M, Freney J, Mettenleiter TC, Vos A (2015) Terrestrial rabies control in the European Union: historical achievements and challenges ahead. Vet J 203:10–17

Osborn DJ, Osbornova J (1998) The mammals of ancient Egypt, vol IV, The natural history of Egypt. Aris & Phillips Ltd, Warminster

Ouchaou B, Amani F (2002) Les carnivores des gisements néolithiques et protohistoriques du nord du Maroc. Quaternaire 13(1):79–87

Palomares F (2013) *Herpestes ichneumon* Egyptian Mongoose (Ichneumon). In: Kingdon JS, Hoffmann M (eds) The mammals of Africa, vol 5, Carnivores, Pangolins, Equids and Rhinoceroses. Bloomsbury, London, pp 306–310

Palomares F, Caro TM (1999) Interspecific killing among mammalian Carnivores. Am Nat 153(5):492–508

Papeş M, Cuzin F, Gaubert P (2015) Niche dynamics in the European ranges of two African carnivores reflect their dispersal and demographic histories. Biol J Linn Soc 114:737–751

Pinho C, Harris DJ, Ferrand N (2007) Contrasting patterns of population subdivision and historical demography in three western Mediterranean lizard species inferred from mitochondrial DNA variation. Mol Ecol 16(6):1191–1205

Recio MR, Virgos E (2010) Predictive niche modelling to identify potential areas of conflicts between human activities and expanding predator populations: a case study of game management and the grey mongoose, *Herpestes ichneumon*, in Spain. Wild Res 37(4):343–354. doi:10.1071/WR09096

Riquelme-Cantal JA, Simón-Vallejo MD, Palmqvist P, Cortés-Sánchez M (2008) The oldest mongoose of Europe. J Archeol Sci 35(9):2471–2473

Rosalino LM, Santos-Reis M (2009) Fruit consumption by carnivores in Mediterranean Europe. Mammal Rev 39(1):67–78

Rosalino LM, Santos MJ, Pereira I, Santos-Reis M (2009) Sex-driven differences in Egyptian mongoose's (*Herpestes ichneumon*) diet in its northwestern European range. Eur J Wildl Res 55(3):293–299. doi:10.1007/s10344-008-0248-x

Santiapillai C, De Silva M, Dissanayake S (2000) The status of mongooses (family: Herpestidae) in Ruhuna National Park, Sri Lanka. J Bombay Nat Hist Soc 97:208–214

Santos MJ, Pinto BM, Santos-Reis M (2007) Trophic niche partitioning between two native and two exotic carnivores in SW Portugal. Web Ecol 7(1):53–62. doi:10.5194/we-7-53-2007

Santos N, Almendra C, Tavares L (2009) Serologic survey for canine distemper virus and canine parvovirus in free-ranging wild carnivores from Portugal. J Wildl Dis 45(1):221–226. doi:10.7589/0090-3558-45.1.221

Santos-Reis M, Santos M, Lourenço S, Marques J, Pereira I, Pinto B (2005) Relationships between stone martens, genets and cork oak woodlands in Portugal. In: Harrison D, Fuller A, Proulx G (eds) Martens and fishers (*Martes*) in human-altered environments. Springer, New York, pp 147–172. doi:10.1007/0-387-22691-5_7

Shi Z, Hu Z (2008) A review of studies on animal reservoirs of the SARS coronavirus. Virus Res 133(1):74–87

Stuart CT (1988) Mongoose of the Gods: the ichneumon. Afr Wildl 42(5):254

Thulin C-G, Simberloff D, Barun A, McCracken G, Pascal M, Islam MA (2006) Genetic divergence in the small Indian mongoose (*Herpestes auropunctatus*), a widely distributed invasive species. Mol Ecol 15(13):3947–3956. doi:10.1111/j.1365-294X.2006.03084.x

Tresic Pavicic A (1936) Mungos na otoku Bracu. Priroda Zagreb 26:60–61

Tvrtkovic N, Krystufek B (1990) Small Indian mongoose *Herpestes auropunctatus* (Hodgson, 1836) on the Adriatic Islands of Yugoslavia. Bonn Zool Beitr 41(1):3–8

Uerpmann M (1995) Early mongooses from Bahrain. In: Buitenhuis H, Uerpmann H-P (eds) Archaeozoology of the Near East II. Backhuys, Leiden, pp 64–71

Veron G, Patou M-L, Pothet G, Simberloff D, Jennings AP (2007) Systematic status and biogeography of the Javan and small Indian mongooses (Herpestidae, Carnivora). Zool Scr 36(1):1–10. doi:10.1111/j.1463-6409.2006.00261.x

Vigne J-D, Zazzo A, Saliège J-F, Poplin F, Guilaine J, Simmons A (2009) Pre-Neolithic wild boar management and introduction to Cyprus more than 11,400 years ago. Proc Natl Acad Sci U S A 106(38):16135–16138. doi:10.1073/pnas.0905015106

Viota M, Rodríguez A, López-Bao J, Palomares F (2012) Shift in microhabitat use as a mechanism allowing the coexistence of victim and killer carnivore predators. Open J Ecol 2:115–120. doi:10.4236/oje.2012.23014

Virgós E, Llorente M, Cortes Y (1999) Geographical variation in genet (*Genetta genetta* L.) diet: a literature review. Mammal Rev 29(2):119–128

Watari Y, Nagata J, Funakoshi K (2011) New detection of a 30-year-old population of introduced mongoose *Herpestes auropunctatus* on Kyushu Island, Japan. Biol Invasions 13(2):269–276. doi:10.1007/s10530-010-9809-5

Wilson DE, Mittermeier RA (2009) Handbook of the mammals of the world, vol 1, Carnivores. Lynx Edicions, Barcelona.

Wozencraft C, Duckworth JW, Choudury A, Muddapa D, Yonzon P, Kanchanasaka B, Jennings A, Veron G (2008) *Herpestes javanicus*. In: The IUCN red list of threatened species, v. 2014.3. www.iucnredlist.org. Downloaded on 18 Feb 2015

Zabala J, Zuberogoitia I, Martinez-Climent JA (2009) Testing for niche segregation between two abundant carnivores using presence-only data. Folia Zool 58(4):385–395

Zapata S, Travaini A, Ferreras P, Delibes M (2007) Analysis of trophic structure of two carnivore assemblages by means of guild identification. Eur J Wildl Res 53(4):276–286. doi:10.1007/s10344-007-0095-1

Zeder MA (2008) Domestication and early agriculture in the Mediterranean Basin: origins, diffusion, and impact. Proc Natl Acad Sci U S A 105(33):11597–11604. doi:10.1073/pnas.0801317105

Chapter 15
Conservation of a New Flagship Species: The Galápagos Pink Land Iguana (*Conolophus marthae* Gentile and Snell, 2009)

Gabriele Gentile, Cruz Marquez, Howard L. Snell, Washington Tapia, and Arturo Izurieta

A New Species of Land Iguana in Galápagos

The Galápagos Pink Land Iguana (*Conolophus marthae*, Fig. 15.1) was first seen on Volcán Wolf (Isabela island) in 1986, when a group formed by Galápagos National Park rangers and technical personnel of the Charles Darwin Foundation accidentally encountered it during a field trip to the remote northwestern slope of Volcán Wolf (Márquez et al. 2010). After this, pink iguanas were only spotted on a few occasions. However, it is only recently that *C. marthae* has been brought to the attention of science as a new species (Tzika et al. 2008; Gentile et al. 2009).

G. Gentile (✉)
Department of Biology, Tor Vergata University, Via della Ricerca Scientifica, Rome 00133, Italy
e-mail: gabriele.gentile@uniroma2.it

C. Marquez
Charles Darwin Foundation, Puerto Ayora, Santa Cruz, Galápagos, Ecuador
e-mail: cruzkallw@gmail.com

H.L. Snell
Department of Biology, University of New Mexico, Albuquerque, NM, USA
e-mail: snell@unm.edu

W. Tapia
Galápagos Conservancy, Puerto Ayora, Santa Cruz, Galápagos, Ecuador

Galápagos National Park Directorate, Puerto Ayora, Santa Cruz, Galápagos, Ecuador

Department of Biology, University of Malaga, Malaga, Spain
e-mail: wtapiaa@gmail.com

A. Izurieta
Galápagos National Park Directorate, Puerto Ayora, Santa Cruz, Galápagos, Ecuador
e-mail: arturo.izurieta@fcdarwin.org.ec

F.M. Angelici (ed.), *Problematic Wildlife*, DOI 10.1007/978-3-319-22246-2_15

Fig. 15.1 A large adult male of *Conolophus marthae* (Photo G. Gentile)

Threats and Actions

A number of threats endanger *C. marthae*. To preserve the species, some of these threats have been tackled with planned actions, discussed in this section. Some of these threats and actions have been the subject of the 2014 meeting of the Iguana Specialist Group of the IUCN, held in Puerto Ayora, Santa Cruz (Galápagos).

Taxonomic Definition and Risk Assessment

Although it is not considered as a threat per se, lack of formal taxonomic status can hinder conservation planning and funding for undescribed species, even though they show unique morphological, ecological, behavioral, and evolutionary traits (Robertson et al. 2011).

The Galápagos Pink Land Iguana was arguably first recognized by the common name of "rosada," meaning "pink" in Spanish (Tzika et al. 2008; Gentile et al. 2009). However, the species was not formally described and internationally recognized as *Conolophus marthae* until 2009 (Gentile and Snell 2009). Despite Gentile and Snell (2009) agreeing with the importance of establishing preserved holotypes when describing a new species, they considered the combination of several threats impacting the only known population of *C. marthae* as incompatible with the sacrifice of an individual to be fixed as the holotype and preserved as a museum specimen. The decision not to sacrifice an individual for the purpose of a holotype fueled further debate on the protocol for describing and recognizing new species that are clearly critically endangered (Donegan 2009; Dubois 2009; Minelli 2009; Nemésio 2009).

The description of *C. marthae* was based on morphological, genetic, and behavioral diagnostic traits. The holotype was branded a free-living adult male and tagged with a passive integrated transponder (PIT) tag, a subcutaneous electronic marker. Genetic data were also deposited in Genbank (http://www.ncbi.nlm.nih.gov/genbank), and exhaustive photo-video documentation was deposited in MorphoBank (Project n. 241; http://www.morphobank.org). To maximize the chances of the holotype being eventually placed as a fixed specimen in a museum collection, the Galápagos National Park Directorate—the legal authority governing Galápagos Protected Areas and its biological diversity—will collect the holotype, or another individual, from the population. They will then move it to an existing captive facility at the Galápagos National Park Center. Here, it will be maintained until its natural death. Subsequently, it will be preserved and deposited in the Governmental Galápagos collection, maintained by the Charles Darwin Foundation in Puerto Ayora, Galápagos. This action will be enforced if continued monitoring of pink iguanas indicates that their population is increasing. A naturally, freshly dead or dying individual can also be considered. However, since 2005 when formal empirical study of the species initiated, no pink iguana cadavers or dying individuals have been encountered in the field.

A delay in taking actions for the conservation of a threatened species may also arise from the lack of proper evaluation of that species' risk of extinction. Increasingly, the IUCN Red List of Threatened Species has been used for conservation policy and planning purposes (Mace et al. 2008), even if it has been suggested that conservation prioritization should take into account the risk of extinction and other aspects, such as measures of the evolutionary relevance of a species (Redding and Mooers 2006; Drummond et al. 2010).

For the purpose of its inclusion in the IUCN Red List, the first evaluation of the risk status of *C. marthae* was completed in 2012. Currently, the species is listed in the IUCN Red List as "Critically Endangered" primarily for its distribution, geographically restricted and limited to a single location, and for the small number of mature individuals (http://www.iucnredlist.org/details/174472/0). It has to be noted that having described and named *C. marthae* was beneficial for the purposes of the inclusion of the species in the IUCN Red List. In fact, despite the inclusion of undescribed species in the IUCN Red List is exceptionally allowed, although discouraged, the permanence of undescribed species in the IUCN Red List is conditional on the publication of the description of the new species. The description must be completed within the subsequent four years or the assessment will be removed (IUCN Standards and Petitions Subcommittee 2014).

Currently, all species of iguanas belonging to the genus *Conolophus* are included under Appendix II of "Convention on International Trade in Endangered Species of Wild Fauna and Flora" (CITES). Upon its discovery, *C. marthae* was included in the "Conservation and Restoration of Ecosystems" program. This was part of the Galápagos Protected Areas management plan.

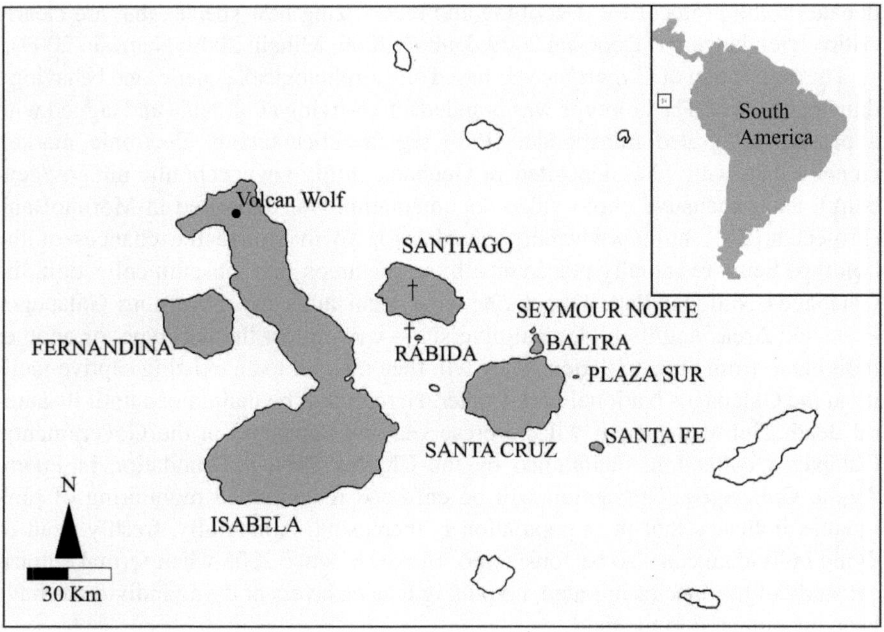

Fig. 15.2 Galápagos archipelago. *Grey* indicates islands where land iguana species occur or have occurred in historic times. *Crosses* indicate extinction in whole island. A *dot* indicates Volcán Wolf (redrawn after Gentile and Snell 2009)

Distribution

C. marthae is endemic to Volcán Wolf on Isabela island (Galápagos, Ecuador). Despite the species first being encountered along the northwestern slopes of the volcano it is most common along the northern slopes of the volcano up to its rim, at altitudes ranging from 600 to 1700 m a.s.l. (Fig. 15.2). The geographic area within which the species can be found is not larger than 25 km^2. However, the largest polygon, obtained by unifying the geographic capture points, measures only 10.9 km^2. Additionally, more than 95 % of observations cluster in a much smaller area (Gentile 2012). Pink iguanas have never been seen inside the caldera, where, in turn, *C. subcristatus* iguanas were nesting in June 2012 and June 2014 (Gentile, personal observation).

Population Size and Recruitment

The single existing population of *C. marthae* approximates a closed population. In stark contrast to *C. subcristatus*, *C. marthae* iguanas were never observed on Volcán Darwin or Volcán Ecuador—the two volcanos on Isabela closest to Volcán Wolf.

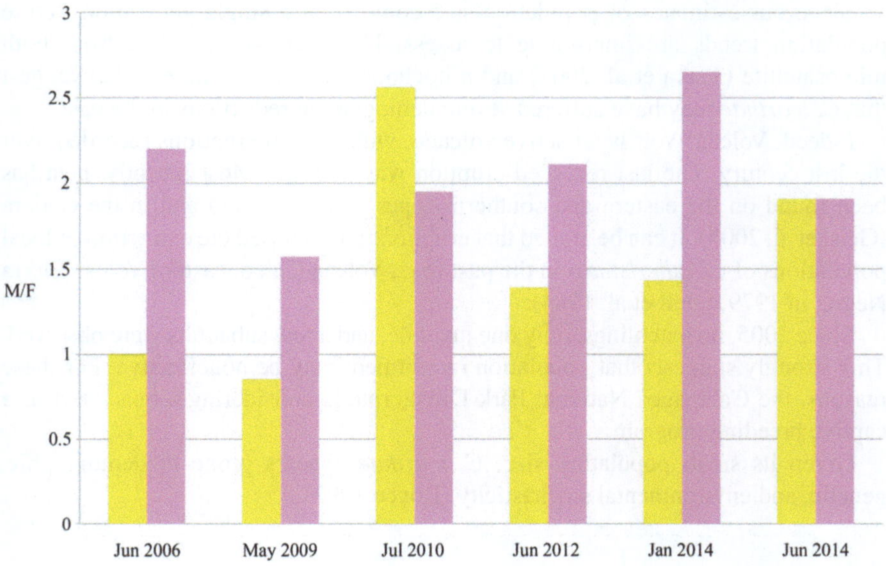

Fig. 15.3 Sex ratio in *C. marthae* (*pink*) and *C. subcristatus* (*yellow*) in Volcán Wolf, Isabela Island

Early mark–recapture data indicated 192 individuals (95 % CL=155–260). This was calculated by applying the Lincoln–Petersen method from two contiguous temporal samples in 2009 and 2010 (Gentile 2012). Figure 15.3 plots the sex ratio of the two populations. Here, the ratio is indicated as the simple proportion of males and females (M/F) captured and marked on each sampling year (2006, 2009, 2010, 2012, 2014). In 2014, M/F was estimated in January and June. Notably, *C. marthae* consistently shows higher M/F values than *C. subcristatus*.

As *Conolophus* species are iteroparous with overlapping generations we pooled temporal samples of adults of mixed cohorts captured from 2005 to 2009 to estimate N_e, taking advantage of single sample methods based on linkage disequilibrium (Hill 1981; Palstra et al. 2009). The effective population size (N_e) of *C. marthae* was estimated to be as large as 90.7 (95 % CL=62.9–148.7). This was calculated by using LDNe (Waples and Do 2008) and 20 microsatellite loci data from a sample of 61 individuals. The analogous estimate, obtained from a sample of 64 *C. subcristatus*, returned N_e=357.7 (95 % CL=200.2–1342.1). Thus, the effective population size of *C. marthae* is four times smaller than the effective size of *C. subcristatus* population in Volcán Wolf alone. Under the assumption that the population size remained constant over the sampling years, the N_e/N ratio for *C. marthae* would be equal to 0.47. It should be noted that, in our estimations, N_e reflects the size of an ideal population experiencing the same rate of random genetic change over time as the real population while N reflects the total census population size including adults, subadults, and juveniles.

Since our estimates of population size come from a single generation, future population trends are impossible to assess. However, genetic data from both microsatellite (Tzika et al. 2008) and mitochondrial DNA (Gentile 2012) suggest that *C. marthae* may have suffered severe demographic reductions in the past.

Indeed, Volcán Wolf is an active volcano, with several eruptions recorded over the last century. The last recorded eruption was in 1982. Most recently, lava has been found on the eastern and southern slopes of the volcano and in the caldera (Geist et al. 2005). It can be argued that eruptions have caused the extinction of local populations of *C. subcristatus* in the past (e.g., Volcán Chico, eastern Volcán Sierra Negra, in 1979, Snell et al. 1984).

Since 2005, no hatchlings, only one juvenile, and a few subadults were observed. This strongly suggests that population recruitment may be noneffective. For these reasons, the Galápagos National Park Directorate is considering a head-start or a captive breeding program.

Given its small population size, *C. marthae* appears prone to demographic, genetic, and environmental stochasticity (Boyce 1992).

Possible Competition with C. subcristatus

As *C. marthae* is syntopic (sensu Rivas 1964) with a population of *C. subcristatus* on Volcán Wolf, there could be competition between the two populations. Competition between iguana species may have negative effects: the introductions of *Iguana iguana* in the area of *I. delicatissima* (Lesser Antilles) resulted in population declines for *I. delicatissima* throughout much of its range (Knapp et al. 2014).

Given the difficult logistics of the site, which limits the duration of field studies, current knowledge of the reproductive biology of the two species in Volcán Wolf is based on circumstantial evidence. In mid-July 2010, the reproductive status of 19 *C. marthae* females was investigated. As ultrasonography proves a reliable, non-destructive method to obtain life history data such as egg and clutch size in wild populations of reptiles (Gilmar and Wolf 2007), we used a Nanomax portable ultrasound system with a 5–8 MHz bandwidth and 10 cm scan depth transducer (FUJIFILM SonoSite, Inc.). The depth settings and screen contrast on the instrument were varied to optimize visualization of the internal anatomy. We applied 0.25–0.5 cm of gel to reduce interference in the signal transmission. We performed transverse, sagittal, and coronal scans of each individual. Continuous video and still images were recorded for future analysis. Five out of the 19 females (26.3 %) showed large shelled eggs (Fig. 15.4), carrying on average 4.4 (±1.5 SD) eggs. No eggs were observed in any of the eight *C. subcristatus* females tested in the same period. In June 2012, out of the 29 *C. subcristatus* females tested, 26 (92.9 %) carried on average 8.4 (±3 SD) shelled eggs. At the same time, one of the 23 (4.3 %) *C. marthae* females carried one egg. In June 2014, nine of 20 (45 %) tested *C. subcristatus* females showed on average 8.5 (±3.3 SD) shelled eggs. Furthermore, a few more females showed follicular eggs (smaller and more spherical than shelled eggs),

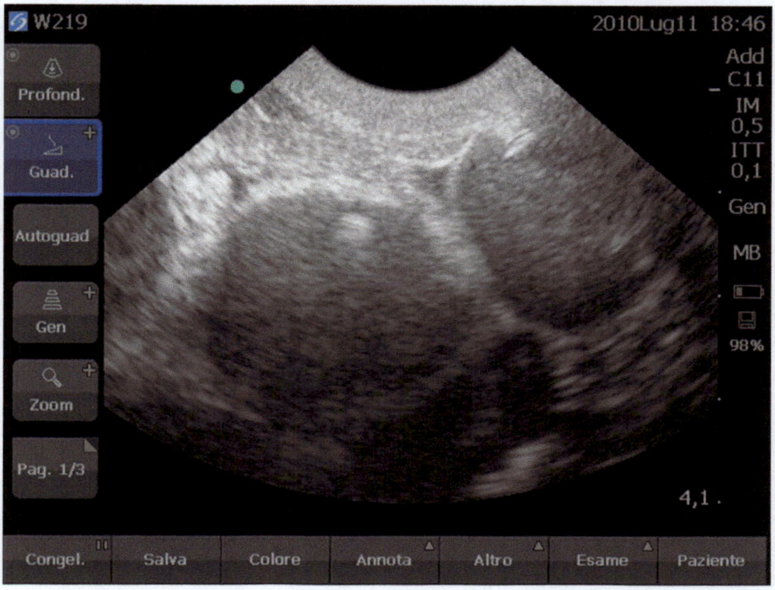

Fig. 15.4 Shelled eggs in a *C. marthae* female. The shell is located around the circumference of the eggs and is more or less echogenic depending on the degree of calcification

while three of 18 (16.6 %) *C. marthae* females carried eggs. Of these, one *C. marthae* female carried three shelled eggs and two carried four to five follicular eggs. While no difference was observed between the average number of shelled eggs carried by *C. subcristatus* in the same sampling periods (June) of 2012 and 2014, the number of *C. subcristatus* females carrying shelled eggs in 2012 and 2014 was significantly different (*Z*-score = 3.678; two-tailed $p \ll 0.01$). The lower percentage was observed in 2014.

Homo-specific pairs of the two species were observed in mid-May 2009 (Fig. 15.5), as well as in sampling trips executed in 2010, 2012, and June 2014. In those periods, both species were sexually active, as indicated sperm at the cloaca of several females and males.

Available data suggest that in the years 2010, 2012, and June 2014 (no ultrasound data are available prior to 2010), only a few *C. marthae* females reproduced, with *C. marthae* showing a smaller clutch size than *C. subcristatus*. This is consistent with the size of *C. marthae*, smaller than the average size of *C. subcristatus* individuals from Volcán Wolf (Gentile 2012). Given that observations were conducted a month later in 2010 (in respect to 2012 and 2014), we can speculate that by mid-July 2010, most *C. subcristatus* females that had mated had already laid their eggs.

Droughts can impact vegetation at the highest altitudes where pink iguanas forage. Adults obtain water from their largely herbivorous diet and are expected to cope sufficiently well with drought as a result. However, scarcity of food caused by droughts has the potential to lower fecundity for the year. This is due to a higher

Fig. 15.5 Male (*left*) and female (*right*) of *C. marthae*, on the rim of Volcán Wolf (Isabela Island) on May 2009 (Photo G. Gentile)

number of infertile eggs being laid, or a failure of females to nest. Further, droughts increase juvenile mortality, exacerbating the reduction in potential recruitment. Indeed, our data would suggest that on Volcán Wolf, as in other island populations (Werner 1983), nesting season for land iguanas appears to occur at the end of the rainy season. In Galápagos, this occurs between the end of December and end of April. Precipitation increases the abundance of trophic resources for herbivorous land iguanas, allowing allocation of fats to be used for yolk formation. Figure 15.6 shows rainfall during the rainy season in 2010, 2012, and 2014. Precipitation was recorded at sea level in Puerto Ayora, Isla Santa Cruz. Although precipitation can quantitatively vary across the islands it is likely that the broadscale pattern across the years may be similar for all islands. Precipitation was more abundant in the 2010 and 2012 rainy seasons (446.2 mm and 435.1 mm, respectively) than in 2014 (176.9 mm). In particular, in 2010 and 2012 precipitation was abundant in the months anteceding the nesting season, while in 2014 a prolonged absence of precipitation was observed up to May. This is reflected in the lower number of *C. subcristatus* females carrying eggs in 2014.

Conolophus females are less likely than males to select and remain in a defined area. In fact, most females enter several male territories during the pre-mating season, establishing temporary relationships. After mating, females leave male territories to reach nesting sites. Given our sampling scheme, these behaviors can be reflected in the sex ratio pattern of the two populations. In fact, in *C. subcristatus*, when sampling was performed from January to June, our sex ratio estimates were

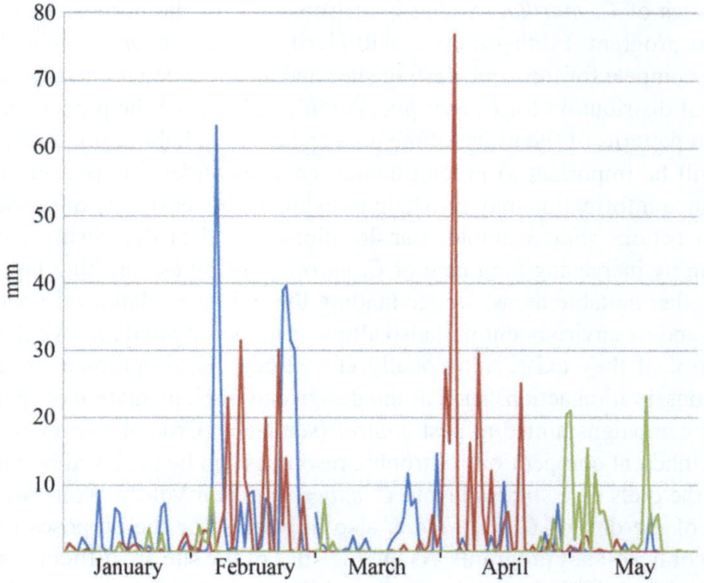

Fig. 15.6 Precipitation in the first semester of years 2010 (*blue*), 2012 (*red*), and 2014 (*green*). Data are from the meteorological station in Puerto Ayora, Santa Cruz Island (Charles Darwin Foundation Meteorological Database: http://www.darwinfoundation.org/datazone/climate/)

closer to 1:1 than when sampling occurred in July 2010, when most females supposedly had already laid eggs. Alarmingly, the sex ratio in *C. marthae* appears to be steadily much higher than 1:1, as would be expected in the case of a small number of females entering male territories for mating.

We need to confirm whether *C. marthae* and the syntopic population of *C. subcristatus* have complete overlapping reproductive seasons as well as whether the two species compete for nesting sites, the location of which is still unknown for *C. marthae*. Competition could also exist for trophic resources. The habitat of *C. marthae* and *C. subcristatus* on Volcán Wolf includes Tropical Dry Shrubland at the top of the volcano and Tropical Dry Forest along the slopes. *C. subcristatus* inhabits areas both surrounding and inside the habitat of *C. marthae*. Preliminary observations (Gentile, unpublished data) suggest that areas surrounding the habitat of *C. marthae* are ecologically distinct from the core area inhabited by *C. marthae*. This suggests that the ecological requirements of *C. marthae* may not be met outside its area. It is clear that uncovering the relationship between area of distribution, habitat characteristics, and its usage will allow us to characterize the ecological niche and formulate a refined habitat suitability model. This could prove crucial in guiding species and implementing habitat conservation actions. This is critical for species that are susceptible and vulnerable to drivers of environmental change. Defining habitat usage for the two species will aid us in identifying the timing and location of competition for resources, should it occur. It will also aid in locating the nesting sites of *C. marthae*, which is instrumental for the purpose of a possible head-start program. Furthermore, it will clarify whether *C. marthae* and *C. subcristatus* compete for the same nesting sites and lead to a better characterization of the area of distribution for *C. marthae*. Additionally, it will help us to understand times and patterns of the usage of the area and develop habitat suitability models, which will be important to predict habitat changes under different climatic scenarios. Such information may also help us evaluate the feasibility of potential conservation actions. For example, translocations aimed at decreasing the risk of extinction by increasing the range of *C. marthae*, or by establishing new populations in other suitable areas. Understanding the intimate relationship between *C. marthae* and its environment will also allow us to identify and remove the barriers to dispersal, if they exist. Additionally, it will help us to optimize strategies for habitat conservation actions such as the design and implementation of focused and effective campaigns aimed at pest control (see later). Crucial information on the possible inherent competition for trophic resources can be achieved by the assessment of the diets of *C. marthae* and *C. subcristatus* on Volcán Wolf. An in-depth analysis of the diet of *C. marthae* is also important for the purposes of captive breeding or head-start programs. As the logistics of the site are difficult, prolonged study on the volcano is prohibitive. Thus, long-lasting direct observations cannot be carried out. For the purposes of diet assessment, the analysis of stomach-content remains can be obtained through stomach flushing. However, this technique is invasive, logistically difficult and, ultimately, not recommendable as it may affect the survival of lizards (Luiselli et al. 2011). Indirect and completely noninvasive methods, by both traditional scat inspection and more recent PCR-based methods, prove very useful when animals are predominantly herbivores (Soininen et al. 2009). These latter methods are particularly well suited to the iguanas on Volcán Wolf. As the scats are morphologically identical in the two species, the correct species assignment of each scat is impossible without the use of molecular tools. An additional advantage of using molecular tools is that they supply indirect information about the geographic occurrence of the species, providing further data with respect to direct capture or sighting. Recently, Di Giambattista et al. (submitted) optimized a protocol for the molecular species assignment of *Conolophus* scats. Currently, a program aimed at the analysis of the feeding behaviors of *C. marthae* and *C. subcristatus* from Volcán Wolf has started as a collaboration between the University Tor Vergata, the Galápagos National Park Directorate and the Charles Darwin Foundation (Jaramillo et al. 2014). Here, a collection of stool samples is processed. The seeds contained in the feces are identified, classified, and then germinated. Germination confirms their identification and tests whether feces contribute to seed dispersal throughout the area of distribution of iguana's species. This approach, although very important, only provides information on the frugivory component of the iguanas' diet. The folivory component is also important and deserves attention by using a different and complementary approach that includes the use of molecular tools.

Possible Hybridization Between C. marthae *and* C. subcristatus

In recent years, it has been documented that introgressive hybridization is a powerful evolutionary factor through which new trajectories may be established (Schwenk et al. 2008), even via lineage fusion, as shown in some Galápagos vertebrates (Garrick et al. 2014; Kleindorfer et al. 2014). In Galápagos, iguanas hybridization is known to occur between the marine iguana *Amblyrhynchus cristatus*—the sister taxon of *Conolophus* (Wiens and Hollingsworth 2000)—and *C. subcristatus* at Plaza Sur island, with introgression being limited to the F1 generation (Rassmann et al. 1997). More recently, Mac Leod et al. (2015) suggested that introgressive hybridization could mask speciation in the marine iguana. However, hybridization may also cause genetic introgression and contamination of pure populations. For example, hybridization between *I. iguana* and *I. delicatissima* throughout the Guadeloupe Archipelago and French West Indies represents a serious threat for *I. delicatissima* (Knapp et al. 2014).

Due to their overlapping ranges and possible overlapping breeding seasons, hybridization could occur, mediating introgression between *C. marthae* and *C. subcristatus* on Volcán Wolf. The head nodding behavior of *C. marthae* is very distinctive and completely different from *C. subcristatus* (Gentile and Snell 2009). This may act as a barrier to hybridization, with other possible unknown mechanisms of mate choice also existing. Gentile et al. (2009) found no evidence of living F1 hybrids. Instead, they discussed the possibility that one *C. subcristatus* individual exhibited a genotype consistent with it being a second generation backcross. Ultimately, introgressive hybridization appeared to be rare and insufficiently strong to prevent genetic differentiation between the two species. However, it must be considered that the *C. marthae* sample used in Gentile et al. (2009) comprised only 15 individuals and nine microsatellite loci. Clarifying the frequency of hybridization and level of genetic introgression between *C. marthae* and *C. subcristatus* is crucial for the purpose of a possible captive breeding program. This hypothesis has been further investigated by Di Giambattista and Gentile (2014) by genotyping 108 *C. marthae* and 162 *C. subcristatus* from Volcán Wolf at 20 microsatellite loci. Their results provide little evidence of successful hybridization with most individuals revealing a proportion of admixture lower than 1 %, and 2–3 % for a small number of iguanas.

Introduced Species

Introduced species such as cattle, donkeys, pigs, goats, dogs, cats, and rats may have a very strong impact on Galápagos wildlife, including iguanas. Nonnative herbivorous species such as donkeys and goats may generate strong competition for food with iguanas and cause habitat destruction. Pigs can destroy nesting habitat while searching for iguana eggs and possibly hatchlings. Feral dogs and cats are predators of hatchlings, juveniles, and adult females, while black rats (*Rattus rattus*) are predators of hatchlings and juveniles.

In mid-1970s, dogs almost caused the extinction of two populations of *C. subcristatus* in Santa Cruz island and in Bahia Cartago (southern Isabela) where *C. subcristatus* has been successfully repatriated after dogs' eradication (Fabiani et al. 2011). Feral dogs can also attack marine iguanas (Burnett and Rudd 1983). In February 2005 a few dogs attacked and bit hundreds of marine iguanas along the coast flanking the town San Cristóbal in San Cristóbal island. Gentile and collaborators counted 147 cadavers of marine iguanas along a 1-km long transect. Land iguanas had disappeared from Baltra by 1954 after the construction and operation of an American airbase during the Second World War. A combination of habitat destruction, human predation, and feral cats proved fatal to the iguana population (Woram 1991).

Such negative effects are observed also in other areas of the world. For example, introduced ungulates put at risk some populations of the rock iguana *Cyclura pinguis* on Anegada in the British Virgin Islands (Mitchell et al. 2000) while pigs have strongly impacted on nests of *Cyclura stejnegeri* (Wiewandt 1977), endemic to Mona Island (Puerto Rico). In the 1970s, a large population of *Cyclura carinata* in Pine Cay (Turks and Caicos Islands) was almost completely destroyed within five years by a few dogs and cats introduced by hotel workers (Iverson 1978). The Jamaican Iguana *Cyclura collei* is currently severely affected by feral cats which occur throughout the area and are known predators of juvenile iguanas (Wilson 2008). Despite some evidence showing coexistence of invasive rats and mice with apparently healthy iguana populations (Mitchell et al. 2000), invasive rodents, particularly rats, may have a great impact on endangered insular populations of iguanas. Recently, Hayes et al. (2012) investigated tail damage (including tail-break and tail-furcation) in 3537 individuals of three species of West Indian rock iguanas (genus *Cyclura*) in the Bahamian Archipelago, including the Turks and Caicos Islands. They found that such damages are significantly higher in populations coexisting with invasive rodents, supporting that they result from failed attempts of predation. Besides the potential effect on the fitness that such damages may cause in injured individuals, it is clear that successful predation has a detrimental effect on the population.

The Galápagos National Park Directorate runs major campaigns to control and eradicate exotic species in Galápagos, including Volcán Wolf. So far, such campaigns have successfully prevented habitat disturbance by non-autochthonous goats in northern Isabela, as well as promoted habitat restoration in southern Isabela.

The Galápagos Hawk (*Buteo galapagoensis*) is the only native predator of *C. marthae* on Volcán Wolf. However, feral cats and black rats also occur in Isabela, including Volcán Wolf. Consequently, given the small population size of *C. marthae*, such pests may pose a threat to the *C. marthae* population's recruitment. On this island, the control of feral cats is particularly challenging. Its large area (4588 km^2) impedes the complete elimination of feral cats (Nogales et al. 2004) and control actions can only aim at mitigating the impact.

In the early 2000s, a 3-year program to eradicate feral cats from the Baltra island (where *C. subcristatus* was repatriated from 1991 onwards) was initially effective. It involved poisoning the cats with sodium monofluoroacetate (compound 1080) and then trapping or shooting them (Phillips et al. 2005). However, the applicability of such a protocol at the Volcán Wolf site is still to be evaluated. Further evaluation of the feral cat population on Volcán Wolf is needed for the purpose of implementing a program for their control.

A strategic plan was implemented by the Galápagos National Park Directorate, aimed at the eradication or mitigation of the introduced population of rodents, such as black rats and house mice (*Mus musculus*), from a medium-small area of the Galápagos islands. North Seymour (1.84 km^2) was used as a pilot to test the plan and train personnel. In the case of North Seymour, the campaign to eradicate black rat was conducted using the anticoagulant rodenticide Brodifacoum (Sevilla-Paredes and Rueda Cordova 2014).

Since the early 1980s, the house mouse has been found on South Plaza Island (0.12 km^2). Since then, a decline in the arborescent prickly pear (*Opuntia echios echios*) has occurred. This cactus is a major food item and source of water for the South Plaza land iguana (*C. subcritatus*) population. Mice undermine the root system of the cactus, ultimately resulting in the death of the cactus. The anticoagulant rodenticide Brodifacoum was also used on this island. This was based on prior research of green iguanas (*Iguana iguana*) that indicated a low risk of toxicity in this species and, presumably, in other iguana species. During a rodent eradication campaign, a plan to safeguard the population of iguanas on South Plaza was developed (Tapia et al. 2014). The applicability of these or similar protocols on Volcán Wolf has not yet been evaluated.

Although it is very unlikely that rats and cats may be completely eliminated from Isabela, mitigation actions and monitoring are in order. Additionally, as in Isabela feral cats and rats coexist, integrated actions should be carried out, aimed at the contemporary mitigation of both cat and rat populations (Rayner et al. 2007; Nogales et al. 2013).

Parasites

Ectoparasite load is high on both *C. marthae* and *C. subcristatus* on Volcán Wolf. In fact, the location is characterized by a massive occurrence of ticks (*Amblyomma usingeri* and *A. macfarlandi*). Ticks are much more abundant on Volcán Wolf than elsewhere in the Galápagos archipelago. Cost of infestation by ticks has been estimated for *A. cristatus* and species' behavioral adaptations to reduce the impact of infestation have been discussed (Wikelski 1999). Ticks infesting reptiles can transmit viruses, bacteria, and hemoparasites (Labuda and Nuttall 2004; Kho et al. 2015; Telford 2009). The hemogregarin *Hepatozoon* (Apicomplexa) appears as the only hemoparasite occurring in Galápagos land iguanas (Fulvo 2010). Both the Volcán Wolf populations of *C. marthae* and *C. subcristatus* show a high prevalence of

Hepatozoon infection, as well as a different leukocyte count (WBC), compared to other populations of land iguanas from the whole archipelago. In particular, the heterophils/lymphocytes (H/L) ratio is higher in infected than in noninfected individuals (Onorati and Gentile 2014). It is interesting to note that in iguanas, the H/L ratio is considered to be an effective indicator of stress (Davis et al. 2008). Whether ectoparasite and hemoparasite load affect the fitness of the two populations is currently under investigation.

Illegal Trade

According to TRAFFIC, the wildlife trade monitoring network (www.traffic.org), the wildlife trade involves hundreds of millions of individual plants and animals from tens of thousands of species. Such illicit trade is now estimated to be worth between US$8 and $10 billion per year globally (Haken 2011).

Illegal wildlife trade in Galápagos is a serious issue. Between 2010 and 2012, the authorities of the Galápagos National Park Directorate recorded four cases of illegal collection of Galápagos reptiles. In July 2012, a German tourist was arrested in Galápagos for trying to illegally transport four land iguanas out of the province. The use of appropriate molecular tools, in combination with a previous genetic characterization of a large number of Galápagos iguanas, proved crucial for the purposes of taxonomic identification and rapid repatriation of confiscated iguanas (Gentile et al. 2013). The work was also used in the case of forensics, following the arrest of the tourist, who was then sentenced to 4 years in prison, the maximum penalty. In fact, according to Ecuadorian law, any attempt to remove wildlife from the Galápagos Islands is a serious environmental crime, punishable under Articles of the Ecuadorian Criminal Code. Cases as the one described above are not accidental. In 2011, the same German tourist was arrested in Fiji for attempting to smuggle local reptiles out of that country. Recently, the Galápagos National Park Directorate uncovered a complex network of people involved in illicit trafficking of Galápagos iguanas (Angermeyer 2014). To clarify, there is no legal trade allowed for Galápagos iguanas protected under CITES. Consequently, any specimen from private captive breeding is illicitly reared as resulting from contraband. Authorities of the Galápagos National Park Directorate are strongly committed to the continuous effort required to prevent the illegal trade of Galápaganian wildlife.

The Galápagos Pink Land Iguana as a New Flagship Species

Over the last 20 years, several definitions of the flagship species concept have been proposed and used. For example, the World Wildlife Fund (WWF) currently highlights priority species as those ecologically or socially important (http://wwf.panda.org/what_we_do/endangered_species/). Among these, flagship species are defined as "iconic animals that provide a focus for raising awareness and stimulating action and funding for broader conservation efforts." In this view, focusing efforts on these species will also help conserve other species that share their habitats and/or are vulnerable to the same threats. While general agreement is still lacking, the socio-economic and strategic significance of the concept has been increasingly emphasized in recent years. In a recent review, a flagship species was defined as "the focus of a broader conservation marketing campaign based on its possession of one or more traits that appeal to the target audience" (Veríssimo et al. 2011). Based on this concept, which no longer has any biological or ecological implications, a methodology was proposed to evaluate existing conservation flagships. Such methodology evaluates the success of a particular flagship by assessing its visibility and recognition (Veríssimo et al. 2014).

It is not the purpose of this chapter to further contribute to the general debate on the flagship species concept. Nevertheless, we will provide some additional considerations of the evolution of *C. marthae*, as well as the social impact the species has had since its discovery. These considerations may help to identify the potential role of *C. marthae* as a flagship species.

Evolutionary Importance

The discovery of a new species of megafauna in one of the most remote, yet greatly investigated, areas of the world was surprising. Indeed, *C. marthae* was found in the Galápagos Islands, one of the most important locations in the world for the development of evolutionary thinking. Charles Darwin visited the Galápagos in 1835 and his 5-week stay in the archipelago proved crucial for the development of his theory of evolution by natural selection. Darwin visited Isabela island but he did not explore Volcán Wolf. Thus, although he noted and commented on both the marine and land iguanas (Darwin 1839), he did not encounter the Galápagos Pink Land Iguana. In a sense, he missed it.

Gentile et al. (2009) performed genetic analyses indicating that the most recent common ancestor between *C. marthae* and the other two congeneric species (*C. subcristatus* and *C. pallidus*, also endemic to Galápagos) existed long ago. They suggested that such an ancestor could have existed between five and six million years (Myr) ago, when the Galápagos did not have their current appearance and none of the existing islands had yet emerged (Geist et al. 2005). Currently, the age of the original split between *C. marthae* and other congeneric species is being revisited, with a more recent proposed origin between one and two million years (Mac Leod et al. (2015)). It is clear that *C. marthae* is a real biogeographic conundrum. It only occurs on Volcán Wolf, which is considered younger than Volcán Sierra Negra, at 0.53 million years Sierra Negra is the oldest volcano of Isabela (Nordlie 1973). Volcán Wolf is almost as old as Volcán Cerro Azul, which is 0.35 million years (Naumann and Geist 2000). The species carries substantial evolutionary information. In fact, it is the only representative of a separate ancient lineage that testifies to the divergence events occurring within *Conolophus* since its original splitting from the marine iguana lineage (Rassmann 1997).

For these reasons, *C. marthae* was featured as the closing scene of the recent series of documentary movies "GALÁPAGOS," written and presented by Sir David Attenborough and produced by Colossus Productions. While advertising the series, Attenborough publicly commented on the species:

> I used to collect stamps and this [pink iguana] was a Penny Black of the natural world in a very big way.

It is interesting to note that the UK's 1840 issued Penny Black was the first adhesive postage stamp ever to be produced. In the series, evolutionary questions regarding *C. marthae* were posed, including the origin and possible selective significance of its color.

The evolutionary importance of the species was also recognized at academic level. In fact, in 2010, the International Institute for Species Exploration at Arizona State University nominated *C. marthae* to be among the most important species newly discovered for the year. In 2012, the same institute included *C. marthae* among the most important new species of the century.

The Galápagos Pink Land Iguana and Society

After the vast mass-media coverage upon its discovery, *C. marthae* became the focus of several educational publications. The species entered the Guinness Book of World Records as the "newest species of iguana" (http://www.guinnessworldrecords.com/world-records/newest-species-of-iguana), although it was reported with a wrong scientific name. The name "Pink Iguana" and logos referring to the species have also started being used for commercial purposes in the Galápagos (Fig. 15.7). Unexpectedly, *C. marthae* was also inspirational to several different initiatives. The link between the animal and evolution was adopted in different ways and contexts to express unicity and capability to evolve in a changing environment. For example, in an economic context, newly founded private business companies were named after the Galápagos Pink Land Iguana to express excellence and ability to perform in a changing market.

The concept was also developed in an artistic context. In 2011, *C. marthae* became the subject of the INPUT Journal, an aesthetic journal on contemporary currents and cultural conditions published by an international ensemble of editors. The species served as a metaphor, considering that the form of the book is near extinction and the art book is distributed via a threatened environment—the art and architecture book store. INPUT asked a community of 12 emerging artists to contemplate *C. marthae* as they considered the art book as an evolving and adapting medium of artistic engagement. The work "INPUT #3—The Pink Iguana" was presented on August 4, 2011 in New York City, http://vimeo.com/26775603.

Fig. 15.7 A signboard portraying the Pink Iguana name and logo in Villamil (Isabela Island) (Photo: L. Di Giambattista)

Remarkably, for its delicate color, *C. marthae* also became the subject of storybooks for children to convey concepts of love and friendship (Fig. 15.8).

The Galápagos Pink Land Iguana and Conservation

In 2009, the WWF-Italy identified a project that led to the discovery of *C. marthae* as among the most relevant Italian projects contributing to the safeguard of biodiversity. It also received attention in 2013, when the Species Survival Commission of the IUCN featured the species in the "Amazing Species"; an initiative to promote popular understanding of threatened species (http://jr.iucnredlist.org/documents/amazingspecies/conolophus-marthae.pdf). Currently, such an initiative has over 21,000 followers on Twitter and over 33,000 on Facebook. These two examples show how *C. marthae* has a role in focusing the attention of the general public on themes of biodiversity and conservation.

Furthermore, *C. marthae* also plays a role as a strategic element in the conservation of the Galápagos Islands. An excellent example of this is provided by Lonesome

Fig. 15.8 Illustration from "The Blue Footed Booby Brothers and The Pink Iguanas," written and illustrated by Gabrielle Shamsey, Pennington, NJ (gabrielleshamsey@verizon.net). The image is courtesy of the Author

George, the last documented member of *Chelonoidis abingdoni* and one of the endemic species of giant tortoises inhabiting the Galápagos. Lonesome George was the most effective icon of Galápagos conservation and a global conservation symbol. Even after passing away in 2012, Lonesome George keeps its significance. Efforts are currently being made to ensure the proper preservation of the specimen and its exposure in a museum, allowing a large number of people to experience the species (http://www.amnh.org/explore/preserving-lonesome-george).

In the Galápagos National Park Directorate's vision, environmental education is crucial as it encourages a change in attitude and behavior in Galápagos citizens. Moreover, it promotes the harmonious relationship between man and nature. *C. marthae* provides an important opportunity for environmental education and the promotion of Galápagos identity in the local population. Through the subjects of natural sciences, locals can learn about the exclusivity of the different species of the archipelago, their interrelationships in the ecosystem, and their lifestyle. In particular, the peculiarity and uniqueness of the Galápagos species increases the pride of Galápagos citizens. They are proud to be a part of the Galápagos delicate ecosystem, internalizing the need to care and magnifying their sense of belonging (De la Rosa 2014).

The "Management Plan for the Protected Areas of Galápagos for the Good Living" establishes principles and criteria for the selection of focal species for the conservation Galápagos biodiversity. Several types of species are identified, which justify a selective administration of biodiversity. Such species are used as management tools on which available resources and actions are focused (http://www.galapagospark.org/noph.php?page=institucion_plan_de_manejo). Among those species, flagships are considered strategic species which, for popular acceptance, are used as flags of programs contributing to global conservation funding. Certainly *C. marthae* is one of them. The idea of taking advantage of more than one unique and evocative species for conservation campaigns is in line with recent evidence (Veríssimo et al. 2014) that shows it is advantageous for organizations to create flagship fleets. Such fleets can broaden the appeal of conservation campaigns, rather than simply preferring the use of a few long-standing flagships.

Acknowledgments Data presented here result from a long-term program in collaboration between the Galápagos National Park and the University of Rome "Tor Vergata," aimed at the conservation of Galápagos land iguanas. We thank the Mohamed bin Zayed Species Conservation Fund for funding. We wish to acknowledge M. Ciambotta, G. Colosimo, M. Cruz, L. Di Giambattista, L. Migliore, M. Onorati, R. Palozzi, and G. Riccarducci for their crucial assistance in the field. We thank the Charles Darwin Foundation, FUJIFILM SonoSite, Inc., and Concepto Azul S.A. for collaborative support. We thank the IUCN ISG for contributing discussions and suggestions during the 2014 IUCN ISG meeting. We wish to thank Sir D. Attenborough and the Colossus Production for being respectful of the pink iguanas' site, during the making of the documentary movie "GALÁPAGOS." We thank K. Cole for reading an early version of the manuscript. We also thank the two anonymous reviewers whose constructive criticism improved the quality of this chapter. Many more colleagues, students, and others have contributed and are contributing to this research; we thank them all. Finally, we are indebted to the Galápagos National Park rangers for their incommensurably valuable participation in this project and friendship.

References

Angermeyer E (2014) Traffic of iguanas. Paper presented at the Galápagos Land and Marine Iguanas workshop, IUCN SSC Iguana specialist group meeting, Puerto Ayora, 28–29 October 2014

Boyce MS (1992) Population viability analysis. Annu Rev Ecol Syst 23:481–506

Burnett BD, Rudd RL (1983) Feral dogs of the Galápagos Islands: impact and control. Int J Stud Anim Probl 4:44–58

Darwin C (1839) The voyage of the Beagle. Dutton, New York

Davis AK, Maney DL, Maerz JC (2008) The use of leukocyte profiles to measure stress in vertebrates: a review for ecologists. Funct Ecol 22:760–772

De la Rosa V (2014) Pink Iguanas: a boost to the environmental education of the archipelago. Paper presented at the Galápagos land and marine Iguanas workshop, IUCN SSC Iguana specialist group meeting, Puerto Ayora, 28–29 October 2014

Di Giambattista L, Gentile G (2014) Assessing hybridization and introgression levels between *Conolophus marthae* and *C. subcristatus*. Paper presented at the Galápagos land and marine Iguanas workshop, IUCN SSC Iguana specialist group meeting, Puerto Ayora, 28–29 October 2014

Donegan TM (2009) Type specimens, samples of live individuals and the Galápagos Pink Land Iguana. Zootaxa 2201:12–20

Drummond SP et al (2010) Influence of a threatened-species focus on conservation planning. Conserv Biol 24:441–449

Dubois A (2009) Endangered species and endangered knowledge. Zootaxa 2201:26–29

Fabiani A et al (2011) Conservation of Galápagos land iguanas: genetic monitoring and predictions of a long-term program on Isla Santa Cruz. Anim Conserv 14:419–429

Fulvo A (2010) Genetic characterization of haemoparasites (*Hepatozoon*) and evaluation of their impact on the populations of land iguana from Galápagos (*Conolophus*). PhD Dissertation, Tor Vergata University, Rome, Italy

Garrick RC et al (2014) Lineage fusion in Galápagos giant tortoises. Mol Ecol 23:5276–5290

Geist DJ et al (2005) Wolf volcano, Galápagos Archipelago: melting and magmatic evolution at the margins of a mantle plume. J Petrol 46:2197–2224

Gentile G (2012) *Conolophus marthae*. http://www.iucnredlist.org/details/174472/0. Accessed 15 Jan 2015

Gentile G, Snell H (2009) *Conolophus marthae* sp. nov. (Squamata, Iguanidae), a new species of land iguana from the Galápagos archipelago. Zootaxa 2201:1–10

Gentile G et al (2009) An overlooked pink species of land iguana in the Galápagos. Proc Natl Acad Sci U S A 106:507–511

Gentile G, Ciambotta M, Tapia W (2013) Illegal wildlife trade in Galápagos: molecular tools help the taxonomic identification of confiscated iguanas and guide their rapid repatriation. Conserv Genet Resour 5:867–872

Gilmar CA, Wolf BO (2007) Use of portable ultrasonography as a nondestructive method for estimating reproductive effort in lizards. J Exp Biol 210:1859–1867

Haken J (2011) Transnational crime in the developing world. Global Financial Integrity. www.gfip.org/storage/gfip/documents/reports/transcrime/gfi_transnational_crime_web.pdf. Accessed 15 Jan 2015

Hayes WK et al (2012) Do invasive rodents impact endangered insular iguana populations? Biodivers Conserv 21:1893–1899

Hill WG (1981) Estimation of effective population size from data on linkage disequilibrium. Genet Res 38:209–216

Iverson JB (1978) The impact of feral cats and dogs on populations of the West Indian rock iguana. *Cyclura carinata*. Biol Conserv 14:63–73

Jaramillo P et al (2014) Preliminary data on the diet of *Conolophus marthae* (Pink Iguana). Can feces contribute to plant dispersal? Paper presented at the Galápagos land and marine Iguanas workshop, IUCN SSC Iguana specialist group meeting, Puerto Ayora, 28–29 October 2014

Kho KL, Koh FX, Tay ST (2015) Molecular evidence of potential novel spotted fever group rickettsiae, *Anaplasma* and *Ehrlichia* species in *Amblyomma* ticks parasitizing wild snakes. Parasit Vector 8:112

Kleindorfer S et al (2014) Species collapse via hybridization in Darwin's tree finches. Am Nat 183:325–341

Knapp C et al (eds) (2014) Lesser Antillean Iguana, *Iguana delicatissima*: conservation action plan, 2014–2016. IUCN Iguana Specialist Group, Gland

Labuda M, Nuttall PA (2004) Tick-borne viruses. Parasitology 129:S221–S245

Luiselli L et al (2011) Stomach flushing affects survival/emigration in wild lizards: a study case with rainbow lizards (*Agama agama*) in Nigeria. Amphibia-Reptilia 32:253–260

Mac Leod A et al (2015) Hybridization masks speciation in the evolutionary history of the Galápagos marine iguana. Proc R Soc Lond Biol 282 doi:10.1098/rspb.2015.0425

Mace GM et al (2008) Quantification of extinction risk: IUCN's system for classifying threatened species. Conserv Biol 22:1424–1442

Márquez CM et al (2010) Estado poblacional de las iguanas terrestres (*Conolophus subcristatus*, *C. pallidus* y *C. marthae*: Squamata, Iguanidae), Islas Galápagos. Bol Téc Lab IASA I 9:25–43

Minelli A (2009) Commentaries on Gentile & Snell (2009): an introduction. Zootaxa 2201:11–11

Mitchell N, Fulford M, Haeffner R (2000) Turks and Caicos iguana (*Cyclura carinata carinata*). WIISG News 3:7–13

Naumann T, Geist D (2000) Physical volcanology and structural development of Cerro Azul Volcano, Isabela Island, Galápagos: implications for the development of Galápagos-type shield volcanoes. Bull Volcanol 61:497–514

Nemésio A (2009) On the live holotype of the Galápagos Pink Land Iguana, *Conolophus marthae* Gentile & Snell, 2009 (Squamata: Iguanidae): is it an acceptable exception? Zootaxa 2201:21–25

Nogales M et al (2004) A review of feral cat eradication on islands. Conserv Biol 18:310–319

Nogales M et al (2013) Feral cats and biodiversity conservation: the urgent prioritization of island management. Bioscience 63:47–53

Nordlie BE (1973) Morphology and structure of the western Galápagos volcanoes and a model for their origin. Geol Soc Am Bull 84:2931–2956

Onorati M, Gentile G (2014) Impact of ecto- and endoparasites on two populations of *Conolophus marthae* and *Conolophus subcristatus* on the Vólcan Wolf (Isabela): Analysis of haematological markers. Paper presented at the Galápagos land and marine Iguanas workshop, IUCN SSC Iguana specialist group meeting, Puerto Ayora, 28–29 October 2014

Palstra FP, O'Connell MF, Ruzzante DE (2009) Age structure, changing demography and effective population size in Atlantic salmon (*Salmo salar*). Genetics 182:1233–1249

IUCN Standards and Petitions Subcommittee (2014) Guidelines for using the IUCN red list categories and criteria. Version 11. Prepared by the Standards and Petitions Subcommittee. http://www.iucnredlist.org/documents/RedListGuidelines.pdf. Accessed 5 Apr 2015

Phillips RB et al (2005) Eradicating feral cats to protect Galápagos Land Iguanas: methods and strategies. Pac Conserv Biol 11:257–267

Rassmann K (1997) Evolutionary age of the Galápagos iguanas predates the age of the present Galápagos islands. Mol Phylogenet Evol 7:158–172

Rassmann K, Trillmich F, Tautz D (1997) Hybridization between land and marine iguana (*Conolophus subcristatus* and *Amblyrhynchus cristatus*) on Plaza Sur. J Zool (Lond) 242:729–739

Rayner MJ et al (2007) Spatial heterogeneity of mesopredator release within an oceanic island system. Proc Natl Acad Sci U S A 104:20862–20865

Redding DW, Mooers AO (2006) Incorporating evolutionary measures into conservation prioritization. Conserv Biol 20:1670–1678

Rivas LR (1964) A Reinterpretation of the concepts "Sympatric" and "Allopatric" with proposal of the additional terms "Syntopic" and "Allotopic". Syst Biol 13:42–43

Robertson BC, Stephenson BM, Goldstien SJ (2011) When rediscovery is not enough: taxonomic uncertainty hinders conservation of a critically endangered bird. Mol Phylogenet Evol 61:949–952

Schwenk K, Brede N, Streit B (2008) Introduction. Extent, processes and evolutionary impact of interspecific hybridization in animals. Phil Trans R Soc B 363:2805–2811

Sevilla-Paredes C, Rueda Cordova D (2014) Restauración Ecológica de la isla Seymour Norte: El caso de la erradicación de la rata negra (*Rattus rattus*). Paper presented at the Galápagos land and marine Iguanas workshop, IUCN SSC Iguana specialist group meeting, Puerto Ayora, 28–29 October 2014

Snell HL, Snell HM, Tracy CR (1984) Variation among populations of Galápagos land iguanas (*Conolophus*): contrasts of phylogeny and ecology. Biol J Linn Soc 21:185–207

Soininen EM et al (2009) Analysing diet of small herbivores: the efficiency of DNA barcoding coupled with high-throughput pyrosequencing for deciphering the composition of complex plant mixtures. Front Zool 6:16. doi:10.1186/1742-9994-6-16

Tapia W et al (2014) Safeguarding the South Plaza Galápagos Land Iguana during rodent eradication using Brodifacoum. Paper presented at the Galápagos land and marine Iguanas workshop, IUCN SSC Iguana specialist group meeting, Puerto Ayora, 28–29 October 2014

Telford SR Jr (2009) Hemoparasites of the reptilia: color atlas and text. CRC Press, Taylor and Francis Group, Boca Raton, 394p

Tzika AC et al (2008) Population genetics of Galápagos land iguana (genus Conolophus) remnant populations. Mol Ecol 17:4943–4952

Veríssimo D, MacMillan DC, Smith RJ (2011) Toward a systematic approach for identifying conservation flagships. Conserv Lett 4:1–8

Veríssimo D et al (2014) Evaluating conservation flagships and flagship fleets. Conserv Lett 7:263–270

Waples RS, Do C (2008) LDNe: a program for estimating effective population size from data on linkage disequilibrium. Mol Ecol Res 8:753–756

Werner DI (1983) Reproduction in the iguana *Conolophus subcristatus* on Fernandina Island, Galápagos: Clutch size and migration costs. Am Nat 121:757–775

Wiens JJ, Hollingsworth BD (2000) War of the Iguanas: conflicting molecular and morphological phylogenies and long-branch attraction in iguanid lizards. Syst Biol 49:143–159

Wiewandt TA (1977) Ecology, behavior, and management of the Mona Island ground iguana, *Cyclura stejnegeri*. PhD Dissertation, Cornell University, Ithaca, New York

Wikelski M (1999) Influences of parasites and thermoregulation on grouping tendencies in marine iguanas. Behav Ecol 10:22–29

Wilson BS (2008) Battling invasive predators to save the Jamaican iguana. Aliens Xamayca 1:2

Woram JM (1991) Who killed the iguanas? Noticias Galápagos 50:2–17

Part VI
Species that Are a Danger to Humans, Man-Eating Wildlife, etc.

This part covers cases in which problematic wildlife species not only have an impact on human activity, or compete with it, but also pose a great threat, sometimes both directly (such as species that are somewhat regularly considered man-eating species) and indirectly, to the activities and very existence of humans (Löe and Röskaft 2004; Lamarque et al. 2009).

This topic is extremely relevant because, in many parts of the world, man's population is booming; therefore, man is searching for more and more space to inhabit. So closer cohabitation is inevitable, if not overlapping, with areas inhabited by wildlife, often including potentially dangerous species. These situations create more and more human–wildlife interactions, resulting in various significant problems that need to be solved.

The first chapter (Dickman and Hazzah 2016) provides a general examination of the reasons why human–wildlife conflict arises, starting with the myths, superstitions, taboos and religions, which often determine and influence the relationship between people and wildlife species. The chapter highlights how absolutely necessary a multidisciplinary approach is to try to resolve the various problems to achieve the sustainable coexistence of humans and wildlife.

The second chapter (Linnell and Alleau 2016) is a review of carnivores that kill man and consider man a possible prey. There are many infrequent cases, but some species prey on man somewhat regularly. This chapter provides a detailed analysis of the case of the grey wolf (*Canis lupus*), which is a predator of humans in North America and Europe. This has recently become a very delicate issue and requires appropriate, urgent measures to mitigate and resolve the conflicts.

The third chapter (McLennan and Hockings 2016) analyses all the cases registered so far of humans being attacked by apes. It highlights the differences between the species, attempts to try to identify the causes and reasons for this behaviour and proposes measures to solve them.

Finally, in the chapter by Le Bel et al. (2016), a very simple and inexpensive means is proposed to improve communication and links between people involved in situations of human–wildlife interaction. This method, using some cases in the southern African region as an example, is based on the use of mobile phones and a network of automated messages, making intervention measures and even conflict prevention much more effective.

References

Dickman AJ, Hazzah L (2016) Money, myths and man-eaters: complexities of human-wildlife conflict. In: Angelici FM (ed) Problematic wildlife—a cross-disciplinary approach. Springer, New York, p 339–356

Lamarque F, Anderson J, Fergusson R, Lagrange M, Osei-Owusu Y, Bakker L (2009) Human-wildlife conflict in Africa. Causes, consequences and management strategies. FAO Forestry Paper 157:1–98

Linnell JDC, Alleau J (2016) Predators that kill humans: myth, reality, context and the politics of wolf attacks on people. In: Angelici FM (ed) Problematic wildlife—a cross-disciplinary approach. Springer, New York, p 357–371

Le Bel S, Chavernac D, Stansfield F (2016) Promoting a mobile data collection system to improve HWC incident recording: a simple and handy solution for controlling problem animals in southern Africa. In: Angelici FM (ed) Problematic wildlife—a cross-disciplinary approach. Springer, New York, p 395–411

Löe J, Röskaft E (2004) Large carnivores and human safety: a review. Ambio 33:283–288

McLennan MR, Hockings KJ (2016) The aggressive apes? Causes and contexts of great ape attacks on humans. In: Angelici FM (ed) Problematic wildlife—a cross-disciplinary approach. Springer, New York, p 373–394

Chapter 16
Money, Myths and Man-Eaters: Complexities of Human–Wildlife Conflict

Amy J. Dickman and Leela Hazzah

Introduction

Human–wildlife conflict—defined at the World Parks Congress as occurring '*when the needs and behaviour of wildlife impact negatively on the goals of humans or when the goals of humans negatively impact the needs of wildlife*'—is a phenomenon that has existed throughout human evolution. Early hominids are thought to have been predated on by leopards (*Panthera pardus*), spotted hyaenas (*Crocuta crocuta*) and sabre-toothed cats (Lee-Thorp et al. 2000), while tales of threatening species have permeated human culture for millennia, appearing in countless myths, songs, stories and works of art (Kruuk 2002; Quammen 2003). Such threats are not merely historical artefacts, though—wild animals attack and kill many hundreds of people a year (Dhanwatey et al. 2013; Loe and Roskaft 2004) and commonly destroy peoples' livelihoods and severely impact their quality of life (Jadhav and Barua 2012; Thirgood et al. 2005). On the other side of the coin, people have had a devastating impact on wildlife, with humans implicated in the extinction of over 300 terrestrial vertebrate species over the past 500 years (Dirzo et al. 2014).

A huge variety of species create conflict with people, including invertebrates, snakes, birds, rodents and other small mammals, and in many cases, those conflicts are resolved by enacting pest control (Marchini 2014). What comprises a 'pest' is a subject worthy of debate—certainly, a hungry leopard prowling around someone's hut could justifiably be considered far more of a pest than a family of rats (*Rattus*

A.J. Dickman (✉)
Wildlife Conservation Research Unit, Department of Zoology, The Recanati-Kaplan Centre, Tubney House, University of Oxford, Tubney, Oxford OX13 5QL, UK
e-mail: amy.dickman@zoo.ox.ac.uk

L. Hazzah
Lion Guardians, P.O. Box 15550-00502, Nairobi, Kenya
e-mail: Lhazzah@gmail.com

F.M. Angelici (ed.), *Problematic Wildlife*, DOI 10.1007/978-3-319-22246-2_16

rattus) taking up residence in a UK home, and yet wildlife killing only tends to incite much controversy in the first case. Although this is often linked to rarity, it is not always the case—poisoning a leopard, even where they are locally common, is likely to generate more debate than poisoning rats, even though persecution of rats has led to them being named as one of the ten most threatened rodent species in Europe (Entwistle and Stephenson 2000). Regardless of differences in how it is viewed depending on the species, it is clear that conflict imposes very serious costs on both humans and wildlife across the globe (Barua et al. 2013; Marchini 2014; Woodroffe et al. 2005). Indeed, it is now one of the most pressing issues in modern biodiversity conservation, as the world's burgeoning human population means that people and wildlife come into contact ever more frequently (Conover 2002), often with harsh consequences for both sides. Lions (*Panthera leo*), elephants (*Loxodonta africana*), and orang-utans (*Pongo* sp.) are just some of the iconic species for which conflict with humans poses a major threat to their continued persistence across much of their range (IUCN 2006; Meijaard et al. 2011; Naughton et al. 1999).

Perhaps the most obvious human–wildlife conflict situation is one where a wild animal destroys someone's property or takes a human life, which may lead to retaliatory action on the part of the human. This kind of direct wildlife damage is undoubtedly important and can have extremely significant impacts on local people and their livelihoods (Barua et al. 2013; Thirgood et al. 2005). For instance, around Zimbabwe's Sengwe Wildlife Research Area, livestock attacks by baboons (*Papio ursinus*), lions, leopards and other species cost householders an average of 12 % of their net annual income (Butler 2000), while around Gishwati forest in Rwanda, crop-raiding by chimpanzees (*Pan troglodytes*) and *Cercopithecus* monkeys incurred a food replacement cost of 10–20 % for local households (McGuinness and Taylor 2014). These costs are not limited to the developing world—in Wisconsin, crop damage by white-tailed deer (*Odocoileus virginianus*) costs over US$34 million a year (Naughton-Treves and Treves 2005). In traditional rural societies, wildlife-related damage can incur cultural costs as well as economic ones, as livestock in particular are often vital sociocultural assets, and their loss affects social standing and status (Dickman 2009).

Human–wildlife conflict can also have more subtle or 'hidden' impacts (Barua et al. 2013). In areas with problematic wildlife, there are opportunity costs where people have to spend time, energy and money protecting their assets, which could be invested in more valuable alternatives such as attending school, generating revenue or engaging in culturally valued activities (Barua et al. 2013; Thirgood et al. 2005). In some cases, wildlife damage forces people to relocate, leading to significant social impacts (Barua et al. 2013). There can also be significant wider scale opportunity costs, from setting aside land for wildlife—a 1995 analysis suggested that Kenya's parks, reserves and forests could generate US$203 million if put to other use, meaning that the US$42 million generated instead by conservation activities was a huge net loss for the country (Norton-Griffiths and Southey 1995). Furthermore, when incidents such as man-eating occur, there are huge behavioural and psycho-social impacts in addition to economic ones, especially as in rural communities the victim is often a male, and therefore a principal wage-earner (Barua et al. 2013; Jadhav and Barua 2012).

However, most of these issues are still linked to wildlife damage as the central problem. Because wildlife damage is usually cited (often vociferously) as the key reason for conflict, it is unsurprising that many mitigation strategies, such as the use of communal herding, guardian animals, thunder-flashes, fladry, chilli or chilli-tobacco fences (Chelliah et al. 2010; McManus et al. 2014; Musiani et al. 2003; Sitati and Walpole 2006) are centred around reducing that damage. These approaches are often successful, and reducing attacks can have a demonstrable effect on wildlife populations—in the Phinda area of South Africa, conservation initiatives such as improving livestock husbandry and response to conflicts were linked to reduced leopard mortality and a leopard population growth rate of 14–16 % (Balme et al. 2009).

While reducing wildlife damage therefore plays an important role in easing human–wildlife coexistence, it will often only address part of the problem. It might seem logical that the different elements of conflict, such as the wildlife damage incurred, the degree of conflict reported (in terms of negativity towards the species concerned), and the response taken are relatively simply related. However, in reality, the situation is usually more complex than initially envisioned, with multiple factors affecting the relationships between the different components (Dickman 2010). Here, we examine factors affecting two key aspects of conflict, namely (1) the relationship between the extent of wildlife damage and attitudes towards wildlife; and (2) the relationship between reported attitudes and the response to conflict (Fig. 16.1). We provide examples from a

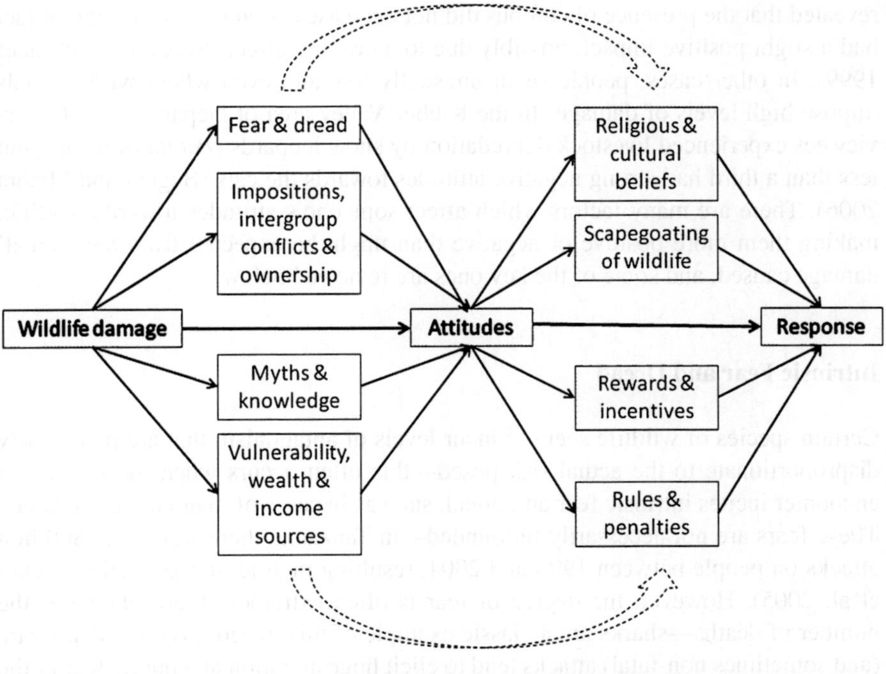

Fig. 16.1 Conceptual model showing some of the key factors likely to influence the relationships between (**a**) wildlife damage and attitudes, and (**b**) attitudes and the response taken

range of study sites and species which demonstrate that issues as varied as religion, economics, cultural beliefs, rules, fear and rewards all influence the complexity of human–wildlife conflict.

Relationship between Level of Damage and Attitudes towards Wildlife

In some cases, the relationship between damage caused by a species, and negativity towards it, seems proportional. In the Pantanal region of Brazil, 82 % of ranchers suffered depredation from jaguars (*Panthera onca*), and unsurprisingly, those people considered them a greater threat than others did (Zimmerman et al. 2005). However, there are many situations where people report very negative views towards a particular species, ostensibly due to the fact that it causes substantial damage, but where closer examination reveals that the value of actual wildlife damage caused by that animal is very low or even absent. For example, Maasai respondents in southern Kenya reported high levels of conflict with lions and negative attitudes toward them; however, actual rates of conflict was quite low, with less than 5 % of all depredation events attributed to lions (Hazzah et al. 2009) (although lions do tend to take relatively valuable stock in the form of cattle). In Zanzibar, the endangered red colobus (*Procolobus kirkii*) is considered by farmers to be one of the most serious local pest species, mainly due to their consumption of coconuts (*Cocos nucifera*), but research revealed that the presence of colobus did not decrease coconut harvests, and in fact had a slight positive impact, possibly due to a pruning effect (Siex and Struhsaker 1999). In other cases, people seem unusually tolerant, even where wild animals impose high levels of damage. In the Kibber Valley area of Nepal, 43 % of interviewees experienced livestock depredation by snow leopards (*Panthera uncia*), but less than a third had strong negative attitudes towards the cats (Bagchi and Mishra 2006). There are many factors which affect someone's attitudes towards wildlife, making them more positive or negative than might be expected from the 'actual' damage caused, and some of the key ones are reviewed below.

Intrinsic Fear and Dread

Certain species of wildlife seem to incur levels of antagonism that are particularly disproportionate to the actual risk posed—this often occurs when the idea of an encounter incites intrinsic fear and dread, such as in cases of man-eating predators. These fears are not necessarily unfounded—in Tanzania, there were over 800 lion attacks on people between 1990 and 2004, resulting in at least 563 deaths (Packer et al. 2005). However, the degree of fear is often extremely high relative to the number of deaths—sharks are a classic example of this, where even relatively rare (and sometimes non-fatal) attacks tend to elicit huge attention and public fear of the species concerned (Neff 2012). In upland Japan, an attack where a brown bear

(*Ursus arctos*) killed seven people (known as the Hokkaido Incident or Sankebetsu Brown Bear Incident) became infamous and contributed to a widespread fear of bears which persists to this day, despite the attack happening a century ago (Knight 2000). Deep-seated fear and dread can also be felt even for species which pose little or no threat to humans, such as spiders in Europe (Prokop and Tunnicliffe 2008). The fear of catastrophic loss, such as when species engage in surplus killing, can also compound fear of and hostility towards wild animals. This has been seen in multiple locations, such as in the western US, where grey wolves (*Canis lupus*) killed up to 98 sheep per attack, reducing local tolerance for them (Muhly and Musiani 2009), and in Chile, where the widespread reporting of surplus killing by pumas (*Puma concolor*) led to intensely negative attitudes, with the cats being perceived as bloodthirsty killers (Murphy and Macdonald 2010). In reality, these incidents of mass killing are rare, but the attention given to these occasional events results in a disproportionately high sense of fear and risk.

Impositions, Inter-Group Conflicts and Ownership

People are far more willing to deal with risks that they undertake voluntarily compared to those which are imposed upon them (Starr 1969), and the unwillingness to deal with risks is exacerbated further if they are imposed by a disliked external group. In the United States, Sweden, Norway and elsewhere, conflicts with grey wolves are heightened by perceptions that the animals are imposed upon rural people by other groups, in these cases remote, urban governments who are unconcerned about the costs incurred by farming communities (Kaltenborn and Bjerke 2002; Knight 2000; Lindquist 2000; Wilson 1997). Similar inter-group conflicts over predator presence emerge in many locations worldwide (Knight 2000)—for instance, in Tanzania's Ruaha landscape, carnivore conflict was heightened because people felt that lions and hyaenas were sent by rival tribes to cause problems and kill their stock (Dickman, pers. obs). In Chile's Araucania Lake Region, 55 % of people surveyed believed (erroneously) that pumas had been released deliberately by wildlife managers into the area, leading to antagonism that the cats' presence was being forced upon them by authorities (Murphy and Macdonald 2010). Similarly, focus groups in Wisconsin revealed negative attitudes towards recolonizing grey wolves, with a commonly cited suspicion that wolves had been reintroduced covertly to the area by the Wisconsin Department of Natural Resources (Browne-Nunez et al. 2015). In Yellowstone National Park, where grey wolves actually were reintroduced, people were resentful that the actions of the 'controlling, domineering, intrusive' federal government had overridden the freedom and self-determination of local people—and even of the wolves themselves—by actively bringing them into the local area (Scarce 1998). The reintroduction of the wolf into Yellowstone was seen by some as a method of exerting social control over peoples' private property, and the wolves were denounced during a speech by activists to local people as 'only a tool being used by those who don't want you around' (Wilson 1997).

Ownership towards a species is also often important: people tend to kill wildlife, even if conflict is low, when they perceive they have no ownership over the resource. Concerns over lack of ownership are often centred around protected areas, which can significantly restrict local peoples' livelihoods—for instance, villagers have reported that the presence of Masaola National Park in Madagascar has limited their options for inter-generational growth and stability so much that they feel 'defeated in the very purpose of life' (Keller 2009). When Namibia obtained independence in 1990, the Ovambo people broke down the fences around Etosha National Park and killed Park animals for meat, as they had previously been banned from hunting in the protected area (IIED 1994). Increasing access and ownership can have marked positive effects—(Hazzah et al. 2013) found that access to protected areas during time of crisis, such as droughts, had a stronger positive influence on Maasai attitudes towards lions, and their likelihood of killing, than more conventional predictors such as rates of conflict. Meanwhile, in the Balikpapan area of Indonesian Borneo, the sun bear (*Helarctos malayanus*) creates substantial conflict, particularly over its destruction of mature coconut trees (Fredriksson 2005). However, the selection of the sun bear as the official mascot for the Balikpapan district in 2001 seems to have improved local attitudes towards the species, with people now having a sense of ownership and pride in the species (Fredriksson 2005).

Myths and Knowledge

Often, the dread and fear mentioned above are linked to local myths regarding a species. In rural Madagascar, the aye-aye (*Daubentonia madagascariensis*) is traditionally believed to be a harbinger of sickness and death, so they are often killed on sight and entire villages have been abandoned after aye-ayes were seen in the vicinity (Simons and Meyers 2001). Fears of relatively small, seemingly innocuous creatures are not restricted to remote cultures—across much of the world, bats are feared, with a persistent belief that they will become tangled in human hair. This myth is so common that researchers have experimented with placing bats on peoples' heads and wrapping them in hair (whereupon the bats merely freed themselves, presumably rather perplexed), but persistent myths can strongly affect views towards a species, even if it causes no apparent damage at all.

Myths can sometimes reflect the kinds of inter-group conflicts mentioned above. This is often the case where animals are believed to be 'shape-shifters', where their body can be inhabited or bewitched by a human spirit, usually as an aggressive act by a rival. In Mozambique and Tanzania, there are suspicions that some people use witchcraft to create 'people-lions' or 'spirit lions', which are used as agents to kill rivals (Dickman 2006; West 2001). In Sierra Leone, chimpanzees sometimes attack young people, and in some cases villagers believe that such attacks are the result of elite outsiders shape-shifting into chimpanzees and stealing body parts from their local victims (Richards 2000). Shape-shifting and bewitching often seems to have an association with species thought to be able to transgress accepted boundaries—this is the case with the spotted hyaena, as its unusual genitalia (where the female's clitoris resembles a penis) has led to the species being viewed as a hermaphroditic deviant which can be bewitched by rivals (Dickman 2009). Primates are also often judged and feared due to their fact that they resemble humans but often transgress accepted boundaries or act in ways that are perceived as immoral. In south-east Asia, orangutans are viewed as 'wild, rude and uncultured' human counterparts (Knight 1999; Rijksen 1995), chimpanzees have been described as 'thieves' and 'rapists' in Uganda (Naughton-Treves 1997) and as having 'low' morality in Sierra Leone (Richards 2000), while in Japan, monkeys are rumoured to have sexually molested women out in the forest (Knight 1999).

Better knowledge about a species can be linked to improved attitudes—Slovakian students who did not believe in myths about bats and who knew more about their biology were significantly more positive towards bats than other pupils (Prokop et al. 2009; Prokop and Tunnicliffe 2008). However, there is not always a positive relationship between knowledge and attitudes—(Simons and Meyers 2001) found that even relatively well-informed people like forest agents held the beliefs described above about aye-ayes. There can also be an interaction between knowledge and personal experience—the most knowledgeable people regarding a species are often those who frequently come into contact with them (such as hunters or pastoralists), and are therefore more likely to have negative experiences (such as predators killing hunting dogs or attacking livestock) and report greater antagonism (Dickman et al. 2014; Heberlein and Ericsson 2008).

Vulnerability, Wealth and Income Sources

Unsurprisingly, people react particularly negatively towards the presence of a species if they are especially vulnerable to its impacts. Vulnerability is often linked to wealth, because wealthier people can afford to invest in asset protection strategies such as employing herders, feeding guarding dogs, protecting crops and building well-constructed livestock enclosures (Naughton-Treves and Treves 2005). Even if wildlife damage still occurs despite these efforts, then an event in a wealthy household is less likely to be catastrophic (Dickman et al. 2013). This means that poverty-stricken households (who are often located in areas rich in biodiversity, including dangerous species; (Loveridge et al. 2010) tend to suffer from 'compounding vulnerability' due to their inability to either prevent wildlife damage or cope with its impact (Naughton-Treves 1997).

Wildlife-related activities can generate huge revenues, at least at a national scale (MTK 2008), and it is often assumed that if people receive some wealth from wildlife, they will be more positive towards the presence of wildlife and areas associated with them. Wildlife-related activities can generate substantial revenues, at least at a national scale (MTK 2008), and it is often assumed that if people receive some wealth from wildlife, they will be more positive towards its presence. This can be the case—in Botswana (Hemson et al. 2009) found that the dislike of living alongside lions and the National Park was less commonly reported amongst tourism employees than other people. In Western Uganda, tourism revenue sharing initiatives around three parks (Kibale, Bwindi Impenetrable, and Mgahinga Gorilla National Park) generated US$83,000 for local communities from 1995 to 1998, which was used to build 21 schools, four clinics, a bridge and a road (Archabald and Naughton-Treves 2001). This appeared to have an important effect in terms of attitudes, with 72 % of respondents saying that the initiatives had improved their attitudes towards the protected area (Archabald and Naughton-Treves 2001). However, a later study around Mgahinga National Park revealed that although financial flows to local communities from mountain gorilla (*Gorilla gorilla beringei*) tourism did reduce local negativity regarding the Parks' creation, they were insufficient to compensate them for the costs of park creation in terms of foregone agricultural production (Adams and Infield 2003). It is important to recognize that improved attitudes towards protected areas do not necessarily translate into improved attitudes towards wildlife species, especially conflict-causing ones. However, this can be the case—around the Community Baboon Sanctuary in Belize, people reported very positive attitudes towards both the protected area and its resident black howler monkeys (*Alouatta nigra*), despite some crop-raiding and disquiet over the level and distribution of benefits from the sanctuary (Alexander 2000). Moreover, studies have shown that people who report negative attitudes and experiences towards protected areas are more likely to negatively perceive the wildlife within the protected area and react accordingly (Chardonnet 2002; Mukherjee 2009; Western 1982).

Around Indonesia's Komodo National Park, research revealed that despite a generally high level of local support for tourism and the Park, people who received income from tourism were actually significantly less likely to support conservation of the Park, possibly because of negative interactions with the Park or its authorities (Walpole and Goodwin 2001). Inequitable distribution of benefits is always a concern regarding revenue-sharing initiatives (Archabald 2000) and can have a marked impact in terms of attitudes towards the park and its wildlife. Marginalized groups are often less likely to be in positions to receive benefits from activities such as tourism, so tend to be more negative—this has been found in locations as diverse as Nepal's Chitwan district (Carter et al. 2014) and Tanzania's Ruaha landscape (Dickman 2009).

Relationship between Reported Attitudes and Response to Conflict

Attitudes can be a good barometer of action taken in response to conflict—amongst Kenyan Maasai, the strongest predictor of actual lion-killing behaviour was attitudes towards lions (Hazzah et al. under review). However, there are many cases where there seems to be a mismatch between the reported attitudes of a species and the action taken (or lack thereof) in response. Frequently, people seem to respond more harshly than seems justified—in Namibia, farmers reported removing an average of 14 cheetahs (*Acinonyx jubatus*) annually, even where they were not considered problematic (Marker et al. 2003). In China's Sichuan Province, researchers found that although attitudes towards Asiatic black bears (*Ursus thibetanus*) were influenced by problems with them, the killing of bears was actually more common in areas without reported conflict (Liu et al. 2011). Interestingly, in Kalimantan, only 7 % of people who reported that they had killed orang-utans said that they did so for self-defence or because they considered the animal a pest, with 41 % saying they did it for an 'unknown reason' (Meijaard et al. 2011).

Conversely, there are instances where relatively few people admit to killing conflict-causing animals, compared to the number who cite problems with them—in Tanzania's Ruaha landscape, nearly all villagers surveyed (94 %) viewed large carnivores as problematic, but only 7.3 % reported having killed one (Dickman et al. 2014). There is often likely to be a marked under-reporting of real killing levels, due to fears of consequences from the authorities or conservation agencies (Fredriksson 2005; St John et al. 2011), but in some cases, the tolerance for conflict-causing species appears to be genuine. In Indonesia's Lore Lindu National Park, Tonkean macaques (*Macaca tonkeana*) are often a serious pest to farmers, but there is marked reluctance to confront or kill them (Riley 2010). Similarly, in rural Japan, monkeys can cause significant economic damage to farmers, but research revealed that a quarter of people believed that damage imposed by monkeys had to be tolerated to some extent (Knight 1999).

Many of the factors mentioned in the section above, such as fear, myths and inter-group conflicts, can also have substantial bearings on whether or not people actually take action in response to conflict, as well as the level of that response. Fear often tempers the desire to act against conflict-causing species—in Tanzania, fear was one of the major reasons why people did not engage in lion hunts, even though almost 90 % viewed lions as problematic (Dickman 2009). In Indonesia, villagers reported an unwillingness to harm crop-raiding monkeys because of a fear that it would incite retribution from the primates (Riley 2010). Similarly, inter-group conflicts can lead to the 'scapegoating' of wildlife described below. However, additional factors also play an important role in determining the response to conflict, such as religious and cultural beliefs, and the relative costs and benefits of, for instance, killing wildlife in response to conflict. These issues are often interwoven with one another, but we have tried to tease out some of the key issues below.

Religious and Cultural Beliefs

There are many instances where wildlife imposes substantial costs on local people, and yet the response seems unusually muted. This might be because of independent personal beliefs held by the individual concerned, or because of rules and tenets imposed by religions which someone adheres to. The relationship between religious beliefs and actions towards wildlife can be complex. White (1967) suggested that Christianity undermines conservation through its doctrine of man's dominion over nature, while in Indonesia, Lee et al. (2009) found that Christians have a higher propensity to hunt wildlife than do Muslims. It has been suggested that the stronger people's orientation towards domination of wildlife, the more likely their attitudes and actions will prioritize human well-being over wildlife, often resulting in wildlife killing (Teel et al. 2010). This was supported by a study in Kenya, which revealed that Maasai who were evangelical Christians were much more likely to report a higher propensity to kill predators then those who attended other churches or none at all (Hazzah et al. 2009).

However, in other cases, religious and cultural beliefs can reduce the chances of wildlife killing. For example, a long list of primate species, including chimpanzees, redtail monkeys (*Cercopithecus ascanius*), rhesus monkeys (*Macaca mulatta*) and baboons (*Papio* sp.), are often serious pests in rural agricultural communities across Asia (Knight 1999). Despite the damage caused, various local beliefs can have protective effects for primates—in several places, including Sulawesi and Thailand, there are beliefs that people can turn into monkeys, and that the kinship between the two groups entitles the primates to protection, even if they cause problems (Riley 2010; Tambiah 1969). Interestingly, the basis of this protection (the similarities between humans and primates) is the same one that underlies the fear in other locations of primates as transgressive shape-shifters (see section 'Myths and Knowledge').

Cultural taboos also prohibit the killing of snow leopards in areas of rural Nepal, where the cat is viewed by local Buddhists as associated with the 'mountain god' (Ale 1998). In those places, retaliatory killing for snow leopard depredation is relatively uncommon, because attacks are thought to be divine retribution for bad actions on the part of the herder, so they are to blame instead of the predator (Ale 1998). However, religious beliefs do not always prevent the killing of wildlife. Around the Kibber Wildlife Sanctuary in the India trans-Himalaya, local households lost 18 % of their livestock to predators over an 18-month period, which amounted to half their annual average yearly income (Mishra 1997). Almost all of these losses were thought to be due to snow leopards, with a few due to other predators like wolves. Both the snow leopard and the Tibetan wolf are protected under Indian law, but the responses towards the species were markedly different—while snow leopards were almost never killed, wolves were killed every year in a dramatic fashion. Despite their Buddhist beliefs, villagers located wolf dens, removed the pups and paraded them around the villages before killing them, often with dynamite (Mishra 1997). Other than the cultural reasoning for this targeted killing, wolves could potentially have been easier to find and kill than the elusive snow leopard. Nevertheless, this kind of 'contagious' conflict, where one species gets blamed for the actions of another, has been documented elsewhere (Dickman et al. 2014) and might be of particular concern where one species is protected by cultural beliefs.

Scapegoating of Wildlife

The kinds of inter-group conflicts mentioned in section 1b can not only influence attitudes towards certain species, but also markedly intensify the responses taken against them. In Kenya, Maasai communities around Nairobi National Park perceived that the Government prioritized wildlife over their cultural values and needs, and this anger eventually resulted in Maasai warriors killing over half the lions in Nairobi National Park (Anonymous 2003). In Japan, macaques are often killed by local people, and while they do indeed raid crops, this monkey culling has been described not as an effective pest control measure, but rather a 'sacrifice' which has the aim of making farmers feel better (Koganezawa 1991), cited in (Knight 1999). The monkeys are also thought to be scapegoats for rural village declines—people are encouraged to vent their frustration and anger by killing monkeys, rather than focusing on the state's role in the breakdown of nationally sponsored rural development initiatives (Maita 1989) cited in Knight (1999).

In some cases, the 'scapegoating' of wildlife can be quite extreme—until 1998, an annual pigeon shoot was held in Hegins, Pennsylvania, where thousands of pigeons were killed in a single day (Hoon Song 2000). Ostensibly, the shoot was a form of problem animal control due to crop losses caused by pigeons. However, research revealed that crop damage was negligible and that thousands of pigeons were in fact brought in and released specifically for the shoot (Hoon Song 2000). It emerged that the pigeons had come to represent the spread of urbanity and moral decay in rural areas, so killing them was an important symbolic act (Hoon Song 2000). These cases highlight that human responses to wildlife are sometimes far more to do with human–human conflict than with the actions of the actual species concerned.

Rewards and Incentives

Whether or not someone acts on their attitudes, and takes action against a species, is likely to be strongly influenced by what that person would gain or lose by doing so. In some cases, there are direct economic rewards for killing wildlife, as exemplified by lion-killing in Tanzania, which is the most important country in the world for lions (with perhaps 40 % of the remaining population; (Riggio et al. 2013). The Sukuma are Tanzania's most populous cattle-raising tribe and yet traditionally, they rewarded young men with valuable gifts (usually of cattle) in return for killing lions which threatened their livelihood and today, even in areas where there is virtually no livestock loss to lions, the community rewards still provide sufficient economic incentive to drive lion-killing (Fitzherbert et al. 2014). In southern Tanzania's Ruaha landscape, the situation is very similar—young men from the Barabaig tribe are rewarded with wealth (in the form of gifts of cattle from other Barabaig households) for killing lions, and this helped drive very high rates of lion-killing around Ruaha National Park (Dickman, pers.obs.).

These community rewards are not limited to African tribes—in Nepal, people who killed wolves were rewarded with money (Mishra 1997), while in Japan, killing black bears still results in bounty payments, which helped incentivize the killing of over 70,000 bears there between 1946 and 1994 (Knight 2000). Similarly, Liu et al. (2011) concluded that in China, although human–wildlife conflicts shaped people's attitudes towards bears, it was the presence of strong economic incentives (illegal trade in bear parts) rather than attitudes which prompted illegal killing.

Rewards are often not purely economic, though—with the Barabaig and other groups such as the Maasai, lion killers received cultural rewards as well as wealth, such as community accolades, attention from women and elevated social status (Hazzah 2011). Many thousands of miles away, in the Pantanal region of Brazil, similar cultural rewards incentivize jaguar hunting, where killing the cats is viewed as an innate component of 'panteineiro' culture and social identity (Marchini and Macdonald 2012).

However, there can also be valuable incentives for tolerating wildlife, which are often economic. Incentive-based schemes have gained popularity over the years as an attempt to increase local attitudes towards predators and provide conservation benefits to the affected communities. Direct 'performance payments' are the clearest example of this, where payments are made to individuals or groups contingent upon specific conservation outcomes, such as the maintenance of a species in a particular area (Dickman et al. 2011; Zabel and Holm-Muller 2008). In Sonora, Mexico, where private ranchland is important for the local jaguar population, the Northern Jaguar Project placed camera-traps on ranchland. They rewarded ranchers with cash payments of between US$50 and US$300 for photographs of jaguar, puma, ocelot (*Leopardus pardalis*) or bobcat (*Lynx rufus*), providing a direct economic incentive to tolerate the presence of these species on private land (Nelson 2009; Nistler 2007). At a larger scale, the Swedish government initiated a performance-payment scheme in 1996 in order to help conserve national populations of wolverines (*Gulo gulo*), lynx (*Lynx lynx*) and wolves, which create substantial conflict with Sami people due to depredation upon reindeer (*Rangifer tarandus*). The payment, made by the state, was calculated depending upon the number of certified carnivore reproduction events on Sami villages' reindeer grazing land, and in 2007, the payment for each certified reproduction of wolverine or lynx was US$29,000. The number of wolverine reproductions in the reindeer area has now exceeded the target of 90 per year, and although it is hard to prove a direct causal relationship with the performance payment initiative, it does suggest success (Zabel and Engel 2010). Of course with any incentive-based scheme, one of the major challenges is ensuring financial sustainability.

Rules and Penalties

The rules governing action taken towards a species (such as killing them) and the resulting penalties can play an important role in determining behaviour. These penalties can be diverse, with just some examples including imprisonment and/or fines (Hazzah et al. 2013; Murphy and Macdonald 2010; St John et al. 2011), community exclusion (Lingard et al. 2003), retaliation by either the species concerned or the ancestral or spirit world (Knight 1999; Kohler 2000; Riley 2010). There can be marked differences in adherence to rules depending on their origin, in particular whether they come from within the community concerned, or are imposed externally. In Madagascar, the endemic radiated tortoise (*Geochelone radiata*) has been legally protected since 1960, but nonetheless been in rapid decline across much of its range, with people killing them for food and trade (Lingard et al. 2003). However, in approximately half the tortoise's range, the Androy people have taboos against eating them, mainly due a perception that they are 'unclean' (although there is also some suggestion of a link to ancestors), and this has had a marked protective effect (Lingard et al. 2003). Violations of the taboo are rare and usually result in community exclusion, with this threat appearing to be far more effective in regulating behaviour than the rules imposed by remote authorities (Lingard et al. 2003). Customary laws do not always guarantee more adherence than national laws—in Indonesia, respondents who reported that the orang-utan was protected by customary law were actually more likely to kill them than people who were uncertain or said they were not protected by such rules, while people who knew that they were protected by national law were less likely to kill them than other people (Meijaard et al. 2011). Similarly, even strong religious rules do not always guarantee adherence, as seen with the killing of wolves (and even occasionally snow leopards) by traditional Buddhist herders (Mishra 1997).

Even though people might be aware of the rules concerning a species, adherence to them will usually be influenced by the likelihood of getting caught. St John et al. (2011) found that approximately one in five farmers in north-eastern South Africa killed leopards despite their protection under the country's Biodiversity Act, suggesting that the national rules did not extend to actual protection on private farms. In the United States, the illegal killing of wolves has been termed the 'shoot, shovel and shut up' approach, with the chance of a transgression being detected relatively unlikely on remote ranches. In comparison, spiritual or customary laws in tightly-knit communities are presumably likely to be far more effective, as there is less opportunity for transgressions to go unnoticed.

There can be other types of penalties for action as well—in East African pastoralist landscapes, lion hunting is an important way of acquiring status and wealth, but it undoubtedly carries significant risk, with people being seriously injured and killed on hunts every year (Dickman, pers.obs.). The potential risks of engaging in lion hunts are sometimes enough to prevent people taking action, even in response to the depredation of highly valued cattle, as people cannot risk the personal and economic impacts that serious injury would incur.

Conclusions for Conflict Mitigation in a Complex World

The examples above demonstrate that conflict is not a simple, linear relationship between damage, attitudes and actions—it is influenced by multiple diverse factors (Fig. 16.1), and means that there is no 'one size fits all' strategy for effective conflict mitigation. However, it is vital to investigate and understand which particular factors drive any specific situation, as that understanding is crucial for determining the most effective action. In some cases, conflict can be significantly reduced by lessening the damage imposed by wild animals, but in many scenarios, other issues such as inter-group conflict, local beliefs and the cost-benefit ratio of wildlife killing are even more important than the 'actual' risk posed by the species. Furthermore, the case studies highlighted in this chapter show that even where a similar factor emerges in different contexts, its impacts might not always be the same, so solutions would need to be culturally and site-appropriate. For example, in some cases where rules seem important determinants of actions, it might be beneficial to utilize customary law, while in others educating people about national law might be more effective. Developing a deep understanding of the drivers of conflict can lead to successful strategies, as has been seen with the transformation of pastoralist warriors killing lions in East Africa (Hazzah et al. 2014). Up until very recently, young men killed lions to gain status, wealth and maintain their social role as community protectors by reducing the threat that lions posed to livestock. In this instance, human–lion conflict was driven not only by wildlife damage, but also by wealth (or lack thereof), local beliefs and community incentives. A conflict mitigation strategy called 'Lion Guardians' was developed to address all these factors—young warriors were employed to track lions and safeguard the community from attacks, enabling them to gain wealth and social status, and to fulfil their traditional roles in a different way (e.g. they help protect against stock attacks by warning of lion presence and helping reinforce enclosures, rather than by killing lions) (Hazzah et al. 2014). This approach, which provides a culturally appropriate platform for warriors to participate in actively conserving lions, has shown to reduce lion-killing by 99 % (Hazzah et al. 2014).

There is no one silver bullet to achieving long-term coexistence between people and wildlife. Unfortunately, conflict is an inevitable reality with an increasing human population and the loss of natural habitat. It is likely that in most conflict scenarios, a multitude of different factors will come into play, making conflict a very complex issue to resolve. However, being aware of relevant drivers in any specific situation will help conservationists develop multifaceted and culturally appropriate mitigation initiatives to help encourage coexistence with wildlife in today's ever more human-dominated world.

References

Adams WM, Infield M (2003) Who is on the gorilla's payroll? Claims on tourist revenue from a Ugandan National Park. World Dev 31:177–190

Ale S (1998) Culture and conservation: the snow leopard in Nepal. Int Snow Leopard Trust Newsl 16:10

Alexander SE (2000) Resident attitudes towards conservation and black howler monkeys in Belize: the Community Baboon Sanctuary. Environ Conserv 27:341–350

Anonymous (2003) Maasai kill half the lions in Nairobi National Park. Cat News 39:5

Archabald K (2000) Can revenue-sharing save wildlife? A case study of Jozani Chwaka Bay Conservation Area, Zanzibar, Tanzania. University of Wisconsin, p 95

Archabald K, Naughton-Treves L (2001) Tourism revenue-sharing around national parks in Western Uganda: early efforts to identify and reward local communities. Environ Conserv 28:135–149

Bagchi S, Mishra C (2006) Living with large carnivores: predation on livestock by the snow leopard (*Uncia uncia*). J Zool 268:217–224

Balme GA, Slotow R, Hunter LT (2009) Impact of conservation interventions on the dynamics and persistence of a persecuted leopard (*Panthera pardus*) population. Biol Conserv 142: 2681–2690

Barua M, Bhagwat SA, Jadhav S (2013) The hidden dimensions of human–wildlife conflict: health impacts, opportunity and transaction costs. Biol Conserv 157:309–316

Browne-Nunez C, Treves A, MacFarland D, Voyles Z, Turng C (2015) Tolerance of wolves in Wisconsin: a mixed-methods examination of policy effects on attitudes and behavioral inclinations. Biol Conserv 189:59–71, http://dx.doi.org/10.1016/j.biocon.2014.12.016

Butler JRA (2000) The economic costs of wildlife predation on livestock in Gokwe communal land, Zimbabwe. Afr J Ecol 38:23–30

Carter NH, Riley SJ, Shortridge A, Shrestha BK, Liu J (2014) Spatial assessment of attitudes toward tigers in Nepal. Ambio 43:125–137

Chardonnet P (2002) Conservation of the African lion: contributions to a status survey. International Foundation for the Conservation of Wildlife, France

Chelliah K, Kannan G, Kundu S, Abilash N, Madhusudan A, Baskaran N, Sukumar R (2010) Testing the efficacy of a chilli-tobacco rope fence as a deterrent against crop-raiding elephants. Curr Sci (Bangalore) 99:1239–1243

Conover M (2002) Resolving human-wildlife conflicts: the science of wildlife damage management. CRC Press, Boca Raton

Dhanwatey HS, Crawford JC, Abade LA, Dhanwatey PH, Nielsen CK, Sillero-Zubiri C (2013) Large carnivore attacks on humans in central India: a case study from the Tadoba-Andhari Tiger Reserve. Oryx 47:221–227

Dickman AJ (2006) Reported conflict between pastoralists and wildlife around the Ruaha National Park, Tanzania. In: Siegrist D, Clivaz C, Hunziker M, Iten S (eds) Exploring the nature of management. Proceedings of the third international conference on monitoring and management of visitor flows in recreational and protected areas. University of Applied Sciences Rapperswil, Rapperswil, pp 405–406

Dickman AJ (2009) Key determinants of conflict between people and wildlife, particularly large carnivores, around Ruaha National Park, Tanzania. Biological anthropology. University College London, London

Dickman AJ (2010) Complexities of conflict: the importance of considering social factors for effectively resolving human–wildlife conflict. Anim Conserv 13:458–466

Dickman AJ, Macdonald EA, Macdonald DW (2011) A review of financial instruments to pay for predator conservation and encourage human-carnivore coexistence. Proc Natl Acad Sci USA 108:13937–13944

Dickman AJ, Marchini S, Manfredo M (2013) The importance of the human dimension in addressing conflict with large carnivores. In: Macdonald DW, Willis K (eds). Key topics in conservation biology, vol 2. Oxford University Press, Oxford, pp 110–126

Dickman AJ, Hazzah L, Carbone C, Durant S (2014) Carnivores, culture and 'contagious conflict': multiple factors influence perceived problems with carnivores in Tanzania's Ruaha landscape. Biol Conserv 178:19–27

Dirzo R, Young HS, Galetti M, Ceballos G, Isaac NJ, Collen B (2014) Defaunation in the Anthropocene. Science 345:401–406

Entwistle AC, Stephenson PJ (2000) Small mammals and the conservation agenda. Conservation biology series. Cambridge University Press, Cambridge, pp 119–140

Fitzherbert E, Caro T, Johnson P, Macdonald DW, Mulder MB (2014) From avengers to hunters: leveraging collective action for the conservation of endangered lions. Biol Conserv 174:84–92

Fredriksson G (2005) Human-sun bear conflicts in East Kalimantan, Indonesian Borneo. Ursus 16:130–137

Hazzah L (2011) Exploring attitudes, behaviors, and potential solutions to lion (*Panthera leo*) killing in Maasailand, Kenya. Nelson Institute of Environmental Studies. University of Wisconsin-Madison, Madison, Madison

Hazzah L, Borgerhoff Mulder M, Frank L (2009) Lions and warriors: social factors underlying declining African lion populations and the effect of incentive-based management in Kenya. Biol Conserv 142:2428–2437

Hazzah L, Dolrenry S, Kaplan D, Frank L (2013) Access to parks during drought influences attitudes and behavior toward lion conservation in Maasailand, Kenya. Environ Conserv 40:266–276

Hazzah L, Dolrenry S, Naughton L, Edwards CT, Mwebi O, Kearney F, Frank L (2014) Efficacy of two lion conservation programs in Maasailand, Kenya. Conserv Biol 28:851–860

Hazzah L, Dolrenry S, Dickman AJ, Bath A (under review) From attitudes to actions: predictors of lion killing by Maasai warriors. Conserv Soc

Heberlein TA, Ericsson G (2008) Public attitudes and the future of wolves *Canis lupus* in Sweden. Wildl Biol 14:391–394

Hemson G, Maclennan S, Mills G, Johnson P, Macdonald DW (2009) Community, lions, livestock and money: a spatial and social analysis of attitudes to wildlife and the conservation value of tourism in a human-carnivore conflict in Botswana. Biol Conserv 142:2718–2725

Hoon Song S (2000) The great pigeon massacre in a deindustralizing American region. In: Knight J (ed) Natural enemies: people-wildlife conflicts in anthropological perspective. Routledge, London, pp 212–228

International Institute for Environment and Development (IIED) (1994) Whose eden? An overview of community approaches to wildlife management. IIED, London

IUCN (2006) Regional conservation strategy for the lion *Panthera leo* in Eastern and Southern Africa. IUCN SSC Cat Specialist Group, Gland, p 60

Jadhav S, Barua M (2012) The elephant vanishes: impact of human–elephant conflict on people's wellbeing. Health Place 18:1356–1365

Kaltenborn BP, Bjerke T (2002) The relationship of general life values to attitudes toward large carnivores. Res Hum Ecol 9:55–61

Keller E (2009) The danger of misunderstanding 'culture'. Madagascar Conserv Dev 4:82–85

Knight J (1999) Monkeys on the move: the natural symbolism of people-macaque conflict in Japan. J Asian Stud 58:622–647

Knight J (2000) Culling demons: the problem of bears in Japan. In: Knight J (ed) Natural enemies: people-wildlife conflicts in anthropological perspective. Routledge, London

Koganezawa M (1991) Nihonzaru no bunpu to hogo no genjo oyobi sono mondaiten—Nikko o chushin ni [The present state and distribution of the Japanese Monkey and related problems: focussing on Nikko]. Yasei dobutsu hogo: 21 seiki e no teigen [Wild animal protection: a proposal for the 21st century]. Nihon Shizen Hogo Kyokai, Tokyo

Kohler A (2000) Half-man, half-elephant: shapeshifting among the Baka of Congo. In: Knight J (ed) Natural enemies: people-wildlife conflicts in anthropological perspective. Routledge, London, pp 50–77

Kruuk H (2002) Hunter and hunted: relationships between carnivores and people. Cambridge University Press, Cambridge

Lee T, Sodhi N, Prawiradilaga D (2009) Determinants of local people's attitude toward conservation and the consequential effects on illegal resource harvesting in the protected areas of Sulawesi (Indonesia). Environ Conserv 36:157–170

Lee-Thorp J, Thackeray JF, van der Merwe N (2000) The hunters and the hunted revisited. J Hum Evol 39:565–576

Lindquist G (2000) The wolf, the Saami and the urban shaman: predator symbolism in Sweden. In: Knight J (ed) Natural enemies: people-wildlife conflicts in anthropological perspective. Routledge, London, pp 170–188

Lingard M, Raharison N, Rabakonandrianina E, Rakotoarisoa J-A, Elmqvist T (2003) The role of local taboos in conservation and management of species: the radiated tortoise in southern Madagascar. Conserv Soc 1:223

Liu F, McShea WJ, Garshelis DL, Zhu X, Wang D, Shao L (2011) Human-wildlife conflicts influence attitudes but not necessarily behaviors: factors driving the poaching of bears in China. Biol Conserv 144:538–547

Loe J, Roskaft E (2004) Large carnivores and human safety—a review. Ambio 33:283–288
Loveridge AJ, Wang SW, Frank LG, Seidensticker J (2010) People and wild felids: conservation of cats and management of conflicts. In: Macdonald DW, Loveridge AJ (eds) Biology and conservation of wild felids. Oxford University Press, Oxford, pp 161–195
Maita A (1989) Sengo sanson Shakai no 'mura kuzushi' to shinrin kanri. In: Setsu U (ed) 'Shinrin shakaigaku' sengen (Declaration on 'Forest sociology'). Yuikaku Sensho, Tokyo
Marchini S (2014) Who's in conflict with whom? Human dimensions of the conflicts involving wildlife. In: Verdade LM, Lyra-Jorge MC, Piña CI (eds) Applied ecology and human dimensions in biological conservation. Springer, New York, pp 189–209
Marchini S, Macdonald DW (2012) Predicting ranchers' intention to kill jaguars: case studies in Amazonia and Pantanal. Biol Conserv 147:213–221
Marker LL, Dickman AJ, Mills MGL, Macdonald DW (2003) Aspects of the management of cheetahs, Acinonyx jubatus jubatus, trapped on Namibian farmlands. Biol Conserv 114:401–412
McGuinness S, Taylor D (2014) Farmers' perceptions and actions to decrease crop raiding by forest-dwelling primates around a Rwandan forest fragment. Hum Dimens Wildl 19:179–190
McManus JS, Dickman AJ, Gaynor D, Smuts BH, Macdonald DW (2014) Dead or alive? Comparing costs and benefits of lethal and non-lethal human-wildlife conflict mitigation on livestock farms. Oryx 49(4):687–695
Meijaard E, Buchori D, Hadiprakarsa Y, Utami-Atmoko SS, Nurcahyo A, Tjiu A, Prasetyo D, Christie L, Ancrenaz M, Abadi F (2011) Quantifying killing of orangutans and human-orangutan conflict in Kalimantan, Indonesia. PLoS One 6:e27491
Mishra C (1997) Livestock depredation by large carnivores in the Indian trans-Himalaya: conflict perceptions and conservation prospects. Environ Conserv 24:338–343
MTK (2008) Ministry of Tourism, Kenya. http://www.tourism.go.ke/ministry.nsf/pages/facts_figures
Muhly TB, Musiani M (2009) Livestock depredation by wolves and the ranching economy in the Northwestern US. Ecol Econ 68:2439–2450
Mukherjee A (2009) Conflict and coexistence in a national park. Econ Pol Wkly 44:52–59
Murphy T, Macdonald DW (2010) Pumas and people: lessons in the landscape of tolerance from a widely distributed felid. In: Macdonald DW, Loveridge AJ (eds) Biology and conservation of wild felids. Oxford University Press, Oxford, pp 431–452
Musiani M, Mamo C, Boitani L, Callaghan C, Gates CC, Mattei L, Visalberghi E, Breck S, Volpi G (2003) Wolf depredation trends and the use of fladry barriers to protect livestock in western North America. Conserv Biol 17:1538–1547
Naughton L, Rose R, Treves A (1999) The social dimensions of human-elephant conflict in Africa: a literature review and case studies from Uganda and Cameroon. African Elephant Specialist, human-elephant task conflict task force of IUCN, Gland
Naughton-Treves L (1997) Farming the forest edge: vulnerable places and people around Kibale National Park. Uganda Geogr Rev 87:27–46
Naughton-Treves L, Treves A (2005) Socio-ecological factors shaping local support for wildlife: crop-raiding by elephants and other wildlife in Africa. In: Woodroffe R, Thirgood S, Rabinowitz A (eds) People and wildlife: conflict or coexistence? Cambridge University Press, Cambridge, pp 252–277
Neff C (2012) Australian beach safety and the politics of shark attacks. Coast Manag 40:88–106
Nelson F (2009) Developing payments for ecosystem service approaches to carnivore conservation. Hum Dimens Wildl 14:381–392
Nistler C (2007) Seeing spots: the return of the jaguar. PERC reports 25. http://www.perc.org/articles/article1016.php
Norton-Griffiths M, Southey C (1995) The opportunity costs of biodiversity conservation in Kenya. Environ Econ 12:125–139
Packer C, Ikanda D, Kissui B, Kushnir H (2005) Lion attacks on humans in Tanzania. Nature 436:927–928
Prokop P, Tunnicliffe SD (2008) Disgusting animals: primary school children's attitudes and myths of bats and spiders. Eurasia J Math Sci Technol Educ 4:87–97

Prokop P, Fancovicova J, Kubiatko M (2009) Vampires are still alive: Slovakian students' attitudes toward bats. Anthrozoos 22:19–30
Quammen D (2003) Monster of God: the man-eating predator in the jungles of history and the mind. Norton, New York
Richards P (2000) Chimpanzees as political animals in Sierra Leone. In: Knight J (ed) Natural enemies: people-wildlife conflict in anthropological perspective. Routledge, London, pp 78–103
Riggio J, Jacobson A, Dollar L, Bauer H, Becker M, Dickman AJ, Funston P, Groom R, Henschel P, de Iongh H, Lichtenfeld L, Pimm S (2013) The size of savannah Africa: a lion's (*Panthera leo*) view. Biodivers Conserv 22:17–35
Rijksen HD (1995) Forest men and man. In: Corbey R, Theunissen B (eds) Ape, man, apeman: changing views since 1600. Department of Prehistory, Leiden University, Leiden
Riley EP (2010) The importance of human–macaque folklore for conservation in Lore Lindu National Park, Sulawesi, Indonesia. Oryx 44:235–240
Scarce R (1998) What do wolves mean? Conflicting social constructions of *Canis lupus* in "bordertown". Hum Dimens Wildl 3:26–45
Siex KS, Struhsaker TT (1999) Colobus monkeys and coconuts: a study of perceived human-wildlife conflicts. J Appl Ecol 36:1009–1020
Simons EL, Meyers DM (2001) Folklore and beliefs about the aye aye (*Daubentonia madagascariensis*). Lemur News 6:11–16
Sitati NW, Walpole MJ (2006) Assessing farm-based measures for mitigating human-elephant conflict in Transmara District, Kenya. Oryx 40:279–286
St John FA, Keane AM, Edwards-Jones G, Jones L, Yarnell RW, Jones JP (2011) Identifying indicators of illegal behaviour: carnivore killing in human-managed landscapes. Proc R Soc B Biol Sci 279:804–812
Starr C (1969) Social benefits vs. technological risks. Science 165:1232–1238
Tambiah SJ (1969) Animals are good to think and good to prohibit. Ethnology 8:423–459
Teel TL, Manfredo MJ, Jensen FS, Buijs AE, Fischer A, Riepe C, Arlinghaus R, Jacobs MH (2010) Understanding the cognitive basis for human-wildlife relationships as a key to successful protected-area management. Int J Sociol 40:104–123
Thirgood S, Woodroffe R, Rabinowitz A (2005) The impact of human-wildlife conflict on human lives and livelihoods. In: Woodroffe R, Thirgood S, Rabinowitz A (eds) People and wildlife: conflict or coexistence? Cambridge University Press, Cambridge, pp 13–26
Walpole MJ, Goodwin HJ (2001) Local attitudes towards conservation and tourism around Komodo National Park, Indonesia. Environ Conserv 28:160–166
West H (2001) Sorcery of construction and socialist modernisation: ways of understanding power in postcolonial Mozambique. Am Ethnol 28:119–150
Western D (1982) Amboseli National Park: enlisting landowners to conserve migratory wildlife. Ambio 11:302–308
White LJ (1967) The historic roots of our ecologic crisis. Science 155:1203–1207
Wilson MA (1997) The wolf in Yellowstone: science, symbol, or politics? Deconstructing the conflict between environmentalism and wise use. Soc Nat Resour 10:453–468
Woodroffe R, Thirgood S, Rabinowitz A (2005) The impact of human-wildlife conflict on natural systems. In: Woodroffe R, Thirgood S, Rabinowitz A (eds) People and wildlife: conflict or coexistence? Cambridge University Press, Cambridge, pp 1–12
Zabel A, Engel S (2010) Performance payments: a new strategy to conserve large carnivores in the tropics? Institute for Environmental Decisions, Zurich
Zabel A, Holm-Muller K (2008) Conservation performance payments for carnivore conservation in Sweden. Conserv Biol 22:247–251
Zimmerman A, Walpole MJ, Leader-Williams N (2005) Cattle ranchers' attitudes to conflicts with jaguar *Panthera once* in the Pantanal of Brazil. Oryx 39:406–412

Chapter 17
Predators That Kill Humans: Myth, Reality, Context and the Politics of Wolf Attacks on People

John D.C. Linnell and Julien Alleau

Introduction: The Ultimate Conflict

Large predators are associated with a wide range of conflicts with human interests. These range from competition with hunters for game, depredation on domestic livestock and/or pets or destruction of property. In addition, many large carnivore species are also periodically associated with a far more serious conflict, the killing of humans. The name "man-eater" and the idea of being killed (and potentially eaten) by a large predator can quite naturally induce fear into people sharing habitat with these species. However, the level of fear that is displayed for any species in any specific area rarely seems to be related to the relevant objective risk of attack. In other words, fear of man-eating is filtered through complex social filters, and may even be cynically manipulated for political goals. In this chapter, we firstly provide an overview of those carnivore species that are associated with killing humans before focusing in detail on the case study of wolves (*Canis lupus*). No other case is currently so controversial or provides a richer body of historical and contemporary material to explore the myth, the reality and the politics of fear.

J.D.C. Linnell (✉) • J. Alleau
Norwegian Institute for Nature Research, PO Box 5685 Sluppen, NO-7485, Trondheim, Norway
e-mail: john.linnell@nina.no; julien.alleau@wanadoo.fr

F.M. Angelici (ed.), *Problematic Wildlife*, DOI 10.1007/978-3-319-22246-2_17

Which Predators Kill People?

There have been two recent reviews of large carnivores that kill humans (Löe and Røskaft 2004; Quigley and Herrero 2005). These reviews identified 12 species that have made multiple predatory attacks on humans, and another 5 species that are associated with anecdotal cases or where they have killed people when provoked. The former include three species of canids (dingo *Canis dingo*, coyote *Canis latrans*, grey wolf, Linnell et al. 2002; Fritts et al. 2003), five cat species (tigers *Panthera tigris*, lion *Panthera leo*, leopard *Panthera pardus*, puma *Puma concolor*, jaguar *Panthera onca*, Neto et al. 2011; Mattson et al. 2011; Bloomgaard 2001; Dhanwatey et al. 2013; Chomba et al. 2012; Yamazaki and Bwalya 1999; Packer et al. 2005; Athreya et al. 2011; Goyal 2001) and four bear species (polar bear *Ursus maritimus*, brown bear *U. arctos*, American black bear *U. americanus*, sloth bear *Melursus ursinus*, Bargali et al. 2005; Rajpurohit and Krausman 2000; Fleck and Herrero 1989; Clark et al. 2012; Chestin 1993; Gunther and Hoekstra 1998). However, only tigers, leopards, sloth bears, lions and brown bears kill people on a regular basis (i.e. annually). There are also isolated and anecdotal reports of striped hyaena (*Hyaena hyaena*) and spotted hyaena (*Crocuta crocuta*), Asiatic black bear (*Ursus thibetanus*), sun bear (*Helarctos malayanus*) and spectacled bear (*Tremarctos ornatus*) killing people. Spotted hyaenas are reputed to have killed many people in Africa although documentation and reliable statistics are rare. It is important to consider that domestic dogs are also associated with many serious and fatal attacks on humans (Avis 1999; Sacks et al. 2000).

Documentation of large carnivore attacks on people is highly variable, as is the level of understanding of their ecology. Although their populations have been greatly reduced in previous centuries, there are still large areas of North America and Europe where large carnivores can come in contact with people. Attacks by brown bears, black bears and mountain lions in North America are well documented and have been subject to a great deal of analysis with respect to identifying patterns, trends and mechanisms (e.g. Herrero 1985; Mattson et al. 2011). In Europe, brown bears and polar bears have attacked people, and these incidents have been studied and summarized (Swenson et al. 1996; Amundsen 2014). The Indian subcontinent is a hot spot for large carnivore attacks on people and there are a number of case studies from specific periods and places (e.g. Barlow et al. 2013; Gurung et al. 2008 for tigers, Bargali et al. 2005; Rajpurohit and Krausman 2000 for sloth bears, Athreya et al. 2011 for leopards) although currently there is no overview of the total situation. The situation from Africa is even more poorly documented, with almost no systematic data compiled from any periods or regions (exceptions are Treves and Naughton-Treves 1999; Kushnir et al. 2014). Thus, it is not possible to develop a conceptual cross-site or cross-species global analysis to assess possible explanations for the existence of man-eaters that have been raised in the popular literature (Corbett 1944; Quammen 2003; Vaillant 2011).

For each species and site, it appears that a different combination of human behaviour, ecological context, landscape characteristics and large carnivore behavior seems to be involved. Large carnivores are such complex animals that individual experience and personality can potentially play major roles in shaping their behavior (Fagen and Fagen 1996; Linnell et al. 1999) although the role of these traits with regard to the propensity of some individuals to kill people has not been rigorously tested. However, given the relatively defenselessness of humans as compared to the killing capability of these species the question should probably be reversed—why do so few people get killed (Quigley and Herrero 2005)? Fatal attacks on humans are exceptional events, and even in regions with a chronic history of attacks such events represent a miniscule proportion of the actual outcomes of human–predator encounters. That being said, it represents a very serious conflict that has the potential to have dramatic impacts on the well-being of rural communities, raises some serious issues associated with environmental justice and can undermine local support for conservation activities (Bhatia et al. 2013; Siemer et al. 2014). It is therefore very important that more attention be paid to this phenomenon and that effective mitigation be devised. The problem at present is that data on attacks on people, like many such practical conflict issues (Butler et al. 2014), rarely make it beyond public databases or management agency reports into the academic literature and are therefore often overlooked.

Wolf Attacks on Humans: Contested Realities

Although there is a great deal that is not understood about the ecology of attacks on people by species like lions, leopards, tigers and bears, at least there is no dispute regarding the fact that these events occur. The situation is very different for the canids. One of the most famous cases surrounds the supposed killing of a baby by a dingo in central Australia in 1980. Initially, it was believed that the mother had killed her own child because it was believed that dingoes did not attack people. However, later evidence cleared her name, and subsequent incidences with dingoes on Fraser Island have established that dingoes can attack people and kill children (Anonymous 2001; Franklin 2012). During the past 100 years, there has also been considerable controversy in Western Europe and North America regarding the danger of wolves to people (Fritts et al. 2003). Historically, western attitudes towards wolves have been negative and have contained elements of hatred and fear, with rumours, legends and myths providing the knowledge basis for public opinion. That wolves were dangerous to people was a common element of these discourses, as epitomized by the role of wolves in folk tales such as Red Riding Hood (Dundes 1989), which was written in a period of frequent wolf attacks on people (Moriceau 2007). Stories like these helped to propagate a mainstreamed and collective belief in the dangers posed by wolves.

The first scientists (North Americans) that began to collect objective data about wolf ecology based on original fieldwork and the critical analysis of sources were the first to begin questioning whether the traditional view of wolves actually reflected reality (Young and Goldman 1944; Mech 1970). Based on their experience and examination of contemporary sources, a new narrative emerged that claimed that there were no documented cases of healthy wolves killing adult humans in North America in the twentieth century. This narrative was widely propagated by the media, not least because of the lasting impact of the fictional work *Never Cry Wolf* by Farley Mowat. This claim may well have been accurate at the time of writing and when considering all the qualifiers in the statements, but it was quickly interpreted by advocates of wolf conservation as an assertion that wolves simply do not kill people. The late twentieth century saw a dramatic change in conservation policy in many countries that facilitated wolf recovery across Europe and North America (Chapron et al. 2014). The message that wolves were not dangerous to people was a standard part of most propaganda presented to the public. However, as wolf recovery has progressed there has been a commensurate increase in negative economic impacts and social conflicts associated with the species (Nie 2003). The potential danger that wolves pose to humans has long been part of this discourse with many stakeholders that have an anti-wolf agenda actively promoting an image of wolves as being potentially dangerous to humans. Against this background of claims and counter claims, there is a real need to clarify the actual evidence for and against wolf attacks on humans. This chapter summarizes the results of two reviews (Linnell et al. 2002; McNay 2002) and more recent literature on this topic. We split the material into two periods, pre- and post-1950, seeing as this represents a transition between major historical periods and the point where scientific publication and modern day ecological studies begin to appear.

Historical Wolf Attacks (Pre-1950)

History is repeatedly used by competing factions to legitimize various positions in politics, and wolf conservation is no exception. The relationship between wolves and humans has a rich cultural history with written source stretching back over 2000 years (Lopez 1978; Carbone 1991; Marvin 2012). Accounts of wolf attacks on humans are widespread throughout history. However, a considerable challenge occurs when trying to assess the accuracy of various sources. There is a whole genre of folktales that portray cases of wolves chasing horse-drawn sleights in winter, or attacking soldiers and postal workers as they travel alone through the woods (Carbone 1991; Snerte 2000). Although widespread in local oral traditions, these tales turned up repeatedly across Europe from the nineteenth century and onwards following the rapid expansion in printed materials (precursors to today's tabloids) in that period which used images and text to generate emotional responses in readers (Sangiovanni 2012). Despite the stories often being portrayed as having occurred locally (in each of many localities), there is no indication that they are based on factual events. In recent years, European historians have studied the administrative archives to gather more robust information about the historical relationships between people and wolves.

The most systematic historical studies have been conducted in France, northern Italy, Finland, Estonia and western Russia (Alleau 2011; Comincini 2002; Lappalainen 2005; Moriceau 2007; Rootsi 2003), with additional local studies having been conducted in countries like Germany, Spain and Sweden (Butzeck 1987; Pousette 2000; Teruelo and Valverde 1992). These studies have mainly focused on the period from the fifteenth to early twentieth centuries from which there are fairly complete administrative archives, especially from the eighteenth to nineteenth centuries. Accounts of wolf attacks on people have been found in many of these searches. In fact, when adding up all the accounts, there are several thousand cases scattered across the centuries for hundreds of wolf attack sequences. The veracity of historical sources can always be questioned, but there is nothing to indicate that there is not a real event behind many, or most, of the cases that have been uncovered (Alleau and Linnell 2015). Historians are trained in interpreting the context of the archival material they examine, and in many cases this material is rich, detailed and comes from multiple parallel sources involving administrative sources written by religious (e.g. parish registers of births and deaths) and state authorities (reports, letters, compensation payments, etc.), as well as contemporary newspaper reports (Alleau and Linnell 2015). Unlike folk tales, these records provide multiple concrete details of places, people, dates and events.

Several patterns emerge from these historical studies. Firstly, despite the total number of cases being quite high, when considering the long time scales (centuries) and large spatial scales (multiple countries) it is clear that being reported as killed by a wolf was not a common event in historical Europe. It is also important to bear in mind how many countries have no such records although this could indicate either that such events did not occur there or that there was an insufficient effort to search for available historical material. Secondly, most of the wolf attacks tend to fall into two categories: rabid or predatory attacks (Linnell et al. 2002). Rabies was a prevalent and well-known disease in Europe prior to the mid-twentieth century, having occurred at least since antiquity (Théodoridès 1986), and rural people would have been very familiar with it. The historical material contains many detailed accounts of rabid wolves biting multiple people within a few hours inside a limited area. Characteristics of such attacks include the behavior of the wolf, the biting of multiple people/animals and the failure to consume the victims. The ferocity of the initial attacks and subsequent infections typically killed many people because of the lack of a treatment for rabies prior to the end of the nineteenth century. Attacks by rabid wolves in historical times have been documented across central, southern and Eastern Europe, including Spain, France, Germany and the European part of Russia (Alleau 2011; Butzeck 1987; Comincini 2002; Moriceau 2007; Rootsi 2003; Teruelo and Valverde 1992).

The second category of attacks concerns multiple cases of individual attacks on people within a limited area over a period of months or years. In these cases, children tended to be selected and the bodies were normally dragged away and consumed. Usually, the episodes ended after some years, or after intensive wolf hunting killed many local wolves. However, in some regions such episodes tended to reoccur after several decades. These cases received a very different treatment in historical records, as they were viewed as being something extreme and outside the normal (of how wolves were meant to behave). Based on the accounts, it appears that these were predatory attacks where individual wolves/wolf packs learnt that it was possible to kill young humans. It must be born in mind that the landscape context and agricultural practices in which these attacks occurred were very different from today's Europe. It was a period with an intense human pressure on the landscape, with relatively little forest and little wild prey. Livestock (in addition to carrion and garbage) would have been the main prey of wolves, and the only thing standing between the wolves and this prey would have been unarmed child shepherds in fragmented landscapes with dispersed settlements (Alleau 2011). Firearms were also generally absent. These clusters of predatory attacks have been described from historical sources from Russia, Estonia, Finland, Sweden, France, Spain and Italy (Alleau 2011; Comincini 2002; Lappalainen 2005; Moriceau 2007; Pavlov 1982; Pousette 2000; Rootsi 2003; Teruelo and Valverde 1992).

There has been a great deal of discussion about the veracity these historical records, especially among wolf advocates. Popular and uncritical works by amateur historians (e.g. Furuseth 2005; Snerte 2000) run the potential of sowing confusion. Some exceptional cases, such as the "Beast of Gévaudan" in France (1764–1767), have become so infused with popular interpretations that it is particularly difficult to separate myth and reality (Clarke 1971). It is obviously impossible to investigate retrospectively historical cases to exclude confusion between wolf attacks and those of feral dogs or wolf-dog hybrids, or indeed any one of many other causes of mortality that could be mistakenly or deliberately misclassified as wolf caused. However, the richness and detail of the archival material and the fact that similar patterns emerge from many different parts of Europe (with different religious, cultural and political traditions) across so many centuries would all tend to indicate that something real was lying behind these reports. This is especially true when we consider the rabies cases where we can find clear parallels with modern cases (see below). Furthermore, recent reports and studies concerning predatory attacks by wolves also give legitimacy to the historical events. There is little doubt that language has been a substantial barrier to creating a wider awareness of this material before now because almost none of the historical material has been published in English.

Recent Wolf Attacks (After 1950)

The role of wolves as a vector of rabies to humans is well documented in the modern medical and veterinary technical literature. This documentation stems from a wide range of Eurasian countries including the former Soviet Union (Cherkasskiy 1988; Kuzmin 2001; Pavlov 1982; Selimov et al. 1978), Turkey (Turkmen et al. 2012), Iran (Bahmanyar et al. 1976; Baltazard and Ghodssi 1954; Beran 1994; Gholami et al. 2014), India (Kumar and Rahmani 1997; Rathod et al. 1997; Shah and Jaswal 1976; Isloor et al. 2014) and China (Fangtao et al. 1988). In addition, there are a number of documented cases of rabid wolves biting people in Canada and Alaska

(McNay 2002). These case studies are so well documented that they leave little doubt as to the potential risks that rabid wolves represent for humans. Key characteristics of these modern day cases include (1) the furious behaviour of the wolf that leads to very severe injuries, often involving bites to head and neck, (2) the involvement of single wolves that bite multiple people in a localized area in a very short time and (3) no attempt to eat the people that are killed or attacked. The consistency of these accounts with the historic accounts gives much credibility to the historical observers. The existence of post-exposure treatment for rabies now saves the lives of most people bitten by wolves, although some people die directly from the attacks and others that are bitten in the head and neck region may die of rabies infection before the post-exposure treatment has time to take effect (Turkmen et al. 2012) or via indirect pathways such as infected wounds or organ transplants (Simani et al. 2012).

In modern times, there have been relatively few predatory attacks on people, so that there is comparatively little material to study. In Western and Northern Europe, wolf populations were greatly reduced, although larger populations persisted in the south and east. The only cases known from Europe are from northwest Spain in the 1950s and 1970s. In three separate episodes, four children were killed and four injured. These cases were investigated by biologists, and it seems possible that wolves were responsible for the attacks (Teruelo and Valverde 1992). More cases are known from India in the 1980s and 1990s where at least three different episodes of predatory attacks on children have been relatively well documented in Bihar (Shahi 1982; Rajpurohit 1999) and Uttar Pradesh (Jhala and Sharma 1997). These cases consisted of clusters of attacks spread over several years that would indicate that one or a few packs developed this specific behavioural pattern. The series of attacks ended when the responsible pack was eradicated. Most recently, a series of cases of predatory attacks have been reported within a localized region of western Iran (Behdarvand and Kaboli 2015; Behdarvand et al. 2014). While the patterns described are consistent with previous reports, the veracity of the underlying data is unknown as not enough details of the individual attacks has been adequately described to permit critical evaluation of the data quality.

North America is striking for its paucity of documented attacks during historical times (Casey and Clark 1996; McNay 2002). However, in the last half century there has been an emerging pattern of attacks where wolves have been showing bold (fearless) behavior around people, sometimes, but not always, in association with food conditioning associated with their use of anthropogenic food sources. In multiple cases, these have led to situations where wolves bite people (McNay 2002; McNay and Mooney 2005). Although not all these cases come from protected areas, several national parks like Denali and Yellowstone have introduced guidelines on how to manage wolves that become fearless or habituated (Anonymous 2003).

Until 2005, none of these attacks had been fatal. However, an adult man was killed at a remote mining exploration site in northern Saskatchewan in November 2005 (McNay 2007). It was clear from the site that he had been killed by a large predator, but the crime scene investigation and autopsy were not handled very efficiently leading to some doubt as to the predator species responsible. Early conclusions that he was killed by a black bear were overturned in a later reanalysis of the evidence (McNay 2007). The new analysis concluded that multiple indices all pointed to the most likely explanation being that he had been killed and fed on by wolves. There was also no evidence for the presence of a black bear in the vicinity. Food-conditioned wolves were known to occur around the mining camp and had behaved aggressively towards other workers in previous days. This was the first documented case of a person being killed by non-rabid wolves in North America in a century, although many wolf-advocates have tried to cast doubt on this finding. Another episode occurred in Alaska in March 2010 when an adult female teacher was killed by wolves while jogging outside a remote village. In this case, the forensic evidence from the scene of the kill, from the autopsy of the woman, and of a number of wolves shot after the event provided conclusive proof of this being an unprovoked attack by healthy wolves (Butler et al. 2011). It is important to point out that the victim was short in stature, jogging at the time and wearing headphones playing music so she may not have even been aware of the wolves and could not take any defensive actions. Between them, these two cases have confirmed that non-rabid wolves can kill adult humans, even in North America. Combining these recent events with the ever-expanding body of historical research should certainly force the more extreme wolf-advocates to reconsider their often uncritical and naïve view of the wolf–human relationship. The evidence now clearly points to the fact that both rabid and non-rabid wolves have killed many people during the centuries.

The Dangers of Wolf Attacks in the Twenty-First Century

The existence of evidence that wolves have killed people during both historical and recent times does not mean that they represent a great danger. It is very important to view the contexts within which attacks occurred.

1. Rabid wolf attacks only occur in areas where the disease is endemic. Rabies has been eliminated from most of Europe in recent years, and subsequently the only cases of rabid wolf attacks in western Eurasia in recent decades have occurred along the European borders with Russia/Belarus, where rabies is still prevalent. The situation is however, worse in the Middle East, central Asia and India, where rabies is widespread. There is very little information concerning the transmission routes between wolves, other wildlife and domestic dogs, making it hard to identify appropriate responses.
2. The major cases of predatory attacks from historical Europe, Spain in the 1950s and 1970s and India in recent decades are all associated with a very specific set of circumstances. They are linked to landscapes with very fragmented habitat, low densities of wild prey, wolf dependence on livestock and anthropogenic foods and high human densities living poor rural lifestyles. They are also all focused on defenseless children placed in vulnerable situations. It can also be related to specific sociopolitical circumstances such as famine, the pandemics such as bubonic plague ("black death") or war which created social and economic

disorganization with unburied bodies (Alleau 2011; Moriceau 2007). These circumstances change the relationships between wolves and humans and increase the probability of undesirable behavior occurring (Alleau and Linnell 2015). From the perspective of present day Europe and North America, it implies a very low risk of such behaviours emerging. These areas currently have very abundant wild prey populations, increasing areas of forest, and rural populations that are generally not conducting activities that would constitute risks (e.g. using children as shepherds). However, certain risks may still remain in other parts of Eurasia or the Indian subcontinent.

3. Recently, there has been a widespread focus on cases of wolves that are developing fearless behavior (i.e. not showing fear when directly confronted with humans). Although there are some historical precedents (Pousette 2000) it is widely reported as an emerging issue from recent decades (e.g. Lescureux and Linnell 2013). There is a lack of robust scientific data on the topic, and there is certainly a need for more research to understand the behavioural mechanisms and processes that lead to the development of risky situations. Wildlife managers dealing with urban coyotes that attack domestic dogs and occasionally people have come a long way in developing conceptual models to understand the habituation process (Schmidt and Timm 2007; Timm and Baker 2007), and these could make a useful starting point to examine the question for wolves. Experience from domestic dog attacks can also provide valuable insights (Avis 1999; Sacks et al. 2000). Developing a more detailed theoretical and empirical understanding, the process is essential before it is possible to determine whether methods like hazing or lethal control are likely to reduce risks.

Despite the need to recognize that the potential for wolf attacks on people is greater than zero and management plans and procedures should take these into account, it is still so small that it is impossible to calculate in a meaningful manner. To put it into context, there are currently >12,000 wolves in Europe and >50,000 wolves in North America, many of which are living in proximity to millions of humans, and yet we only find evidence for a handful of attacks in recent decades.

The Politics of Fear

Despite the low level of objective risk that wolves represent for people, large segments of the public express fear of wolves (Røskaft et al. 2003; Zimmermann et al. 2001). Fear is also frequently used as an argument by anti-wolf advocates in efforts to undermine conservation legislation and reduce the current level of legal protection offered to wolves as well as frequently being used in the media and mentioned in social media discourses. Understanding the reason for this mismatch between objective assessment of risk and the public discourse around risk requires an understanding of the psychology and sociology of human–large carnivore conflicts

(Bjerke et al. 2002). Central to this is the idea that many of the conflicts associated with large carnivores are more related to social conflicts between different groups of humans rather than the direct impacts that carnivores per se may have on people (Redpath et al. 2013). With respect to wolves, these conflicts include those between rural and urban areas, traditional vs modern lifestyles and values, hunting vs non-hunting interests and experience-based vs academic knowledge systems (Skogen et al. 2013). Wolves have become potent symbols of these wider conflicts in Europe and North America, such that opposition to the wolf has become a form of social opposition to the changes influencing rural life (Nie 2003; Skogen and Krange 2003). Within this context, fear has just become one more element in a wider political struggle. Unfortunately, it is not just the fear based on individual experience (which would actually confirm that the risk from wolves is low) which is entering the debate. In contrast, fear is being actively used by certain leading individuals within the anti-wolf advocacy world to build up an opposition to wolves (Geist 2014). These individuals are very active in spreading what at best is a highly biased and very selective discourse around wolves, but which also contains a large amount of misinformation and fabrication. The popular media is also picking-up on these messages and giving them wide coverage. On the other hand, the widespread denial of wolf attacks that is voiced by many wolf advocates increases the knowledge gap. Given the wider social and political conflicts for which wolves have become symbolic, this combination of fear mongering and denial falls onto fertile ground within social groups that are respectively united in their opposition to, or support for, wolves. The result has been an escalation in social conflicts over wolves (Brox 2000) and a reduction in the extent to which scientific knowledge serves as a basis for decision-making.

Managing the Fear of Wolves

Although the risk of people being attacked by wolves is incredibly low in the modern world, the risk is not zero. Global efforts to combat rabies will clearly help to reduce the incidences of rabid wolf attacks, and improvements in health care that make post-exposure treatment more widely available will clearly help to save lives. Rabies is still a major global disease, responsible for over 50,000 deaths annually, with domestic dogs being responsible for most cases (e.g. Gholami et al. 2014). Wolves are the responsible vector in only a tiny proportion of these deaths; however, there is little doubt that the specific aggressive nature of these cases has greatly shaped human attitudes towards wolves during historical times. The situations where predatory attacks occur are not widespread in Europe and North America although they persist in areas like India and Central Asia. There is a clear need to better understand the mechanisms and processes that leads a few individual wolves to develop the unusual behavior of killing people; however, the fact that this so rarely happens makes it very hard to study. In reality, the most important strategy is probably to be able to respond rapidly and remove individuals who develop unacceptable behavioural patterns. In this context, it is important to underline the need for agreeing on what are considered acceptable and unacceptable behaviours. Wolves living in proximity to people are not automatically a danger. The difference is only evident in direct encounters between wolves and people and is most obvious when wolves begin to actively approach people. There would be considerable benefits from an expert process to develop guidelines for dealing with such wolves. A holistic strategy must also contain a focus on putting wolf attacks into context against other more familiar risks, such as that posed by domestic dogs. This example is also useful because the recommended responses are broadly similar.

While the previous actions focus on dealing with the real risks posed by wolves, the far greater challenge lies in managing the fear of wolves. As long as wolves remain politicized symbols for wider conflicts it is going to be hard to bring observed levels of fear down to levels more in line with the objective risk they pose. However, a first step at least will be to build a broader agreement about the knowledge base on wolf attacks. For many decades, wolf-advocates have tried to deny or trivialize the facts that wolves have killed people in both historical and modern times. This has led to a situation where many stakeholders have felt that there is a battle-of-knowledge between their own experience, what they hear within their own social circles, and what the so-called external experts claim, creating a credibility-gap and an active opposition to much academic knowledge. Openly admitting that wolves have killed people in specific settings and explicitly coming up with procedures to manage these situations will at least be a first step towards building trust and approaching a common knowledge base for further discussions. Hopefully, this will be based on an objective analysis of the situation rather than one on cynical fear mongering and myth (Bjerke et al. 2002). Unfortunately, this manner of explanation, underlining the complexity of contexts in which attacks happens is far less media-friendly than the more simplified, sensationalized and polarized versions that currently dominate in the press.

Acknowledgements The author was financed by the Research Council of Norway and the Norwegian Environment Agency during the production of this chapter.

References

Alleau J (2011) Garder ses distances: une histoire des relations homes-loups dans les Alpes occidentales (XVIe–XVIIIe siècle). PhD thesis, Caen University, France

Alleau J, Linnell JDC (2015) The story of a man-eating beast in Dauphiné, France (1746–1756). In: Masius P, Sprenger J (eds) A fairytale in question: historical interactions between humans and wolves. White Horse Press, Isle of Harris, pp. 79–100

Amundsen B (2014) Uten nåde: isbjørn og mennesker på Svalbard. Vigmostad and Bjørke, Oslo

Anonymous (2001) Risk assessment: risk to humans posed by the dingo population on Fraser Island. Queensland Environmental Protection Agency. www.env.qld.gov.au/cgi-bin/w3-msql/environment/park/fraser/msqwelcome.html?page=dingo_risk.pdf

Anonymous (2003) Management of habituated wolves in Yellowstone National Park. Yellowstone National Park, Yellowstone

Athreya V, Odden M, Linnell JDC, Karanth KU (2011) Translocation as a tool for mitigating conflict with leopards in human-dominated landscapes of India. Conserv Biol 25(1):133–141

Avis SP (1999) Dog pack attack: hunting humans. Am J Forensic Med Pathol 20(3):243–246

Bahmanyar M, Fayaz A, Nour-Salehi S, Mohammadi M, Koprowski H (1976) Successful protection of humans exposed to rabies infection. JAMA 236(24):2751–2754

Baltazard M, Ghodssi M (1954) Prevention of human rabies: treatment of persons bitten by rabid wolves in Iran. Bull World Health Organ 10:797–803

Bargali HS, Naim Akhtar A, Chauhan NPS (2005) Characteristics of sloth bear attacks and human casualties in North Bilaspur Forest Division, Chhattisgarh, India. Ursus 16(2):263–267

Barlow ACD, Ahmad I, Smith JLD (2013) Profiling tigers (Panthera tigris) to formulate management responses to human-killing in the Bangladesh Sundarbans. Wildl Biol Pract 9(2):30–39

Behdarvand N, Kaboli M (2015) Characteristics of gray wolf attacks on humans in an altered landscape in the west of Iran. Hum Dimens Wildl 20(2):112–122

Behdarvand N, Kaboli M, Ahmadi M, Nourani E, Mahini AS, Aghbolaghi MA (2014) Spatial risk model and mitigation implications for wolf-human conflict in a highly modified agroecosystem in western Iran. Biol Conserv 177:156–164

Beran GW (1994) Rabies and infections by rabies related viruses. In: Beran GW (ed) Handbook of zoonoses. Section V: Viral. 2nd ed. CRC Press, London, pp 307–358

Bhatia S, Athreya V, Grenyer R, Macdonald DW (2013) Understanding the role of representations of human-leopard conflict in Mumbai through media-content analysis. Conserv Biol 27(3):588–594

Bjerke T, Røskaft E, Skogen K, Torgersen S (2002) Hva påvirker folks frykt for ulven? In: Linnell JDC, Bjerke T (eds) Frykten for ulven: en tverrfaglig utredning, vol 722, NINA Oppdragsmelding., pp 75–100

Bloomgaard P (2001) Frontiers of fear: tigers and people in the Malay world, 1600–1950. Yale University Press, London

Brox O (2000) Schismogenesis in the wilderness: the reintroduction of predators in Norwegian forests. Ethnos 65(3):387–404

Butler L, Dale B, Beckmen K, Farley S (2011) Findings related to the March 2010 fatal wolf attack near Chignik Lake, Alaska, ADF&G/DWC/WSP-2011-2. Wildlife Special Publication

Butler JRA, Linnell JDC, Morrant D, Athreya V, Lescureux N, McKeown A (2014) Dog eat dog, cat eat dog: social-ecological dimensions and implications of dog predation by wild carnivores. In: Gompper M (ed) Free-ranging dogs and wildlife conservation. Oxford University Press, Oxford, pp 117–143

Butzeck S (1987) The wolf Canis lupus L. as a mediator of rabies to the German population in the 16th and 17th century. Z Gesamte Hyg 33(12):666–669

Carbone G (1991) La peur du loup. Decouvertes Gallimard Histoires Naturelles. Kapp Lahore, Jombart

Casey D, Clark TW (1996) Tales of the wolf: fifty-one stories of wolf encounters in the wild. Homestead, Moose

Chapron G, Kaczensky P, Linnell JDC, von Arx M, Huber D, Andrén H, López-Bao JV, Adamec M, Álvares F, Anders O, Balčiauskas L, Balys V, Bedő P, Bego F, Blanco JC, Breitenmoser U, Brøseth H, Bufka L, Bunikyte R, Ciucci P, Dutsov A, Engleder T, Fuxjäger C, Groff C, Holmala K, Hoxha B, Iliopoulos Y, Ionescu O, Jeremić J, Jerina K, Kluth G, Knauer F, Kojola I, Kos I, Krofel M, Kubala J, Kunovac S, Kusak J, Kutal M, Liberg O, Majić A, Männil P, Manz R, Marboutin E, Marucco F, Melovski D, Mersini K, Mertzanis Y, Mysłajek RW, Nowak S, Odden J, Ozolins J, Palomero G, Paunović M, Persson J, Potočnik H, Quenette P-Y, Rauer G, Reinhardt I, Rigg R, Ryser A, Salvatori V, Skrbinšek T, Stojanov A, Swenson JE, Szemethy L, Trajçe A, Tsingarska-Sedefcheva E, Váňa M, Veeroja R, Wabakken P, Wölfl M, Wölfl S, Zimmermann F, Zlatanova D, Boitani L (2014) Recovery of large carnivores in Europe's modern human-dominated landscapes. Science 346(6216):1517–1519

Cherkasskiy BL (1988) Roles of the wolf and the racoon dog in the ecology and epidemiology of rabies in the USSR. Rev Infect Dis 10(Suppl 4):S634–S636

Chestin IE (1993) Brown bear aggressiveness towards humans in Russia. In: Bears of Russia and adjacent countries—state of populations. Proceedings of the 6th conference of specialists studying bears, Moscow. Central Forest Reserve, pp 89–94

Chomba C, Senzota R, Chabwela H, Mwitwa J, Nyirenda V (2012) Patterns of human-wildlife conflicts in Zambia, causes, consequences and management responses. J Ecol Nat Environ 4(12):303–312

Clark DA, van Beest FM, Brook RK (2012) Polar bear—human conflicts: state of knowledge and research needs. Can Wildl Biol Manage 1(1):1–9

Clarke CHD (1971) The beast of Gévaudan. Nat Hist 80:44–51, 66–73

Comincini M (2002) L'uomo e la "bestia antropofaga": storia del lupo nell'Italia settentrionale dal XV al XIX secolo. Unicopli, Italy

Corbett J (1944) Man-eaters of Kumaon. Oxford University Press, London

Dhanwatey HS, Crawford JC, Abade LAS, Dhanwatey PH, Nielsen CK, Sillero-Zubiri C (2013) Large carnivore attacks on humans in central India: a case study from the Tadoba-Andhari Tiger Reserve. Oryx 47(2):221–227

Dundes AE (1989) Little Red Riding Hood: a casebook. University of Wisconsin Press, Madison

Fagen R, Fagen JM (1996) Individual distinctiveness in brown bear, Ursus arctos L. Ethology 102:212–226

Fangtao L, Shubeng C, Yinzhon W, Chenzhe S, Fanzhen Z, Guanfu W (1988) Use of serum and vaccine in combination for Prophylaxis following exposure to rabies. Rev Infect Dis 10(Suppl 4):S766–S770

Fleck S, Herrero S (1989) Polar bear conflicts with humans. In: Bromley M (ed) Bear-people conflicts: proceedings of a symposium on management strategies. Northwest Territories Department of Renewable Resources, Yellowknife, Northwest Territories, pp 201–202

Franklin AS (2012) Dingoes in the dock. New Sci 213(2852):28–29

Fritts SH, Stephenson RO, Hayes RD, Boitani L (2003) Wolves and humans. In: Mech LD, Boitani L (eds) Wolves: behavior, ecology, and conservation. University of Chicago Press, Chicago, pp 289–316

Furuseth A (2005) Drept av bjørn og ulv: en historisk oversikt over mennesker drept og skadet av rovdyr i Norge de siste 400 år. Landbruksforlaget, Oslo

Geist V (2008) Death by Wolves and the power of Myths: the Kenton Carnegie Tragedy. Fair Chase 23(4):29–33

Gholami A, Fayaz A, Farahtaj F (2014) Rabies in Iran: past, present and future. J Med Microbiol Infect Dis 2(1):1–10

Goyal PS (2001) Man-eating leopards—status and ecology of leopard in Pauri Garhwal, India. Carnivore Damage Prev News 3:9–10

Gunther KA, Hoekstra HE (1998) Bear-inflicted injuries in Yellowstone National Park, 1970–1994. Ursus 10:377–384

Gurung B, Smith JLD, McDougal C (2008) Factors associated with human-killing tigers in Chitwan National Park, Nepal. Biol Conserv 141:3069–3078

Herrero S (1985) Bear attacks: their causes and avoidance. Nick Lyons Books, New York

Isloor S, Marissen WE, Veeresh BH, NithinPrabhu K, Kuzmin IV, Rupprecht CE, Satyanarayana ML, Deepti BR, Sharada R, Neelufer MS, Yathiraj S, Abdul Rahman S (2014) First case report of rabies in a wolf (Canis lupus pallipes) from India. J Vet Med Res 1(3):1012

Jhala YV, Sharma DK (1997) Childlifting by wolves in eastern Uttar Pradesh, India. J Wildl Res 2(2):94–101

Kumar S, Rahmani AR (1997) Status of Indian grey wolf Canis lupus pallipes and its conservation in marginal agricultural areas of Solapur District, Maharashtra. J Bombay Nat Hist Soc 94(3):466–472

Kushnir H, Weisberg S, Olson E, Juntunen T, Ikanda D, Packer C (2014) Using landscape characteristics to predict risk of lion attacks on humans in south-eastern Tanzania. Afr J Ecol 52(4):524–532

Kuzmin I (2001) Rabies in Russia 1960–1998. RABNET. www.who.int/emc/diseases/zoo/Russia_data/russiarabies index.html
Lappalainen A (2005) Sudet jäljet. Metsäkustannus, Hämeenlinna
Lescureux N, Linnell JDC (2013) The effect of rapid social changes during post-communist transition on perceptions of the human—wolf relationships in Macedonia and Kyrgyzstan. Pastoralism 3:4
Linnell JDC, Odden J, Smith ME, Aanes R, Swenson JE (1999) Large carnivores that kill livestock: do "problem individuals" really exist? Wildl Soc Bull 27(3):698–705
Linnell JDC, Løe J, Okarma H, Blancos JC, Andersone Z, Valdmann H, Balciauskas L, Promberger C, Brainerd S, Wabakken P, Kojola I, Andersen R, Liberg O, Sand H, Solberg EJ, Pedersen HC, Boitani L, Breitenmoser U (2002) The fear of wolves: a review of wolf attacks on humans. NINA Oppdragsmelding 731:1–65
Löe J, Røskaft E (2004) Large carnivores and human safety: a review. Ambio 33(6):283–288
Lopez BH (1978) Of wolves and men. Charles Scribner's Sons, New York
Marvin G (2012) Wolf. Reaktion Books, London
Mattson D, Logan K, Sweanor L (2011) Factors governing risk of cougar attacks on humans. Hum Wildl Interact 5(1):135–158
McNay ME (2002) Wolf–human interactions in Alaska and Canada: a review of case history. Wildl Soc Bull 30:831–843
McNay ME (2007) A review of evidence and findings related to the death of Kenton Carnegie on November 8, 2005 near points north, Saskatchewan. Alaska Department of Fish and Game, Fairbanks
McNay ME, Mooney PW (2005) Attempted predation of a child by a gray wolf, Canis lupus, near Icy Bay, Alaska. Can Field Nat 119:197–201
Mech LD (1970) The wolf: the ecology and behavior of an endangered species. American Museum of Natural History, New York
Moriceau JM (2007) Histoire du méchant loup : 3000 attaques sur l'homme en France (XVe–XXe siècle). Fayard, Paris
Neto MF, Garrone Neto D, Haddad Junior V (2011) Attacks by jaguars (Panthera onca) on humans in central Brazil: report of three cases, with observation of a death. Wilderness Environ Med 22(2):130–135
Nie MA (2003) Beyond wolves: the politics of wolf recovery and management. University of Minnesota Press, London
Packer C, Ikanda D, Kissui B, Kushnir H (2005) Lion attacks on humans in Tanzania. Nature 436:927–928
Pavlov M (1982) [Wolf]. Moscow
Pousette E (2000) De måanniskoätande vargarna. Bjørkelangen Bok and Papir, Bjørkelangen
Quammen D (2003) Monster of god: the man-eating predator in the jungles of history and the mind. W.W. Norton, London
Quigley H, Herrero S (2005) Characteristics and prevention of attacks on humans. In: Woodroffe R, Thirgood S, Rabinowitz A (eds) People and wildlife: conflict or coexistence? Cambridge University Press, Cambridge, pp 27–48
Rajpurohit KS (1999) Child lifting: wolves in Hazaribagh, India. Ambio 28(2):162–166
Rajpurohit KS, Krausman PR (2000) Human-sloth bear conflicts in Madhya Pradesh, India. Wildl Soc Bull 28(2):393–399
Rathod NJ, Salunke S, Bawiskar VR (1997) Clinical profile of wolf bite cases in Jalgaon District. J Assoc Phys India 45(11):866–867
Redpath SM, Young J, Evely A, Adams WM, Sutherland WJ, Whitehouse A, Amar A, Lambert RA, Linnell JDC, Watt A, Gutierrez RJ (2013) Understanding and managing conservation conflicts. Trends Ecol Evol 28:100–109
Rootsi I (2003) Rabid wolves and man in Estonia of the 18th and 19th centuries. Acta Zool Lith 13:72–77

Røskaft E, Bjerke T, Kaltenborn BP, Linnell JDC (2003) Patterns of self reported fear towards large carnivores among the Norwegian public. Evol Human Behav 24:184–198

Sacks JJ, Sinclair L, Gilchrist J, Golab GC, Lockwood R (2000) Breeds of dogs involved in fatal human attacks in the United States between 1979 and 1998. J Am Vet Med Assoc 217(6):836–840

Sangiovanni A (2012) Le parole e le figure: storia dei media in Italia dall'età liberale alla Seconda Guerra mondiale. Donzelli Editore, Italy

Schmidt RH, Timm RM (2007) Bad dogs: why do coyotes and other canids become unruly? Proc Wildl Damage Manage Conf 12:287–302

Selimov MA, Klyueva EV, Aksenova TA, Lebedeva IR, Gribencha LF (1978) Treatment of patients bitten by rabid or suspected rabid wolves with inactivated tissue culture rabies vaccine and rabies gammaglobulin. In: IABS W (ed) Symposium on the standardization of rabies vaccines for human use produced in tissue cultures (rabies III). S. Karger, Basel, pp 141–146

Shah U, Jaswal GS (1976) Victims of a rabid wolf in India: effect of severity and location of bites on development of rabies. J Infect Dis 134(1):25–29

Shahi SP (1982) Status of the grey wolf (Canis lupus pallipes Sykes) in India—a preliminary survey. J Bombay Nat Hist Soc 79(3):493–502

Siemer WF, Decker DJ, Shanahan JE, Wieczorek Hudenko HA (2014) How do suburban coyote attacks affect residents' perceptions: insights from a New York case study. Cities Environ 7(2):7

Simani S, Fayaz A, Rahimi P, Eslami N, Howeizi N, Biglari P (2012) Six fatal cases of classical rabies virus without biting incidents, Iran 1990–2010. J Clin Virol 54(3):251–254

Skogen K, Krange O (2003) A wolf at the gate: the anti-carnivore alliance and the symbolic construction of community. Sociol Ruralis 43(3):309–325

Skogen K, Krange O, Figari H (2013) Ulvekonflikter: en sosiologisk studie. Akademika forlag, Oslo

Snerte K (2000) Ulvehistorier. Samlaget, Oslo

Swenson JE, Sandegren F, Heim M, Brunberg S, Sørensen OJ, Söderberg A, Bjärvall A, Franzén R, Wikan S, Wabakken P, Overskaug K (1996) Er den skandinavisk bjørnen farlig? NINA Oppdragsmelding 404:1–26

Teruelo S, Valverde JA (1992) Los lobos de Morla. Circulo de Bibliotilia Venatoria, Madrid

Théodoridès J (1986) Histoire de la rage. Masson, Paris

Timm RM, Baker RO (2007) A history of urban coyote problems. Proc Wildl Damage Manage Conf 12:272–286

Treves A, Naughton-Treves L (1999) Risk and opportunity for humans coexisting with large carnivores. J Hum Evol 36:275–282

Turkmen S, Sahin A, Gunaydin M, Tatli O, Karaca Y, Turedi S, Gunduz A (2012) A wild wolf attack and its unfortunate outcome: rabies and death. Wilderness Environ Med 23(3):248–250

Vaillant J (2011) The tiger: a true story of vengeance and survival. Vintage Depatures, New York

Yamazaki K, Bwalya T (1999) Fatal lion attacks on local people in the Luangwa Valley, Eastern Zambia. S Afr J Wildl Res 29(1):19–21

Young SP, Goldman EA (1944) The wolves of North America. Dover, New York

Zimmermann B, Wabakken P, Dötterer M (2001) Human-carnivore interactions in Norway: how does the re-appearance of large carnivores affect people's attitudes and levels of fear? For Snow Landsc Res 76(1/2):137–154

Chapter 18
The Aggressive Apes? Causes and Contexts of Great Ape Attacks on Local Persons

Matthew R. McLennan and Kimberley J. Hockings

Attacks and Human-Directed Aggression by Wild Animals

One of the main challenges to biodiversity conservation globally is the rising level of interaction between humans and wild animals, and the resulting conflicts that can emerge (Conover 2002; Hill et al. 2002; Woodroffe et al. 2005). Living alongside wildlife can impose substantial costs upon local people that are frequently cited as the drivers of human–wildlife 'conflict' (see section 'Understanding Great Ape Attacks in the Context of Human–Wildlife "Conflict"' for discussion of conflict definitions), including financial and social costs associated with crop losses and livestock depredation, and risks to human health and well-being from wildlife (Hill 2004; Thirgood et al. 2005; Wang and Macdonald 2006; Mackenzie and Ahabyona 2012; Barua et al. 2013). Human-directed aggression by wild animals—and physical attacks by large mammals in particular—are an especially serious cause of negative human–wildlife interactions. We define an attack as an aggressive interaction involving physical contact, potentially leading to human injury or loss of life. While usually rare, a single attack by a wild animal can elicit more hostility and panic than less immediately severe, but persistent problems associated with human–wildlife sympatry, such as wildlife feeding on agricultural crops (commonly referred

M.R. McLennan (✉)
Anthropology Centre for Conservation, Environment and Development,
Oxford Brookes University, Oxford, UK
e-mail: mmclennan@brookes.ac.uk

K.J. Hockings
Anthropology Centre for Conservation, Environment and Development,
Oxford Brookes University, Oxford, UK

Departmento de Antropologia, Universidade Nova de Lisboa, Lisbon, Portugal

Centre for Research in Anthropology (CRIA), Lisbon, Portugal
e-mail: khockings@brookes.ac.uk

F.M. Angelici (ed.), *Problematic Wildlife*, DOI 10.1007/978-3-319-22246-2_18

to as ‘crop raiding’) or livestock depredation (Hockings et al. 2010). Wildlife attacks—including fear of attack (Kaltenborn et al. 2006)—directly impact the willingness of local communities to tolerate wild animals in their environment, thus reducing their support for conservation (Hill et al. 2002). Even so, and despite their intrinsic negative impact on human well-being, detailed and systematic records of animal attacks are rare, and little is often known about the circumstances surrounding cases (Quigley and Herrero 2005; Thirgood et al. 2005). It is therefore imperative to seek a better understanding of the frequency, causes and circumstances of human-directed aggression by different wildlife groups and species, in order to inform appropriate mitigation strategies and facilitate human–wildlife coexistence. Here, our focus is on large mammal attacks only.

Large mammals attack hundreds of people globally each year, with attacks by big cats (e.g. tigers, *Panthera tigris*), bears (e.g. sloth bear *Melursus ursinus*) and large herbivores such as elephants (*Elephas maximus* and *Loxodonta africana*) receiving most attention (Rajpurohit and Krausman 2000; Löe and Röskaft 2004; Packer et al. 2005; Dunham et al. 2010). Attacks by wild large mammals are particularly problematic because they tend to elicit strong, often ‘negative’, responses (Quigley and Herrero 2005) and are difficult to prevent entirely where humans and wild animals share an environment (Kushnir et al. 2010; Dhanwatey et al. 2013). Attacks occur for diverse reasons and classifying them is often not straightforward. Quigley and Herrero (2005) used the broad categories of ‘provoked’ and ‘unprovoked’ to characterise attacks by terrestrial large carnivores. Although the circumstances surrounding provoked attacks vary, they occur most commonly when a person enters an animal’s ‘personal space’ (i.e. the area around an animal in which it reacts to human presence, which depends on the specific conditions of the situation; Quigley and Herrero 2005). Entering an animal’s personal space need not imply intentionality or aggression by the person(s). For example, attacks by sloth bears in India commonly occurred when the bears encountered unsuspecting humans who were engaged in activities such as defecation outdoors (Bargali et al. 2005). In other situations, however, an animal attacks when a person(s) enters its personal space and purposefully attempts to approach, touch, capture, injure or kill it. For example, Neto et al. (2011) report a nonfatal attack on a man by a jaguar (*Panthera onca*), which reportedly occurred after his dogs cornered the cat; the man was attacked when he approached to help his dogs fight the jaguar. Regardless of whether purposive human aggression is involved, most provoked attacks are ultimately defensive with the animal responding to a perceived human threat (Quigley and Herrero 2005).

Provoked attacks also occur when a person(s) has food or garbage attractants that draw an animal near the person to within its personal space, as characterises some attacks by brown bears (*Ursus arctos*) (Herrero and Higgins 2003). Feeding wild animals, or using food to lure an animal closer, are commonly cited factors in wildlife attacks on tourists (Orams 2002). More generally, attacks by large mammals become increasingly likely where natural habitats are degraded through human activities that reduce or alter the food choices available, potentially causing animals to range nearer to human settlements (Bargali et al. 2005).

Unprovoked attacks occur when the animal approaches and attacks a person(s) who is the principle attractant, for example, predation on humans by large carnivores (Quigley and Herrero 2005). While predatory attacks are opportunistic in many cases, big cats can become dedicated 'man-eaters' in some situations, repeatedly targeting humans as prey (e.g. lions *Panthero leo*, Kerbis-Peterhans and Gnoske 2001; tigers, Gurung et al. 2008). Aside from predation, unprovoked attacks can also result when the person and animal are intent on using the same space, and the animal attacks when it is not given 'right of way' and the person(s) is unable to scare it off (Quigley and Herrero 2005); no food or other attractant is involved. However, some attacks of this type might also be motivated by defensive instincts, as discussed above. Finally, disease plays a part in some unprovoked attacks, as when rabid wolves (*Canis lupus*) attack humans (Linnell et al. 2002). Wildlife attacks inevitably generate fear and hostility towards the species concerned, which can lead to retaliatory killing of threatened species (e.g. tigers, Inskip et al. 2014). Even among persons who rarely or never encounter a potentially dangerous wild animal, fear of an attack is sufficient to promote negative attitudes towards wildlife (Kaltenborn et al. 2006).

To date, great apes have featured only rarely in published reports of wild animal attacks on local persons, and few attempts have been made to evaluate characteristics of ape attacks (but see Hockings et al. 2010). Like many large mammals, however, great apes increasingly occupy disturbed habitats amid expanding human populations, and consequently come into contact with people more frequently in some areas than previously (Hockings et al. 2015). This increased proximity has led to growing reports of aggressive interactions between humans and great apes, including cases of apes attacking people (Hockings and Humle 2009). The survival of many ape populations requires finding ways for humans and apes to coexist together in shared landscapes. Given the declining conservation status of all great apes (IUCN 2014), and the potential for ape aggression towards people to reduce support for conservation efforts, a greater understanding of ape attacks on humans is needed to inform appropriate conflict mitigation strategies.

Wild great apes sometimes direct aggressive behaviour towards human researchers, often during early phases of habituation. Such aggression does not usually involve physical contact (e.g. charging displays by adult male chimpanzees; Grieser-Johns 1996; McLennan and Hill 2010). However, gorillas undergoing habituation have been known to attack researchers physically, for example, by grabbing their legs and/or biting (Doran-Sheehy et al. 2007; Ando et al. 2008). More serious attacks (i.e. involving biting or beating causing substantial injury or potential loss of life) on researchers by wild great apes have been reported only very rarely (see Kutsukake and Matsusaka 2002 for an incident involving an unhabituated female chimpanzee; and see White and Edwards 2000 p. 60 for an attack on a researcher by a lone silverback gorilla). Attacks on persons by rehabilitated ex-captive great apes have occurred following their release into natural environments (e.g. orangutans: Yeager 1997; Dellatore 2007), including serious attacks by rehabilitant chimpanzees (Borner 1985; Agoramoorthy and Hsu 1999). However, attacks by ex-captive apes are likely related to their loss of fear of humans, while serious attacks on researchers by wild apes appear precipitated by unusual circumstances (see above references). Therefore, in this chapter we focus exclusively on aggressive interactions between wild great apes and local persons. First, we review reports of aggression towards humans, including physical attacks, and evaluate their likely causes and contexts. We consider potential differences among great ape taxa in their propensity to attack humans and possible reasons for variation. Finally, we consider great ape attacks in relation to recent developments in the human–wildlife conflict literature, and discuss how a detailed understanding of the contexts of attacks, along with a better understanding of human cultural attitudes and practises and human–human conflict, are important to inform appropriate strategies to reduce likelihood and impact of attacks.

Great Apes and Humans

The nonhuman great apes (hereafter 'great apes')—chimpanzees (*Pan troglodytes*), bonobos (*Pan paniscus*), gorillas (*Gorilla* spp.) and orangutans (*Pongo* spp.)—are widely used as charismatic mega-fauna for conservation. All great ape species and subspecies are listed as Endangered or Critically Endangered by the International Union for the Conservation of Nature (IUCN 2014). A mere estimated 300,000–400,000 individuals of all great ape species remain in the wild with most populations declining due to habitat loss, hunting and disease (IUCN 2014; Rainer et al. 2014). With rapid human population growth in ape range counties, these threats show no sign of slowing down. It has been predicted that by 2030, less than 10 % of African and 1 % of Asian great ape habitat will remain undisturbed from human activities such as logging, mining, agriculture and infrastructural development (GLOBIO model analysis, Nelleman and Newton 2002; see also Junker et al. 2012 for African apes and Struebig et al. 2015 for orangutans). Today, most long-term great ape research sites are impacted by humans and their activities in one way or another (Hockings et al. 2015). Great apes require large spaces yet are increasingly forced into fragmented and restricted ranges, often outside of protected areas (e.g. West African chimpanzees *P. t. verus*: Kormos et al. 2003; Bornean orangutans *P. pygmaeus*: Wich et al. 2012). Consequently, over the coming years and decades we can predict rising levels of contact between humans and great apes, with inevitable increases in negative interactions arising from competition for space and resources, as well as conflicts among different human groups over management of great apes (see section 'Mitigating Conflicts Caused by Attacks').

While persistence of great apes in heavily modified human-dominated habitats is largely a contemporary phenomenon, humans and great apes have a long history of coexistence in some regions. For example, at Lopé, Gabon, humans have coexisted with chimpanzees and gorillas for at least 60,000 years (Tutin and Oslisly 1995). As our closest living relatives, great apes share with us a suite of morphological, behavioural and cognitive similarities, which are widely recognised by local human communities familiar with these animals (e.g. Sept and Brooks 1994;

Sicotte and Uwengeli 2002; Lingomo and Kimura 2009; Sousa et al. 2014). In areas where apes are heavily hunted, they are unlikely to persist in very close proximity to human settlements. However, in some human cultures great apes are afforded special significance that promotes tolerant attitudes towards them; for example, apes are sometimes offered protection from hunting or persecution due to religious beliefs (orangutans: Abram et al. 2015) or local taboos and/or totemic beliefs that consider them ancestors (chimpanzees: Yamakoshi 2005; gorillas: Etiendem et al. 2011). Perhaps not surprisingly, great apes feature prominently in stories and folktales of some human societies that have traditionally coexisted with apes. Notably, such stories often allude to the formidable strength and potential dangerousness of sympatric apes (Richards 1995; Köhler 2005; Giles-Vernick and Rupp 2006; Thompson et al. 2008; Oishi 2013).

Variation in Ape Attacks on Humans

Chimpanzees

Considerably more data are available on wild chimpanzee–human interactions than for other great apes. Chimpanzees are sympatric with humans throughout much of their range in equatorial Africa. In areas where human population density is relatively low and encroachment on natural habitat is not extensive, interactions between people and chimpanzees are largely neutral (e.g. parts of West Africa: Dunnett et al. 1970; Leciak et al. 2005; Duvall 2008; Hockings and Sousa 2013). But where human population densities are higher and chimpanzees are not hunted for food, they can persist in highly fragmented and degraded habitat alongside human farming communities (e.g. Uganda: Reynolds et al. 2003, McLennan 2008; Guinea: Hockings 2009; Sierra Leone: Halloran et al. 2014; see also Fig. 1 in Hockings and McLennan 2012). In such circumstances, people and chimpanzees may utilise the same space and resources and encounter one another frequently, inevitably leading to competition and conflict (Hockings and Humle 2009). For example, chimpanzees exposed to agriculture readily learn to exploit some human crops, which can become important items in their diet (Hockings and McLennan 2012; McLennan and Hockings 2014). 'Conflicts' over natural resources, particularly agricultural crops, reportedly occur throughout the chimpanzees' geographical range (Hockings and McLennan 2012).

Available data imply that a 'habitat disturbance threshold' exists beyond which the frequency of human–chimpanzee contact rises and interactions become increasingly hostile, with both chimpanzees and people directing aggressive behaviour towards the other (McLennan 2008). For example, eastern chimpanzees (*P. t. schweinfurthii*) at Bulindi in the Hoima District of western Uganda inhabit shrinking forest fragments surrounded by farmland and villages. Farmers in this region are traditionally tolerant of chimpanzees, perceiving them to have a 'good' character (Hill and Webber 2010). However, during the past decade forest fragments were extensively logged and cleared for farming, particularly tobacco cash cropping. Chimpanzee behaviour is perceived by local villagers to have undergone recent negative changes, concurrent with the widespread conversion of forest to farmland, including persistent crop 'raiding' and aggression towards people (McLennan and Hill 2012). Adult chimpanzees directed frequent threatening behaviour towards local persons encountered on farmland and village paths, as well as inside forest, such as charging, mobbing and pursuing them (McLennan and Hill 2010), and showed willingness to engage in prolonged agonistic interactions with humans (for an example, see McLennan 2010a). At the same time, some farmers responded to the rise in encounters with chimpanzees and more frequent crop losses with harassment including shouting, stone-throwing and chasing with dogs (McLennan and Hill 2010). In such circumstances, the likelihood of a chimpanzee attack increases.

Chimpanzee attacks are reported from regions of close human–chimpanzee sympatry in West, East and Central Africa, but details of the surrounding circumstances are often vague or absent (e.g. Mutombo et al. 1983; Richards 1995; McLennan 2008; Halloran et al. 2014). It can be difficult to obtain facts about human or chimpanzee behaviour prior to an attack, in part because people are reluctant to admit any wrongdoing. For example, in Uganda's Hoima District many villagers are aware of the legal status of chimpanzees. Local accounts suggested that two recent attacks in one subcounty 'involved a chimpanzee first being speared or attacked with *pangas* (machetes), or set upon by dogs. In these cases, an attempt may have been made to take an infant chimpanzee from its mother or otherwise confront a crop raiding ape' (McLennan 2008, p. 50). Thus, there was a strong indication that these were 'provoked' attacks. Even so, it is unclear how reliable local reports were.

More detailed accounts are available from several other sites in East and West Africa. Hockings et al. (2010) reported 11 attacks on local persons by chimpanzees (*P. t. verus*) in the heavily human-influenced habitat of Bossou, Guinea, between 1995 and 2009. In 10 of 11 cases, attacks were directed towards children between 18 months and 12 years old. These varied in severity. In three cases, local onlookers reported that injuries were sustained when a chimpanzee dragged the child along the ground during a social display (which functions to intimidate a subordinate or gain rank); when the child's skin was scratched by the ape's sharp fingernails; or when the chimpanzee bit the child directly (Fig. 18.1). Although one child sustained life-threatening injuries, none of the attacks were fatal and none were predatory, which is typically characterised by eating from the victim (cf. Wrangham et al. 2000). Attacks took place on a road and on narrow paths bordering forest, or in cultivated fields and orchards, where opportunities for human–chimpanzee encounters were high. All attacks coincided with wild fruit scarcity, increased levels of crop foraging by chimpanzees, and cropping seasons that likely increased human use of paths. Only one incident was witnessed by researchers. On this occasion, an adult male chimpanzee approached and attempted to take bananas (i.e. an attractant) from a child, but was chased back to the forest by a researcher to prevent a potentially more serious attack. This individual was known to be nearby when other attacks occurred, and due to his confrontational behaviour and general lack of fear of people (chimpanzees at Bossou are well habituated), researchers

Fig. 18.1 Adult male chimpanzee at Bulindi, Uganda. Chimpanzees have large canine teeth that can inflict severe injuries to victims of biting attacks (Photo: Georgia Lorenti)

strongly suspect this same male was responsible for several other attacks on local persons. In at least some cases, the chimpanzee(s) was probably provoked before attacking (e.g. children throwing stones or carrying food that attracted the chimpanzee). Still, in 8 of 11 cases, it was not possible to confirm exactly why the attack took place. Reasons might include unreported provocation by people, hunger motivating the chimpanzees to feed in agricultural areas thus increasing the likelihood of close range encounters with humans, and adult male chimpanzees asserting their dominance, either to impress conspecifics or assert dominance over humans (Hockings et al. 2010).

Five chimpanzee attacks on children aged between c. 6 months and 6–7 years old are known to have occurred at Bulindi since 2006. These incidents followed extensive clearance of local forests and rise in human–chimpanzee interactions, as detailed above; according to villagers no physical attacks occurred prior to these. While no attack was fatal, in three cases the child sustained serious injuries requiring hospitalisation (McLennan 2010b, and unpublished data). For example, a 4-year-old boy was attacked at a well in a narrow strip of forest. Although local reports varied it appears chimpanzees were travelling past the well when they were 'disturbed' by a group of children; several villagers suggested the children most likely threw stones at the apes. As the children ran away, the youngest child fell down and was grabbed and bitten on the head, foot and below the armpits by a chimpanzee (McLennan 2010b). In at least two other cases, children or young men were alleged to have harassed chimpanzees prior to an attack; thus, these cases were probably 'provoked'. In one case, however, a chimpanzee apparently grabbed and badly mauled a baby which the mother had placed under shade while digging in her field. In at least three cases, the attacking ape was reportedly 'very big' and not closely associated with younger chimpanzees, suggesting an adult male. As at Bossou, no attack at Bulindi to-date appears motivated by predatory instincts since in no case did the attacking ape eat from the victim or inflict a fatal bite. Reynolds (2005) reported four chimpanzee attacks on children in village areas surrounding Budongo Forest, north of Bulindi, including one which was fatal. No part of the body was eaten in this latter case; however, victims were badly bitten in all cases. Unfortunately, information about human behaviour prior to these incidents is insufficient to reliably categorise them as provoked or unprovoked.

Nevertheless, unequivocal incidents of chimpanzees preying on human children have been documented. Wrangham et al. (2000) reported eight cases over 4 years in which a wild chimpanzee—thought to be a single adult male (Wrangham 2001)—caused severe injury or death to children between the ages of 6 months and 5 years from villages bordering Kibale National Park, Uganda. In all cases, victims were either alone or accompanied only by other children or women. The chimpanzee(s) ate from all victims that could be carried off to an undisturbed site; three children were eviscerated. Injuries to other victims were similar to those seen in monkeys preyed on by chimpanzees (e.g. hands and feet bitten off) (Wrangham et al. 2000). Attacks occurred in a habitat matrix composed of scattered villages, cultivated fields and secondary forest. The chimpanzee(s) exhibited bold behaviour by travelling up to 180 m from the forest edge to capture victims; twice a baby was removed from the doorway of a village house (Wrangham 2001). Two further confirmed incidents of predation occurred in and around Gombe National Park, Tanzania. The first occurred outside the park before 1960 when a male chimpanzee seized a baby from a woman's back, injuring the woman, and killed and partially ate the baby (Thomas 1961 cited in Goodall 1986). A more recent case in 2002 involved a well-habituated and particularly fearless adult male chimpanzee from the main study group in Gombe. This individual approached two women walking through the park and grabbed a 14-month-old baby from one of them; the baby was carried off, killed and partially eaten (Kamenya 2002). Another attack occurred within the park during which a 6-year-old boy was bitten by a chimpanzee and suffered substantial injuries to his face, but it is unclear if this attack was predatory motivated (Goodall 1986). McLennan (2008) reported a fatal attack on a child at a sugarcane plantation near the sparsely forested Kasongoire Forest Reserve, near Budongo, in what might have been a predatory incident. Retaliatory killings of chimpanzees by local villagers are known to have occurred in response to attacks around both Kibale and Budongo (Wrangham 2001; Reynolds 2005).

Gorillas

Western gorillas (*G. gorilla*) coexist with humans in regions across their distribution in the lowland forests of Central Africa (e.g. Etiendem et al. 2011; Oishi 2013). Sabater Pi (1966) reported seven attacks by western lowland gorillas (*G. g. gorilla*) on local persons over a 10-year period at Rio Muni (mainland Equatorial Guinea). Few details of human behaviour prior to these incidents were given; however, most victims were hunters and most attacks seem to have been committed by injured male gorillas. However, in the most detailed account, a hunter startled a small gorilla group at close range (<5 m) and the (uninjured) dominant male charged and attacked, presumably in defence of the group. Thus, there is no indication that attacks were 'unprovoked' according to Quigley and Herrero's (2005) criteria. Attacking gorillas inflicted severe injuries on victims with their teeth and hands. Sabater Pi concluded that attacks on humans were overall rare considering the frequent opportunities for contact between people and gorillas in local forests and plantations. In a further report, a solitary Cross River (*G. g. diehli*) silverback male attacked an adult man who was setting traps in the Kagwene Gorilla Sanctuary, Cameroon. The man attempted to run and was attacked from behind (A. Nicholas pers. comm. in Hockings and Humle 2009); whether this incident was unprovoked or not is unknown. Elsewhere in southeastern Cameroon, hunters recounted incidents of persons being attacked by gorillas (Oishi 2013). Most attacks seem to have occurred in the context of hunting the apes; however, hunters also claimed gorillas sometimes ambush people unexpectedly in the forest (Oishi 2013).

At a small number of sites in East Africa populations of eastern gorillas (*G. beringei*) live alongside high-density farming communities. Most published data on human–gorilla 'conflicts' are from Bwindi Impenetrable National Park, Uganda, where nearly half of the World's mountain gorillas (*G. b. beringei*) live. Some of Bwindi's gorillas, including groups habituated for tourism, spend a substantial amount of time outside park boundaries where the landscape is dominated by smallholder agriculture and villages, with one habituated group ranging more than 1 km from the park border (Goldsmith et al. 2006). Outside the park, gorillas cause frequent damage to local farmers' crops (e.g. to banana plantations). Additionally, there have been incidents of gorillas attacking people on agricultural land outside the park that left local persons (mostly men) seriously injured. Madden (2006) reported at least eight attacks in two parishes bordering the park during 1996–1998. The attacking individuals were thought to be mature males in all cases, with one individual likely involved in the four cases in one parish. Madden (2006, p. 182) further reported that 'in most cases the offending individuals have been habituated to human presence for the purposes of facilitating ecotourism'. Insufficient information is available to determine whether these attacks were provoked or unprovoked. However, Madden suggested that some attacks may follow from a gorilla being surprised, or being a surprise to, local people.

Bonobos

The geographic range of bonobos is restricted to the central Congo Basin, south of the Congo River. Although in some areas bonobos live in proximity to villages (e.g. at Wamba: Idani et al. 2008; Salonga National Park: Thompson et al. 2008), and reportedly consume agricultural crops at some sites (e.g. Lake Tumba: Inogwabini and Bewa 2009), little published data exist on the nature and extent of human–bonobo interactions. Thompson et al. (2008) reported that the Iyaelima people who live alongside bonobos within Salonga National Park told folk stories of men fighting physically with bonobos. The Iyaelima claimed to avoid encounters with bonobos, believing the apes can beat up or kill people. However, aside from anecdotes reported in the media (e.g. Catholic World News 2011), no detailed or reliable published data on wild bonobo attacks on humans seem to be available.

Orangutans

Orangutan populations are plummeting throughout their remaining distribution in northern Sumatra, Borneo and Malaysia. They are primarily forest dwellers and were traditionally thought to lack the resilience and adaptability to cope with major habitat disturbance. However, recent studies show that orangutans can persist in degraded habitat including logged forest, plantation landscapes and agroforest systems, indicating greater ecological resilience than previously thought (*P. abelii*: Campbell-Smith et al. 2011; *P. pygmaeus*: Meijaard et al. 2010; Ancrenaz et al. 2010; Ancrenaz et al. 2015). Like African apes surviving in human-dominated habitats, orangutans may enter farmland to feed on cultivated foods (Salafsky 1993; Marchal and Hill 2009; Campbell-Smith et al. 2010); in such areas, they likely have frequent interactions with local human communities. Although ex-captive orangutans are known to attack humans, we are unaware of any published accounts detailing an attack by a wild orangutan on a person. However, Yuwono et al. (2007, p. 21) note that 'this species will not attack people unless it is first provoked, or cornered or otherwise feels threatened'. One case of serious aggression by a wild orangutan against a human occurred in the Sekonyer River area of Tanjung Puting in Central Borneo. A local man was hunting deer when his dogs encountered an adult male orangutan on the ground and chased him. When the man also approached, the orangutan attacked him (Galdikas, pers. comm. 2015). Reports suggest that local people who live in proximity to these great apes consider them dangerous and may fear an attack by wild orangutans (Campbell-Smith et al. 2010). Indeed, in an extensive survey of local knowledge about orangutan killings in Kalimantan, 'self-defence' was the most frequently reported reason for orangutan killings in village areas after hunting for food (Meijaard et al. 2011). And in a related survey, 15 % of respondents who had personally killed an orangutan gave 'fear' or 'self-defence' as the reason for killing compared to 8 % who reported killing one because of crop damage (Davis et al. 2013).

Why Do Chimpanzees Feature Most Prominently in Reports of Great Ape Attacks?

Available data suggest that physical attacks on humans by wild great apes in Africa and Asia are a rare occurrence. For example, at both Bossou and Bulindi where chimpanzee–human interactions have been studied in most detail, attacks occurred less than once per year, even though chimpanzees at both sites encounter local people daily (Hockings 2009; McLennan and Hill 2010). Despite the rarity of physical attacks by apes, local communities who share landscapes closely with these animals perceive a threat to their personal safety and often fear them (Madden 2006; Campbell-Smith et al. 2010; McLennan and Hill 2012). Evidently, far more published information is available on attacks by wild chimpanzees and, to a lesser extent, wild gorillas compared to the other great apes, bonobos and orangutans. The lack of published accounts of wild orangutan attacks is notable given that, like chimpanzees, orangutans seem able to hang-on in highly disturbed habitats near people, unless persecuted. This indicates that wild orangutan attacks are exceptionally rare, probably because of the relatively peaceful nature of these apes. We found no published information about attacks by wild bonobos, which have been less-studied compared to other great apes. Research on human–bonobo interactions, for example, in areas where bonobos are known to feed on agricultural crops, would be informative in this regard.

The perpetrators of attacks on humans by wild chimpanzees and gorillas seem to be mature males in most cases. However, attack characteristics differ between these African apes. Reports indicate that the target of gorilla aggression is usually an adult (e.g. a hunter) while victims of chimpanzee attacks are overwhelmingly children. Probably, the chimpanzees' smaller body size makes attacks on adult humans risky; unlike captive apes, wild chimpanzees are presumably unaware of their greater strength relative to adult humans (cf. Kabasawa et al. 2008 who report a fatal attack on a man by escaped sanctuary chimpanzees). Gorilla attacks appear most often motivated by defensive instincts, such as when gorillas are wounded by hunters or otherwise perceive the human(s) as a threat. In chimpanzees, the picture is more mixed, with attacks falling into both provoked (i.e. in response to harassment) and unprovoked (e.g. predation or competition for 'right of way') categories. It has been suggested that some wildlife attacks occur when animals lose their fear of humans (e.g. bears *Ursus* spp.; Conover 2002). The perpetrators of some ape attacks were well habituated for tourism or research (e.g. Bwindi gorillas, Bossou chimpanzees). While habituation might make apes less reticent about directing physical aggression towards humans in some situations, it is important to note that many recorded attacks involved semi-habituated (Bulindi chimpanzees) or entirely unhabituated individuals (e.g. predatory chimpanzee(s) at Kibale).

Both gorillas and chimpanzees can inflict severe injuries on victims of attacks (see Khalil et al. 2011 for an assessment of injuries sustained by an adult woman following an attack by a captive male chimpanzee) (Fig. 18.1). However, unlike attacks by some other large mammals (e.g. big cats), attacks by apes rarely cause fatalities. Among great apes, only chimpanzees prey on human children as food. Some bonobo and orangutan populations hunt mammals, including sympatric primates (Surbeck and Hohmann 2008; Hardus et al. 2012). However, hunting and meat-eating—particularly monkey hunting—is overall more prevalent in chimpanzees (e.g. Watts and Mitani 2002). Chimpanzee predation on children represents the most extreme illustration of great ape aggression towards humans. Even so, unequivocal evidence of chimpanzees targeting children as prey comes from two sites only (Kibale and Gombe); no evidence indicates that attacks at Bossou and

Bulindi, among other sites (e.g. around Budongo) were predatory motivated. Therefore chimpanzees' aggressive behaviour may not necessarily lead to predatory attacks, where they encounter children frequently. Hunting frequencies vary among chimpanzee populations as do the prey species available (Newton-Fisher 2007). Notably, neither Bulindi nor Bossou chimpanzees regularly eat meat; at Bossou diurnal primates are absent due to past human hunting activities (Hockings et al. 2012), whereas at Bulindi suitable prey occur (e.g. black and white colobus monkeys, *Colobus guereza*) but no evidence suggests the chimpanzees eat them (McLennan 2010b and unpublished data). Therefore, the risk of predatory attacks against children might be greatest at human-impacted sites where chimpanzees regularly hunt and consume sympatric monkeys. The suggestion that chimpanzee predatory attacks at Kibale were carried out by a single adult male (Wrangham 2001) raises the intriguing possibility that, like some large carnivores, chimpanzees can become 'man-eaters' in exceptional situations.

Why might chimpanzees show a greater propensity to attack humans than other great apes? Opportunity may be part of the answer. Compared to gorillas, bonobos and orangutans, chimpanzees have a broader geographical distribution and exploit a wide variety of habitat types including dense lowland rainforest, dry savanna woodland, and montane forest. This ecological flexibility may equip them with the resilience to cope with human disturbance better than other great apes (McLennan and Hockings 2014). Even so, most reports of chimpanzee attacks come from areas in East Africa where apes aren't traditionally eaten or parts of West Africa where cultural taboos limit hunting, enabling chimpanzees to live in close proximity to local human communities (Hockings and McLennan 2012) (Fig. 18.2). By contrast,

Fig. 18.2 An adult male chimpanzee at Bossou in Guinea crossing a village homestead having 'raided' a papaya fruit. Such close interactions between chimpanzees and people are not uncommon in areas where the apes are not hunted, and where human settlements and agriculture encroach on chimpanzee habitat (Photo: Kimberley J. Hockings)

bonobos and western gorillas are found in Central Africa where human societies tend to occur at lower densities and are more likely to hunt apes, limiting opportunities for close coexistence. In Sumatra, Borneo and Malaysia large-scale commercial agriculture, particularly of oil palm, is the predominant land-use activity around many orangutan habitats (Nantha and Tisdell 2009; Struebig et al. 2015). Guards and workers in these plantations are often men. In contrast, small-scale agriculture characterises most areas of human–chimpanzee sympatry in Africa. In these landscapes, young children commonly guard crop fields and collect water from forest wells, elevating the risk of aggressive encounters with chimpanzees.

From a comparative behavioural perspective, chimpanzees display more frequent aggression to one another as a normal aspect of social life compared to other great apes (Wrangham and Peterson 1996; Wilson et al. 2014). Thus, they can be considered more aggressive generally. Moreover, chimpanzees kept in captivity can be notoriously dangerous to humans (Kabasawa et al. 2008; Khalil et al. 2011). Wild chimpanzee social systems are characterised by strong male bonds, and aggressive confrontations with local people might be a means for adult males to display boldness and assert dominance in front of other males; at the same time, group males might also provide coalitionary support during aggressive interactions with humans (for an example, see McLennan 2010a). This contrasts with other great apes in which males are either more solitary (orangutans), social groups include a single or small number of adult males (gorillas) or males show comparatively weak bonds (bonobos).

Mitigating Conflicts Caused by Attacks

Understanding Great Ape Attacks in the Context of Human–Wildlife 'Conflict'

Hockings and Humle (2009, p. 1) defined human–great ape conflicts as 'any human–great ape interaction which results in negative effects on human social, economic or cultural life, great ape social, ecological or cultural life or the conservation of great apes and their environment'. Definitions of human–wildlife conflict vary but are often constructed in a way that sets animals in conflict with people (Redpath et al. 2014). Redpath et al. (2014) have questioned whether such definitions encourage the misidentification of the antagonists in conflict situations, thereby reducing the likelihood of finding effective mitigation solutions. While the term 'human–wildlife conflict' has become commonplace in discussions of human–wildlife interactions, it obscures the underlying 'human–human conflicts' that stem from differential goals, perceptions and power relations among the human groups concerned (Madden and McQuinn 2014; Marchini 2014; Hill 2015). By directing focus wholly on the proximate factors of a conflict (e.g. wildlife attacks), we risk ignoring underlying social factors that might be driving it (Dickman 2010; Redpath et al. 2013). For example, Redpath et al. (2014) reviewed recent articles dealing with 'human–wildlife conflicts'. Most animal species involved were predators or large herbivores, mostly of conservation concern, and almost all underlying conflicts were between those who sought to defend conservation objectives and those defending other objectives, mainly livelihoods and human safety.

To avoid misunderstandings in discussions about wildlife attacks, we should differentiate between (1) the direct impacts of attacks on humans, and (2) conflicts among those humans defending the wildlife, including conservation practitioners, and those defending other positions, such as local community members, and be open about the different interests involved in the conflict (Young et al. 2010; Redpath et al. 2014). Gorilla attacks around Bwindi provide an illustrative example. Fear of attack was reported by villagers to impede their activities and movements; for example, some farmers were afraid to work in their gardens and children were afraid to walk to school. Household economics were also affected when individuals injured by gorilla attacks, usually men, were unable to work (Madden 2006). The perceived lack of support from park authorities inflated feelings of ill-will among some community members towards the gorillas and the park generally. As Madden (2006, p. 184) points out, 'the threat of attacks and [crop] raids is also a constant reminder to local people that they lack empowerment under existing government wildlife laws, and that many individuals and families are continually at risk of suffering harm due to gorillas that far outweighs any benefits they may receive from the park's community-oriented revenue sharing programme' (for similar arguments, see also Laudati 2010; Tumusiime and Svarstad 2011). Such issues raise difficult ethical questions regarding promoting human–ape coexistence, where they encounter each other regularly (see McLennan and Hill 2013 for detailed discussions of the ethics of great ape conservation in human-dominated landscapes).

Mitigating Human–Ape Impacts

Mitigating conflicts involving humans and great apes presents challenges because some interest groups seek to conserve great apes while others who are negatively affected by the conflict (e.g. local people) may want them removed or eliminated. Like other large-bodied mammals, due to the size and strong physical presence of apes, fear of physical harm can be a strong driver of conflict (McLennan and Hill 2012; Hockings et al. 2014). Culturally sensitive conservation actions to promote coexistence should therefore aim to foster tolerance and reduce fear, through promoting or developing existing positive attitudes towards apes, while working with people to develop practical measures to reduce encounters and prevent them from escalating into aggressive events (see below). Conflicts can escalate when local people feel that the needs or values of wildlife and/or other human groups, such as the government or tourists, are given priority over their own needs. As illustrated by the Bwindi example above, this is especially applicable when people believe they have little to gain and much to lose by living alongside protected and high-profile wildlife including great apes (McLennan and Hill 2013).

Selective removal of 'problem' apes, i.e. aggressive adult males, as occasionally advocated in conflict situations for other species (e.g. male Asian elephants;

Sukumar 1991) is not appropriate in most situations because of the Endangered or Critically Endangered status of all great apes (IUCN 2014). At Kibale, where the predatory attacks were thought to be committed by a single adult male chimpanzee, removal was considered the best option; however, the male responsible for the final attack was ultimately tracked down and killed by local people before the authorities could intervene (Wrangham 2001). Translocation of problem apes, as has been attempted with mixed success with problem tigers (Goodrich and Miquelle 2005) and leopards *Panthera pardus* (Athreya et al. 2011), is unlikely to be a feasible option for a variety of reasons including the cost and practicalities involved in relocating wild great apes, the difficulty of finding alternative suitable habitats away from humans, and the impact that removal of individuals (e.g. adult males) might have on group social dynamics. Providing financial compensation to victims of ape attacks is likewise complicated. Unless administered carefully, compensation for problematic ape behaviour (including crop damage) can create the perception of 'ownership' of wild apes, for example, by conservation organisations or research teams. Following chimpanzee attacks on children at Bossou, immediate medical expenses were provided to the victim's families as a sign of good will, although it was emphasised that the researchers were not responsible for the chimpanzees or their actions (Hockings et al. 2010).

Land-use changes: The regular cutting back of vegetation along edges of fields, paths and trails frequented by both humans and apes might decrease the incidence of surprise encounters in potentially high-risk areas, thus reducing the likelihood of an animal attacking in response to a perceived human threat. If local people encounter apes most frequently in specific locations to which the apes are drawn, such as a fruit tree, removing or repositioning the attractant outside of the apes' ranging area can reduce the likelihood of encounters. For example, removal of papaya trees in proximity to houses at Bossou successfully reduced chimpanzee forays into people's homesteads, and hence the likelihood that chimpanzees would encounter children (Hockings 2007). Most traditional barriers such as trenches or fences do not deter apes from crossing into agricultural areas or homesteads, and electric fences are expensive and difficult to maintain, and impractical in forest–farm mosaics where crop fields are scattered among forest patches (cf. Honda et al. 2009). Establishment of buffer zones of open land or difficult-to-cross vegetation (e.g. tea) might block or restrict ape travel routes through agricultural land and reduce human and ape encounters (Hockings and Humle 2009). However, where there is limited land for farming and little remaining forest, spare land to create such buffers may be unavailable and apes might have no choice but to cross farmland when travelling among fragmented forest patches (e.g. chimpanzees at Bulindi; McLennan 2010b). Cash crops (e.g. tea, tobacco) and food crops (e.g. vegetables such as potato, onion, cabbage) that are seemingly seldom or never exploited by chimpanzees, including crops they are either unpalatable or inedible when raw (e.g. chilli), could potentially act as economically viable buffers to lessen ape movements into fields and villages (see Hockings and McLennan 2012). Nevertheless, some such crops are associated with increased deforestation of chimpanzee habitats (e.g. tobacco in Uganda;

McLennan 2008). Therefore, the utility of particular crops to help reduce encounters between people and great apes must be carefully balanced against their environmental impact (Hockings and McLennan 2012).

Changing human behaviour: Education programmes should provide informed advice to local people on 'best practice' when encountering great apes to avoid the likelihood of encounters escalating into aggression. In general, people (especially children) should never deliberately provoke apes by throwing sticks or stones, firing gunshots or slingshots, or chasing them with dogs. People often panic upon encountering an ape, and human infants and young children are sometimes left behind by fleeing adults or older children (Hockings et al. 2010; McLennan 2010b), leaving younger children vulnerable to attack. For gorillas, standing up while avoiding eye contact, while holding onto another person, can prevent a charge (Doran-Sheehy et al. 2007). The worst reaction is for a person to run away, as this can provoke an ape to charge, potentially leading to grabbing and biting (Hockings and Humle 2009). Adult male chimpanzees are less likely to confront men than women and children (Wrangham 2001; Hockings et al. 2010; McLennan and Hill 2012), so where possible, adult men should take a lead position when walking through forest or along paths. Children living in proximity to great apes should not be left unsupervised near forest edges, including on agricultural land, and should not venture alone into the forest.

Measures employed by different groups of people to deter apes from entering villages or agricultural areas vary, as do their effectiveness. Around Budongo, guarding of fields, involving regular patrolling of field perimeters by a male guard armed with a stick, was highly effective (albeit time consuming) for deterring chimpanzees (Hill and Wallace 2012). However, other commonly used techniques including shouting, banging objects, throwing stones and using dogs can provoke ape aggression, as noted above—especially if used by children. We acknowledge that these ideals may be hard for some households to put into practice since having children guard crops or fetch water (while adults engage in other tasks) is common in rural Africa.

Concluding Remarks

Physical aggression towards humans by wild great apes in Africa and Asia is overall rare, especially considering that people and apes encounter each other daily at some sites. Compared to other great apes, however, attacks by chimpanzees in areas of high human–ape spatial overlap stand out as a relatively predictable, if infrequent, occurrence. In addition, chimpanzee attacks are more likely to be unprovoked and may involve predation on human children. Unfortunately, detailed information on the circumstances surrounding ape attacks is often lacking. Owing to ongoing human encroachment and modification of great ape habitats, we predict rising levels of contact between humans and great apes in the future. To establish appropriate mitigation strategies that facilitate sustainable human–great ape coexistence, attacks need to be documented thoroughly and objectively. These data should be combined with further research into the relative effectiveness of different mitigation actions, information on the cultural value of great apes to local people and an increased understanding of underlying conflicts among the different human stakeholder groups.

Acknowledgements We are grateful to the anonymous reviewers whose comments helped us to improve the manuscript. During the writing of this chapter, the authors were supported by fellowships from the Leverhulme Trust (to M.M.) and Fundação para a Ciência e a Tecnologia (to K.H.; ref: IF/01128/2014).

References

Abram NK, Meijaard E, Wells JA et al (2015) Mapping perception of species' threats and population trends to inform conservation efforts: the Bornean orangutan case study. Divers Distrib 21:487–499

Agoramoorthy G, Hsu MJ (1999) Rehabilitation and release of chimpanzees on a natural island. J Wildl Rehabil 22:3–7

Ancrenaz M, Ambu L, Sunjoto I et al (2010) Recent surveys in the forests of Ulu Segama Malua, Sabah, Malaysia, show that orang-utans (*P. p. morio*) can be maintained in slightly logged forests. PLoS One 5, e11510

Ancrenaz M, Oram F, Ambu L (2015) Of Pongo, palms and perceptions: a multidisciplinary assessment of Bornean orang-utans *Pongo pygmaeus* in an oil palm context. Oryx 49:465–472

Ando C, Iwata Y, Yamagiwa J (2008) Progress of habituation of western lowland gorillas and their reaction to observers in Moukalaba-Doudou National Park, Gabon. Afr Study Monogr 39:55–69

Athreya V, Odden M, Linnell JD et al (2011) Translocation as a tool for mitigating conflict with leopards in human-dominated landscapes of India. Conserv Biol 25:133–141

Bargali HS, Akhtar N, Chauhan NPS (2005) Characteristics of sloth bear attacks and human casualties in North Bilaspur Forest Division, Chhattisgarh, India. Ursus 16:263–267

Barua M, Bhagwat SA, Jadhav S (2013) The hidden dimensions of human–wildlife conflict: health impacts, opportunity and transaction costs. Biol Conserv 157:309–316

Borner M (1985) The rehabilitated chimpanzees of Rubondo Island. Oryx 19:151–154

Campbell-Smith G, Simanjorang HV, Leader-Williams N et al (2010) Local attitudes and perceptions towards crop-raiding by Sumatran orangutans (*Pongo abelii*) and other non-human primates in Northern Sumatra, Indonesia. Am J Primatol 72:866–876

Campbell-Smith G, Campbell-Smith M, Singleton I et al (2011) Apes in space: saving an imperilled orangutan population in Sumatra. PLoS One 6, e17210

Catholic World News (2011) Protect humans, not just animals, bishop in Congo urges. Catholic World News. http://www.catholicculture.org. Accessed Jan 2015

Conover MR (2002) Resolving human–wildlife conflicts: the science of wildlife damage management. Lewis, Boca Raton

Davis JT, Mengersen K, Abram NK et al (2013) It's not just conflict that motivates killing of orangutans. PLoS One 8, e75373

Dellatore DF (2007) Behavioural health of reintroduced orangutans (*Pongo abelii*) in Bukit Lawang, Sumatra Indonesia. MSc thesis, Oxford Brookes University, Oxford

Dhanwatey HS, Crawford JC, Abade LA et al (2013) Large carnivore attacks on humans in central India: a case study from the Tadoba-Andhari Tiger Reserve. Oryx 47:221–227

Dickman AJ (2010) Complexities of conflict: the importance of considering social factors for effectively resolving human–wildlife conflict. Anim Conserv 13:458–466

Doran-Sheehy DM, Derby AM, Greer D et al (2007) Habituation of western gorillas: the process and factors that influence it. Am J Primatol 69:1354–1369

Dunham KM, Ghiurghi A, Cumbi R et al (2010) Human–wildlife conflict in Mozambique: a national perspective, with emphasis on wildlife attacks on humans. Oryx 44:185–193

Dunnett S, van Orshoven J, Albrecht H (1970) Peaceful co-existence between chimpanzee and man in West Africa. Bijdr Dierkd 40:148–153

Duvall CS (2008) Human settlement ecology and chimpanzee habitat selection in Mali. Landsc Ecol 23:699–716

Etiendem DN, Hens L, Pereboom Z (2011) Traditional knowledge systems and the conservation of Cross River gorillas: a case study of Bechati, Fossimondi, Besali, Cameroon. Ecol Soc 16:22

Giles-Vernick T, Rupp S (2006) Visions of apes, reflections on change: telling tales of great apes in Equatorial Africa. Afr Stud Rev 49:51–73

Goldsmith ML, Glick J, Ngabirano E (2006) Gorillas living on the edge: literally and figuratively. In: Newton-Fisher NE, Notman H, Paterson JD, Reynolds V (eds) Primates of Western Uganda. Springer, New York, pp 405–422

Goodall J (1986) The chimpanzees of Gombe: patterns of behaviour. Harvard University Press, Cambridge

Goodrich JM, Miquelle DG (2005) Translocation of problem Amur tigers *Panthera tigris altaica* to alleviate tiger-human conflicts. Oryx 39:454–457

Grieser-Johns B (1996) Responses of chimpanzees to habituation and tourism in the Kibale Forest, Uganda. Biol Conserv 78:257–262

Gurung B, Smith JLD, McDougal C et al (2008) Factors associated with human-killing tigers in Chitwan National Park, Nepal. Biol Conserv 141:3069–3078

Halloran AR, Cloutier CT, Monde S et al (2014) The Tonkolili Chimpanzee Project in Sierra Leone: implications for chimpanzee conservation strategies in anthropogenic landscapes. Afr Primates 9:15–22

Hardus ME, Lameira AR, Zulfa A et al (2012) Behavioral, ecological, and evolutionary aspects of meat-eating by Sumatran orangutans (*Pongo abelii*). Int J Primatol 33:287–304

Herrero S, Higgins A (2003) Human injuries inflicted by bears in Alberta: 1960–98. Ursus 14:44–54

Hill CM (2004) Farmers' perspectives of conflict at the wildlife–agriculture boundary: Some lessons learned from African subsistence farmers. Hum Dimens Wildl 9:279–286

Hill CM (2015) Perspectives of 'conflict' at the wildlife–agricultural boundary: 10 years on. Hum Dimens Wildl. 20:296–301

Hill CM, Wallace GE (2012) Crop protection and conflict mitigation: reducing the costs of living alongside non-human primates. Biodivers Conserv 21:2569–2587

Hill CM, Webber AD (2010) Perceptions of nonhuman primates in human–wildlife conflict scenarios. Am J Primatol 72:919–924

Hill CM, Osborn FV, Plumptre AJ (2002) Human–wildlife conflict: Identifying the problem and possible solutions. Albertine Rift Technical Report Series, vol 1. Wildlife Conservation Society, New York

Hockings KJ (2007) Human–chimpanzee coexistence at Bossou, The Republic of Guinea: a chimpanzee perspective. PhD thesis, University of Stirling, Stirling

Hockings KJ (2009) Living at the interface: human–chimpanzee competition, coexistence and conflict in Africa. Interact Stud 10:183–205

Hockings KJ, Humle T (2009) Best practice guidelines for the prevention and mitigation of conflict between humans and great apes. IUCN/SSC Primate Specialist Group, Gland

Hockings KJ, McLennan MR (2012) From forest to farm: systematic review of cultivar feeding by chimpanzees—management implications for wildlife in anthropogenic landscapes. PLoS One 7, e33391

Hockings KJ, Sousa C (2013) Human–chimpanzee sympatry and interactions in Cantanhez National Park, Guinea-Bissau: current research and future directions. Primate Conserv 26:57–65

Hockings KJ, Yamakoshi G, Kabasawa A et al (2010) Attacks on local persons by chimpanzees in Bossou, Republic of Guinea: long-term perspectives. Am J Primatol 72:887–896

Hockings KJ, Humle T, Carvalho S et al (2012) Chimpanzee interactions with nonhuman species in an anthropogenic habitat. Behaviour 149:299–324

Hockings KJ, McLennan MR, Hill CM (2014) Fear beyond predators. Science 344:981

Hockings KJ, McLennan MR, Carvalho S et al (2015) Apes in the Anthropocene: flexibility and survival. Trends Ecol Evol 30:215–222

Honda T, Miyagawa Y, Ueda H et al (2009) Effectiveness of newly-designed electric fences in reducing crop damage by medium and large mammals. Mamm Stud 34:13–17

Idani GI, Mwanza N, Ihobe H et al (2008) Changes in the status of bonobos, their habitat, and the situation of humans at Wamba in the Luo Scientific Reserve, Democratic Republic of Congo. In: Furuichi T, Thompson J (eds) The Bonobos: behavior, ecology and conservation. Springer, New York, pp 291–302

Inogwabini BI, Bewa M (2009) Bonobo food items, food availability and bonobo distribution in the Lake Tumba Swampy forests, Democratic Republic of Congo. Open Conserv Biol J 3:1–10

Inskip C, Fahad Z, Tully R et al (2014) Understanding carnivore killing behaviour: exploring the motivations for tiger killing in the Sundarbans, Bangladesh. Biol Conserv 180:42–50

IUCN (2014) IUCN Red list of threatened species. Version 2014.3. www.iucnredlist.org. Accessed Dec 2014

Junker J, Blake S, Boesch C et al (2012) Recent decline in suitable environmental conditions for African great apes. Divers Distrib 18:1077–1091

Kabasawa A, Garriga RM, Amarasekaran B (2008) Human fatality by escaped *Pan troglodytes* in Sierra Leone. Int J Primatol 29:1671–1685

Kaltenborn BRP, Bjerke T, Nyahongo J (2006) Living with problem animals—self-reported fear of potentially dangerous species in the Serengeti Region, Tanzania. Hum Dimens Wildl 11:397–409

Kamenya S (2002) Human baby killed by Gombe chimpanzee. Pan Afr News 9:26

Kerbis-Peterhans JC, Gnoske TP (2001) The science of 'man-eating' among lions *Panthera leo* with a reconstruction of the natural history of the 'man-eaters of Tsavo'. J East Afr Nat Hist 90:1–40

Khalil A, Spiotta AM, Barnett GH (2011) Difficulties with the neurological assessment of humans following a chimpanzee attack: case report. J Neurosurg 115:140–144

Köhler A (2005) Of Apes and men: Baka and Bantu attitudes to wildlife and the making of eco-goodies and baddies. Conserv Soc 3:407–435

Kormos R, Boesch C, Bakarr MI et al (2003) West African chimpanzees: status survey and conservation action plan. IUCN, Gland, www.primate-sg.org/action.plans.htm

Kushnir H, Leitner H, Ikanda D, Packer C (2010) Human and ecological risk factors for unprovoked lion attacks on humans in southeastern Tanzania. Hum Dimens Wildl 15:315–331

Kutsukake N, Matsusaka T (2002) Incident of intense aggression by chimpanzees against an infant from another group in Mahale Mountains National Park, Tanzania. Am J Primatol 58:175–180

Laudati AA (2010) The encroaching forest: struggles over land and resources on the boundary of Bwindi Impenetrable National Park, Uganda. Soc Nat Resour 23:776–789

Leciak E, Hladik A, Hladik CM (2005) Le palmier à huile *(Elaeis guineensis)* et les noyaux de biodiversité des forêts-galeries de Guinée maritime: à propos du commensalisme de l'homme et du chimpanzee. Rev Ecol Terre Vie 60:179–184

Lingomo B, Kimura D (2009) Taboo of eating bonobo among the Bongando people in the Wamba Region, Democratic Republic of Congo. Afr Study Monogr 30:209–225

Linnell JDC, Andersen R, Andersone Z et al (2002) The fear of wolves: a review of wolf attacks on people. NINA Oppdragsmelding 731:65

Löe J, Röskaft E (2004) Large carnivores and human safety: a review. Ambio 33:283–288

Mackenzie CA, Ahabyona P (2012) Elephants in the garden: financial and social costs of crop raiding. Ecol Econ 75:72–82

Madden F (2006) Gorillas in the garden: human–wildlife conflict at Bwindi Impenetrable National Park. Policy Matters 14:180–190

Madden F, McQuinn B (2014) Conservation's blind spot: the case for conflict transformation in wildlife conservation. Biol Conserv 178:97–106

Marchal V, Hill CM (2009) Primate crop-raiding: a study of local perceptions in four villages in North Sumatra, Indonesia. Primate Conserv 24:107–116

Marchini S (2014) Who's in conflict with whom? Human dimensions of the conflicts involving wildlife. In: Verdade LM, Lyra-Jorge MC, Piña CI (eds) Applied ecology and human dimensions in biological conservation. Springer, Berlin, pp 189–209

McLennan MR (2008) Beleaguered chimpanzees in the agricultural district of Hoima, western Uganda. Primate Conserv 23:45–54

McLennan MR (2010a) Case study of an unusual human–chimpanzee conflict at Bulindi, Uganda. Pan Afr News 17:1–4

McLennan MR (2010b) Chimpanzee ecology and interactions with people in an unprotected human-dominated landscape at Bulindi, western Uganda. PhD thesis, Oxford Brookes University, UK

McLennan MR, Hill CM (2010) Chimpanzee responses to researchers in a disturbed forest–farm mosaic at Bulindi, western Uganda. Am J Primatol 72:907–918

McLennan MR, Hill CM (2012) Troublesome neighbours: changing attitudes towards chimpanzees (*Pan troglodytes*) in a human-dominated landscape in Uganda. J Nat Conserv 20:219–227

McLennan MR, Hill CM (2013) Ethical issues in the study and conservation of an African great ape in an unprotected, human-dominated landscape in western Uganda. In: MacClancy J, Fuentes A (eds) Ethics in the field: contemporary challenges. Berghahn, New York, pp 42–66

McLennan MR, Hockings KJ (2014) Wild chimpanzees show group differences in selection of agricultural crops. Sci Rep 4:5956

Meijaard E, Albar G, Rayadin Y et al (2010) Unexpected ecological resilience in Bornean orangutans and implications for pulp and paper plantation management. PLoS One 5, e12813

Meijaard E, Buchori D, Hadiprakarsa Y et al (2011) Quantifying killing of orangutans and human-orangutan conflict in Kalimantan, Indonesia. PLoS One 6, e27491

Mutombo M, Jezek Z, Arita I et al (1983) Human monkeypox transmitted by a chimpanzee in a tropical rain-forest area of Zaire. Lancet 321:735–737

Nantha HS, Tisdell C (2009) The orangutan–oil palm conflict: economic constraints and opportunities for conservation. Biodivers Conserv 18:487–502

Nelleman C, Newton A (2002) Great apes—the road ahead: An analysis of great ape habitat, using GLOBIO methodology. United Nations Environment Programme, Nairobi

Neto MFC, Garrone Neto D, Haddad V Jr (2011) Attacks by Jaguars (*Panthera onca*) on humans in Central Brazil: report of three cases, with observation of a death. Wilderness Environ Med 22:130–135

Newton-Fisher NE (2007) Chimpanzee hunting behaviour. In: Henke W, Rothe H, Tattersall I (eds) Handbook of physical anthropology. Springer, New York, pp 1295–1320

Oishi T (2013) Human–gorilla and gorilla–human: dynamics of human–animal boundaries and interethnic relationships in the central African rainforest. Rev Primatol 5:63

Orams MB (2002) Feeding wildlife as a tourism attraction: a review of issues and impacts. Tour Manage 23:281–293

Packer C, Ikanda D, Kissui B, Kushnir H (2005) Lion attacks on humans in Tanzania. Nature 436:927–928

Quigley H, Herrero S (2005) Characterization and prevention of attacks on humans. In: Woodroffe R, Thirgood S, Rabinowitz A (eds) People and wildlife: conflict or coexistence? Cambridge University Press, Cambridge, pp 27–48

Rainer H, White A, Lanjouw A (eds) (2014) State of the Apes: extractive industries and Ape conservation. Cambridge University Press, Cambridge

Rajpurohit KS, Krausman PR (2000) Human–sloth-bear conflicts in Madhya Pradesh, India. Wildl Soc Bull 28:393–399

Redpath SM, Young J, Evely A et al (2013) Understanding and managing conservation conflicts. Trends Ecol Evol 28:100–109

Redpath SM, Bhati S, Young J (2014) Tilting at wildlife: reconsidering human–wildlife conflict. Oryx 49:222–225

Reynolds V (2005) The chimpanzees of the Budongo Forest. Oxford University Press, Oxford

Reynolds V, Wallis J, Kyamanywa R (2003) Fragments, sugar, and chimpanzees in Masindi District, western Uganda. In: Marsh LK (ed) Primates in fragments. Springer, New York, pp 309–320

Richards P (1995) Local understanding of primates and evolution: some Mende beliefs concerning chimpanzees. In: Corbey R, Theunissen B (eds) Ape, man, apeman: changing views since 1600. Leiden University, Leiden, pp 262–273

Sabater Pi J (1966) Gorilla attacks against humans in Rio Muni, West Africa. J Mammal 47:123–124

Salafsky N (1993) Mammalian use of a buffer zone agroforestry system bordering Gunung Palung National Park, West Kalimantan, Indonesia. Conserv Biol 7:928–933

Sept JM, Brooks GE (1994) Reports of chimpanzee natural history, including tool use, in 16th- and 17th-century Sierra Leone. Int J Primatol 15:867–878

Sicotte P, Uwengeli P (2002) Reflections on the concept of nature and gorillas in Rwanda: implications for conservation. In: Fuentes A, Wolfe L (eds) Primates face to face. Cambridge University Press, Cambridge, pp 163–182

Sousa J, Vicente L, Gippoliti S et al (2014) Local knowledge and perceptions of chimpanzees in Cantanhez National Park, Guinea-Bissau. Am J Primatol 76:122–134

Struebig MJ, Fischer M, Gaveau DL et al (2015) Anticipated climate and land-cover changes reveal refuge areas for Borneo's orang-utans. Glob Chang Biol 21:2891–2904

Sukumar R (1991) The management of large mammals in relation to male strategies and conflict with people. Biol Conserv 55:93–102

Surbeck M, Hohmann G (2008) Primate hunting by bonobos at LuiKotale, Salonga National Park. Curr Biol 18:R906–R907

Thirgood S, Woodroffe R, Rabinowitz A (2005) The impact of human–wildlife conflict on human lives and livelihoods. In: Woodroffe R, Thirgood S, Rabinowitz A (eds) People and wildlife: conflict or coexistence? Cambridge University Press, Cambridge, pp 13–26

Thompson JM, Nestor LM, Kabanda RB (2008) Traditional land-use practices for bonobo conservation. In: Furuichi T, Thompson J (eds) The Bonobos: behavior, ecology and conservation. Springer, New York, pp 227–244

Tumusiime DM, Svarstad H (2011) A local counter-narrative on the conservation of mountain gorillas. Forum Dev Stud 38:239–265

Tutin CEG, Oslisly R (1995) *Homo*, *Pan* and *Gorilla*: co-existence over 60 000 years at Lopé in central Gabon. J Hum Evol 28:597–602

Wang SW, Macdonald DW (2006) Livestock predation by carnivores in Jigme Singye Wangchuck National Park, Bhutan. Biol Conserv 129:558–565

Watts DP, Mitani JC (2002) Hunting behavior of chimpanzees at Ngogo, Kibale National Park, Uganda. Int J Primatol 23:1–28

White L, Edwards A (eds) (2000) Conservation research in the African rain forests: a technical handbook. Wildlife Conservation Society, New York

Wich SA, Gaveau D, Abram N et al (2012) Understanding the impacts of land-use policies on a threatened species: is there a future for the Bornean orang-utan? PLoS One 7, e49142

Wilson ML, Boesch C, Fruth B et al (2014) Lethal aggression in *Pan* is better explained by adaptive strategies than human impacts. Nature 513:414–417

Woodroffe R, Thirgood S, Rabinowitz A (eds) (2005) People and wildlife: conflict or co-existence? Cambridge University Press, Cambridge

Wrangham R (2001) Moral decisions about wild chimpanzees. In: Beck BB, Stoinski TS, Hutchins M et al (eds) Great apes and humans: the ethics of coexistence. Smithsonian Institution Press, Washington, DC, pp 230–244

Wrangham RW, Peterson D (1996) Demonic males: apes and the origins of human violence. Houghton Mifflin Harcourt, Boston

Wrangham RW, Wilson ML, Hare BA et al (2000) Chimpanzee predation and the ecology of microbial exchange. Microb Ecol Health Dis 12:186–188

Yamakoshi G (2005) What is happening on the border between humans and chimpanzees? Wildlife conservation in West African rural landscapes. In: Hiramatsu K (ed) Coexistence with nature in a 'glocalizing' world: field science perspectives. Kyoto University, Kyoto, pp 91–97

Yeager CP (1997) Orangutan rehabilitation in Tanjung Puting National Park, Indonesia. Conserv Biol 11:802–805

Young JC, Marzano M, White RM et al (2010) The emergence of biodiversity conflicts from biodiversity impacts: characteristics and management strategies. Biodivers Conserv 19:3973–3990

Yuwono EH, Susanto P, Saleh C et al (2007) Guidelines for the better management practices on avoidance, mitigation and management of human–orangutan conflict in and around oil palm plantations. WWF–Indonesia, Indonesia

Chapter 19
Promoting a Mobile Data Collection System to Improve HWC Incident Recording: A Simple and Handy Solution for Controlling Problem Animals in Southern Africa

Sébastien Le Bel, David Chavernac, and Fiona Stansfield

A management consultant once said; "If you can't measure something, you can't manage it." (Drucker 2009)

Wildlife is a renewable natural resource with multifaceted values (Chardonnet et al. 2002; Kojwang 2010), but for many local rural Africans its hostility (Rosa and Joubert 2009) still overshadows any expected outcomes from conservation initiatives. The need to address the dysfunction of this wildlife–domestic interaction is highlighted by the escalating problem of poaching and elephant crop raiding that appears to be linked to illegal elephant killing (Hart and Smith 2001).

Mitigation packages are offering some technical solutions applicable to the communities involved, but HWC reporting remains poor, thus undermining a sound understanding of this phenomenon. Given the situation, this chapter explores how new communication technologies, based on the booming mobile phone sector in Sub-Saharan Africa, might best lead to a clearer understanding of HWC and thereby contribute to reducing revenge killings of charismatic species such as the elephant.

Human–Wildlife Conflict, an Escalating Problem

Human–Wildlife Conflict (HWC) can be defined as *Any interaction between humans and wildlife that results in negative impacts on human social, economic or cultural life, on the conservation of wildlife populations or on the environment* (WWF 2005). Inevitably, when wildlife and humans share the same landscape and resources

S. Le Bel (✉) • D. Chavernac • F. Stansfield
ES Department, BSEF Research Unit, CIRAD, 34398 Montpellier, Cedex 5, France

BIOS Department, CMAE Research Unit, Elephant Research and Conservation Unit, CIRAD, Chiredzi, Zimbabwe
e-mail: sebastien.le_bel@cirad.fr; david.chavernac@cirad.fr; fionajane28@gmail.com

F.M. Angelici (ed.), *Problematic Wildlife*, DOI 10.1007/978-3-319-22246-2_19

Fig. 19.1 An elephant bull busy foraging in the town of Victoria Falls © PPP

(Distefano 2004), HWC cannot be reduced to the biological control of overabundant populations or to management of a human–wildlife interface without taking into account its social impact; in other words, the capacity of a community to support and locally manage a certain level of conflict (Woodroffe et al. 2005).

HWC is not restricted to a particular geographical region or climate situation and is a common phenomenon in urban areas (Breitenmoser et al. 2005; McGinnis 2008) even in southern Africa, as illustrated by this picture of an elephant bull roaming in the town of Victoria Falls in Zimbabwe (Fig. 19.1). Hot spots still remain around conservation areas; the underlying causes across regions have been attributed to land-use changes and high human population growth (Woodroffe et al. 2005; Araman 2009), which is expected to double in Africa from 0.8 to 1.8 billion people in the next 40 years (ILRI 2009). Africans will be packed more tightly into cities and their crops will increasingly impinge upon territories populated by wildlife, thus giving rise to more frequent and severe incidences of HWC (Jeke 2014).

A political issue: The social impact of HWC has become a major political factor; this issue is regularly raised by communities when the governmental authorities venture into the field (Lamarque 2010), especially when the situation is exacerbated by the media reporting on the negative perceptions of the general public regarding those species that cause the most conflict.

The statement "*Next time elephants will be the ones voting for you*" is a classic warning to local leaders when the local people consider that the pressure of wildlife (elephant) conflicts is not being properly addressed; HWC has often been a manifestation of underlying human–human conflict (Dickman 2010). On a regional scale, HWC has become such a concern that its importance was openly recognized in February 2010 by the Southern African Development Community (SADC)

Fig. 19.2 A few examples of the direct costs of sharing space and resources with wildlife: a maize field raided by baboons, a granary that has been damaged by an elephant, and a farmer injured by a leopard © PPP

Technical Committee on Wildlife, which designated HWC as one of the main problems for Africa's rural populations in terms of personal security and economic loss (LeBel et al. 2010).

Losses and fears: For local rural communities, human–wildlife interaction means losses and fears, with disruptions to livelihoods and food insecurity, which are undermining conservation and mitigation strategies (Barua et al. 2013). The impact of HWC can be summarised as follows according to the two categories of costs (Goredema 2009; WWF 2005).

The obvious costs of HWC are the direct ones (Fig. 19.2), with damage to food crops being grown for on-farm consumption (e.g. maize, millet, fruit, and vegetables), damage to cash crops causing a reduction in household income (e.g. cotton, tobacco, fruits, and vegetables), and damage to water storage, water reticulation, and water pumping equipment (reservoirs, pipes, and pumps), and food stores (grain and other dried products). In Namibia, for example, a rough estimation of the combined costs of HWC to communal area farmers is about USD 1 million annually, with the drop in net income for crop enterprises varying from 28 to 202 % depending on the distance of the crop enterprise from a wildlife habitat (WWF 2008). In line with this damage, other adverse effects of HWC include livestock loss or injury, the transmission of diseases from wildlife to livestock, and competition between livestock and wildlife for grazing and water. At national level, losses are barely significant, but for the individual stock owner they can be catastrophic and can make a difference between economic independence and poverty.

The indirect costs of HWC are much more difficult to estimate, but have a significant impact on the willingness to accept the presence of wildlife (Decker and Purdy 1988). It includes the constant fear of the potential harm caused by wild animals, the need to guard property to prevent damage, restrictions on people's movements at night, and access to water and non-timber forest products (WWF 2005). Living with wildlife also has other costs, such as the increased risk of contracting diseases such as malaria during night-time guarding.

Conservation challenges: Human–wildlife interactions have been detrimental to wild mammals; many species have been reduced in numbers due to hunting, pastoralism, habitat modification, disease control, fence mending, or problem animal control (Happold 1995; Taylor and Martin 1987). With the restricted use of lethal control measures and the success of conservation programmes, southern Africa is facing a paradoxical situation of an overabundance of wildlife and thus increased occurrence of HWC. For emblematic species such as elephants, conflicts with neighbouring human communities in southern Africa have become a major constraint for the sustainable management and conservation of free-ranging populations (Hoare 2001; Lee and Graham 2006). With the increase in elephant populations in southern Africa of 5 % per annum (Cumming and Jones 2005) and the expansion of human settlements into wildlife areas, local communities living on marginal land adjacent to protected areas are faced with an increasing occurrence of human–elephant conflict (HEC) (Nelson et al. 2003). For example, between 2002 and 2006, more than 5000 cases of HEC were recorded in Zimbabwe, which resulted in the killing of 774 elephants during subsequent problem animal control operations (Campfire 2007).

The Challenge of Monitoring HWC

Preventing and mitigating HWCs are top conservation priorities, particularly in areas where wildlife and people cohabit. Modern approaches seek to reduce the magnitude of HWC and deal with problem animals that cause conflicts using lethal and non-lethal measures, while increasing the level of tolerance in the affected human populations (Sillero-Zubiri et al. 2006). Mitigation measures and HWC management strategies (WWF 2005; FAO 2009) offer a range of solutions based on gaining greater insight into the behaviour of problem animals, testing various deterrent measures and adopting policies to reduce the extent of conflicts. The reduction in HWC is also expected to change human negative perceptions of wildlife.

HWC has been a complex problem and a combination of approaches is required to manage conflicts, including wildlife barriers, property protection, traditional methods, and the removal of specific problem animals (Nelson et al. 2003; WWF 2005). For any HWC management strategy to succeed, it must be sustainable and should therefore be administered by the local community (WWF 2005). To achieve this, mitigation measures were recently compiled for southern Africa as a set of handy solutions in an HWC tool box developed by FAO (LeBel et al. 2010; Czudek and LeBel 2011). For ease of use, they were classified according to the five types of conflict that local communities are facing: human threat, crop raiding, infrastructure damage, water competition, and livestock threat. Depending on the expected outcomes, they can also be grouped into four categories: awareness measures, preventing access, chasing away, and removing problem animals.

Reporting and assessing the local impact of HWC is part of this mitigation package, with the development of a decision-support system and forms to be filled (WWF 2005), with a specific design for HEC (Hoare 2002; Osborn and Parker 2002). The objective is to provide information for land-use and development planning, to assist in developing appropriate HWC management strategies, and to adapt strategies and actions over time as data indicates what works and why.

In practice, the lack of communication and trust between wildlife authorities and people concerned by HWC makes the effectiveness of the reporting poor. It was perceived as an ineffectual exercise in Zimbabwe because of the intervention delay (LeBel et al. 2011). In Mozambique, HWC cases are under-reported despite the loss of human lives and the number of revenge killings (240 people and 304 wild animals killed between 1997 and 2004) (Anderson and Pariela 2005). The same was reported in Ghana where only a few records were available due to decreased reporting by farmers rather than low incidents over those years, with most incidents being related to elephant damage (Haricha et al. 2013). On the other hand, in Zambia over-reporting of HWC has been observed, especially when it concerns large or mega herbivores; the lethal control of those problem animals becomes a handy source of free meat for the local communities (Chomba et al. 2012). This last example illustrates the situation where authorities have chosen not to understand the problem, but rather to remove it using lethal methods, even if this shortcut does not overcome the problem (Hoare 2001). The case study of the reporting system in Mozambique and in Zimbabwe will give an idea of how complicated the system can be when the question arises as to what information needs to be recorded and to whom it needs to be sent (LeBel 2011b, c).

Understanding the flow of information: The reporting of an HWC incident can follow a complex route, as the information has to be provided to different categories of people who need either to be informed or to provide a sound response to the incident. In Zimbabwe, reporting follows the administrative layers of the Rural District Council (Fig. 19.3), whereas in Mozambique co-management of the wildlife resource by two different directorates (DNTF and DNAC[1]) adds a further layer of complexity.

Choosing explanatory variables: Accurate and consistent data are critical for good decision-making regarding HWC management. A workable balance therefore has

[1]DNAC: National Directorate of Conservation Areas (Direcção Nacional das Áreas de Conservação). DNTF: National Directorate of Lands and Forests (Direcção Nacional de Terras e Florestas).

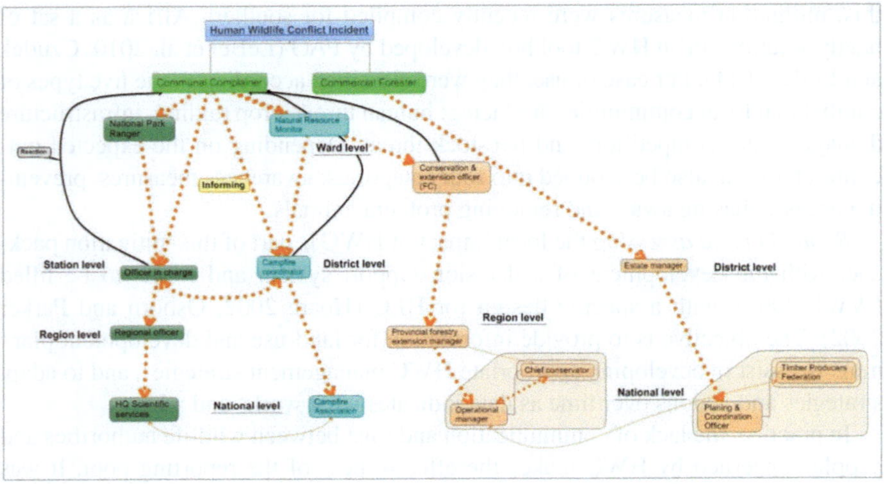

Fig. 19.3 Flow of information generated after an HWC incident in Zimbabwe on communal lands and in a forestry concession

to be found between the amount of data that decision makers need for assessing the situation, and the local capacity for recording the events. If too much information is to be recorded, this may result in producing HWC forms that are partially or wrongly filled out, such as the problem animal control form used in Zimbabwe. If too little information is requested, there will be insufficient data to correctly analyse the situation, which was the situation encountered in Mozambique (LeBel 2011a). In an attempt to address this problem, the "5W's and an H" method (Apte et al. 2001) was chosen to guide us in the process of developing data trees of the key information needed to understand an HWC problem. More than 30 variables (grouped as shown in Table 19.1) were selected by local stakeholders; this increased the challenge of rapid on-site analysis and reporting to improve speed and the capacity to respond. Although game guards have a sound knowledge of the landscapes in which they work and of the wildlife living alongside them, they cannot be expected to provide the same kind of data as researchers.

Starting with MOMS (Management Oriented Monitoring System), a user-friendly management tool developed in Namibia for the new-born Conservancies in the 2000s (Diggle 2006), could be a first step to improving the recording of HWC events. The advantages of MOMS are multiple: no 'data mining' process (i.e. spending a lot of time collecting data for others); developing a sense of 'ownership' of the data and motivating game scouts; no sophisticated or expensive equipment required; little technical input required. The Event Book system utilized for HWC monitoring allows immediate visualization of the consequences of certain management decisions, which makes them an excellent tool for conflict mitigation involving stakeholders and decision-makers. This was introduced in Zimbabwe (Taylor 2010) and in Mozambique to monitor illegal offtake and natural resource management (Costa 2007).

Table 19.1 HWC data tree guidelines

"5W+H"	Questions concerning the HWC incident	Information recorded
When	When did it happen?	Date of the incident
		Time of day
Where	Where did it happen?	Location (type of habitat)
		Coordinates (GPS or reference grid)
		Designated HWC hotspot or not
Who	Who does it concern?	ID of the complainant and of the affected people
		ID of the animal involved (species, age, sex, and behaviour)
What and Why	What was the impact? Why did it happen?	Description of the incident: human casualty, crop damage, equipment or infrastructure destruction, livestock predation, etc.
How	How was it mitigated?	Action taken by the complainant and the wildlife authority to prevent HWC and to control problem animals

MOMS and its Event Book system is considered to be the most robust and systematic system for monitoring HWC incidents. However, the absence of spatial perception of the impact of HWC and the need for rapid on-site reporting led us to seek some solutions to upgrade this recording system with a real-time communication system.

Investing in Mobile Data Collection Systems

The mobile phone is a basic piece of equipment owned by just about everyone in today's fast-moving world. Since 2000, this booming sector has grown by 44 % in terms of the number of connections in Sub-Saharan Africa (SSA), the fastest-growing mobile market in the world, as compared to an average 34 % for developing regions and 10 % for developed regions overall. There are currently more than 454 million connections in SSA and this trend is expected to continue growing from 60 % in 2012 to reach 75 % of the population and 700 million connections in 2016 (Vital Wawe Consulting 2009).

Mobile communication systems are opening up new areas for improvement based on an assessment of the impact of real-time communication systems. Applications are multiple, making it possible to bypass poor infrastructure and connect remote areas directly with the tools and services that enrich our lives; for the underserved population, text message-based (SMS) services appear to be an essential tool for start-up (Rao 2012). The UN Foundation has categorized the opportunities for mobile technology in six types: education and awareness, remote data collection, remote monitoring, communication and training, disease and epidemic outbreak tracking, and diagnosis and treatment support (Vital Wawe Consulting 2009). These systems have a greater ability to influence behaviour than radio and television and have proven particularly effective in targeting hard-to-reach populations and rural areas (Vital Wawe Consulting 2009).

The following example provides an idea of what are now the integral 'touch points' in most Africans' lives. In less than a decade, almost a hundred SMS applications have been developed for disease control interventions (Déglise et al. 2012), disease surveillance (Robertson and Nelson 2010), mobile money platforms, or market information for small farmers (Rao 2012). In the banking sector, the leading cell phone company in Kenya (Safaricom) launched M-PESA, a popular SMS-based money transfer system that allows individuals to deposit, send, and withdraw funds using their cell phone (Jack and Suri 2010). In the agriculture sector, mobile phones also have a substantial impact by improving pricing, reducing wastage, and increasing efficiency (Deloitte 2012). For the health sector, from simple reminders for vaccinations to grassroots information gathering on diseases, mobile phones are becoming a key cornerstone of health programmes in a growing number of African countries, offering an innovative and potentially effective means of creating a real-time reporting system (Vital Wawe Consulting 2009; Rosewell et al. 2013).

Conservation projects have much to gain from engaging with mobile phone technology (Graham et al. 2011). A first step could be payment via mobile phone, offering a direct and immediate way of rewarding wildlife informers for good intelligence services.

Wildlife applications: Mobile phone services are developing critical roles in the arena of wildlife management and conservation; the number of wildlife applications available for smartphones has increased tremendously in the last few years, with dozens of applications to find parks and trails, for wildlife sightings, wildlife references, bird-specific applications, and plant identification (SPNL 2014; Brigida 2011; White and White 2011). Some were developed to improve the rescue of injured or orphaned wildlife in Australia with the Wildlife Rescue App (IFAW 2015) or by offering a new service provider such as the Wildlife Mobile in the UK where 10 % of the cost of each call goes towards saving the natural world (WWF 2014). Specific applications for illegal trade and the sighting of poaching evidence are supported by the UN (Cressa and Zommers 2014) such as 'Wildlife Witness', a TRAFFIC application to encourage the public to report illegal activity involving wildlife (TRAFFIC 2014), or the 'apeAPP' mobile phone application launched by the Great Apes Survival Partnership (GRASP) in 2014 to halt the illegal trade in great apes (GRASP 2014).

Mobile phones were recently tested for HEC mitigation in the vicinity of Laikipia, Kenya (Graham et al. 2011), alerting farmers or rangers if elephants were about to cross the boundary of the protected areas for crop raiding (BBC 2005; Turrettini 2008). In India, an SMS service informs people about elephant movements and operates red-flashing LED lights in strategic locations when elephants are around (Saju 2012). The absence of mobile phone applications for HWC mitigation raises the question of selecting the most appropriate technology for a real-time monitoring scheme with the capacity to inform decision-makers and improve the understanding of conflicts.

Investing in Mobile Data Collection Systems (MDCS) offers a further step with the processing of georeferenced information in a timely manner. The NOMAD (HumanitariaN Operations Mobile Acquisition of Data) project classified MDCS mobile information technologies in humanitarian contexts according to three important considerations for the usability of these tools: active development and user communities, the technical complexity and ease of setup, and the appropriateness for rapid assessments (Chris Jung 2011). This places a heavy burden on software used for spatio-temporal surveillance (Robertson and Nelson 2010) if the overall objective is to monitor wildlife populations, maintain them at adequate levels, and restore natural habitats and the balance between predator and prey species.

To explore the feasibility of HWC monitoring, a series of tests were conducted in southern and central Africa (LeBel et al. 2014a, b). We started with FrontlineSMS (http://www.frontlinesms.com), a free open source software package used by a variety of organizations to distribute and collect information via text messages (Banks 2007a, b). The software does not require an internet connection and can work with only a GSM (Global System for Mobile Communications) or a GRPS (General Packet Radio Service) modem and a computer (Gow and Waidyanatha 2010). It can send and receive text messages, manage group contacts, respond to messages, and trigger other events. At each demonstration site, a series of keywords were created to log into the new local system for training purposes and to record HWC incidents. Three types of automatic functions were developed: an auto reply message directed to the message sender, message transfer to decision-makers, and the uploading of a database (Fig. 19.4).

Producing an HWC SMS involves typing a keyword with a space after it, then copying the value of each variable (numbers and abbreviation) captured on the form. It functions as a handy way of alerting decision-makers; however, the integrity of the message transferred to the database appears to be more challenging in the

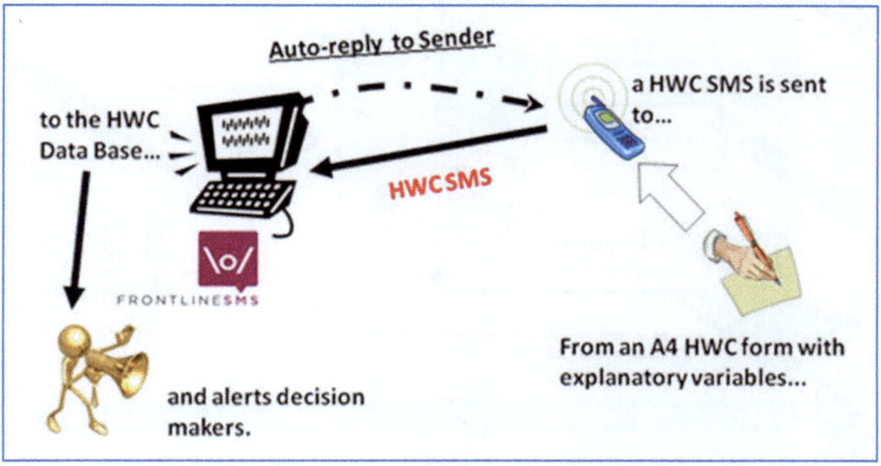

Fig. 19.4 Collecting, managing, and transferring HWC information with FrontlineSMS

long run. From a demo site in central Africa, more than a hundred SMSs were sent per month, but only 47 % were valid, the remainder being partially transmitted or without content (LeBel et al. 2014c). Besides the technical problems, we struggled to build a strong appropriation of this communication system; this highlights the importance, when implementing any plan, of considering the time needed to garner support for novel surveillance methods among users and stakeholders (Robertson et al. 2010). Three improvements could make the use of this MDCS more user-friendly: the use of a GPRS modem supporting a more complete set of AT commands, and with a higher data transmission speed, facilitating the typing of text messages with the use of templates pre-stored in the informants' mobile phones, and linking the use of FrontlineSMS with a reward system to improve informants' livelihoods. To cope with the need for georeferenced information and for sharing wildlife-based information in a timely manner (Fig. 19.5), we moved from solutions based on SMS services for open-source Android applications. Such applications, which were tested during the development of health information systems, combine the attributes of availability, standardization, timeliness, user-friendliness, reliability, security, and shareability (Chavernac et al. 2015) and have the ability to increase the effectiveness of a national HWC mitigating strategy.

Current HWC mitigation monitoring is undertaken with KoBoCollect, an application from the KoBoToolbox. KoBoToolbox (2013), developed by the Harvard Humanitarian Initiative, is an open source of tools for data collection and analysis

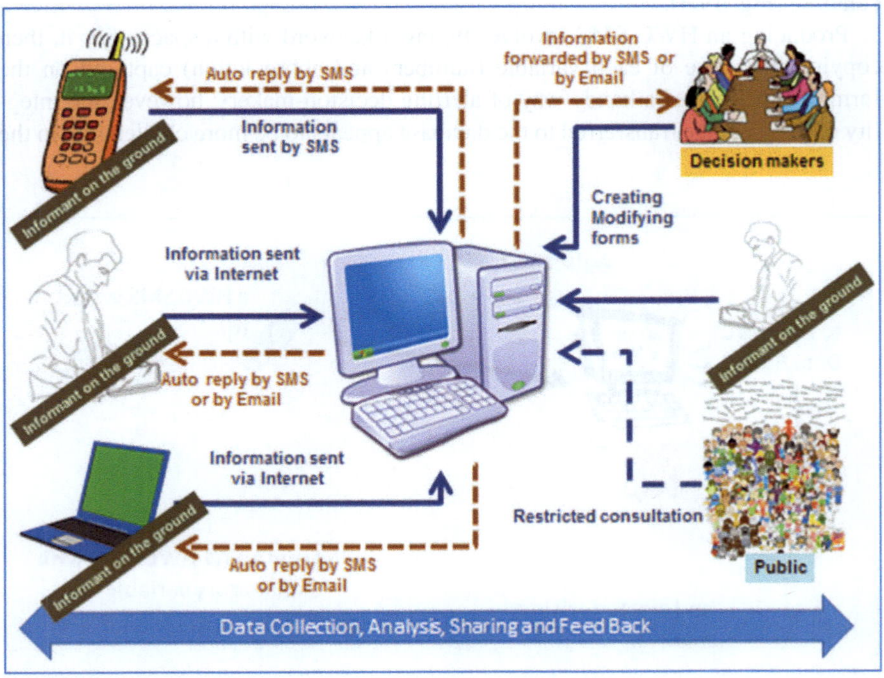

Fig. 19.5 Conceptual design of a real-time wildlife-based information system

in humanitarian emergencies (Kreutzer 2014). It is based on the OpenDataKit and is an Android application for smartphones, laptops, or any other device. With this application, you can collect data on- and off-line that can be synchronized into a database. Use of this tool requires the following three steps: (1) building a form, (2) downloading using a mobile device and collecting data, (3) analysing and managing data.

Building an HWC form: The form builder for KoBoToolkox was developed to be simple and easy for anyone to use. It does not require any technical knowledge or advanced training and forms can be developed on-screen or imported from Excel. For the development of an HWC form, apart from the general information of the location and date of the event, we propose the provision of six categories of information (Fig. 19.6): the wildlife species involved in the conflict, the victim or informer reporting the incident, the type and extent of crop damage, the type and extent of livestock predation, the type and extent of infrastructure damage, and the control and mitigation measures taken by the community or by the wildlife authority.

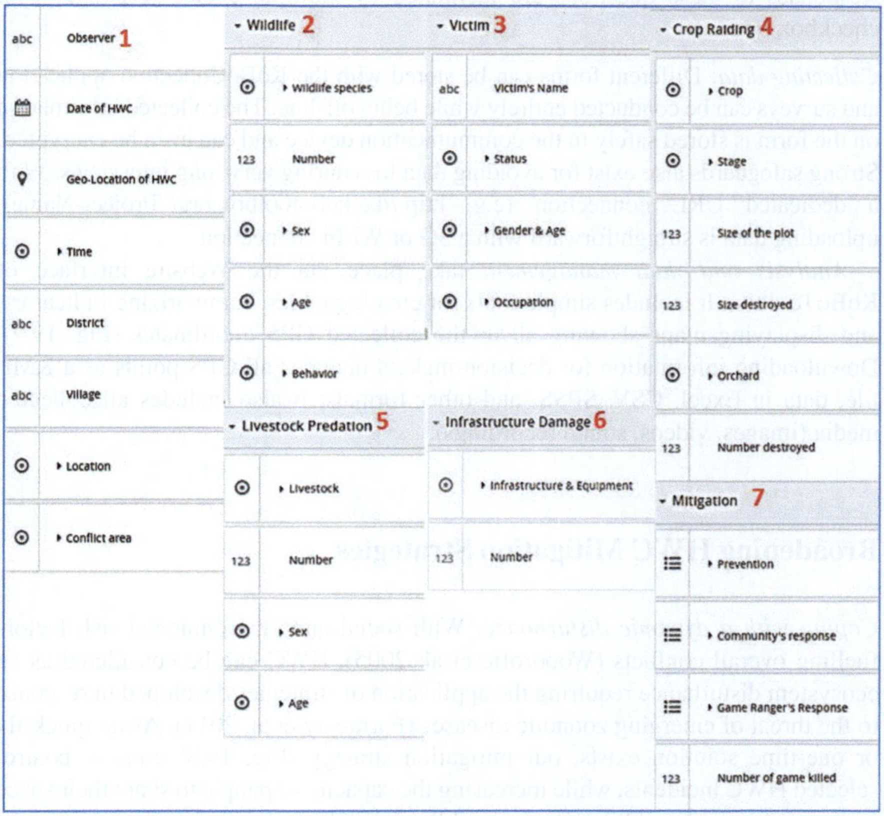

Fig. 19.6 Headings of the HWC form related to (1) the observer, (2) the wildlife, (3) the victim, (4) crop destruction, (5) livestock predation, (6) infrastructure destruction, and (7) mitigation measures

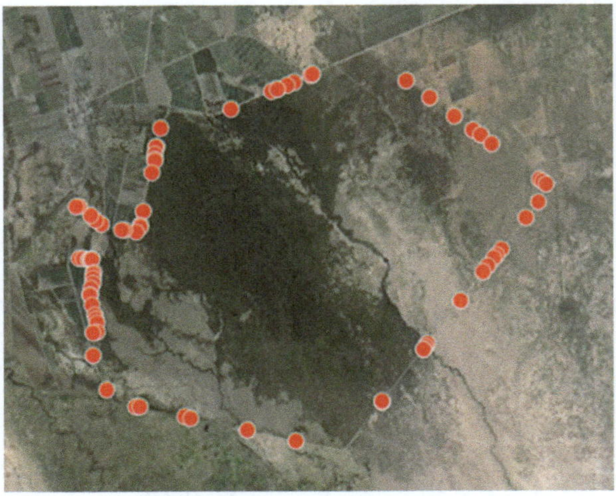

Fig. 19.7 Display of the breakages of a game fence by wildlife collected using KoBoCollect in Tanzania

Responses to these questions are facilitated by multiple choice responses with checkbox options.

Collecting data: Different forms can be stored with the KoBoCollection application and surveys can be conducted entirely while being off-line. The collected information on the form is stored safely in the communication device and can even be encrypted. Strong safeguards also exist for avoiding data loss during very long interviews. With a dedicated URL connection (e.g. http://kc.kobotoolbox.org/"Project-Name), uploading data is straightforward with a 3G or Wi-Fi connection.

Analysis and data management take place via the Website interface of KoBoToolbox. It includes simple tools for creating tables, summarizing indicators, and displaying maps showing all of the collected GPS coordinates (Fig. 19.7). Downloading information for decision-makers includes all GPS points as a KML file, data in Excel, CSV, SPSS, and other formats; it also includes all collected media (images, videos, sound recordings).

Broadening HWC Mitigation Strategies

Coping with a dynamic disturbance: With social and environmental risk factors fuelling overall conflicts (Woodroffe et al. 2005), HWC can be considered as an ecosystem disturbance requiring the application of strategies developed in response to the threat of emerging zoonotic diseases (Formenty et al. 2011). As no quick-fix or one-time solution exists, our mitigation strategy (Fig. 19.8) aims to control selected HWC incidents, while increasing the capacity of people to share their space and resources with wildlife (Decker and Purdy 1988).

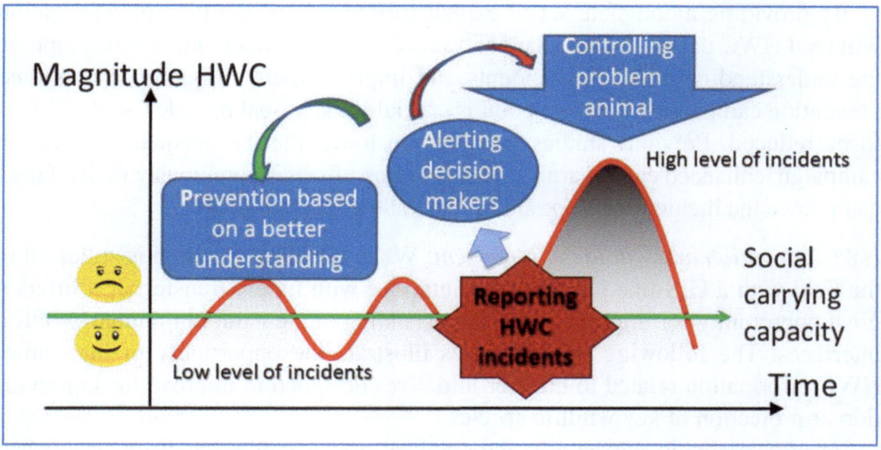

Fig. 19.8 Mitigation strategic model whereby HWC is treated as an environmental disease

Using information technology provided by MDCS, a suggested coping strategy should provide a complete set of prevention measures, an early warning mechanism alerting decision-makers, and the application of control solutions for problem animals, with a view to learning from past experience (LeBel and Czudek 2011).

The cornerstone of this strategy is the early detection of HWC incidents, which presupposes the deployment and training of informers. While reporting can be improved with proper communication technology (Android application combined with ad-hoc MDCS), the success of such strategies will depend on the motivation of informers. An initial option could be to reward them for good intelligence services using their mobile phone to credit airtime. Another type of incentive could be to offer real-time and quick assistance in the event of, particularly, human casualties or major crop destruction/livestock predation. Such a set-up would be more than relevant if it was linked to a compensation or insurance scheme for animal-induced damage (WWF 2008).

Early HWC detection has the capacity to alert decision-makers in good time, thereby allowing a quick and adequate reaction to control dangerous animals. Bringing wildlife authorities to the front line of HWC mitigation presupposes that they are properly equipped and trained not only to react quickly, but also to assess from a distance the urgency of the situation, based on the "Eisenhower Decision Matrix" (Krogerus and Tschäppeler 2008). Another difficulty wildlife authorities will be facing is the choice of adequate mitigation measures on arrival at the site where they will be urged to use the handy shortcut of lethal control of the so-called problem animal. Not only does this not solve the problem (Hoare 2001), even if it does lessen the HWC stress for the community (Chomba et al. 2012), such an option will quickly pave the way for uncontrolled bushmeat production or create chaotic situations such as the trading or removal of problem animals as part of a sport-hunting activity. Moreover, the success of such an approach, particularly for senior management, will be to accept the transparency that comes with real-time information.

By providing a complete set of explanatory variables, the flow of information will feed HWC databases. Regular analysis of that information will help to improve the understanding of HWC incidents and improve their mitigation by targeted prevention campaigns. This last point is crucial if the global prevalence of HWC is to be reduced. Previous studies have demonstrated that an adequate prevention campaign, enhanced early warning, and guarding efforts on previously raided farms can reduce the incident of crop raiding by 90 % (Sitatai et al. 2005).

HWC mitigation and wildlife enforcement: When HWC information is collected in the field with a GIS-like incident map interface with timely transferral, it offers a great opportunity for improving the understanding of the nature of human–wildlife interfaces. The following two examples illustrate the opportunity of integrating HWC information related to broader initiatives designed to improve the conservation or protection of key wildlife species.

For a charismatic species like the elephant, a correlation has been established between the level of conflict and illegal killing. HEC has become a priority objective of elephant management at many sites for CITES (2010). Some elephant populations have been negatively impacted by the scale of removal of problem animals. Proper recording of information related to elephant conflicts could be integrated in the MIKE (Monitoring of Illegal Killing of Elephants) programme to improve understanding of the causes of the problem and of illegal elephant killing (Hart and Smith 2001).

Another option worth exploring might be to link up with SMART, which stands for Spatial Monitoring and Reporting Tool (Bhammar and Wooten 2014). Based on an open-source software program SMART was created by, and for, the conservation community in order to engage them on the frontline of fighting wildlife crime in Africa and elsewhere (SMART 2014). It is a step-up from MIST (Management Information System), which has fewer data conglomeration capabilities, but is still used for field work in some parks.

References

Anderson J, Pariela F (2005) Strategies to mitigate human-wildlife conflicts in Mozambique. Wildlife Management Working Paper No. 8. FAO, Rome

Apte PR, Shah H, Mann D (2001) "5W's and an H" of TRIZ Innovation. TRIZ Journal (September 2001)

Araman A (2009) Human-elephant conflict in northern Mozambique. Paper presented at the Human Wildlife Conflict Training of Trainers Workshop, 9–11 Dec 2009, Harare

Banks K (2007a) And then came the Nigerian elections: The story of FrontlineSMS. Stanford J African Studies 4

Banks K (2007b) FrontlineSMS—a text messaging system for NGOs. Oryx 40(1):17

Barua M, Bhagwat SA, Jadhav S (2013) The hidden dimensions of human–wildlife conflict: health impacts, opportunity and transaction costs. Biol Conserv 157:309–316

BBC (2005) Texting to save Kenyan elephants. http://news.bbc.co.uk/cbbcnews/hi/newsid_4340000/newsid_4347300/4347363.stm

Bhammar H, Wooten N (2014) 20 new and powerful conservation tools. Yale School of Forestry & Environmental Studies Blog. http://environment.yale.edu/blog/2014/12/20-new-and-powerful-conservation-tools/20. New and Powerful Conservation Tools. Accessed 9 Feb 2015

Breitenmoser U, Angst C, Landry J-M, Breitenmoser-Würsten C, Linnell JDC, Weber JM (2005) Non-lethal techniques for reducing depredation. In: Rosie Woodroffe STAR (ed) People and wildlife. Conflict or coexistence? vol 9, Conservation biology. Cambridge University Press, Cambridge, pp 49–71

Brigida D (2011) 25+ Nature and Wildlife Mobile Apps. Wildlife Promise. http://blog.nwf.org/2011/08/the-best-wildlife-and-nature-iphone-and-android-apps/. Accessed 12 Feb 2015

Campfire (2007) PAC statistics 2002-2006. Campfire Association, Harare

Chardonnet P, des Clers B, Fischer J, Gerhold R, Jori F, Lamarque F (2002) The value of wildlife. Revscitech Off Int Epiz 21(1):15–51

Chavernac D, Hendrikx P, LeBel S, Lancelot R (2015) Les nouvelles technologies de l'information appliquées à la vigilance en santé animale. Bulletin épidémiologique, santé animale et alimentation (No. 66/Spécial vigilance vis-à-vis des maladies exotiques):51–54

Chomba C, Senzota R, Chabwela H, Mwitwa J, Nyirenda V (2012) Patterns of human-wildlife conflicts in Zambia, causes, consequences and management responses. J Ecol Nat Environ 4(12):303–313

CITES (2010) African elephant action plan. Fifteenth meeting of the conference of the parties Doha (Qatar), 13–25 March 2010

Costa A (2007) Management oriented monitoring system in Bazaruto National Park. Results from 2006. Marine programme. WWF

Cressa D, Zommers Z (2014) Emerging Technologies: Smarter ways to fight wildlife crime. UNEP Global Environmental Alert Service. http://na.unep.net/geas/archive/pdfs/GEAS_Jun2014_EmergingTechnologies_illegalwildlife.pdf. Accessed 12 Feb 2015

Cumming D, Jones B (2005) Elephants in Southern Africa: Management issues and options. WWF/SARPO Occasional Paper

Czudek R, LeBel S (2011) FAO/BIO-HUB Human-Wildlife Conflict Mitigation Toolkit: helping people to prevent and mitigate conflict with wild animals. Paper presented at the Southern African Wildlife Management Association Symposium, Hartenbos Holidays Resort, Southern Cape, 18–21 September 2011

Decker DJ, Purdy KG (1988) Toward a concept of wildlife acceptance capacity in wildlife management. Wildl Soc Bull 16(1):53–57

Déglise C, Suggs LS, Odermatt P (2012) SMS for disease control in developing countries: a systematic review of mobile health application. J Telemed Telecare 18:273–281

Deloitte LLP (2012) Sub-Saharan Africa mobile observatory 2012. GSM, London

Dickman AJ (2010) Complexities of conflict: the importance of considering social factors for effectively resolving human–wildlife conflict. Animal Conservation (13): 458–466. doi:10.1111/j.1469-1795.2010.00368.x

Diggle R (2006) Feasibility for a community-based wildlife monitoring system for the Srepok wilderness area, Cambodia. Report on the feasibility for a community-based wildlife monitoring system for the Srepok wilderness area. WWF, IIED, London

Distefano E (2004) Human-wildlife conflict worldwide: collection of case studies, analysis of management strategies and good practices. FAO, Rome

Drucker P (2009) "If you can't measure it, you can't manage it." MarketCulture Blog, vol 2015

FAO (2009) Human-wildlife conflict in Africa: causes, consequences and management strategies, vol 157. Forestry paper. FAO, Rome

Formenty P, Odugleh-Kolev A, Grein T, Abela-Rider B, Merianos A, Drager-Dayal R, Drury P, Eremin S, Pessoa-Da-Silva C, Slattery R, Ryan M (2011) From forecasting to control of emerging infectious diseases of zoonotic origin: linking animal and human health systems. Paper presented at the Animal Health and Biodiversity—Preparing for the Future Compendium of the OIE Global Conference on Wildlife 23–25 February 2011, Paris

Goredema L (2009) Relevance of HWC to CBNRM. WWF, Regional CBNRM Project

Gow GA, Waidyanatha N (2010) Using mobile phones in a real-time biosurveillance program: Lessons from the frontlines in Sri Lanka and India. In: IEEE International Symposium on Technology and Society (ISTAS'10), Wollongong, NSW, 16 Dec 2009

Graham MD, Adams WM, Kahiro GN (2011) Mobile phone communication in effective human-elephant conflict management in Laikipia, Kenya. Oryx 46(1):137–144

GRASP (2014) Mobile conservation on the go! (GRASP). Great Apes Survival Partnership. http://www.apeapp.org/. Accessed 12 Feb 2015
Happold DCD (1995) The interactions between humans and mammals in Africa in relation to conservation: a review. Biodivers Conserv 4:395–414
Haricha FK, Treydtea AC, Sauerborna J, Owusu EH (2013) People and wildlife: conflicts arising around the Bia conservation area in Ghana. J Nat Conserv 21(5):342–349
Hart JA, Smith KH (2001) Monitoring Illegal Killing of Elephants (MIKE) Central African Pilot Project. Technical report no 3 monitoring of elephant poaching, anti-poaching effort, and law enforcement in central Africa. CITES
Hoare RE (2001) Management implications of new research on problem elephants. Pachyderm 30:44–48
Hoare RE (2002) Towards a standardized data collection protocol for human-elephant conflict situation in Africa. In: Human-wildlife conflict: identifying the problem and possible solutions. Albertine Rift Technical Report Series, vol 1. WCS
IFAW (2015) Help rescue wildlife in NSW: download our app. http://www.ifaw.org/australia/our-work/animal-rescue/help-rescue-wildlife-nsw-download-our-app. Accessed 12 Feb 2015
ILRI (2009) Climate livestock and poverty—challenges at the interface. International Livestock Research Institute, Nairobi
Jack W, Suri T (2010) The Economics of M-PESA. August 2010 edn. MIT, Cambridge
Jeke L (2014) Human-wildlife coexistence in Omay communal land, Nyaminyami Rural District Council in Zimbabwe. Mediterr J Soc Sci 5 (20):809, 818
Jung C (2011) Mobile data collection systems: a review of the current state of the field. CNES, NOMAD: HumanitariaN Operations Mobile Acquisition of Data: 25
Kojwang HO (2010) Estimating and reflecting the economic values of natural resources in national accounts: a focus on non-consumptive use of wildlife, including examples from the forest sector. FAO & CIC, Budapest
Kreutzer T (2014) KoBoToolbox used by OCHA in Central African Republic. OCHA. http://www.kobotoolbox.org/blog/kobotoolbox-used-ocha-central-african-republic. Accessed 15 Feb 2015
Krogerus M, Tschäppeler R (2008) The decision book: 50 models for strategic thinking. Kein & Aber, Zurich
Lamarque F (2010) Memorandum of the consultancy mission carried out in Maputo, Mozambique by Francois Lamarque. Programm Officer—IGF Foundation, IGF/FAO
LeBel S (2011a) International consultant on Human-Wildlife Conflict mitigation and HWC database development—Mission report. FAO—Projects TCP/MOZ/3301 and TCP/ZIM/3301, Maputo, 14–23 March 2011
LeBel S (2011b) Mission Report In Mozambique. Project Number: TCP/MOZ/3301. International consultant on Human-Wildlife Conflict mitigation and HWC database development. FAO/DNTF, Maputo
LeBel S (2011c) Mission report in Zimbabwe (TCP/ZIM/3301) International consultant on Human-Wildlife Conflict mitigation and HWC database development. Preliminary Report Chiredzi, Harare, Hwange, Mbire. FAO, Harare, June 2011
LeBel S, Czudek R (2011) Promoting SMS to manage Human-Wildlife Conflicts, a first tentative to improve the control and understanding of such a problem in Southern Africa. Paper presented at the Southern African Wildlife Management Association Symposium, Hartenbos Holidays Resort, Southern Cape, 18–21 September 2011,
LeBel S, Mapuvire G, Czudek R (2010) Human-Wildlife Conflict Toolkit: comprehensive solutions for farmers and communities. Unasylva 236(61):12–13
LeBel S, Murwira A, Mukamuri B, Czudek R, Taylor R, LaGrange M (2011) Human Wildlife Conflicts in Southern Africa: Riding the Whirl Wind in Mozambique and in Zimbabwe. The importance of biological interactions in the study of biodiversity. InTech, Rijeka
LeBel S, Chavernac D, Czudek R, Mapuvire G (2014a) Data mobile data collection system as an option to foster grass root wildlife management: The case study of Human-Wildlife Conflict Mitigation In Africa. Paper presented at the 6th IUCN World Parks Congress, Sydney, 12–19 November 2014
LeBel S, Chavernac D, Mapuvire G, Cornu G (2014b) FrontlinesSMS as an early warning network for human-wildlife mitigation: Lessons learned from tests conducted in Mozambique and Zimbabwe. EJISDC 60 (6):1–13

LeBel S, Daniel C, Gaidet N, Vliet Nv, Nguinguiri JC (2014c) Addressing information gaps on hunting sustainability in Central Africa: opportunities and limits of innovative technologies. In: 51st Annual meeting of the association for tropical biology and conservation, Cairns, 20–24 July 2014

Lee PC, Graham MD (2006) African elephants Loxodonta africana and human–elephant interactions: implications for conservation. International Zoo Yearbook 40:9–19

McGinnis P (2008) Background on Urban Deer Management. Urban Deer Management Plan, Grand Haven

Nelson A, Bidwell P, Sillero-Zubiri C (2003) A review of human-elephant conflict management strategies. Wildlife Conservation Research Unit, Oxford University, People and Wildlife Initiative

Osborn FV, Parker GE (2002) Living with elephants II: A manual for implementing an integrated programme to reduce crop loss to elephants and to improve livelihood security of small-scale farmers. www.elephantpepper.org/downloads/manual/%202.2.pdf Mid Zambezi Elephant Project, Harare

Rao M (2012) Mobile Africa report 2012. Sustainable innovation ecosystems. MobileMonday

Robertson C, Nelson TA (2010) Review of software for space-time disease surveillance. Int J Health Geogr 9(16):8

Robertson C, Sawford K, Daniel SLA, Nelson TA, Stephen C (2010) Mobile phone–based infectious disease surveillance system, Sri Lanka. Emerg Infect Dis 16(10):1524–1531

Rosa P, Joubert P (2009) Wildlife exploitation in Sub-Saharan Africa: an entrepreneurship analysis. Paper presented at the AGSE 2009

Rosewell A, Ropa B, Randall H, Dagina R, Hurim S, Bieb S, Datta S, Ramamurthy S, Mola G, Zwi AB, Ray P, MacIntyre R (2013) Mobile phone-based syndromic surveillance system, Papua New Guinea. Emerg Infect Dis 19(11):1811–1818

Saju MT (2012) How SMS saves humans from elephants in the tea meadows of Valparai. The Time of India. http://timesofindia.indiatimes.com/city/coimbatore/How-SMS-saves-humans-from--elephants-in-the-tea-meadows-of-Valparai/articleshow/16921832.cms. Accessed 12 Feb 2015

Sillero-Zubiri C, Sukumar R, Treves A (2006) Living with wildlife: the roots of conflict and the solutions. In: Macdonald D, Service K (eds) Key topics in conservation biology. Wiley-Blackwell, Oxford, pp 255–272

Sitatai NW, Walpole MJ, Leader-Williams N (2005) Factors affecting susceptibility of farms to crop raiding by African elephants: using a predictive model to mitigate conflict. J Appl Ecol 42:1175–1182

SPNL (2014) The Top 9 Wildlife Apps For Your Mobile Phone. BirdLife International. http://www.spnl.org/the-top-9-wildlife-apps-for-your-mobile-phone/. Accessed 12 Feb 2015

Taylor RD (2010) Save Valley Conservancy—MOMS user's manual. Module1: anti-poaching. Patrol report summary and illegal activity report. Version 1 SVC

Taylor RD, Martin RB (1987) Effects of veterinary fences on wildlife conservation in Zimbabwe. Department of National Parks and Wild Life Management, Harare

TRAFFIC (2014) New app to build awareness and information on illegal wildlife trade in South-East Asia. Traffic. http://www.traffic.org/home/2014/4/9/new-app-to-build-awareness-and-information-on-illegal-wildli.html. Accessed 12 Feb 2015

Turrettini E (2008) Text messages save elephants' lives, villagers' crops. Textually.org. http://www.textually.org/textually/archives/2008/10/021426.htm. Accessed 12 Feb 2015

Vital Wawe Consulting (2009) mHealth for development: the opportunity of mobile technology for healthcare in the developing world. UN Foundation-Vodafone Foundation Partnership, Washington, DC

White J, White N (2011) Tracking the wild. http://www.trackingthewild.com/. Accessed 12 Feb 2015

Woodroffe R, Thirgood S, Rabinowitz A (eds) (2005) People and wildlife: conflicts or coexistence? vol 9, Conservation Biology. Cambridge University Press, Cambridge

WWF (2005) Human wildlife conflict manual. WWF-SARPO, Harare

WWF (2008) Common ground—reducing human wildlife conflict. Species Programme, Gland

WWF (2014) Wildlife mobile. http://www.wildlifemobile.co.uk/our-vision. Accessed 12 Feb 2015

Part VII
Some Special Cases: Wind Farms and Fauna, Bird-Strikes, Electrocution, etc.

This part explores some issues which have been around for years and are becoming more numerous and frequent, i.e. the impact on fauna (birds and bats in these cases) of fixed or mobile structures built by man, or bird strikes. This phenomenon includes cases of collisions between birds and airplanes and other aircrafts (helicopters, etc.), collisions between birds or bats and the turbine blades on wind farms, and birds that perch on high-tension electric lines, towers, solar panels and other similar structures (Manville 2005).

Collisions between birds and aircraft can not only cause bird death but also cause serious damage to the aircraft, and in some cases, even fatal accidents for humans (Thorpe 2012).

There are four chapters in this part. The first (Manville 2016) looks at the situation of endangered American migratory birds and bats (Microchiroptera), the numbers of which are decreasing due to collisions at tall structures such as communication towers, structural lines, panels and on-shore wind farms. The chapter emphasises not only the immediate lethal effects but also the negative impact (radiation, etc.) and long-term effects on populations and habitats. There is also a review of some of the most modern and effective measures being taken to mitigate these negative effects.

The second chapter (Thorpe 2016) is a comprehensive review, spanning more than a century, of bird strikes (strictly speaking), i.e. collisions between birds and aircrafts, military aircrafts, helicopters and other flying vehicles. The approach of the chapter is intentionally technical, including considerations and tips to mitigate the damage frequently reported by aircrafts and to avoid causing human deaths.

The third chapter (McKee et al. 2016) is also an exhaustive discussion of collisions between birds and aircrafts. First, it provides a historical introduction to the phenomenon; however, it then focuses on an approach to containing the phenomenon, which should include the adoption of various measures, both at airports and in the air, as well as the structural strengthening of the aircrafts themselves.

The last chapter addresses the impact land-based wind farms may have on birds and bats (Zwart et al. 2016). Although wind energy is a clean energy source that has been rapidly developing, especially over the last 20 years and is now widespread throughout the world (Musgrove 2010), the chapter underlines that knowledge of the long-term, direct impact on birds and bats is still considered insufficient (see Saidur et al. 2011). Finally, many possible solutions are highlighted that can minimise the impact of wind power facilities on wildlife, habitat and the landscape.

References

Manville AM II (2005) Bird strikes and electrocutions at power lines, communication towers, and wind turbines: state of the art and state of the science—next steps toward mitigation. In: Ralph CJ, Rich TD (eds) Bird conservation implementation in the Americas: Proc. 3rd International partners flight conference, USDA Forest Service Gen Tech Rept PSW-GTR-191, pp 1051–1064

Manville AM II (2016) Impacts to birds and bats due to collisions and electrocutions from some tall structures in the United States—wires, towers, turbines, and solar arrays: state of the art in addressing the problems. In: Angelici FM (ed) Problematic wildlife—a cross-disciplinary approach. Springer, New York, p 415–442

McKee J, Shaw P, Dekker A, Patrick K (2016) Approaches to wildlife management in aviation. In: Angelici FM (ed) Problematic wildlife—a cross-disciplinary approach. Springer, New York, p 465–488

Musgrove P (2010) Wind power. Cambridge University Press, Cambridge

Saidur R, Rahim RA, Islam MR, Solangi KH (2011) Environmental impact of wind energy. Renew Sustain Energy Rev 15:2423–2430

Thorpe J (2012) 100 Years of fatalities and destroyed civil aircraft due to bird strikes. World Birdstrike Association, Amsterdam. http://worldbirdstrike.com/Stavanger/100%20years%20of%20fatalities%20and%20destroyed%20civil%20aircfaft%20due%20to%20bird%20strikes%20Paper.pdf

Thorpe J (2016) Conflict of wings—birds versus aircraft. In: Angelici FM (ed) Problematic wildlife—a cross-disciplinary approach. Springer, New York, p 443–463

Zwart MC, McKenzie AJ, Minderman J, Whittingham MJ (2016) Conflicts between birds and onshore wind farms. In: Angelici FM (ed) Problematic wildlife—a cross-disciplinary approach. Springer, New York, p 489–504

Chapter 20
Impacts to Birds and Bats Due to Collisions and Electrocutions from Some Tall Structures in the United States: Wires, Towers, Turbines, and Solar Arrays—State of the Art in Addressing the Problems

Albert M. Manville II

Introduction

Air and airspace as habitats are relatively new concepts (Kunz et al. 2008; Diehl 2013) for many individuals, academics, scientists, and agencies, including federal agencies such as the U.S. Fish and Wildlife Service (hereafter FWS); action agencies that implement FWS guidelines, rules and regulations such as the Bureau of Land Management and the U.S. Forest Service; and state agencies. Tall structures such as communication towers, power transmission lines, commercial wind turbines, solar power towers, and buildings extend into the airspace, in some cases to great heights (e.g., 229 m above ground level [AGL; 750 ft] for some wind turbine rotor swept areas, 610 m AGL [2000 ft] for some digital television (DTV) communication towers, and 442 m AGL [1451 ft] for Chicago's Willis high-rise tower). These tall structures can have deleterious direct effects and impacts to flying wildlife, not to mention indirect effects caused by air and facility disturbance from infrasound noise and lighting, barriers, and fragmented habitats. The overall goal for developers of tall structures and the agencies that regulate them should be to do no harm to protected wildlife species and minimize impacts to their habitats such as the U.S. Interior Department's "smart from the start" initiative (2011 doi.gov) for renewable energy development calling for minimal impacts from development. Attention is focused here toward that overall goal. Several industries whose efforts

A.M. Manville II (✉)
Advanced Academic Programs, Krieger School of Arts and Sciences, Johns Hopkins University, Washington, DC, USA

Wildlife and Habitat Conservation Solutions LLC, Falls Church, VA, 22043, USA

Division of Migratory Bird Management, U.S. Fish and Wildlife Service, Arlington, VA, USA
e-mail: whcsllc006@verizon.net

F.M. Angelici (ed.), *Problematic Wildlife*, DOI 10.1007/978-3-319-22246-2_20

have recently been implemented to minimize harm to birds and to a lesser extent to bats are also assessed. These include the electric utility and the communication tower industries. Several other industries that could significantly reduce harm and impact to both bird and bat species and their habitats are discussed, but the majority of companies are not doing so, in major part based on the assessment of this author due to lack of regulations. These include the commercial, land-based wind industry in the U.S. and the industrial solar energy industry, currently in the Southwest U.S.

Status of and Impacts to Avifauna and Bats in North America

Avian Status and Legal Protections

Migratory birds—i.e., by federal legislative definition those that migrate across U.S., Canadian and/or Mexican borders, of which 1027 species are currently protected in the United States (50 Code of Federal Regulations [C.F.R.] 10.13), are a public trust resource, meaning they belong to everyone. Almost all North American continental birds are protected by the Migratory Bird Treaty Act of 1918, as amended (MBTA;16. U.S.C. 703 et seq.), which implements and regulates bilateral protocols with Canada, Mexico, Japan, and Russia. The Act is a strict liability statute; proof of criminal intent in the injury or killing of birds is not required by authorities for cases to be made.

The Statute and its regulations protect migratory birds, their parts, eggs, feathers, and nests from un-permitted "take" (migratory bird nests are protected during the breeding season while eagle nests are protected year-round), although efforts are currently underway by FWS to develop a permit where "take" could be allowed under MBTA. A Federal permit is required to possess a migratory bird and its parts, and the MBTA currently provides no provision for the accidental or incidental "take" (causing injury or death) of a protected migratory bird, even when otherwise normal, legal business practices or personal activities are involved. The U.S. Congress noted the "take" of even one protected migratory bird to be a violation of the Statute, with fines and criminal penalties that can be extensive. For example, Moon Lake Electric Cooperative was fined $100,000 (U.S.) in 1999 for electrocuting migratory birds; and PacifiCorp was fined $10,500,000 (U.S.) for electrocuting birds in 2009 (the final 2014 settlement agreement included $400,000 (U.S.) in fines, $200,000 restitution to the State of Wyoming, and $1,900,000 to the National Fish and Wildlife Foundation for eagle conservation). A Duke Energy Wind Facility was fined $1,000,000 (U.S.) in 2013 for killing protected birds in wind turbine blade collisions. All the cases involved several years probation for the company executives and all required significant improvements and upgrades to facilities. Companies can also be fined under the criminal misdemeanor provisions of MBTA which can occur when steps to avoid or minimize "take" are not implemented and "take" subsequently results. This occurs after field staff and agents from the FWS's Office of

Law Enforcement have advised a proponent of concerns and suggested measures to avoid or minimize “take” and such recommendations have been ignored or only minimally implemented. It is important to note that the vast majority of “take” by industry goes un-investigated let alone unenforced due to lack of funding, staff, and other priorities.

Bald (*Haliaeetus leucocephalus*) and Golden Eagles (*Aquila chrysaetos*) are also protected by the Bald and Golden Eagle Protection Act (BGEPA; 50 C.F.R. 22.3, 22.26 and 22.27). “Take” under BGEPA is more expansive than under MBTA and includes pursuit, shooting, poisoning, capturing, killing, trapping, collecting, molesting, and disturbing both species (50 C.F.R. 22.3). Permits are required for disturbance take and take resulting in mortality (50 C.F.R. 22.26), and for take of nests (50 C.F.R. 22.27).

The overall objective of the FWS is to maintain bird populations at stable or increasing numbers. This is a daunting challenge due to the direct and indirect impacts of all of the structural issues discussed in this chapter, plus many others briefly mentioned below. As a result, there are growing numbers of Birds of Conservation Concern (BCCs; USFWS 2008)—species in decline but not yet ready for federal listing as threatened or endangered. Currently, there are 273 species and subspecies on the national BCC, Service Regional BCC and Bird Conservation Region BCC lists (USFWS 2008), providing an early warning of likely peril unless the population trends are reversed. These BCC lists require periodic reviews and updates under provisions of the Fish and Wildlife Conservation Act (16 U.S.C. 2901–2912).

Federally listed bird species are those designated and protected under the Endangered Species Act (ESA; 7 U.S.C. 136, 16 U.S.C. 1531 et seq.). Listed species include 78 endangered and 15 threatened bird species on the List of Threatened and Endangered Species. An endangered species faces a significant risk of extinction in the near, foreseeable future throughout all or a significant portion of its range. A threatened species is at risk of becoming endangered in the near future. Collectively, BCC and ESA-listed birds represent at least 366 bird species (36 %) in decline, some seriously, with numbers of both listed and BCC species growing (Manville 2013a). Additionally, the FWS is also tasked to maintain stable or increasing breeding populations of Bald and Golden Eagles under implementing regulations of BGEPA and compliance with the National Environmental Protection Act (NEPA, 42 U.S.C. 4321 et seq.).

Birds are critically important to us all. Birds provide key ecosystem services that fuel a multi-billion dollar (U.S.) industry through pollination, insect, and weed-seed control efforts in the agribusiness and forest products industries. Without migratory birds, there would be untold additional problems requiring more pesticide, herbicide, and other chemical use. Feeding, photographing, and watching migratory birds also fuel a $32 billion/year (U.S.) recreation industry in the U.S., representing an estimated 20 % of the U.S. adult population involved in these endeavors. It is asserted that more adults in the U.S. feed, photograph, and watch birds than play golf (Carter 2013; MountainNature.com 2015).

A number of migratory bird species—notably Bald and Golden Eagles, Common Ravens (*Corvus corax*), American Crows (*C. brachyrhynchos*), hawks, falcons, doves, owls, and hummingbirds—are revered by and protected by Tribal law of some Native American Tribes and Canadian First Nations Peoples. Some of these very species are also at considerable risk from habitat disturbance, habitat fragmentation, injury, and death from land-based wind turbine blade collisions (Erickson et al. 2014), communication tower and guy wire collisions (Gehring et al. 2009), and heating/array impacts with solar facilities (Kagan et al. 2013).

Problems and Challenges for Migratory Birds

In an attempt to roughly assess the annual status of breeding bird populations in North America, several FWS biologists estimated a minimum of ten billion breeding landbirds in the United States exclusive of Alaska and Hawaii, and a minimum fall population of 20 billion migratory birds in North America north of Mexico based on Breeding Bird Survey data (Manville 2005, citing Aldrich et al. 1975; Banks 1979; J. Trapp 2001 pers. comm.). It is difficult to reliably quantify the total annual spring and fall breeding landbird populations in North America. The number of imperiled/declining North American birds continues to increase, the number of imperiled populations continues to grow continent-wide, and the numbers of birds on bird conservation, species of concern, watch lists, state-endangered, and federal-endangered species lists are growing in North America—in some cases at troubling, rapidly declining population rates (Manville 2013a).

The large, estimated annual loss of birds is due to a number of factors. Natural mortality can decimate some bird populations (e.g., starvation, disease, predation, parasitism, stress, nutrient deficiencies, and accidents), recognizing that some of these factors can also be human-related. Additionally, the direct and indirect impacts from humans are extensive. According to the theory, natural mortality tends to decrease to compensate for reduced density, but when mortality such as from structures exceeds a threshold, it can become additive to natural mortality, becoming exploitive (Allen et al. 2006). The mortality factors related to our human footprint include collisions with structures (e.g., building windows, power lines, communication towers and guy wires, wind turbine blades, solar power towers and mirrors, monuments, and bridges)—several of which are discussed in this chapter. Birds are also killed or injured by domestic and feral cats, illegal shootings, collisions with vehicles and aircraft, poisoning from pesticides and contaminants, drowning in oil and wastewater pits, impacts from oil and chemical spills, electrocutions at power line infrastructure, entanglement and drowning in fishing gear, drowning in stock tanks, "take" from hunting and crippling loss (i.e., birds injured but not killed by licensed hunters which subsequently die), poaching, poisoning from lead and other metals, direct loss of breeding habitat, and documented impacts to birds from climate change, among others (Manville 2013a, b). Individually and collectively, these impacts may become additive and all should be assessed cumulatively.

Frequently, proponents from one industry sector, concerned citizens, politicians, and conservationists supporting a specific type of industry will compare estimated levels of mortality from their sector of industry to another. For example, building windows are estimated to kill upwards of 1 billion birds/year in the U.S. (Klem and Saenger 2013; Loss et al. 2013b)—probably the greatest single source of structurally caused bird mortality in the U.S. Compare this to the estimated impacts to birds from power line collisions in the U.S., which may number from 8 to 57 million bird deaths annually based on sensitivity analysis and a meta-review of studies (Loss et al. 2014). Electrocutions, meanwhile, may kill from 0.9 to 11.6 million birds annually in the U.S. (Loss et al. 2014). However, collisions with communication towers may "take" *only* 6.8 million birds/year in North America, most of which are in the U.S. (Longcore et al. 2012). Proponents of the communication tower and cellular telephone industries will frequently make these comparisons to favor their own sector from further scrutiny as does the wind generation industry.

A recent estimate by Loss et al. (2013a) suggests a median estimate of 2.4 billion birds killed annually in the U.S. by domestic and feral cats—the largest projected source of human-related mortality to birds yet published in North America. Using this estimate for comparison is misleading since cats tend to concentrate on smaller birds. By comparing mortality from cats to the most recent estimates of mortality caused by commercial land-based wind turbines, the wind energy estimates are several orders of magnitude smaller, resulting in what might at face value be interpreted as insignificant. For several reasons, this comparison is very misleading. Some birds may have evolved adaptions to cat predation (e.g., sparrows and starlings), but behaviors for avoiding rotating blades and structures that appear as water have not evolved (USFWS 2015 pers. comm). Mortality must be cumulatively assessed for all known and projected causes, including for wind generation. Arguing that wind-generation-caused bird mortality is small by comparison may fail to include it among cumulative effects. Some bird species are more vulnerable to "take" which was acknowledged by Erickson et al. (2014) when concerns were raised about the mortality to 13 species of BCC (USFWS 2008) by the wind industry based on available data.

Collisions with land-based, wind energy turbine blades were recently estimated to kill 440,000 birds/year based on a 2008 estimate of some 22,000 operating turbines (Manville 2009) and have more recently been estimated to kill 573,000 birds/year in the U.S., of which an estimated 83,000 are raptors, based on a 2012 estimate of some 34,400 operating monopole and lattice-constructed turbines (Smallwood 2013). Loss et al. (2013c) attempted to estimate bird mortality at monopole-constructed turbines in the U.S., projecting an average of 234,000 bird deaths/year. Erickson et al. (2014) conservatively estimated annual bird mortality in the U.S. and Canada at 368,000 for all bird species killed. In the opinion of this author and some FWS biologists, field staff, wind energy leads, and law enforcement agents (FWS 2014 and 2015 pers. comm., FWS 2014 confidential internal memos), there continues to be a problem with the transparency, reliability, consistency, and rigor of many of the reports evaluated and subsequent mortality estimates published. These concerns are discussed beyond. Loss et al. (2013c) acknowledged the need for the public release of industry reports and a further evaluation of risk to birds before proceeding with a widespread shift to taller and larger turbines. Those recommendations are essentially being ignored. However, as wind generation grows exponentially, impacts to birds and bats are elevated. As of December 31, 2014, 65,879 megawatts (MW) of installed capacity (more than 48,000 utility scale turbines) were operating in the U.S. (DOE WINDExchange 2015, American Wind Energy Association 2015).

From the perspective of commercial, land-based wind energy, there is yet another problem with these mortality comparisons. The relatively low level of estimated wind energy mortality does not account for the current disproportionate take of Golden Eagles (GOEAs) by wind turbines in the Western U.S. Of approximately 67–75 GOEAs killed/year at Altamont Pass Wind Resource Area, California (Smallwood 2013), there are additional records of more than 79 GOEAs and six Bald Eagles (BAEAs) that have been documented killed in the West at other commercial wind energy facilities from 1997 to 2012 (Pagel et al. 2013), contrary to assertions by some wind energy proponents that eagle mortality is only a problem at Altamont Pass, California. These figures represent a substantial underestimation of the number of GOEAs killed at wind facilities in the Western U.S. (Pagel et al. 2013) since records continue to be collected by FWS staff detailing more eagle mortalities (FWS 2014 and 2015 confidential unpublished data). The Pagel et al. (2013) discoveries were not based on any systematic mortality or monitoring surveys. The growing "take" of eagles and the effects to eagle territories and eagle use areas are growing concerns as more wind facilities are built and become operational. Additionally, there is a growing—but still low—level of take of BAEAs nationwide at wind energy facilities, but more records exist of eagle fatalities from both species at wind energy facilities which have not been released by wildlife agencies since the publication of Pagel et al. (2013; FWS 2015 pers. comm., FWS 2014 and 2015 confidential unpublished data).

There is also a disproportionately large but still poorly substantiated level of take of passerines at wind facilities nationwide (Smallwood 2013; Erickson et al. 2014). A proportion of the migratory birds killed at wind facilities which are Birds of Conservation Concern (BCCs; USFWS 2008) continues to grow (Manville 2009, 2013a; Erickson et al. 2014). These BCC species are already in decline and in some cases in significant peril, but not yet listed under the Endangered Species Act. The current status of BCC species is a growing concern and not easily rectified by lack of federal and state agency resources to address these issues. Yet proponents of the wind generation industry will frequently cite other larger estimated sources of mortality to estimated mortality from wind turbines (AWEA 2015) rather than focusing on addressing the problems of wind turbines indiscriminately killing multiple bird species.

The bottom line, when trying to understand the dynamics of bird (and for that matter bat) populations, all impacts of tall structures and alternate energy sources should be assessed through cumulative effects analyses under the National Environmental Policy Act (NEPA). However, not all projects (i.e., from single turbines to large wind facilities) require NEPA review unless proponents want and apply for a BGEPA or ESA "take" permit, are located on public/federal property, or are receiving federal funding (Manville 2013a). Performing a NEPA review can be challenging, especially given data gaps, unknowns, and uncertainties. However, cumulative effects analysis can best be performed by coordination between the project proponent's consultant and the FWS NEPA specialist/coordinator for the FWS Region where the project is being proposed. This will help determine the need for a NEPA Environmental Assessment, an Environmental Impact Statement, or possible categorical exclusion.

In addition to the impacts from causes due to natural mortality, additive mortality, or a continuum between compensatory mortality and additivity (Peron 2013), project proponents should also include cumulative impacts from cats, windows, power lines, wind turbines, solar facilities, lighting, communication towers, and all other anthropogenic structures including bridges and airports. The impacts should be assessed over the lifetime of all the structures and other impact sources. Additionally, the growing effects of climate change should be incorporated in any cumulative effects analysis (Manville 2013a).

The situation makes for a complicated review with many dynamics involved in assessing the status of bird and other populations. The good news: as scientifically validated, peer-reviewed, and published best-management practices, best available technologies, proven conservation measures, and other tools become publicly available, they should be systematically and consistently implemented. This approach makes the best conservation sense, provides the most bang for the buck, and may help reverse declining populations trends.

Status and Impacts to Bats in North America

Among some of the most maligned yet important animals in the world, insectivorous bats (Microchiroptera) play critical roles and provide key ecosystem services to humanity. Unfortunately, the roles bats play are hugely misunderstood by the public. In the U.S., bats alone save billions of dollars each year by protecting the forest products and agricultural industries. The estimated savings range from \$4 billion–\$53 billion/year (U.S. dollars, averaging \$22.9 billion; Boyles et al. 2011). For example, a single big brown bat (*Eptesicus fuscus*) can consume from 3000 to 7000 mosquitoes/night, some of which may be carrying West Nile virus, malaria, and chikungunya virus, among other diseases. A colony of 20 million Mexican free-tailed bats (*Tadarida brasiliensis*) in Central Texas can consume ≥113,398 kg (0.25 million pounds) of insects/night (Cryan et al. 2014). Insectivorous bats consume June beetles (subfamily Melolonthinae), leafhoppers (family Cicadellidae), spotted cucumber beetles (*Diabrotica undecimpunctata*), green stink bugs (*Chinavia hilaris*), corn ear worm larvae (*Helicoverpa zea*), gypsy moths (*Lymantria dispar dispar*), spotted budworms (*Heliothis* spp.), and many other pests.

Of the 45 species of bats found in the contiguous 48 United States, six are federally listed under the ESA (FWS.gov). These include the gray (*Myotis grisescens*),

Indiana (*M. sodalis*), Ozark big-eared (*Corynorhinus townsendii ingens*), Virginia big-eared (*C. t. virginianus*), lesser long-nosed (*Leptonycteris yerbabuenae*), and the Mexican long-nosed (*L. navies*) bats. Highly troubling are recent deleterious impacts to cave-dwelling bats, especially those in the genus *Myotis* (e.g., little brown [*M. lucifugus*] and Indiana bat), from the fungal disease known as White-nosed Syndrome (WNS; *Pseudogymnoascus destructans*). To date, WNS is conservatively estimated to have killed more than seven million hibernating bats in 25 U.S. States and six Canadian Provinces. Population declines of >80 % of the bats in the Northeastern United States have recently been reported (Reynolds et al. 2015). All efforts to protect bats and reverse population declines are critically important and any efforts that can reduce or eliminate additional compensatory and/or additive mortality should be employed.

Addressing Problems Through Stressor Management

One approach being used by wildlife agencies, specifically the FWS in addressing direct, indirect, and cumulative impacts to migratory birds—and other fauna including bats—is through stressor management. A stressor is defined as any alteration or addition to the environment that when applied to a resource becomes a threat to the individual bird and/or its population. Stressors can be both anthropogenic and natural. For example, dissecting a project's construction and operational schedule can delineate each stressor. Common avian stressors that impact breeding, foraging, migration, migration corridors, and wintering areas include artificial lighting, noise, human/habitat disturbance, the addition of structures to the landscape, and the removal and manipulation of vegetation. The principle behind stressor management is to focus on the *cause* of the impact (e.g., installation of lighting) rather than its *effect* (e.g., nighttime bird attraction). Previously, managing project effects had focused on fixing the consequences of an action such as marking communication tower guy-support wires with bird deterrent devices to reduce bird collisions—admittedly costly, often difficult, and not necessarily effective. By constructing an un-guyed, monopole, or lattice-support tower, guy wire collisions are avoided. Stressor management today aims to deconstruct a project, providing a more tangible impact analysis by identifying the full spectrum of avian stressors associated with the lifecycle of a project. The stressors produced by each individual activity (e.g., brush clearing, dredging, using heavy machinery, or installing structural lighting), within each phase of a project (i.e., pre-construction, construction, post-construction/operation, and decommissioning), helps the project proponent realistically anticipate the problems that might be associated with their project and identify cost-effective ways to avoid or minimize the individual stressors at their source before they become realized threats to migratory birds (Morris and Kershner 2013; E. Kershner 2013 pers. comm.).

Discussion: Projected Impacts to Birds and Bats from Specific Industry Sectors

Direct and Indirect Effects of Transmission and Distribution Powerline Collisions and Electrocutions

The impacts of transmission and distribution powerlines on migratory birds have not been carefully or systematically monitored, even though dozens of peer-reviewed studies have been published in scientific journals assessing impacts to birds from powerless (e.g., APLIC 2006, 2012). This is in part due to the millions of kilometers (miles; APLIC 2012; Manville 2013a) of distribution lines and nearly 1.207 million km (0.75 M miles; APLIC 2012; Manville 2013a) of transmission lines in the U.S.; lack of adequate utility and agency staff to systematically survey them for dead birds; lack of pressure by the regulatory agencies on the industry; lack of recognition of the problem; and lack of adequate agency funding (Manville 2009, 2011). For purposes of comparison, distribution lines in rural and urban areas generally carry from 2.4 kilovolts (kV) up to 60 kV of electricity, using transformers to step down the voltage going into homes, offices, and other structures. Distribution lines are often placed above ground as undergrounding increases the cost. High voltage transmission lines carry from 60 to >700 kV and are generally located on tall pylon power towers, or other platforms. Transmission lines can be placed underground, but the challenges to maintain them can be significant, plus the costs range from three to 20 times that of above-ground placement, which are significant increases (APLIC 2006; B. Bolin 2013 pers. comm.).

Collisions and electrocutions are both important avian problems, but each has different impacts and rates of mortality vary between species (Manville 2013a). Although different species have different vulnerabilities, other than BAEA, GOEAs, and buteos (i.e., soaring hawks; APLIC 2006), there generally are not enough data to generate a clear quantitative picture of how vulnerable different species are to electrocutions. Vulnerability, time of day/night, weather conditions, visual acuity, disturbance, and issues still not well understood about avian vision all affect collision impacts (Martin 2011, 2014), but all need further quantitative testing, peer review, and publication.

Bird collisions occur primarily with energized transmission wires and the smaller, static (lighting arresting) wires generally located on top of the transmission towers which are not as visible to birds in flight (APLIC 2012). Visual acuity can be critically important since birds must depend on eyesight to see and avoid obstacles such as static wires close-up (Martin 2011, 2014).

Electrocutions, however, occur primarily at distribution lines and their infrastructures, although flashovers (contact between two energized wires, or an energized and grounded structure) have been occasionally documented from raptor "streamers" (streams of liquid fecal waste) which contact energized transmission wires (APLIC 2006). Distribution power lines supplying alternating current are frequently constructed in three, energized (hot) phases, with an additional ground wire separate from them. Because each energized phase is different, electrocutions can occur between them, or between a hot and the ground wire. For birds which touch phased distribution lines placed too close together, electrocutions can result from phase-to-phase line contact (often between fleshy parts of a bird's anatomy, e.g., wrist to foot, or wrist-to-wrist); phase-to-ground contact; or when feathers are wet (resulting in electrocutions and not infrequently power outages). Uninsulated power pole infrastructure can cause bird electrocutions by touching equipment such as exposed wire bushings, bare jumper wires, unprotected fused cutouts, unprotected switches, and by other means. Even small birds such as passerines can be at risk of electrocution (APLIC 2006).

In addition to direct impacts (e.g., Bevanger and Broseth 2004—in an empirical study in Norway), birds, bats, and other fauna are also impacted by the indirect effects of transmission and distribution lines, powerline utility poles, solar power towers and solar mirrors, and their infrastructure. These include the introduction of barriers to movement, habitat fragmentation, site avoidance/abandonment, disturbance, loss of population vigor, behavioral modification, creation of sub-optimal or marginal habitats, loss of refugia, and intraspecific and interspecific competition for resources (Manville 2013a). It is important to note that most of these indirect effects are difficult to quantify, difficult to separate from other impacts, and for the most part have not been quantitatively tested, critically reviewed, and published in refereed journals.

To better understand and address these issues, considerable research has and continues to be conducted on understanding the indirect effects of transmission and distribution lines, among other tall structures. Power lines, wind energy facilities, communication towers, and oil pumping facilities have been suspected of causing negative effects to some bird species, notably some species of grouse (Manville 2004). The imperiled status of many of these species better explains the research focus. For example, the Attwater's Prairie-chicken (*Tympanuchus cupido attwater*) is Federally ESA-listed as endangered, the Gunnison Sage-grouse (*Centrocercus minimus*) is threatened, the Lesser Prairie-chicken (*T. pallidicinctus*) is threatened, and the Greater Prairie-chicken (*T. cupido*) has been petitioned for federal listing. Research on the direct and indirect effects of tall structures on prairie-chickens, sage-grouse, and Sharptail-grouse (*T. phasianellus*) has been extensive (e.g., Connelly et al. 2000; Braun et al. 2002; Hagen 2003; Wolfe et al. 2003a, b; Pitman 2003; Hagen et al. 2004; Patten et al. 2004; Connelly et al. 2004—all summarized in Manville 2004). Research and studies continue with more recent advances discussed in APLIC (2012). Winder et al. (2014) and Winder et al. (2015 in press) empirically tested the recommendation by FWS (Manville 2004) for avoiding development within an 8-km (five mile) buffer from leks by wind energy facilities affecting Greater Prairie-chickens. Both studies showed negative effects on both males and females of this species within eight km, supporting FWS's previous buffer recommendation. Evaluation and proper power line routing continue to be assessed and implemented to address direct and indirect effects on federally endangered Whooping Cranes (*Grus americana*; APLIC 2012).

Bats have been found incidentally in bird mortality searches in both transmission and distribution powerline corridors. While the recommendations from the Avian Power Line Interaction Committee (APLIC 2006, 2012) have been primarily focused on avoiding and minimizing impacts to protected migratory birds, the recommendations and best practices may also benefit bats, especially where bird-wire marking devices are installed. However, until research is conducted on the etiology of bat-wire collisions, the benefits of APLIC recommendations for bats will continue to remain speculative.

Addressing Problems and Attempting to Resolve Impacts to Birds from Powerline Collisions and Electrocutions: An Electric Utility-FWS Partnership

The North American partnership between members of the electric utility industry, including investor-owned utilities, electric cooperatives, electric administrations, several federal agencies, the Edison Electric Institute, Electric Power Research Institute, FWS, and some Canadian (e.g., Canadian Wildlife Service and Environment Canada) and Mexican partners (e.g., Semarnat and the Mexican Institute of Ecology), is noteworthy and deserves closer examination. Called the Avian Power Line Interaction Committee (APLIC), the group's proactive approach in addressing effects from avian impacts as well as dealing with threats associated with electric utility infrastructure has become well-known.

Begun as an ad hoc collaborative in the early 1970s to specifically address Whooping Crane-powerline collisions and GOEA electrocutions at distribution line infrastructure, the APLIC partnership has been significantly expanded and was codified in 1989 with the creation of the committee housed within and managed by the Edison Electric Institute where records are maintained. It has grown to more than 55 members today (www.aplic.org).

While APLIC's initial and early focus centered on avoiding raptor electrocutions and Whooping Crane collisions, its orientation has expanded to all birds, including much more involvement among company members, other stakeholders including vendors, members of academic and research communities, and the interested general public. Similarly, the FWS's involvement with electric utilities—as well as other industries which it regulates—has focused, in descending order of priority, on education, exchange of information, and lastly enforcement—the three "E's" (J. Birchell 2012 pers. comm.). While APLIC has been touted as one of the longest and possibly most productive partnerships FWS has had with any industry sector to date, the partnership between the electric utility industry and FWS has not been without some controversy. FWS law enforcement agents and prosecuting attorneys at the Department of Justice made two criminal cases against the industry, with multi-million dollar (U.S.) penalties, including against the Moon Lake Electric Cooperative in 1999 and PacifiCorp in 2009—previously referenced. While APLIC members are sensitive to the cases and the media surrounding them, in the opinion of this author the cases have served to garner the undivided attention of some of the industry, resulting in more proactive cooperation with FWS and the other regulators. The same cannot be said for the wind generation industry where only one criminal case, previously referenced, has been prosecuted.

APLIC has set the industry standard for a proactive approach to addressing stressors *prior* to wire and infrastructure placement and operation. These include the development and release of APLIC's 2005 *Avian Protection Plan (APP) Guidance* (APLIC 2005), a collaborative effort between APLIC and FWS.[1] The *APP Guidance* lays out 12 principles for companies, cooperatives, public service and utility districts, and electric administrations to follow, while developing and implementing a proactive plan to address potential impacts from wire collisions and electrocutions. By developing and implementing an APP, a utility is ideally focused on the *cause* of a problem (e.g., wire collision and infrastructure electrocution, disturbance to nesting GOEAs due to excessive noise, or removal of vegetation negatively affecting birds) and taking steps to address it proactively, including throughout any new construction. As a result, the APP becomes a business and operational tool and better protects the utility against prosecution from FWS. There are, to date, more than 100 APPs already developed or under development by electric utilities and cooperatives, exclusive of any additional APPs required under court order (e.g., Moon Lake and PacifiCorp).

To proactively deal with stressors as well as deal with existing threats, APLIC periodically publishes best management practices and best operational technologies based primarily on peer-reviewed, published scientific studies to address electrocutions (most recently, *Suggested Practices for Avian Protection on Power Lines: the State of the Art in 2006*)[2] and collisions (most recently, *Reducing Avian Collisions with Power Lines: the State of the Art in 2012*).[3] These documents and their recommendations are designed for use on existing power line infrastructure (e.g., retrofits—focused on addressing threats) and for all new construction (i.e., anticipating and avoiding potential stressors, where possible). Both documents, in part, deconstruct the powerline/infrastructure projects, focusing on the true problems, helping to identify other activities that may produce stressors, and suggesting cost-effective ways to identify and avoid or minimize the stressor component of an activity while still allowing the activity to proceed. Included in the APLIC (2006) document are chapters on regulations and compliance, biological aspects of avian electrocution, power line design and avian safety (in considerable detail), and the development of an APP, among others. Similarly, in APLIC (2012), there are chapters on progress in dealing with collision issues (in North America, internationally, with the need for future research priorities), avian regulations and compliance, understanding bird collisions, minimizing collision risks, powerline marking to reduce collisions, and APPs.

[1] A document this author helped craft and negotiate.

[2] Coauthored by this author.

[3] Coauthored by this author.

APLIC also teaches short courses and other training modules dealing with avian-wire interactions, funds bird-utility research, and holds bi-annual meetings open to the public—including 1.5-day avian interaction workshops. The work of APLIC and its members has resonated in Canada, Mexico, Europe, Asia, Australia, and elsewhere. Fundamentally, APLIC has set the benchmark for other industries to follow in enabling a means to proactively address two significant threats to birds by identifying, avoiding, and minimizing the primary avian stressors associated with that activity. This still allows the activity to proceed in an effective and efficient way by enhancing reliable electrical energy delivery. In June 2014, APLIC and FWS celebrated their 25th anniversary working collaboratively since the committee was formed, while previously working in an ad hoc capacity since the 1970s (aplic.org).

While Loss et al. (2014) attempted to refine nationwide estimates for wire collisions and electrocutions, they did not attempt to summarize the overall efficacy of APLIC recommendations. Instead, they called for more information on the proportion of utilities implementing new best practices and retrofits, the degree with which these practices are reducing mortality, and the need for a consistent, peer-reviewed monitoring protocol. APLIC has yet to publish a nationwide meta-review of how best practices and suggested mitigation measures have worked to date. However, both APLIC documents (2006, 2012) do summarize empirical findings of mortality reduction based on some specific studies reported in these documents. FWS agents and field biologists routinely request the use of APLIC standards (2006, 2012) as benchmarks for addressing wire collisions and electrocutions, even though the recommendations are voluntary (FWS 2014 pers. comm.). In this author's opinion, one notable example of success should be credited to Puget Sound Energy, in western Washington. Where collision issues are identified as problems, this company has reduced to near-zero additional distribution wire collisions from Trumpeter Swans (*Cygnus buccinator*) by marking wires with bird diverter devices where birds are feeding at adjacent potato fields and may collide with the lines (M. Walters 2014 pers. comm.; pse.org/environment).

Collisions and Radiation Effects from Communication Towers: Addressing Problems to Birds

Tower Collision Mortality

Communication towers, which vary from short (<61 m AGL [200 ft]) monopole cellular telephone towers and antenna arrays to tall (>610 m AGL [2000 ft]) radio, television, and emergency broadcast towers, have two impacts on migratory birds, and to a lesser extent on bats since mortalities are reported only anecdotally to bird deaths. Information was first published in the late 1940s of a large, single night bird collision with a radio tower in Baltimore, Maryland (Aronoff 1949). More recently, information has been published on the suspected etiology of avian-tower collisions.

Frequently during nighttime migrations, birds are overwhelmed by inclement weather events, forcing bird fall-out, significant reductions in flight heights, and resultant attraction to lighted structures and confusion (Manville 2007, 2009, 2014a). Mortality has previously been conservatively estimated at 4–5 million birds killed in the U.S. annually (Manville 2002, 2005, 2009) based on limited, empirical data, and extrapolation from Banks' (1979) estimate. Current estimates of 6.8 million birds/year in the U.S. and Canada (Longcore et al. 2012) are based on a meta-review of 38 studies for which mortality data were available and corrected for sampling error, searcher efficiency, and scavenging. The vast majority of these bird deaths are in the U.S. (Longcore et al. 2012). In another review, at least 13 species of Birds of Conservation Concern were estimated to suffer annual mortality of 1–9 % of their estimated total population based solely on tower collisions in the U.S. or Canada (Longcore et al. 2013). These include estimated annual mortality of >2 % for the Yellow Rail (*Cocturnicops noveboracensis*), Swainson's Warbler (*Limnothlypis swainsonii*), Pied-bill Grebe (*Podilymbus podiceps*), Bay-breasted Warbler (*Setophaga castanea*), Golden-winged Warbler (*Vermivora* chrysoptera), Worm-eating Warbler (*S. discolor*), Prairie Warbler (*S. discolor*), and Ovenbird (*Seiurus aurocapilla*). Up to 350 species of birds have been documented killed at communication towers (Manville 2007, 2014a).

Radiation Effects

The much less documented but growing concern to birds and other wildlife involves effects of non-thermal, nonionizing microwave (and other) radiation from communication towers on nesting and roosting wild birds, an impact yet unstudied in the U.S. In Europe, impacts have been well-documented. Balmori (2005) found strong negative correlations between levels of tower-emitted microwave radiation and bird breeding, nesting, and roosting in the vicinity of electromagnetic fields in Spain. He documented nest and site abandonment, plumage deterioration, locomotion problems, and death in House Sparrows (*Passer domesticus*), White Storks (*Ciconia ciconia*), Rock Doves (*Columba livia*), Magpies (*Pica pica*), Collared Doves (*Streptopelia decaocto*), and other species. While these species had historically been documented to roost and nest in these areas, Balmori (2005) did not observe these symptoms prior to construction of the cellular phone towers. Balmori and Hallberg (2007) and Everaert and Bauwens (2007) found similar strong negative correlations among male House Sparrows. Under laboratory conditions in the U.S., T. Litovitz (2000 pers. comm.) and DiCarlo et al. (2002) raised troubling concerns about impacts of low-level, non-thermal radiation from the standard 915 MHz cell phone frequency on domestic chicken embryos (*Gallus gallus*)—with lethal results (www.healthandenvironment.org/wg_emf_news/6143). Given the findings of the studies mentioned above, and an extensive meta-review of the published studies by Panagopoulos and Margaritis (2008), field studies should be conducted in North America by third-party, independent research entities with no vested interest in the outcomes to validate potential impacts of communication tower radiation—both direct and indirect—to birds and other animals. However, to date, these have yet to be performed.

Efforts to Reduce Bird Collisions at Communication Towers

The FWS's Division of Migratory Bird Management became actively involved in the avian-tower collision issue in early 1998 with a large, single-night bird kill of up to 10,000 mostly Lapland Longspurs (*Calcarius lapponicus*) at a lighted, gas pumping facility and three surrounding communication towers in western Kansas (Manville 2001). To begin addressing the issue, the FWS published *Voluntary Guidelines for Communication Tower Design, Siting, Construction, Operation, and Decommissioning* in September 2000.[4] It developed and chaired the Communication Tower Working Group, focusing on the science surrounding bird attraction to lights, the dynamics of bird collisions, and efforts focused on dealing with stressors and their threats. The interim, voluntary *Guidelines* published in 2000 were updated in 2013 based on FWS recommendations provided on the record to the Federal Communications Commission (FCC) in 2007, 2011, 2012, and 2013 (Manville 2013a, b, 2014a). Changes in lighting and reductions in tower height and guy-support wires (Manville 2007; Gehring et al. 2009, 2011; Longcore et al. 2012) appear to preliminarily be reducing bird deaths, but a systematic review of these changes is recommended to determine empirically if the FWS guidelines, FCC licensing, and Federal Aviation Administration (FAA) lighting updates are reducing bird mortality. The FAA is finalizing updates to their 2007 lighting circular (FAA 2007), which incorporates new changes to steady-burning, red pilot warning obstruction lights generally placed on tall structures >61 m AGL (200 ft) in height (Manville 2013a; J. Gehring 2015 pers. comm.). Birds are particularly sensitive to the color red at night, especially if the red lights burn continuously rather than flashing or strobed (Gehring et al. 2009).

This development is highly noteworthy given the coordination, research, and work done by J. Gehring (Gehring et al. 2009, 2011). Specifically, new breakthroughs in better understanding the roles of lighting (especially steady-burning, red incandescent L-810 lights), tower height, and the use of guy support wires could—once fully implemented by the FCC and the FAA—reduce bird attraction and collision mortality by more than 50 % based on recent research and meta-reviews (Gehring et al. 2009, 2011; Longcore et al. 2012, 2013). That projected reduction in mortality still needs to be empirically assessed and verified, strongly suggesting the need in the opinion of this author for systematic mortality monitoring based on accepted monitoring protocols (e.g., Gehring et al. 2009).

Meanwhile, the vast majority of the FWS's voluntary recommendations are intended to proactively address the effects of stressors and their threats *before* tower

[4] Coauthored by this author.

siting and construction occur. These includes recommendations for collocation of antennas, use of a lattice or monopole construction, avoiding wetlands and other important bird areas, building in already degraded sites, eliminating L-810 lighting, keeping towers unlit and unguyed, following APLIC (2006, 2012) recommended standards for wire infrastructure, minimizing habitat footprints, down-shielding security lighting using only motion or heat-sensitive types, decommissioning inactive towers, and other steps (Manville 2013b). The efficacy of each of these recommendations will need, in the opinion of this author, to be systematically monitored and assessed to see how well each is working and modified or adapted as necessary to make them most effective. Since lighting changes will ultimately result in energy cost savings for tower owners and lessees, it is hoped that the majority of communication tower construction projects will comply with the suggested lighting practices and other best practice recommendations, and that re-licensing, existing retrofits, and new construction will collectively result in significant reductions in both "take" and habitat alteration and fragmentation. While no similar partnership like APLIC exists among the communication tower operators and FWS, that industry is represented by a consortium of trade associations. These include CTIA, PCIA, the National Tower Erectors Association, and the National Association of Broadcasters. Members of the consortium are beginning to acknowledge, appreciate, and address the benefits of constructing and maintaining bird-friendly communication towers.

The impacts of tower radiation, especially on nesting birds, are still unstudied in the U.S. Until independent, third-party research can be conducted and results analyzed, no recommendations can yet be provided on this issue—other than to proceed using the precautionary approach and to keep emissions as low as reasonably achievable. The precautionary approach, based in part on Article #15 of the 1992 Rio Conference (unep.org), recommends that where serious harm may result, lack of scientific certainly is not a reason for postponing implementation of cost-effective measures. Aside from the field and laboratory studies referenced above, there remains much uncertainty about effects from nonionizing radiation on migratory birds and other wildlife.

Collisions and Habitat Impacts from Commercial, Land-Based Wind Turbines: Addressing Bird and Bat Impacts

The Effects

Land-based commercial wind energy electrical-generating facilities are relatively new structures on the landscape, only operating in the U.S. since the 1980s at Altamont Pass Wind Resource Area, California (Righter 1996; Smallwood and Thelander 2004). However, from the 1980s to the present, commercial wind generation in the U.S. has grown explosively (DOE 2015). The U.S. Department of

Energy's 2015 WINDExchange (DOE 2015) indicates that 65,879 MW of installed capacity (more than 48,000 utility-scale turbines) were operating by the end of 2014. It is not at all surprising that estimated bird mortality has grown from what was first presented as an average of 34,000 bird deaths/year in 2000 (Erickson et al. 2001, estimating mortality based on a review of only 12 projects). In 2008, as the industry continued to grow exponentially and mortality monitoring protocols by consultants remained inconsistent between nearly every project, Manville (2009) estimated 440,000 bird deaths/year by correcting for six major biases inadequately addressed in then existing project review. These included in decreasing order of bias concern (1) variability in the duration and intensity of carcass searches (including observer bias and lack of credible levels of detection), (2) failure to address carcass searches during some migration and most nesting, (3) effects of inclement weather, (4) size of the search areas, (5) unaccounted crippling loss incidents, and (6) impacts from wind wake and blade wake turbulence. Manville (2009) did not include the formula and actual calculations he used to develop his estimate, in major part due to a lack of space in the peer-reviewed Proceedings. He took the industry's 2008 estimate of 58,000 annual bird deaths, attempting to update it reflective of biases still inadequately addressed by industry consultants. Using conceptual models developed by Huso (2008, later published in 2010), he attempted to address concerns over estimators (Huso 2008), especially where biases remained very large between projects and continued to be unaddressed by many industry consultants. Finally, Manville (2009) weighted the inconsistencies addressed by Huso (2008) in a decreasing order of bias concerns listed above. By selecting decreasingly weighted percentages for the six biases, he roughly calculated a range of annual bird mortality from 440,000 to 690,000, selecting the lowest estimate. Due to the numerous biases in the industry's 2008 cumulative mortality estimate, Manville made no attempt to apply any statistical rigor to his estimate (Manville 2012). By 2012, Smallwood (2013) estimated 573,000 bird deaths, of which some 83,000 were raptors, from wind facilities nationwide based on closer review and analysis. His estimate included a correction for inadequate survey and assessment of passerines killed based on approximately 34,400 then operating turbines across the U.S. in 2012. Loss et al. (2013c) estimated 234,000 birds killed at monopole-constructed wind turbines in the U.S. (excluding lattice turbine structures), while Erickson et al. (2014) estimated 368,000 birds killed at turbines in the U.S. and Canada. There continues to be some disagreement regarding the methodologies and rigor used to assess mortality.

Others (e.g., Sovacool 2009) have published comparisons of bird mortality from wind energy to fossil fuel, nuclear energy, and other sources. While these comparisons can be instructive, the analytical methods used to develop the estimates are often highly variable, duration and intensity of monitoring may differ greatly, scientific peer review may not have been conducted (Ferrer et al. 2012; Smallwood 2013), and reporting mortality in the aggregate (i.e., number of birds estimated killed) fails to detect species-level effects necessary to make conservation assessments and decisions (Longcore et al. 2013).

Impacts especially to Golden Eagles continue to be especially troubling. To date, only the Shiloh IV Wind Project, Solano Country, California, a 102-MW facility, has a pending eagle "take" (50 C.F.R. 22.26) permit to injure and/or kill up to five GOEAs over a 5 year period (http://www.fws.gov/cno/press/release.cfm?rid=628). The pending permit is not without controversy as at least two retired FWS law enforcement agents have spoken out against the project and its permit (Wiegand 2014) as have several environmental groups (Associated Press 2014).

Smallwood (2013) estimated at least 888,000 insectivorous bats killed/year at U.S. commercial wind energy facilities, which was based on 51,630 MW of installed wind capacity in 2012, now at more than 65,879 MW by late December 2014, and growing (DOE 2015). Bats are currently being lost in unprecedented numbers from blade collisions and barotrauma, most susceptible of which are the tree roosting bats including the hoary (*Lasiurus cinereus*), Eastern red (*L. borealis*), and silver-haired bats (*Lasionycteris noctivagans*; Cryan et al. 2014). Why these bats remain more susceptible to collisions with turbine blades, especially at low blade speeds, remains yet unknown. It appears that bat behaviors that evolved at tall trees are now proving maladaptive to flying around turbine blades (Cryan et al. 2014).

Like the impacts from other industry sectors, commercial wind energy projects cause direct and indirect effects on birds and bats. Due, however, to the massive footprint of some of these projects—i.e., hundreds of km^2—effects can be accentuated. The direct effects of turbines and their projects include bird and bat collision mortality, and barotrauma in bats and anecdotally reported in small birds (Manville 2009). Direct habitat loss, creation of barriers, loss of grasslands, direct fragmentation of habitat, increase in habitat edge, increase in nest parasitism and predation, and impacts on water quality can also be problematic (e.g., Sovacool 2009). From the perspective of indirect effects, numerous concerns have also been raised. These include reduced nesting and breeding densities, loss of population vigor and overall densities, habitat and site abandonment, loss of refugia, attraction to modified habitats including suboptimal ones, effects on behavior (e.g., stress, interruption, and modification), displacement, avoidance, and habitat unsuitability (Manville 2004; Gillespie 2013; Winder et al. 2014, 2015 in press). Indirect effects can be incredibly difficult to quantify, with further difficulties teasing out specific effects from others.

Beginning to Address the Problems

The FWS went through a long and detailed, multi-year process (2007–2010), coincident with the process to develop an eagle "take" permit mechanism, working through the Wind Energy Federal Advisory Committee (FAC) to develop and update the FWS's 2003 interim, voluntary land-based wind energy guidelines. This author served as one of two technical scientific advisors to the FAC. The 2003 document[5]

[5] Cowritten by this author

was open to 2 years of public comment. The resultant product was the *2012 Service Wind Energy Guidelines* (WEG) available on the FWS's website at www.fws.gov. While the specific guidelines are not prescriptive and only provide recommendations, they do recommend a detailed, tiered process for addressing stressors and their threats—notably Tiers 1, 2, and 3 focused on pre-construction landscape and site review. *If* a wind developer does perform its due diligence and properly sites wind facilities in bird, bat, and habitat-friendly locations, the project is unlikely to impact trust resources including birds in a significant way—i.e., negatively affecting their populations. However, there still is no permitting mechanism for "take" of migratory birds, and the permitting mechanism for eagle "take" requires important data on adult survivorship, territorial and foraging range integrity, adult breeding viability, recruitment, and disturbance to justify proposed levels of "take." The permitting process continues to remain a work in progress within FWS.

However, other than proper site location—i.e., siting turbines in low risk, degraded habitats, developed sites, or other locations where birds and bats will be minimally impacted—options are very limited. These low-risk sites still need to be clearly documented using accepted, scientific protocols that that can tie in low risk to factors that reduce rates of bird collision and minimize impacts from habitat alteration. These efforts continue to be a work in progress. There are no best practices or best available technologies for birds yet available for large-scale, wind energy developers. Such practices and technologies need to be independently peer-reviewed, scientifically validated, and acknowledged by independent experts as accepted tools to avoid or minimize "take" and/or affect habitats. In short, no silver bullet exists. Blade feathering (i.e., changing the pitch of the blades so they no longer cut into the wind), seasonal shutdowns, and electronic monitoring with automated Supervisory Control and Data Acquisition (SCADA) radar systems tied to feathering—which incidentally emit large quantities of radio frequency radiation—have only been reported to show limited success. Additionally, setbacks from ridge edges and turbine alignment have also shown some promise, but only with limited success (e.g., Smallwood and Thelander 2004). SCADA, for example, is very expensive to operate and companies using the system are finding it to be ineffective due to issues of sensitivity, response time to feathering, and verification of approaching targets (FWS 2015 pers. comm.). Mortality data are generally not shared with FWS or other agencies, or made available for third party data collection or independent peer review. This makes the efficacy of mitigation measures unclear, unknown, and difficult to verify (e.g., Wiegand 2014; Associated Press 2014). The smaller and shorter, vertical axis helix, flow-through turbines are far more efficient but more expensive than current technologies. They do have some promise in being more bird- and bat-friendly (FWS 2015 pers. comm.). Economies of scale suggest that higher blade heights with larger rotor swept areas are more efficient, overall less expensive per megawatt produced, but at a growing cost to wildlife and their habitats (Loss et al. 2013c). Rotor-swept areas now exceed 2.8 ha (seven acres) in area, larger than the entire area of three modern 747 jets. This is a situation quite different from what APLIC published through its 2006 and 2012 *Suggested Practices* documents that contain quantified and scientifically validated best practices and best available technologies. Many of these practices have been shown to significantly reduce wire collisions, electrocutions, and habitat alterations.

Hoary, Eastern red, silver-haired, and little brown bats are being heavily impacted by turbine blades. Whether these impacts are compensatory, additive, or represent a continuum between compensation and additivity (Peron 2013) still remains unclear and needs much more assessment. However, for insectivorous bats, there may be a conservation measure that could significantly deter blade collisions. Insectivorous bats tend to forage for insects when wind speeds are low (e.g., ~0.5 to 3.5 m/s) and the insects are present and readily available. Insectivorous bats remain highly susceptible to collisions and even barotrauma at these low wind speeds. By increasing the cut-in speed of turbine blades—i.e., the speed of the wind at which the blades begin to rotate—from ~3.0 to 6.0 or 6.5 m/s, bat mortality in a Pennsylvania study was reduced by up to 93 % (Arnett et al. 2011). While this change results in a loss of only a small fraction of energy production, it could significantly reduce bat mortality and therefore deserves careful consideration (Arnett et al. 2011; Arnett and Baerwald 2013). However, because the recommendation in the FWS's WEG is only voluntary, few companies are currently implementing this or other useful mitigation measure (Williams 2014; Manville 2014b).

Based on public comment, review, and internal assessment, the FWS published its updated, *Eagle Conservation Plan Guidance, Module 1, Land-based Wind Energy, Version 2* (ECPG), in April 2013. Like the WEG, it recommends approaches to avoiding and minimizing eagle "take" and impacts to eagle territories and eagle use areas based on a tiered protocol using the stressor management approach—i.e., identifying the stressors, their threats, and the consequences. While following the ECPG is voluntary, where disturbance "take" and/or "take" resulting in mortality are likely to occur, a permit (50 C.F.R. 22.26 or 22.27) is strongly recommended as un-permitted "take" may have legal consequences (Associated Press 2014). The goal of the ECPG is to ensure that the breeding population of both species of eagles remains stable or increasing. While the FWS published the authorization for the take permits in 2009 (50 C.F.R. 22.26 for eagle "take" and 22.27 for nest "take") along with the required NEPA documentation, the implementation of the regulations and permitting are a work in progress.

Studies are beginning to be published on the indirect effects of commercial wind energy facilities including on grassland bird density, nest survival, bird avoidance and attraction, and bat presence at turbines, turbine pads, and the generation facilities in Iowa (Gillespie 2013). As previously discussed, Winder et al. (2014) and Winder et al. (2015 in press) are validating a FWS recommendation (Manville 2004) of an 8-km (five-mile) buffer between Greater Prairie-Chicken leks and wind facilities. Research into indirect effects continues.

For numerous reasons, it has become increasingly clear that independent, third-party monitoring of wind facilities and site studies, and solar facilities briefly discussed next, must also be implemented. Unfortunately, with FWS's voluntary WEG guidance, that currently seems unlikely. Instances of data falsification and obfuscation of data; data release limitations through confidentiality agreements signed by project biologists, contractors, and cooperators; submission of fraudulent reporting;

and inadequate monitoring have been reported to FWS's Office of Law Enforcement (e.g., Wiegand 2014). Also reported were concerns about vested consultant interests, spotty reporting, proprietary data, and an unwillingness to work with FWS (FWS 2014 and 2015 pers. comm.)—unlike many of the companies in the electric utility industry. As Williams (2014:67) reminds us, "…some wildlife mortality is inevitable with even the best projects. But nothing will do more harm to the industry than excusing or tolerating wildlife-stupid projects that give it a bad name." If the public remains concerned, their voices need to be heard, and in turn, the industry needs to proactively address these concerns.

Beginning to Address Problems to Birds from Collisions and Heat Impacts at Industrial Solar Facilities in the Southwest

Problems to Birds and Other Wildlife

Industrial-scale solar development is relatively new to the U.S. Not until 1979 was the first industrial solar facility installed and operated in the U.S. in the Mojave Desert, which used a heliostat-power tower-solar receiver boiler generation system. Named Solar One, it had a tower of 86 m AGL (282 ft) in height, and a heliostat field of 765 m (2510 ft) in diameter—small by current power tower standards. At Solar One, McCrary et al. (1986) collected and reported 70 bird fatalities involving 26 species, 57 birds of which died from collisions while 13 died from burning. More recently, Leitner (2009) raised additional concerns and made suggestions for the proper selection of solar sites, including more research and mitigation. However, based on preliminary discoveries, a recent publication with troubling results (Kagan et al. 2013), and specific new recommendations by researchers, the environmental project review for the current solar technologies continues to be sorely inadequate.

There are three types of solar-generating facilities: (1) photovoltaic systems, (2) trough systems, and (3) solar power towers.

(1) Photovoltaics directly convert sunlight into energy (e.g., Desert Sunlight—at 1619+ ha [4000+ acres], with more than eight million panels, is probably the largest solar facility in the world). These flat panel systems can each cover enormous areas, displacing foraging habitats for GOEAs (a species of concern for FWS), their prey, and other species. In California's Imperial County alone, 91 km^2 (35 mi^2) of flat panel photovoltaics have already been and are being proposed for development. In a recent 2013 opportunistic survey conducted by staff of FWS and reported by the National Fish and Wildlife Forensics Laboratory (NFWFL; Kagan et al. 2013), where no pre-determined carcass sampling protocol was used, 61 bird carcasses retrieved from Desert Sunlight were transported to NFWFL to determine cause of death. Birds apparently mistook the shiny mirrored surfaces of the cells for water, resulting in blunt force trauma, predation, and unknown causes. Bird carcasses have

also incidentally been found at other flat panel projects in California's Central Valley, Imperial Valley, and in Nevada. These reports are only incidental to facility operations, not based on systematic surveys—which is a quandary.

(2) Trough systems consist of parabolic mirrors which are about 9m (30 ft) tall and can be hundreds of meters long. They focus sunlight onto tubes which convert heat to electricity (e.g., Genesis Solar Energy). From the Genesis site, 31 bird carcasses were opportunistically evaluated by NFWFL for cause of death. The results included impact trauma, predation, and unknown causes (Kagan et al. 2013). It is important to note that the number of carcasses found to date far outnumber the 31 reported several years ago by Kagan et al. (2013; FWS 2015 pers. comm.). These carcasses were found opportunistically, with no research study design, based on no third-party monitoring.

(3) Solar power towers are by far the most complex of industrial solar generation and also the most deadly to both birds and bats—based on the preliminary evidence. They consist of thousands of mirrors (e.g., Ivanpah with more than 300,000—the largest industrial solar steam generating system in the world). The mirrors intensely reflect solar energy to a power-generating tower (for Ivanpah, 140 m AGL [459 ft]), producing steam at temperatures of up to 427 °C (800 °F). This, in turn, runs a turbine and has an air-cooled condenser. Ivanpah has been characterized as a "mega-trap" for wildlife by the NFWFL (Kagan et al. 2013). In addition to significant bat and monarch butterfly (*Danaus plexippus*) mortality, the facility has attracted other insects, which in turn have attracted insect-eating birds, which were incapacitated by the solar energy flux, in turn attracting avian and mammalian predators. This has created an entire food chain vulnerable to injury and death. Carcasses collected opportunistically at Ivanpah included 141 birds which died from solar flux ($N=47$), impact trauma ($N=24$), predation ($N=5$), undetermined trauma ($N=14$), and "unknown" ($N=46$; Kagan et al. 2013). Even more troubling is a very recent, preliminary report (FWS 2015 unpublished data) by third-party monitors of 130 birds killed during a 4-h observation period at Crescent Dunes solar steam power project, Nye County, Nevada. Virtually all the birds were vaporized (FWS 2015 pers. comm.).

If just three commercial solar energy facilities are killing $N=233$ protected migratory birds based only on opportunistic and incidental monitoring during a few visits—i.e., information not gathered via pre-determined, robust, and peer-reviewed protocols for mortality monitoring—then how many birds, bats, and imperiled insects (e.g., monarchs) are actually being killed/year? It must be emphasized that the $N=233$ number represents only what FWS opportunistic visits discovered several years ago. Current FWS Special Purpose-Utility (Avian Take Monitoring) Annual Reports (SPUT; FWS Form 3-202-17) indicate that for Desert Sunlight, Genesis, and Ivanpah alone, more than 1000 birds killed representing almost 160 different species have been reported to FWS (2015 unpublished FWS data; also reported on www.kcet.org). This is far greater than the Kagan et al. (2013) preliminary reporting. While no GOEA carcasses have yet been found, solar facilities are displacing thousands of hectares of breeding and foraging habitat. One estimate suggests that up to 28,000 birds, including rapidly declining populations of Western Grebes (*Aechmophorus occidentals*; a BCC species), Common Loons (*Gavia mimer*), Peregrine Falcons (*Falco peregrinus*), Burrowing Owls (*Athene cunicularia*), Short-eared Owls (*Asio flames*), and others, are being killed each year in commercial solar arrays now operating only in Southern California, with a focus on Ivanpah (Center Biological Diversity 2014). However, until reporting is consistent, systematic, robust, and scientifically credible, the direct, indirect, and cumulative effects of industrial solar development on resident and wintering/migrant birds will remain uncertain. The lack of peer-reviewed data and a push by the current administration to fast-track renewable energy only complicates the situation.

These developments clearly do not bode well for industrial solar development. Apparently a number of FWS biologists raised major concerns before projects were even approved, let alone constructed, but their concerns did not resonate (FWS 2014 and 2015 pers. comm. and internal communications).

Beginning to Address the Problems

It is time to go back to the basics, using sound science and accepted protocols for monitoring as the drivers for developing industrial solar energy. These protocols should be scientifically credible, sufficiently robust, field tested, peer-reviewed, and accepted as valid by the scientific community—e.g., Gehring et al. 2009, as modified to apply to solar monitoring. Agencies need to maintain the leadership willing to stand up to the powerful industries and not be swayed by "green washing" (i.e., industry touting its actions as environmentally friendly and responsible, when in fact they can be very impactful). Because it is so challenging, enacting change within the agencies can be incredibly difficult. For example, on Bureau of Land Management public lands where the focus is on the development of solar facilities, thorough pre-construction risk assessment must be implemented, along with a full NEPA review of proposed projects, including citizen participation in the process (e.g. testimony, peer review, and litigation). Meanwhile, here is a preliminary list of some suggested mitigation for wildlife impacts at industrial solar facilities—which is far from exhaustive. All should be further tested using empirical field studies and published in refereed scientific journals, indicating which techniques are most effective. Bird and bat mortality can be reduced through fencing, nets, perch deterrents, exclusionary measures, UV-reflective glass, suspended operations during peak bird presence, use of video cameras and trained dogs for detection of carcasses, at least 2 years of daily bird and bat mortality searches—adjusting for scavenger removal including by Common Ravens, and addressing observer bias—and other measures as suggested by Kagan et al. (2013). Independent peer review of the agencies and contractors' statistics is also critical. How these projects were approved without sufficient oversight is very troubling. In this author's opinion, this same concern also applies to land-based wind development.

Conclusion

The issues discussed above present huge challenges, especially since we still know so little about the overall, cumulative impacts of powerlines, communication towers, commercial wind projects, and commercial solar arrays on birds, bats, and their habitats. If electric transmission, electronic communication, and renewable energy development are to be bird-, bat-, and habitat-friendly, changes must take place. This suggests a complete paradigm shift in assessing sites, adequately predicting pre-construction risks, validating risks during post-construction monitoring and assessment, and reversing ongoing very troubling trends.

To begin making this shift, this author recommends the development of an accepted monitoring protocol for each industry sector. Each protocol should be empirically based, scientifically valid, sufficiently robust—of the appropriate duration and intensity, with a consistent study design, field tested, peer-reviewed, and published in a refereed scientific journal. Post-construction monitoring should ideally include empirically driven, field-tested, and validated conservation and mitigation measures. Where such measures currently do not exist (e.g., industrial solar arrays and wind energy projects), research should continue to try to find them. Mitigation replacement/compensation measures for "take" and impacts to wildlife habitats should also be developed, empirically evaluated, peer-reviewed, published, and adopted, where most effective.

The guidelines for avoiding or minimizing impacts to migratory birds at communication towers, electric utilities, and commercial wind turbines have, for the most part, been voluntary—generally left up to the discretion of the industry proponents. This has often resulted in huge inconsistencies in monitoring (e.g., this author recounts a consultant providing four days of bird monitoring data at a proposed wind energy site to represent an entire migratory season of three months). As a result, a regulatory (e.g., implemented through the U.S. Code of Federal Regulations) versus voluntary approach has been suggested, including by this author, but under the current political climate in the U.S., that is highly unlikely. If regulations were developed, the suggested, empirically based monitoring protocols mentioned above should be incorporated as part of them. Also important, the agencies required by law and statute to manage wildlife and wildlife habitats need to acknowledge and implement their trust and statutory responsibilities regarding the wildlife they are entrusted to protect and conserve. Based on this author's experiences, politics rather than sound science seem to drive many current decisions. The Department of Interior and Department of Energy might be good places to begin the shift.

Based on the experiences of this author, there is some good news. With collaborative efforts such as those of APLIC long in place—and generally working well—the bar has been set high for other industries and agencies to follow. Where companies and their consultants are working with FWS, other agencies, and the public to better understand and minimize the impacts from human structures, their efforts should be applauded. This is a very good, but still too rare a thing.

Acknowledgments The author wishes to thank B. Blake Levitt, medical/science journalist, author, and former contributor to the *New York Times;* J. Kalina, Ph.D., author and conservation biologist; J.E. Pagel, Ph.D., raptor ecologist and wildlife biologist; and two anonymous referees for their critical review, constructive comments, and suggested changes to this document.

References

Aldrich JW, Banks RC, Cade TJ et al (1975) Report of the American Ornithologists' Union ad hoc Committee on Scientific and Educational Use of Birds. Auk 92(3 Suppl):1A–27A

Allen MS et al (2006) Implications of compensatory and additive mortality to the management of selected sport fish populations. Lakes Reservoirs Res Mgt 3(1):67–79. doi:10.1111/j.1440-1770.1998.tb00033.x

American Wind Energy Association (2015) Get the facts. www.awea.org

Arnett EB, Baerwald EF (2013) Impacts of wind energy development on bats: implications for conservation. In: Adams RA, Pedersen SC (eds) Bat evolution, ecology and conservation. Springer, New York, pp 435–456

Arnett EB et al (2011) Altering turbine speed reduces bat mortality at wind energy facilities. Front Ecol Envir 9(4):209–214

Aronoff A (1949) The September migration tragedy. Linnaean NewsLett 3(1):2

Associated Press (2014) California wind farm avoids prosecution for eagle deaths, June 26

Avian Power Line Interaction Committee (2006) Suggested practices for avian protection on power lines: the state of the art in 2006. Edison Elec. Inst., APLIC, and CA Energy Comm., p 1- 207

Avian Power Line Interaction Committee (2012) Reducing avian collisions with power lines: the state of the art in 2012. Edison Elec. Inst. and APLIC, p 1-159

Balmori A (2005) Possible effects of electromagnetic fields from phone masts on a population of White Stork (*Ciconia ciconia*). Electromag Biol and Med 24:109–119

Balmori A, Hallberg O (2007) The urban decline of the House Sparrow (*Passer domestics*): a possible link with electromagnetic radiation. Electromag Biol Med 26:141–151

Banks RC (1979) Human related mortality of birds in the United States. U.S. Fish & Wildlife Service, Natl. Fish & Wildlife Lab., Spec. Sci Rept—Wildlife No. 215:1-16

Bevanger K, Broseth H (2004) Impact of power lines on bird mortality in a subalpine area. An Biodiv Cons 27(2):67–77

Boyles JG, Cryan PM et al (2011) Economic importance of bats in atriculture. Science 332(6025):41. doi:10.1126/science. 1201366

Braun CE et al (2002) Oil and gas development in Western North America: effects of sagebrush steppe avifauna with particular emphasis on sage grouse. Trans. 67th N Am Wildl Nat Resour Conf, pp 337–349

Carter E (2013) Birding in the United States: demographic and economic analyses. USFWS Rep 2011–1:1–16

Center for Biological Diversity (2014) Annual estimate of bird mortality at BrightSource Energy's Ivanpah Solar Electric Generation System

Avian Power Line Interaction Committee (2005) Avian Protection Plan (APP) Guidelines, Edison Elec Inst's Avian Power Line Interaction Cmt and USFWS, pp 1–84

Connelly JW, Schroeder MA et al (2000) Guidelines to manage sage grouse populations and their habitats. Wildl Soc Bull 28(4):967–985

Connelly JW, Knick ST et al (2004) Conservation assessment of greater sage-grouse and sagebrush habitats. W Assoc Fish Wildl Agencies, Cheyenne, pp 1–610

Cryan PM, Gorresen PM et al (2014) Behavior of bats at wind turbines. Proc Natl Acad Sci U S A 111(42):15126–15131. doi:10.1073/pnas.1406672111

Department of Energy WINDExchange (2015) Installed wind capacity as of 31 Dec 2014. apps2.eere.energy.gov
DiCarlo A, White N et al (2002) Chronic electromagnetic field exposure decreases HSP70 levels and lowers cytoprotection. J Cell Biochem 84:447–454
Diehl RH (2013) The airspace is habitat. Trends Ecol Evol 28(7):377–379
Erickson WP, Johnson GD et al (2001) Avian collisions with wind turbines: a summary of existing studies and comparisons to other sources of avian collision mortality in the United States. WEST, Cheyenne, pp 1–60
Erickson WP et al (2014) A comprehensive analysis of small-passerine fatalities from collisions with turbines at wind energy facilities. PLoS One 9(9):e107491. doi:10.1371/journal.pone.0107491
Everaert J, Bauwens D (2007) A possible effect of electromagnetic radiation from mobile phone base stations on the number of breeding House Sparrows (*Passer domesticus*). Electromag. Biol Med 26:63–72
Federal Aviation Administration (2007) Obstruction marking and lighting, Advisory Circ AC 70/7460-1K. U.S. Department of Transportation, Washington, DC
Ferrer M et al (2012) Weak relationship between risk assessment studies and recorded mortality in wind farms. J Appl Ecol 49:38–46
Gehring J, Kerlinger P, Manville AM II (2009) Communication towers, lights and birds: successful methods of reducing the frequency of avian collisions. Ecol Appl 19(2):505–514
Gehring J, Kerlinger P, Manville AM II (2011) The role of tower height and guy wires on avian collisions with communication towers. J Wildl Manage 75(4):848–855
Gillespie MK (2013) Bird and bat responses to wind energy development in Iowa. Master's Thesis. Iowa State University, Ames, pp 1–82
Hagen CA (2003) A demographic analysis of lesser prairie-chicken populations in southwestern Kansas: survival, population viability, and habitat use. Ph.D. Dissertation, Division of Biology, College Arts Sciences, Kansas State University, pp 1–199
Hagen CA, Jamison BE et al (2004) Guidelines for managing lesser prairie-chicken populations and their habitats. Wildl Soc Bull 32(1):69–82
Huso M (2008) A comparison of estimators of bat (and bird) fatality at wind power generation facilities. Oct 28 PowerPoint Pres. National Wind Wildlife Collaborative Reseach VII Meeting, Milwaukee, pp 1–48
Huso MMP (2010) An estimator of wildlife fatality from observed carcasses. Environmetrics 22(3):318–329. doi:10.1002/env.1052
Kagan RA, Viner TC et al (2013) Avian mortality at solar energy facilities in Southern California—a preliminary analysis. National Fish and Forensics Laboratory, Ashland, pp 1–28
Klem D Jr, Saenger PG (2013) Evaluating the effectiveness of select visual signals to prevent bird-window collisions. Wilson J Ornithol 125(2):406–411
Kunz TH et al (2008) Aeroecology: probing and modelling the aerosphere. Integr Comp Biol 48:1–11
Leitner P (2009) The promise and peril of solar power—as solar facilities spread, desert wildlife faces risks. Wildlife Professional 3(1):48–53
Longcore T, Rich C, Mineau P et al (2012) An estimate of avian mortality at communication towers in the United States and Canada. PLoS One 7(4):1–17, Open Access
Longcore T, Rich C, Mineau P et al (2013) Avian mortality at communication towers in the United States and Canada: which species, how many, and where? Biol Conserv 158:410–419
Loss S, Will T, Marra P (2012) Direct human-caused mortality of birds: improving quantification of magnitude and assessment of population impact. Front Ecol 10(7):357–364
Loss SR, Will T, Marra PP (2013a) The impact of free-ranging domestic cats on wildlife of the United States. Nat Commun 4(1396):1–7. doi:10.1038/ncomms2380
Loss SR, Will T, Loss SS, Marra PP (2013b) Bird-building collisions in the United States: data-driven estimates of annual mortality and species vulnerability. Briefing memo for April 18, 2013, stakeholder information meeting at USFWS HQ Office, Arlington, 1 pp.

Loss SR, Will T, Marra PP (2013c) Estimates of Bird Collision Mortality at Wind Facilities in the Contiguous United States. Biol Conserv 168:201–209

Loss SR, Will T, Marra PP (2014) Refining estimates of bird collision and electrocution mortality at power lines in the United States. PLoS One 9(7):e101565. doi:10.1371/journal.pone.0101565

Manville AM II (2002) The ABC's of avoiding bird collisions at communication towers: the next steps. Proc. Workshp. Avian Interactions Utility and Communication Structures, Elec Power Res Inst, pp 1–16

Manville AM II (2007) Comments of the U.S. Fish and Wildlife Service submitted electronically to the FCC on 47 CFR Parts 1 and 17, WT Docket No. 03-187, FCC 06-164, Notice of proposed rulemaking, "Effects of communication towers on migratory birds," Feb 2, pp 1–32

Manville AM II (2011) Estimates of annual human-caused mortality to North American birds (with literature citations). Div Mig Bird Mgt, USFWS, Anchorage, pp 1–12

Manville AM II (2012) Wind Turbine Fatality Number—Media Statement and Background Information. Internal Interior Dept Briefing Memo, Div Mig Bird Mgt, USFWS, March 15, pp 1–12

Manville AM II (2014a) Status of U.S. Fish and Wildlife Service developments with communication towers with a focus on migratory birds: updates to Service staff involved with tower issues—a peer reviewed webinar. Talking Points and Literature Citations, March 7, pp 1–13

Manville AM II (2014b) The impacts on birds, bats and their habitats from commercial wind and solar energy development: unintended consequences. Faculty Speaker Series, Johns Hopkins Univ., Krieger Schl Arts & Sci, Oct 10, pp 1–37

Manville AM II (2013b) U.S. Fish and Wildlife Service revised guidelines for communication tower design, siting, construction, operation, retrofitting, and decommissioning—Suggestions based on previous USFWS recommendations to FCC regarding WT Docket No. 03–187, FCC 06–164, Notice of Proposed Rulemaking, "Effects of Communication Towers on Migratory Birds," Docket No. 08–61, FCC's Antenna Structure Registration Program, and Service (2012) Wind Energy Guidelines. Div Mig Bird Mgt, Arlington, pp 1–5

Manville AM II (2001) Avian mortality at communication towers: steps to alleviate a growing problem. In: Levitt BB (ed.). Proc. "Cell Towers Forum" State of Science/State of Law, ISBN 1-884820-62-X, pp 75–86, 227–228

Manville AM II (2004) Prairie grouse leks and wind turbines: U.S. Fish and Wildlife Service justification for a 5-mile buffer from leks; additional grassland songbird recommendations. DMBM, USFWS, pp 1–17

Manville AM II (2005) Bird strikes and electrocutions at power lines, communication towers, and wind turbines: state of the art and state of the science—next steps toward mitigation. In: Ralph CJ, Rich TD (eds), Bird Conservation Implementation in the Americas: Proc. 3rd Internatl Partners Flight Conf, USDA Forest Service Gen Tech Rept PSW-GTR-191, pp 1051–1064

Manville AM II (2009) Towers, turbines, power lines and buildings—steps being taken by the U.S. Fish and Wildlife Service to avoid or minimize take of migratory birds at these structures. In: Rich TD, Arizmendi C, Demarest DW, Thompson C (eds.). Tundra to Tropics: Connecting Birds, Habitats and People. Proc. 4th Internatl Partners Flight Conf, pp 262–272

Manville AM II (2013a) Anthropogenic-related bird mortality focusing on steps to address human-caused problems. Invited, peer-reviewed white paper for Anthropogenic Panel 5th Internatl Partners in Flight Conf. August 27, Snowbird, UT. Div Mig Bird Mgt, USFWS, pp 1–16

Martin GR (2011) Understanding bird collisions with man-made objects: a sensory ecology approach. Ibis 153(2):239–254

Martin GR (2014) The subtlety of simple eyes: the tuning of visual fields to perceptual challenges for birds. Philos Trans R Soc Lond B Biol Sci 369(1636):20130040

McCrary MD, McKernan RL et al (1986) Avian mortality at a solar energy power plant. J Field Ornithol 57(2):135–141

Morris C, Kershner E (2013) Migratory bird conservation for Federal partners. National Conservation Training Center, Div Mig Bird Mgt Training Module, USFWS

MountainNature.com (2015) Birding in North America—the Field Guide for the Next Millennium

Pagel JE, Kritz KJ et al (2013) Bald eagle and golden eagle mortalities at wind energy facilities in the contiguous United States. J Raptor Res 47(3):311–315

Panagopoulos DJ, Margaritis LH (2008) Mobile telephony radiation effects on living organisms (Chap. 3). In: Harper AC, Buress RV (eds) Mobile telephones. Nova Science, New York, pp 107–149. ISBN 978-1-60456-436-5

Patten MA, Wolfe DH et al (2004) Habitat fragmentation, rapid evolution, and population persistence. Evol Ecol Res 7:1–29

Peron G (2013) Compensation and additivity of anthropogenic mortality: life-history effects and review of methods. J Anim Ecol 82(2):408–417

Pitman JC (2003) Lesser prairie-chicken nest site selection and nest success, juvenile gender determination and growth, and juvenile survival and dispersal in southwestern Kansas. M.Sc. Thesis, Div Biology, College Arts and Sciences, Kansas State University, pp 1–169

Reynolds HT, Ingersoll T, Barton HA (2115) Modeling the environmental growth of *Pseudogymnoascus destructans* and its impact on the white-nose syndrome epidemic. J Wildl Disease doi 10.7589/2014-06-157

Righter RW (1996) Wind energy in America: a history. University of Oklahoma Press, Norman, pp 1–361

Smallwood KS (2013) Comparing bird and bat fatality-rate estimates among North American wind energy projects. Wildl Soc Bull 37(1):19–33

Smallwood KS, Thelander CG (2004) Developing methods to reduce bird mortality in the Altamont Pass Wind Resource Area. Final Report by BioResources Consultants, Ojai, CA to California Energy Commission, Public Interest Energy Research-Environmental Area, Contract No. 500-01-019, Spiegel L (Progr Mngr), pp 1–363

Sovacool BK (2009) Contextualizing avian mortality: a preliminary appraisal of bird and bat fatalities from wind, fossil-fuel, and nuclear electricity. Energy Policy 37:2241–2248

U.S. Fish and Wildlife Service (2008) Birds of conservation concern 2008. U.S. Dept Interior, Fish and Wildlife Service, Div Mig Bird Mgt, Arlington, pp 1–85. http://www.fws.gov/migratorybirds/

Wiegand J (2014) Wind power slaughter: ex-USFWS agent speaks out on Shiloh IV (California), Master resource—Free Market Energy Mag, July 16

Williams T (2014) A mighty wind. Audubon Magazine, March–April, pp 32, 34, 64–67

Winder VL, McNew LB et al (2014) Space use by female Greater Prairie-Chickens in response to wind energy development. Ecosphere 5:art3

Winder VL, Gregory AJ et al (2015) Responses of male Greater Prairie-Chickens to wind energy development. Condor 117. do:1650/CONDOR-14-98.1 (in press)

Wolfe DH, Patten MA, Sherrod SK (2003a) Factors affecting nesting success and mortality of Lesser Prairie-Chickens in Oklahoma. ODWC Federal Aid in Wildlife Restoration Project W-146-R Final Rept. OK Department of Wildlife Conservation, pp 1–23

Wolfe DH, Patten MA, Sherrod SK (2003b) Causes and patterns of mortality in Lesser Prairie-Chickens. Poster presented at meetings of The Wildlife Society, Burlington, VT, and Prairie Chicken Technical Committee, OK Department of Wildlife Conservation. OK Biol Surv and George M. Sutton Avian Res Ctr

Chapter 21
Conflict of Wings: Birds Versus Aircraft

John Thorpe

Introduction

1.1 It is hard to believe that soft feathery birds are a wildlife hazard, yet they are a very real threat to aircraft. They have caused the deaths of more than 450 people and the destruction of over 500 civil and military aircraft. This means that birds–aircraft collisions are high in the list of wildlife hazards. Surprisingly, the first fatality occurred over 100 years ago, only 9 years after powered flying began. Celebrity pilot 'Cal' Rogers, the first person to fly across the USA, was piloting his Wright Flyer Model B in an exhibition flight along the shore at Long Beach, California, in April 1912 when he went through a flock of Gulls (*Larus* sp.) and one of them jammed the flying controls causing the aircraft to crash into the sea breaking Cal's neck (Fig. 21.1).

1.2 In the next 35 years, there were a couple of civil aircraft accidents due to collisions with birds, both in India and involving birds of prey. It is known that during World War 2, for example, the Royal Air Force Far East Command lost at least nine aircraft and 11 aircrew during a 15 month period as a result of collisions with birds, while the American Far East Air Force stated it 'suffered more damage from birds than from enemy action'.

1.3 In recent years, the subject has been expanded to include wildlife ranging from dogs, deer, foxes, and other animals that can get onto airports. These comprise a small proportion of incidents and are most likely to damage an aircraft's landing gear, whilst it is likely that only one engine would be at risk.

Note: In this Chapter, the units used are the standard aviation convention of feet (0.3048 m) for height/altitude and knots, nautical miles per hour (1.85 km/h) for speed.

J. Thorpe (✉)
F.R.Ae.S., International Bird Strike Committee, Brackens, Chapel Road,
Smallfield, Surrey RH6 9JH, UK
e-mail: johnthorpe@flysafe.fsnet.co.uk

F.M. Angelici (ed.), *Problematic Wildlife*, DOI 10.1007/978-3-319-22246-2_21

Fig. 21.1 The Wright Flyer wreckage on the shore at Long Beach after it collided with Gulls (*Larus* sp.) (photo via J Thorpe)

Accidents to Airliners and Executive Jets (Thorpe 2012)

2.1 In spite of the steady expansion of civil aviation, it wasn't until 1960 that any real notice was taken of the threat posed to aircraft by birds sharing the same airspace. On 4th October, 1960, a 4-engined turboprop Lockheed Electra was taking off from Boston Logan Airport, USA, and immediately collided with a large flock of European Starlings (*Sturnus vulgaris*, 80 g) (bird weights Brough 1983). One engine had to be shut down and two others lost power resulting in loss of control and the aircraft spinning into the sea. Sixty two people were killed with ten seriously injured survivors. Suddenly, birds were recognised as a serious hazard to aviation. This accident remains the worst bird strike accident in aviation history.

2.2 Globally, a total of 28 airliners and 19 executive jets have been destroyed or damaged beyond economic repair as a result of collision with birds. Of these, 18 resulted in fatalities killing 207 people including 7 on the ground. Airliner losses range from Boeing 737s in Ethiopia and Spain to a Boeing 747 in Belgium and an Airbus A300 in India. Executive jet losses include accidents in France, Italy, and the UK, but the majority are in the USA where a high proportion of such flights take place. Surprisingly, there has only been one fatal accident to a jet-powered civil airliner in many million flying hours. This was in September 1988 when an early Boeing 737 was taking off from Bahir Dar, Ethiopia, and collided with a flock of Speckled Pigeons (*Columba guinea*, 320 g), damaging both engines which lost power; 35 were killed and 21 injured in the ensuing forced landing. The pigeons had been feeding on grass seed that had flourished in soil brought from another site to fill trenches dug for cable laying. Some further cases are in the following paragraphs (Fig. 21.2).

2.2.1 In November 1962, a Vickers Viscount turboprop en-route at night from Washington DC to Newark, New York, collided with a Whistling swan (*Cygnus columbianus*, 6 kg) *at 6000 ft* over Maryland. The impact tore

Fig. 21.2 The burnt out remains of the Boeing 737 following loss of both engines due to ingestion of Speckled Pigeons (*Columba guinea*, 320 g). 35 of the 56 on board were killed (photo via J. Thorpe)

off the left-hand tailplane, rendering the aircraft uncontrollable and it crashed killing the 17 on board.

2.2.2 The 139 occupants of a McDonnell Douglas DC10 had a very lucky escape when on 12th November 1975 at Kennedy Airport New York. During the first take-off of the day on the particular runway, at a speed of 156 kts, which was just below the speed at which the aircraft can be stopped on the remaining runway, several Gulls species comprising Great Black-backed (*Larus marinus*, 1.7 kg), Ring-billed (*Larus delawarensis*, 385 g), and Herring (*Larus argentatus*, 1.1 kg), rose from the runway. The aircraft struck many birds and the take-off was abandoned. Number 3 engine disintegrated and fan blades ruptured a wing fuel tank starting a massive fire. On the wet runway and without the assistance of engine reverse thrust, the aircraft was finally stopped on a taxiway with collapsed landing gear and was destroyed by fire. All the occupants were airline employees who escaped successfully although two received serious injuries. They were thoroughly familiar with the emergency evacuation procedures – if it had been a normal passenger flight the result would have been very different. Subsequently the engine fan casings were modified (Fig. 21.3).

2.2.3 A British Aerospace 125 Executive jet was taking off at dusk on 20th November, 1975 from a manufacturer's airfield at Dunsfold, UK, when at about 75 ft just after becoming airborne, the aircraft encountered a flock of Lapwings (*Vanellus vanellus*, 215 g). Both engines lost power

Fig. 21.3 The DC10 burning fiercely after an engine disintegrated as the result of an engine ingested a number of the larger varieties of Gulls (*Larus* sp.). All 139 on board escaped (photo Port of New York Authority)

and the pilot attempted to land back on the runway but it over-ran and crossed a road into a field before being destroyed by fire. The two pilots were slightly injured, the seven passengers were unhurt. As it crossed the road, it had struck a car killing the lady driver and five children. Subsequently, traffic lights were installed to stop vehicles when aircraft were taking off. The Lapwing distress call tape used in the pre-digital age had broken and was away for repair.

2.2.4 In April 1978, a trainee pilot supervised by an instructor was doing touch and go landings in an early version of the Boeing 737 at Gossellies in Belgium. As the aircraft was about to lift-off, a flock of Woodpigeons (*Columba palumbus*, 465 g) were seen ahead. Because he believed both engines might have ingested birds, the pilot abandoned the take-off even though the speed was beyond the decision speed V1, below which the aircraft can stop within the length of the runway. The aircraft over-ran into an industrial estate with a collapsed right landing gear and the right engine torn from the wing and was destroyed by fire. It was found that the left engine had only ingested a single bird, which had caused severe damage. The sudden halving of acceleration had convinced the pilot that both engines had been damaged. It demonstrated that ingestion of a single bird can trigger a chain of events leading to the destruction of an aircraft (Fig. 21.4).

2.2.5 During a morning departure from Lunken Airport, Cincinnati, USA, on 7th April, 1981, a Lear 23 was climbing through 4000 ft when a Common Loon (*Gavia immer*, 3.7 kg) penetrated the right windshield killing the

Fig. 21.4 The wreckage of the Boeing 737 at Gossellies, Belgium, after a single Woodpigeon (*Columba palumbus*, 465 g) was ingested by an engine. The three crew escaped (photo via J Thorpe)

co-pilot and seriously injuring the pilot. Windshield debris damaged engine No. 2 which had to be shut down. Loss of hydraulics meant that the flaps and wheel brakes were inoperative and wind-blast resulted in communication difficulties. The injured pilot demonstrated great skill in successfully returning to Lunken Airport.

2.2.6 On 20th January, 1995, just as a Dassault Falcon 20 executive jet rotated on take off from Paris Le Bourget, it encountered a flock of Lapwings (*Vanellus vanellus*, 215 g), a number were ingested in No. 1 engine. The pilot reported he was returning due to an engine fire, witnesses saw the rear of the aircraft engulfed in flames. A tight left hand circuit was flown at a height of about 500 ft above ground level, (AGL), in an attempt to land back on the runway but it was about 30° off the runway heading, landed heavily with left bank in a nose-down attitude just to the right of the runway. It was destroyed by impact and fire killing all ten on board. About 15 dead birds were found on the runway close to the point where the aircraft lifted off. The engine fan disc with most of the blades sheared off at the root was found about 500 m to the side of the runway. It had separated and shrapnel had penetrated the rear fuselage puncturing the engine fuel feeder tank. The cockpit voice tape revealed that while taxiing the pilots had remarked 'look at those birds there'. The person responsible for airport bird control had gone off duty due to illness. There is considerable on-going litigation involving the Airport Authority, and the aircraft and engine manufacturers (Fig. 21.5).

2.2.7 While below 500 ft on short finals to land at Ciampino Airport, Rome, a Boeing 737-800 encountered an enormous European Starling flock (*Sturnus*

Fig. 21.5 The remains of the Falcon 20 which attempted to force land at Paris Le Bourget after an engine disintegrated during take-off rupturing a fuel tank and setting the aircraft on fire. All 10 on board were killed. 15 dead Lapwings (*Vanellus vanellus*, 215 g) were found on the runway (photo BEA Report f-In950120)

vulgaris, 80 g). Although an immediate go-around was attempted there were multiple strikes on the nose, wings, windshield and both engines lost power with burning smell and vibration. The engines did not respond and remained at 40 % N1 so a skilfully executed glide landing was made. The aircraft impacted hard forcing the left hand landing gear into the wing structure writing off the aircraft; it slid to a halt with one engine resting on the runway. One person of the 172 on board was injured during the emergency evacuation. The accident happened on 10th November 2008, less than three months before the accident described below.

2.2.8 It was the 15th January, 2009 accident to an Airbus A320 that hit the headlines worldwide. Having taken off from La Guardia Airport, New York, it was climbing through 3200 ft when it collided with a flock of Canada Geese (*Branta canadensis*, 3.6 kg) and both engines lost most of their power. Without enough height to return or reach another airport, it was very skilfully ditched in the Hudson River, New York. All 155 on board were successfully rescued by ferries and other boats.

2.3 In the 47 fatal and destroyed aeroplanes in this group of accidents, the main identified species causing them were gulls—*Larus* sp. at 27 %, followed by various water birds (including Whistling Swan—*Cygnus columbianus*, cranes—*Grus* sp., Lapwings—*Vanellus vanellus*, loon—*Gavia immer*, White Pelican—*Pelecanus erythrorhynchos,* and Canada Goose—*Branta canadensis*) at 15 %, various birds of prey (including vultures—*Gyps* sp., Black Kites—*Milvus migrans,* and Sparrow Hawk—*Falco sparverius*) at 12 %, Pigeons—*Columba* sp. 10 %, and European Starlings—*Sturnus vulgaris* at 7 % (Thorpe 2012) (Fig. 21.6).

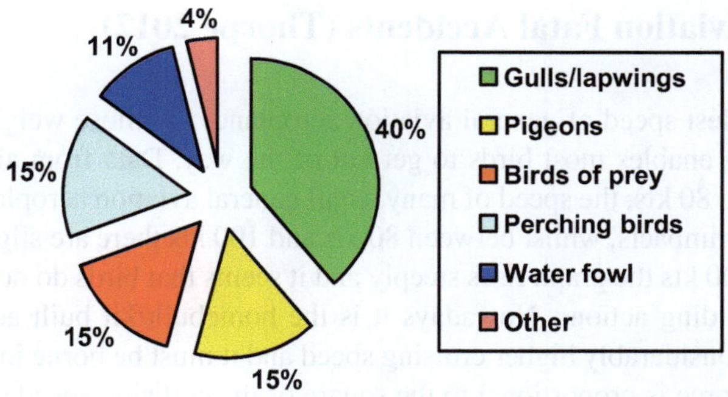

Fig. 21.6 The bird species, where known, in the 47 airliner and executive jet accidents

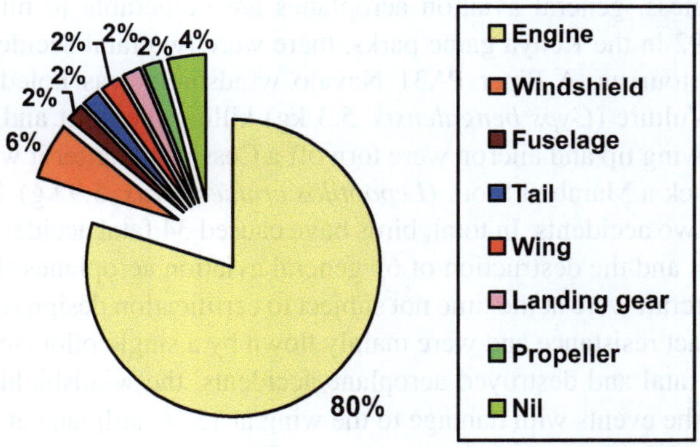

Fig. 21.7 Part struck, where known, in the above 47 accidents

2.4 The main cause, in 80 % of cases, was damage to engines after ingestion of a bird or birds, followed by the windshield with 8 %. A number of these were early Russian turboprop airliners operating as freight aircraft in parts of the world where measures to reduce the risk were minimal or nonexistent. Executive jets often operate from aerodromes with little or nothing in the way of measures to reduce the risk and in many cases their engines are of an age that pre-dates any requirement for them to be tested for bird impact resistance (Fig. 21.7).

2.5 The reasonable record for fatalities of jet airliners may in part be due to improved awareness of the problem, implementation of better airport measures in many parts of the world, and tougher airworthiness criteria for all but the oldest aircraft and engines (see para. 7.2.1).

General Aviation Fatal Accidents (Thorpe 2012)

3.1 The modest speed of general aviation aeroplanes i.e. those weighing less than 5,700 kg enables most birds to get out of the way. Data from airliners shows that up to 80 kts, the speed of many small general aviation aeroplanes, there are very few impacts, whilst between 80 kts and 100 kts there are slightly more but above 100 kts the graph rises steeply and it seems that birds do not have time to take avoiding action. Nowadays it is the homebuilt/kit built aeroplanes that have a considerably higher cruising speed and it must be borne in mind that the impact force is proportional to the square of the collision speed so that a small increase in speed results in a big increase in the impact force. In a number of the general aviation accidents, the pilot was attempting to avoid birds by taking evasive action and either lost control or collided with obstructions or the ground.

3.2 Nevertheless, general aviation aeroplanes are vulnerable to bird hazards. In 1991/1992 in the Kenya game parks, there were two fatal accidents to aircraft carrying tourists. A Piper PA31 Navajo windshield was holed by a White-backed Vulture (*Gyps bengalensis*, 5.3 kg) killing the pilot and shortly afterwards a wing tip and aileron were torn off a Cessna 402 after it was believed to have struck a Marabou Stork (*Lepoptilos crumeniferus*, 5.9 kg). In all, 16 died in these two accidents. In total, birds have caused 34 fatal accidents resulting in 71 deaths and the destruction of 61 general aviation aeroplanes (Fig. 21.8).

3.3 These aircraft were at the time not subject to certification design requirement for bird impact resistance and were mainly flown by a single pilot (see para. 7.2.4).

3.4 In these fatal and destroyed aeroplane accidents, the windshield was holed in 44 % of the events with damage to the wing at 13 %, tailplane at 10 %, and the engine and flying controls both with 8 %. The 13 % of nil cases was when the pilot crashed while attempting to avoid birds (Figs. 21.9 and 21.10).

3.5 The birds causing these accidents are very different from those involved in airliner accidents, the cause in 50 % of them were Birds of Prey—*Falconiformes* with a combination of Geese—*Anser* sp., Ducks—*Anas* sp. and Swans—*Cygnus* sp. in 21 % and Gulls—*Larus* sp. 15 % (Fig. 21.11).

3.6 Although it has not yet caused an accident, on the open space of an airfield where nesting sites may be scarce, birds can quickly build a nest in a variety of places in an aircraft. This includes inside the engine cowling presenting a fire hazard or in the fuselage among the control cables. This is more prevalent in general aviation aircraft which do not fly as frequently as airliners earning their keep.

Helicopter Fatalities (Thorpe 2012)

There have been eight fatal accidents causing 19 deaths and the destruction of 12 helicopters. The total is small considering most helicopters operate low-down where birds fly most frequently. In the past, helicopters had a relatively slow cruising speed, coupled with rotor noise, giving sufficient warning for birds to get out of the

Fig. 21.8 The Cessna 402 wing tip and aileron together with part of the Marabou Stork (*Lepoptilos crumeniferus*, 5.9 kg) embedded in it. All seven on board were killed after the pilot was unable to maintain control (photo via J Thorpe)

Fig. 21.9 Part of general aviation aircraft struck, where known, in the fatal and destroyed aeroplane accidents

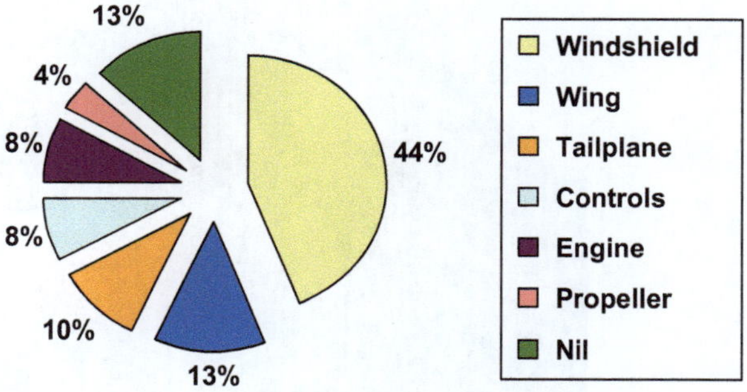

Fig. 21.10 This African white-backed vulture (*Gyps africanus*, 1.7 kg) smashed through the windshield of the Cessna 206 while it was flying at 2500 ft over The Okavango nature reserve in Botswana. The pilot and four passengers escaped with minor injuries (photo via J Thorpe)

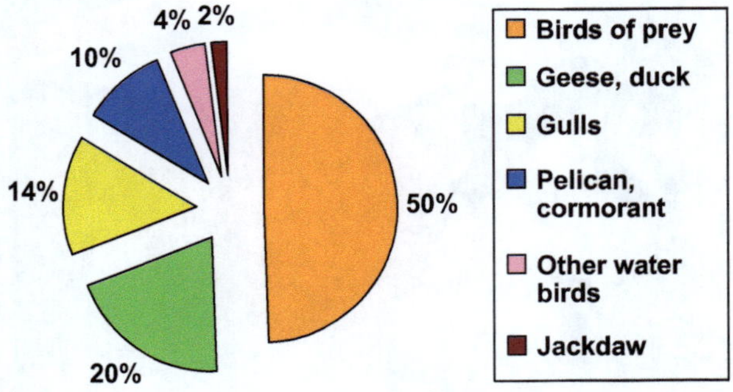

Fig. 21.11 The bird species, where known, causing the general aviation accidents

Fig. 21.12 A typical case of a shattered helicopter windshield (photo via J. Thorpe)

way. However, the trend towards faster and environmentally quieter helicopters may well result in increased problems, especially as vulnerable windshields were holed in 50 % of the helicopter accidents, particularly after collision with heavy birds (Raven—*Corvus corax*, 1.1 kg, Buzzard—*Buteo* sp. and Red-tailed Hawk—*Buteo jamaicensis*, 1.1 kg) (Fig. 21.12).

Military Aviation Losses (Transport Canada 2001)

5.1 Information on military aircraft losses is harder to obtain but it is known that since 1950 there have been over 350 aircraft lost of which at least 63 were fatal to 141 occupants and to four on the ground. Military aircraft frequently operate at low level at high speed, 420 to 450 kts, which is where birds are most likely to be encountered, resulting in a large increase in impact force. Most single and twin-engined fighters and training aircraft usually have ejector seats which have saved many pilots lives. At least 40 twin-engined military aircraft have been destroyed and surprisingly five or more of them had four engines.

5.1.1 On 17th November, 1980, a 4-engined RAF Nimrod MR2 maritime reconnaissance aircraft was making a dawn take-off from RAF Kinloss in Scotland when at 20 ft it collided with a flock of mixed Gulls (*Larus* sp.) badly damaging three engines and crashed in woodland killing the two pilots, fortunately the other 17 on board escaped.

5.1.2 In September 1987, a USAF Rockwell B-1B bomber was flying at 600 ft and 560 kts on a low level training route in Colorado, USA, when an American White Pelican (*Pelecanus erythrerhynchos*, 7 kg) struck the wing just above an engine nacelle starting a fire that damaged the hydraulic control system causing the aircraft to roll uncontrollably. Three of the six crew were unable to eject and were killed. The aircraft had not been designed for a major bird strike and the fleet was subsequently modified to reduce its vulnerability.

5.1.3 Early in the morning of 22nd September, 1995, a fully loaded USAF E-3B AWACS, which is based on the Boeing 707 airliner, was taking off from Elmendorf Base in Anchorage, Alaska. At lift off, 153 kts, it struck a flock of Canada Geese (*Branta canadensis*, 3.6 kg) and some were ingested by both left hand engines which lost power such that it could not climb and became uncontrollable. It crashed on a forested hillside killing all 24 crew members.

5.1.4 Another E-3 AWACS was destroyed on 14th July, 1996, when taking off from Aktion Air Base in Greece when multiple bird strikes caused the crew to abandon the take-off at approx. 140 kts. It ran off the end of the runway into the sea. The 14 crew were lucky to escape without serious injury.

5.1.5 A Belgian Air Force 4 turboprop-engined Lockheed C-130H Hercules with 41 on board was on final approach to the Dutch Air Force base at Eindhoven, Netherlands, when it flew through a massive flock of Lapwings (*Vanellus vanellus*, 215 g) and European Starlings (*Sturnus vulgaris*, 80 g), dozens were ingested by the engines causing three to lose power. It crashed beside the runway and fire erupted. Air Traffic Control had not informed the fire service that there were passengers on board and they concentrated on extinguishing the fire to rescue the cockpit crew before opening the rear door by which time 34 of the 41 on board had died from the effects of toxic fumes (Fig. 21.13).

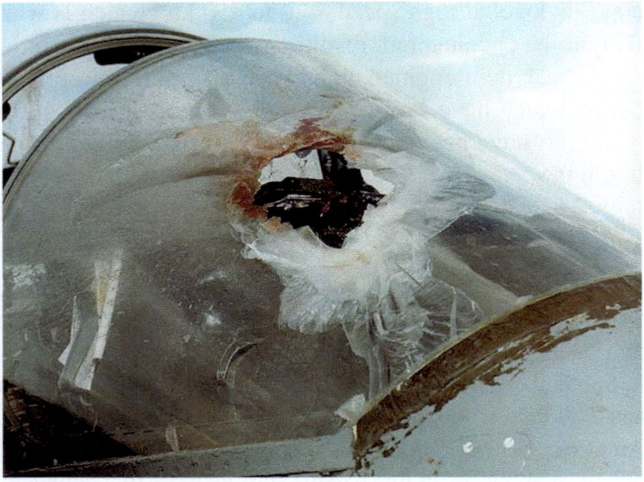

Fig. 21.13 Typical damage to the windshield of a military fast jet following impact with a bird which has the same effect as a bullet (photo via J. Thorpe)

5.1.6 In January 2014, a US Air Force Sikorsky HH-60G Pave Hawk helicopter was on a routine night low level exercise which passed close to a nature reserve in North Norfolk, UK. While flying at 310 ft above the ground at 110 kts, it impacted with a flock of Geese; (reportedly weighing between 2.7 and 5.4 kg) three shattered the windshield knocking the pilot and co-pilot unconscious and disabling the trim and flight path stabilisation system. The helicopter crashed killing the four crew members. The geese were probably Pink-footed Geese (*Anser brachyrhynchus*, av. wt. 2.45 kg), which differs from the reported weight. Whatever the exact weight, the windshield would not have been able to withstand that sort of impact.

Nonfatal Bird Strikes

6.1 Bird strikes occur about once in every 2000 flights with serious damage in about 1 in 15,000 flights, a rare event but which cannot be ignored. There have been many cases of multiple engine damage, fortunately either with enough runway length to abandon take-off, or sufficient power available to return. European airlines continue to experience about 20 cases per year where more than one engine ingests birds. Forty years ago, a high proportion of airline flights used 4-engined aeroplanes, whereas nowadays most flights use twin-engined aeroplanes thus increasing the serious consequences that can result from a double engine strike. It has been estimated that bird strikes worldwide cost the aviation industry over US$1 billion per annum in engine and airframe

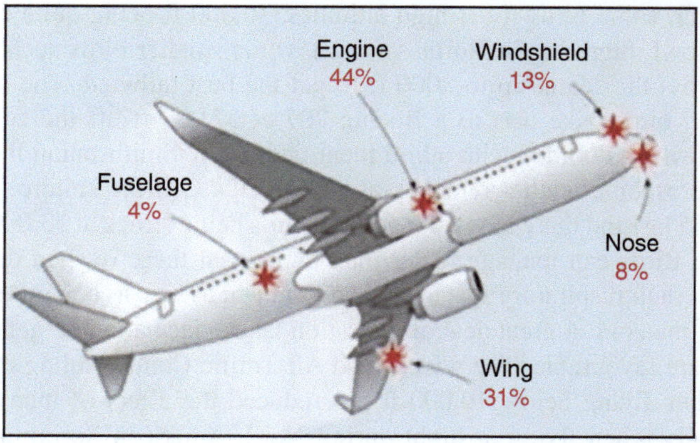

Fig. 21.14 Data from the European Aviation Safety Agency on the Aircraft part struck in both damaging and non-damaging collisions with birds (figure from EASA report)

damage and the associated cost of delays. It should be borne in mind that, for example, a Rolls Royce Trent engine, which is in widespread use, costs nearly US$30 million.

6.2 In one 5-year period, European airlines reported over 7500 strikes, some causing damage, but the majority had no effect. Gulls were involved in almost 40 % of the strikes followed by Lapwings (14 %), various Swift, Swallow, and Martin species (11 %), which are harmless to airliners and various Birds of Prey at 10 %. Lapwing and European Starling strikes have reduced in recent years due to the declining populations (Fig. 21.14).

6.3 In the USA, there were 138,257 reported bird strikes to civil aircraft between 1990 and 2013. Due to better reporting, during the 24-year period, there was a sixfold increase in reported bird strikes, both damaging and non-damaging, while the number of commercial flights only trebled. About 9 % of the reports resulted in damage; although the percentage of damaging strikes has fallen in commercial aviation, it has not in general aviation. Fifty-two percent of bird strikes occurred between July and October. Birds are slightly more likely to be struck during the landing (i.e. descent, approach, or landing roll) phase of flight compared with take-off and climb when the aircraft climbs rapidly to get to its cruising altitude as fast as Air Traffic Control permits it, thus getting out of the bird-rich environment quicker than during the more protracted approach and landing phase. For commercial and GA aircraft, 71 and 74 % of bird strikes, respectively, occurred at or below 500 ft above ground level (AGL). Above 500 ft AGL, the number of strikes declined by 34 % for each 1000-ft gain in height for commercial aircraft, and by 43 % for GA aircraft. Strikes occurring above 500 ft, when speeds are higher, were more likely to cause damage than strikes at or below 500 ft. During this 23-year period, the highest reported US bird strike was 31,300 ft (Federal Aviation Administration 2014).

6.4 However, some birds fly at high altitudes, 10,000 ft being quite common for water fowl migrating in North America, while smaller birds such as thrushes may cross the sea at up to 8000 ft to get the best tailwind. The highest ever reported bird strike was to a Boeing 707 at 37,000 ft off the coast of West Africa, which collided with a bird identified by the Smithsonian Institute from the microscopic feather remains, as a Ruppell's Griffon Vulture (*Gyps rueppelli*, 7.5 kg) and there are two reports of birds being struck at 33,000 ft over the Sahara. Birds can manage better than humans at these oxygen depleted altitudes as their respiratory system is more efficient as well as being well-insulated against the cold. A great deal of migration takes place at night, generally when winds are favourable. The widespread Air Traffic Control ruling of 250 kts or less when flying below 10,000 ft has reduced the effect of many collisions, especially when flying at night during North American Spring and Autumn waterfowl migration.

Reducing the Threat

Scope of Paragraph

The information in this paragraph provides a brief overview of some ways in which the threat can be reduced; it is not intended to be a comprehensive guide to all aspects and readers should refer to appropriate documents in the References.

Civil Aircraft and Engines (Eurocontrol Skybrary 2014)

7.2.1. One solution to reducing the hazard is to make the aircraft and engines better able to resist bird impacts. This adds to the weight and makes the aircraft less economic to operate. Airliner engines have been made progressively stronger and the latest can now withstand a single bird of up to 3.65 kg without hazard to the aircraft, although thrust will be lost. Jet engines can also withstand ingestion of a certain mass of birds, depending on area of the intake, while still retaining an adequate proportion of the thrust. This capability must be demonstrated by tests on an operating engine. (Full details are contained in reference documents.) A number of relatively recent aeroplanes and engines meet the improved ingestion criteria, but it is not feasible to apply new standards retrospectively to previously produced engines which are likely to be in service for many years to come. Some may wonder why a metal mesh or guard is not fitted over the intake. Unfortunately, this solution has not been practicable as the guard would tend to reduce engine efficiency and may be more susceptible to icing. Also, the guard would have to be very strong to withstand the forces from bird strikes or damage from other debris (for example, a burst tyre) resulting in pieces of the guard being ingested by the

Fig. 21.15 Typical bird damage to an older jet engine, in this case a Pratt & Whitney JT8D (photo via J Thorpe)

engine. However, research is still ongoing to develop an effective system and materials to deflect birds and other foreign objects from jet engines without degrading performance (Fig. 21.15).

7.2.2. Under long-standing International rules civil airliner windshields must be able to withstand, without hazard, impact with a bird of 1.8 kg at the maximum allowed speed at sea level, Vc, or 0.85 Vc at 8000 ft whichever is the most critical. In 1989, an A320 aircraft at 2500 ft and 250 knots IAS collided with a Vulture (*Gyps* sp., 4.5 kg) just above the cockpit windscreen. Although the windshields were not penetrated, the impact destroyed four of the six cockpit display units (CRT's) and shock loaded an engine fire button in the roof panel causing one engine to shut itself down. Penetration of jet or turboprop airliner multi-layer windshields by a bird of any size is extremely rare, even though in some cases the outer layer has been shattered. As a result of the Vickers Viscount accident described in para. 2.2.1, the US Federal Aviation Administration brought in a rule relating to a 3.6 kg bird impacting with an airliner tail area

7.2.3. The lighter weight Executive jets of less than 5700 kg (e.g. Lear 23 and some 24s) were Certificated to US Part 23 Requirements for general aviation aircraft which did not require bird impact resistance.

7.2.4. In the past, there was no bird strike legislation for general aviation aeroplanes, and helicopters, it has been known for a Swift (*Apus apus*, 40 g) to penetrate a helicopter windshield. However, commuter aeroplanes with between 10 and 19 seats have had a 1 kg bird proof windshield. There are now requirements and conditions for newly certificated small aeroplanes (at the maximum approach speed using landing flaps) and for helicopters. Some aeroplanes may have an option for a thicker windscreen, which will also provide a quieter cockpit environment. Although the new rules will not prevent penetration by a large bird, it will absorb some of the impact force and reduce pilot injury. Furthermore, designers of kit/homebuilt aircraft which are in many cases much faster than older factory produced aeroplanes, should consider the windscreen thickness and its structure in relation to the aircraft cruise speed (UK Civil Aviation Authority 2013).

Military Aircraft and Engines

The strength of military aircraft is a more complex matter and is almost always shielded from public view by the need for Security. Each country that manufactures military aircraft will have its own airworthiness standards, although co-operation with prospective customer countries is likely to influence what each aircraft can cope with.

Aerodrome Environment

7.4.1 The main area where measures can be taken to reduce the threat to aircraft from birds is the airfield which will be affected by the nature of the surrounding area and the local problem and solutions are generally specific to the particular site. The aerodrome area must be kept unattractive to birds which are looking for food, water, and shelter. Garbage dumps and landfill sites are a major attraction for many bird species and must not be allowed where close to an airfield or in a position where birds may 'commute' across the airfield to their roost sites or to open water. International Rules suggest 13 km from an airport (International Civil Aviation Organisation 2014).

7.4.2 Many years ago, scientific trials involving bird-counts on a number of airfields over a couple of years showed that for most species, including Gulls (*Larus* sp.), mowing the grass to a height of between 15 and 20 cm restricted the birds view of threatening predators so that the birds went elsewhere such that the number using the airfield was reduced by an average of 80 %. This may not be possible at some airports where they are unable to grow this 'long grass' due to the lack of rainfall, unsuitable soil, etc. (International Bird Strike Committee 2006).

7.4.3 If the growing of crops is permitted on an airport they must be a variety that does not attract birds, while grass should not be allowed to go to seed as it may attract large flocks of smaller birds.

7.4.4 It must not be forgotten that birds fly at night and that their nocturnal activity can be affected by the tide if near the coast, by the weather conditions, and by the brightness of moonlight. They will also fly in poor visibility.

7.4.5 When birds are present, a variety of scaring techniques can be used, the main one being the use of distress calls for the species causing the problem. These are now digitised and broadcast through a loudspeaker on top of an airport vehicle. The bird's response is instinctive and they show little sign of getting used to these. Some species such as Pigeons (*Columba* sp.) and many Birds of Prey (*Falconiformes*) do not have distress calls and other means have to be used (Wildlife Management at Aerodromes, CAP 772, UK Civil Aviation Authority 2014).

7.4.6 Shell crackers which leave a smoke trail before exploding are very effective. A range of other methods includes gas cannon, radio-controlled model aircraft, lasers, trained dogs, effigies of dead birds, falconry, and lasers. The effectiveness of these varies depending upon the problem species and the particular site. There is rarely one simple solution as birds learn quickly what is a real threat which is why effigies, gas cannon, and fixed installations can quickly become ineffective. If used, they need to be moved to a different position on an almost daily basis. They do not get used to distress calls. All means should be regarded as 'tools in a toolbox'.

7.4.7 Mass killing of birds is not usually a realistic measure as it leaves an ecological hole that other birds will fill. Occasional killing may be necessary as a means of reinforcing the threat from other dispersal methods.

7.4.8 Proper identification of bird remains is important in determining what the problem is, even small downy feathers are enough for identification under a microscope and DNA techniques are becoming more widely available.

7.4.9 The most effective measure is the use on aerodromes of properly trained and equipped, motivated staff who enjoy their job and understand birds and their habits. The old saying should be borne in mind 'if you think safety is expensive—try an accident'. If all else fails, it may be necessary to reschedule aircraft movements; perhaps only appropriate during the migration season.

Use of Radar

During the Second World War, it was realised that the slow moving 'angels' that appeared on early radar screens were birds. In some parts of the world, migrating birds follow set routes used by countless previous generations. Some countries including in particular Israel, Germany, Netherlands, and the USA have had great success in using specially adapted radar, assisted in some cases by ground observers to track these migrations, thereby reducing bird strike damage by curtailing or delaying military training flights, particularly those at low-level, such that routes and times avoid the worst periods. This is not practical for much of civil aviation where airliners follow set routes, are flying to a timetable with departure and arrival slots and at many airports noise curfews or restrictions. In any case, they are intent as soon as possible climbing to their cruise altitude. On a local basis, birds may 'commute' across an aerodrome when moving to feeding grounds from lakes, coastal marshes, and roosts. At some airfields, specially adapted radar is used to track these local movements. Nowadays, most Air Traffic radars interrogate a transponder in the aircraft and the basic primary radar has had slow moving objects removed otherwise vehicles would clutter the screen. Considerable radar development is underway and future airborne radar systems could be adapted to 'see' birds, while the aircraft is flying that could provide aircrew with 'real time' warning of the hazard. A parallel is the development that has already taken place of airborne warning systems to minimise mid-air collisions between aircraft.

Flight Crew Actions

Parked aircraft may provide an attractive Spring-time nest site. During pre-flight inspection of the aircraft, pilots should check apertures for signs of bird entry. A nest may be a fire hazard or can obstruct the flying controls, etc. If when taxiing for take-off, flight crew see birds on, near, or flying over the runway, they should delay their departure and alert Air Traffic Control so that aerodrome personnel can be called to remove the potential hazard. After take-off the risk of a strike decreases rapidly as the aircraft gets higher, 70 % of strikes occur at less than 200 ft above the ground. When airborne, even when flying slowly, if a bird is seen, it is almost impossible to avoid it in time. In any case, birds are much more agile than aircraft and when threatened birds take evasive action by diving; this author has personal experience on two occasions when flying a general aviation aeroplane. Near the ground there have been cases of pilots losing control when trying to avoid birds and crashing during the attempt. Pilots who engage in low flying tasks should wear a protective helmet, or at the very least, goggles in case of windscreen damage or penetration. When approaching to land if birds are visible in the landing area, pilots may need to go-around and alert ATC to call ground staff to remove the hazard. Following a bird strike, any remains should be identified by an expert, even small feathers can be sufficient when examined under a microscope.

Legislation, Reporting, and Liability

8.1 ICAO, the International Civil Aviation Organisation, has for many years implemented in Annex 14 para. 9.5 measures on Bird Hazard Reduction together with ICAO Airport Services Manual Part 3 'Bird Control and Reduction' (ICAO 2014). The standards detailed are reasonably well-applied in the developed world, particularly Europe, North America, the Antipodes, and parts of the Far East. In Europe, the European Aviation Safety Agency has endorsed EASA, ADR.OPS.B.020 Wildlife strike hazard reduction.

8.2 Everyone in aviation including pilots, airport personnel, Air Traffic, and engineers are encouraged to report ALL bird strikes, whether they cause damage or not. Reports should be sent to the Aviation Authority in each country and forwarded to the International Civil Aviation Organisation (ICAO) in Montreal for inclusion in IBIS (the International Birdstrike Information System). It is only by knowing what and where the problems are that the international threats can be reduced. In some countries, reporting is mandatory in others only voluntary, while those with limited resources are likely to ignore the problem.

8.3 There have been a number of court cases where airfield personnel have been found guilty of negligence and jailed. Furthermore, aircraft operators have taken legal action against an airport. One of the most important involved the nonfatal destruction in December 1973 of a Falcon 20 executive jet at Norwich Airport in the UK. The aircraft owners brought a case in the High Court of Justice against the airport for their loss. In summary, the Judgement was that the airport had failed to take 'reasonable precautions' as described in the then current Civil Aviation Authority documentation on the subject. Cases for restitution for loss or damage have been bought at other airports including Amsterdam, Chicago, New York Kennedy, and Paris Le Bourget. Where they were able to prove that they have complied with relevant material, they were found not liable for damages (Fig. 21.16).

Fig. 21.16 The Norwegian registered Falcon 20 that was the subject of The court case at Norwich Airport, UK. It was deemed beyond economic repair by the insurers (photo Eastern Counties Newspapers)

Final Remarks

9.1 Although *not* a major cause of aircraft accidents, bird strikes are nevertheless a serious safety and economic hazard. Aircraft continue to be destroyed and occupants killed or injured in accidents due to:

- Striking birds
- Attempting to avoid birds
- Birds being the start of a chain of events

Aerodrome remedial measures and tougher aircraft/engines appear to have improved transport aircraft safety. However, twin-engined airliners have by and large replaced those with three or four engines and there is a risk of both engines being damaged. Engine damage is the major risk for this group of aircraft, with flocking Gulls (*Larus* sp.) the main threat causing 33 % of the accidents. This underlines the importance of the thorough application of aerodrome bird control measures by dedicated staff.

9.2 Business jets appear to be particularly vulnerable especially when operated from aerodromes with little or no bird control measures. In recent years, a significant number of accidents involve early Russian aircraft operating from 'remote' or third world areas where bird control measures are unlikely.

9.3 'General aviation' aeroplanes are most vulnerable to the windshield being holed, the cause of 50 % of the accidents. Birds of Prey (*Accipitriformes*) are generally heavy and were responsible for half of the accidents. This group of aircraft mostly fly at heights where almost-impossible-to-spot-birds are most prevalent.

9.4 Exactly half of helicopter accidents were due to the windshield being holed, sometimes by heavy birds. Again, helicopters mainly operate low down where most birds fly and the trend is towards faster, quieter helicopters, providing less time for birds to take evasive action.

9.5 Bird strike accidents are a rare event that occur out-of-the-blue even at airports which may consider that adequate measures are in place to minimise the risk. All should bear in mind that 'complacency is the enemy of safety'.

Acknowledgments With thanks to W. John Richardson LGL Ltd, Canada, the late Bruce McKinnon of Transport Canada, Dr. Richard Dolbeer and Sandra E. Wright of US Dept of Agriculture and in particular to the un-known and un-credited photographers of the illustrations used in this chapter.

References

Aircraft Certification for Bird Strike Risk, Eurocontrol Skybrary (2014) Brussels, Belgium. Available via http://www.skybrary.aero/index.php/Aircraft_Certification_for_Bird_Strike_Risk

Bird Avoidance, General Aviation Safety Sense Leaflet 10c. UK Civil Aviation Authority (2013). London, England. Available via http://www.caa.co.uk/application.aspx?catid=33&pagetype=65&appid=11&mode=detail&id=1164

Brough T (1983) Average weight of birds'. UK Agricultural Science Service, Guilford
Bird Control and Reduction, International Civil Aviation Organisation, ICAO, Airport services manual, Part 3, Montreal Canada.
International Bird Strike Committee (2006) Recommended practices—standards for aerodrome bird/wildlife control. Available via http://www.int-birdstrike.org/Standards_for_Aerodrome_bird_wildlife%20control.pdf
Sharing the Skies (2001) TP13549 Transport Canada, Montreal, Canada. Available via http://www.tc.gc.ca/eng/civilaviation/publications/tp13549-menu-2163.htm
Thorpe J (2012) 100 Years of fatalities and destroyed civil aircraft due to bird strikes. World Birdstrike Association, Amsterdam. Available via http://worldbirdstrike.com/Stavanger/100%20years%20of%20fatalities%20and%20destroyed%20civil%20aircfaft%20due%20to%20bird%20strikes%20Paper.pdf
Wildlife Management at Aerodromes, CAP 772 (2014). UK Civil Aviation Authority London, England. Available via http://www.caa.co.uk/application.aspx?appid=11&mode=detail&id=2726
Wildlife Strikes to Civil Aircraft in the United States 1990 to 2013, Federal Aviation Administration (2014). Washington DC. Available via http://wildlife.faa.gov/downloads/Wildlife-Strike-Report-1990-2013-USDA-FAA.pdf

Chapter 22
Approaches to Wildlife Management in Aviation

Jeffrey McKee, Phillip Shaw, Arie Dekker, and Kylie Patrick

Abbreviations

AGL	Above Ground Level
AHAS	Avian Hazard Advisory Service
AIREP	Airborne Report
ATC	Air Traffic Control
ATIS	Automatic Terminal Information Service
ATSB	Australian Transport Safety Bureau Area Forecast
ARFOR	Area Forecast
AUSALPA	Australian Airline Pilots Association
BAM	Bird Avoidance Model
BASH	Bird Aircraft Strike Hazard
BCAS	Bird Collision Avoidance System
BIRDTAM	Bird Notice to Airmen
CAA	Civil Aviation Authority
CAP	Civil Aviation Publication
CASA	Civil Aviation Safety Authority
EASA	European Aviation Safety Agency
ERAU	Embry—Riddle Aeronautical University
ENRAM	European Network for the Radar surveillance of Animal Movement

J. McKee (✉) • P. Shaw
Avisure Pty Ltd, Burleigh Heads, QLD, Australia
e-mail: jmckee@avisure.com

A. Dekker
Royal Netherlands Air Force (Ret), PO Box 8762, 4820 BB Breda, The Netherlands

Const. Huygensstraat 4, 1901 JA Castricum, The Netherlands

K. Patrick
Avisure Pty Ltd, Brasilia, Brazil

F.M. Angelici (ed.), *Problematic Wildlife*, DOI 10.1007/978-3-319-22246-2_22

ESA	European Space Agency
FAA	Federal Aviation Administration
FOD	Foreign Object DebrisGIS Geographic Information System
GIS	Geographic Information System
IAF	Israeli Air Force
IATA	International Airline Transport Association
ICAO	International Civil Aviation Organisation
IFALPA	International Federation of Airline Pilots Associations
IFATCA	International Federation of Air Traffic Controllers Associations
KSIA	King Shaka International Airport
METAR	Meteorological Terminal Aviation Routine Weather Report
NBAA	National Business Aviation Association
NEXRAD	Next Generation Radar
NOTAM	Notice to Airmen
OPERA	Operational Program for the Exchange of Weather Radar Information
RAAF	Royal Australian Air Force
RAF	Royal Air Force, United Kingdom
RNLAF	Royal Netherlands Air Force
SEATAC	Seattle—Tacoma Airport
SMR	Small Mobile Radar
SMS	Safety Management System
TAF	Terminal Area Forecast
USAF	United States Air Force
USDA	United States Department of Agriculture
WBA	World Birdstrike Association
WHMP	Wildlife Hazard Management Plan

The Conflict Begins

In 1904, the Wright brothers made 105 flights averaging only about 30 s per flight before control issues, particularly pitch instability, abruptly terminated each sortie. In July of that year, a serious crash convinced them that unless their Wright Flyer was significantly redesigned, powered flight would remain a curiosity with little practical application (Centennial of Flight 2014). Over the next year, they redesigned and built the Wright Flyer III increasing its roll stability and pitch and yaw authority. The improvements worked well, and on October 5th, 1905, they achieved controlled sustainable flight, flying 24 miles in 40 min and exhausting their fuel reserves (Wright Bros Aeroplane Company 2014). Suddenly, the Wright brothers were in the realms of practical flight and one can imagine their exuberance during the preliminary trials in September when they regularly achieved controlled flights lasting more than 5 min without crashing. That exuberance may well have prompted Orville Wright to "chase a flock of birds for two rounds." on September 7, killing one of the birds, recording the first "wildlife strike," and so beginning the conflict between aviation and wildlife

Fig. 22.1 Eugene Gilbert (1889–1918). During the 1911 Paris to Madrid air race, an eagle attacked his aircraft and he shot at it from the cockpit while in flight. Apparently, he deliberately aimed away from the bird attempting to scare it rather than trying to kill it. Eugene could be considered the father of aviation wildlife management. He is the first to document aviation related wildlife dispersal and he instinctively demonstrated three fundamental principles that remain valid today: firstly, pilot-in-command is ultimately responsible for managing the aircraft's collision risk; secondly, management should aim to separate the wildlife's flight path from the aircraft's flight path; and finally, it is not always necessary to kill the wildlife to achieve this aim. *Photo taken by Enrique Guinea Maquibar, Vitoria-Gasteiz, Euskal Herria in March 1913. Reproduced courtesy of Javier Berasaluce Bajo, Municipal Archives of Vitoria-Gasteiz, Euskal Herria, Spain*

(Wright Stories 2014). That conflict remains essentially unresolved and one that plagues us today both as an air safety and conservation issue.

By 1911, powered flight had become a practical, if not precarious, exercise and application had expanded into cargo payloads, aerial bombing exercises, parachute jumps, amphibious operations, and limited passenger transport. It had also become a competitive sport; the French aviator, Eugene Gilbert (Fig. 22.1), during the 1911 Madrid to Paris air race, reported shooting at an eagle from the cockpit to deter it from harassing his aircraft (Early Aviators 2014).

Not long after Gilbert's bizarre experience, the inevitable happened; on April 3rd, 1912, the first human fatality due to wildlife strike was reported in the USA. In September of the previous year, Calbraith Rodgers miraculously survived after his Wright Flyer Model B collided with a chicken coop (killing several chickens) after take-off on the second leg of his trans-continental crossing. This may be the first recorded incident of liability as a result of animal strike; Calbraith was required to immediately pay compensation to the chicken farmer. Only 7 months later, having successfully completed his epic east to west flight across the USA, he was killed when his aircraft crashed into the sea off Long Beach California. He was conducting an exhibition flight when his aircraft struck a gull that entangled the controls. Reports from the time suggest that, like Orville Wright in 1905, Calbraith was deliberately "buzzing" the flock when he collided with the bird (Wikisource 2014, Lienhard 2003). Ironically, he had survived innumerable crashes and personal injury during the transcontinental flight that took a gruelling 84 days and required 70 legs (Fiddlers Green 2014). At that time, undercarriage was a vague concept and airports were unheard of, so each landing was essentially a controlled crash requiring the aircraft to be rebuilt before the next leg. When he finally arrived in California, only the rudder and a wing strut were original components of his aircraft. He had skirted thunderstorms, had "a run in with an eagle," and replaced the engine twice (Aerofiles 2014).

The Early Years

Before World War II, wildlife strike remained a relatively benign aviation safety issue. Between 1923 and 1940, the UK Royal Air Force (RAF), reported six hull-loses and no fatalities as a result of wildlife strike. However, in the 1940s aircraft movement rates surged and flight speeds dramatically increased. Consequently, RAF reported 33 hull loses and 16 fatalities resulting from wildlife strike over this decade (Richardson and West 2005). Then, in the 1950s, the advent of larger and faster jet aircraft and a global increase in air passenger traffic sparked the emergence of wildlife strike as a modern day air safety issue.

In the following 50 years while aircraft, air operations, and air safety graphically advanced into the Space Age and the era of streamlined high capacity transport, wildlife strike management in civil aviation remained conceptually and technically static. Back in 1911, opportunistically and no doubt in desperation, Eugene Gilbert pioneered the practice of shooting at birds to scare them away from aircraft. Curiously, this practice remains to this day one of the mainstays of wildlife strike mitigation, but since Gilbert's initial efforts the practice has devolved in focus; management emphasis shifted from scaring birds away from aircraft to scaring them away from aerodromes.

Wildlife Strike Today

Since the turn of the millennium, the rate at which wildlife and aircraft collide has increased. Between 2000 and 2010, civil wildlife strike rates (collisions per 10^4 aircraft movements) steadily increased in all but one of nine developed countries surveyed (McKee et al. 2012). Wildlife strike is almost always fatal to the wildlife involved and sometimes it has serious consequence for aircrew and passengers. Strikes causing damage or having an adverse effect on flight constitute roughly 5–10 % of total strikes, but strikes causing hull loss or human fatality are relatively rare. Absolute numbers (per year) of compromised flights, fatalities, and hull loss resulting from strike have increased over the last 40 years, but it is difficult to obtain data normalised for aircraft movements and therefore derive a meaningful trend for serious wildlife-related accidents. For more in-depth discussion of damage and serious accidents, see Thorpe (2015), Dolbeer (2013), Dolbeer et al. (2013) and Eschenfelder (2009). An excellent archive and discussion of serious wildlife strike incidents from 2007 to the present can be found at the Italian Birdstrike Consulting & Training web site (BC&T News and Events 2015). In 2000, total costs to the aviation industry resulting from wildlife strike were estimated at US$1.2B/annum, and the average cost of a strike to high capacity aircraft was estimated at US$39,000 (Allan 2000). Downstream effects such as delay, cancellation, or aborted procedures incurred three quarters of these costs. These industry cost estimates are likely to be gross underestimates, as many countries do not maintain reliable or mandatory wildlife strike-reporting procedures.

The conservation and animal welfare costs of strike are harder to reliably quantify. Tens of thousands of animals die each year as a direct result of collisions with aircraft, and arguably, ten times that number are culled as part of aerodrome wildlife strike management programs. To date the air safety implications of strike have been the sole focus of collision consequence and the wildlife attrition has been ignored; there has been little discussion on the effect of wildlife strike on conservation, endangered species management, and biodiversity. Similarly, the ethics and efficacy of culling programs aimed at preventing wildlife strike are rarely addressed with rigor. As a result, some communities, particularly those in first world countries, are becoming less inclined to blithely accept cavalier calls for lethal wildlife control, particularly if the rationale for those measures are not evidence-based (Bridger 2013; Uhlfelder 2013). It appears that large scale culling programs are initiated more because the approach graphically imparts confidence that "something is being done" rather than because there is any evidence that the approach is effective in reducing strike rates. A recent excellent review of this issue by the Swiss Ornithological Institute should be available by early 2016 (Rey and Liechti 2015 in print). While air safety must remain a paramount priority, it is clear that the time for the application of more effective and less invasive wildlife control measures is long overdue within the civil aviation industry.

Factors Contributing to Changing Strike Risk

Several factors are thought to be driving the widespread increase in wildlife strike rates. Improved surveillance and reporting procedures have caused apparent increases in strike rates in some countries such as the UK and Australia, where, over the last 10 years, wildlife strike reporting changed from voluntary to mandatory. There are also biological, air-operational, and cultural factors that are thought to be driving real strike rate increases. Urbanisation and conservation practice in Europe and North America have led to population resurgences of high strike risk species such as Canada Geese *(Branta canadensis)* bringing them into more frequent conflict with aircraft in both terminal and en-route airspace (Buurma 1996; Dolbeer and Eschenfelder 2003; Dolbeer 2013; Eschenfelder and DeFusco 2010; Dolbeer 2011; Moller 2009; Maragakis 2009). Similarly, complex factors have contributed to range shifts and urban redistribution of Australian White Ibis (*Threskiornis molucca*), and Flying Foxes (*Pteropus spp.*) in Australia (Smith 2009; Roberts et al. 2011). In turn, the increasing abundance of these species in urban landscapes has brought them into more frequent conflict with aircraft in terminal airspace (Patrick et al. 2008; Parsons et al. 2009; McKee et al. 2010).

In addition, over the last 20 years design evolutions in commercial aircraft have incidentally resulted in them becoming more susceptible to wildlife strike. Passenger aircraft have larger frontal areas and greater weight limits and they are becoming quieter while maintaining or increasing reference speeds. Closing speeds in an air-borne strike to a modern high capacity airliner would rarely be less than 120 kts. At these speeds, collision detection, recognition, and avoidance is often beyond the response time of either the aircrew or the wildlife (Dolbeer 2013; Kelly et al. 1999b). In contrast over the same period, improvements in materials, hull, and engine design have made aircraft less susceptible to catastrophic failure as a result of wildlife strike (Dolbeer 2013). This engineering-based approach has reduced wildlife strike risk by reducing strike consequence, but has done little to address the incident rate or the rate of wildlife attrition.

Although wildlife strike rate is, by consensus, standardised as strikes per 10^4 aircraft movements, different countries use different input parameters and assumptions in deriving their wildlife strike rates. In individual cases, it can be difficult to discern whether reporting surveillance, biological, or operational factors are most contributory to observed rate increases. Consequently, there remains some debate as to whether the global strike rate increase is real or artefact. However even given the obvious limitations of definitions, data, and analysis, it remains clear that strike rates to civil aircraft have not and are not significantly decreasing. The corollary to this observation is that traditional approaches to wildlife hazard management and strike mitigation in civil aviation have been ineffective. Despite 50 years of "management," civil strike rates have at best remained static and at worst are steadily increasing to the detriment of air safety and the environment.

The Traditional Wildlife Management Paradigm

Managing aviation-wildlife conflicts is problematic because flying animals operate in an open system and collision between wildlife and aircraft is a dynamic and seemingly unpredictable event. The obvious complexity of the system has led to the issue either being relegated to the "too hard basket" or over-simplified to a point where the management paradigm derives from several logical misconceptions. Some maintain the belief that wildlife strike is an irrelevant and essentially insoluble problem, an attitude of studied professional neglect that tends to depower cooperative management efforts. Many promote the rationale that since 90 % of strikes occur in the vicinity of aerodromes (ICAO 2012), it is therefore an aerodrome problem. This proximity argument was and remains a convenient misdirection; by implication, it assigns responsibility for a complex dynamic airspace problem solely to aerodrome operators and tacitly exonerates the bulk of the industry from contributing to a solution. As a result, the default management approach has become aerodrome-centric and in most countries today the expectation is almost entirely on aerodrome operators to prevent wildlife strike. The aim of this traditional approach is to prevent collision between wildlife and aircraft by attempting to create a wildlife exclusion bubble around airports.

Aerodromes use both active and passive management techniques to try and achieve this aim. Active techniques include the use of noise, light, predator simulations, and trained predators to scare wildlife away from the aerodrome proper. Active techniques also include trapping, relocation, selective euthanasia, broad scale culling, and reproductive control. Passive means include fencing to exclude incursions by terrestrial species and landscape modification to make the aerodrome less attractive to both resident and transient species. More recently, passive management has extended into attempts at managing land and wildlife populations adjacent to the aerodrome; however, this can be fraught with major legal and practical considerations as aerodrome operators normally have no jurisdiction outside their airport boundary. This mechanism is essentially an attempt to extend the exclusion bubble outwards from the airport, but in most situations where it has been tried, the practice is reminiscent of "trying to hold the tide back" and the policy is not backed by enforceable legislation. A broad overview of the principles and requirements of current aerodrome wildlife hazard practice is located on the World Birdstrike Association (WBA) web site and in the International Civil Aviation Organization's (ICAO) Airport Services Manual Document 9137. More in-depth descriptions of the active and passive methods commonly used for aerodrome strike management can be found in the Federal Aviation Administration (FAA) Wildlife Hazard Manual, the United States Air Force Bird Aircraft Strike Hazard (BASH) Guidelines, and the Transport Canada publication, 'Sharing the Skies'.

There are several limitations to the airport-centric approach. Firstly, it is an indirect and static approach to a dynamic problem. Rather than keeping wildlife away from aerodromes, management should be primarily aimed at keeping wildlife and aircraft separated during both the planning and execution phases of flight. To do this effectively, it is necessary to understand and manage the flight paths of both the wildlife and the aircraft.

The observation that approximately 90 % of civil wildlife strikes occur within the vicinity of an airport probably holds true globally, but it is a superficial rationale for the premise that airports alone should hold responsibility for wildlife strike prevention. When examined more closely, this apparent geographic segregation of strike probability is a function of height above ground level. The majority of bird movements occur within 3000’ AGL and it is only an unfortunate association that most civil aircraft movements within this altitude block occur during approach and departure “in the vicinity” of aerodromes. In reality, wildlife strike risk is not an aerodrome problem. It is a below 3000' AGL airspace problem and thus the primary responsibility for managing the problem should reside with those managing and using the airspace. Airport operators have no authority over airspace or aircraft flight paths, and only a very limited capacity to manage the flight paths of wildlife and only then inside the aerodrome boundary. Airport operators are highly skilled at maintaining a safe *static* environment for take-off, landing, and taxi, but normally they have no mandate or experience in managing aircraft collision avoidance even when aircraft are inside the aerodrome boundary. Given that wildlife strike is primarily a flight collision avoidance issue, the level of mitigation responsibility held by airport operators should be adjunctive only. The current skewed balance between authority and responsibility for this issue can best be illustrated by *reductio ad absurdum*; for example, 90 % of all wind shear incidents and 100 % of all crosswind landing incidents occur in the vicinity of an aerodrome. Under the twisted aegis of the current wildlife strike paradigm, these observations mandate that airport operators should be held accountable for these incidents because they failed to stop the wind blowing. Similarly, wildlife are not a fixed component of an airport’s environment, and in most cases, they cannot be effectively managed as such. For the purposes of aviation hazard management, wildlife are better classed as autonomous air space meteors or non-controlled traffic.

While there is no doubt that contemporary aerodrome wildlife management is both necessary and useful, truly effective wildlife collision avoidance can only be managed at an integrated operational level by applying the same conceptual models used to mitigate other dynamic hazards such as traffic separation and weather avoidance. This in turn implies that the operational sectors that hold authority over aircraft flight paths need to positively engage in strike mitigation.

Secondly, the current wildlife management approach (Fig. 22.2) is based on ecologically and operationally flawed premises. The concept that managing a small area within an airport boundary can significantly reduce the rate at which wildlife will infringe the airspace and conflict with aircraft may apply in select circumstances, but it is exceptional rather than generally applicable. Aircraft and flying wildlife operate in an open system, and while aircraft movements around an airport are somewhat predictable, wildlife movement patterns currently are not. In most cases, the way in which wildlife use the airspace is dependent on regional and mesoscale factors such as fluctuations in resource availability, ambient conditions, season, and climate and most of these determinants are outside the control scope of aerodrome operators. The expectation for airport operators to manage their airfield to help reduce strike risk is entirely reasonable and practical. However, the extended expectation that they alone can provide effective management at the multiple spatial and temporal scales required (Martin et al. 2011) to significantly reduce strike rates is impractical and unjust. A good recent example is the emergent conflict with flying-foxes (*Pteropus* spp.); as a result of large-scale habitat changes across their range, these animals are redistributing into urban areas creating an increasing strike

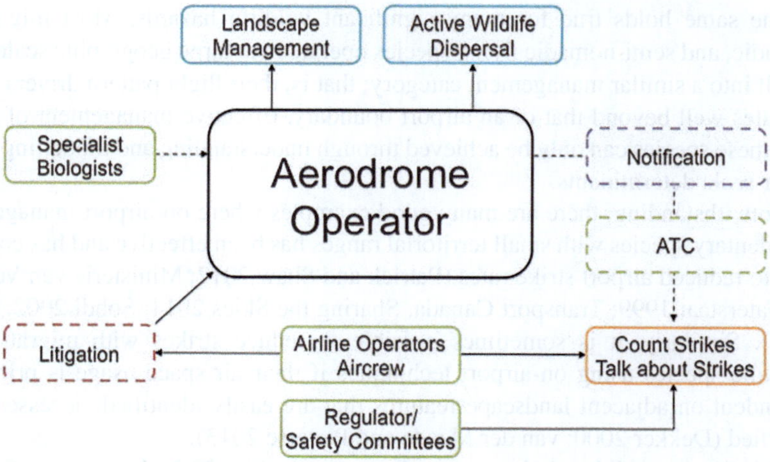

Fig. 22.2 A simplified model summarising the components of the traditional airport-centric wildlife management approach. Most of the expertise necessary to understand the biological component of wildlife strike risk and mitigation resides with specialist biologists. Risk mitigation actions are the responsibility of airport operators and include landscape management, active wildlife dispersal and rudimentary hazard notification by NOTAM. ATC, aircrew and airline operators have no formal responsibilities and very little active input in this system other than occasionally trying to recover strike damage costs from airports. Notification of wildlife hazard by NOTAM is often ineffective. NOTAMS that include precise information about expected wildlife hazards may help with flight planning. However, if they are unsupported by regular reports providing real time hazard updates, they cannot inform of dynamic changes in threat status and cannot provide aircrew with any meaningful information about immediate wildlife threats

risk (Patrick et al. 2008; Parsons et al. 2009). Flying-foxes forage over regional scale areas (Roberts et al. 2012) and the geographic relationship between their roost sites and spatio-temporally patchy food resources determine their choice of flight paths and thus their potential conflict rates with aircraft. No traditional airport-based active or passive mitigation techniques are known to alter their flight tracks. Therefore, the only practical way to mitigate the collision risk is by operational separation; i.e. observe, understand, and if possible anticipate flying-fox movement patterns then relay relevant information to pilots, so they can amend flight accordingly. Other than to ensure that nothing on the airport is an attractant for flying foxes, there is very little an aerodrome operator can or should contribute in this scenario. It is a fundamental airspace-flight path management issue and remains well outside the skill set and jurisdiction of aerodromes.

This conceptual approach works very well to mitigate air traffic conflict and to avoid hazardous weather. Meteorologists, air traffic controllers, and pilots manage these issues dynamically, while aerodrome operators retain only a secondary responsibility for maintaining some of the infrastructure necessary for the procedures to work. It would be ecologically and operationally untenable to expect that aerodrome operators bear the main responsibility for assessing and mitigating dynamic weather and traffic hazards.

The same holds true for many significant wildlife hazards. Most migratory, nomadic, and semi-nomadic flying species operate over large geographic scales and so fall into a similar management category; that is, their flight pattern drivers occur at scales well beyond that of an airport boundary. Effective management of strike with these species can only be achieved through understanding and modelling those larger scale determinants.

Notwithstanding, there are many good examples where on-airport management of sedentary species with small territorial ranges has been effective and has contributed to reduced airport strike rates (Patrick and Shaw 2012; Ministerie van Verkeer en Waterstaat 1999; Transport Canada, Sharing the Skies 2014; Sohdi 2002; Shaw 2008). Similarly, it is sometimes possible to reduce strikes with migratory or nomadic species using on-airport techniques if their air space usage is primarily dependent on adjacent landscape features that are easily identified, accessed, and modified (Dekker 2000; van der Meade and Pieterse 2013).

Thirdly, the traditional airport-centric approach (Fig. 22.2) does not conform to the requirements of an integrated aviation safety management system. Most of the biological skills and knowledge required to understand wildlife movements and thereby reduce the prevalence of wildlife in the airspace resides with external ornithologists and ecologists who assist airport management to assess and control the site strike risk. Some of this knowledge and understanding may filter through to on-ground staff responsible for airside wildlife management, but very little reaches any other industry sector.

Aircraft are most at risk from wildlife strike and aircrew and Air Traffic Control (ATC) are best placed to control aircraft movements thereby avoiding wildlife conflict. Yet in civil aviation today, despite the fact that wildlife strike is the most common cause of in-flight collision, civil aircrews and air traffic controllers remain essentially untrained in wildlife hazard recognition and strike risk management. Consequently, wildlife strike management in civil aviation remains isolated, relatively impotent, and lacking operational relevance. In 2006, a scheduled Boeing 767 ingested wildlife into one engine on departure, but elected to continue to the destination regardless. The investigation report identified that "…The crew had no training regarding wildlife strikes, nor was any required. The operator had no wildlife strike policy other than to report strikes, nor was any required…" (ATSB 2007). This is at odds with integrated management approaches used for all other dynamic environmental hazards to aviation where formal training, examination, and currency practice in hazard identification, prioritisation, and response are mandatory.

Over the last 50 years, the aircraft design and engineering sector together with regulators have also contributed to wildlife strike mitigation, albeit wildlife strike consequence mitigation. Following the horrific fatal strike accident involving a Lockheed Electra in Boston in 1960, the industry adopted "minimum engine failure after ingestion" specifications and subsequently minimum ratings on hull and windscreen strengths (Federal Aviation Regulations 2014; for overview see Demers & McVey 2015) This engineering approach to wildlife strike mitigation has no doubt helped reduce post-strike catastrophic failure rates; however, it does not address the core issue of primary collision avoidance.

Finally, a concerning limitation of the current airport-centric approach in civil aviation is that it is inherently adversarial. There is an increasing trend for airlines to litigate against aerodrome operators in an attempt to recoup strike damage costs to their aircraft and in some cases these attempts have been successful (Dale 2009; Dolbeer 2006; Battistoni 2009). Most of these successes are prefaced on the idea that airports must provide a safe environment for aircraft operations; but they ignore the fact that birds, like thunderstorms, are dynamic environmental phenomena that come and go as they please through the airspace and airports cannot subsume the command responsibility of aircrew to avoid collision with these phenomena. Airports might rightly be held liable for strikes occurring because they failed to modify a wildlife attractant within their boundary, but that should be the limit of their liability because that is the limit of their authority. At a more low key level, we note anecdotes where, after a run of strikes at a particular port, carriers have threatened the aerodrome operator with unscheduled wildlife management audits or with moves to cease operations unless they "do more" to prevent strikes. Ironically, when approached in return to outline their active contributions to reducing strike risk, most carriers have no answer. This trend is at odds with current air safety practice that is more constructively directed at identifying both the proximate and ultimate causes of an incident and finding cooperative and cross-disciplinary solutions to prevent recurrence.

The skewed aerodrome focus of traditional strike management is further entrenched by the scope of global and national wildlife hazard statutory guidelines. ICAO wildlife strike mitigation guidelines are listed under Annex 14, *Aerodromes*; in Australia, they are detailed in the Civil Aviation Safety Regulations Part 139, Manual of Standards (*Aerodromes*); in the UK, they are defined in the Civil Aviation Authority CAPs 168 & 772 Licensing of *Aerodromes*, Wildlife Strike Management for *Aerodromes*; and the International Air Transport Association (IATA) present their wildlife strike guidelines within Safety and Security, OH&S policy for maintenance crews. Given that wildlife strikes are common real-time in-flight collisions, it seems almost absurd that wildlife strike risk assessment and mitigation are not described or legislated for in any flight operational or air traffic management context. The exception is that some countries have guidelines specifying a nebulous role for ATC in strike prevention. This usually consists of a requirement for controllers to notify aircraft of reported bird hazards in the vicinity.

Management Approaches Based on Operational Separation

Since the 1970s, wildlife strike management in the military aviation arena evolved in a different direction. Most birds operate well below 10,000 ft AGL and, except on approach and departure, modern civil passenger aircraft normally operate well above this level. However, military operations often require sustained low-level flight below this altitude and therefore incur substantially higher wildlife exposure than their civil counterparts. Military low-level operations are also often conducted at high speed and as impact forces vary with the square of closing speed, the consequences of strike to military aircraft are more frequently catastrophic. Thus for military operations, the spurious "aerodrome proximity" argument never fulminated; in fact, most military aviation authorities report the opposite spatial risk profile with most damaging strikes occurring en route rather than in the vicinity of aerodromes. For example, between 1976 and 2008, the Royal Netherlands Air Force (RNLAF) reported a total of 3352 strikes, 67 % of which occurred en route, with the remaining 23 % in terminal airspace. Twenty-two per cent of en route strikes resulted in damage, while 13 % of strikes in terminal airspace resulted in damage (Dekker, European Space Agency Web site 2009). Similarly in Australia between 1960 and 2014, there have only been three serious hull loss accidents, two of which were fatal and all of these involved strikes to military jets at low altitude (Richardson 1994).

The United States Air Force (USAF) alone currently reports over 4000 strikes per year at estimated costs averaging US$24M per year over the last 12 years (USAF BASH Statistics 2014), with 34 crew killed and 17 aircraft lost to wildlife strike between 1974 and 2000 (DeFusco 2000). However, in response to rigorous and integrated hazard management initiatives, their strike rate has steadily declined since 2005.

Globally, 66 serious military strike incidents resulting in hull loss, ejection, or fatality were documented between 1990 and 2008 from 21 air forces (Project Get Out and Walk 2014), and a more formal review of military wildlife strikes between 1950 and 1999 estimated at least 283 aircraft lost from 27 air forces (Richardson and West 2000). These figures are likely to underestimate the real attrition rate, as only a small minority of countries detail military aircraft losses.

Hence, necessity motivated military aviation authorities to confront the reality of wildlife strike as a dynamic in-flight collision problem and to develop procedural approaches to strike prevention. Conventional airport-based management remains an integral component of strike prevention in most military arenas, but it has been supplemented and overshadowed by more operationally relevant dynamic separation techniques. These techniques are predicated on accurate wildlife detection and movement modelling, wildlife hazard forecasting, flight planning, and avoidance (Bird Avoidance Model, BAM). The primary aim of this approach is to provide aircrew with specific position, altitude, and tracking information about wildlife hazards so that they can plan and conduct their flight around high-risk airspace and times.

The Israeli Air Force (IAF) pioneered this approach in the 1970s. At that time, the IAF suffered several fatal accidents and degraded capability as a result of wildlife strikes. In response, they mapped the main bird migration routes through Israel's airspace and developed a workable BAM to provide rudimentary risk forecasts for airspace over time. The IAF applied this model, planned their flight tracks and altitudes accordingly, and in the subsequent 17 years, recorded a 76 % reduction in the serious strike incident rate. It's estimated that this program has saved many IAF aircrew lives and at least US$500M (Leshem et al. 2005; Ovadia 2005). These initial efforts at separation-based procedures were solely strategic in that they modelled expected bird movement patterns to allow better flight planning. More recently,

Fig. 22.3 A medium power (MPR) Air Defence Radar in the Netherlands (*left*). The raw signal from these sensors can be processed to selectively display primary returns from birds. Ultimately, this data contributes to integrated bird hazard information for aircrews. The information can be displayed to analysts in various forms including on Google Earth® (*right*). In this case, the primary map is also overlayed with ambient surface wind vector arrows (*yellow*). The two *red* annuli represent bird returns from two separate MPRs, one in Holland and one in Belgium. The *blank* areas on the inside of each annulus (close to each station) show no signal as there are no birds intersecting the beam swathe at that range–altitude combination. The outside fringe of each annulus represents the range-altitude combination where the beam swathe exits the migration level of the flock. The dual compass inset diagram displays a vector interpolation (relative to the ground) from the bird returns of each station: the average vector is 18 m/s to the south–west for the returns around the northerly station and 24 m/s to the south–west for the southern station. Fusion of this data with overlapping information from regional weather radars, SMRs and other remote sensors can provide a comprehensive picture of the wildlife hazard risk across the region. *Photos courtesy of Hans van Gasteren of the Royal Netherlands Air Force*

IAF have developed a long range radar tracking system that can provide real-time bird movement data allowing aircraft in flight to avoid high-risk areas or levels (Ovadia 2012).

Also in the late 1970s, the Royal Netherlands Air Force (RNLAF) began experimenting with similar approaches to strike mitigation. They used alternately processed raw signal data from extant military Air Defence Radars (Fig. 22.3) to detect and model large-scale bird migration movements at ranges out to 150 km (van Belle et al. 2007). Aircraft could then avoid low-level operations during peak migration times. Between 1989 and 2007, RNLAF further refined their detection capability and procedures recording a sustained reduction in en-route strike rate from around 40 strikes per 10,000 flight hours flown to 3 strikes per 10,000 flight hours (Dekker, European Space Agency Web site 2009). Quite remarkably, this project has evolved into a trans-national consortium providing short-term bird movements forecasts at 1 square degree resolution for four air forces in northern Europe. This project, Flysafe (2014), is coordinated by the European Space Agency (ESA) under their integrated applications program and resolves biological data together with information from satellite, air defence, and the OPERA (Eumetnet 2014) weather radar networks to produce near real-time bird forecasts. The main operational output is the European Bird Notice to Airmen (BIRDTAM), which is a short-term area forecast of bird density available to air force personnel for use in flight planning (Dekker et al.

2008). Flysafe has recently completed its proof of concept phase and is now being extended in scope with the aims of further reducing en route strike rates and providing a workable platform for real time separation procedures (Flysafe 2014).

The attrition rate and loss of capability resulting from wildlife strike over the last 50 years led the USAF down a similar path, developing a BAM in the 1980s for use in North America. The original BAM was based on historical bird migratory data; however in the late 1990s, it was supplemented with weather forecast information and radar return data from the continental NEXRAD radar network (Gauthreaux and Schmidt 2013). Calibration of raw returns showed that NEXRAD could reliably detect at least 7 of the top 11 risk species found across the USA (Kelly et al. 1999a). Today, this project has evolved into the USAF Avian Hazard Advisory System (AHAS), which provides short-term forecasts of relative bird risk by area, route, and location across continental United States. AHAS bird severity forecasts are accessible online to both military and civil pilots (AHAS 2014).

The success of the AHAS system in reducing strike rate is not as apparent as it is with the Israeli or European systems probably reflecting the program's shorter time in service. Anecdotal reports suggest that USAF Air Combat Command registered a 50 % reduction in strikes in the first year that AHAS was mandated for operational service, although this degree of success is not evident in the broad strike statistics available directly from the USAF BASH web site (USAF BASH 2014). However, it is evident from these statistics that absolute strike numbers have consistently decreased after peaking at just over 5000 in 2005.

Since 2003, the USAF has also been developing a Small Mobile Radar (SMR) strike mitigation program in parallel with regional approaches. The main aim of this system is to provide real-time hazard information to pilots so that they can avoid birds in flight. In contrast to the broad scale used in forecasting with the AHAS NEXRAD system, SMRs are designed to detect targets over much smaller airspace (e.g. out to 5 nm and up to 5000' AGL), but can provide useable bird track, velocity, and altitude information. They are ideal for hazard detection around airports and in low-level operations areas such as bombing ranges. The USAF implemented a trial of SMR and real time separation procedures at Dare Range Colorado in 2003 prior to the trial aircraft using the range averaged a Class A/B (damage greater than USD$0.5M) strike incident every 18 months. With the system operational between 2004 and 2012, the USAF reported no Class A/B incidents at the range and that the system has increased range availability (Merritt et al. 2012). However, it is not the SMR per se that is pivotal for effective mitigation. The key factors are the procedures that prescribe timely threat detection, notification, and appropriate flight path modification. The same process can be achieved using trained observers instead of SMR; the SMR makes threat detection more spatially efficient and more accurate over a greater range of visibility conditions. For a summary of the relative efficacy of different airborne wildlife detection procedures, see Brand et al. 2011.

The USAF has since deployed and trialled SMR for assessment at several US air force facilities and units have been tested for deployment in combat theatres (Le Boeuf et al. 2008). The proper application of airport radar or other remote-sensing systems overcome some of the detection and analytical limitations of traditional wildlife management and the evidence suggests that if they are integrated

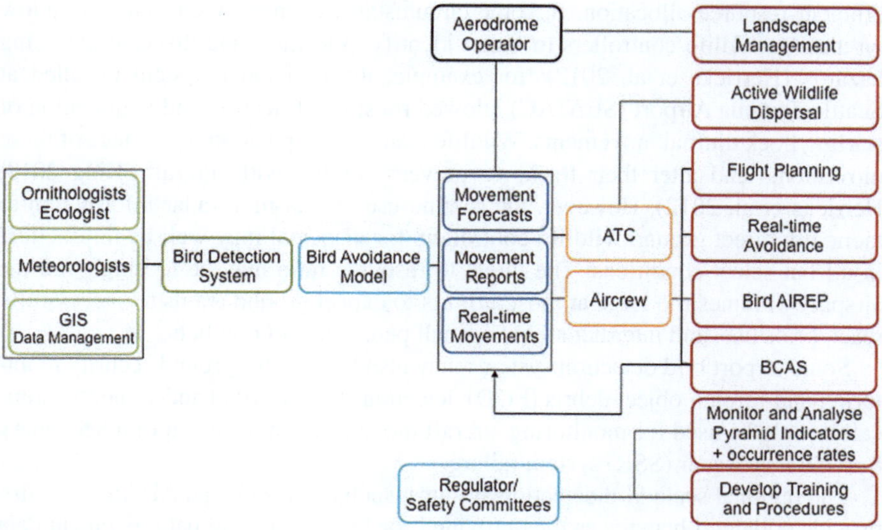

Fig. 22.4 A conceptual model summarising the components of a dynamic wildlife separation system. The main engine for this system is a constantly updated BAM supported by biological and environmental specialists and remote-sensing technologies. The main outputs to aviation are a series of interlocked hazard forecasts and real time situation reports that are relayed to aircrew for flight planning and flight execution, similar to the current ARFOR-TAF-METAR-ATIS system currently in use for weather hazard avoidance. With this approach, aircrew and ATC are the primary collision mitigation elements. Airport mitigation measures are used with on-airport landscape modification remaining a responsibility of the aerodrome operator, but in this model active wildlife controllers are managed directly by ATC. In this model, ATC and aircrew receive formal training in wildlife recognition and risk assessment. Thus, both sectors are continually iterating informed risk assessments and considered operational decisions about the real-time hazard level and both sectors are constantly supplementing the data stream. The BAM at the centre of this model will also output useable information to Public Health, Agriculture, Environmental and Conservation authorities

properly with flight procedures, they will result in sustained strike rate reductions (Dekker et al. 2008; Merritt et al. 2012; Leshem et al. 2005). The main elements of an ideal separation based mitigation model are outlined in Fig. 22.4.

Collateral Benefits of Avian Remote Sensing Systems

The remote-sensing technologies used to support wildlife hazard detection and collision avoidance include radar, thermal imaging, satellite and airborne optical and multispectral imaging systems, and ground-based fusion systems. These technologies may provide potentially useful fringe benefits beyond notifying aircrew of imminent wildlife collision. Firstly, they may allow more precisely targeted passive and active on-airport wildlife management, which in turn translates into more efficient resource allocation. In some circumstances, timely radar data can allow on-airport wildlife controllers to better identify, prioritise, and disperse incoming hazards (Herricks et al. 2012): for example, the trial radar system installed at Seattle-Tacoma Airport (SEATAC) allowed roost identification and anticipation of starling flock diurnal movements. Wildlife teams could position in advance of these movements and alter their tracks to prevent conflict with aircraft (King 2013; Herricks et al. 2012). However, the routine use of real-time radar information to tactically direct ground wildlife controllers for dispersal may well be impractical in all but select conditions. The threat to resource ratio may be too high and the airspace volumes too large at most airports to expect ground teams to successfully react, prioritise, and interdict all but a small percentage of conflicts.

Some airport bird detection systems may also be used for ground security monitoring and foreign object debris (FOD) detection (Carter 2012) and some, theoretically, could be used for monitoring aircraft movements in the event of a Secondary Surveillance Radar (SSR) system failure.

The regional scale bird detection systems, such as Flysafe and AHAS, may also provide collateral benefits as they provide broad scale bird and bat movement data that in turn may be useful for global health, agriculture, and conservation management (Ginati et al. 2010). Bird and bat movement models, and the movement forecasts developed with these programs, could be applied to risk assessment and mitigation for emergent human diseases such as Ebola virus, Avian influenza, and Henipah virus diseases and similarly for significant animal production pathogens such as Newcastle disease virus (McKee et al. 2011). In the conservation arena, these models may also help resolve the ecology of long distance bird and bat seed dispersers, improving our understanding of forest recruitment and ultimately contributing to carbon management. For recent reviews of the more general applications of radar and remote sensing in these arenas, see Gauthreaux and Belser (2003), Xiao et al. (2007), van Gasteren et al. (2008), Tran et al. (2010), Shaumon-Baranes et al. (2014), and ENRAM (2014).

Impediments to Progress

While there are technical and procedural issues to be addressed before the adoption of radar-based wildlife separation procedures can be realised in civil aviation (Nohara 2009; Bunch and Herricks 2010; Nohara et al. 2012; King 2013; Beason et al. 2013), significant cultural and commercial limitations prevail. Firstly, integration of these procedures will be long-term, capital- and data-intensive projects with a significant lag time before useable operational information is available. Consequently, the cost-benefit of adapting these methods to civil aviation remains in question. Secondly, the information derived from remote sensors and BAMs is only of value if successfully used to alter the flight vectors of both aircraft and wildlife to prevent collision. Thus to be operationally viable, these approaches require the active engagement of those who have the authority to alter aircraft flight paths;

that is, air traffic control (ATC) and aircrew. In turn, aircrew and air traffic controllers need background and procedural training in aviation wildlife hazard management analogous to the way they are currently trained in meteorology (McKee et al. 2012; Greeves 2013).

However, there currently exists a strong reluctance in both pilot and ATC professional groups to engage at this level with the wildlife strike issue. Despite consensus that wildlife strike is a genuine safety issue, many believe that adopting wildlife-aircraft separation procedures would be overly complex, would not be adequately resourced, and may eventually compromise cockpit and tower workflow and safety. In addition, there are concerns by airline and airport operators that such procedures may reduce movement rates and incur financial loss and there exists a negative perception, particularly expressed by ATC, that if they actively and routinely engage in strike mitigation practices, they may be then held liable for strikes that still occur despite their best efforts. Ironically, under ICAO 4444 section 7.3 and 7.4, ATC are currently required to notify aircraft of wildlife hazards in such a way that there is enough specific information and time for the pilot to take useful mitigation action. Although this requirement is essentially impractical in the current air traffic control context, the requirement and thus the liability already exist and in Europe that liability has been successfully prosecuted as a result of a serious strike followed by engine failure (Battistoni 2009).

These fears are also well-illustrated in the equivocal International Federation of Air Line Pilots Associations (IFALPA) draft policy statement on bird detection techniques (IFALPA AGE 2011), which at the outset acknowledges the worrying trend in civil strike rates and the need for new management approaches. On the other hand, the policy presumes conflict detection by airports and advocates transfer of collision avoidance responsibility to ground personnel, but fails to prescribe anything other than spectator roles for flight crew or ATC. The policy, rightly, is very specific about the advisory nature of wildlife notifications and maintaining final pilot command authority in wildlife avoidance situations, but then abrogates any command responsibility for contributing to a solution. The authors cite the complexity of application, commercial considerations, safety and legal responsibility of aircrew and air traffic controllers all as reasons for retreating from the issue.

On face value, real time wildlife separation procedures do appear quite complex. Implementation requires new training and new notification systems, which would need to be globally standardised and in turn would need to be supported by wildlife movement forecasting and reporting models. Furthermore, given the global diversity in biome, climate, and animal behaviour, standardising critical parameters such as bird hazard levels for risk models and notification systems will be challenging. The perception of complexity in applying operational wildlife separation to high movement rate civil operations is further confounded by the well-entrenched pre-judgement that they must result in chaotic traffic sequence disruption and significant commercial penalty. In reality, the level of disruption or penalty is more likely to depend on how the procedures and technology are staged into a complex traffic environment. Once set in motion, new technologies and procedures rapidly embed and evolve; in practice, implementation should be no more daunting and no more disruptive than the introduction of new weather avoidance or aircraft separation procedures. In the past, these issues have been well-managed by careful planning, staged implementation, and well-supported execution that circumvent apparently insurmountable complexities—in other words, by approaching the issue in precisely the same pragmatic way aviators have been doing since Orville Wright first "chased a flock [of Starlings] around Beard's corn field".

It is certainly possible to begin introducing both the technology and the procedures for strategic and tactical wildlife separation in civil aviation (Sowden and Eschenfelder 2009), particularly since the process is supported by 50 years of proof of concept from military aviation. In military aviation, the high attrition rate associated with strike during low-level operations was a critical motivation for implementing separation-based mitigation approaches. However, at the moment the same rationale does not exist in the civil arena. It is harder for civil operators to accept the need to alter the strike mitigation paradigm when there are only sporadic reminders of the catastrophic consequences of wildlife strike to humans and essentially no ethic within the industry that concedes any significance to the continuum of fatal consequence to wildlife.

We note that many of the cultural and psychosocial restraints discussed above in relation to civil aviation are not relevant to military aviation where the imperative is on efficiency, safety, and maintaining capability rather than profit and fear of liability. With the exception of combat operations, lower movement rates and less stringent scheduling imperatives with military aviation allow more latitude for aircrews to prioritise wildlife strike avoidance. The significant point here is that *aircrew* should proactively prioritise and drive wildlife strike mitigation; it should not be a primary responsibility of aerodrome operators, as pilot-in-command is always the final authority for the disposition of the aircraft and therefore always ultimately responsible for what the aircraft collides with.

Structural and cultural differences between military and civil systems also contribute to the divergence in attitudes towards operationally integrated wildlife management. In military aviation ATC, aircrew, aircraft operator, and aerodrome management generally all fall under the same command structure and all work to common goals; in civil aviation, these elements are fragmented and often, to some extent, commercially and culturally competitive. Hence, implementation of operational change is easier to coordinate in military settings. Competition between civil aviation sectors remains one of the main reasons why responsibility for wildlife strike mitigation is still entrenched and confined to the aerodrome sector. It is easier to continue letting aerodromes attempt to manage wildlife strike alone than it is for other industry sectors to engage and incur the nominal extra cost and liability.

The Way Ahead

Notwithstanding the challenges discussed above, there is increasing recognition in civil aviation that a dynamic wildlife hazard mitigation approach is necessary and integrated operational separation procedures will likely be adopted in some form despite their perceived limitations. The Hudson River event of 2009 and several similar near catastrophes in Europe have motivated a search for approaches that will demonstrably and sustainably reduce strikes rates. There are now several commercial companies manufacturing terminal bird radar systems for civil use and several civil airports are trialling these units initially as adjuncts to on-airport management and as a means of developing local bird movement models (King 2013). More importantly, the advent of radar and other accurate detection technologies has encouraged adoption of more contemporary risk assessment approaches based on monitoring pyramid indicators rather than the collision events themselves (Klope et al. 2009).

Recently, robust biological and ecological survey work together with an SMR and co-ordinated cross-disciplinary cooperation resulted in the first successful introduction of radar-based bird hazard advisory procedures in civil aviation. These procedures apply to airspace in the vicinity of King Shaka International Airport (KSIA), Republic of South Africa, and help prevent aircraft conflicting with swallows (*Hirundo rustica*) that egress a roost-site on the runway 06 approach (Marshall 2010; Merritt et al. 2012). Several factors seem to have contributed to the success of this program, including the fact that KSIA is a relatively low movement-rate port and thus the airspace is amenable to procedural experimentation. In addition, the primary threat target (a large flock of swallows) can be unambiguously identified by radar and the flock flight paths and flight timings are semi-random allowing some degree of block time and airspace predictability to the threat. Most importantly, the process involved cross-disciplinary cooperation by airport, airspace, airline, and conservation authorities, providing good outcomes for both aircraft and birds and dramatically underscoring the legitimacy of the separation approach in civil operations. Of related relevance is a novel program, recently devised in Central Europe, which is a conceptually different but nevertheless an effective operational approach to strike mitigation. The low-cost carrier, Wizzair, was elected to independently implement a Safety Management System (SMS)-based wildlife hazard management plan (WHMP), which to our knowledge is the first significant *airline*-based WHMP to be deployed (Pekk 2012). This plan included lobbying for better regulatory oversight, integrated cockpit, and communication protocols in addition to the carrier itself taking responsibility for coordinating and driving wildlife management compliance initiatives across all sectors of its flight range. Given the fact that it is aircraft, not aerodromes, which collide with wildlife, this aggressive usurpation of responsibility by a carrier is a refreshingly logical, albeit long overdue, change to the mitigation paradigm. It is also apparently effective; in the first year of deployment the program resulted in a 20 % reduction in strike rate, a 10 % reduction in damaging strike rate, and a 40 % reduction in total delay time to the Wizzair fleet (Pekk 2012). To our minds, a fusion approach derived from both the King Shaka and Wizzair experiences is certainly one way forward to achieve just and effective management of the problematic aircraft strike issue.

Another positive sign is that national and international regulators are becoming more engaged with wildlife avoidance. The US FAA recently released guidelines (FAA AC 2010) for the selection, deployment, and use of avian radar systems at aerodromes. Similarly, recent international wildlife strike and safety meetings are seeing increasing representations from ICAO and the International Airline Transport Association (IATA). Most encouraging is the observation that the cultural and fear barriers discussed above are beginning to dissipate as a result of extensive technology validations (Brand et al. 2011), good communications and multi-stakeholder discussions (Nohara et al. 2012; Hale and Koros 2014), and also as a result of strong advocacy from progressive professional associations, particularly the National Business Aviation Association (NBAA), Embry-Riddle Aeronautical University (ERAU), and the Australian Airline Pilots Association (AusALPA). Similarly, significant efforts have been directed at defining how wildlife threat information can be standardised, organised, and simply communicated in operational settings (Nohara et al. 2012). It also should be noted that both Airbus Industries and Boeing are actively engaged in strike mitigation research (Nicholson and Reed 2011; Papin 2012). Finally, there is some evidence that strike liability is starting to be partitioned more equitably. Prior to 2006, most successful litigation held aerodrome operators alone accountable for strike damage. However, a recent post-strike finding in favour of an aircraft operator split the liability for the aircraft damage between the aerodrome operator, ATC, and the regulator (Battistoni 2009), finally highlighting the reality that mitigation of in-flight collisions devolves to all operational sectors.

Two of the most significant challenges to further progress include: an imperative to clearly demonstrate that radar-based separation approaches will provide a cost benefit as well as an obvious safety benefit to civil operations; and the necessity to secure positive and practical engagement in the process from the International Federation of Air Traffic Controllers Associations (IFATCA). Given the positive experience with military strike management over the past 50 years, we are optimistic that these hurdles will be overcome and we anticipate a time when accurate bird forecasting and co-ordinated real-time management will reduce aircraft and wildlife conflict.

Acknowledgments The authors extend their gratitude for the ideas, discussions, sound advice, and sometimes outright disagreement provided by many dedicated colleagues. These include but are not limited to: Squadron Leader (Ret) Peter "Bong" Nelson AFC, Royal Australian Air Force, Australia; Colonel Jean Paul Apon, Royal Netherlands Air Force, the Netherlands; Tenente-Coronel Aviador Henrique Rubens Balto de Oliveira, Brazilian Air Force, Centro de Investigação e Prevenção de Acidentes Aeronáuticos, Brazil; Dr. Ian Faulconbridge, Managing Director, Magpie Technologies, Australia; Captain Gary Cooke, National Business Aviation Association, USA; Will Jamieson, Avisure, Canada; Jeff Follett, Martin Ziviani and Ronel Jit, Avisure, Australia; Hans van Gasteren, Luit Buurma and Albert de Hoon, Ministry of Defense, the Netherlands; Captain Alex Winslow, Qatar Airways, Qatar; Joep Kievits, KLM Royal Dutch Airlines, the Netherlands; Siete Hamminger and Addy Borst, Robin Radar, the Netherlands; Pal Ranestad, Avinor, Norway; Captain Paul Eschenfelder, Embry-Riddle Aeronautical University, USA; Dr. Tim Nohara and Dr. Bob Beason, Accipiter Radar, Canada; Emeritus Professor Edwin Herricks, Centre of Excellence for Airport Technology, University of Illinois, USA; Captain Brian Greeves, Managing Director, Aviation Solutions, Australia; Professor Amnon Ginati, and Davide Coppola, the European Space Agency, Europe; Professor Willem Bouten and Dr. Judy Shamoun-Baranes, Amsterdam University, the Netherlands.

References

Allan J (2000) The costs of bird strike and bird strike prevention. USDA National Wildlife Research Center Symposia, human conflicts with wildlife, economic considerations. Paper 18
ATSB (2007) Air safety investigation reports, Investigation 200605807
Battistoni V (2009) Bird strikes and the courts: the Antonov Case. Proc. Joint USA/Canada Birdstrike Committee meeting, Victoria, BC, September 2009
Beason R, Nohara T, Weber P (2013) Beware the Boojum: caveats and strengths of avian radar. Human-Wildlife Interactions 7:16–46
Brand M, Key G, Nohara T, Beason R, Herricks E, King R, Gauthreaux S, Begier M, Bowser C, Swift M, Klope M. Griese H, Dotur C (2011) Integration and validation of avian radars, ESTCP Project RC-200723
Bridger R (2013) Plane Truth. Aviation's real impact on people and the environment. Pluto Press 9th October 2013, ISBN: 9780745330327
Bunch W, Herricks E (2010) Observational validation of avian radar systems. Proc. FAA worldwide airport technology transfer conference, Atlantic City, April 2010
Buurma LS (1996) Superabundance of bird; trends, wetlands and aviation. Proc. Bird Strike Committee Meeting Europe 23:43–50
Carter N (2012) A real-time bird detection system for near-airfield airspace utilizing a thermo-optical scanning system. Proc. Int. Bird Strike Committee Meeting, Stavanger, June 2012
Dekker A (2000)poor long grass: low density ground cover for the runway environment. Proc. Int. Bird Strike Committee Meeting, Amsterdam, April 2000
Dale L (2009) Personal and corporate liability in the aftermath of bird strikes: a costly consideration. Hum Wildl Conflicts 3:216–225
DeFusco R (2000) Current status of the USAF bird avoidance model (BAM) Proc. Int. Bird Strike Committee Meeting, Amsterdam, April 2000
Dekker A, van Gasteren H, Bouten W, Shamoun-Barnes J, Borst A, Holleman I, Dokter A, Ginati A, Garafalo G (2008) The European Space Agency's Flysafe project, looking at the bird strike problem from another perspective. Proc. Int. Bird Strike Committee Meeting, Brazil, November 2008
Demers C, McVey L (2015) An inside look at revising bird ingestion certification standards for commercial and business jet aircraft. In: Proceedings of the joint USA/Canada Birdstrike Committee meeting, Montreal, Sept 2015
Dolbeer RA (2006) Birds and other wildlife hazards at airports: liability issues for airport managers. USDA National Wildlife Research Center, Staff publications No. 142
Dolbeer RA, Eschenfelder P (2003) Amplified bird strike risk related to population increases of large birds in North America. Proc. Int. Bird Strike Committee Conference 26:49–67
Dolbeer RA (2011) Increasing trend of damaging bird strikes with aircraft outside the airport boundary: implications for mitigation measures. Hum Wildl Interact 5:235–248
Dolbeer RA, Wright SE, Weller J, Begier MJ (2013) Wildlife strikes to civil aircraft in the United States 1990-2012 FAA National Wildlife Strike database Report 19, September 2013
Dolbeer RA (2013) The history of wildlife strikes and management at airports. In: DeVault T, Blackwell B, Belant J (eds) Wildlife in airport environments: preventing animal-aircraft collisions through science-based management. The Johns Hopkins University Press in association with The Wildlife Society, Baltimore, pp 1–6
Eschenfelder P (2009) Successful strategies for aviation wildlife mitigation. Proc Joint USA/Canada Birdstrike Committee Meeting, Victoria, BC, September 2009
Eschenfelder P, DeFusco R (2010) Bird strike mitigation beyond the airport, pilots must be prepared for bird strike avoidance and damage control. Aerosafety World, August 2010, pp 44–47
Federal Aviation Regulations (2014) Regulations regarding minimal structural requirements to prevent catastrophic failure after birdstrike. FAR Parts, 23.775, 25.571 25.631, 25.775, 33.7
Gauthreaux S, Belser C (2003) Radar Ornithology and biological conservation. The Auk 120:266–277

Gauthreaux S, Schmidt P (2013) Radar technology to monitor hazardous birds at airports. In: DeVault T, Blackwell B, Belant J (eds) Wildlife in airport environments: preventing animal-aircraft collisions through science-based management. The Johns Hopkins University Press in association with The Wildlife Society, Baltimore, pp 141–151
Ginati, A. Coppola D, Garofalo G, Shamoun-Baranes J, Bouten W, van Gasteren H, Dekker A, Sorbi S. (2010) FlySafe: an early warning system to reduce risk of bird strikes. ESA Bull 144, November 2010
Greeves B (2013) Wildlife hazard education and training for pilots. Proc. Australian Aviation Wildlife Hazard Group Forum, Brisbane, Australia 2013
Hale M, Koros A (2014) Bird radar knowledge elicitation activity. Final Report for Federal Aviation Authority, Concept Development and Validation Branch, Wildlife Surveillance Concept (WiSC) initiative
Herricks E, Bunch W, Lazar P, Woodworth E (2012) Characterising bird movement on and around airports using avian radar systems Proc. Int. Bird Strike Committee Meeting, Stavanger, June 2012
Kelly TA, Merritt R, Donalds T, White R (1999a) The avian hazard advisory service. Proc. Joint USA/Canada Birdstrike Committee meeting, Victoria, BC, 1999
Kelly TC, Bolger R, O'Callaghan MJ (1999b) The behavioural response of birds to commercial aircraft. Proc. Joint USA/Canada Birdstrike Committee meeting, Victoria, BC, 1999
King R (2013) Interim report. Research on bird detecting radar. FAA Report DOT/FAA/TC-13/3
Klope M, Beason R, Nohara T, Begier M (2009) Role of near-miss bird strikes in assessing hazards. Human-Wildlife Conflicts 39(2):208–215
Le Boeuf E, Wilkens T, Merritt R (2008) The USAF Small Mobile Radar (SMR) program: an overview and status report on the operational use of dedicated avian radar systems at USAF airfields and ranges. Proc. Bird Strike Committee USA/Canada meeting, Orlando, FL, August 2008
Leshem Y, Ovadia O, Dinevich L, Raz O. (2005) A national network of bird and weather radars in Israel—from vision to reality. Proc. Int. Bird Strike Committee Meeting, Athens, May 2005
Maragakis I (2009) Bird Population trends and their impact on Aviation Safety. EASA Report Safety Analysis and Research Department, 9th January 2009
Martin J, Belant J, DeVault T, Blackwell B, Burger L, Riffel S, Wang G (2011) Wildlife risk to aviation: a multi-scale issue requires a multi-scale solution. Hum Wildl Interact 5(2):198–203
McKee J, Shaw P, Patrick K (2010) Bird Strike—perched out on a limb. Proc. Australia & NZ Association of Air Safety Investigators conference, Canberra, May 2010
McKee J, Patrick K, Shaw P (2011) Satellites, aircraft and emerging wildlife zoonoses: a message from the European Space Agency. Proc. Australasian Wildlife Diseases Conference, Coorong SA, September 2011
McKee J, Shaw P, Patrick K, Winslow A, Eschenfelder P (2012) An aviation Wildlife hazard management syllabus for pilots: what would it look like? Proc. Int. Bird Strike Committee Meeting, Stavanger, Norway, June 2012
Marshall L (2010) Bird arrivals, departures a priority at new airport in Africa. National Geographic, March 24, 2010
Merritt R, Zakrajsek E, Smith A. (2012) Integrating bird detection radar systems into airport operations. Proc Int. Bird Strike Committee Meeting, Stavanger, June 2012
Moller A (2009) Successful city dwellers: a comparative study of the ecological characteristics of urban birds in the Western Palearctic. Oecologia 159:849–858
Nicholson R, Reed W. (2011) Strategies for prevention of bird-strike events, Boeing Company, Aero Magazine 3rd quarter 2011
Nohara T (2009) Could avian radar have prevented US Airways Flight 1549's bird strike? Proc. Joint USA/Canada Birdstrike Committee meeting, Victoria, BC, September 2009
Nohara T, Sowden R, Perkins R, Beason R (2012) Aviation Stakeholder management of bird strike risks-enhancing communication processes to pilots and air traffic controllers for information derived from avian radar. Proc. Int. Bird Strike Committee Meeting, Stavanger, June 2012
Ovadia O (2005) Ten years of birdstrikes in the Israeli air force. Proc. Int. Bird Strike Committee Meeting, Athens, May 2005

Ovadia O (2012) Real time warning system in the IAF. Authority and responsibility. Proc. Int. Bird Strike Committee meeting, Stavanger, June 2012

Papin N (2012) Use of embedded systems to reduce birdstrike likelihood—research on pertinent acoustic/visual signals. Proc. Int. Bird Strike Committee meeting, Stavanger, Norway, June 2012

Parsons J, Blair D, Luly J, Robson S (2009) Bat strikes in the Australian aviation industry. J Wildl Manage 73(4):526–529

Patrick K, McKee J, Shaw P (2008) A practical approach to managing the flying fox hazard at a tropical Australian airport: an example of successful proactive risk management. 2008 Proc. Int. Bird Strike Committee Meeting, Brazil, November 2008

Patrick K, Shaw P, (2012) Bird Strike Hazard Management Programs at Airports: what works? Proc. 5th Symposium of Flight Safety, Department of Science and Aerospatiale Technology, 28–30 August 2012, São José dos Campos/SP, Brazil

Pekk A (2012) Managing the unmanageable: views on the role airline companies can play. Proc. Int. Bird Strike Committee Meeting, Stavanger, Norway, June 2012

Rey L, Liechti F (2015) Lethal control on airports: an effective tool to mitigate the risk of bird strikes? A literature review on behalf of the Swiss Federal Office of Civil Aviiation (FOCA), *in print*

Richardson WJ (1994) Serious birdstrike-related accidents to military aircraft of ten countries: preliminary analysis of circumstances. Proc. Bird Strike Committee of Europe, Vienna, August 1994

Richardson WJ, West T (2000) Serious birdstrike accidents to military aircraft: update list and summary. Proc Int. Bird Strike Committee meeting, Amsterdam, April 2000

Richardson WJ, West T (2005) Serious birdstrike accidents to UK military aircraft 1923 to 2004: numbers and circumstances. Proc. Int. Bird Strike Committee Meeting, Athens, May 2005

Roberts B, Catterall C, Eby P, Kanowski J (2011) Latitudinal range shifts in Australian flying-foxes: a re-evaluation. Austral Ecol 37:12–22

Roberts B, Catterall C, Eby P, Kanowski J (2012) Long distance and frequent movements of the flying fiox *Pteropus poliocephalus*. Implications for management. PLoS One 7(8), e42532. doi:10.1371/journal.pone.0042532

Shamoun-Baranes J, Alves J, Bauer S, Dokter A, Huppop O, Koistinen J, Leijnse H, Liechti F, van Gasteren Hand Chapman J (2014) Continental scale radar monitoring of the aerial movements of animals. Movement Ecol 2:9–15

Shaw P (2008) A practical and cost effective approach to airport wildlife management in the context of local landuse: two Australian case studies. Proc. Bird Strike Committee USA/Canada meeting, Orlando, FL, August 2008

Smith A (2009) Population ecology of the Australian White Ibis, *Threskiornis molucca,* in the urban environment. Ph.D. Thesis, University of Technology, Sydney, 2009

Sodhi N (2002) Perspectives in Ornithology. Competition in the air: birds versus aircraft. The Auk 119:587–595

Sowden R, Eschenfelder P (2009) Integrating avian radar into the aviation operating environment. Proc. Joint USA/Canada Birdstrike Committee meeting, Victoria, BC, September 2009

Thorpe J (2015) This publication

Tran A, Goutard F, Chamaille L, Baghdadi N, Lo D (2010) Remote sensing and avian influenza: a review of image processing methods for extracting key variables affecting avian influenza virus survival in water from Earth Observation satellites. Int J Appl Earth Observ Geoinf 12:1–8

Uhlfelder E (2013) Bloody Skies: the fight to reduce deadly bird-plane collisions. News, Environment. National Geographic, November 7, 2013

Van Belle J, Shamoun-Baranes J, van Loon E, Bouten W (2007) An operational model predicting bird migration intensities for flight safety. J Appl Ecol 44:864–874

Van der Meade M, Pieterse E (2013) A four track approach to reduce the risk of wildlife strikes at Amsterdam Schiphol Airport. J Airport Manage 7:165–172

Van Gasteren H, Shamoun-Baranes J, Ginati A, Garofalo G (2008) Avian Alert: a bird migration early warning system. Proc. 58th International Astronomical Congress, Glasgow, September 2008

Xiao X, Gilbert M, Slingenberg J, Lei F, Boles S (2007) Remote sensing, ecological variables and wild bird migration related to outbreaks of highly pathogenic H5N1 avian influenza. J Wildl Dis 43:40–47

Web Sites

Aerofiles (2014) http://www.aerofiles.com/vinfiz.html
AHAS Avian Hazard Advisory Service (2014) http://www.usahas.com/home/
BC&T News and Events (2015) http://www.birdstrike.it/en/index.php?News_and_Events
Centennial of Flight (2014) http://centennialofflight.net/essay/Wright_Bros/1905_Flyer/WR8.htm
Dekker, European Space Agency (2009) Advanced research in Telecommunications http://artes.esa.int/sites/default/files/Bird%20strikes%20in%20military%20aviation_Arie%20Dekker.pdf
Early Aviators (2014) http://www.earlyaviators.com/egilbert.htm
ENRAM European Network for the Radar Surveillance of Animal Movement (2014) http://www.enram.eu
Eumetnet (2014) OPERA Operational Program on the Exchange of weather RAdar information http://www.eumetnet.eu/opera
FAA AC (2010) Airport avian radar systems AC 150/5220-25 http://www.faa.gov/airports/resources/advisory_circulars/index.cfm/go/document.current/documentNumber/150_5220-25
FAA Wildlife Hazard Manual http://digitalcommons.unl.edu/cgi/viewcontent.cgi?article=1127&context=icwdm_usdanwrc
Fiddlers Green (2014) http://www.fiddlersgreen.net/models/Aircraft/Vin-Fiz.html
Flysafe ESA Bird avoidance model (2014) http://artes-apps.esa.int/projects/flysafe
ICAO Airport Services Manual Document 9137 (2012) http://www.birdstrike.org/wp-content/uploads/2014/10/ICAO-AirportServicesManual-Part3-FourthEdition-20121.pdf
IFALP AGE (2011) Position statement 12POS04, Bird detection techniques. http://www.ifalpa.org/downloads/Level1/IFALPA%20Statements/Airport%20Issues/12POS04%20-%20Bird%20detection%20techniques.pdf
Lienhard J (2014) http://www.uh.edu/engines/epi1914.htm
Ministerie van Verkeer en Waterstaat (1999) 1999http://www2.vlieghinder.nl/naslagdocs/CDrom/REGELS_SCHIPHOL/2.3_TNL/4.3.4.1_Bird_control_at_airports.pdf
Project Get Out and Walk (2014) Ejections and bird strikes http://www.ejection-history.org.uk/birds/birdstrikes.htm
Transport Canada (2014) Sharing the skies http://www.tc.gc.ca/publications/en/tp13549/pdf/hr/tp13549e.pdf
US Air Force BASH guidelines (2014) http://static.e-publishing.af.mil/production/1/af_se/publication/afpam91-212/afpam91-212.pdf
US Air Force BASH Statistics (2014) http://www.afsec.af.mil/organizations/bash/
Wikisource (2014) http://en.wikisource.org/wiki/Aviator_C.P._Rodgers_Almost_Instantly_Killed._His_Biplane_Falls_Distance_of_200_Feet
World Birdstrike Association Manual of best practice. http://worldbirdstrike.com/index.php/resources/publications/ibsc-best-practices-manual
Wright Bothers Aeroplane Company (2014) http://www.wright-brothers.org/Information_Desk/Just_the_Facts/Airplanes/Flyer_III.htm
Wright Stories (2014) http://wrightstories.com/bird-strikes/

Chapter 23
Conflicts Between Birds and On-Shore Wind Farms

Mieke C. Zwart, Ailsa J. McKenzie, Jeroen Minderman, and Mark J. Whittingham

Background

Wind power is an important source of renewable energy, providing around 2.1 % of electricity worldwide during 2011 (Table 23.1). This figure may rise to 20 % by 2050, according to some projections (IPCC 2012). The use of on-shore wind farms has increased dramatically over the last decade (GWEC 2013) (Fig. 23.1). While the exploitation of renewable energy sources will be fundamental to combating climate change, this rapid expansion of wind farm development has raised issues about potential harmful effects on wildlife. Birds are one of the key groups of concern (IPCC 2012) and may be affected by wind farms both through direct collision with turbines and through habitat and ecosystem modifications associated with wind farm developments (Drewitt and Langston 2006). In this chapter, we will first review these effects and the mechanisms by which they may occur. We will then outline possible mitigation strategies against any potential adverse effects on wildlife.

Current Evidence on the Effects of Wind Farms on Birds

Wind turbines can affect bird populations in two main ways—directly, via mortality after collision with wind farm infrastructure, or indirectly, via disturbance and/or displacement effects caused by the presence or operation of turbines. First, we will focus on collision effects and then look into disturbance effects in a later section.

M.C. Zwart (✉) • A.J. McKenzie • M.J. Whittingham
School of Biology, Newcastle University, Ridley Building, Newcastle-Upon-Tyne NE1 7RU, UK
e-mail: mieke.c.zwart@gmail.com

J. Minderman
School of Biology, University of St Andrews, Dyers Brae House, Greenside Place, St Andrews, Fife, KY16 9TH, UK

F.M. Angelici (ed.), *Problematic Wildlife*, DOI 10.1007/978-3-319-22246-2_23

Table 23.1 Electricity generation by wind power contributing to total electricity generated from all sources and from renewable sources

Compared to	Region	% Contribution 2007	2008	2009	2010	2011
All sources	North America	0.75	1.19	1.69	2.12	2.82
	Central and South America	0.11	0.13	0.20	0.32	0.43
	Europe	2.94	3.34	3.91	4.22	5.11
	Eurasia	0.02	0.03	0.03	0.04	0.08
	Middle East	0.02	0.03	0.03	0.02	0.03
	Africa	0.21	0.22	0.29	0.37	0.42
	Asia and Oceania	0.37	0.54	0.78	1.00	1.36
	World	0.90	1.15	1.45	1.69	2.12
Renewable sources	North America	4.85	7.21	9.59	12.35	14.55
	Central and South America	0.17	0.19	0.29	0.47	0.63
	Europe	13.80	14.81	15.95	16.11	19.47
	Eurasia	0.12	0.16	0.19	0.25	0.50
	Middle East	0.56	1.76	1.81	0.98	1.38
	Africa	1.22	1.31	1.63	2.05	2.41
	Asia and Oceania	2.64	3.53	5.07	6.11	8.54
	World	4.82	5.90	7.13	8.18	10.14

Data from U.S. Energy Information Administration (2014)
This table shows the percentage contribution of wind power to total electricity generated (*top*) and to renewable electricity generated (*bottom*) for eight international regions for the years 2007–2011

Collision

Every year, hundreds of millions of birds are killed due to collisions with a variety of human-made structures, for example vehicles, building and windows, power lines, communication towers, and wind turbines (Erickson et al. 2001). Some authors suggest that bird fatalities due to collisions with wind turbines or associated structures are low compared with other causes of mortality (e. g. Erickson et al. 2001). However, the potential effect that mortality caused by collisions with wind turbines may have on certain bird populations should not be underestimated (Hunt 2002; Madders and Whitfield 2006). High mortality rates have been reported at some wind farms, for example at the Altamont Pass in California, a large wind farm with 5400 turbines, where an estimated 1127 raptors are killed each year (Smallwood and Thelander 2008); an estimated mortality rate of 0.21 raptor/turbine/year. At Tarifa in Southern Spain, the estimated mortality rate was 0.15 birds/turbine/year for griffon vulture (*Gyps fulvus*) and 0.19 birds/turbine/year for common kestrel (*Falco tinnunculus*) (Barrios and Rodríguez 2004). It is not only raptors that have been reported to collide with wind turbines as an estimated seven little terns (*Sterna albifrons*), 238 common terns (*Sterna hirundo*), and 84 sandwich terns (*Sterna sandvicensis*) are thought to have collided at a wind farm in Zeebrugge,

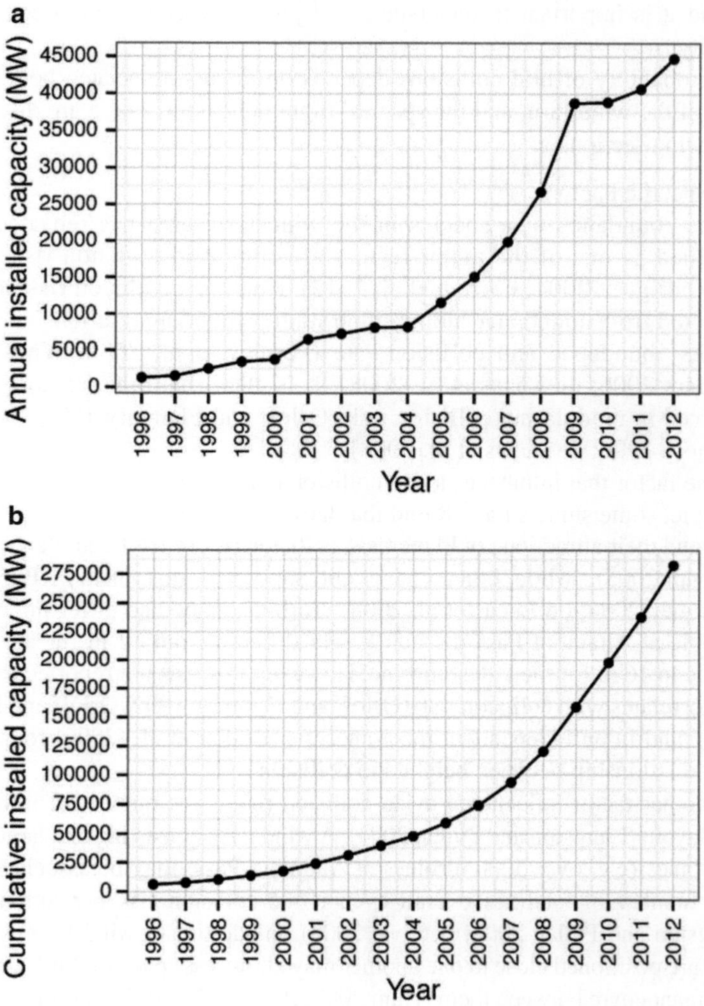

Fig. 23.1 Global-installed wind capacity from 1996 to 2012. (**a**) Global annual installed wind capacity in MW and (**b**) Cumulative global-installed wind capacity in MW. Data from GWEC (2013)

Belgium, during two breeding seasons (Everaert and Stienen 2006). By contrast, at a wind farm in Malaga in Southern Spain, only one collision of a common kestrel was recorded during the study period of two years and no collisions were identified for any other species, which included raptors, passerines, and non-passerines (Farfán et al. 2009). Furthermore, Rothery et al. (2009) reported that possibly two gannets (*Morus bassanus*) had collided at a wind farm at Blyth in England over a post-construction study period of 3 years and no collisions were reported for eight other seabird species. Thus, it is clear that collision mortality varies in time and space and it is important to understand and predict what factors affect collision probability.

The probability of bird collisions depends on factors associated both with the location of the wind farm and the species' flight behaviour. These are discussed in the following sections.

1. Site-Specific Factors

   The abundance of a species in the wind farm area has repeatedly been identified as one of the major factors affecting bird collision risk (Barrios and Rodríguez 2004; Carrete et al. 2012). However, collision risk cannot be predicted from abundance alone: some studies found no relationship between species abundance and collision rate (Fernley et al. 2006; Whitfield and Madders 2006; de Lucas et al. 2008). It is clear that other factors must be involved in modulating collision risk (Orloff and Flannery 1992; Orloff and Flannery 1996; de Lucas et al. 2008).

   One factor that influences avian collision risk is the type of turbine used. For example, some studies have found that lattice towers could provide perches for birds and their attraction could increase collision risk (Orloff and Flannery 1992; Percival 2005), while others have not supported this theory (Barrios and Rodríguez 2004; de Lucas et al. 2008). Turbines may also differ in height, for example at Altamont Pass the hub height differed from 12 m for the smallest turbine to 46 m for the tallest turbine. However, this was not reported as a factor that correlates with collision risk (Orloff and Flannery 1992; Orloff and Flannery 1996), and in an independent meta-analysis, Hötker et al. (2006) found only a weak relationship between height and collision risk.

   Another factor influencing avian collision risk is the number of turbines and their layout (Langston and Pullan 2003). A wind farm consisting of a large number of turbines (e.g. over 5000 turbines at Altamont Pass) may be associated with a large number of fatalities overall even if the collision risk per turbine is low (Langston and Pullan 2003; Percival 2003). In addition, a wind farm where turbines are positioned close to one another may allow less space for birds to successfully manoeuvre between them (Hunt 2002; Percival 2005). Furthermore, turbines located at the end of a row were reported to have higher collision rates at Altamont Pass (Orloff and Flannery 1992; Orloff and Flannery 1996; Smallwood and Thelander 2004), but the underlying causes for this difference are unknown and similar differences were not reported at Tarifa (de Lucas et al. 2008).

   Topographical features have also been suggested to influence collision risk in birds. Vultures, for example, require more lift to successfully evade turbines at higher altitudes, which might not always be available. Furthermore, many raptors use updrafts to aid their flight and thus areas with weaker updrafts can have higher mortality rates (de Lucas et al. 2008). For example, at a site in Spain, vulture mortality per turbine was higher in areas with gentle slopes producing weaker updrafts (de Lucas et al. 2008). In addition, wind farms located near features such as a sharp change in relief (e.g. plateau edges) and/or on mountain ridges resulted in higher raptor mortality (Percival 2005; Hötker et al. 2006).

Hunt (2002), in contrast, found that slope was not related to the number of fatalities, but the authors suggested that this may be because other factors (such as turbine spacing) were more important in this instance.

Finally, some studies showed correlations between prey availability and golden eagle (*Aquila chrysaetos*) collision mortality (Hunt 2002; Smallwood and Thelander 2004), suggesting that hunting raptors may not notice the turbines as they search for potential prey (Martin 2011).

2. Species-Specific Factors

In addition to the site-specific factors discussed above, collision risk can also be affected by interspecific variation in behaviour and physiology. While birds in flight tend to focus their attention on what is below them to allow for effective detection of foraging opportunities, they also focus on conspecifics or roost sites (Martin 2011). They may have learnt to expect the open airspace above vegetation to be highly predictable and largely free of hazards (Martin 2011). In addition, the visual system of birds provides high resolution vision in the lateral fields but not in the frontal field. As a result, birds (particularly in flight) may have limited awareness of what is in front of them (Martin 2011), increasing collision risk with 'unexpected' objects such as wind turbines.

Variation in flight maneuverability, which depends largely on morphology (Drewitt and Langston 2008), is another factor affecting collision risk. Although it is unknown which morphology factors influence collision risk, some suggestions have been made. For example, larger, relatively heavier species tend to have lower flight maneuverability and are thus less able to avoid wind turbines when necessary (Garthe and Hüppop 2004). In addition, many soaring birds are also less maneuverable as they have a weak-powered flight and use updrafts or thermals to power their flight (Tucker 1971; Pennycuick 1975; de Lucas et al. 2008). The number of flights, their duration, and height also influence collision risk (Garthe and Hüppop 2004; Drewitt and Langston 2008). For example, many passerines making local movements, as opposed to those during migration, tend to fly lower than the rotor swept area of larger turbines reducing the risk of collision (Hötker et al. 2006).

3. Other Factors

A number of additional factors affecting bird collision risk that are not directly related either to the properties of the wind farm or bird biology and ecology have been identified. Certain weather conditions can influence flight ability. For example, heavy winds will affect flight maneuverability (Langston and Pullan 2003). Furthermore, fog and heavy rain will impede vision and thus also affect collision risk (Larsen and Guillemette 2007).

It has been shown that collision risk changes with different seasons. In winter, lower temperatures mean that thermal updrafts are less common, affecting the flight ability of soaring birds. Indeed, de Lucas et al. (2008) found higher collision rates during winter than other seasons. Another study concerning little, common, and sandwich terns showed that collision risk was higher during chick provisioning (Everaert and Stienen 2006).

Disturbance

Substantial amounts of infrastructure (e.g. access tracks) are created during construction of wind farms. This, combined with the 'footprint' of the turbines themselves, causes a certain amount of direct habitat loss and/or fragmentation. While this loss of habitat is negligible for smaller wind farms, when the development consists of hundreds or even thousands of turbines, this loss can be considerable. Furthermore, birds may also avoid the area surrounding the wind farm, causing indirect habitat loss. This has been reported for different species in different seasons, primarily raptors, geese, ducks, and waders (Hötker et al. 2006) (Table 23.2).

However, birds do not always avoid turbine sites. For example, no disturbance effects have been found for most passerine species (Devereux et al. 2008; Farfán et al. 2009), or a range of other species, e.g. willow ptarmigan (*Lagopus lagopus*) (Bevanger et al. 2010; Douglas et al. 2011). Thus, disturbance behaviour appears to be species-specific, and it is unclear why certain species are affected while others are not. In addition, avoidance behaviour can be season-specific as a recent study found that black grouse (*Tetrao tetrix*) were avoiding wind farms during the breeding season, but there was no indication of avoidance during the winter (Zwart et al. 2015a). While we do not understand all the mechanisms driving avoidance behaviour, a range of contributing factors have been identified which we will outline below.

Firstly, noise produced by turbines could affect bird communication or foraging efficiency and birds might therefore perceive areas close to wind farms as of lower habitat quality. Most noise studies on animals have focused on the effects of urban or traffic noise. For example, great tits (*Parus major*) adjust the pitch of their song in response to urban noise (Slabbekoorn and Peet 2003) and traffic noise is correlated with a reduction in reproductive performance (Reijnen et al. 1996; Halfwerk et al. 2011). In another study, noise lowered foraging efficiency in chaffinches (*Fringilla coelebs*) (Quinn et al. 2006). There is limited information currently published on the impacts of wind turbine noise. Recent work has suggested that anti-predator behaviour in ground squirrels (*Spermophilus beecheyi*) and territorial behaviour in European robins (*Erithacus rubecula*) are affected by wind turbine noise (Rabin et al. 2006 Zwart et al. 2015b), but whether such effects can be generalized to other species is currently unclear. Zeiler and Grünschachner-Berger (2009) suggested that black grouse (*Tetrao tetrix*) may have left a wind farm site because of song disruption. However, the impacts of wind farm noise on bird distribution have not been directly addressed.

Secondly, increased human activity associated with wind farms could affect bird populations (Langston and Pullan 2003; Madders and Whitfield 2006; Zeiler and Grünschachner-Berger 2009). Such an increase would most likely be due to wind farm maintenance, but could also result from increases in tourism. After the construction of a wind farm in Norway, hiker activity increased as access to the area was improved through the newly created tracks that accompanied the wind farm development (Bevanger et al. 2010). Human disturbance is known to affect birds in a number of ways including reduced intake rates (de Boer and Longamane 1996;

Table 23.2 List of examples of disturbance effects by wind farms

Species	Scientific name	Country	Disturbance	Season	Reference
American Kestrel	*Falco sparverius*	US	Yes	Summer	Garvin et al. (2011)
Bewick's Swan	*Cygnus bewickii*	Netherlands	Yes	Winter	Fijn et al. (2007)
Black Grouse	*Tetrao tetrix*	Austria	Yes	Breeding	Zeiler and Grünschachner-Berger (2009)
Common Eider	*Somateria mollissima*	Denmark	Yes	Winter	Larsen and Guillemette (2007)
Cormorant	*Phalacrocorax carbo*	UK	Yes	Breeding	Rothery et al. (2009)
Corvids	*Corvidae*	UK	No	Winter	Devereux et al. (2008)
Dunlin	*Caldris alpina*	Norway	Yes	Breeding	Bevanger et al. (2010)
Eurasian skylark	*Alauda arvensis*	UK	No	Winter	Devereux et al. (2008)
Gamebirds		UK	No	Winter	Devereux et al. (2008)
Golden Plover	*Pluvialis apricaria*	UK	Yes		Pearce-Higgins et al. (2009)
Golden Plover	*Pluvialis apricaria*	Norway	Yes	Breeding	Bevanger et al. (2010)
Golden Plover	*Pluvialis apricaria*	UK	No	Breeding	Douglas et al. (2011)
Granivores		UK	No	Winter	Devereux et al. (2008)
Great Black-backed Gull	*Larus marinus*	UK	No	Breeding	Rothery et al. (2009)
Northern Harrier	*Circus cyaneus*	US	Yes	Summer	Garvin et al. (2011)
Pheasant	*Phasianus colchicus*	UK	Yes	Winter	Devereux et al. (2008)
Pink-footed Goose	*Anser brachyrhynchus*	Denmark	Yes		Larsen and Madsen (2000), Madsen and Boertmann (2008)
Red-tailed Hawk	*Buteo jamaicensis*	US	Yes	Summer	Garvin et al. (2011)
Sandwich Tern	*Sterna sandvicensis*	UK	No	Breeding	Rothery et al. (2009)
Tundra Bean Goose	*Anser serrirostris*	Netherlands	Yes	Winter	Fijn et al. (2007)
Turkey Vulture	*Cathartes aura*	US	Yes	Summer	Garvin et al. (2011)
Wheatear	*Oenanthe oenanthe*	Norway	Yes	Breeding	Bevanger et al. (2010)

(continued)

Table 23.2 (continued)

Species	Scientific name	Country	Disturbance	Season	Reference
White tailed eagles	*Haliaeetus albicilla*	Norway	Yes	Breeding	Bevanger et al. (2010), Dahl et al. (2012)
Willow Ptarmigan	*Lagopus lagopus*	Norway	No	Breeding	Bevanger et al. (2010)
Willow Ptarmigan	*Lagopus lagopus scotica*	UK	No	Breeding	Douglas et al. (2011)

This list was constructed via a literature search using "wind farms" AND "disturbance" AND "birds" OR "wind farms" AND "avoidance" AND "birds" as key words at the Web of Science™. This is not an exhaustive list

Goss-Custard et al. 2006) and increased nest predation (Lord et al. 2001) when they flee their foraging or nesting grounds due to an approaching person.

Third, physical properties of the wind farm such as turbine size and layout may alter bird distributions. Larger turbines can have a greater effect on nesting birds than smaller turbines (Hötker et al. 2006; Madsen and Boertmann 2008), which could be because larger turbines are more spaced out and thus cover a larger area. For example, birds were found not to actively avoid small wind turbines (micro-turbines or domestic turbines, 6–18 m hub height and often installed singly) (Minderman et al. 2012). In contrast, breeding birds, particularly songbirds, have been shown to be less affected by larger turbines (Hötker et al. 2006). Within wind farms, turbines can be positioned in a number of layouts, for example in clusters or rows. One study suggested that clusters might lead to a greater disturbance of pink-footed geese (*Anser brachyrhynchus*), as a cluster layout often coincides with their preferred habitat of open landscape (Larsen and Madsen 2000).

Finally, the construction of the wind farm might in fact cause more of an effect than the operational state (Douglas et al. 2011; Pearce-Higgins et al. 2012). If this is the case, it would be expected that the birds will return to the site over time after construction is completed. This has only been reported in a few cases (e.g. Madsen and Boertmann 2008) and some studies have reported that there is no habituation (Hötker et al. 2006; Stewart et al. 2007; de Lucas et al. 2008), but further longer-term studies are necessary to test this hypothesis.

In addition to indirect habitat loss, avoidance may lead to habitat fragmentation—the turbines lowering habitat quality in the surrounding area and thus breaking up a single patch of habitat into several smaller patches.

In conclusion, further studies are needed to fully understand the disturbance effects of wind farms on birds. In particular, while raptors are a key group that have been shown to be at risk of collision, studying population-level impacts of turbines is challenging due to the low breeding densities of these species (Newton 1979). It is worth adding one final note of caution: some of the effects of turbines on birds may have gone unnoticed as studies might not have been long enough for an effect to be detected (Garvin et al. 2011) or due to a lack of Before-After Control-Impact

(BACI) studies (Madders and Whitfield 2006). The before-after design involves data collection at a wind farm site prior to construction and compares it with data after construction. Collecting data before and after construction from a wind farm site *and* a control site is known as a BACI design.

Population-Level Effects

Both direct collision mortality and disturbance effects may have population-level consequences. These effects are likely to be highly species-specific and we discuss the potential impacts at a population-level below.

Consequences of Direct Collision Mortality

In contrast to disturbance effects, population-level consequences of collision mortality are thought to be more direct. Mortality from collisions could have a major impact on the population level of a species (Langston and Pullan 2003), particularly for long-lived species with low productivity (Langston and Pullan 2003; Hötker et al. 2006). Species with a small global range or population size might be particularly vulnerable. It is therefore important to consider the status of the birds that are using the proposed wind farm site in order to determine the potential effects. It is important to note that population effects may not be immediately apparent; for example, recruitment from other populations can replace the local nesting population, despite the number of birds being killed by the wind farm. The area would thus have become an ecological sink as more adults are coming into the area than leaving it (Smallwood and Thelander 2008).

Consequence of Disturbance Effects

The population-level consequences of disturbance effects are difficult to quantify and few studies have done so (Pearce-Higgins et al. 2012). Habitat loss caused by turbines is expected to cause a decrease in the overall quality of remaining habitat (Larsen and Madsen 2000; Madders and Whitfield 2006). The population-level response to this decrease in habitat quality depends on whether alternative habitat is available (Langston and Pullan 2003). For example, geese and swans moved from control areas to the wind farm area only when food availability in the control area was depleted (Fijn et al. 2007). Furthermore, birds might be displaced into less suitable habitat because optimal habitat might already support the maximum number of that species (e.g. insufficient availability of nesting locations or food resources), which may reduce their ability to survive and reproduce (Madders and Whitfield 2006; Dahl et al. 2012). This drop in productivity affects long-lived species with low annual productivity and slow maturation more than short-lived species with higher annual productivity (Langston and Pullan 2003; Hötker et al. 2006).

Alternatively, avoidance of turbine development areas may cause flights (e.g. between breeding and foraging grounds or migration flights) to be altered: the so-called barrier effect. Changes in flight paths may incur extra energy costs as travelling distances are increased. These increased energy costs could adversely affect survival or breeding success. For example, while flight lines of breeding little, common, and sandwich terns feeding young passed through a wind farm area, the same site was avoided during the non-breeding season, suggesting that they could not afford the extra flight time during the breeding season (Everaert and Stienen 2006). Migrating common eiders (*Somateria mollissima*) and geese have been reported to fly around an offshore wind farm in Denmark (Desholm and Kahlert 2005) and in England (Plonczkier and Simms 2012). The population-level consequences of the barrier effect for migratory populations are unclear, although they are expected to be limited if increases in flight time are relatively small (Desholm 2003).

Prevention and Mitigation

From looking at the factors that affect disturbance or collision caused by wind farms, it is clear that the impact on birds can be minimised by careful wind farm placement. Wind farm construction on sites where particularly sensitive species are present or where collision risk is high, as predicted from factors discussed above, should be avoided.

Therefore, the following is recommended:

1. Wind farms should avoid areas that are highly used by species sensitive to collision or disturbance. These include areas that are important to raptors, such as mountain ridges, and important foraging sites. Furthermore, wind farms should not be built in areas where there are large numbers of flights of sensitive species, such as migration crossing points or between nesting and feeding areas (Langston and Pullan 2003; Percival 2005; Hötker et al. 2006).
2. Wind farms should avoid areas that are designated as, or qualify for, sites of international or national nature conservation (Langston and Pullan 2003).
3. Wind farms should be placed so that they are parallel to the main flight direction (Hötker et al. 2006). For example, they could be placed parallel to migration route or flights between roosting and feeding areas.
4. Wind farms should have corridors so that birds can fly easily between them (Hötker et al. 2006).
5. Wind turbines should not have perching opportunities or other features that could attract birds (Hötker et al. 2006).
6. The height of the mast should be chosen so that the collision risk is low and/or any disturbance is minimal (Hötker et al. 2006).

In Europe, Competent Authorities require Environmental Impact Assessments (EIAs) to be carried out prior to any wind farm developments taking place.

These aim to ensure that the development is placed in a suitable location which minimises adverse impacts on wildlife (Directive 2011/92/EU). In brief, EIAs require a range of ecological surveys to be carried out, including those to determine which bird populations might be affected by any development. In addition, the sensitivity of those populations to any impact is determined and the scale of any potential effects is assessed. Finally, recommendations are made as to the acceptability of the predicted effects of the proposed development (Percival 2003). Outside Europe, there is little information on requirements that is easily accessible. Canada (Kingsley and Whittam 2005) and Mexico (Martínez 2008) have similar guidelines in place to those in Europe. In the United States, survey requirements vary extensively by state; some states have very detailed guidelines regarding the placement of wind farms, while others do not have any (Jodi Stemler Consulting 2007).

Given the possibility of bird mortality as outlined in the previous section, a key element of many EIAs is the estimation of likely bird mortality due to collision. To this end, numerical models are constructed that predict the number of bird fatalities per turbine per unit time, given the characteristics of the proposed turbines and bird activity in the area. The most widely used collision risk model was developed by Band et al. (2007) (Fig. 23.2). In this model, the collision rate is a product of a range of variables, including: (1) the size (wingspan and length) and flight speed of the given bird species; (2) the dimensions of the rotors and the speed of rotation; and (3) the avoidance rate of given species. The number of birds flying through this danger zone is calculated from vantage point surveys (Fig. 23.2) and is then multiplied by the collision risk rate to predict the number of collisions. One weakness of this model, among others, lies in the difficulty of estimating avoidance rate. The authors of the Band model tentatively suggest the use of a 95 % avoidance rate when data

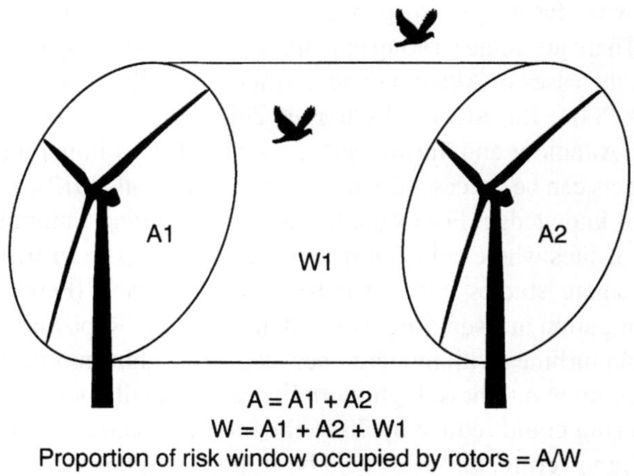

Fig. 23.2 An illustration of how to calculate the number of birds flying through the collision risk window (referred to as 'W' in (Band et al. 2007)). W is the sum of all birds observed flying through the areas A1, A2, and W1 during a period of time, e.g. 1 month, divided by the number of hours recording in that period (to calculate an hourly collision risk estimate). Thus, the lower bird would be classed as passing through the risk window, whereas the upper bird would not

are lacking, but recognise that this figure is somewhat arbitrary and advise against using it (Band et al. 2007). Despite this, the model remains the most used model in the UK. However, many studies have taken this avoidance rate figure as an absolute value for which it was never designed. Other studies suggest that avoidance rates may be a considerably higher than the original figure proposed by Band et al. (Desholm and Kahlert 2005; Chamberlain et al. 2006), suggesting that the Band model, when the 95 % avoidance rate is assumed, may overestimate collision rates. Conversely, a recent study by Ferrer et al. (2011) found that actual collision rates were in fact considerably higher than those predicted using the Band model, further highlighting the shortcomings of current collision risk modelling. Crucially, these risk models do not take into account many of the factors that were discussed earlier in this chapter, and no other models that do are currently used in practice (at least in Europe). It will be interesting to investigate, as the field develops, what effect the incorporation of these additional factors has on model performance.

Sensitivity maps can be used to visualise the suitability of potential sites for wind farm development. To date, maps have been created for Scotland (Bright et al. 2008) and the national waters of Germany (Garthe and Hüppop 2004). The map for Scotland is based on Special Protected Areas and the distribution of 16 bird species, although some sensitive species have not been taken into account (Bright et al. 2008), while the German map is based on densities of bird species occurring in the area and their sensitivity to wind farm development (Garthe and Hüppop 2004).

Tools like collision risk modelling and sensitivity mapping provide an additional tool for use in the assessment of the effects of wind farms on birds. Although there remain many unknowns in the interactions between birds and wind farms, we should make use of all available tools and use them with the best data available to date, in order to minimise the effect of wind farms to the best of our ability. However, we must do so carefully, acknowledging all the necessary caveats.

Effects of wind farms on bird populations are only possible to measure post-construction. There are some striking examples of significant impacts of wind farms on birds, as in the cases of Altamont and Tarifa (for details see above) (Smallwood and Thelander 2004; Barrios and Rodríguez 2004). Although no EIAs were performed before Altamont and Tarifa were constructed, it is important to recognise that not all effects can be successfully predicted (Ferrer et al. 2012), at least with our current level of knowledge. For example, some of the highest mortality rates have been reported at sites where collision risk was estimated to be sufficiently low during risk assessment studies conducted before construction (Ferrer et al. 2012). Alternative mitigation measures are required in such case. Repowering wind farms, by replacing old turbines with modern ones, can reduce bird mortality by avoiding areas which are known to have high mortality rates. Smallwood et al. (2009) suggested repowering could reduce by 70 % the mortality caused by the wind farm. In addition to repowering, turbines could be stopped at times when collision risk is highest. For example, in a recent case study, mortality was halved when turbines were stopped when griffon vultures were observed near them, while only 0.07 % of energy production was lost (de Lucas et al. 2012).

Conclusion

In many cases, effects of wind farms on bird populations are limited to species on the site, although substantial problems have been reported at some sites. Factors that contribute to collision risk include flight behaviour and the topography surrounding the wind farm. The studies reviewed in this chapter suggest that some adverse effects maybe prevented by appropriate placement of wind farms, and EIAs and sensitivity maps provide vital means to do so. Unexpected effects post-construction may be mitigated in variety of ways, including shutting down turbines during times of high collision risk or repowering of old turbines.

Currently, we do not fully understand the interaction between birds and wind farms and thus our predictions of potential effects may be inaccurate. Further research is therefore needed to improve understanding of both causes and consequences of collision mortality and displacement effects, and additional data are needed to more accurately estimate model parameters (e.g. avoidance rates). As some studies lack pre-monitoring data and could therefore have missed some disturbance effects, data from both pre- and post-construction EIA surveys could benefit new research. However, many of these are not publicly available due to commercial client confidentiality. In spite of such issues, collaborative studies between academics, consultants and NGO partners will be most likely to make genuine contributions to improving our understanding of conflicts between birds and wind turbines.

References

Band W, Madders M, Whitfield D (2007) Developing field and analytical methods to assess avian collision risk at wind farms. In: de Lucas M, Janss GFE, Ferrer M (eds) Birds wind farms risk assess. Mitig, Quercus, pp 259–275

Barrios L, Rodríguez A (2004) Behavioural and environmental correlates of soaring-bird mortality at on-shore wind turbines. J Appl Ecol 41:72–81. doi:10.1111/j.1365-2664.2004.00876.x

Bevanger K, Berntsen F, Clausen S et al (2010) Pre- and post-construction studies of conflicts between birds and wind turbines in coastal Norway (BirdWind). Report on findings 2007-2010. NINA Rep 620:1–152

Bright J, Langston R, Bullman R et al (2008) Map of bird sensitivities to wind farms in Scotland: a tool to aid planning and conservation. Biol Conserv 141:2342–2356. doi:10.1016/j.biocon.2008.06.029

Carrete M, Sánchez-Zapata JA, Benítez JR et al (2012) Mortality at wind-farms is positively related to large-scale distribution and aggregation in griffon vultures. Biol Conserv 145:102–108. doi:10.1016/j.biocon.2011.10.017

Chamberlain DE, Rehfisch MR, Fox AD et al (2006) The effect of avoidance rates on bird mortality predictions made by wind turbine collision risk models. Ibis (Lond 1859) 148:198–202. doi:10.1111/j.1474-919X.2006.00507.x

Dahl EL, Bevanger K, Nygård T et al (2012) Reduced breeding success in white-tailed eagles at Smøla windfarm, western Norway, is caused by mortality and displacement. Biol Conserv 145:79–85. doi:10.1016/j.biocon.2011.10.012

De Boer WF, Longamane FA (1996) The exploitation of intertidal food recources in Inhaca bay, Mozambique, by shorebirds and humans. Biol Conserv 78:295–303

De Lucas M, Janss GFE, Whitfield DP, Ferrer M (2008) Collision fatality of raptors in wind farms does not depend on raptor abundance. J Appl Ecol 45:1695–1703. doi:10.1111/j.1365-2664.2008.01549.x

De Lucas M, Ferrer M, Bechard MJ, Muñoz AR (2012) Griffon vulture mortality at wind farms in southern Spain: Distribution of fatalities and active mitigation measures. Biol Conserv 147:184–189. doi:10.1016/j.biocon.2011.12.029

Desholm M (2003) How much do small-scale changes in flight direction increase overall migration distance? J Avian Biol 34:155–158

Desholm M, Kahlert J (2005) Avian collision risk at an offshore wind farm. Biol Lett 1:296–298. doi:10.1098/rsbl.2005.0336

Devereux CL, Denny MJH, Whittingham MJ (2008) Minimal effects of wind turbines on the distribution of wintering farmland birds. J Appl Ecol 45:1689–1694. doi:10.1111/j.1365-2664.2008.01560.x

Directive 2011/92/EU (2012) Directive 2011/92/EU of the European parliament and of the council of 13 December 2011 on the assessment of the effects of certain public and private projects on the environment. Off J Eur Union L 26:1–21

Douglas DJT, Bellamy PE, Pearce-Higgins JW (2011) Changes in the abundance and distribution of upland breeding birds at an operational wind farm. Bird Study 58:37–43. doi:10.1080/00063657.2010.524914

Drewitt AL, Langston RHW (2006) Assessing the impacts of wind farms on birds. Ibis (Lond 1859) 148:29–42

Drewitt AL, Langston RHW (2008) Collision effects of wind-power generators and other obstacles on birds. Ann N Y Acad Sci 1134:233–266. doi:10.1196/annals.1439.015

Erickson WP, Johnson GD, Strickland MD et al (2001) Avian collisions with wind turbines: A summary of existing studies and comparisons to other sources of avian collision mortality in the United States. National Wind Coordinating Committee, c/o RESOLVE, Inc., Washington, DC, pp 1–62

Everaert J, Stienen EWM (2006) Impact of wind turbines on birds in Zeebrugge (Belgium). Biodivers Conserv 16:3345–3359. doi:10.1007/s10531-006-9082-1

Farfán MA, Vargas JM, Duarte J, Real R (2009) What is the impact of wind farms on birds? A case study in southern Spain. Biodivers Conserv 18:3743–3758. doi:10.1007/s10531-009-9677-4

Fernley J, Lowther S, Whitfield P (2006) A review of goose collisions at operating wind farms and estimation of goose avoidance rate. British Trust for Ornithology, The Nunnery, Thetford, pp 1–18

Ferrer M, de Lucas M, Janss GFE et al (2012) Weak relationship between risk assessment studies and recorded mortality in wind farms. J Appl Ecol 49:38–46. doi:10.1111/j.1365-2664.2011.02054.x

Fijn RC, Krijgsveld KL, Prinsen HAM, et al. (2007) Effecten op zwanen en ganzen van het ECN windturbine testpark in de Wieringermeer. Aanvaringsrisico's en verstoring van foeragerende vogels. 1–100

Garthe S, Hüppop O (2004) Scaling possible adverse effects of marine wind farms on seabirds: developing and applying a vulnerability index. J Appl Ecol 41:724–734

Garvin JC, Jennelle CS, Drake D, Grodsky SM (2011) Response of raptors to a windfarm. J Appl Ecol 48:199–209. doi:10.1111/j.1365-2664.2010.01912.x

Goss-Custard JD, Triplet P, Sueur F, West AD (2006) Critical thresholds of disturbance by people and raptors in foraging wading birds. Biol Conserv 127:88–97. doi:10.1016/j.biocon.2005.07.015

GWEC (2013) Global wind statistics 2012. 1–4.

Halfwerk W, Holleman LJM, Lessells CM, Slabbekoorn H (2011) Negative impact of traffic noise on avian reproductive success. J Appl Ecol 48:210–219. doi:10.1111/j.1365-2664.2010.01914.x

Hötker H, Thomsen K-M, Jeromin H (2006) Impacts on biodiversity of exploitation of renewable energy sources: the example of birds and bats—facts, gaps in knowledge, demands for further research, and ornithological guidelines for the development of renewable energy exploitation. NABU 1–65

Hunt G (2002) Golden eagles in a perilous landscape: predicting the effects of mitigation for wind turbine blade-strike mortality. California Energy Commission, Sacramento

IPCC (2012) Renewable energy sources and climate change mitigation. Special report of the intergovernmental panel on climate change. Clim Policy, 1–1076

Jodi Stemler Consulting (2007) Wind power siting regulations and wildlife guidelines in the United States. Association of Fish & Wildlife Agencies & U. S. Fish & Wildlife Service, Denver

Kingsley A, Whittam B (2005) Wind turbines and birds: a background review for environmental assessment. 1–81.

Langston RHW, Pullan JD (2003) Windfarms and birds: An analysis of the effects of windfarms on birds, and guidance on environmental assessment criteria and site selection issues. RSPB/ BirdLife Int Rep 1–58

Larsen JK, Guillemette M (2007) Effects of wind turbines on flight behaviour of wintering common eiders: implications for habitat use and collision risk. J Appl Ecol 1:516–522. doi:10.1111/j.1365-2664.2007.1303.x

Larsen JK, Madsen J (2000) Effects of wind turbines and other physical elements on field utilization by pink-footed geese (Anser brachyrhynchus): A landscape perspective. Landsc Ecol 15:755–764. doi:10.1023/A:1008127702944

Lord A, Waas JR, Innes J, Whittingham MJ (2001) Effects of human approaches to nests of northern New Zealand dotterels. Biol Conserv 98:233–240

Madders M, Whitfield DP (2006) Upland raptors and the assessment of wind farm impacts. Ibis (Lond 1859) 148:43–56. doi:10.1111/j.1474-919X.2006.00506.x

Madsen J, Boertmann D (2008) Animal behavioral adaptation to changing landscapes: spring-staging geese habituate to wind farms. Landsc Ecol 23:1007–1011

Martin GR (2011) Understanding bird collisions with man-made objects: a sensory ecology approach. Ibis (Lond 1859) 153:239–254

Martínez C (2008) Proyecto de Norma Oficial Mexicana NOM-151-SEMARNAT, Evaluaci6n de los impactos de la generaci6n de la energía oleoeléctrica sobre la vida silvestre, Abril 2008. CONABIO, Mexico D.F

Minderman J, Pendlebury CJ, Pearce-Higgins JW, Park KJ (2012) Experimental evidence for the effect of small wind turbine proximity and operation on bird and bat activity. PLoS One 7, e41177. doi:10.1371/journal.pone.0041177

Newton I (1979) Population ecology of raptors. Poyser, Birkhamstead

Orloff S, Flannery A (1992) Wind turbine effects on avian activity, habitat use, and mortality in Altamont Pass and Solano Country wind resource areas. California Energy Commission, Sacramento

Orloff S, Flannery A (1996) A continued examination of avian mortality in the Altamont Pass wind resource area. California Energy Commission, Sacramento

Pearce-Higgins JW, Stephen L, Langston RHW et al (2009) The distribution of breeding birds around upland wind farms. J Appl Ecol 46:1323–1331. doi:10.1111/j.1365-2664.2009.01715.x

Pearce-Higgins JW, Stephen L, Douse A, Langston RHW (2012) Greater impacts of wind farms on bird populations during construction than subsequent operation: results of a multi-site and multi-species analysis. J Appl Ecol 49:386–394. doi:10.1111/j.1365-2664.2012.02110.x

Pennycuick CJ (1975) Mechanics of flight. Avian Biol 5:1–75

Percival SM (2003) Birds and wind farms in Ireland: a review of potential issues and impact assessment. Ecology Consulting, Durham

Percival SM (2005) Birds and windfarms: what are the real issues? Br Birds 98:194–204

Plonczkier P, Simms IC (2012) Radar monitoring of migrating pink-footed geese: behavioural responses to offshore wind farm development. J Appl Ecol 49:1187–1194. doi:10.1111/j.1365-2664.2012.02181.x

Quinn JL, Whittingham MJ, Butler SJ, Cresswell W (2006) Noise, predation risk compensation and vigilance in the chaffinch Fringilla coelebs. J Avian Biol 37:601–608. doi:10.1111/j.2006.0908-8857.03781.x

Rabin LA, Coss RG, Owings DH (2006) The effects of wind turbines on antipredator behavior in California ground squirrels (Spermophilus beecheyi). Biol Conserv 131:410–420. doi:10.1016/j.biocon.2006.02.016

Reijnen R, Foppen R, Meeuwsen H (1996) The effects of traffic on the density of breeding birds in Dutch agricultural grasslands. Biol Conserv 75:255–260

Rothery P, Newton I, Little B (2009) Observations of seabirds at offshore wind turbines near Blyth in northeast England. Bird Study 56:1–14. doi:10.1080/00063650802648093

Slabbekoorn H, Peet M (2003) Birds sing at a higher pitch in urban noise. Nature 424:267

Smallwood KS, Thelander CG (2004) Developing methods to reduce bird mortality in the altamont pass wind resource area. Progr Manag 1–6

Smallwood KS, Thelander CG (2008) Bird mortality in the Altamont Pass Wind Resource Area, California. J Wildl Manage 72:215–223. doi:10.2193/2007-032

Smallwood KS, Neher L, Energy C (2009) Map-based repowering of the Altamont Pass Wind Resource Area based on burrowing owl burrows, raptor flights, and collisions with wind turbines. Energy 1–49

Stewart GB, Pullin AS, Coles CF (2007) Poor evidence-base for assessment of windfarm impacts on birds. Environ Conserv 34:1–11. doi:10.1017/S0376892907003554

Tucker VA (1971) Flight energetics in birds. Am Zool 11:115–124

Whitfield DP, Madders M (2006) A review of the impacts of wind farms on hen harriers circus cyaneus and an estimation of collision avoidance rates. Natural Research Information Note 1 (revised). 1:1–32.

Zeiler HP, Grünschachner-Berger V (2009) Impact of wind power plants on black grouse, Lyrurus tetrix in Alpine regions. Folia Zool 58:173–182

Zwart MC, Robson P, Rankin S, et al. (2015a) Using Environmental Impact Assessment and post-construction monitoring data to inform wind energy developments. Ecosphere 6(2):26. http://dx.doi.org/10.1890/ES14-00331.

Zwart MC, Dunn JC, McGowan PJK, Whittingham MJ (2015b) Wind farm noise suppresses territorial defense behavior in a songbird. Behav Ecol arv128. doi: 10.1093/beheco/arv128

Part VIII
Bushmeat: A Socioecological Problem. The Overexploitation of Wildlife for Nutritional and Traditional Purposes. World Animal Trade, Extinction Risk and Socio-economic Issues

This part examines some extremely important issues regarding wildlife exploitation, especially that of mammals and birds, although the phenomenon does include other animal *taxa*. For the purposes of this book, it, of course, focuses more specifically on warm-blooded vertebrates.

Bushmeat, or trade in the meat of wildlife, is extremely widespread in West and Central Africa, as well as in other areas of the world such as tropical Asia and South America, albeit to a lesser extent (Nasi et al. 2008). The main reason behind the trade in bushmeat is to augment the protein requirements of many agro-forest-pastoral populations with low living standards (Nasi et al. 2008). However, reasons also exist that are related to superstitions, local religions and traditional medicines. There are many problems related to this phenomenon, from the severe impact on many species that are at risk of extinction (Nasi et al. 2008) to the serious health problems caused by the spread of serious animal diseases and epidemics (Karesh and Noble 2009).

Regarding the aforementioned, the first part provides a complete review of the issues related to the transmission of diseases and zoonoses linked to the 'bushmeat' phenomenon (Kurpiers et al. 2016), from which important information may be gleaned.

The second chapter (Sollund 2016) discusses the immense problem of the world wildlife trade, much of which is illegal (Rosen and Smith 2010). This trade, which is certainly one of the most lucrative in the world (secondary only to drug and weapons trafficking), has, as one can easily imagine, an absolutely devastating impact on rare species and species in danger of extinction (Rosen and Smith 2010). In fact, many species have already become extremely rare or almost extinct due to this business (e.g. Yi-Ming et al. 2000). Sollund's (2016) chapter is a case in point for the illegal wildlife trade (IWT) and provides the example of Norway, which is addressing the legal and practical aspects of countering the sale of objects of animal origin, such as ivory objects sold in web auctions, which are popular all over the world and frequently attended by thousands of collectors and dealers.

References

Karesh WB, Noble E (2009) The bushmeat trade: increased opportunities for transmission of zoonotic disease. Mt Sinai J Med 76:429–434

Kurpiers LA, Schulte-Herbrüggen B, Ejotre I, Reeder DAM (2016) Bushmeat and emerging infectious diseases—lessons from Africa. In: Angelici FM (ed) Problematic Wildlife- A cross-disciplinary approach. Springer, New York, p 507–551

Nasi R, Brown,D, Wilkie D, Bennett E, Tutin C, van Tol G, Christophersen T (2008) Conservation and use of wildlife-based resources: the bushmeat crisis. Secretariat of the Convention on Biological Diversity, Montreal, and Center for International Forestry Research (CIFOR), Bogor. Technical Series 33, pp 1–50

Rosen GE, Smith KF (2010) Summarizing the evidence on the international trade in illegal wildlife. Ecohealth 7:24–32

Sollund R (2016) Wildlife trafficking in a globalized world: an example of motivations and *modus operandi* from a Norwegian case study. In: Angelici FM (ed) Problematic wildlife—a cross-disciplinary approach. Springer, New York, p 553–570

Yi-Ming L, Zenxiang G, Xinhai L, Sung W, Niemelä J (2000) Illegal wildlife trade in the Himalayan region of China. Biodivers Conserv 9:901–918

Chapter 24
Bushmeat and Emerging Infectious Diseases: Lessons from Africa

Laura A. Kurpiers, Björn Schulte-Herbrüggen, Imran Ejotre, and DeeAnn M. Reeder

Introduction

Emerging infectious diseases (EIDs) are human diseases that are either newly discovered or are increasing in incidence or geographical range. Some diseases, such as measles, sleeping sickness, and bubonic plague, emerged in prehistoric or ancient times (Babbott and Gordon 1954; Hays 2005; Steverding 2008), whereas others, such as Ebola virus, Nipah virus, and SARS, emerged more recently (World Health Organization 1978; Chua et al. 2000; Guan et al. 2003). The trend of EID emergence is accelerating: over 300 distinct emerging disease events have been recorded in the last six decades and more than 35 new infectious diseases have emerged in humans since 1980 (Lederberg et al. 2003; Jones et al. 2008).

Upwards of 75 % of EIDs in humans are of zoonotic origin, which means the pathogen originates in animals and is transmitted to humans (Taylor et al. 2001; Jones et al. 2008; Karesh and Noble 2009). Although many zoonotic pathogen spillovers arise in domestic animals, including livestock, the majority (71.8 %) of zoonotic EIDs arise from wildlife species (Jones et al. 2008). In many developing countries, domesticated animals live in close proximity to wildlife. This facilitates the movement of pathogens between them and to humans through interactions with sylvatic disease cycles or through two-step wildlife-to-domestic animal-to-human emergences. Examples include rabies infections, which move between wildlife and domestic dogs, with recurring spillovers to humans; and the Henipah viruses, in which *Pteropus* flying foxes are the reservoir host and domestic pigs or horses are

L.A. Kurpiers • I. Ejotre • D.M. Reeder (✉)
Department of Biology, Bucknell University, Lewisburg, PA 17837, USA
e-mail: Laura.Kurpiers@bucknell.edu; Imran.Ejotre@bucknell.edu; DeeAnn.Reeder@bucknell.edu

B. Schulte-Herbrüggen
Stockholm Resilience Centre, Stockholm University, 10691 Stockholm, Sweden
e-mail: bjorn.schulte-herbruggen@su.se

F.M. Angelici (ed.), *Problematic Wildlife*, DOI 10.1007/978-3-319-22246-2_24

amplifier hosts from which spillovers to humans have been documented (Childs et al. 2007; Daszak et al. 2007). Not surprisingly, the most devastating pandemics in human history, the Black Death, Spanish influenza, and HIV/AIDS, were all caused by zoonoses from wildlife (Morens et al. 2008).

Zoonotic diseases can spill between animal hosts and humans in a variety of ways, including through (a) shared vectors, such as mosquitoes for malaria, (b) indirect contact, such as exposure to rodent feces in a peridomestic setting, or (c) direct contact with an animal through consumption, animal bites, scratches, body fluids, tissues, and excrement (Wolfe et al. 2005a). Most pathogens infecting animals fail to make the jump into humans, but 33 % of zoonotic pathogens (~286 out of 868 zoonotic pathogen species studied) that have spilled over are known to be transmissible between humans (Taylor et al. 2001). Of all EIDs, zoonotic spillovers from wildlife have been identified as the most significant, growing threat to global health (Cleaveland et al. 2007; Jones et al. 2008).

Recent evidence highlights the link between infectious diseases and biodiversity loss, land use changes, and habitat fragmentation (Cleaveland et al. 2007; Maganga et al. 2014; Gottdenker et al. 2014). Although additional research on the relationship between habitat degradation and EIDs is needed, Gottdenker et al. (2014) reviewed 305 studies incorporating a broad variety of diseases and found that the most common land use change types related to zoonotic disease transmission were deforestation, habitat fragmentation, agricultural development, irrigation, and urbanization. Functionally, the mechanisms influencing disease spillover include disruption of food web structures, changes in host–pathogen interactions, and mixing of pathogen gene pools resulting in increased pathogen genetic diversity (Jones et al. 2013). Many studies have shown that habitat fragmentation and biodiversity loss correspond to an increase in disease and pathogen abundance and diversity within a host species (Allan et al. 2003; Gillespie et al. 2005; Keesing et al. 2006; Salzer et al. 2007; Cottontail et al. 2009; Young et al. 2014). Specifically, the emergence or re-emergence of many zoonotic diseases including yellow fever, Lyme disease, hantavirus pulmonary syndrome, Nipah virus encephalitis, influenza, rabies, malaria, and human African trypanosomiasis have been linked to anthropogenic habitat changes (Jones et al. 2013).

Many of these human environmental changes are occurring in sub-Saharan Africa where human bushmeat activities have been linked to numerous virulent disease outbreaks, including Ebola (Leroy et al. 2004a), HIV (Van Heuverswyn and Peeters 2007), and monkeypox (Rimoin et al. 2010). Pathogen spillover from bushmeat can occur through consumption; however, the main risks are associated with exposure to body fluids and feces during handling and butchering (Kilonzo et al. 2014; Paige et al. 2014). Historically, when a spillover occurred, the likelihood of an epidemic was limited because hunter-gatherer tribes were generally small and widely dispersed, hampering disease transmission between groups of people. Once agricultural expansion occurred, human population densities increased, and people became better connected, diseases could spread more easily. As a result, transmissions of infectious diseases from animals to humans have led to devastating outcomes across the globe (LeBreton et al. 2006). EIDs cause hundreds of thousands of deaths annually (Bogich et al. 2012). Some outbreaks have spread across large regions and became pandemics, costing the global economy tens of billions of dollars (e.g., SARS, H5N1, the 2014–2015 West African Ebola outbreak) and bringing entire nations to the brink of economic collapse.

In this review, we explore the links between bushmeat-related activities and EIDs in sub-Saharan Africa, where the vast majority of African emerging infectious zoonotic diseases occur (Jones et al. 2008). The recent Ebola outbreaks have highlighted the potential role of bushmeat as a source of pathogens, but a comprehensive review of the different pathogens that may emerge from wildlife through bushmeat-related activities is lacking. Although we are in no way suggesting that this issue is more important than other pressing health crises in sub-Saharan Africa (such as malaria prevention/treatment and improving healthcare infrastructure), we argue that a better assessment of the public health threats associated with this human-wildlife interaction is warranted and necessary to improve management of future disease outbreaks.

Bushmeat

The term "bushmeat" refers to the meat derived from wild animals for human consumption (Milner-Gulland and Bennett 2003) (Fig. 24.1). It includes a wide range of animals, such as invertebrates, amphibians, insects, fish, reptiles, birds, and mammals, including as many as 500 species in sub-Saharan Africa (Ape Alliance 2006). Although research has focused largely on mammals and, to a lesser extent, birds, theoretically any wildlife species harvested for bushmeat could be a potential source of zoonotic disease that can spillover during the hunting, butchering, and preparation process (Wolfe et al. 2000; Karesh and Noble 2009). Hunters face risk of injury from live animals, which might allow animal blood to enter the hunter's bloodstream through open wounds. While small animals can be carried in bags, large animals are commonly carried on the shoulder or back, bringing the hunter in close contact with the animal and facilitating transfer of body fluids (LeBreton et al. 2006). The highest risk of disease transmission occurs during the butchering of animals, e.g. skinning, opening of the body cavity, removal of organs, and cutting of meat. More people butcher than hunt animals (83 % and 42 %, respectively, LeBreton et al. 2006) and butchering involves the use of sharp tools, which may lead to cuts during the process. Subramanian (2012) found that 38 % of respondents cut themselves on a regular basis during butchering. Women are especially at risk of disease transmission as they engage more often in butchering and in food preparation than men. In discussing the links between bushmeat and disease, we refer to this all-encompassing suite of risky behaviors as "bushmeat-related activities."

Nonhuman primates, rodents, and bats have all been linked to the spillover of zoonotic diseases into humans (Cleaveland et al. 2007; Jones et al. 2008; Kilonzo

Fig. 24.1 Bushmeat being smoked in rural South Sudan; photo credit Adrian Garside

et al. 2014). A review of the West and Central African bushmeat literature including market, offtake, and consumption surveys documented a total of 177 species from 25 orders that were harvested for bushmeat, including 134 (76 %) mammal species, 24 (14 %) bird species, 18 (10 %) reptile species, and 1 (<1 %) amphibian species (Taylor et al. 2015). Among mammals, the largest group was primates (48 species) including western gorillas (*Gorilla gorilla*), bonobos (*Pan paniscus*), and common chimpanzees (*Pan troglodytes*), followed by ungulates (34 species), carnivores (22 species), and rodents (16 species). In terms of biomass offtake, however, ungulates are generally the most prominent group. Although the Taylor et al. (2015) study is very comprehensive, it only included studies that: (1) provided a quantitative measure of bushmeat offtake, consumption, and/or market availability/sales; (2) used non-biased data collection methods and systematically sampled settlements/hunters to prevent selection bias; (3) identified carcasses to the species level; and (4) recorded either the number of carcasses or the total biomass (kg). For a more inclusive and general review of existing Central African bushmeat studies, see Wilkie and Carpenter (1999), and for West African studies, see Schulte-Herbrüggen (2011). Fa et al. (2006) found that of the approximately one million carcasses traded in the Cross-Sanaga region of Nigeria and Cameroon, 99 % were mammals; of which around 40 % were ungulates, 30 % rodents, and nearly 15 % primates. However, as wildlife populations become depleted, such as near urban areas and intensively used agricultural landscapes, smaller bodied mammals comprise a larger share of hunters' offtake (Bowen-Jones et al. 2003; Schulte-Herbrüggen et al. 2013a).

Livelihood Importance

Humans have hunted wild animals for consumption and to protect their crops for millennia (Shipman et al. 1981; Grubb et al. 1998; Davies et al. 2007), and it remains an important source of food and income security among rural communities today (de Merode et al. 2004; Brashares et al. 2011). Bushmeat is an important source of animal protein in many West and Central African countries, with up to 90 % of total animal protein consumption coming from wild animals (Fa et al. 2003). Overall, the contribution of bushmeat to protein and food security is generally lower in urban than rural areas and is highest among remote rural communities (Brashares et al. 2011). For example, the relative importance of bushmeat in the diet of rural Gabonese households ranged from 13 % of total household consumption value in a village near a town to 25 % in a remote community (Starkey 2004). Similarly, for rural Equatorial Guinea, Allebone-Webb (2008) showed that bushmeat consumption contributed 43 % to total protein consumption in a village with poor transport links, but only 18 % in a village with good connections. In remote Cameroonian communities with very few opportunities for purchasing alternative protein sources, bushmeat comprised 80–98 % of animal protein consumption (Muchaal and Ngandjui 1999). In rural communities with relatively good market access and low levels of bushmeat consumption, the importance of bushmeat for food has been shown to increase seasonally during the agricultural lean season (e.g. the planting season between harvests) when farming households receive little income (Dei 1989, de Merode et al. 2004, Schulte-Herbrüggen et al. 2013b) and during the dry season when fish is not available (Poulsen et al. 2009). Bushmeat is also an important source of nutrients, especially among children. Evidence from rural Madagascar shows that removing bushmeat consumption would result in a 29 % increase in the number of children suffering from anemia and triple the cases of anemia among children in the poorest households (Golden et al. 2011).

Most hunters sell at least part of their harvest making it an important source of income, especially where alternative income-generating activities are lacking. The importance of bushmeat in household economies varies across sites and individual hunting households, ranging from 38 % to more than 90 % of the total cash income earned (reviewed in Schulte-Herbrüggen 2011). In rural Gabon, hunting accounts for up to 72 % of household incomes, with the proportion rising in poorer, more remote communities (Starkey 2004). Hunters are also more likely to sell large animals and keep small animals for their own consumption, because the latter fetch a lower price per animal and may be less marketable (van Vliet and Nasi 2008). Finally, households facing income shortages during the agricultural lean season and requiring cash income to pay for urgent expenditures, such as hospital bills, are more likely to sell bushmeat than keep it for own consumption (de Merode et al. 2004).

Overall, income from bushmeat sales can be lucrative and compare favorably with alternative work in many rural places. Vega et al. (2013) found that commercial hunters in Equatorial Guinea generated a mean of US$2000 per year from bushmeat sales.

Hunters supplying markets in Central African logging concessions earned twice the income of junior technicians working at a logging company (Tieguhong and Zwolinski 2009). Rural Kenya hunters can earn 2.5 times the average salary in the area (Fitzgibbon et al. 1995), and Ghanaian hunters can earn income similar to that of a graduate entering Wildlife Service, and up to 3.5 times the government minimum wage (Ntiamoa-Baidu 1998). Very successful Zambian hunters have been reported earning just below the mean annual income in a single hunting trip (Brown 2007).

The sale of bushmeat historically occurred at a local level, but with increased transportation routes and globalization, the bushmeat trade is expanding to supply urban and international demand. In the past, novel pathogens entering the rural communities may not have spread beyond the community, but this is no longer the case as remote rural areas are connected to urban areas, and increased global trade networks and air travel increases the risk of disease transmission worldwide (Brashares et al. 2011). This expanding trade network links hunters to consumers, and with many people along this commodity chain coming into contact with bushmeat, the opportunity for disease spillover can occur at many points. For example, the commodity chain supplying bushmeat to an urban market in Ghana includes hunters, wholesalers, market traders, restaurant owners, and consumers (Mendelson et al. 2003). The bushmeat commodity chain supplying an urban market in Democratic Republic of the Congo is comprised of hunters, porters who carry the meat to the road, the bicycle traders who transport the meat into town, and the market-stall owners who sell the bushmeat to consumers (de Merode and Cowlishaw 2006). A recent study from Ghana estimates that a minimum of 128,000 bats are sold each year through a commodity chain that stretches up to 400 km and involves multiple vendors (Kamins et al. 2011a). In Zambia, Mozambique, and Malawi, well-developed and complex rural-urban trade supply networks link rural hunters to urban consumers who are willing to pay high prices for bushmeat (Barnett 1997). Understanding commodity chains is important, as pathogens likely remain viable for some period after an animal is killed. For example, Prescott et al. (2015) demonstrated that Ebola virus remains viable on monkey carcasses for at least seven days, with viral RNA detectable for 10 weeks.

Scale of Bushmeat Harvest in Sub-Saharan Africa

Bushmeat has become a multi-million dollar business due to a growing human population and is now serving both subsistence and trade objectives. Harvest volumes have been estimated at 12,000 tones per year in the Cross-Sanaga rivers region of Nigeria and Cameroon (Fa et al. 2006), 120,000 tones per year in Côte d'Ivoire (Caspary 1999), 385,000 tons per year in Ghana (Ntiamoa-Baidu 1998), and at total of 1–4.9 million tons per year in Central African forests (Wilkie and Carpenter 1999; Fa et al. 2002).

However, it is important to recognize that our understanding of the scale of bushmeat harvest is limited by the availability of information and hence current regional harvest estimates might underestimate actual harvest volumes. Despite substantial effort in recent years, our knowledge is still site-specific and data are lacking from many regions. Most surveys have been restricted to relatively small areas or market catchments from which national estimates were extrapolated. Research efforts have focused on Central Africa with some data available for 60 % of countries compared to 30 % of West African countries (Taylor et al. 2015). A large number of sites with detailed bushmeat data are concentrated in the Cross-Sanaga region of Nigeria and Cameroon, where Fa et al. (2006) collected market data at 86 sites, hence presenting a geographical bias in our understanding of bushmeat harvest. Furthermore, the majority of available data samples (79.3 % and 53.6 %, in West and Central Africa, respectively) identified by Taylor et al. (2015) come from market surveys with poorly defined catchment areas, compared to offtake and consumption surveys. Strong variation between individual estimates highlights the problems with extrapolation of survey data to national or regional levels and the effects of sampling strategies (hunter versus market surveys), timing of survey (open season versus lean season), survey location, and extrapolation methods. Individual figures should therefore be treated with caution, but the overall message remains: bushmeat is harvested at an enormous scale exposing those involved in the bushmeat commodity chain to zoonotic diseases.

Drivers of Increased Bushmeat Hunting and Disease Risks

The current scale of bushmeat hunting is primarily the result of socio-demographic changes (Wilkie and Carpenter 1999). Africa's human population has risen from 0.2 billion in 1950 to 0.9 billion in 2013 and is expected to rise to 2.2 billion by 2050 (United Nations 2013). Where alternative sources of animal protein and income are scarce, human population growth has been linked to increasing hunting intensity (Brashares et al. 2001).

Bushmeat has been and remains a staple source of animal protein among the rural poor, yet recent attention has focused on urban consumers of bushmeat as a driver of increased hunting. Urban consumers generally have a range of meat sources from which to choose, but value bushmeat for its taste, cultural connotations, and as a luxury food item (Fa et al. 2009). While urban consumers generally consume less bushmeat than rural consumers (Brashares et al. 2011), urban populations in Africa have increased dramatically from about 15 % of the total population in 1950 to 40 % in 2014 (United Nations 2014) and have created a strong demand for bushmeat and hence market for rural hunters.

The increasing demand for bushmeat has been accompanied by changes in hunting technology and improvements in hunting efficiency. Traditional hunting tools, such as nets and bow and arrow, have been replaced with more modern tools of guns and snares. Modern guns have an up to 25-times higher rate of return compared to traditional weapons (Wilkie and Curran 1991), substantially increasing the ease and cost-effectiveness of hunting (Alvard 1995). This enables hunters to catch more animals and sell a larger part of their catch (Bowen-Jones and Pendry 1999; Bowen-Jones et al. 2003; Nasi et al. 2008).

Hunting efficiency has also improved as remote forests have become more accessible through the construction of logging roads and improved transportation (Wilkie et al. 1992; Auzel and Wilkie 2000). For example, after the construction of 140 km of logging roads in northern Congo, the average time for a hunting trip was reduced from 12 to 2 hours (Wilkie et al. 2001). Development of rural businesses, such as timber companies, attracts workers and their families to remote locations, increasing bushmeat demand, especially when no hunting regulations are in place and alternative protein sources are not provided (Auzel and Wilkie 2000; Bennett and Gumal 2001; Poulsen et al. 2009). The effect of logging company presence on hunting pressure was documented in Gabon where ape populations decreased 50 % between 1983 and 2000 as a result of hunting (Walsh et al. 2003). In addition, agricultural expansion and mining have exerted a strong force in changing the African landscape and influencing human migration patterns (Norris et al. 2010). Due to increased access, people are brought into closer contact with wildlife, which facilitates accessibility to bushmeat hunting and makes transportation of bushmeat from rural to urban areas easier and more cost-effective (Wolfe et al. 2005a).

Along with increased ease of transportation comes the opportunity for bushmeat to be traded on the international market. The international trade in bushmeat has recently gained attention as both a driver of bushmeat hunting and the cross-border spread of zoonotic diseases. Illegal wildlife trade is the second-largest black market worldwide, involving millions of animals and estimated to be worth US$50–150 billion per year (United Nations Environment Programme 2014). Case studies at airports screening passenger luggage for bushmeat estimated that approximately 5 tons of bushmeat per week arrive at Paris Roissy-Charles de Gaulle airport (Chaber et al. 2010) and 8.6 tons per year at Zurich and Geneva airports (Falk et al. 2013). As bushmeat hunting, globalization, and human interconnectedness increase, the potential for zoonoses leading to EIDs also increases. This risk was highlighted when retroviruses (e.g., simian foamy virus) and herpesviruses (cytomegalovirus and lymphocryptovirus) were found in confiscated primates at US airports (Smith et al. 2012).

Bushmeat as a Source of Zoonotic Diseases in Sub-Saharan Africa

Indisputable evidence of the transmission of pathogens from wildlife to humans exists only for relatively few cases because the standard of proof is very high. Nevertheless, the evidence for spillovers is very strong and many pathogens can be classified as very likely to spillover (Jones et al. 2008; Kilonzo et al. 2014). Furthermore, countless pathogen species of zoonotic potential will likely be discovered as surveillance increases (Taylor et al. 2001; Jones et al. 2008). Our close phylogenetic relationship with nonhuman primates increases the likelihood that pathogen spillover from these animals to humans will cause infection (Childs et al. 2007).

Moreover, it is not surprising that many studies have focused on spillover events from nonhuman primates to humans given the high prevalence of these largely diurnal mammals in the bushmeat trade (Taylor et al. 2015). For instance, nonhuman primates of the family Hominidae include the Gorillinae and Paninae, which show a genetic difference of only 2 % or less with humans (Gonzalez et al. 2013), and members of these subfamilies share many morphological, physiological, and ecological features that may have a direct role in the transmission of infectious diseases (Davies and Pedersen 2008). Cleaveland et al. (2007), in their assessment of the risk of disease emergence by taxa, found that the relative risk of disease emergence was highest for bats, followed closely by primates, then ungulates and rodents. There have been surprisingly few studies of the connection between hunting of birds or other vertebrates and EIDs, especially in Africa, but surveillance for zoonotic pathogens in African birds is strongly needed (e.g., for avian influenza tracking see Simulundu et al. 2011, 2014).

The characteristics of different species may render them more or less susceptible to hunting. Behavioral traits such as communal nesting, large-group living, loud acoustic performances, and a diurnal lifestyle—which are found in many primate species—may facilitate the detection and harvesting of several individuals at one time (Bodmer 1995). Taste preferences for certain species influence hunters' decisions as do attempts to maximize returns by preferring large-bodied animals that provide more food or fetch a higher price when sold than small-bodied species (Bodmer 1995). Bats, especially the larger fruit bats popular in the bushmeat trade, are susceptible to hunting because they are often found in large, sometimes vocal groups that are visible during the day or in high concentrations in caves (Mickleburgh et al. 2009). Increased human encroachment in recent decades (Kamins et al. 2011b) has driven some bat species to become peridomestic (O'Shea et al. 2011; Plowright et al. 2011), which renders them easy targets for hunting. Finally, sick animals may be less successful in evading hunters and hence more easily hunted, thereby increasing the risk of disease transmission to hunters.

In addition to the behavioral traits that may influence which species are hunted, physiological traits of these species may make them more likely to harbor and transmit diseases. For example, bats, which are present in the bushmeat trade and comprise the highest risk among all wildlife for harboring emerging diseases (Cleaveland et al. 2007), present unique traits that suit them to hosting pathogens. These traits include: (1) relatively long lifespans for their body size (Munshi-South and Wilkinson 2010), which may facilitate pathogen persistence for chronic infections; (2) flight, which allows movement and dispersal over long distances and which creates high body temperatures that may select for co-evolution with viruses that can live at febrile temperatures and are therefore highly virulent in humans (O'Shea et al. 2014); (3) physiological similarity across sympatric species that roost together in high densities enabling pathogens adapted to any of the sympatric species to spillover to others (Streicker et al. 2010); and (4) regulation of their immune systems in such a way as to make them more likely to host, but remain unaffected by viral pathogens, serving as the reservoir host for emerging and highly virulent viruses (Baker et al. 2013).

Despite the fact that pathogens are common and often occur in high numbers in basically all animals, only a relatively small proportion of these pathogens will spillover to humans (Cleaveland et al. 2007). That said, when spillover events do occur, they can be not only deadly but costly. For example, the United Nations Development Program (2015) has estimated that West Africa as a whole may lose US$3.6 billion per year between 2014 and 2017 due to the 2014–2015 Ebola outbreak. This loss stems from the cumulative effects of closed borders, decreased trade, decreased foreign direct investment, and decreased tourism, resulting in increased poverty levels and food insecurity.

To understand the dynamics of spillover events and risks in relation to the pathogen, a number of factors must be considered, including: (1) the evolutionary history of the pathogen, (2) how the pathogen is maintained among its wildlife host(s), (3) how the pathogen is transmitted across a species barrier, (4) whether a productive infection is produced in the new host, (5) whether that infection produces significant disease in that host, and (6) whether morbidity and/or mortality levels in the secondary host are sufficient to be considered significant (Childs et al. 2007). From this, it follows that emerging pathogens are not an arbitrary selection of all pathogens. Becoming established in a human host typically requires adaptations, often for increased virulence, as has been documented in HIV (Wain et al. 2007; Etienne et al. 2013). Generalist pathogens have the ability to infect more than one host species and have higher relative emergence risk than pathogens that are very host-specific (Cleaveland et al. 2007); this is especially true for pathogens that can infect species in more than one taxonomic order. One example of this generalist "broad" host range is found in the newly described African henipavirus, which can enter and infect cells of nonhuman primates, bats, and humans (Lawrence et al. 2014).

Of particular importance for understanding bushmeat-related spillover events is whether a wildlife species is a natural or incidental pathogen host. Natural or reservoir hosts are a natural part of the pathogen life cycle and may maintain the infectious pathogen for prolonged periods of time, often without showing symptoms. In contrast, an incidental or dead-end host may be infected by the pathogen and may even transmit it, but it is not a part of the normal maintenance cycle of the pathogen and is more likely to be affected by it than natural hosts. For example, contact with sick common chimpanzees and western gorillas has been tightly linked to Ebola virus spillover in several outbreaks (Leroy et al. 2004b). Like their human cousins, these great apes are largely considered incidental or dead-end hosts for this virus and do not maintain it long-term in nature. In the case of this deadly filovirus, understanding what species are true reservoirs (likely fruit bats in the family Pteropodidae; Pourrut et al. 2007, 2009; Hayman et al. 2010, 2012) and the spillover events between these reservoirs and other mammals (including apes, carnivores, and ungulates; Leroy et al. 2004a) will prove critical to mitigating the components of disease transmission that are due to bushmeat-related activities. Unfortunately, it is often difficult to definitively determine the natural host(s) of a particular pathogen as it requires, in descending order of importance, isolation of the agent from individuals of the target species, detection of pathogen-specific nucleic acid sequences from individuals, and serological evidence that an individual has been exposed previously. Indeed, the study of reservoir systems and how infectious agents move between and within them can be complex, requiring rigorous and sophisticated analyses of multiple interrelated variables (Gray and Salemi 2012; Viana et al. 2014).

Descriptions of the types of pathogens potentially encountered through bushmeat-related activities can be found below, with several important and well-studied examples described in more detail. In their review of global trends in EIDs, in which they separately listed each antimicrobial pathogen strain that has recently emerged, Jones et al. (2008) report that the vast majority of pathogens involved in EIDs are bacterial or rickettsial, followed by viral or prion, then protozoa, fungi, and helminths. Other studies have ranked viruses as more prevalent (Taylor et al. 2001; Woolhouse et al. 2005; Cleaveland et al. 2007). In Jones et al.'s (2008) analysis of 335 EID events between 1940 and 2004, only four EIDs list bushmeat as the driver; other significant drivers were socioeconomic factors such as human population density. These four bushmeat-related EID events were all significant events; all due to viruses (Ebolavirus, human immunodeficiency virus-1, monkeypox virus, and SARS), suggesting that viruses are the most important pathogens in regard to spillover due to bushmeat-related activities (see also Kilonzo et al. 2014). We review the literature from sub-Saharan Africa in relation to bushmeat species by pathogen type (viruses, bacteria, helminths, protozoa, fungi, and prions), noting the significant potential for pathogens not yet associated with bushmeat-related activities to also be involved. Very few studies have considered all of the potential zoonotics in a region or in a taxonomic group. Magwedere et al.'s (2011) comprehensive study of zoonotics in Namibia is an exception.

Overview of Pathogens Related to Bushmeat Activities

Table 24.1 summarizes these pathogens by bushmeat host taxonomic group, conservatively listing only those species/pathogen combinations that have been tied strongly to spillovers from wildlife to humans via bushmeat-related activities and recognizing that this link is often putative and difficult to establish. Thus, Table 24.1 does not include some of the potential but not demonstrated spillover risks of poorly studied groups such as helminths and protozoans. Furthermore, due to their close genetic relationship with humans, common chimpanzees and western gorillas may share many pathogens of all varieties with humans, but the direction of spillover is not always clear (e.g. tourist interactions may spread disease from humans to apes) and much of these data are not discussed herein. Also not included in the table are studies where pathogens are not determined to species and, consequently, the bushmeat host–human link is unclear, or where exposure would be via an insect vector, which could be encountered when handling bushmeat. While we have attempted a very thorough treatment of pathogens that meet our criteria for inclusion in the table, it is possible that some relevant studies have been missed.

Table 24.1 Bushmeat species and zoonotic pathogens for which strong evidence for spillovers via bushmeat-related activities exists (see criteria for inclusion in text)

Bushmeat species	Pathogen	Location	References
Great Apes (Chimpanzee, Bonobo, Gorilla)[a]			
Pan troglodytes (Common chimpanzee)	Zaire Ebolavirus (V)	Cameroon, Gabon, Republic of Congo	Leroy et al. (2004a, b)
	Tai Forest Ebolavirus (V)	Côte d'Ivoire	Le Guenno et al. (1995), Wyers et al. (1999)
	HIV-1/SIVcpz (V)	Cameroon, Democratic Republic of the Congo, Tanzania	Santiago et al. (2002), Worobey et al. (2004), Van Heuverswyn et al. (2007)
	HLTV/SLTV-1 (V)	Central & Eastern Africa	Gao et al. (1999), reviewed in Sharp and Hahn (2010), Peeters et al. (2013)
	Simian Foamy Virus (V)	Cameroon, Côte d'Ivoire, Gabon, Republic of Congo, Tanzania	Calattini et al. (2006), Liu et al. (2008)
	Strongyloides fulleborni (H)	Gabon	Mouinga-Ondémé et al. (2012)
	Entamoeba histolytica (P)	Tanzania	Gillespie et al. (2010)
	Balantidium coli (P)	Central African Republic, Tanzania	Lilly et al. (2002), Gillespie et al. (2010)
	Giardia intestinalis (P)	Guinea Bissau	Sak et al. (2013)
	Bacillus anthracis (B)	Côte d'Ivoire	Leendertz et al. (2004)
Pan paniscus (Bonobo)	HTLV/STLV-2, HTLV/STLV-3 (V)	Democratic Republic of the Congo	Ahuka-Mundeke et al. (2011), Van Brussel et al. (1998)
Gorilla gorilla (Western gorilla)	Zaire Ebolavirus (V)	Cameroon, Gabon	Leroy et al. (2004a, b)
	HIV-1/SIVgor (V)	Cameroon	Takehisa et al. (2009)
	HTLV/STLV-1 (V)	Cameroon	Courgnaud et al. (2004), Nerrienet et al. (2004)
	Simian Foamy Virus (V)	Cameroon, Gabon	Wolfe et al. (2004b), Mouinga-Ondémé et al. (2012)
	Rabies (V)	Central African Republic, Kenya	Karugah (1997)
	Strongyloides fulleborni (H)	Central African Republic	Lilly et al. (2002)
	Entamoeba histolytica (P)	Central African Republic	Lilly et al. (2002)
	Balantidium coli (P)	Central African Republic	Lilly et al. (2002)
	Giardia intestinalis (P)	Central African Republic, Rwanda	Sak et al. (2013), Hogan et al. (2014)

Other nonhuman primates			
Colobus angolensis (Angola colobus)	HTLV/STLV-3 (V)	Democratic Republic of the Congo	Ahuka-Mundeke et al. (2012)
	Strongyloides fulleborni (H)	Uganda	Gillespie et al. (2005)
	Entamoeba histolytica (P)	Uganda	Gillespie et al. (2005)
Colobus guereza (Mantled guereza)	*Strongyloides fulleborni* (H)	Cameroon, Uganda	Gillespie et al. (2005), Pourrut et al. (2011)
	Entamoeba histolytica (P)	Uganda	Gillespie et al. (2005)
Piliocolobus badius (Western red colobus)	HTLV/STLV-1 (V)	Côte d'Ivoire	Leendertz et al. (2010)
Piliocolobus tephrosceles (Ugandan red colobus)	HTLV/STLV-1 (V)	Uganda	Goldberg et al. (2009)
	Strongyloides fulleborni (H)	Uganda	Gillespie et al. (2005)
	Entamoeba histolytica (P)	Uganda	Gillespie et al. (2005)
Piliocolobus tholloni (Thollon's red colobus)	HTLV/STLV-1, HTLV/STLV-3 (V)	Democratic Republic of the Congo	Ahuka-Mundeke et al. (2012).
Lophocebus albigena (Gray-cheeked mangabey)	HTLV/STLV-1, HTLV/STLV-3 (V)	Cameroon	Liégeois et al. (2012), Locatelli and Peeters (2012)
	Strongyloides fulleborni (H)	Cameroon	Pourrut et al. (2011)
Lophocebus aterrimus (Black crested mangabey)	HTLV/STLV-3 (V)	Democratic Republic of the Congo	Ahuka-Mundeke et al. (2012)
Papio anubis (Olive baboon)	Zaire Ebolavirus (V)	Cameroon	Leroy et al. (2004b)
	HTLV/STLV-1 (V)	Ethiopia, Kenya	Mahieux et al. (1998), Meertens et al. (2001), Takemura et al. (2002), Locatelli and Peeters (2012)
Papio cynocephalus (Yellow baboon)	HTLV/STLV-3 (V)	Tanzania	Voevodin et al. (1997)
Papio hamadryas (Hamadryas baboon)	HTLV/STLV-3 (V)	Eritrea, Ethiopia, Senegal	Goubau et al. (1994), Takemura et al. (2002), Meertens and Gessain (2003)
Papio ursinus (Chacma baboon)	HTLV/STLV-1 (V)	South Africa	Mahieux et al. (1998)
	Bacillus anthracis (B)	Namibia	Magwedere et al. (2012)

(continued)

Table 24.1 (continued)

Bushmeat species	Pathogen	Location	References
Papio sp. (Baboon sp.)	Rabies (V)	Kenya, Namibia, Zambia	Munang'andu (1995), Karugah (1997), Magwedere et al. (2012)
Theropithecus gelada (Gelada)	HTLV/STLV-3 (V)	Ethiopia	Van Dooren et al. (2004)
Cercocebus agilis (Agile mangabey)	HTLV/STLV-1, HTLV/STLV-3 (V)	Cameroon	Nerrienet et al. (2001), Courgnaud et al. (2004), Liégeois et al. (2008), Sintasath et al. (2009a), Locatelli and Peeters (2012)
	Strongyloides fulleborni (H)	Cameroon	Pourrut et al. (2011)
	Entamoeba histolytica (P)	Central African Republic	Lilly et al. (2002)
	Balantidium coli (P)	Central African Republic	Lilly et al. (2002)
Cercocebus atys (Sooty mangabey)	HIV-2/SIVsm (V)	West Africa	Hirsch et al. (1989), reviewed in Sharp and Hahn (2010), Peeters et al. (2013)
	HTLV/STLV-1 (V)	Sierra Leone	Traina-Dorge et al. (2005)
Cercocebus torquatus (Collared mangabey)	HTLV/STLV-1, HTLV/STLV-3 (V)	Cameroon	Meertens et al. (2001, 2002), Liégeois et al. (2008, 2012)
Mandrillus leucophaeus (Drill)	Zaire Ebolavirus (V)	Cameroon	Leroy et al. (2004b)
Mandrillus sphinx (Mandrill)	Zaire Ebolavirus (V)	Cameroon	Leroy et al. (2004b)
	HTLV/STLV-1 (V)	Cameroon	Nerrienet et al. (2001), Courgnaud et al. (2004), Liégeois et al. (2012)
	Simian foamy virus (V)	Cameroon, Gabon	Wolfe et al. (2004b), Mouinga-Ondémé et al. (2010)
Allenopithecus nigroviridis (Allen's swamp monkey)	HTLV/STLV-1 (V)	Democratic Republic of the Congo	Meertens et al. (2001)
Miopithecus ogouensis (Gabon talapoin)	HTLV/STLV-1 (V)	Cameroon	Courgnaud et al. (2004)
	Strongyloides fulleborni (H)	Cameroon	Pourrut et al. (2011)

Erythrocebus patas (Patas monkey)	HTLV/STLV-1 (V)	Cameroon, Central African Republic, Senegal	Ishikawa et al. (1987), Saksena et al. (1994)
Chlorocebus aethiops (Grivet)	Marburg virus (V)	Uganda	Smith (1982)
	HTLV/STLV-1 (V)	Ethiopia, Senegal	Meertens et al. (2001), Takemura et al. (2002)
Chlorocebus pygerythrus (Vervet monkey)	HTLV/STLV-1 (V)	Kenya	Meertens et al. (2001)
	Leptospira (B)	Botswana	Jobbins and Alexander (2015)
Chlorocebus sabaeus (Green monkey)	HTLV/STLV-1 (V)	Senegal	Meertens et al. (2001), Locatelli and Peeters (2012)
Chlorocebus tantalus (Tantalus monkey)	HTLV/STLV-1 (V)	Kenya	Meertens et al. (2001), Locatelli and Peeters (2012)
Cercopithecus albogularis (Sykes' monkey)	HTLV/STLV-1 (V)	Kenya	Mwenda et al. (1999)
Cercopithecus ascanius (Red-tailed monkey)	HTLV/STLV-1 (V)	Democratic Republic of the Congo	Ahuka-Mundeke et al. (2012)
	Strongyloides fulleborni (H)	Uganda	Gillespie et al.(2004)
Cercopithecus cephus (Moustached guenon)	HTLV/STLV-1, HTLV/STLV-3 (V)	Cameroon	Courgnaud et al. (2004), Liégeois et al. (2008, 2012), Locatelli and Peeters (2012)
	Strongyloidesfulleborni (H)	Cameroon	Pourrut et al. (2011)
Cercopithecus lhoesti (L'Hoest's monkey)	*Strongyloides fulleborni* (H)	Uganda	Gillespie et al. (2004)
Cercopithecus mitis (Blue monkey)	*Strongyloides fulleborni* (H)	Kenya, Uganda	Munene et al. (1998), Gillespie et al. (2004)
Cercopithecus mona (Mona monkey)	HTLV/STLV-1, HTLV/STLV-3 (V)	Cameroon	Meertens et al. (2001), Sintasath et al. (2009b), Locatelli and Peeters (2012)
	Strongyloides fulleborni (H)	Cameroon	Pourrut et al. (2011)
Cercopithecus neglectus (De Brazza's monkey)	Zaire Ebolavirus (V)	Cameroon	Leroy et al. (2004b)
	HTLV/STLV-1 (V)	Democratic Republic of the Congo	Ahuka-Mundeke et al. (2012)
	Simian Foamy Virus (V)	Cameroon	Wolfe et al. (2004b)
	Strongyloides fulleborni (H)	Cameroon	Pourrut et al. (2011)

(continued)

Table 24.1 (continued)

Bushmeat species	Pathogen	Location	References
Cercopithecus nictitans (Greater spot-nosed monkey)	HTLV/STLV-1, HTLV/STLV-3 (V)	Cameroon	Meertens et al. (2001), Courgnaud et al. (2004), Liégeois et al. (2008, 2012), Sintasath et al. (2009b)
	Strongyloides fulleborni (H)	Cameroon	Pourrut et al. (2011)
Cercopithecus pogonias (Crested mona monkey)	HTLV/STLV-1 (V)	Cameroon	Courgnaud et al. (2004), Liégeois et al. (2008)
	Strongyloides fulleborni (H)	Cameroon	Pourrut et al. (2011)
Cercopithecus wolfi (Wolf's mona monkey)	HTLV/STLV-1 (V)	Democratic Republic of the Congo	Locatelli and Peeters (2012)
"Vervet monkey"	Rabies (V)	Zambia	Munang'andu (1995)
	Strongyloides fulleborni (H)	Uganda	Gillespie et al. (2004)
Unspecified primate sp.	Rabies (V)	Ethiopia, Ghana, Kenya, Malawi, Mozambique; Namibia, Sudan, Uganda, Zimbabwe	See summary and discussion in Gautret et al. (2014)
Bats			
Eidolon helvum (African straw-colored fruit bat)	Zaire Ebolavirus (V)	Ghana	Hayman et al. (2010)
	Lagos bat virus (V)	Ghana, Kenya, Nigeria, Senegal	Boulger and Porterfield (1958), Institut Pasteur (1985), Hayman et al. (2008, 2010), Kuzmin et al. (2008)
	Henipaviruses (V)	Cameroon, Ghana, Republic of Congo, Zambia	Hayman et al. (2008), Drexler et al. (2009), Baker et al. (2012), Weiss et al. (2012), Muleya et al. (2014), Pernet et al. (2014)
Hypsignathus monstrosus (Hammer-headed fruit bat)	Marburgvirus (V)	Gabon, Republic of Congo	Pourrut et al. (2009)
	Zaire Ebolavirus (V)	Ghana, Gabon, Republic of Congo	Pourrut et al. (2007, 2009), Hayman et al. (2012)
Epomops franqueti (Franquet's epauletted fruit bat)	Marburgvirus (V)	Gabon, Republic of Congo	Pourrut et al. (2009)
	Zaire Ebolavirus (V)	Ghana, Gabon, Republic of Congo	Pourrut et al. (2007, 2009), Hayman et al. (2012)
Epomophorus gambianus (Gambian epauletted fruit bat)	Zaire Ebolavirus (V)	Ghana	Hayman et al. (2012)

Epomophorus wahlbergi (Wahlberg's epauletted fruit bat)	Lagos Bat Virus (V)	South Africa	Markotter et al. (2006)
Micropteropus pusillus (Peters's lesser epauletted fruit bat)	Marburgvirus (V)	Gabon, Republic of Congo	Pourrut et al. (2009)
	Zaire Ebolavirus (V)	Gabon, Republic of Congo	Pourrut et al. (2009)
	Lagos Bat Virus (V)	Central African Republic	Sureau et al. (1980)
Rousettus aegyptiacus (Egyptian rousette)	Marburgvirus (V)	Democratic Republic of the Congo, Gabon, Kenya, Republic of Congo, Uganda	Swanepoel et al. (2007), Towner et al. (2007, 2009), Pourrut et al. (2009), Kuzmin et al. (2010b), Amman et al. (2012)
	Zaire Ebolavirus (V)	Gabon, Republic of Congo	Pourrut et al. (2009)
	Lagos Bat Virus (V)	Kenya	Kuzmin et al. (2008)
Myonycteris torquata (Little collared fruit bat)	Zaire Ebolavirus (V)	Gabon, Republic of Congo	Pourrut et al. (2007, 2009)
Nanonycteris veldkampii (Veldkamp's dwarf epauletted fruit bat)	Zaire Ebolavirus (V)	Ghana	Hayman et al. (2012)
Rhinolophus eloquens (Eloquent horseshoe bat)	Marburgvirus (V)	Democratic Republic of the Congo	Swanepoel et al. (2007)
Miniopterus inflatus (Greater long-fingered bat)	Marburgvirus (V)	Democratic Republic of the Congo	Swanepoel et al. (2007)
Miniopterus schreibersii (Schreibers's long-fingered bat)	Duvengahe virus (V)	South Africa	Paweska et al. (2006)
Nycteris gambiensis (Gambian slit-faced bat)	Lagos Bat Virus	Senegal	Institut Pasteur (1985)
Mops condylurus (Angolan free-tailed bat)	Zaire Ebolavirus (V)	Gabon	Pourrut et al. (2009)
Unspecified bat sp.	Duvengahe virus (V)	Kenya	van Thiel et al. (2009)

(continued)

Table 24.1 (continued)

Bushmeat species	Pathogen	Location	References
Rodents			
Funisciurus anerythrus (Thomas's rope squirrel)	Monkeypox virus (V)	Democratic Republic of the Congo	Khodakevich et al. (1986)
Funisciurus sp. (African striped squirrel sp.)	Monkeypox virus (V)	Ghana	Reynolds et al. (2010)
Heliosciurus gambianus (Gambian sun squirrel)	Monkeypox virus (V)	Ghana	Reynolds et al. (2010)
Paraxerus cepapi (Smith's bush squirrel)	*Leptospira* (B)	Botswana	Jobbins and Alexander (2015)
Xerus sp. (African ground squirrel sp.)	Monkeypox virus (V)	Ghana	Reynolds et al. (2010)
Unspecified squirrel sp.	Rabies (V)	Namibia, Zimbabwe	Pfukenyi et al. (2009), Magwedere et al. (2012)
Lophuromys sikapusi (Rusty-bellied brush-furred rat)	Mokola virus (V)	Central African Republic	Saluzzo et al. (1984)
Mastomys natalensis (Natal mastomys)	Lassa Fever (V)	Guinea	Ter Meulen et al. (1996)
Rattus norvegicus (Brown rat)	*Leptospira* (B)	Botswana	Jobbins and Alexander (2015)
Unspecified "rat species"	Rabies (V)	Namibia	Magwedere et al. (2012)
Cricetomys sp. (Giant pouched rat sp.)	Monkeypox virus (V)	Ghana	Reynolds et al. (2010)
Atherurus africanus (African brush-tailed porcupine)	Salmonella (B)	Gabon	Bachand et al. (2012)
Aardvark			
Orycteropus afer (Aardvark)	*Leptospira* (B)	Botswana	Jobbins and Alexander (2015)

Ungulates			
Equus burchellii (Burchell's zebra)	*Bacillus anthracis* (B)	Namibia	Magwedere et al. (2012)
Phacochoerus aethiopicus (Desert warthog)	Rabies (V)	Namibia	Magwedere et al. (2012)
Phacochoerus africanus (Common warthog)	*Leptospira* (B)	Botswana	Jobbins and Alexander (2015)
Alcelaphus buselaphus (Hartebeest)	Rabies (V)	Namibia	Magwedere et al. (2012)
	Bacillus anthracis (B)	Namibia	Magwedere et al. (2012)
Connochaetes taurinus (Blue wildebeest)	Rabies (V)	Namibia	Magwedere et al. (2012)
	Bacillus anthracis (B)	Namibia, Tanzania	Lembo et al. (2011), Magwedere et al. (2012)
Antidorcas marsupialis (Springbok)	Rabies (V)	Namibia	Magwedere et al. (2012)
	Bacillus anthracis (B)	Namibia	Magwedere et al. (2012)
	Brucella (B)	Namibia	Magwedere et al. (2011)
Syncerus caffer (African buffalo)	*Bacillus anthracis* (B)	Tanzania	Lembo et al. (2011)
	Brucella (B)	Botswana, Mozambique	Alexander et al. (2012), Tanner et al. (2014)
Taurotragus oryx (Common eland)	Rabies (V)	Namibia, Zimbabwe	Pfukenyi et al. (2009), Magwedere et al. (2012)
	Bacillus anthracis (B)	Namibia	Magwedere et al. (2012)
Tragelaphus strepsiceros (Greater kudu)	Rabies (V)	Namibia, Zimbabwe	Pfukenyi et al. (2009), Magwedere et al. (2012)
Cephalophus sp. (Duiker sp.)	Zaire Ebolavirus (V)	Gabon	Leroy et al. (2004a)
Sylvicapra grimmia (Bush duiker)	Rabies (V)	Zimbabwe	Pfukenyi et al. (2009)
Hippotragus niger (Sable antelope)	Rabies (V)	Zimbabwe	Pfukenyi et al. (2009)
Oryx gazelle (Gemsbok)	Rabies (V)	Namibia	Magwedere et al. (2012)
Unspecified "oryx," "antelope," "duiker"	Rabies (V)	Namibia	Magwedere et al. (2012)

V virus, *H* helminth, *P* protozoan, *B* bacteria

[a]See text for further discussions of pathogens in great apes, including uncertainty as to whether apes or humans are the source of spillovers

Viral Pathogens

Viruses are obligatory intracellular parasites characterized primarily by the nature of their nucleic acids (DNA or RNA; single or double stranded, etc.). They are the most abundant form of life on earth; many viruses are recognized as important disease-causing agents, and they are subject to frequent mutation and thus evolution. The advent of modern molecular techniques has advanced our understanding of viral diversity and pathogenesis in both animal and human hosts. For example, in relation to bushmeat, it is now clear that many virus variants are present in hunted nonhuman primate species, which have received most of the research attention, and that these variants have crossed between nonhuman primates and humans on multiple occasions (Peeters and Delaporte 2012; Table 24.1). Bats and rodents are also major zoonotic virus carriers (Meerburg et al. 2009; Baker et al. 2013); other taxonomic groups are less studied, at least in sub-Saharan Africa. Several sub-Saharan African viruses of importance are vector-borne, including Rift Valley Fever and Crimean-Congo hemorrhagic fever. While one presumes that this would make them unlikely to spread via bushmeat-related activities, the possibility remains that animal handling could present a risk (Magwedere et al. 2012). However, no significant links between vector-borne viruses and bushmeat hunting have been made, and we will not include a discussion of these viruses here.

HIV/SIV: The most notable virus to emerge from the bushmeat interface is human immunodeficiency virus (HIV). While the origin of HIV was long obscured, Human HIV-1 and HIV-2 are believed to have evolved from strains of simian immunodeficiency virus (SIV) (Hahn et al. 2000; Lemey et al. 2003; Van Heuverswyn and Peeters 2007; Sharp and Hahn 2010; Peeters and Delaporte 2012; Peeters et al. 2013; Kazanji et al. 2015). Evidence suggests that SIV crossed over to humans by blood contact when hunters had an exposed open wound or injured themselves during the butchering of nonhuman primates (Hahn et al. 2000; Wolfe et al. 2004a, b; Karesh and Noble 2009). The closest relatives of HIV-1 found among nonhuman primates are SIVcpz and SIVgor, from common chimpanzees and western gorillas in west central Africa (Gao et al. 1999; Sharp et al. 2005; Keele et al. 2006; Van Heuverswyn et al. 2006, 2007; Takehisa et al. 2009) and at least four separate spillovers have occurred (Peeters et al. 2013). HIV-2 is derived from SIVsmm from sooty mangabeys (*Cercocebus atys*) in West Africa (Apetrei et al. 2005; Hirsch et al. 1989; Gao et al. 1992; Ayouba et al. 2013), where high viral genetic diversity exists and where transmission is believed to have occurred at least eight times.

The potential for future and continued spillovers from SIVs is high, and multiple species-specific variants exist. For example, Peeters et al. (2002) and Peeters (2004) estimated that more than 20 % of nonhuman primates hunted for food are infected with a variant of SIV; Locatelli and Peeters (2012) and Peeters et al. (2013) noted that at least 45 species-specific variants of SIV from at least 45 primate species are currently recognized. Aghokeng et al. (2010) sampled 1856 nonhuman primate carcasses from 11 species found in bushmeat markets in Cameroon. They documented low overall prevalence of SIV (only 2.93 % of carcasses), with the lowest prevalence found among the most common species in the market. However, they did find SIV variants in about 70 % of the tested primate species. In total, serological evidence of SIV infection has been documented for at least 40 different primate species (Aghokeng et al. 2010; Liégeois et al. 2011, 2012). Cross-species transmission of strains and co-infection with more than one strain have been documented, sometimes followed by genetic recombination (Hahn et al. 2000; Bibollet-Ruche et al. 2004; Aghokeng et al. 2007; Gogarten et al. 2014), a recipe for future spillovers into humans (Locatelli and Peeters 2012).

Human T-Cell Lymphotropic Virus (HTLV): Similar to HIV, human T-lymphotropic viruses (HTLV) are related to simian viral lineages in which significant diversity has been found (Ahuka-Mundeke et al. 2012; Peeters and Delaporte 2012). All three sub-Saharan great apes and 30 additional nonhuman primates have been documented to have STLV/HTLV variants and a variety of HTLV viruses have been documented in wildlife and in central African hunters (Calattini et al. 2009, 2011; Courgnaud et al. 2004; Sintasath et al. 2009a, b; Wolfe et al. 2005b; Zheng et al. 2010; Locatelli and Peeters 2012). Similar to HIV/SIV, dual infections with more than one variant have been documented in nonhuman primates (Agile mangabeys, *Cercocebus agilis*; Courgnaud et al. 2004) and in humans (Calattini et al. 2011; Wolfe et al. 2005b).

Simian Foamy Virus: Simian foamy retroviruses (SFV) are endemic in most African primates (Hussain et al. 2003; Switzer et al. 2005; Peeters and Delaporte 2012) and are known to transmit to humans (Sandstrom et al. 2000; Switzer et al. 2004; Calattini et al. 2007; Mouinga-Ondémé et al. 2010, 2012). Like the other retroviruses discussed above (HIV and HTLV), SFV is genetically diverse and relatively host species-specific. In Cameroon, Wolfe et al. (2004b) documented three geographically independent SFV infections, which could be traced to De Brazza's monkey (*Cercopithecus neglectus*), mandrill (*Mandrillus sphinx*), and western gorilla. Likewise, in Gabon, Mouinga-Ondémé et al. (2010, 2012) documented human spillover events involving multiple strains of SFV, with infected humans having been bitten by common chimpanzees, western gorillas, or mandrills infected with their respective variant of SFV.

Ebola and Marburg Viruses: There are seven species of filoviruses currently identified, five of which occur in sub-Saharan Africa—Genus *Ebolavirus*: Tai forest ebolavirus (TAFV), Sudan ebolavirus (SUDV), Zaire ebolavirus (EBOV), Bundibugyo virus (BDBV); Genus *Marburgvirus:* Marburg virus (MARV). These pathogens are periodically emerging viruses, typically from single spillover events, which cause hemorrhagic fevers (reviewed by Olival and Hayman 2014; Rougeron et al. 2015 (but note that Rougeron's listing for a single case of SUDV in Sudan in 2011 is erroneous)). The 2014–2015 West Africa outbreak of EBOV is still ongoing at the time of this writing (Labouba and Leroy 2015). While the zoonotic source of this outbreak is unknown, three initial outbreaks of the Ebola virus in the Democratic Republic of the Congo from 1976 to 1979 involved victims who were reported to have handled western gorilla or common chimpanzee carcasses or to have had physical contact with people who touched the animals (Leroy et al. 2004a, b). Similarly,

Marburgvirus was first identified in laboratory workers who had dissected imported grivet (*Chlorocebus aethiops*) (Martini et al. 1968; Siegert et al. 1968). Both western gorillas and common chimpanzees have suffered significant mortality from filovirus outbreaks (Walsh et al. 2003; Leroy et al. 2004a, b; Bermejo et al. 2006; Rizkalla et al. 2007) and antibodies to EBOV were documented in several other primate species by Leroy et al. (2004b). The single case of TAFV occurred in an ethnologist likely infected while performing a necropsy of a dead common chimpanzee following a rash of common chimpanzee deaths in the Tai National Park in Côte d'Ivoire (Le Guenno et al. 1995; Wyers et al. 1999). Beyond primates, other incidental hosts in the wild are possible, as was demonstrated for duikers (*Cephalophus* spp.) (Leroy et al. 2004a; Rouquet et al. 2005). As reviewed by Weingartl et al. (2013), both dogs (naturally) and pigs (at least experimentally) can also be infected. During the 2001–2002 EBOV outbreak in Gabon, Allela et al. (2005) found over 30 % seroprevalence in dogs living in villages with EBOV human and animal cases. Those dogs appeared to be asymptomatic and were presumed to be exposed by scavenging wild animals.

Although incidental hosts likely play important roles in the ecology of these viruses, especially when moribund or dead animals are consumed, strong evidence suggests that bats are the natural reservoir hosts for at least Marburgvirus and EBOV. For Marburgvirus, the cave dwelling and densely packed Egyptian rousette fruit bat (*Rousettus aegyptiacus*) is now well-documented as a reservoir host (Towner et al. 2009; Amman et al. 2012), but antibodies against the virus and/or the presence of viral RNA have been found in several other species (see Table 24.1). The strong association of Marburgvirus with the Egyptian rousette makes sense in light of the outbreaks of this virus in people visiting tourist caves or working in mines (Adjemian et al. 2011; Timen et al. 2009; Towner et al. 2009; Amman et al. 2012). The picture for EBOV is less clear, but evidence of infection has been found in at least eight sub-Saharan bat species (Pourrut et al. 2007, 2009; Hayman et al. 2010, 2012; Table 24.1). Of the ten bat species listed in Table 24.1 for Marburgvirus and EBOV, seven are fruit bats, which are relatively larger and more visible, and thus targets of bushmeat hunters. That said, bushmeat hunting of these bats is not ubiquitous throughout their range and cannot solely explain filovirus spillovers. Mari Saéz et al. (2015) unconvincingly suggested the non-fruit bat, *Mops condylurus,* might have been the source of the 2014–2015 West African Ebola outbreak. Pourrut et al. (2009) found evidence of antibodies against ZEBOV in this species, but there is no real evidence that this free-tailed bat played a role in the 2014–2015 outbreak. To date, no bat host has been identified for BDBV, SUDV, or TAFV and broader surveillance for indications of these viruses in bats and other hosts should be conducted.

Henipaviruses and Other Paramyxoviruses: Hendra virus and Nipah virus (HNVs) are paramyxoviruses in the genus *Henipavirus* that emerged in Australia and southeast Asia, respectively, with fruit bats in the genus *Pteropus* (family Pteropodidae) as reservoir hosts (reviewed by Croser and Marsh 2013). However, recent studies have identified *Henipavirus* and Henipa-like viruses in sub-Saharan African fruit bats, which are a phylogenetically distinct clade of pteropodid bats that do not overlap distributionally with any *Pteropus* species. Documentation of *Henipavirus* and related RNA (Drexler et al. 2009; Muleya et al. 2014; Baker et al. 2012) and anti-*Henipavirus* antibodies (Hayman et al. 2008; Pernet et al. 2014) in the African straw-colored fruit bat (*Eidolon helvum*) clearly show that this deadly and diverse viral group is present in sub-Saharan Africa. This bat species is a frequent target of hunters and a significant protein source where it is found (Kamins et al. 2011b). Weiss et al. (2012) documented the presence of this group of viruses in these bats found live in bushmeat markets. Strong evidence of spillover to humans was documented by Pernet et al. (2014) who found antibodies against HNVs in human samples from Cameroon. These seropositive human samples were found almost exclusively in individuals who reported butchering these bats. This bat is also a long-distance migrator with significant panmixia across the continent, which could facilitate viral transmission between bats (Peel et al. 2013).

The paramyxovirus story in sub-Saharan Africa is still unfolding. Both Drexler et al. (2012) and Baker et al. (2012) describe great diversity in paramyxoviruses from sub-Saharan bats. In their comprehensive study of the evolutionary history of this virus family, Drexler et al. (2012) found that the *Henipavirus* lineage originated in Africa and identified bats as the likely origin of this large family of viruses. A precautionary tale from sub-Saharan Africa comes from the recent discovery and naming of the Sosuga virus from a wildlife researcher who became very ill after handling and dissecting hundreds of bats and rodents in Uganda and South Sudan (Albariño et al. 2014). This virus is most closely related to Tuhoko virus 3, a rubula-like virus recently isolated from the Leschenault's Rousette fruit bat (*Rousettus leschenaultii*) in southern China. Amman et al. (2015) subsequently found Sosuga virus in *R. aegyptiacus* captured from multiple locations in Uganda; the researcher infected by this virus handled this species extensively in her studies.

Rabies and Other Lyssaviruses: Rabies is the oldest known zoonotic EID, recorded as early as the twenty-third century BC (Steele and Fernandez 1991). An estimated 25,000 people die in Africa each year from rabies (Dodet et al. 2015), some portion of which may be from exposure that occurs in bushmeat-related activities, although most human cases can be attributed to domestic dogs. Rabies virus (RABV) is in the *Lyssavirus* genus. It is joined in Africa by at least five additional species: Lagos bat virus (LBV), Mokola virus (MOKV), Duvenhage virus (DUVV), Shimoni bat virus (SHIBV), and the newly proposed Ikoma lyssavirus (IKOV). These viruses have bat(s) as their reservoir host (Banyard et al. 2014) with two exceptions. The Mokola virus is found in shrews (*Crocidura* spp.), rusty-bellied brush-furred rat (*Lophuromys sikapusi*; Saluzzo et al. 1984), and companion animals (Delmas et al. 2008; Kgaladi et al. 2013). The Ikoma virus has thus far only been documented in African civets (*Civettictis civetta*; Table 24.1, Marston et al. 2012). A variety of wildlife species can be secondary hosts of rabies (e.g., in Botswana, see Moagabo et al. 2009) and rabies has been documented to occur in a number of nonhuman primate species, including those encountered in the bushmeat trade (Gautret et al. 2014). Lyssaviruses are found worldwide, but the greatest genetic diversity is in Africa and Lagos bat virus may be more than one species (Delmas et al. 2008; Markotter et al. 2008; Kuzmin et al. 2010a). While most human cases are due to rabies virus, Duvenhage virus has been documented in human fatalities associated with bat scratches that likely transmitted the virus (van Thiel et al. 2009; Paweska et al. 2006). Mokolo virus has been detected in two human cases without mortality (Kgaladi et al. 2013).

The lyssavirus story in Africa will continue to emerge due to increased surveillance and improved molecular techniques. The 2012 discovery of Ikoma virus in an African civet in Serengeti National Park in Tanzania, where domestic dogs are largely absent and detection in bat hosts is nonexistent (Marston et al. 2012; Horton et al. 2014), highlights the likelihood that many more lyssaviruses exist in a variety of host species. The true diversity of lyssaviruses in Africa, and the potential for human spillover via bushmeat-related activities, remains to be discovered.

Lassa and Other Arenaviruses: Arenaviruses include a number of zoonotic species, typically transmitted from rodents to humans. Lassa virus is the best known of the viral hemorrhagic arenaviruses in Africa and is well-documented in West Africa, especially Guinea, Sierra Leone, Nigeria, and Liberia. As with some of the bacterial pathogens described below, the primary risk comes from peridomestic exposure to the rodent host, the natal mastomys (*Mastomys natalensis*), via exposure to urine or fecal materials. However, Ter Meulen et al. (1996) found a strong association between hunting of peridomestic rodents and antibodies to and symptoms of Lassa virus, tying bushmeat-related activities to the spillover of this virus to humans.

Human Monkeypox Virus: Contrary to its moniker, the reservoir hosts of human monkeypox virus (MPX) are neither monkeys nor humans, but rather rodents. The first case of human monkeypox was identified in 1970 in the Democratic Republic of the Congo, with subsequent outbreaks in Liberia, Sierra Leone, Côte d'Ivoire, Nigeria, and Democratic Republic of the Congo (reviewed by Reynolds et al. 2010; Rimoin et al. 2010). Recent MPX increases in the Democratic Republic of the Congo and elsewhere have been attributed to cessation of the human smallpox vaccine, which conferred some immunity to other pox viruses (Rimoin et al. 2010). Human and nonhuman primate infections are suspected to result from wildlife exposure such as would occur in bushmeat-related activities; infected species include squirrels (e.g., Thomas's rope squirrel, *Funisciurus anerythrus*; Khodakevich et al. 1986; African ground squirrels; *Xerus* sp.; Reynolds et al. 2010), dormice (*Graphiurus* sp.; Reynolds et al. 2010), and giant pouched rats (*Cricetomys* sp.; Reynolds et al. 2010). The outbreak that occurred in the USA in 2007 after exposure to rodents in the illegal pet trade also linked human monkeypox to rope squirrels, dormice, and pouched rats (Hutson et al. 2007). While dormice are small and not likely to be the target of hunting, the diurnal and highly visible squirrels and the giant pouched rats are routinely hunted (Taylor et al. 2015), making the spillover to humans highly plausible.

Bacteria

Jones et al. (2008) list 54.3 % of EID events as being caused by bacteria and there is good evidence to suggest that bacterial pathogens have the potential to be just as important as viruses when it comes to those that may spillover due to bushmeat-related activities, but in this capacity they have received far less attention (Cantas and Suer 2014). Transmission pathways for bacterial pathogens can occur through direct exposure to body fluids or feces, but they can also possibly be transferred indirectly through exposure to disease vectors such as fleas and ticks when handling animals. In a rare survey of bacterial pathogens that might spillover via bushmeat-related activities, Bachand et al. (2012) sampled muscle from 128 bushmeat carcasses from multiple species at markets in Gabon for the presence of *Campylobacter*, *Salmonella*, and *Shigella*. While they only recorded the presence of *Salmonella*, the potential for contamination and thus spillover of enteric pathogens from carcass handling remains high, especially in the days after purchase when pathogens continue to replicate. Bacteria in the genus *Leptospira* are endemic sub-Saharan African pathogens that have a high risk of spillover during bushmeat-related activities as they are shed in urine. Jobbins and Alexander (2015) documented their widespread presence in wild mammals, birds, and reptiles, highlighting the role that wildlife may play in leptospirosis. The bushmeat interface may also play a role in human cases of anthrax, caused by *Bacillis anthracis*, which is largely a disease of grazing herbivorous mammals, but to which common chimpanzees are also susceptible (Leendertz et al. 2004). If bushmeat includes not only the hunting of apparently healthy animals but also sick animals or salvage of contaminated carcasses, the risk of human outbreaks increases (Hang'ombe et al. 2012).

A number of bacterial pathogens are vector-borne, which at face value would make them unlikely to spread via bushmeat-related activities. However, especially for bacteria with flea or tick as vectors, as opposed to mosquitoes for example, one can envision that animal handling could present a risk. The most frightening among the vector-borne bacterial pathogens is plague, caused by the bacteria *Yersinia pestis* and transmitted through the infected fleas of rodents. Africa remains an endemic region of importance for this pathogen (World Health Organization 2005; Davis et al. 2006). Fleas and ticks are also responsible for transmitting rickettsial pathogens, such as *Rickettsia africae*, which causes African tick-bite fever (ATBF). Mediannikov et al. (2012) collected ticks from duikers and a pangolin that were living in close proximity to humans in Guinea and found *R. africae* in 10 % of ticks collected from the tree pangolin (*Manis tricuspis*), suggesting the potential for spillover with the close handling of these animals. Further research is clearly and urgently needed to fully assess the potential for bacterial disease spillovers via bushmeat-related activities.

Helminths

The helminths or "worm-like" animals include many parasites of zoonotic potential, although Taylor et al. (2001) found helminthes less likely to cause EIDs. Humans engaging in bushmeat-related activities are likely exposed to these pathogens via exposure to fecal material in which eggs are shed, from transcutaneous exposure to infectious larvae, or from consumption of uncooked meat (McCarthy and Moore 2000).

Several studies have examined the prevalence of helminths in animals from bushmeat markets and found high rates of multiple species. For example, Adejinmi and Emikpe (2011) collected fecal samples from greater cane rats (*Thryonomys swinderianus*) and bush duikers (*Sylvicapra grimmia*) in bushmeat markets in Nigeria and documented high prevalence rates (83.3 % and 53.8 %, respectively) of helminth ova in feces as well as larvae from fecal cultures. Likewise, Magwedere et al. (2012) and Mukaratirwa et al. (2013) reviewed the evidence for *Trichinella* infection in humans, livestock, and wildlife in sub-Saharan Africa and noted that bush-pigs (*Potamochoerus* spp.) and desert warthogs (*Phacochoerus aethiopicus*) are a source for human infection. As is the case with many other pathogens, humans and nonhuman primates share susceptibility to many parasitic helminth species (Pedersen et al. 2005; Pourrut et al. 2011). Pourrut et al. (2011) sampled gastrointestinal parasites from 78 wild monkeys of 9 species collected from bushmeat markets in Cameroon and documented high helminth loads, including species known to infect humans. Gillespie et al. (2010) had similar findings from common chimpanzee fecal samples. Overall, the available evidence suggests that spillover of many of these pathogens during bushmeat-related activities is likely.

Protozoan

Protozoans are a paraphyletic group of eukaryotic organisms that are neither animals, plants, nor fungi and include amoebas and giardia. The risk of protozoan spillover from bushmeat-related activities is similar to that for helminths and bacteria in that exposure to feces, bodily fluids, and even potentially to meat could transmit disease to a permissive human host (Pourrut et al. 2011). A number of protozoans are important pathogens with zoonotic potential (Taylor et al. 2001). Perhaps the best example are the amoebozoa, which cause diarrheal disease and which are documented in a variety of animals, including bushmeat species such as nonhuman primates (Gillespie et al. 2010; Pourrut et al. 2011). Gillespie et al. (2010) documented the amoeba *Entamoeba histolytica* and the ciliated protozoan *Balantidium coli* in common chimpanzees; both are human pathogens (although the direction of spillover is uncertain, as common chimpanzees and other primates may have obtained this parasite from humans). Indeed, Lilly et al. (2002) documented both protozoans in common chimpanzees, western gorillas, agile mangabeys, and humans living in the same region in Central African Republic. A number of other nonhuman primates have had documented *E. histolytica* infections as well (see Table 24.1). Other protozoan examples include *Toxoplasma gondii*, which causes the disease toxoplasmosis, but could not be detected during a recent, albeit small scale, survey of bushmeat (Prangé et al. 2009) and water/foodborne parasites such as *Giardia.* Recent studies have documented *Giardia* in a variety of species that exist in the bushmeat trade, including western gorilla and African buffalo (*Syncerus caffer*) (Hogan et al. 2014).

Fungi

Fungi are increasingly being recognized as important pathogens that may emerge, even in humans (Jones et al. 2008; Fisher et al. 2012), and a number of fungi are considered medically important. In particular, fungal infections are problematic for people who are immunosuppressed (e.g., from HIV infection), in which case their immune systems are unable to adequately fight the infection. Nonetheless, we have uncovered no examples of EIDs in Africa caused by fungal pathogens not related to human immunosuppression, as even the 1950s outbreak of cryptococcal meningitis in the Democratic Republic of the Congo has been likely linked to co-infection by HIV (Molez 1998; Jones et al. 2008).

Prions

Only 5 % of prion diseases are acquired (as opposed to inherited), but these include the well-publicized outbreaks of scrapie, bovine spongiform encephalopathy (BSE, or "mad cow disease"), and chronic wasting disease (CWD) in ungulates from Europe and North America. Of these, only BSE has been detected in humans and in captive-held primates (Imran and Mahmood 2011a, b; Bons et al. 1999; Lee et al. 2013), likely due to consumption of contaminated meat products. The authors have found no descriptions of infectious prion diseases in Africa, but this poorly studied pathogen type may well be present in the world's second largest continent. As it relates to bushmeat-related practices, prions can be found in nearly all tissues and are resistant to degradation, even by cooking, rendering them a potential pathogen worth watching.

Local Knowledge and Perception of Disease Risk

The risk of disease spillover from bushmeat to hunters is highest during butchering and especially if no precautions are taken. Whether hunters take precautions may depend on their knowledge and perception of disease risk. There is increasing evidence that the perception of and knowledge about zoonotic diseases is generally low but varies strongly between sites. A survey among rural bushmeat hunters and traders in Sierra Leone showed that 24 % reported knowledge of disease transmission from animals to humans (Subramanian 2012). Similarly, 23 % of rural–urban hunters and traders in Ghana perceived a disease risk from a bat-bushmeat activity, with significantly more respondents associating risk with bat consumption than bat preparation or hunting (Kamins et al. 2014). Individuals who participate in butchering wild animals typically associate less risk to meat preparation and consumption than those who do not participate in butchering (Kamins et al. 2014) (Fig. 24.2).

Fig. 24.2 A pangolin being prepared in rural Ghana; photo credit Laura Kurpiers

LeBreton et al. (2006) found that hunters and butchers who perceived personal risks were significantly less likely to butcher wild animals, but that risk perception was not associated with hunting and eating bushmeat. Thirty-three percent of bushmeat consumers in a Ghanaian market were not aware that zoonotic diseases could be transmitted from bushmeat to humans. Those who were aware gave Ebola (48 %) and anthrax (16 %) as examples of zoonotic diseases (Kuukyi et al. 2014). In contrast, a large-scale survey among rural Central African population showed that the majority (74 %) of respondents perceived contact with bushmeat blood or body fluids as dangerous (LeBreton et al. 2006). Unfortunately, studies in this field can be challenging, as reported perceptions may differ from actual or 'revealed' behaviors and beliefs (Wilkie 2006).

Although there seems to be some level of risk awareness in certain human populations, several studies report a distinct lack of precautionary behavior, resulting in hunters, butchers, and consumers exposing themselves to zoonotic diseases. LeBreton et al. (2006) found that only 4 % of hunters and 2 % of people reporting butchering indicated that they took precautions against contact with animal blood and fluids while hunting and butchering. Furthermore, the few that took precautions may not have protected themselves adequately, as the most common response was "generally being careful." This was followed by "washing hands," and the least number of participants reporting "avoiding contact with blood, draining blood from carcasses and wearing suitable clothing." Paige et al. (2014) examined human–animal interactions in western Uganda and found that nearly 20 % of participants reported either being injured by an animal or having contact with a primate. The most commonly reported animal injuries were bites (72.9 %) and scratches (23.2 %). In a separate study, it was also shown that although Ghanaian hunters generally handle live bats, they do not typically use protective measures such as gloves, and thereby come into contact with blood through scratches and bites (Kamins et al. 2014). Given the lack of awareness and precautionary measures taken among people who come into contact with bushmeat, the opportunity for new zoonotic pathogens to spillover into humans remains high (LeBreton et al. 2006). This is especially true, since the current rate of hunting wild animals will likely continue—at least until domestic animal production increases and can support the protein needs of the local people.

The Way Forward

Current global disease control efforts focus almost exclusively on responding long after a spillover event has occurred, which increases the risk of a single spillover event causing an epidemic or pandemic. This retroactive response to emerging disease outbreaks is often costly economically and in terms of human well-being (Childs and Gordon 2009; United Nations Development Program 2015). Increased pre-spillover surveillance measures along with quantification of spillover risk is critically needed. For example, Wolfe et al. (2004b, 2005b) found that 1 % of rural Cameroonians are infected with wild primate variants of T-lymphotropic viruses and another 1 % are infected with wild primate variants of simian foamy virus. These sorts of data are simply lacking for most emergent disease systems. Here we will discuss the regulatory and educational measures that could be taken to mitigate the risk of a zoonotic spillover event and spread. Such efforts should be undertaken as a part of a comprehensive response to other sub-Saharan public health crises so as to not divert scarce resources. For example, increases in EID surveillance efforts and in post-emergence management go hand in hand with the improved healthcare infrastructure that must become a priority for sub-Saharan Africa.

At face value, the risk of disease transmission would be reduced if people stopped harvesting bushmeat; however, this scenario is not realistic given the importance of bushmeat in many communities in Africa for which there is limited affordable access to alternate protein sources (Pike et al. 2010; Gebreyes et al. 2014). A more practical option may be to restrict hunting of nonhuman primates, as many zoonotic EIDs have come from them, and instead allow communities to hunt smaller-bodied mammals with higher reproductive rates. Any intervention aiming to restrict access to wildlife should involve community leaders and stakeholders during public outreach to reduce the risk of alienating communities (Monroe and Willcox 2006). The education and enforcement necessary to implement such a restriction must consider the cultural and economic contexts surrounding individual communities. Consider, for example, the problems with enforcement of access restrictions and the history of antagonistic relationships due to exclusion from protected areas between conservationists and local communities. Without proper educational outreach, this could result in backlash from local communities. Furthermore, using zoonotic diseases to enforce hunting restrictions runs the risk of demonizing species considered to be the main disease carriers. Nonhuman primates could then become targets and their populations could be decimated (Pooley et al. 2015).

A more realistic strategy may be to concentrate on preventing future zoonotic spillover events through culturally appropriate education and preventing the spread of diseases through better disease surveillance. In that effort, it would also be important to incorporate collaborative and interdisciplinary approaches between veterinary researchers, ecologists, microbiologists, public health researchers, and anthropologists to develop surveillance and research approaches that will be both culturally appropriate and improve detection of zoonotic diseases tied to bushmeat hunting (Kilonzo et al. 2014).

Education

The risk of disease transmission could be reduced through community education that focuses on people with high levels of exposure to wild animals (Wolfe et al. 2007). Communicating with hunters and butchers about the risks associated with bushmeat and promoting awareness of safer techniques may reduce current levels of pathogen exposure and transmission. To enhance the effectiveness of prevention campaigns, it is particularly important to reinforce the potential for infections during hunting and butchering as this may be overlooked by some hunters (LeBreton et al. 2006). Because the risk perception of hunters and those engaging in butchering wild animals has a negative association with the level of participation in meat preparation and consumption (Kamins et al. 2014), this may reduce current levels of pathogen exposure and transmission, if not by discouraging individuals to participate in preparation and consumption, then by encouraging those individuals to more proactively consider safety and preventative measures.

Global Viral Forecasting (GVF; now “Global Viral” and “Metabiota”) has been pivotal in educating vulnerable populations in rural central Africa by providing information on the risk of zoonotic disease transmission from hunting wild animals (LeBreton et al. 2012). Hunters are informed about disease risks associated with different species, what steps can be taken to avoid infections, and how they can reduce their contact with blood and body fluids of wild animals. Hunters are urged to redirect hunting efforts away from apes and monkeys and towards less risky species such as antelope and rodents, while also being discouraged from butchering animals when there are cuts or injuries on their hands and limbs.

Of course, a common aspect of such attempts at social outreach and education is that even when it is possible to promote awareness, individuals may not believe the hazard is important or that it could affect them. Some authors have even found that when people do believe the risk is real and relevant, there is often little evidence that this knowledge promotes a change in behavior (McCaffrey 2004). For example, a pilot education program among Ghanaian hunters resulted in substantially improved understanding of disease risk, yet largely failed to change peoples' behavior (Kamins et al. 2014). When asked about what would change their behavior, participants responded; becoming ill from zoonotic disease followed by alternative livelihoods and stricter laws. Because awareness is not directly related to behavior, Monroe and Willcox (2006) suggest that campaigns should not rely on the threat of infection to change behavior, but should rather use community leaders to change cultural norms associated with hunting and educate people involved in butchering about best practices of how to protect themselves.

Surveillance

With the increasing prevalence of zoonotic disease emergence and the associated risk for public health, we have to improve our understanding of the dynamics of spillover events of pathogens from animal to human hosts (Rostal et al. 2012) and improve systematic global monitoring efforts. This could help detect, define, and control local human emergence while it is still locally confined and before it has a chance to spread globally. Improved detection and surveillance will lead to a better prioritization of public health efforts. One of the most effective strategies in terms of early detection of an emergent pathogenic threat would be to focus surveillance efforts among people who are highly exposed to at-risk animals and on the animal populations to which they are exposed (LeBreton et al. 2012). Bushmeat hunters would be an important target group, as they are in contact with bodily fluids from animals and are at risk for transmission and infection from novel pathogens.

As an example, GVF has established monitoring programs at multiple sites throughout Central Africa to detect the moment of a pathogen spillover, which can then be used to predict and ultimately prevent zoonotic disease emergences (Evans and Wolfe 2013). In order to track and provide data for EIDs, this effort coordinates the collection of filter-paper blood samples from both hunted animals and people who hunt and butcher wild animals. Early results have shown that this type of surveillance can assist in early detection of new diseases by offering insight into pathogen origin. It would also help describe the spillover dynamics of new or existing diseases. Such data are valuable for developing a detailed, mechanistic understanding of the processes that drive disease emergence and prevent spillovers from spreading in early stages of an outbreak. Contextualizing the relative or actual risks of spillovers would be vital for the preferential allocation of resources to high-risk regions or humans who perform high-risk activities (Daszak et al. 2007). As part of these efforts, improved knowledge of how anthropogenic environmental changes and sociological or demographic factors affect the risk of disease emergence will likely be a cost-effective and sustainable mechanism to reduce or control disease spillover risks (Daszak et al. 2007).

Call for Research

The social and environmental issues surrounding bushmeat represent a complex problem for conservation, global public health, and sustainable development, as it is often the poorest and most vulnerable populations who depend on bushmeat for income or food security. Accordingly, the challenge should be addressed in a holistic manner, by integrating multiple efforts to achieve common objectives. Although much progress has been made not only in addressing the problems concerning bushmeat harvest and zoonotic disease spillover, there is much work to be done. Research that would pave the way for future efforts would include the quantification of social response to environmental policy change (e.g., in the context of harvest restriction), development of a more representative picture of bushmeat consumption in Africa, a broader exploration of the many classes of pathogens within wildlife, and more thorough understanding and quantification of the dynamics behind spillover events and the risks to humans. Such efforts could facilitate the development of policy and infrastructure that would help curb the dependency on bushmeat, reduce risks associated with bushmeat harvest, and help understand in what circumstances zoonotic disease spillover events occur.

There is still uncertainty as to how education should be implemented in different regions and what features of such education would be most valuable for local people. Such an effort might consist of surveying rural bushmeat-harvesting populations across Africa and using the resulting data to contextualize priorities and goals in a way that could help standardize education approaches. While some locations in Africa have had extensive research in the scope and impact of bushmeat harvest, much of Africa has been neglected in those efforts. A more developed understanding of the location, scale, and structure of bushmeat harvest throughout the continent would help researchers and policy-makers prioritize efforts related to disease surveillance, education, or aid. The study of zoonotic spillover events related to viruses, while not completely developed, has received far more attention than the related fields of spillover from bacterial or other non-virus pathogens. There is significant interest in pursuing these lines, as they represent an underdeveloped body of knowledge that could have significant impacts related to human health and disease ecology.

References

Adejinmi JO, Emikpe GE (2011) Helminth parasites of some wildlife in Asejire Game Reserve, Nigeria. South Afr J Wildl Res 41:214–217

Adjemian J, Farnon EC, Tschioko F, Wamala JF, Byaruhanga E, Bwire GS, Kansiime E et al (2011) Outbreak of Marburg hemorrhagic fever among miners in Kamwenge and Ibanda Districts, Uganda, 2007. J Infect Dis 204(Suppl 3):S796–S799

Aghokeng AF, Bailes E, Loul S, Courgnaud V, Mpoudi-Ngolle E, Sharp PM, Delaporte E et al (2007) Full-length sequence analysis of SIVmus in wild populations of mustached monkeys (*Cercopithecus cephus*) from Cameroon provides evidence for two co-circulating SIVmus lineages. Virology 360:407–418

Aghokeng AF, Ayouba A, Mpoudi-Ngole E, Loul S, Liégeois F, Delaporte E, Peeters M (2010) Extensive survey on the prevalence and genetic diversity of SIVs in primate bushmeat provides insights into risks for potential new cross-species transmissions. Infect Genet Evol 10:386–396

Ahuka-Mundeke S, Liégeois F, Lunguya O, Mbenzo V, Mbende M, Inogwabini B, Ayouba A et al (2011) Evidence of STLV 2 and STLV 3 infections in wild living bonobos (*P. paniscus*) from the Democratic Republic of Congo. Retrovirology 8:A93

Ahuka-Mundeke S, Mbala-Kingebeni P, Liégeois F, Ayouba A, Lunguya-Metila O, Demba D, Bilulu G et al (2012) Identification and molecular characterization of new simian T cell lymphotropic viruses in nonhuman primates bushmeat from the Democratic Republic of Congo. AIDS Res Hum Retroviruses 28:628–635

Albariño CG, Foltzer M, Towner JS, Rowe LA, Campbell S, Jaramillo CM, Bird BH et al (2014) Novel paramyxovirus associated with severe acute febrile disease, South Sudan and Uganda, 2012. Emerg Infect Dis 20:211–216

Alexander KA, Blackburn JK, Vandewalle ME, Pesapane R, Baipoledi EK, Elzer PH (2012) Buffalo, bush meat, and the zoonotic threat of brucellosis in Botswana. PLoS One 7, e32842

Allan BF, Keesing F, Ostfeld RS (2003) Effect of forest fragmentation on Lyme disease risk. Conserv Biol 17:267–272

Allebone-Webb S (2008) Evaluating dependence on wildlife products in Equatorial Guinea. Dissertation. Imperial College, London

Allela L, Bourry O, Pouillot R, Délicat A, Yaba P, Kumulungui B, Rouquet P et al (2005) Ebola virus antibody prevalence in dogs and human risk. Emerg Infect Dis 11:385–390

Alvard M (1995) Shotguns and sustainable hunting in the Neotropics. Oryx 29:58

Amman BR, Carroll SA, Reed ZD, Sealy TK, Balinandi S, Swanepoel R, Kemp A et al (2012) Seasonal pulses of Marburg virus circulation in juvenile *Rousettus aegyptiacus* bats coincide with periods of increased risk of human infection. PLoS Pathog 8, e1002877

Amman BR, Albariño CG, Bird BH, Nyakarahuka L, Sealy TK, Balinandi S, Schuh AJ et al (2015) A recently discovered pathogenic Paramyxovirus, Sosuga Virus, is present in *Rousettus aegyptiacus* fruit bats at multiple locations in Uganda. J Wildl Dis 51(3):774–779

Ape Alliance (2006) Recipes for survival: controlling the bushmeat trade. WSPA, London

Apetrei C, Metzger MJ, Richardson D, Ling B, Telfer PT, Reed P, Robertson DL et al (2005) Detection and partial characterization of simian immunodeficiency virus SIVsm strains from bush meat samples from rural Sierra Leone. J Virol 79:2631–2636

Auzel P, Wilkie D (2000) Wildlife use in Northern Congo: hunting in a commercial logging concession. In: Robinson JG, Bennett JG (eds) Hunting for sustainability in tropical forests. Columbia University Press, New York, pp 413–426

Ayouba A, Akoua-Koffi C, Calvignac-Spencer S, Esteban A, Locatelli S, Li H, Li Y et al (2013) Evidence for continuing cross-species transmission of SIVsmm to humans: characterization of a new HIV-2 lineage in rural Cote d'Ivoire. AIDS 27:2488–2491

Babbott FLJ, Gordon JE (1954) Modern measles. Am J Med Sci 228:334–361

Bachand N, Ravel A, Onanga R, Arsenault J, Gonzalez J (2012) Public health significance of zoonotic bacterial pathogens from bushmeat sold in urban markets of Gabon, Central Africa. J Wildl Dis 48:785–789

Baker KS, Todd S, Marsh G, Fernandez-Loras A, Suu-Ire R, Wood JLN, Wang L et al (2012) Co-circulation of diverse paramyxoviruses in an urban African fruit bat population. J Gen Virol 93:850–856

Baker ML, Schountz T, Wang LF (2013) Antiviral immune responses of bats: a review. Zoonoses Public Health 60:104–116

Banyard AC, Evans JS, Luo TR, Fooks AR (2014) Lyssaviruses and bats: emergence and zoonotic threat. Viruses 6:2974–2990

Barnett R (1997) Food for thought: the utilization of wild meat in eastern and southern Africa. Traffic East/Southern Africa

Bennett EL, Gumal MT (2001) The interrelationships of commercial logging, hunting, and wildlife in Sarawak. In: Fimbel RA, Grajal A, Robinson JG (eds) The cutting edge - wildlife conservation in logged tropical forests. Columbia University Press, New York, pp 359–374

Bermejo M, Rodríguez-Teijeiro JD, Illera G, Barroso A, Vilà C, Walsh PD (2006) Ebola outbreak killed 5000 gorillas. Science 314:1564

Bibollet-Ruche F, Bailes E, Gao F, Pourrut X, Barlow KL, Clewley JP, Mwenda JM et al (2004) New simian immunodeficiency virus infecting De Brazza's monkeys (*Cercopithecus neglectus*): evidence for a *Cercopithecus* monkey virus clade. J Virol 78:7748–7762

Bodmer R (1995) Managing Amazonian wildlife: biological correlation of game choice by detribalized hunters. Ecol Appl 5:872–877

Bogich TL, Olival KJ, Hosseini PR et al (2012) Using mathematical models in a unified approach to predicting the next emerging infectious disease. In: Aguirre AA, Ostfeld RS, Daszak P (eds) New directions in conservation medicine: applied cases of ecological health. Oxford University Press, Oxford, pp 607–618

Bons N, Mestre-Frances N, Belli P, Cathala F, Gajdusek DC, Brown P (1999) Natural and experimental oral infection of nonhuman primates by bovine spongiform encephalopathy agents. Proc Natl Acad Sci U S A 96:4046–4051

Boulger L, Porterfield J (1958) Isolation of a virus from Nigerian fruit bats. Trans R Soc Trop Med Hyg 52:421–424

Bowen-Jones E, Pendry S (1999) The threat to primates and other mammals from the bushmeat trade in Africa, and how this threat could be diminished 1. Oryx 33:233–246

Bowen-Jones E, Brown D, Robinson EJZ (2003) Economic commodity or environmental crisis? An interdisciplinary approach to analysing the bushmeat trade in central and west Africa. Area 35:390–402

Brashares JS, Arcese P, Sam MK (2001) Human demography and reserve size predict wildlife extinction in West Africa. Proc Biol Sci 268:2473–2478

Brashares JS, Golden CD, Weinbaum KZ, Barrett CB, Okello GV et al (2011) Economic and geographic drivers of wildlife consumption in rural Africa. Proc Natl Acad Sci U S A 108:13931–13936

Brown D (2007) Is the best the enemy of the good? Livelihoods perspectives on bushmeat harvesting and tradesome issues and challenges. In: Davies G, Brown D (eds) Bushmeat and livelihoods: wildlife management and poverty reduction. Blackwell Publishing, Oxford, pp 111–124

Calattini S, Nerrienet E, Mauclère P, Georges-Courbot MC, Saib A, Gessain A (2006) Detection and molecular characterization of foamy viruses in Central African chimpanzees of the *Pan troglodytes troglodytes* and *Pan troglodytes vellerosus* subspecies. J Med Primatol 35:59–66

Calattini S, Betsem EBA, Froment A, Mauclère P, Tortevoyen P, Schmitt C, Njouom R et al (2007) Simian foamy virus transmission from apes to humans, rural Cameroon. Emerg Infect Dis 13:1314–1320

Calattini S, Betsem E, Bassot S, Chevalier SA, Mahieux R, Froment A, Gessain A (2009) New strain of human T lymphotropic virus (HTLV) type 3 in a Pygmy from Cameroon with peculiar HTLV serologic results. J Infect Dis 199:561–564

Calattini S, Betsem E, Bassot S, Chevaliera SA, Tortevoyea P, Njouomc R, Mahieuxa R et al (2011) Multiple retroviral infection by HTLV type 1, 2, 3 and simian foamy virus in a family of Pygmies from Cameroon. Virology 410:48–55

Cantas L, Suer K (2014) Review: the important bacterial zoonoses in "one health" concept. Front Public Health 2:144

Caspary H (1999) Wildlife utilisation in Cote d'Ivoire and West Africa—potential and constraints for development cooperation. GTZ, Eschborn

Chaber AL, Allebone-Webb S, Lignereux Y, Cunningham AA, Rowcliffe JM (2010) The scale of illegal meat importation from Africa to Europe via Paris. Conserv Lett 3:317–321

Childs JE, Gordon ER (2009) Surveillance and control of zoonotic agents prior to disease detection in humans. Mt Sinai J Med 76:421–428

Childs JE, Richt JA, Mackenzie JS (2007) Introduction: conceptualizing and partitioning the emergence process of zoonotic viruses from wildlife to humans. In: Childs JE, Mackenzie JS, Richt JA (eds) Wildlife and emerging zoonotic diseases: the biology, circumstances and consequences of cross-species transmission. Springer, Berlin, pp 1–31

Chua KB, Bellini WJ, Rota PA, Harcourt BH, Tamin A, Lam SK, Ksiazek TG (2000) Nipah virus: a recently emergent deadly paramyxovirus. Science 288:1432–1435
Cleaveland S, Haydon DT, Taylor L (2007) Overviews of pathogen emergence: which pathogens emerge, when and why? In: Childs JE, Mackenzie JS, Richt JA (eds) Wildlife and emerging zoonotic diseases: the biology, circumstances and consequences of cross-species transmission. Springer, Berlin, pp 85–111
Cottontail V, Wellinghausen N, Kalko E (2009) Habitat fragmentation and haemoparasites in the common fruit bat, *Artibeus jamaicensis* (Phyllostomidae) in a tropical lowland forest in Panamá. Parasitology 136:1133–1145
Courgnaud V, Van Dooren S, Liégeois F, Pourrut X, Abela B, Loul S, Mpoudi-Ngole E et al (2004) Simian T-cell leukemia virus (STLV) infection in wild primate populations in Cameroon: evidence for dual STLV type 1 and type 3 infection in agile mangabeys (*Cercocebus agilis*). J Virol 78:4700–4709
Croser EL, Marsh GA (2013) The changing face of the henipaviruses. Vet Microbiol 167:151–158
Daszak P, Epstein JH, Kilpatrick AM, Aguirre AA, Karesh WB, Cunningham AA (2007) Collaborative research approaches to the role of wildlife in zoonotic disease emergence. Curr Top Microbiol Immunol 315:463–475
Davies TJ, Pedersen AB (2008) Phylogeny and geography predict pathogen community similarity in wild primates and humans. Proc Biol Sci 275:1695–1701
Davies AG, Schulte-Herbrüggen B, Kümpel NF, Mendelson S (2007) Hunting and trapping in Gola Forests, south-eastern Sierra Leone: Bushmeat from farm, fallow and forest. In: Davies AG, Brown D (eds) Bushmeat and livelihoods: wildlife management and poverty reduction. Blackwell, Oxford, pp 15–29
Davis S, Makundi R, Machang'u R, Leirs H (2006) Demographic and spatio-temporal variation in human plague at a persistent focus in Tanzania. Acta Trop 100:133–141
De Merode E, Cowlishaw G (2006) Species protection, the changing informal economy, and the politics of access to the bushmeat trade in the Democratic Republic of Congo. Conserv Biol 20:1262–1271
De Merode E, Homewood K, Cowlishaw G (2004) The value of bushmeat and other wild foods to rural households living in extreme poverty in Democratic Republic of Congo. Biol Conserv 118:573–581
Dei GJS (1989) Hunting and gathering in a Ghanaian rain forest community. Ecol Food Nutr 22:225–243
Delmas O, Holmes EC, Talbi C, Larrous F, Dacheux L, Bouchier C, Bourhy H (2008) Genomic diversity and evolution of the lyssaviruses. PLoS One 3, e2057
Dodet B, Tejiokem MC, Aguemon AR, Bourhy H (2015) Human rabies deaths in Africa: breaking the cycle of indifference. Int Health 7:4–6
Drexler JF, Corman VM, Gloza-Rausch F, Seebens A, Annan A, Ipsen A, Kruppa T et al (2009) Henipavirus RNA in African bats. PLoS One 4, e6367
Drexler JF, Corman VM, Müller MA, Maganga GD, Vallo P, Binger T, Gloza-Rausch F et al (2012) Bats host major mammalian paramyxoviruses. Nat Commun 3:796
Etienne L, Hahn BH, Sharp PM, Matsen FA, Emerman M (2013) Gene loss and adaptation to hominids underlie the ancient origin of HIV-1. Cell Host Microbe 14:85–92
Evans G, Wolfe N (2013) APIC conference: Infection prevention in an era of emerging pandemics. Hospital Infection Control & Prevention 40:7
Fa JE, Juste J, Burn R, Broad G (2002) Bushmeat consumption and preferences of two ethnic groups in Bioko Island, West Africa. Hum Ecol 30:397–416
Fa JE, Currie D, Meeuwig J (2003) Bushmeat and food security in the Congo Basin: linkages between wildlife and people's future. Environ Conserv 30:71–78
Fa JE, Seymour S, Dupain JEF, Amin R, Albrechtsen L, Macdonald D (2006) Getting to grips with the magnitude of exploitation: Bushmeat in the Cross–Sanaga rivers region, Nigeria and Cameroon. Biol Conserv 129:497–510
Fa JE, Albrechtsen L, Johnson PJ, Macdonald DW (2009) Linkages between household wealth, bushmeat and other animal protein consumption are not invariant: evidence from Rio Muni,

Equatorial Guinea: Bushmeat consumption and household wealth. Anim Conserv 12:599–610

Falk H, Dürr S, Hauser R, Wood K, Tenger B, Lortscher M, Schuepbach-Regula G (2013) Illegal import of bushmeat and other meat products into Switzerland on commercial passenger flights. Rev Sci Tech 32:727–739

Fisher MC, Henk DA, Briggs CJ, Brownstein JS, Madoff LC, McCraw SL, Gurr SJ (2012) Emerging fungal threats to animal, plant and ecosystem health. Nature 484:186–194

Fitzgibbon CD, Mogaka H, Fanshawe JH (1995) Subsistence hunting in Arabuko-Sokoke Forest, Kenya, and its effects on mammal populations. Conserv Biol 9:1116–1126

Gao F, Yue L, White AT, Pappas PG, Barchue J, Hanson AP, Greene BM, Sharp PM, Shaw GM, Hahn BH (1992) Human infection by genetically diverse SIVSM-related HIV-2 in West Africa. Nature 358:495–499

Gao F, Bailes E, Robertson DL, Chen Y, Rodenburg CM, Michael SF, Cummins LB et al (1999) Origin of HIV-1 in the chimpanzee *Pan troglodytes troglodytes*. Nature 397:436–441

Gautret P, Blanton J, Dacheux L, Ribadeau-Dumas F, Brouqui P, Parola P, Esposito DH, Bourhy H (2014) Rabies in nonhuman primates and potential for transmission to humans: a literature review and examination of selected French national data. PLoS Negl Trop Dis 8, e2863

Gebreyes WA, Dupouy-Camet J, Newport MJ, Oliveira CJB, Schlesinger LS, Saif YM, Kariuki S et al (2014) The global one health paradigm: challenges and opportunities for tackling infectious diseases at the human, animal, and environment interface in low-resource settings. PLoS Negl Trop Dis 8, e3257

Gillespie TR, Greiner EC, Chapman CA (2004) Gastrointestinal parasites of the guenons of western Uganda. J Parasitol 90:1356–1360

Gillespie TR, Chapman CA, Greiner EC (2005) Effects of logging on gastrointestinal parasite infections and infection risk in African primates. J Appl Ecol 42:699–707

Gillespie TR, Lonsdorf EV, Canfield EP, Meyer DJ, Nadler Y, Raphael J, Pusey AE et al (2010) Demographic and ecological effects on patterns of parasitism in eastern chimpanzees (*Pan troglodytes schweinfurthii*) in Gombe National Park, Tanzania. Am J Phys Anthropol 143:534–544

Gogarten JF, Akoua-Koffi C, Calvignac-Spencer S, Leendertz SAJ, Weiss S, Couacy-Hymann E, Koné I et al (2014) The ecology of primate retroviruses - an assessment of 12 years of retroviral studies in the Tai national park area, Cote d'Ivoire. Virology 460:147–153

Goldberg TL, Sintasath DM, Chapman CA, Cameron KM, Karesh WB, Tang S, Wolfe ND et al (2009) Coinfection of Ugandan red colobus (*Procolobus [Piliocolobus] rufomitratus tephrosceles*) with novel, divergent delta-, lenti-, and spumaretroviruses. J Virol 83:11318–11329

Golden CD, Fernald LCH, Brashares JS, Rasolofoniaina BJR, Kremen C (2011) Benefits of wildlife consumption to child nutrition in a biodiversity hotspot. Proc Natl Acad Sci U S A 108:19653–19656

Gonzalez JP, Prugnolle F, Leroy E (2013) Men, primates, and germs: an ongoing affair. In: Mackenzie JS, Jeggo M, Daszak P, Richt JA (eds) One health: the human-animal-environment interfaces in emerging infectious diseases. Springer, Berlin, pp 337–353

Gottdenker NL, Streicker DG, Faust CL, Carroll C (2014) Anthropogenic land use change and infectious diseases: a review of the evidence. EcoHealth 11(4):619–3210

Goubau P, Van Brussel M, Vandamme AM, Liu HF, Desmyter J (1994) A primate T-lymphotropic virus, PTLV-L, different from human T-lymphotropic viruses types I and II, in a wild-caught baboon (*Papio hamadryas*). Proc Natl Acad Sci U S A 91:2848–2852

Gray RR, Salemi M (2012) Integrative molecular phylogeography in the context of infectious diseases on the human-animal interface. Parasitology 139:1939–1951

Grubb P, Jones TS, Davies AG, Edberg E, Starin ED (eds) (1998) Mammals of Ghana, Sierra Leone and the Gambia. Trendrine, St. Ives

Guan Y, Zheng BJ, He YQ, Liu XL, Zhuang ZX, Cheung CL, Luo SW et al (2003) Isolation and characterization of viruses related to the SARS coronavirus from animals in southern China. Science 302:276–278

Hahn BH, Shaw GM, De Cock KM, Sharp PM (2000) AIDS as a zoonosis: scientific and public health implications. Science 287:607–614

Hang'ombe MB, Mwansa JCL, Muwowo S, Mulenga P, Kapina M, Musenga E, Squarre D et al (2012) Human-animal anthrax outbreak in the Luangwa valley of Zambia in 2011. Trop Doct 42:136–139

Hayman DT, Suu-Ire R, Breed AC et al (2008) Evidence of henipavirus infection in West African fruit bats. PLoS One 3, e2739

Hayman DT, Emmerich P, Yu M, Wang LF, Suu-Ire R, Fooks AR, Cunningham AA et al (2010) Long-term survival of an urban fruit bat seropositive for Ebola and Lagos bat viruses. PLoS One 5, e11978

Hayman DT, Yu M, Crameri G, Wang LF, Suu-Ire R, Wood JLN, Cunningham AA (2012) Ebola virus antibodies in fruit bats, Ghana, West Africa. Emerg Infect Dis 18:1207–1209

Hays JN (2005) Epidemics and pandemics: their impacts on human history. ABC-CLIO, Santa Barbara

Hirsch VM, Olmsted RA, Murphey-Corb M, Purcell RH, Johnson PR (1989) An African primate lentivirus (SIVsm) closely related to HIV-2. Nature 339:389–392

Hogan JN, Miller WA, Cranfield MR, Ramer J, Hassell J, Noheri JB, Conrad PA et al (2014) Giardia in mountain gorillas (*Gorilla beringei beringei*), forest buffalo (*Syncerus caffer*), and domestic cattle in Volcanoes National Park, Rwanda. J Wildl Dis 50:21–30

Horton DL, Banyard AC, Marston DA, Wise E, Selden D, Nunez A, Hicks D et al (2014) Antigenic and genetic characterization of a divergent African virus, Ikoma lyssavirus. J Gen Virol 95:1025–1032

Hussain AI, Shanmugam V, Bhullar VB, Beer BE, Vallet D, Gautier-Hion A, Wolfe ND et al (2003) Screening for simian foamy virus infection by using a combined antigen Western blot assay: evidence for a wide distribution among Old World primates and identification of four new divergent viruses. Virology 309:248–257

Hutson CL, Lee KN, Abel J, Carroll DS, Montgomery JM, Olson VA, Li Y et al (2007) Monkeypox zoonotic associations: insights from laboratory evaluation of animals associated with the multi-state US outbreak. Am J Trop Med Hyg 76:757–768

Imran M, Mahmood S (2011a) An overview of animal prion diseases. Virol J 8:493

Imran M, Mahmood S (2011b) An overview of human prion diseases. Virol J 8:559

Institut Pasteur (1985) Isolations of Lagos bat virus in West Africa. Internal reports of Centre Collaborateur OMS de Reference et Recherche Pour les Arbovirus. http://www.pasteur.fr/recherche/banques/CRORA/bibref/br01600.htm and http://www.pasteur.fr/recherche/banques/CRORA/bibref/br01610.htm. Accessed 26 Feb 2015

Ishikawa KI, Fukasawa M, Tsujimoto H, Else JG, Isahakia M, Ubhi NK, Ishida T et al (1987) Serological survey and virus isolation of simian T-cell leukemia/T-lymphotropic virus type I (STLV-I) in non-human primates in their native countries. Int J Cancer 40:233–239

Jobbins SE, Alexander KA (2015) Evidence of *Leptospira* sp. infection among a diversity of African wildlife species: beyond the usual suspects. Trans R Soc Trop Med Hyg 109(5):349–351

Jones KE, Patel NG, Levy MA, Storeygard A, Balk D, Gittleman JL, Daszak P (2008) Global trends in emerging infectious diseases. Nature 451:990–993

Jones BA, Grace D, Kock R, Alonso S, Rushton J, Said MY, McKeever D et al (2013) Zoonosis emergence linked to agricultural intensification and environmental change. Proc Natl Acad Sci U S A 110:8399–8404

Kamins AO, Restif O, Ntiamoa-Baidu Y, Suu-Ire R, Hayman DT, Cunningham AA, Wood JLN et al (2011a) Uncovering the fruit bat bushmeat commodity chain and the true extent of fruit bat hunting in Ghana, West Africa. Biol Conserv 144:3000–3008

Kamins AO, Restif O, Rowcliffe M, Cunningham A, Wood J (2011b) Use of bats as bushmeat: implications for human health in Ghana, West Africa. Ecohealth 7:S102

Kamins AO, Rowcliffe JM, Ntiamoa-Baidu Y, Cunningham AA, Wood JLN, Restif O (2014) Characteristics and risk perceptions of Ghanaians potentially exposed to bat-borne zoonoses through bushmeat. Ecohealth 1–17

Karesh WB, Noble E (2009) The bushmeat trade: increased opportunities for transmission of zoonotic disease. Mt Sinai J Med 76:429–434

Karugah AK (1997) Rabies in Kenya. In: Proceeding of the SEARG International Symposium. http://searg.info/fichiers/articles/1997025027L.PDF. Accessed 20 Feb 2015

Kazanji M, Mouinga-Ondémé A, Lekana-Douki-Etenna S, Caron M, Makuwa M, Mahieux R, Gessain A (2015) Origin of HTLV-1 in hunters of nonhuman primates in Central Africa. J Infect Dis 211:361–365

Keele BF, Van Heuverswyn F, Li Y, Bailes E, Takehisa J, Santiago ML, Bibollet-Ruche F et al (2006) Chimpanzee reservoirs of pandemic and nonpandemic HIV-1. Science 313:523–526

Keesing F, Holt RD, Ostfeld RS (2006) Effects of species diversity on disease risk. Ecol Lett 9:485–498

Kgaladi J, Wright N, Coertse J, Markotter W, Marston D, Fooks AR, Freuling CM et al (2013) Diversity and epidemiology of Mokola virus. PLoS Negl Trop Dis 7, e2511

Khodakevich L, Jezek Z, Kinzanzka K (1986) Isolation of monkeypox virus from wild squirrel infected in nature. Lancet 1:98–99

Kilonzo C, Stopka TJ, Chomel B (2014) Illegal animal and (bush) meat trade associated risk of spread of viral infections. In: Singh SK (ed) Viral Infections and Global Change. Wiley, Hoboken, pp 179–194

Kuukyi FS, Amfo-Otu R, Wiafe E (2014) Consumer views of bushmeat consumption in two Ghanaian markets. Appl Res J 1:1

Kuzmin IV, Niezgoda M, Franka R, Agwanda B, Markotter W, Beagley JC, Urazova OY et al (2008) Lagos bat virus in Kenya. J Clin Microbiol 46:1451–1461

Kuzmin IV, Niezgoda M, Franka R, Agwanda B, Markotter W, Breiman RF, Shieh WJ et al (2010a) Marburg virus in fruit bat, Kenya. Emerg Infect Dis 16:352–354

Kuzmin IV, Mayer AE, Niezgoda M, Markotter W, Agwanda B, Breiman RF, Rupprecht CE (2010b) Shimoni bat virus, a new representative of the Lyssavirus genus. Virus Res 149:197–210

Labouba I, Leroy EM (2015) Ebola outbreaks in 2014. J Clin Virol 64:109–110

Lawrence P, Pérez BE, Drexler JF, Corman VM, Müller MA, Drosten C, Volchkov V et al (2014) Surface glycoproteins of the recently identified African Henipavirus promote viral entry and cell fusion in a range of human, simian and bat cell lines. Virus Res 181:77–80

Le Guenno B, Formenty P, Wyers M, Gounon P, Walker F, Boesch C (1995) Isolation and partial characterisation of a new strain of Ebola virus. Lancet 345:1271–1274

LeBreton M, Prosser AT, Tamoufe U, Sateren W, Mpoudi-Ngole E, Diffo JLD, Burke DS et al (2006) Patterns of bushmeat hunting and perceptions of disease risk among central African communities. Anim Conserv 9:357–363

LeBreton M, Pike B, Saylors K, le Doux Diffo J, Fair JN, Rimoin AW, Ortiz N, Djoko C, Tamoufe U, Wolfe ND (2012) Bushmeat and infectious disease emergence. In: Aguirre AA, Ostfeld RS, Daszak P (eds) New directions in conservation medicine: applied cases in ecological health. Oxford University Press, New York, pp 164–178

Lederberg J, Hamburg MA, Smolinski MS (eds) (2003) Microbial threats to health: emergence, detection, and response. National Academies Press, Washington, DC

Lee J, Kim SY, Hwang KJ, Ju YR, Woo HJ (2013) Prion diseases as transmissible zoonotic diseases. Osong Public Health Res Perspect 4:57–66

Leendertz FH, Ellerbrok H, Boesch C, Couacy-Hymann E, Mätz-Rensing K, Hakenbeck R, Bergmann C et al (2004) Anthrax kills wild chimpanzees in a tropical rainforest. Nature 430:451–452

Leendertz SAJ, Junglen S, Hedemann C, Goffe A, Calvignac S, Boesch C, Leendertz FH (2010) High prevalence, coinfection rate, and genetic diversity of retroviruses in wild red colobus monkeys (*Piliocolobus badius badius*) in Taï National Park, Côte d'Ivoire. J Virol 84: 7427–7436

Lembo T, Hampson K, Auty H, Beesley CA, Bessell P, Packer C, Halliday J et al (2011) Serologic surveillance of anthrax in the Serengeti ecosystem, Tanzania, 1996-2009. Emerg Infect Dis 17:387–394

Lemey P, Pybus OG, Wang B, Saksena NK, Salemi M, Vandamme AM (2003) Tracing the origin and history of the HIV-2 epidemic. Proc Natl Acad Sci U S A 100:6588–6592
Leroy EM, Rouquet P, Formenty P, Souquiere S, Kilbourne A, Froment JM, Bermejo M et al (2004a) Multiple Ebola virus transmission events and rapid decline of central African wildlife. Science 303:387–390
Leroy EM, Telfer P, Kumulungui B, Yaba P, Rouquet P, Roques P, Gonzalez JP et al (2004b) A serological survey of Ebola virus infection in central African nonhuman primates. J Infect Dis 190:1895–1899
Liégeois F, Lafay B, Switzer WM, Locatelli S, Mpoudi-Ngolé E, Loul S, Heneine W et al (2008) Identification and molecular characterization of new STLV-1 and STLV-3 strains in wild-caught nonhuman primates in Cameroon. Virology 371:405–417
Liégeois F, Butel C, Mouinga-Ondéme A, Verrier D, Motsch P, Gonzalez JP, Peeters M et al (2011) Full-length genome sequence of a simian immunodeficiency virus from a wild-captured sun-tailed monkey in Gabon provides evidence for a species-specific monophyletic SIVsun lineage. AIDS Res Hum Retroviruses 27:1237–1241
Liégeois F, Boué V, Mouacha F, Butel C, Ondo BM, Pourrut X, Leroy E et al (2012) New STLV-3 strains and a divergent SIVmus strain identified in non-human primate bushmeat in Gabon. Retrovirology 9:28
Lilly AA, Mehlman PT, Doran D (2002) Intestinal parasites in gorillas, chimpanzees, and humans at Mondika Research Site, Dzanga-Ndoki National Park, Central African Republic. Int J Primatol 23:555–573
Liu W, Worobey M, Li Y, Keele BF, Bibollet-Ruche F, Guo Y, Goepfert PA et al (2008) Molecular ecology and natural history of simian foamy virus infection in wild-living chimpanzees. PLoS Pathog 4, e1000097
Locatelli S, Peeters M (2012) Cross-species transmission of simian retroviruses: how and why they could lead to the emergence of new diseases in the human population. AIDS 26:659–673
Maganga GD, Bourgarel M, Vallo P, Dallo TD, Ngoagouni C, Drexler JF, Drosten C et al (2014) Bat distribution size or shape as determinant of viral richness in African bats. PLoS One 9, e100172
Magwedere K, Bishi A, Tjipura-Zaire G, Eberle G, Hemberger Y, Hoffman LC, Dziva F (2011) Brucellae through the food chain: the role of sheep, goats and springbok (*Antidorcus marsupialis*) as sources of human infections in Namibia. J S Afr Vet Assoc 82:205–212
Magwedere K, Hemberger MY, Hoffman LC, Dziva F (2012) Zoonoses: a potential obstacle to the growing wildlife industry of Namibia. Infect Ecol Epidemiol 2012:2
Mahieux R, Pecon-Slattery J, Chen GM, Gessain A (1998) Evolutionary inferences of novel simian T lymphotropic virus type 1 from wild-caught chacma (*Papio ursinus*) and olive baboons (*Papio anubis*). Virology 251:71–84
Mari Saéz A, Weiss S, Nowak K, Lapeyre V, Zimmermann F, Düx A, Kühl HS et al (2015) Investigating the zoonotic origin of the West African Ebola epidemic. EMBO Mol Med 7:17–23
Markotter W, Randles J, Rupprecht CE, Sabeta CT, Taylor PJ, Wandeler AI, Nel LH (2006) Lagos bat virus, South Africa. Emerg Infect Dis 12:504–506
Markotter W, Kuzmin I, Rupprecht CE, Nel LH (2008) Phylogeny of Lagos bat virus: challenges for lyssavirus taxonomy. Virus Res 135:10–21
Marston DA, Horton DL, Ngeleja C, Hampson K, McElhinney LM, Banyard AC, Haydon D et al (2012) Ikoma lyssavirus, highly divergent novel lyssavirus in an African civet. Emerg Infect Dis 18:664–667
Martini GA, Knauff HG, Schmidt HA, Mayer G, Baltzer G (1968) A hitherto unknown infectious disease contracted from monkeys. "Marburg-virus" disease. Ger Med Mon 13:457–470
McCaffrey S (2004) Thinking of wildfire as a natural hazard. Soc Nat Res 17:509–516
McCarthy J, Moore TA (2000) Emerging helminth zoonoses. Int J Parasitol 30:1351–1359
Mediannikov O, Diatta G, Zolia Y, Balde MC, Kohar H, Trape JF, Raoult D (2012) Tick-borne rickettsiae in Guinea and Liberia. Ticks Tick-Borne Dis 3:43–48
Meerburg BG, Singleton GR, Kijlstra A (2009) Rodent-borne diseases and their risks for public health. Crit Rev Microbiol 35:221–270

Meertens L, Gessain A (2003) Divergent simian T-cell lymphotropic virus type 3 (STLV-3) in wild-caught *Papio hamadryas papio* from Senegal: widespread distribution of STLV-3 in Africa. J Virol 77:782–789

Meertens L, Rigoulet J, Mauclère P, Van Beveren M, Chen GM, Diop O, Dubreuil G et al (2001) Molecular and phylogenetic analyses of 16 novel simian T cell leukemia virus type 1 from Africa: close relationship of STLV-1 from *Allenopithecus nigroviridis* to HTLV-1 subtype B strains. Virology 287:275–285

Meertens L, Mahieux R, Mauclère P, Lewis J, Gessain A (2002) Complete sequence of a novel highly divergent simian T-cell lymphotropic virus from wild-caught red-capped mangabeys (*Cercocebus torquatus*) from Cameroon: a new primate T-lymphotropic virus type 3 subtype. J Virol 76:259–268

Mendelson S, Cowlishaw G, Rowcliffe JM (2003) Anatomy of a bushmeat commodity chain in Takoradi, Ghana. J Peasant Stud 31:73–100

Mickleburgh S, Waylen K, Racey P (2009) Bats as bushmeat: a global review. Oryx 43:217–234

Milner-Gulland EJ, Bennett EL (2003) Wild meat: the bigger picture. Trends Ecol Evol 18:351–357

Moagabo KT, Monyame KB, Baipoledi EK, Letshwenyo M, Mapitse N, Hyera JMK (2009) A retrospective longitudinal study of animal and human rabies in Botswana 1989-2006. Onderstepoort J Vet Res 76:399–407

Molez JF (1998) The historical question of acquired immunodeficiency syndrome in the 1960s in the Congo River basin area in relation to cryptococcal meningitis. Am J Trop Med Hyg 58:273–276

Monroe MC, Willcox AS (2006) Could risk of disease change bushmeat-butchering behavior? Anim Conserv 9:368–369

Morens DM, Folkers GK, Fauci AS (2008) Emerging infections: a perpetual challenge. Lancet Infect Dis 8:710–719

Mouinga-Ondémé A, Betsem E, Caron M, Makuwa M, Sallé B, Renault N, Saib A et al (2010) Two distinct variants of simian foamy virus in naturally infected mandrills (*Mandrillus sphinx*) and cross-species transmission to humans. Retrovirology 7:105

Mouinga-Ondémé A, Caron M, Nkoghé D, Telfer P, Marx P, Saïb A, Leroy E et al (2012) Cross-species transmission of simian foamy virus to humans in rural Gabon, Central Africa. J Virol 86:1255–1260

Muchaal PK, Ngandjui G (1999) Impact of village hunting on wildlife populations in the western Dja Reserve, Cameroon. Conserv Biol 13:385–396

Mukaratirwa S, Grange LL, Pfukenyi DM (2013) *Trichinella* infections in animals and humans in sub-Saharan Africa: a review. Acta Trop 125:82–89

Muleya W, Sasaki M, Yasuko ORBA, Ishii A, Thomas Y, Nakagawa E, Ogawa H et al (2014) Molecular epidemiology of paramyxoviruses in frugivorous *Eidolon helvum* bats in Zambia. J Vet Med Sci 76:611–614

Munang'andu HM (1995) Rabies in Zambia. In: Proceeding of the SEARG International Symposium. http://searg.info/fichiers/articles/1995036040L.PDF. Accessed 20 Feb 2015

Munene E, Otsyula M, Mbaabu DAN, Mutahi WT, Muriuki SMK, Muchemi GM (1998) Helminth and protozoan gastrointestinal tract parasites in captive and wild-trapped African non-human primates. Vet Parasitol 78:195–201

Munshi-South J, Wilkinson GS (2010) Bats and birds: exceptional longevity despite high metabolic rates. Ageing Res Rev 9:12–19

Mwenda JM, Sichangi MW, Isahakia M, Van Rensburg EJ, Langat DK (1999) The prevalence of antibodies to simian T-cell leukaemia/lymphotropic virus (STLV-I) in non-human primate colonies in Kenya. Ann Trop Med Parasitol 93:289–297

Nasi R, Brown D, Wilkie D, Bennett E, Tutin C, Van Tol G, Christophersen T (2008) Conservation and use of wildlife-based resources: the bushmeat crisis. Secretariat of the Convention on Biological Diversity, Montreal. and Center for International Forestry Research (CIFOR), Bogor Technical Series, p. 50

Nerrienet E, Meertens L, Kfutwah A, Foupouapouognigni Y, Gessain A (2001) Molecular epidemiology of simian T-lymphotropic virus (STLV) in wild-caught monkeys and apes from

Cameroon: a new STLV-1, related to human T-lymphotropic virus subtype F, in a *Cercocebus agilis*. J Gen Virol 82:2973–2977

Nerrienet E, Meertens L, Kfutwah A, Foupouapouognigni Y, Ayouba A, Gessain A (2004) Simian T cell leukaemia virus type I subtype B in a wild-caught gorilla (*Gorilla gorilla gorilla*) and chimpanzee (*Pan troglodytes vellerosus*) from Cameroon. J Gen Virol 85:25–29

Norris K, Asase A, Collen B, Gockowksi J, Mason J, Phalan B, Wade A (2010) Biodiversity in a forest-agriculture mosaic - the changing face of West African rainforests. Biol Conserv 143:2341–2350

Ntiamoa-Baidu Y (1998) Sustainable harvesting, production and use of bushmeat. Wildlife Department, Ghana, Accra

O'Shea TJ, Neubaum DJ, Neubaum MA, Cryan PM, Ellison LE, Stanley TR, Rupprecht CE et al (2011) Bat ecology and public health surveillance for rabies in an urbanizing region of Colorado. Urban Ecosystems 14:665–697

O'Shea TJ, Cryan PM, Cunningham AA, Fooks AR, Hayman DT, Luis AD, Peel AJ et al (2014) Bat flight and zoonotic viruses. Emerg Infect Dis 20:741–745

Olival KJ, Hayman DT (2014) Filoviruses in bats: current knowledge and future directions. Viruses 6:1759–1788

Paige SB, Frost SDW, Gibson MA, Jones JH, Shankar A, Switzer WM, Ting N et al (2014) Beyond bushmeat: animal contact, injury, and zoonotic disease risk in Western Uganda. Ecohealth 11:534–543

Paweska JT, Blumberg LH, Liebenberg C, Hewlett RH, Grobbelaar AA, Leman PA, Croft JE et al (2006) Fatal human infection with rabies-related Duvenhage virus, South Africa. Emerg Infect Dis 12:1965–1967

Pedersen AB, Altizer S, Poss M, Cunningham AA, Nunn CL (2005) Patterns of host specificity and transmission among parasites of wild primates. Int J Parasitol 35:647–657

Peel AJ, Sargan DR, Baker KS, Hayman DT, Barr JA, Crameri G, Suu-Ire R et al (2013) Continent-wide panmixia of an African fruit bat facilitates transmission of potentially zoonotic viruses. Nat Commun 4:2770

Peeters M (2004) Cross-species transmissions of simian retroviruses in Africa and risk for human health. Lancet 363:911–912

Peeters M, Delaporte E (2012) Simian retroviruses in African apes. Clin Microbiol Infect 18:514–520

Peeters M, Courgnaud V, Abela B, Auzel P, Pourrut X, Bibollet-Ruche F, Loul S et al (2002) Risk to human health from a plethora of Simian immunodeficiency viruses in primate bushmeat. Emerg Infect Dis 8:451–457

Peeters M, Jung M, Ayouba A (2013) The origin and molecular epidemiology of HIV. Expert Rev Anti Infect Ther 11:885–896

Pernet O, Schneider BS, Beaty SM, LeBreton M, Yun TE, Park A, Zachariah TT (2014) Evidence for henipavirus spillover into human populations in Africa. Nat Commun 5:5342

Pfukenyi DM, Pawandiwa D, Makaya PV, Ushewokunze-Obatolu U (2009) A retrospective study of wildlife rabies in Zimbabwe, between 1992 and 2003. Tropl Anim Health Prod 41:565–572

Pike BL, Saylors KE, Fair JN, LeBreton M, Tamoufe U, Djoko CF, Rimoin AW et al (2010) The origin and prevention of pandemics. Clin Infect Dis 50:1636–1640

Plowright RK, Foley P, Field HE, Dobson AP, Foley JE, Eby P, Daszak P (2011) Urban habituation, ecological connectivity and epidemic dampening: the emergence of Hendra virus from flying foxes (*Pteropus* spp.). Proc Biol Sci 278:3703–3712

Pooley S, Fa JE, Nasi R (2015) No conservation silver lining to Ebola. Conserv Biol 29(3):965–967

Poulsen JR, Clark CJ, Mavah G, Elkan PW (2009) Bushmeat supply and consumption in a tropical logging concession in Northern Congo. Conserv Biol 23:1597–1608

Pourrut X, Delicat A, Rollin PE, Ksiazek TG, Gonzalez JP, Leroy EM (2007) Spatial and temporal patterns of Zaire ebolavirus antibody prevalence in the possible reservoir bat species. J Infect Dis 196(Suppl 2):S176–S183

Pourrut X, Souris M, Towner JS, Rollin PE, Nichol ST, Gonzalez JP, Leroy E (2009) Large serological survey showing cocirculation of Ebola and Marburg viruses in Gabonese bat populations, and a high seroprevalence of both viruses in *Rousettus aegyptiacus*. BMC Infect Dis 9:159

Pourrut X, Diffo JLD, Somo RM, Bilong Bilong CF, Delaporte E, LeBreton M, GonzaleZ JP (2011) Prevalence of gastrointestinal parasites in primate bushmeat and pets in Cameroon. Vet Parasitol 175:187–191

Prangé A, Perret C, Marié JL, Calvet F, Halos L, Boireau P, Davoust B (2009) Toxoplasmosis: a survey on meat products in Cote d'Ivoire. Med Trop (Mars) 69:629–630

Prescott J, Bushmaker T, Fischer R, Miazgowicz K, Judson S, Munster VJ (2015) Postmortum stability of Ebola virus. Emerg Infect Dis 21(5):856–859

Reynolds MG, Carroll DS, Olson VA, Hughes C, Galley J, Likos A, Montgomery JM et al (2010) A silent enzootic of an Orthopoxvirus in Ghana, West Africa: evidence for multi-species involvement in the absence of widespread human disease. Am J Trop Med Hyg 82:746–754

Rimoin AW, Mulembakani PM, Johnston SC, Lloyd Smith JO, Kisalu NK, Kinkela TL, Blumberg S et al (2010) Major increase in human monkeypox incidence 30 years after smallpox vaccination campaigns cease in the Democratic Republic of Congo. Proc Natl Acad Sci U S A 107:16262–16267

Rizkalla C, Blanco-Silva F, Gruver S (2007) Modeling the impact of Ebola and bushmeat hunting on western lowland gorillas. Ecohealth 4:151–155

Rostal MK, Olival KJ, Loh EH, Karesh WB (2012) Wildlife: the need to better understand the linkages. In: Jeggo M, Daszak P, Rict JA, Mackenzie J (eds) One Health: the human-animal-environment interfaces in emerging infectious diseases. Springer, Berlin, pp 101–125

Rougeron V, Feldmann H, Grard G, Becker S, Leroy EM (2015) Ebola and Marburg haemorrhagic fever. J Clin Virol 64:111–119

Rouquet P, Froment JM, Bermejo M, Kilbourn A, Karesh W, Reed P, Kumulungui B et al (2005) Wild animal mortality monitoring and human Ebola outbreaks, Gabon and Republic of Congo, 2001-2003. Emerg Infect Dis 11:283–290

Sak B, Petrzelkova KJ, Kvetonova D, Mynarova A, Shutt KA, Pomajbikova K, Kalousova B et al (2013) Long-term monitoring of microsporidia, *Cryptosporidium* and *Giardia* infections in western Lowland Gorillas (*Gorilla gorilla gorilla*) at different stages of habituation in Dzanga Sangha Protected Areas, Central African Republic. PLoS One 8, e71840

Saksena NK, Herve V, Durand JP, Leguenno B, Diop OM, Digouette JP, Mathiot C et al (1994) Seroepidemiologic, molecular, and phylogenetic analyses of simian T-cell leukemia viruses (STLV-I) from various naturally infected monkey species from central and western Africa. Virology 198:297–310

Saluzzo JF, Rollin PE, Dauguet C, Digoutte JP, Georges AJ, Sureau P (1984) First isolation of Mokola virus from a rodent *Lophuromys sikapusi*. Ann Inst Pasteur Virol 135:57–61

Salzer JS, Rwego IB, Goldberg TL, Kuhlenschmidt MS, Gillespie TR (2007) *Giardia* sp. and *Cryptosporidium* sp. infections in primates in fragmented and undisturbed forest in western Uganda. J Parasitol 93:439–440

Sandstrom PA, Phan KO, Switzer WM, Fredeking T, Chapman L, Heneine W, Folks TM (2000) Simian foamy virus infection among zoo keepers. Lancet 355:551–552

Santiago ML, Rodenburg CM,Kamenya S, Bibollet-Ruche F, Gao F, Bailes E, Meleth S et al (2002) SIVcpz in wild chimpanzees. Science 295:465

Schulte-Herbrüggen B (2011) The importance of bushmeat in the livelihoods of cocoa farmers living in a wildlife depleted farm-forest landscape, SW Ghana. Dissertation. University College, London

Schulte-Herbrüggen B, Cowlishaw G, Homewood K, Rowcliffe JM (2013a) The importance of bushmeat in the livelihoods of West African cash-crop farmers living in a faunally-depleted landscape. PLoS One 8, e72807

Schulte-Herbrüggen B, Rowcliffe JM, Homewood K, Kurpiers LA, Witham C, Cowlishaw G (2013b) Wildlife depletion in a West African farm-forest mosaic and the implications for hunting across the landscape. Hum Ecol 41:795–806

Sharp PM, Hahn BH (2010) The evolution of HIV-1 and the origin of AIDS. Philos Trans R Soc Lond B Biol Sci 365:2487–2494

Sharp PM, Shaw GM, Hahn BH (2005) Simian immunodeficiency virus infection of chimpanzees. J Virol 79:3891–3902

Shipman P, Bosler W, Davis KL, Behrensmeyer AK, Dunbar RIM, Groves CP, Thackeray F et al (1981) Butchering of giant geladas at an Acheulian site. Curr Anthropol 22:257–268

Siegert R, Shu HL, Slenczka HL, Peters D, Muller G (1968) The aetiology of an unknown human infection transmitted by monkeys (preliminary communication). Ger Med Mon 13:1–2

Simulundu E, Ishii A, Igarashi M, Mweene AS, Suzuki Y, Hang'ombe BM, Namangala B et al (2011) Characterization of influenza A viruses isolated from wild waterfowl in Zambia. J Gen Virol 92:1416–1427

Simulundu E, Nao N, Yabe J, Muto NA, Sithebe T, Sawa H, Manzoor R et al (2014) The zoonotic potential of avian influenza viruses isolated from wild waterfowl in Zambia. Arch Virol 159:2633–2640

Sintasath DM, Wolfe ND, LeBreton M, Jia H, Garcia AD, Diffo JLD, Tamoufe U et al (2009a) Simian T-lymphotropic virus diversity among nonhuman primates, Cameroon. Emerg Infect Dis 15:175–184

Sintasath DM, Wolfe ND, Zheng HQ, LeBreton M, Peeters M, Tamoufe U, Djoko CF et al (2009b) Genetic characterization of the complete genome of a highly divergent simian T-lymphotropic virus (STLV) type 3 from a wild *Cercopithecus mona* monkey. Retrovirology 6:97. doi:10.1186/1742-4690-6-97

Smith M (1982) Field aspects of the Marburg virus outbreak: 1967. Primate Supply 7:11–15

Smith KM, Anthony SJ, Switzer WM, Epstein JH, Seimon T, Jia H, Sanchez MD et al (2012) Zoonotic viruses associated with illegally imported wildlife products. PLoS One 7:71–79

Starkey M (2004) Commerce and subsistence: the hunting, sale and consumption of bushmeat in Gabon. Dissertation. University of Cambridge, Cambridge

Steele JH, Fernandez P (1991) History of rabies and global aspects. In: Baer GM (ed) The natural history of rabies. CRC, Boca Raton, pp 1–26

Steverding D (2008) The history of African trypanosomiasis. Parasite Vectors 1:411

Streicker DG, Turmelle AS, Vonhof MJ, Kuzmin IV, McCracken GF, Rupprecht CE (2010) Host phylogeny constrains cross-species emergence and establishment of rabies virus in bats. Science 329:676–679

Subramanian M (2012) Zoonotic disease risk and the bushmeat trade: assessing awareness among hunters and traders in Sierra Leone. Ecohealth 9:471–482

Sureau P, Tignor G, Smith A (1980) Antigenic characterization of the Bangui strain (ANCB-672D) of Lagos bat virus. Ann Inst Pasteur Virol 131:25–32

Swanepoel R, Smit SB, Rollin PE, Formenty P, Leman PA, Kemp A, Burt F et al (2007) Studies of reservoir hosts for Marburg virus. Emerg Infect Dis 13:1847–1851

Switzer WM, Bhullar V, Shanmugam V, Cong M, Parekh B, Lerche NW, Yee JL et al (2004) Frequent simian foamy virus infection in persons occupationally exposed to nonhuman primates. J Virol 78:2780–2789

Switzer WM, Salemi M, Shanmugam V, Gao F, Cong M, Kuiken C, Bhullar V et al (2005) Ancient co-speciation of simian foamy viruses and primates. Nature 434:376–380

Takehisa J, Kraus MH, Ayouba A, Bailes E, Van Heuverswyn F, Decker JM, Li Y et al (2009) Origin and biology of simian immunodeficiency virus in wild-living western gorillas. J Virol 83:1635–1648

Takemura T, Yamashita M, Shimada MK, Ohkura S, Shotake T, Ikeda M, Miura T et al (2002) High prevalence of simian T-lymphotropic virus type L in wild Ethiopian baboons. J Virol 76:1642–1648

Tanner M, Inlameia O, Michel A, Maxlhuza G, Pondja A, Fafetine J, Macucule B et al. (2014) Bovine tuberculosis and brucellosis in cattle and African buffalo in the Limpopo National Park, Mozambique. Transbound Emerg Dis 62:632–638

Taylor LH, Latham SM, Woolhouse ME (2001) Risk factors for human disease emergence. Philos Trans R Soc Lond B Biol Sci 356:983–989

Taylor G, Scharlemann JPW, Rowcliffe M, Kümpel N, Harfoot MBJ, Fa JE, Melisch R et al (2015) Synthesising bushmeat research effort in West and Central Africa: a new regional database. Biol Conserv 181:199–205

Ter Meulen J, Lukashevich I, Sidibe K, Inapogui A, Marx M, Dorlemann A, Yansane ML et al (1996) Hunting of peridomestic rodents and consumption of their meat as possible risk factors for rodent-to-human transmission of Lassa virus in the Republic of Guinea. Am J Trop Med Hyg 55:661–666

Tieguhong JC, Zwolinski J (2009) Supplies of bushmeat for livelihoods in logging towns in the Congo Basin. J Hortic Forestry 5:65–80

Timen A, Koopmans MP, Vossen AC, van Doornum GJ, Gunther S, van den Berkmortel F, Verduin KM et al (2009) Response to imported case of Marburg hemorrhagic fever, the Netherlands. Emerg Infect Dis 15:1171–1175

Towner JS, Pourrut X, Albariño CG, Nkogue CN, Bird BH, Grard G, Ksiazek TG et al (2007) Marburg virus infection detected in a common African bat. PLoS One 2, e764

Towner JS, Amman BR, Sealy TK, Reeder-Carroll SA, Comer JA, Kemp A, Swanepoel R et al (2009) Isolation of genetically diverse Marburg viruses from Egyptian fruit bats. PLoS Pathog 5, e1000536

Traina-Dorge VL, Lorino R, Gormus BJ, Metzger M, Telfer P, Richardson D, Robertson DL et al (2005) Molecular epidemiology of simian T-cell lymphotropic virus type 1 in wild and captive sooty mangabeys. J Virol 79:2541–2548

United Nations (2013) World population prospects: the 2012 revision. In: United Nations, Department of Economic and Social Affairs, Population Division. http://www.prb.org/pdf13/2013-population-data-sheet_eng.pdf. Accessed 26 Feb 2015

United Nations (2014) World urbanization prospects: the 2014 revision. In: United Nations, Department of Economic and Social Affairs, Population Division. http://esa.un.org/unpd/wup/Highlights/WUP2014-Highlights.pdf. Accessed 26 Feb 2015

United Nations Development Program (2015) Socio-economic impact of Ebola virus disease in West African countries. www.africa.undp.org/content/dam/rba/docs/Reports/ebola-west-africa.pdf. Accessed 22 Mar 2015

United Nations Environment Programme (2014) UNEP year book 2014 emerging issues in our global environment. United Nations, Nairobi

Van Brussel M, Salemi M, Liu HF, Gabriels J, Goubau P, Desmyter J, Vandamme AM (1998) The simian T-lymphotropic virus STLV-PP1664 from *Pan paniscus* is distinctly related to HTLV-2 but differs in genomic organization. Virology 243:366–379

Van Dooren S, Shanmugam V, Bhullar V, Parekh B, Vandamme AM, Heneine W, Switzer WM (2004) Identification in gelada baboons (*Theropithecus gelada*) of a distinct simian T-cell lymphotropic virus type 3 with a broad range of Western blot reactivity. J Gen Virol 85:507–519

Van Heuverswyn F, Peeters M (2007) The origins of HIV and implications for the global epidemic. Curr Infect Dis Rep 9:338–346

Van Heuverswyn F, Li Y, Neel C, Bailes E, Keele BF, Liu W, Loul S et al (2006) Human immunodeficiency viruses: SIV infection in wild gorillas. Nature 444:164

Van Heuverswyn F, Li Y, Bailes E, Neel C, Lafay B, Keele BF, Shaw KS et al (2007) Genetic diversity and phylogeographic clustering of SIVcpzPtt in wild chimpanzees in Cameroon. Virology 368:155–171

van Thiel PP, de Bie RM, Eftimov F, Tepaske R, Zaaijer HL, van Doornum GJ, Schutten M et al (2009) Fatal human rabies due to Duvenhage virus from a bat in Kenya: failure of treatment with coma-induction, ketamine, and antiviral drugs. PLoS Negl Trop Dis 3, e428

van Vliet N, Nasi R (2008) Hunting for livelihood in northeast Gabon: patterns, evolution, and sustainability. Ecol Soc 13(2):33

Vega MG, Carpinetti B, Duarte J, Fa JE (2013) Contrasts in livelihoods and protein intake between commercial and subsistence bushmeat hunters in two villages on Bioko Island, Equatorial Guinea. Conserv Biol 27:576–587

Viana M, Mancy R, Biek R, Cleaveland S, Cross PC, Lloyd-Smith JO, Haydon DT (2014) Assembling evidence for identifying reservoirs of infection. Trends Ecol Evol 29:270–279

Voevodin A, Samilchuk E, Allan J, Rogers J, Broussard S (1997) Simian T-lymphotropic virus type 1 (STLV-1) infection in wild yellow baboons (*Papio hamadryas cynocephalus*) from Mikumi National Park, Tanzania. Virology 228:350–359

Wain LV, Bailes E, Bibollet-Ruche F, Decker JM, Keele BF, Van Heuverswyn F, Li Y et al (2007) Adaptation of HIV-1 to its human host. Mol Biol Evol 24:1853–1860

Walsh PD, Abernethy KA, Bermejo M, Beyers R, De Wachter P, Akou ME, Huijbregts B et al (2003) Catastrophic ape decline in western equatorial Africa. Nature 422:611–614

Weingartl HM, Nfon C, Kobinger G (2013) Review of Ebola virus infections in domestic animals. Dev Biol 135:211–218

Weiss S, Nowak K, Fahr J, Wibbelt G, Mombouli JV, Parra HJ, Wolfe ND et al (2012) Henipavirus-related sequences in fruit bat bushmeat, Republic of Congo. Emerg Infect Dis 18:1536–1537

Wilkie D (2006) Bushmeat: a disease risk worth taking to put food on the table? Anim Conserv 9:370–371

Wilkie DS, Carpenter JF (1999) Bushmeat hunting in the Congo Basin: an assessment of impacts and options for mitigation. Biodivers Conserv 8:927–955

Wilkie DS, Curran B (1991) Why do Mbuti hunters use nets? Ungulate hunting efficiency of archers and net-hunters in the Ituri rain forest. Am Anthropol 93:680–689

Wilkie DS, Sidle JG, Boundzanga GC (1992) Mechanized logging, market hunting, and a bank loan in Congo. Conserv Biol 6:570–580

Wilkie DS, Sidle J, Boundzanga G, Blake S, Auzel P (2001) Defaunation or deforestation: commercial logging and market hunting in northern Congo. Columbia University Press, New York

Wolfe ND, Eitel MN, Gockowski J, Muchaal PK, Nolte C, Prosser AT, Torimiro JN et al (2000) Deforestation, hunting and the ecology of microbial emergence. Global Change Hum Health 1:10–25

Wolfe ND, Prosser TA, Carr JK, Tamoufe U, Mpoudi-Ngole E, Torimiro JN, LeBreton M et al (2004a) Exposure to nonhuman primates in rural Cameroon. Emerg Infect Dis 10:2094–2099

Wolfe ND, Switzer WM, Carr JK, Bhullar VB, Shanmugam V, Tamoufe U, Prosser AT et al (2004b) Naturally acquired simian retrovirus infections in central African hunters. Lancet 363:932–937

Wolfe ND, Daszak P, Kilpatrick AM, Burke DS (2005a) Bushmeat hunting, deforestation, and prediction of zoonotic disease. Emerg Infect Dis 11:1822–1827

Wolfe ND, Heneine W, Carr JK, Garcia AD, Shanmugam V, Tamoufe U, Torimiro JN et al (2005b) Emergence of unique primate T-lymphotropic viruses among central African bushmeat hunters. Proc Natl Acad Sci U S A 102:7994–7999

Wolfe ND, Dunavan CP, Diamond J (2007) Origins of major human infectious diseases. Nature 447:279–283

Woolhouse ME, Haydon DT, Antia R (2005) Emerging pathogens: the epidemiology and evolution of species jumps. Trends Ecol Evol 20:238–244

World Health Organization (1978) Ebola haemorrhagic fever in Sudan, 1976. Bull World Health Organ 56:247–270

World Health Organization (2005) Plague. Wkly Epidemiol Rec 80(80):138–140

Worobey M, Santiago ML, Keele BF, Ndjango JB, Joy JB, Labama BL, Dhed AB et al (2004). Origin of AIDS: Contaminated polio vaccine theory refuted. Nature 428:820

Wyers M, Formenty P, Cherel Y, Guigand L, Fernandez B, Boesch C, Le Guenno B (1999) Histopathological and immunohistochemical studies of lesions associated with Ebola virus in a naturally infected chimpanzee. J Infect Dis 179(Suppl 1):S54–S59

Young HS, Dirzo R, Helgen KM, McCauley DJ, Billeter SA, Kosoy MY, Osikowicz LM et al (2014) Declines in large wildlife increase landscape-level prevalence of rodent-borne disease in Africa. Proc Natl Acad Sci U S A 111:7036–7041

Zheng H, Wolfe ND, Sintasath DM, Tamoufe U, Lebreton M, Djoko CF, Diffo Jle D et al (2010) Emergence of a novel and highly divergent HTLV-3 in a primate hunter in Cameroon. Virology 401:137–145

Chapter 25
Wildlife Trafficking in a Globalized World: An Example of Motivations and Modus Operandi from a Norwegian Case Study

Ragnhild Sollund

Introduction[1]

The illegal wildlife trade (IWT) is one of the fastest growing illegal trades today; it is repeatedly positioned among the trafficking of illegal drugs, arms, and humans in regard to the economy involved in the trade as one of the largest illegal trades worldwide (Zimmerman 2003; Wyatt 2013a; UNODC 2012). Unfortunately, due to the clandestine nature of IWT and weak law enforcement and control of it, the scale of international IWT is difficult to measure. Low estimates put the legal trade in wildlife to be worth from 5 to 50 billion USD annually (Reeve 2002, p. 10), while high estimates place it at 159–160 billion USD annually (Warchol 2007; Duffy cited in White 2011, p. 55; Schneider 2008). More recently, a report by Chatham House places its worth at over 300 billion USD per year (Vines 2014, p. 9). In contrast, the IWT has been valued from 6 billion USD annually (Warchol 2007) to 10 billion USD annually (Schmidt 2004), and more recently, up as high as 20 billion USD annually (Alacs and Georges 2008).

In this chapter, the illegal trade will be examined using a Norwegian penal case still awaiting trial in order to uncover the modus operandi of an offender, his motivations for engaging in the trade, and priorities and problems in control and law enforcement. The term "wildlife" usually includes both flora and fauna, but in the present case study the focus is on animals. I begin this chapter by presenting the theoretical point of departure and briefly discussing IWT, before presenting the methodology used in the research project which is the basis for the chapter. Thereafter, I present the specific penal case. Findings from the exploration of this case are further used to discuss problems in the control and law enforcement of IWT

[1] Part of the chapter's literature review also forms part of the case study report produced for the EFFACE project.

R. Sollund (✉)
Department of Criminology and Sociology of Law, University of Oslo, Oslo, Norway
e-mail: Ragnhild.sollund@jus.uio.no

F.M. Angelici (ed.), *Problematic Wildlife*, DOI 10.1007/978-3-319-22246-2_25

given the motivations for involvement in this crime that are evident in the case and whether efforts made to prevent IWT are adequate.

The theoretical basis for this study is found in green criminology. This particular field of criminology acknowledges that not only criminalized acts which are harmful are of interest to green criminologists, but also harmful acts which are not yet criminalized, harmful acts which are supported by powerful players and structures in society, for example, states and corporations, and harmful acts which are part of daily practice, for example, pollution caused by carbon emissions (e.g. White 2013; Lynch and Stretesky 2014). Of special concern for green criminologists are the concepts of rights and justice. These are delineated as *human rights and environmental justice*—environmental rights are an extension of human or social rights so as to enhance the quality of human life, now and in the future; *ecological citizenship and ecological justice*—humans are merely one component of complex ecosystems that should be preserved in their own right in light of the rights of the environment; and *animal rights and species justice*—environmental harm is constructed in relation to the role nonhuman animals play within environments and their intrinsic right to not suffer abuse, whether this be one-on-one harm, institutionalized harm, or harm arising from human actions that affect climates and environments globally (White 2013, p. 6). The last one is of particular relevance for the present study, which adopts a non-speciesist moral point of departure which acknowledges that nonhuman animals can be regarded as victims because of their capacity to experience suffering (Regan 1983), and that animal abuse can be defined as such no matter whether this takes place in accordance with or in breach of law, whether as one-to-one harm or institutionalized harm (Beirne 1999). Beirne and South (2007, p. xiv) thus argue that "[...]green criminology should be a harm-based discourse that addresses violations of what some have variously termed environmental morality, environmental ethics, and animal rights." From this perspective, wildlife trade is a crime no matter whether or not it is in breach of law because those who are victims of the trade undoubtedly suffer and often die as a consequence of it.

Literature Review

The literature persuasively argues that the escalation in the collection and killing of wildlife is influenced by market forces and thus motivated largely by the potential for substantial economic gain. Because of low detection rates and lenient punishments, the potential rewards for offenders far outweigh the risk of being penalized (Europol 2013; European Commission 2014a, p. 3). IWT is facilitated by globalization which has opened borders and expanded the marketplace. The Internet plays a significant role as an intermediary between supply and demand (IFAW 2008), something this chapter highlights. Practices which involve the use of animals for food (e.g., eating pangolin as a delicacy) and as medicine (e.g., rhino horn, tiger, and leopard powder) are also widespread and probably facilitated by globalization, including human migration and travel, as witnessed in the EU (Van Uhm in press). However, there is also growing evidence that the impact of the IWT is widespread—evidence suggests IWT erodes state authority, fuels civil conflict, threatens national stability and international security, and provokes substantial economic losses internationally (Lawson and Vines 2014; Wyatt 2013b)—and this has led to increased international political attention.

According to Baille et al. (as quoted in Rivalan et al. 2007), the IWT threatens a third of the world's species; among the most critically endangered are rhinoceros and African elephant. During the past 40 years the Living Planet Index (LPI), which measures more than 10,000 representative populations of mammals, birds, reptiles, amphibians, and fish, has identified a 52 % decline of these populations (WWF 2014a). Although habitat loss likely accounts for much of this decline, hunting and "abducting"[2] also constitute a serious threat for many species. Consequently, CITES (*Convention on International Trade in Endangered Species* of Wild Fauna and Flora, n.d.) lists five thousand animal species as threatened, many because of trade or the combination of loss of habitat and trade (WWF 2014b; Reid 1992).

The body of interdisciplinary scientific literature investigating IWT as well as legal wildlife trade (LWT) is growing. For example, biologists are studying the impact IWT has on specific species affected by the trade and on their natural environments. The trafficking of parrots for the pet trade is endangering many parrot species. The methods of abduction are particularly destructive, for example, when entire cohorts of chicks are taken from nests and the nesting trees are cut as part of the procedure. As is the case in the illegal reptile trade, up to 90 % of rare birds die before they reach their destination, largely due to the crude and harsh methods by which animals are trafficked (Warchol et al. 2003, p. 8; see also Guzman et al. 2007). (For studies on parrots and trafficking see Weston and Memon 2009; Gonzales 2003; Herrera and Hennessey 2007; Eniang et al. 2008; Metz 2007; Guzman et al. 2007. For studies on reptiles and trafficking see Ellis 2005; Zhang et al. 2008; Alves et al. 2008). Even so, this trade remains lucrative. As is the case with drug traffickers, animal traffickers can sustain such substantial losses because of the considerable revenue which can be made from even a single transaction. For example, a pair of rare parrots can be sold for 50,000 € in Europe (Van Uhm in press).

A lot of the trafficking goes on in local markets (e.g. Weston and Memon 2009; Lee 1996; Warchol et al. 2003; Zhang et al. 2008), for example, reptiles are used widely for medicinal purposes in Brazil (see Alves et al. 2008) and as food in Southwest China (see Zhang et al. 2008). Furthermore, studies which have been done on tiger and rhinoceros derivatives have established that traditional medicine and also newer beliefs about the healing properties of animals or their parts play an

[2] I regard the concept "poaching" as anthropocentric and implying the acceptance of animals as property and therefore I prefer the terms "abduction," when animals are taken from their habitats while still alive (Sollund 2011) and killed.

important role in the growth of IWT (e.g. Ellis 2005; Minnaar 2013). This is also the case in the use of bear bile (Burgess et al. 2014).

More recently, the focus on the criminal nature of IWT has given rise to numerous studies on the ivory trade (Warchol et al. 2003; Lemieux and Clarke 2009), the parrot trade (Pires and Clarke 2011; Pires and Moreto 2011; Sollund 2011), and IWT more generally (Schneider 2012; Wyatt 2013a). The literature includes studies focusing on trafficking within and to specific geographic locations (Wyatt 2009, 2011; Herbig 2010; Sollund 2011, 2013). While several of these contributions adopt conventional criminological approaches in their theoretical explanations of the IWT and possible responses to it, such as viewing IWT as criminal and a breach of law which must be addressed through conventional means such as situational crime prevention (Wellsmith 2010, 2011; Hill 2015; Lemieux 2014; Pires and Clarke 2011; Pires and Moreto 2011; Schneider 2012), others adopt a broader green criminology perspective, including concern about animal rights and ecological and species justice (Sollund 2011, 2013; Wyatt 2013a, b). Reports produced by NGOs such as TRAFFIC,[3] IFAW, WWF[4], and WSPA provide the most substantial, and often the most current, evidence on the patterns, trends, and nature of the IWT and the ways in which it is counteracted (e.g. WWF/Dalberg 2012; TRAFFIC 2008; Burgess et al. 2014). These NGOs play a central role in responding to the IWT through "policing" (albeit without police enforcement powers), educating, and campaigning. International agencies such as the UN, Interpol/Europol, and the EU have contributed to the literature with key reports providing international and local perspectives on the scale, nature, and impact of the IWT. According to European Commission estimates, the number of African elephants illegally killed has doubled over the last decade, and the quantity of ivory seized has tripled. The illegal killing of rhinoceros has escalated sharply in South Africa, with over 1000 animals killed in 2013, compared to 13 in 2007. In total, since 2010, about 80 % (2500) of the entire South African rhinoceros population has been killed. The attention given to "iconic" species such as the rhinoceros would suggest they are the ones killed and trafficked most prolifically. However, the species most frequently seized in the EU are birds and reptiles (Van Uhm in press). The worldwide reptile trade demands the supply of millions of individuals every year. Reptiles are used medicinally, for example, in Brazil and China (see Ellis 2005; Zhang et al. 2008; Alves et al. 2008; Santana 2008), in the food and fashion industries, and in the pet trade (see Herbig 2010; Sollund 2013; Engler and Parry-Jones 2007). A substantial number die during capture, transportation, and thereafter in captivity, due to lack of proper care and nutrition (RSPCA n.d.).

---

[3] TRAFFIC has existed as an NGO since 1976 and is a monitoring network. It researches IWT (and LWT) and is governed by a steering group consisting of members from partner organizations WWF and IUCN. Its aim is not to prohibit WLT, but to make it sustainable. TRAFFIC employs approximately 120 persons of more than 25 nationalities, based in 30 countries. It is organized in seven regional teams and a headquarters office in the UK, which operates as a registered charity (see TRAFFIC n.d.a).

[4] For TRAFFIC, see TRAFFIC (n.d.b); for WWF, see WWF (2014a).

Despite the substantial loss of animal life and the widespread decline of eco-systems and the number of species, it is important to emphasize that the majority of wildlife trade is taking place in *compliance* with laws and regulations, including the CITES Convention, as trade in a specific species is only banned when that species is critically endangered (see Appendix I for a list of species which are banned from trade).[5] Before I proceed to present the specific and the methodology, I will briefly discuss the Internet as a marketplace for IWT, since this is central to my later discussion.

The Internet as a Marketplace for IWT

A quick google search confirms that eBay is important in connecting buyers and sellers of ivory. eBay has officially taken a stance against commercializing ivory. According to its website, "eBay has decided to institute a global ban on the sale of all types of ivory. This global ban will be effective January 1, 2009" (eBay Inc. 2008). This ban is circumvented through selling ivory under the label "faux ivory," although many items are obviously not faux but genuine, not least antiques (see eBay n.d.).

There are exceptions to the ban on ivory, depending on when the ivory item was crafted. According to an Interpol and IFAW investigation into the illegal ivory trade on the Internet, ivory is considered antique or pre-convention (i.e., the CITES Convention) if the raw ivory was acquired or introduced into the EU before the adoption of the Convention, or if the worked ivory was acquired more than 50 years ago. These ivory items are subject to less strict regulations. Commercial trade within the EU in these antique ivory items may be allowed with certification. For the commercial trade of worked ivory items which are more than 50 years old, no certificate is needed. If the ivory item is not antique, an import and/or export permit must be obtained to prove its legality. Exemptions to the CITES Convention also exist provided the items are considered to be personal effects: The EU CITES legislation includes various categories of personal effects of ivory that are listed in Annex A and Annex B. Importing ivory which is listed in Annex A of the EU CITES legislation as a personal effect requires export and import permits which must be presented to customs before entering the EU. An import permit is not needed for a personal effect of ivory listed under Annex B. In both cases, however, an export permit from the country of origin is necessary (Interpol/IFAW 2013, p. 10). According to The Norwegian Environment Agency, the practice is to be restricted in Norway in order to close the loopholes which exist in defining ivory (or other CITES products) as personal effects because of the massive volume of the Internet trade worldwide and the pressure it puts on many species (Miljødirektoratet n.d.).

The Interpol/IFAW report (2013) establishes that the Internet trade of ivory is a major problem in the EU, especially in western Europe. The time frame studies for the report were only a couple of hours per day in nine countries over a fortnight. A conservative estimate was that 4500 kg of ivory were traded during this brief time

[5] See CITES (2015).

(p. 11). According to the report, significant loopholes in the legislation of the Internet trade of ivory exist: "[There is] no obligation on the side of the seller to prove on the Internet that the ivory item for sale complies with the existing legislation; there is no obligation to inform the buyer of existing regulations; and there is no obligation for the website owner to comply with existing legislation" (p. 15). This implies that buyers may be ignorant of the fact that they are committing a crime when they purchase ivory through Internet auctions. This does not imply they are not committing a crime, because being unaware that you are committing a crime does not absolve an offender in other kinds of crimes, although this would depend on the damage caused and the character and seriousness of the offence.

Analysis of seizures from EU Customs authorities show that 60 % of traded ivory was categorized as personal effects, 22 % came as mail parcels (as in the case discussed later in this chapter), and 9 % came concealed in freight or transport (IFAW 2013, p. 20). Another IFAW report based on a study of Internet trade of CITES products in eight countries (Australia, US, UK, France, Germany, Russia, Canada, and China) found that the most frequently traded wildlife product was ivory, followed by exotic birds: "Of the 7122 auctions, advertisements and communiqués reviewed in this study, 5223 (73.3 %) offered elephant products. Elephant products, predominantly ivory, were the most commonly available products in six of the eight countries profiled, with the United States responsible for ten times more ivory listings than the next closest country (the United Kingdom)" (IFAW 2008, p. 10). An Italian case study found that the Internet boosted IWT in a substantial way and that through the Internet, collectors had entered the IWT market (Lavorgna 2014).

Methodology and Case Study Description

In collecting data for this project, I used a multifaceted qualitative approach including interviews with law enforcement agents and experts in Norway, Colombia, and Brazil, interviews with offenders (Norwegian), and analyses of confiscation reports from Norwegian Customs Directorate [Tolldirektoratet] about CITES seizures. Still pending is an analysis of 7–800 penal cases involving the illegal importation or keeping of (illegal) exotic wildlife or CITES-listed species. The data from interviews with law enforcement agents in Norway are used to support the analysis in the chapter, but the focus here is on one case and accompanying documents. This case consists of Customs confiscation reports, Customs reports to the police, correspondence concerning the verification of species, photographs, reports of police interviews (including quotations from transcripts) with the offender, etc.[6] These reports all pertain to one person who is accused of bringing

[6]The project has been approved by the Norwegian Social Science Data Services and access to penal cases and police reports have been granted by the state attorney in Norway. All ethical issues pertaining to the access to criminal offenders and information about offenders have been addressed and all data which could lead persons to be identified have been anonymized.

products of CITES-listed species to Norway, mainly by means of post and delivery services. The case has been under investigation for approximately two years (as of January 2015, the time of writing) and is still awaiting trial.

Case

Police and Customs reports refer to different dates when parcels addressed to the accused were first checked and seized because they contained illegal wildlife products. The accused is a man who was not in waged labour because of an occupational injury. In one report, the Police state that Customs opened ten parcels from different European countries, all addressed to the accused. Six of these contained ivory. Another police report stated that Customs stopped and opened ten parcels addressed to the accused. Eight of these contained ivory, one a stuffed bird and one a deer's head. This report also stated that there were similar unsettled cases against the same man. Yet another seizure of parcels addressed to the accused revealed two containing 12 CITES-listed products: eight made from ivory, three made from hippo teeth, and one stuffed falcon. Other Customs confiscation reports pertaining to this man deal with Nile crocodile heads (CITES I). All together several elephant tusks, either raw or decorated, have been confiscated. In a report of a confiscation made by Customs in the fall of 2012, items were found to have been sent from a large number of different private addresses in several European countries, as well as from companies specializing in Internet auctions.

Following the seizure of the parcels, the police went to the accused's home with a search warrant. Here they found a chaotic collection of several hundred stuffed animals and birds, many of species which are protected under the Bern and CITES Conventions (Norway is party to both) such as wolverine, lynx and several hawk, eagle, owl, and falcon species. The search and police interviews further revealed that the man had several reptile skins, including a cobra, a stuffed monitor lizard, and a very large number of stuffed wading, predator and song birds, bats, fox, otter, and hare, and pelts from different animals (This list is not exhaustive). There were glass showcases containing stuffed animals, preponderantly birds, everywhere in the house. The ivory items found also were many in number, although the number has not yet been specified.

According to the police interview reports, it appears that the accused had acted in an almost feverish way when purchasing items through Internet auctions. He could have 200 auction bids going at the same time. According to the interview report, "it was almost like an obsession for him. [...] Ten to fifteen parcels could arrive daily." In another interview report, the police state that "it became like an obsession to acquire things cheaply; he learnt the tricks of how to succeed with this. At times it was like a gambler's addiction; he therefore won many auctions." He purchased items from many different auction websites including eBay, Skanfil, and Finn.no. He also travelled to Denmark and England to buy items, always from private persons. In an interview he stated that he panicked when parcels kept arriving; he lost track of what he had ordered and felt he lost control and so hid the parcels unopened in the attic. He did not want his family to see them. He further explained that he had been collecting marine items since childhood—a hobby he was introduced to by a relative—especially from the whale hunting period in Norway,[7] and he thus also had many items in his possession made from whale ivory and whale bone, including narwhale. Because he was not working, he had a lot of time to spend on his stuffed animal collection. In fact, he had bought whole collections of stuffed birds from death estates.

The police interviews with the accused reveal he was aware that labelling ivory as "faux" or naming it "ox bone" is a well-known strategy among ivory collectors who make their purchases on the Internet and among traders (IFAW 2013; see also Davenport 2014). Elephant ivory may also be sold as mammoth or hippo ivory, or detection may be evaded through misspelling the word "ivory," or by omitting it entirely in the ad (IFAW 2008, p. 15).

Motivations: Collections

Collecting may be fundamental for humans and nonhumans and be deeply ingrained in us. According to Rigby and Rigby (1944, cited in Case 2009, p. 739), an early collection of seal impressions is dated to the fifth century BCE, indicating that collecting has a long tradition. Or picture squirrels or spotted nutcrackers that collect and store their nuts for the winter. As gatherers, humans depended on their capacity to forage food found in nature. Collecting mushrooms and berries are still important pastimes for people even when they do not depend on these for food, and the passion with which mushrooms are collected often far exceeds the need and use humans have for the mushrooms. O'Brien (year) states that in the US, one in three people collects (cited in McIntosh 2002, p. 85). Collecting is defined as follows: "If the predominant value is representative or representational, i.e., if said object or idea is valued chiefly for the relation it bears to some other object or idea, or objects, or ideas, such as being one of a series, part of a whole, a specimen of a class, then it is the subject of a collection" (Durost year, cited in Pearce 2003, p. 157). Pearce adds that this definition holds the valuable distinction between objects held *for use*, with a helpfully wide idea of what constitutes 'use', and objects held as *part of a sequence*: it is the idea of series or class which creates the notion of the collection (Pearce 2003, p. 159).

A collection and the activity of collecting can hold many sociological and psychological meanings (McIntosh and Schmeichle 2004). Pearce's definition above might be applied to the ivory collection of the accused; the photographs in

[7] Whale hunting in Norway is still going on. The peak of this activity was in the late 1800s and took place in the Antarctic and in Norway as well as in open seas; many whale species were driven to the brink of extinction until the foundation of the International Whaling Commission and the moratorium introduced against whaling in 1986. Norway opposed the moratorium and more than 10,000 minke whales have been killed in Norway since (Martinsen 2013, pp. 103–105).

the case documents demonstrate that the ivory items collected by the man were decorative, rather than having a user value, which further emphasizes their value as a collection. Collectors may collect for leisure, aesthetics, competition, risk, fantasy, a sense of community, prestige, domination, sensual gratification, sexual foreplay, a desire to reframe objects, the pleasing rhythm of sameness and difference, ambition to achieve perfection, extending the self, reaffirming the body, producing gender identity, and achieving immortality (McIntosh and Schmeichle 2004). McIntosh and Schmeichel point out that all these motivations involve self-fulfilling or self-enhancement needs. Collectors may further be categorized as: (1) passionate collectors, obsessive and emotional, and willing to pay any price for the right item; (2) inquisitive collectors for whom collecting is an investment; (3) the hobbyist who collects purely for enjoyment; and (4) expressive collectors who collect as a statement of who they are (2004, p. 87). The accused may fit into several of these categories, except that of inquisitive collector (#2) as he did not resell what he collected. A study by Formanek (2003) of why people collect revealed there are a number of motivations: the meaning of collecting (1) to the self, (2) to others, collecting (3) as preservation, restoration, history, and a sense of continuity, (4) as financial investment, and (5) as addiction. The last category may be of particular interest in the present case judging by the hundreds of dead animals and animal products in the man's house. For the accused, collecting began as enjoyment, something he was introduced to as a child, but became more than a hobby. In the end it was also a statement of who he was and where he came from; both the man and family members, having worked as sailors and in similar positions, had collected items related to the sea. As he was unable to work, his passion for collecting seems to have gradually increased until it turned into a compulsive disorder. The way in which stuffed animals were piled up around the house indicates he had lost control of these collections.

Ivory can hold attraction for various reasons: it is rare and therefore valuable, it is beautiful and therefore enjoyable, and in the collectors' mindset, it is a collection which the collector can *continue* to pursue because variations of similar objects can be accumulated. In this way, it could give purpose and meaning to the life of the accused, a man who is unemployed and collecting can replace waged labour. As poetically stated by Danet and Katriel: "In our opinion, in its contemporary form, and perhaps in all ages since it first blossomed about 3000 years ago (Alsop 1982; Bazin 1967), collecting is an aesthetic activity that gives expression to the universal experience of the ephemerality of human existence" (2003, p. 236). More prosaically, one can state that consumerism is driving collectors; the accused in this case did not even stop to admire or systemize his collection, he merely purchased items for it. This may be counter to the ordinary drive people have to collect. Clifford describes the collection process as "an excessive, sometimes even rapacious need to *have* is transformed into rule-governed, meaningful desire. Thus the self that must possess but cannot have it all learns to select, order, classify in hierarchies—to make 'good' collections" (Clifford 2003, p. 260). In the present case, the urge to *have* seems to have overruled the urge to *systematize* and thus *refine* and culturalize the collection—the ivory parcels were hidden away. They seem thus not have been something he took pride in; on the contrary, he was ashamed of the way he had purchased them, indicating he could not prevent himself from taking part in the auctions. Although precious items of ivory may serve to ascribe social status to a collector because ivory is highly prized and increasingly rarer, in the present case it seems craving these items or perhaps even more craving the act of purchase through participation in the auctions seems to have been an important motivation. The accused may, however, have had several related motivations for his engagement in IWT. I will treat them separately and then proceed to a discussion of how to prevent people with similar motivation from getting involved in this crime

Internet Auction Addiction?

Collecting ivory may have been a compulsory act of addiction on the part of the accused; the police interviews suggest that the compulsion may have been reinforced by the *mode of purchase*, which can be compared to a gambler's addiction. For gamblers, the Internet provides ample opportunities for indulging in their addiction at any time of day, thus amplifying the situational and structural factors facilitating the activity (Griffiths 2003) and making it harder to resist the temptation. In the same way, Internet auction websites have provided collectors with similar opportunities. Structural *characteristics* are viewed by Griffiths as those which are responsible for the reinforcement of the addiction, and which may satisfy gamblers' needs and actually facilitate excessive gambling. Technological advance can (and will) have a potentially large impact on the development and maintenance of gambling behavior. Factors which facilitate the purchase of illegal ivory on the Internet are the privacy in which the passion may be out-acted, the pressure during the process in which one must act swiftly in order not to lose the bid, the reward being close at hand, and the attraction of the beautiful item, including its rarity and art value, on display in the pictures and described in the text. In relation to gambling addictions, Griffiths states that addiction is essentially about rewards and the immediacy of rewards. Therefore, the more potential rewards there are, the more addictive the activity is likely to be (1999, p. 268). In the present case, rewards came with far more certainty than in gambling and may therefore have had an even greater effect on the motivation to proceed with the activity. The rewards were also highly cherished, at least during the auctions. The accused's description of acting "like [he was] obsessed" may however suggest that the possibility of winning the auction drove him to proceed as much as the desire for the object.

Therapists are beginning to consider online auction addiction as a medical disorder under the rubric of Internet addiction (Peters and Bodkin 2007). An additional element of Internet auctions is the competition between the bidders. Other elements than pure luck, which is more prevalent in ordinary gambling, come into play, such as the capacity of bidders to psychologically foresee the actions of their opponents during the auctions. Consequently, the taste of victory and of having deserved to win the auctions may be sweeter and be an encouragement to proceed. The statements of the accused in the police interrogations, however, indicate that his Internet auction participation was compulsory in nature, indicating an online auction addiction. Such an addiction is defined as: "[...] the compulsive need to participate in online auctions that over time generates harmful consequences for the consumer" (Peters and Bodkin 2007, p. 3). Harmful consequences in this instance are in addition to encouraging the illegal killing of animals of critically endangered species. They also include direct harmful effects to the accused himself—committing criminal offenses and spending his resources on IWT. According to his own statements made during police interviews, the accused was engaged in activities similar to those of Internet auction addicts; the compulsory behavior forcing them to partake in more and more auctions for items they do not need, incurring financial costs, and storing and concealing the items (Peters and Bodkin 2007). The accused was a passionate collector. His was a hobby which resulted in him using the Internet as a medium through which he got access to auctions throughout the world. He could purchase whatever he desired; however, it seems that the auctions themselves became as important as the items which were acquired by means of them.

A third important motivation for his addiction may have been his lack of work and the considerable time he had on his hands because of that. Waged labour provides meaning of life for many (see Jahoda 1982), and when waged labour is absent many will seek for other ways to create meaning and self-realization. By turning into an expert on (dead) wildlife,[8] and through possessing and building his collections, the accused may have achieved a satisfaction he otherwise would have lacked in life due to his situation.

Law Enforcement and Prevention of IWT on the Internet

At this point, it is unknown how the accused will be punished, and this case stands out as unique in the data of this research project. I can therefore not claim that this type of offense is representative of people's motivation for being involved in the IWO as their cases appear; for example, in Norway, people's desire for pets drives the trafficking of live animals, something the accused did not participate in (Sollund 2013). Neither can I claim it is representative of all IWT crimes. However, the items in the accused's collection are similar to many others described in other confiscation reports and penal case files in the data—objects people purchased either online or as souvenirs while on vacation. While reptile keepers also may be passionate about their hobbies, what distinguishes this case is the compulsory drive to purchase dead animals as stuffed or their products, and how this may have led the accused to commit these crimes. Evidently, he had no concern for the fact that his hobby entailed the death of innumerable individuals. The animals involved were alienated as objects; displayed in his house or randomly thrown into boxes as parts.

[8] Although the police reports showed that several of the dead animals in his house were categorized wrongly according to species.

What this case exemplifies in a stark manner is the great number of individuals who may be harmed and killed indirectly through the actions of a single man; the ivory confiscations are connected to several elephants, so if such behavior is indeed representative of collectors purchasing wildlife on the Internet, the speed with which elephant species are driven to extinction becomes comprehensible. In addition, in his collection there were large numbers of eagles, falcons, hawks, reptiles, big carnivores, etc., species which are threatened. Because animals are intricately a part of the ecosystems they live in, the "hobby" of collectors has a mortal component, affecting not only the desired species but also entire ecosystems.

The accused claimed in police interviews that he had tried to ensure that he did not commit a crime in the legal sense.[9] Assuming those who engage in IWT are disinterested in animal welfare and biodiversity, one can also assume that they will pay more attention to the punitive consequences for themselves which their actions may result in, than to the more important, broader consequences of environmental destruction. Ignorance of the possible punitive consequences may thus be important in explaining why offenders commit these crimes. A further criminalization of such acts, and awareness of this fact, may deter potential criminals in this field. Even for people who do care for animal welfare and nature, lack of awareness of the consequences concerning these issues is likely a push factor for IWT.

Those who are highly motivated to commit such crimes may be harder to deter despite their awareness of the consequences—harm to animals/nature or potential punishment—whether they commit such acts to profit from them, or for private rewards/satisfaction, as in the present case. Deterrence also depends on the level of risk and the kind of punishment implied. For egg collectors in the UK, a fine may simply be regarded as an additional cost of the hobby (Nurse 2013, p. 69). The risk of going to prison would likely be more of a detriment, but the statements the accused made in interviews do not imply that he perceived such punishment as probable. His carelessness when buying on the Internet may indicate that in spite of his knowledge that he was doing something illegal, he was unable to refrain from doing it because his drive/passion/compulsion for Internet auctions and collecting was greater than the risk he ran, like for others who purchase WLT products on the Internet (Lavorgna 2014).

There is also contradicting information in the police interview reports; the accused says on one occasion that he would not do anything illegal, yet he says he is aware of the prevalence of ivory sold as ox bone and mammoth and that he bought such items himself. That he was aware of this suggests that he was willing to take a risk because the satisfaction he received from purchasing the items outweighed the risk. The secretive nature of Internet activity and the multiple opportunities to offend it offers, whether buyers and sellers trade in live or dead animals, make it extremely important

[9] See introduction. As mentioned, in green criminology terminology many harms are regarded as crimes even when they are not formally criminalized through laws or regulation (see South 2008; Beirne and South 2007; Sollund 2015; Walters 2010, 2013; White 2013).

for law enforcement agents to target in a strategic, preventative way. CITES[10] suggests that units dedicated to investigating wildlife crime linked to the Internet be established at national levels, that wildlife trade issues be incorporated into existing units that investigate or monitor computer or cyber-crime, and that mechanisms to coordinate the monitoring of Internet-related wildlife trade be established at national levels. These require that far more resources are allotted to prevent and prosecute such crimes than what is currently the situation, not the least in Norway.

As shown, the Internet trade also provides multiple opportunities for laundering, that is, selling ivory as "antique," "ox bone," "mammoth," or "faux ivory." This suggests that while WLT continues to be regulated rather than banned, it will not cease because of the difficulties of regulation. A legal market motivates IWT as it hides, facilitates, legitimates, and encourages the trade in wildlife (Sollund 2011). This is evidenced by the identification of forged CITES import and export certificates, corruption on the part of Customs officials and police, wild-caught animals laundered as captive-bred, and the sale of ivory stockpiles which failed to reduce demand as expected (Shepherd et al. 2012; Warchol et al. 2003; UNODC 2012; Wyatt 2009, p. 153; Van Uhm in press). The way CITES functions (or fails to function) as protection for threatened wildlife has, therefore, been subject to debate (Dickson, 2005; Kievit 2000; Reeve 2002; Sollund 2011). To control (meaning to distinguish similar legal and illegal products/animals that are similar) the trade therefore seems an impossible task. Take Norway as an example. Customs officers in Norway might think confiscating wildlife is complicated and burdensome because they are not competent to distinguish a CITES-listed animal from one which is not listed or to recognize the difference between a forged or genuine CITES certificate.

Still, it must be noted that a ban may also have unexpected and unacceptable consequences. Rivalan et al. (2007) reveal that the timing between the announcement and effectuation of a species ban (e.g. moving a species from CITES II to CITES I appendix) is important, because during this period there is an increase in trafficking which may significantly harm the species, as evidenced by the decline in rhinoceroses. Bans can increase collectors' interest in a species, thereby increasing its value and the incentive to offend (Low 2003 cited in Sollund 2011, p. 445). Lemieux and Clarke (2009), Guzman et al. (2007), and Moyle (2003) also suggest that an international ban may serve to increase the domestic trade of a species (e.g. parrots), a consequence equally harmful to the survival of the species. Guzman et al.'s (2007) study of the parrot trade in Mexico indicates that bans reduce transnational WLT, but not abductions for local markets which is guided by cultural practices and traditions. The ban on ivory, which was first introduced in 1989, resulted in an increase in elephant populations, but civil wars and unregulated markets impacted the effectiveness of the ban (Lemieux and Clarke 2009). On

[10] Consolidated Resolution Conf. 11.3 on Compliance and enforcement of the 11th meeting of the Conference of the Parties to CITES, Gigiri (Kenya), 10–20 April 2000, as amended at the 13th, 14th and 15th meetings of the Conference of the Parties to CITES (http://www.cites.org/eng/res/11/11-03R15.php).

a large scale, however, the regulation of WLT rather than a total ban necessarily prolongs and legitimates the trade in endangered species, until the species reaches a critical state at which time a ban is introduced (Sollund 2011).

Conclusion

IWT is an increasing problem and this chapter has highlighted the role of the Internet in it. It has shown aspects of the trade which serve to reveal the motivations that consumers of wildlife may have for engaging in IWT. It establishes that IWT practices follow the development of the modern world where opportunities to get involved in IWT are increasing. Technological developments that are available in the modern world are used to full advantage in IWT; when travel is cheaper and more accessible, it is easier for people to become involved as consumers. The Internet increases the options people have to partake in such practices without moving from the comfort of their homes (Lavorgna 2014). The Internet also produces an increased distance between the object which is purchased (the animal this once was and the way it was killed) and the buyer, as well as between the actor and the act. Because wildlife products are so easily accessible, the seriousness of such crimes may be downplayed by the offenders. The existence of Internet auctions and the way people get addicted to them further underlines the importance of the need for coordinated action in surveillance of relevant Internet sites such as eBay. This requires that surveillance is a priority both at national and international levels (as suggested by CITES). In Norway, this is, at this point, not a priority of law enforcement agencies; the same way CITES cases are generally not prioritized.

An important reason why IWT is increasing may be the legitimation of WLT through the CITES Convention—CITES regulates rather than bans trade (in general) in order to secure nonhuman species as resources for prolonged trade (Sollund 2011). The Convention perpetuates the idea that nonhuman animals are there for humans to use as they please, whether as decoration, food, pets, medicine, status symbols, or collector items. This emphasizes the material rather than the intrinsic value of nonhuman animals. Many will argue that precisely because of the economic value of WLT, it must be sustained and prolonged for human benefit, thus emphasizing the rights humans have to live in and off of their environment (and other beings in it). Others argue that, given the importance of eco-justice and species justice, human practice of emptying ecosystems of their inhabitants for the purpose of trade is not only unsustainable, but also a crime. Such a perspective acknowledges the rights of individual animals to live unharmed by humans. Costa Rica has taken a step in that direction by banning hunting; it may be an example for other states to follow. While attitudes remain anthropocentric and ignorance of the harm involved in these crimes remains pervasive, control and law enforcement persist as the main instruments to prevent ITW. Creating awareness about the costs of IWT is crucial not the least among consumers in the western world. Given that these crimes are low in the priority of the judicial system, are leniently punished, usually by means of insignificant fines (e.g. Lowther et al. 2002), are often ignored and undetected, and may substantially reward the criminals involved, there is little to prevent these crimes from being committed and to ensure that lives and species are saved.

Acknowledgments I am grateful to the Norwegian Animal Protection Fund for financial contribution to the project. The project is also a part of the EFFACE, funded by the FP7 under the European Commission. See http://efface.eu/. I am further grateful to my proof reader Kerry Fast and two anonymous referees for comments and suggestions for ways to improve the chapter.

References

Alacs E, Georges A (2008) Wildlife across our borders: a review of the illegal trade in Australia. Aust J For Sci 40:147–160

Alves RRD, da Silva Veira WL, Santana GG (2008) Reptiles used in traditional folk medicine: conservation implications. Biodivers Conserv 17:2037–2049

Beirne P (1999) For a nonspeciesist criminology: animal abuse as an object of study. Criminology 37:117–149

Beirne P, South N (2007) Introduction: approaching green criminology. In: Beirne P, South N (eds) Issues in green criminology: confronting harms against environments, humanity and other animals. Devon, Willan, pp xiii–xvi

Burgess EA, Stoner SS, Foley K-E (2014) Brought to bear: an analysis of seizures across Asia (2000–2011). TRAFFIC: Petaling Jaya, Selangor, Malaysia

Case DO (2009) Serial collecting as leisure, and coin collecting in particular. Library Trends 57:729–752

CITES (n.d.) Convention on international trade in endangered species of wild fauna and flora. Retrieved from http://www.cites.org/eng/disc/text.php

CITES (2015) Appendices. Retrieved from http://www.cites.org/eng/app/appendices.php

Clifford J (2003) Collecting ourselves. In: Pearce SM (ed) Interpreting objects and collections. Routledge, London, pp 258–269

Danet B, Katriel T (2003) No two alike: play and aesthetics in collecting. In: Pearce SM (ed) Interpreting objects and collections. Routledge, London, pp 220–240

Davenport J (2014) eBay traders “are flouting ban on ivory by calling it faux.” London Evening Standard, February 10. Retrieved from http://www.standard.co.uk/news/uk/ebay-traders-are-flouting-ban-on-ivory-by-calling-it-faux-9119071.html

Dickson B (2005) What is the goal of regulating wildlife trade? Is regulation a good way to achieve the goal? In: Oldfield S (ed) The trade in wildlife: regulation for conservation. Earthscan, London, pp 23–33

eBay Inc. (n.d.) Faux ivory. Retrieved from http://www.ebay.com/bhp/faux-ivory

eBay Inc. (2008) eBay to institute global ban on ivory sales. October 20. Retrieved from http://blog.ebay.com/ebay-to-institute-global-ban-on-ivory-sales/

Ellis R (2005) Tiger bone and rhino horn: the destruction of wildlife for traditional Chinese medicine. Island Press, Washington, DC

Eniang EA, Akpan CE, Eniang ME (2008) A survey of African grey parrots (*Psittacus erithacus*) trade and trafficking in Ekonganaku area of Ikpan forest block, Nigeria. Ethiopian J Env Stud Manage 1:68–73

Engler M, Parry-Jones R (2007) Opportunity or threat? The roles of the European Union in global wildlife trade. TRAFFIC, Brussels

European Commission (2014a) Communication from the commission to the council and the European parliament on the EU approach against wildlife trafficking., Retrieved from http://ec.europa.eu/environment/cites/pdf/communication_en.pdf

Europol (2013) Threat assessment 2013 environmental crime in the EU. Retrieved from https://www.europol.europa.eu/content/threat-assessment-2013-environmental-crime-eu

Formanek R (2003) Why they collect: collectors reveal their motivations. In: Pearce SM (ed) Interpreting objects and collections. Routledge, London, pp 327–446

Gonzales JA (2003) Harvesting, local trade, and conservation of parrots in the Northeastern Peruvian Amazon. Biol Conserv 114:437–446

Griffiths M (1999) Gambling technologies: prospects for problem gambling. J Gambl Stud 15:265–293

Griffiths M (2003) Internet gambling: issues, concerns, and recommendations. Cyber Psychol Behav 6:557–568

Guzman JCC, Sanchez SME, Grosselet M, Gamez JS (2007) The illegal parrot trade in Mexico: a comprehensive assessment. Retrieved from http://www.defenders.org/publications/the_illegal_parrot_trade_in_mexico.pdf

Herbig J (2010) The illegal reptile trade as a form of conservation crime: a South African criminological investigation. In: White R (ed) Global environmental harm: criminological perspectives. Willan, Devon, pp 110–132

Herrera M, Hennessey B (2007) Quantifying the illegal trade in Santa Cruz de la Sierra, Bolivia, with emphasis on threatened species. Bird Conserv Int 17:295–300

Hill J (2015) A systems thinking perspective on the motivations and mechanisms driving wildlife poaching. In: Sollund R (ed) Critical criminology in a changing world: green harms and crimes. Palgrave, London

IFAW (2008) Killing with keystrokes. An investigation of the illegal wildlife trade on the world wide web. Retrieved from http://www.ifaw.org/sites/default/files/Killing%20with%20Keystrokes.pdf

IFAW (2013) Project web: an investigation into the ivory trade over the Internet within the European Union. Retrieved from http://www.ifaw.org/united-states/resource-centre/project-web-investigation-ivory-trade-over-internet-within-european-union

Jahoda M (1982) Employment and unemployment: a social-psychological analysis. Cambridge University Press, Cambridge

Kievit H (2000) Conservation of the Nile crocodile: has CITES helped or hindered? In: Hutton J, Dickson B (eds) Endangered species threatened convention: the past, present and future of CITES. Earthscan, London, pp 88–97

Lavorgna A (2014) Wildlife trafficking in the Internet age. Crime Sci 3(1):1–12

Lawson K, Vines A (2014) Global impacts of the illegal wildlife trade, The costs of crime, insecurity and institutional erosion. Chatham House, London

Lee J (1996) Poachers, tigers and bears…oh my! Asia's illegal wildlife trade. Northwest J Int Law Business 498:495–515

Lemieux AM (ed) (2014) Situational prevention of poaching. Routledge, London

Lemieux AM, Clarke R (2009) The international ban on ivory sales and its effects on elephant poaching in Africa. Br J Criminol 49:451–471

Lowther J, Cook D, Roberts M (2002) Crime and punishment in the wildlife trade. WWF/TRAFFIC and University of Wolverhampton, Wolverhampton

Lynch MJ, Stretesky PB (2014) Exploring green criminology: toward a green criminological revolution. Ashgate, Farnham

Martinsen S (2013) Dyrs verdi i lys av utvalgte fangsttradissjoner. [The value of animals in light of certain hunting traditions?] In: Sollund R, Tønnessen M, Larsen G (eds) Hvem er villest i landet her? Råskap mot dyr i antropocen—menneskets tidsalder. Spartacus SAP, Oslo, pp 99–121

McIntosh WD, Schmeichle B (2004) Collectors and collecting: a social psychological perspective. Leis Sci 26:85–97

Miljødirektoratet (n.d.) Lovgrunnlaget I Norge. http://www.miljodirektoratet.no/no/Tema/Arter-og-naturtyper/Handel-med-trua-arter-CITES/Lovgrunnlaget-i-Norge/. Accessed on 10 March 2015

Metz S (2007) Rehabilitation of Indonesian parrots from the illegal wild bird trade: early experiences on Seram Island, Indonesia. Association of Avian veterinarians (Australian

Committee), Melbourne. Retrieved from http://www.indonesian-parrot-project.org/pdf_files/AAV_07_6-29-07.PDF
Minnaar A (2013) The poaching of rhino in South Africa: a conservation, organised and economic crime? In: Sorvatzioti D, Antonopolous G, Papanicoulaou G, Sollund R (eds) Critical views on crime, policy and social control. University of Nicosia Press, Nicosia, pp 196–229
Moyle B (2003) Regulation, conservation and incentives. In: Oldfield S (ed) The trade in wildlife: regulation for conservation. Earthscan, London
Nurse A (2013) Animal harm: perspectives on why people harm and kill animals. Ashgate, Farnham
Pearce SM (2003) The urge to collect. In: Pearce SM (ed) Interpreting objects and collections. Routledge, London, pp 157–160
Peters C, Bodkin CP (2007) An exploratory investigation of problematic online auction behaviors: experiences of eBay users. J Retail Cons Serv 14:1–16
Pires S, Clarke R (2011) Are parrots CRAVED? An analysis of parrot poaching in Mexico. J Res Crime Delinq 49:122–146
Pires S, Moreto WB (2011) Preventing wildlife crimes: solutions that can overcome the "tragedy of the commons". Eur J Crime Policy 17:101–123
Reeve R (2002) Policing international trade in endangered species. The CITES treaty and compliance. Earthscan, London
Regan T (1983) The case for animal rights. Berkeley University Press, Berkeley
Reid WR (1992) How many species will there be? In: Whitmore TC, Sayer JA (eds) Tropical deforestation and species extinction. Chapman & Hall, New York
Rivalan P, Delmas V, Angulo E, Bull LS, Hall RJ, Courchamp F et al (2007) Can bans stimulate wildlife trade? Nature 447:529–530
RSPCA (n.d.) Far from home—reptiles that suffer and die in captivity. Retrieved from www.rspca.org.uk/…/LocateAsset%3Fasset=document
Schmidt C (2004) Environmental crimes: profiting at the earth's expense. Environ Health Perspect 112:96–103
Schneider JL (2008) Reducing the illicit trade in endangered wildlife: the market reduction approach. J Contemp Crim Justice 24:274–295
Schneider JL (2012) Sold into extinction: the global trade in endangered species. Praeger, Santa Barbara
Shepherd CR, Stengel CJ, Nijman V (2012) The export and re-export of CITES-listed birds from the Solomon Islands. Southeast Asia, Petaling Jaya, Selangor, Malaysia, TRAFFIC
Sollund R (2011) Expressions of speciesism: the effects of keeping companion animals on animal abuse, animal trafficking and species decline. Crime Law Soc Change 55:437–451
Sollund R (2013) Animal trafficking and trade: abuse and species injustice. In: Walters R, Westerhuis D, Wyatt T (eds) Emerging issues in green criminology: exploring power, justice and harm pp 72-93. Palgrave, Basingstoke, pp 72–93
Sollund R (2015) Critical, green criminology: an agenda for change. In: Sollund R (ed) Critical criminology in a changing world: green harms and crimes. Palgrave, London
South N (2008) Nature, difference and rejection of harm: expanding the agenda for green criminology. In: Sollund R (ed) Global harms. Ecological crime and speciesism. Nova Science, New York, pp 198–201
TRAFFIC (2008) What's driving the wildlife trade? A review of expert opinion on economic and social drivers of the wildlife trade and trade control efforts in Cambodia, Indonesia, Lao PDR and Vietnam. East Asia and Pacific Region sustainable development discussion papers. East Asia and Pacific Region Sustainable Development Department, World Bank, Washington, DC
TRAFFIC (n.d., a) Publications. Retrieved from http://www.traffic.org/search-publications/
TRAFFIC (n.d., b) What we do. Retrieved from http://www.traffic.org/overview/
UNODC (2012) Wildlife and forest crime analytic toolkit. Revised edition. Retrieved from http://www.unodc.org/documents/Wildlife/Toolkit_e.pdf
Van Uhm D (2014) Illegal wildlife trade to the EU and harms to the world. In: Sapens T, White R, Huisman W (eds). Environmental crime and the world (n.p.). Ashgate, London

Vines A (2014) Global impacts of the illegal wildlife trade: the costs of crime, insecurity and institutional erosion. Retrieved from http://www.chathamhouse.org/publications/papers/view/197367#sthash.2O2ErEK7.dpuf
Walters R (2010) Toxic atmospheres, air pollution, trade and the politics of regulation. Crit Criminol 18:307–323
Walters R (2013) Air crimes and athmospheric justice. In South N, Brisman A (eds.) Routledge handbook of green criminology. Routledge, London, 134–150
Warchol GL (2007) The transnational illegal wildlife trade. Criminal Justice Studies: a Critical. Journal of Crime. Law and Society 17:57–73
Warchol GL, Zupan L, Clarke W (2003) Transnational criminality: an analysis of the illegal wildlife market in Southern Africa. Int Crim Justice Rev 13:1–26
Wellsmith M (2010) The applicability of crime prevention to problems of environmental harm: a consideration of illicit trade in endangered species. In: White R (ed) Global environmental harm: criminological perspectives. Willan, Devon, pp 132–150
Wellsmith M (2011) Wildlife crime: the problems of enforcement. Eur J Crime Pol 17:125–148
Weston MK, Memon MA (2009) The illegal parrot trade in Latin America and its consequences to parrot nutrition, health and conservation. Bird Populations 9:76–83
White R (2011) Transnational environmental crime: toward an eco-global criminology. Routledge, London
White R (2013) Environmental harm. An eco-justice perspective. Polity, Croydon
Wyatt T (2009) Exploring the organization in Russia Far East's illegal wildlife trade: two case studies of the illegal fur and illegal falcon trades. Global Crime 10:144–154
Wyatt T (2011) Illegal raptor trade in the Russian Federation. Contemp Justice Rev 14:103–123
Wyatt T (2013a) Illegal wildlife trade: a deconstruction of the crime, the offenders and the victims. Palgrave, Basingstoke
Wyatt T (2013b) The local context of transnational wildlife trafficking: the Heathrow animal reception centre. In: Walters R, Westerhuis D, Wyatt T (eds) Emerging issues in green criminology: exploring power, justice and harm. Palgrave, Basingstoke, pp p128–137
WWF (2014a) Overview. Retrieved from http://www.worldwildlife.org/threats/deforestation
WWF (2014b) Living planet report 2014: species and spaces, people and places. Retrieved from http://wwf.panda.org/about_our_earth/all_publications/living_planet_report/
WWF/Dalberg (2012) Fighting illicit wildlife trafficking: a consultation with governments. WWF International, Gland
Zhang L, Ning HN, Sun S (2008) Wildlife trade, consumption and conservation awareness in southwest China. Biodivers Conserv 17:1493–1516
Zimmerman ME (2003) The black market for wildlife: combating transnational organized crime in the illegal wildlife trade. Vand J Transnat'l L 36:1657–1689

Part IX
Hidden Species: An Appropriate Scientific Approach to Cryptozoology

The addition of this rather peculiar part to the book stems from a belief that the issue of 'cryptic' species, or wildlife species the existence of which is unclear and/or which are often known only to local people, is very important and needs to be addressed in a serious and pragmatic manner. It is known that these situations can often seriously affect the quality of life of many types of people (Turner 2009) and that sizable funds are allocated, sometimes inappropriately, to promote and realise some of the expeditions to search for these unknown species (Hill et al. 2006).

Moreover, cryptozoology (Heuvelmans 1968, 1982) is largely considered a pseudoscience (*inter alia*, Loxton and Prothero 2013); however, it certainly could be considered a specific branch of zoology (Rossi 2012). However, this issue's ambiguous reputation has so far been earned due to the utterly uncontrolled and unjustified proliferation of books, informative articles, films and documentaries which have little or nothing to do with science and/or cryptozoology (e.g. Clark and Coleman, 1978).

The purpose of Rossi's (2016) chapter is precisely to try to show how the proper study of cryptozoology can contribute to scientific research, as has often been the case so far, particularly regarding the discovery of species which are unknown to science and whose existence is seemingly unlikely (Van Roosmalen and Van Roosmalen 2002).

References

Clark J, Coleman L (1978) Creatures of the outer space. Warner Books, New York

Heuvelmans B (1968) In the wake of the sea-serpent. Hill & Wang, New York

Heuvelmans B (1982) What is cryptozoology? Cryptozoology 1:1–12

Hill GE, Mennill DJ, Rolek BW, Hicks TL, Swiston KA (2006) Evidence suggesting that ivory-billed woodpeckers (*Campephilus principalis*) exist in Florida. Avian conservation and ecology 1(3):2. [online]. http://www.ace-eco.org/vol1/iss3/art2/

Loxton D, Prothero DR (2013) Abominable science!: origins of the Yeti, Nessie, and other famous cryptids. Columbia University Press, New York

Rossi L (2012) Criptozoologia—Animali misteriosi tra scienza e leggenda. Photocity, Pozzuoli
Rossi L (2016) A review of Cryptozoology: towards a scientific approach to the study of "hidden animals". In: Angelici FM (ed) Problematic wildlife—a cross-disciplinary approach. Springer, New York, p 573–588
Turner SS (2009) Negotiating nostalgia: the rhetoricity of thylacine representation in Tasmanian Tourism. Soc Anim 17:97–114
Van Roosmalen MGM, Van Roosmalen T (2002) The description of a new marmoset genus, *Callibella* (Callitrichinae, Primates), including its molecular phylogenetic status. Neotrop Primates 11:1–10

Chapter 26
A Review of Cryptozoology: Towards a Scientific Approach to the Study of "Hidden Animals"

Lorenzo Rossi

Introduction

"Cryptozoology" is a term that defines a branch of zoology that is generally considered a pseudoscience (Simpson 1984; Prothero 2007; Dubois and Nemésio 2007; Loxton and Prothero 2013) devoted to the study of animal species whose existence is not supported by empirical evidence, but rather hypothesized via indirect and uncertain information, including oral traditions, eyewitness accounts, and inconclusive physical evidence. Since its first appearance in the literature (Blancou 1959), both the word "cryptozoology" and its meaning have been the subject of heated discussions, so that hitherto a commonly accepted definition has yet to be found, and several authors have proposed a very personal vision of this discipline (Paxton 2002). The first person to use this term in a paper, with the aim to establish a new subdiscipline in the study of animal biology, was French-Belgian zoologist Bernard Heuvelmans (1965), universally known as the "Father of Cryptozoology". However, the first formal attempts to define cryptozoology and its methodology were only published several years later, between 1982 and 1998, in the peer-reviewed journal *Cryptozoology*, which, despite meeting all the requirements of a formal publication, suffered from low circulation. Several papers by Heuvelmans were also published in non-technical or non-English journals (e.g. Heuvelmans 1987a, b, 1997) and likely were little known not only to the majority of the academic world, but also within the small circle of the so-called "cryptozoological community". Although cryptozoology is considered a pseudoscience by several authors, other researchers argue against this interpretation, remarking on some aspects of this discipline (Raynal 1989; Naish 2001; Paxton 2002; Woodley et al. 2008; Rossi 2012). However, in this debate, little reference has been made to the existing scientific

L. Rossi (✉)
Associazione Orango, Via San Cristoforo, 196, 47522 Cesena, FC, Italy
e-mail: lorenzo@criptozoo.com

F.M. Angelici (ed.), *Problematic Wildlife*, DOI 10.1007/978-3-319-22246-2_26

literature on cryptozoology (e.g. Simpson 1984); and furthermore, not all the critics are adequately knowledgeable of it (Conway et al. 2013). The purpose of this chapter is to review the available literature in order to determine the history, definition, aims, and methodology of cryptozoology according to Heuvelmans' ideas, as well as the main criticisms of it. The epistemological aspects of cryptozoology will also be addressed in order to suggest how it may be included among the scientific disciplines, in addition to if and how cryptozoology could actually contribute to the discovery of new animal species and to biodiversity conservation.

Material and Methods

Reviewing the literature was not an easy process: a great deal of interesting information was found in "unorthodox sources" (e.g. magazines, blogs, grey literature, etc.) rather than in technical and scientific papers due to the fact that cryptozoology is, at best, a controversial topic. While perusing such sources is usually strongly discouraged by scientific journals, in this situation it becomes necessary due to the unique nature of the subject covered in this chapter. Furthermore, some significant criticisms have been published outside the peer-reviewed literature. In order to provide a synthesis of the status of cryptozoology as understood by Heuvelmans, and be both coherent and easily understandable, sources do not always appear in chronological order. In fact, Heuvelmans' writings cover a period of about sixty years, during which the author changed his views on cryptozoology and integrated several new ideas. The criticisms of cryptozoology have been subdivided into categories, and special attention has been given to how coherent they are with Heuvelmans' ideas.

Results

History, Beginnings, and Current Status of Cryptozoology

Even though several zoologists and biologists have, in years past, worked on so-called "mysterious animals" (e.g. Oudemans 1892; Krumbiegel 1950), the birth of modern cryptozoology is attributed to Scottish-born American naturalist Ivan T. Sanderson (1911–1973) and to French-Belgian zoologist Bernard Heuvelmans (1916–2001), who independently invented the term "cryptozoology" (Heuvelmans 1968). Sanderson became quite famous in the USA thanks to a series of radio and TV shows, and to several books and magazine articles devoted to zoology, but also to bizarre subjects such as UFOs and paranormal phenomena (Heuvelmans 1997; Conway et al. 2013). Reading one of these odd papers on the alleged survival of dinosaurs in Africa (Sanderson 1948) inspired Heuvelmans to dedicate his life to collecting information on animals that could potentially be discovered (Heuvelmans 1984). Heuvelmans undertook to tackle these alleged zoological mysteries in six books, originally only published in French (Heuvelmans 1955, 1958a, 1965, 1978, 1980; Heuvelmans and Porchnev 1974), raising a variety of diverging opinions from within the academic world (e.g. Johnson 1959; Reed 1959; Hedgpeth 1968). In particular, two of these books (Heuvelmans 1955, 1965) achieved great commercial success and were translated into several languages (e.g. Heuvelmans 1958b, c, 1968). Thus, cryptozoology enjoyed worldwide fame.

On January 8–9, 1982, the International Society of Cryptozoology (ISC) was founded in Washington, DC, at a meeting in the Department of Vertebrate Zoology, National Museum of Natural History, Smithsonian Institution, thanks to the vision of biochemist Roy P. Mackal (1925–2013) from the University of Chicago and agronomist J. Richard Greenwell (1942–2005) from the University of Arizona. The ISC aimed to reach the scientists who were interested in cryptozoology, yet had some reservations due to its controversial nature. Moreover, the ISC promoted cryptozoology as a subdiscipline of zoology and animal biology (Greenwell 1982). For this reason, the ISC published a peer-reviewed yearly journal (*Cryptozoology, The Interdisciplinary Journal of the International Society of Cryptozoology*) with papers, field reports and news, and a quarterly bulletin (*The ISC Newsletter*). Although the Society and its journal received a rather lukewarm response from the academic community (e.g. May 1984; Simpson 1984; Heuvelmans 1997), the journal continued to be published until 1998, when the ISC ceased its activities due to internal dispute and financial problems. One of the main sources of internal dispute was how the public perceived cryptozoology as a discipline. In fact, as pointed out by Arment (2004), in the 1970s and 1980s, several so-called "paranormal investigators" started to collect and publish numerous reports of mysterious animals whose apparently strange nature and features were explained with the supernatural. Heuvelmans (1997) proposed to separate these two different currents into cryptozoology (the "science of the hidden animals", which he conceived of as a branch of zoology) and "crypto-zoology" (meant to be a "hidden" or esoteric zoology not concerned with real, "flesh and blood" animals). Yet, over the years, the latter became the most common interpretation of cryptozoology among the general public, mostly thanks to the publication of commercially successful books on supernatural zoology (e.g. Clark and Coleman 1978; Keel 1970). On the other hand, hardly any further attempts were made to promote a more scientific approach to this field. As a matter of fact, aside from *Cryptozoology*, only three peer-reviewed journals have been devoted to this discipline: *The Cryptozoology Review* (1996–2004), *Kraken: Archives de Cryptozoologie* (2008–2011), and *The Journal of Cryptozoology* (2012–present). Therefore, cryptozoology has been given very little recognition by the academic world and is not considered a scientific discipline (e.g. Carroll 2003).

Cryptozoology According to Bernard Heuvelmans

Heuvelmans (1982) coined the term "cryptozoology" using three Greek words as roots: "Kryptos" (Κρυπτος) (hidden), "Zoon" (Ζoν) (animal), and "Logòs" (Λόγος) (discussion, i.e. science), hence defining it as the "science of hidden animals". According to Heuvelmans, these "hidden animals" are animals whose existence is unknown to science but not to the local people with whom they share a geographical area or animals about which we have some indirect knowledge (such as local stories, sightings, footprints, etc.), which is, however, insufficient to demonstrate their existence (Heuvelmans 1982). Heuvelmans defined and subdivided this indirect knowledge into "circumstantial evidence" and "witness evidence", borrowing these terms from the legal jargon of the forensic sciences (Wigmore 1935). Thus, in cryptozoology, a fundamental requirement is the existence of indirect evidence concerning an alleged, still-unknown to science, animal species that is defined as "ethnoknown" (Greenwell 1985).

Heuvelmans preferred to refer to the animals studied by cryptozoology as "hidden" rather than "unknown" because, in his opinion, this label included not only potential taxa that were not yet formally described, but also belonging to species considered extinct whose survival is hitherto unknown (Heuvelmans 1982). Later, he extended this definition to include populations of animals already known to science, but living in areas where their distribution has yet to be documented (Greenwell 1984; Heuvelmans 1988). Cryptozoology's main contribution to scientific research is to accelerate the completion of the planet's biodiversity inventory allowing new species, which could potentially already be endangered, to receive prompt legal protection as soon as possible (Heuvelmans 1982, 1988). In order to achieve this, once any information about an apparently still unknown animal has been obtained, the cryptozoologist must acquire as much information as possible about this taxon by thoroughly examining a broad range of sources (such as myth, folklore, history, and archaeology), collecting witness statements and analysing all the alleged indirect evidence of its existence (e.g. footprints, skin fragments, hair tufts, pictures, films, and so on) (Heuvelmans 1988). Any thus collected information must then be carefully screened and evaluated in order to discard any unreliable report or non-zoological explanations of the phenomenon. Subsequently, a sort of identikit of the animal is compiled to situate it in its correct taxonomic position with as much accuracy as is possible in order to increase the likelihood of discovering it in nature and, therefore, formally describing it (Heuvelmans 1982, 1988). According to Heuvelmans (1984), the necessity of cryptozoology is demonstrated by the fact that, in the history of zoological discovery, many animals—even large ones—would have been discovered earlier if a cryptozoological approach had been applied. For instance, the giant panda (*Ailuropoda melanoleuca*), formally discovered in 1890, had been described as *bei-shung* (white bear) in Chinese manuscripts as early as 621 A.D. (Morris and Morris 1966), and the West Indian or African coelacanth (*Latimeria chalumnae*), formally discovered in 1938, was already well-known as *kombessa* among the natives of the Comoros Islands (Smith 1953, 1956). When screening resources, Heuvelmans emphasizes the great importance of myths and legends because usually the rarest and lesser-known animals tend to become myths, thereby making it easier to transform them into something quite different from their original zoological status in oral and written tradition (Heuvelmans 1987b, 1990). For instance, some Chinese natural history treatises from the II century A.D. described the *fen-chu* a gigantic hairy rodent weighing about 600 kg and armed with two huge pickaxe-shaped teeth that lived in the area north of China (Siberia). This legendary beast was later identified as mammoths (*Mammuthus primigenius*). Their bodies had been exquisitely preserved in the permafrost and were occasionally found by indigenous Siberian hunters who traded commercially with the Chinese and contributed to creating this legend (Heuvelmans 2007). Another mysterious animal of Chinese folklore, the *mé*, was described as a cross between a bear, an elephant, and a rhinoceros, and proved to be an extreme mythification of the Malayan tapir (*Tapirus indicus*), discovered by Western scientists in 1816 (Heuvelmans 2007). Strong mythification may lead zoologists to believe that some evidence is nothing more than unsubstantiated legend; therefore, cryptozoology should also de-mythify this information by elaborating accurate scientific theories (Heuvelmans 1982). Furthermore, Heuvelmans points out that the possibility of scientifically describing a species even before the capture or collection of a specimen and the depositing of a holotype (in the form of a dead specimen) should be cryptozoology's long-term objective. This is why Heuvelmans hoped that the International Commission on Zoological Nomenclature would come to accept a separate naming system, such as the "parataxa" proposed in paleontology by Moore and Sylvester-Bradley (1957) for paleoichnological systematics, as traces are difficult to attribute to a given species. Finally, after the "hidden animal" is described, it would pass from cryptozoology into zoology proper (Heuvelmans 1982, 1984, 1997).

Criticism

Animals Studied in Cryptozoology Are not Scientifically Plausible

One of the first problems that the newly born ISC had to tackle was the application of a specific technical term that would be used in place of the vague and often out of context "hidden animal", or the all too misleading and decidedly inadequate "monster" (Greenwell 1983). Wall (1983), therefore, proposed to adopt the word "cryptid", which is still universally used today in cryptozoology. Heuvelmans (1986) compiled a first systematic checklist of cryptids (yet without using this term, referring instead to *apparently unknown animals with which cryptozoology is concerned*). On this list, which had 138 entries, he included all the alleged animal forms that appeared in his books, plus others he had learned of from other texts, newspapers, personal communications, and field studies from over more than 35 years of activity. One of the major criticisms of this checklist was that cryptozoology emphasized the search for large animals, to the detriment of the small ones, which does not make much sense in the natural sciences, given that animals with low body mass constitute the vast majority of the species in any ecosystem (van Valen 1983). Groves (1984) and Simpson (1984) argued that the checklist included mainly large terrestrial mammals, judging new species of large mammals very unlikely to be discovered. Later, Loxton and Prothero (2013) emphasized, according to the proposed checklist and several works from "authorities" in the field (e.g. Krantz 1999; Mackal 1976, 1980, 1987), cryptozoology seems to focus almost exclusively on very unlikely creatures from a biological perspective, such as Bigfoot, the Yeti, the Monster of Loch Ness, the Mokele Mbembe, and so on, notwithstanding the fact that any search for unconfirmed animals should by definition belong to cryptozoology itself. Yet, another problem concerning cryptozoological entities is their alleged range of distribution: according to Loxton and Prothero (2013) and Groves (1984), it is in fact entirely possible that future discoveries will and must come from either very scarcely populated areas or museum collections, while some of the most famous cryptids seem to live in areas where human presence is quite strong. Hence, if they were real, then they should have been discovered long before.

Cryptozoological Nomenclature Is not Applicable

van Valen (1983) criticized the possibility of describing a species only on theoretical grounds because the holotype, as Valen pointed out, is the only objective data that can demonstrate the real existence of any organism. Happel (1983) also criticized this point, stating that the possibility of describing a species before it is actually discovered is absolutely irrelevant to the scope of cryptozoology. According to Pauwels and Chérot (1997), cryptozoology was born of a misunderstanding by Heuvelmans: he believed that the International Code of Zoological Nomenclature necessarily called for the registration of the more or less complete remains of an animal in order to make it possible to arrive at a scientific description; but there are no clear rules for this (e.g. Wakeham-Dawson et al. 2002; Polaszek et al. 2005; Donegan 2008). Dubois and Nemésio (2007) indicate that, while the ICZN does not require holotypes, in the form of a preserved specimen of the species in question (onomatophores) to be registered, it also categorically rules out all "hypothetical concepts", i.e. all those animals whose existence—past or present—is formally known only to the mind of the author, whether it is a prediction or not. According to these authors, cryptids are a perfect example of "hypothetical concepts" and add that if registering a holotype as an onomatophore was clearly one of the ICZN's rules, then Latin names proposed only through eyewitness descriptions, footprints, or pictures to describe cryptozoological animals would cease to present any philosophical problems regarding their validity.

Other Pseudo-scientific Aspects

According to Loxton and Prothero (2013), cryptozoology should be considered a pseudo-science because it promotes statements that seem scientific but are not actually guided by the scientific method of verification and falsification of hypothesis. Furthermore, often the so-called cryptozoologists tend to hold on to their ideas even when there is strong evidence against their case and attempt to use ad hoc hypotheses to avoid admitting their errors. Conway et al. (2013) point out that the majority of cryptozoological hypotheses and evaluations have been published in popular literature, thus avoiding peer-review, and that so far there is still no convincing evidence as to the existence of the most famous cryptids. These authors also criticize the excessively literal interpretations used by cryptozoologists when working on cryptids identified mainly in folklore (e.g. lake monsters, African dinosaurs, etc.). As a matter of fact, in many cases the interdisciplinary approach seems to be exclusively used to credit zoological interpretations of a given phenomenon, ignoring other possible causes (Meurger and Gagnon 1998). Even Groves (1984) supports this view, suggesting that cryptozoologists should first ask themselves if a given cryptid could truly exist rather than first asking from which animal a given legend could stem. Yet another criticism shared by several authors is that cryptozoology largely uses eyewitness accounts that, due to their inherent unreliability, cannot be considered valid data (Loxton and Prothero 2013; Shermer 1997, 2003). According to Mckinney (2013), standard zoology already searches for new animal species, thus there is no need to create a separate discipline. Nor can cryptozoology be classified as a scientific discipline because it does not address problems that have not already been dealt with by other known disciplines. Therefore, it is unnecessary and for this reason should be considered a pseudo-science. Even Loxton and Prothero (2013) highlight that the discovery of species such as the coelacanth, okapi (*Okapia johnstoni*), and Komodo monitor lizard (*Varanus komodoensis*), often offered by cryptozoologists as examples of the validity of cryptozoology, actually have no value at all as they all belong to the field of standard zoology. Finally, according to Naish (2007), the overlapping of cryptozoology and zoology is so broad that one wonders if cryptozoology actually exists at all.

Discussion

Animals Studied in Cryptozoology Are not Scientifically Plausible

According to Paxton (2011), even though the zoological community does not use the term "cryptid", it does not mean that it cannot be used, as it was once formally defined. Unfortunately, this is a problem with no simple solution. According to Heuvelmans, if an animal belongs to a potentially unknown species, to a surviving form of a species that is considered extinct, or to a known species living outside its recognized area of distribution, and there is indirect evidence of said animal, then this organism should be situated within the field of cryptozoology, and thus can be considered a cryptid. On the other hand, Loxton and Prothero's 2013 criticism is essentially correct: cryptozoology mainly emphasizes entities whose biology and ethology strongly clash with current scientific knowledge and whose existence is not supported at all. Upon examination of the 119 papers and field reports published in the 13 volumes of *Cryptozoology*, the manuscripts can be subdivided as follows:

- 12 Concerning technical definitions and proposals relating to cryptozoology
- 39 Concerning so-called "relict hominids" (Bigfoot, Yeti, Yowie, etc.)
- 32 Concerning lake monsters (Nessie, Champ, etc.) and sea monsters ("Caddy" and giant octopuses)
- 7 Concerning the possible survival of species considered extinct (3 of which are on Mokele Mbembe, an alleged surviving sauropod dinosaur from central Africa)

- 16 Concerning the existence of possible new, yet undiscovered, species
- 4 Concerning animals with new possible ranges
- 6 Concerning folklore and ethnozoology
- 3 Concerning subjects not necessarily connected to cryptozoology, such as the possible cloning of extinct species from their DNA

Thus, it is absolutely impossible to consider cryptozoology a science as long as it continues to concern itself with impossible creatures. However, although cryptozoological literature seems to be full of bizarre creatures, this is not because there is no valid method to distinguish between the possible and the unreal, but only because, in my opinion, this method has rarely been used—paradoxically, even by its inventor. In fact, Heuvelmans (1987a) points out that if a hitherto unclassified animal form was described by a traveller or native, this would not justify it being studied by a cryptozoologist, as a cryptozoologist must be called upon if there is a "certain implicit plausibility" and if it is "coherent with the most advanced scientific knowledge of our time". If this evaluation method was applied correctly, cryptids such as Bigfoot, lake monsters, and surviving dinosaurs would be utterly excluded from cryptozoology. As for criticisms of the size of cryptids, the majority having a large body size, Heuvelmans never considered size discriminatory. However, since cryptozoology is based on eyewitness reports and local traditions, cryptids generally should have a size that makes it possible to observe them (Heuvelmans 1983). However, small-sized species already locally known before their official discovery clearly demonstrate that testimonial and circumstantial evidence can be used to find "new" animals, notwithstanding body size. For instance, five different new species of New World monkeys whose body weight ranges from 150 to 1200 g (*Callibella humilis, Callithrix manicorensis, Callithrix acariensis, Callicebus bernhardi,* and *Callicebus stephennashi*) were discovered and described by zoologist Marcus van Roosmalen and his co-workers thanks to indications provided by local people, who told van Roosmalen that in different areas similar monkeys exhibited different colours (van Roosmalen et al. 1998, 2000, 2002; van Roosmalen and van Roosmalen 2002; van Roosmalen 2014). Finally, although the discovery of mid- and large-body-sized animals is statistically less likely than that of small-sized ones, and even if we narrow our scope to include only the most recently discovered terrestrial forms, we find the discovery in 2010 of the Northern Sierra Madre forest monitor (*Varanus bitatawa*), a 2-m long lizard, and in 2013, the lowland tapir (*Tapirus kobomani*), weighing 110 kg (Welton et al. 2010; Cozzuol et al. 2013). It should also be pointed out that the latter taxon had already been described by van Roosmalen (van Roosmalen and van Hooft 2013; van Roosmalen 2014) under the name *Tapirus pygmaeus* after its discovery thanks to the indications of local natives.

Cryptozoological Nomenclature Is not Applicable

If, on the one hand, the ICZN does not clearly state that dead type-specimens must be registered as holotypes in order to assign a scientific name to a species (Donegan 2008), attempts to describe cryptozoological species are usually extremely disappointing (Dubois and Nemésio 2007). Textbook cases are: Scott and Rines's description of the Loch Ness Monster (1976) based on alleged underwater pictures of flipper-like appendages that later were revealed to be heavily retouched (Binns 1983; Campbell 1996; Shine 2006); or the famous "Minnesota Iceman", the alleged corpse of an unknown hominid found frozen in a block of ice and exhibited in sideshows around the USA which was judged real by Heuvelmans (1969). Heuvelmans considered it real and described it as *Homo pongoides* via a visual and photographic analysis, but later it proved to be a masterfully produced mannequin (West 2011). However, perhaps the most emblematic case of the risks of describing a cryptid based solely on pictures and testimonial evidence is possibly the so-called "Cadborosaurus", the alleged sea serpent of British Columbia. Mainly based on old pictures of an apparently unusual carcass found in 1937 in the stomach of a sperm whale, LeBlond and Bousfiled (1995) and Bousfiled and LeBlond (1995) described *Cadborosaurus willsi* as a surviving form of sauropterygian plesiosaur. Using the same pictures and testimonial accounts of alleged sightings, Saggese (2009) later offered a different interpretation of the creature, making reference to a highly specialized Sirenian closely related to Steller's sea cow (*Hydrodamalis gigas*), which he named *Cadborotherium willsi.* Setting aside the highly questionable method used by these authors in their tentative description of the alleged sea serpent (see below for a further discussion of this case), it is clear that, from the same sources, two different genera belonging to two different classes are described, proving that the evidence itself is ambiguous at best and cannot be accepted as the basis of a scientific description worthy of its name. As a matter of fact, Donegan (2008) reports several cases of species that have been described without registering onomatophores, yet the quality of the indirect evidence used in these descriptions (clear pictures and observations made by scientists in the field) cannot be compared to that used in the majority of cryptozoological cases (blurred pictures, grainy video, confused descriptions made by alleged eyewitness, etc.). The debate on establishing clearer ICZN guidelines is beyond the scope of this chapter. However, if there is currently sufficient evidence to describe a species, then it can be described, and, in my opinion, a "parallel" nomenclature would only create greater confusion and the result would be void of any scientific criteria. Therefore, I believe that the description of a species on very poor evidence, or perhaps even before its actual discovery, would be none other than an exercise in style that facilitates the demonstration of the existence of certain authors to science more than the demonstration of the existence of certain species—in addition to being completely irrelevant to cryptozoology. Woodley (2011) suggests using an independent classification system in cryptozoology based on the concept of *aequivotaxa* (from the Latin "aequivocus", or ambiguous, uncertain), where "hypothetical concepts" can find a place based on given requirements, such as:

1. The "aequivotaxon" must be supported by a holotype in the form of a detailed description, pictures, images, audio recordings, biological samples, or any other evidence.
2. The description of the "aequivotaxon" must not contain attempts to deduce its biological affinities.
3. The proposed name must be different from the one used to christen the species in case of an effective future discovery of the examined cryptid

According to Woodley (2011), funding an International Commission on Aequivological Nomenclature would be the most important step towards the institution of cryptozoology as a formal discipline. Yet, I think that the application of the scientific method would be even more important so as to avoid "hypothetical concepts" and better safeguard the image of cryptozoology. Therefore, I am unable to support Woodley's unnecessary proposal. In fact, I am of the opinion that such classification, albeit well elaborate and clever, would not be used by professional zoologists and would only confine cryptozoology to a sphere of research that is cut off from the recognized biological sciences. My proposal of a method that would allow for the study of cryptids without needing to provide a possibly premature "scientific" description consists of improving on the checklist method introduced by Heuvelmans (1986). By applying the most up-to-date zoological knowledge, not to mention a good dose of common sense, such a checklist would allow to rule out unreal cryptids (such as "a 30 m long anaconda", "lacustrine plesiosaurs", and so on) and coherently and systematically group together all the reports of potential new *taxa* and sightings of species considered extinct that have been collected by field zoologists and biologists during their research. Such reports often risk suffering from limited circulation because scientific zoological journals are not generally interested in publishing material that may be solely theoretical and devoid of any results. Such a checklist could be regularly published on official media created for this specific purpose and could prove to be of great help to biodiversity conservation: the formal discovery of new species could be helped by knowing that this potentially new species has already been reported in a given area, in the event that such an animal actually existed; or researchers could find out that these reports regard a species that is already known and thus that cryptid could be struck off the list and zoologists (and their colleagues) could focus their research on more potentially interesting cases. Moreover, if such a checklist had official status, it would help circumvent ethical problems, such as in the description of the lowland tapir, encouraging any researcher who has already collected and reported evidence on the discovery of a new *taxon* to be included, or at least cited, in the description.

Other Pseudoscientific Aspects

Criticism which considers cryptozoological theories pseudoscientific is certainly not without grounds. The use of ad hoc theories and data manipulation which only take into consideration that which furthers these theories, ignoring that which is contrary to the given hypothesis is, unfortunately, common practice in cryptozoology.

One remarkable example is the description of nine different species of large unknown marine animals based on the analysis of alleged sightings (Heuvelmans 1965). According to Heuvelmans, this analysis was conducted with scientific rigour and allowed him to discover how the apparently unrelated and confused reports of so-called "sea serpents" were actually obscuring a logical and coherent picture because the description of the sighted creature was strictly connected to the sighting area. Heuvelmans (1965) stated that these nine kinds of sea serpents would have occupied different ranges and niches, thus being biologically believable. However, in Magin's (1996) critical examination of his work, the author pointed out that several sightings considered valid by Heuvelmans were actually hoaxes, and that Heuvelmans himself had piloted the data in order to substantiate his theories. For instance, given that a specific ocean area was inhabited by long-necked sea serpents, any sighting coming from that area was automatically included in the category "long-necked", even though in the description, the neck of the animal was either not mentioned or reported as "not long" (e.g. Heuvelmans 1968, pp. 286, 360, 412, 580–582). Another good example comes from the already mentioned "formal" descriptions of the *Cadborosaurus*. In fact, although witnesses have often described a sort of mane along the neck of the animal, Bousfiled and LeBlond (1995) are not inclined to consider this feature in their description of the creature. On the other hand, Saggese (2009) includes the mane, yet chooses to ignore the many reports of horny projections on the head of the animal, considering them misinterpretations and oversights. Hence, both Bousfiled and LeBlond (1995) and Saggese (2009) were eager to exclude the features that would not fit into their own taxonomical hypothesis on the nature of the alleged cryptid (for a plesiosaur, a mane, and horns for a sirenian). Yet, despite the fact that many claims in cryptozoology are not scientific, this does not imply that the method itself is not scientific (Paxton 2002). In my opinion, the fundamental ideas forming the basis of cryptozoology are empirically demonstrable:

- The inventory of our planet's fauna is largely incomplete. Not only new species are discovered every year, but mathematical models have been developed to estimate the number of potential species yet to be discovered (e.g. Giam et al. 2012).
- It is possible to discover new species previously known only through circumstantial and testimonial evidence, especially those provided by the local population (e.g. Sheil and Lawrance 2004; Cozzuol et al. 2013).
- Species considered extinct may have survived and may be rediscovered. Sheffers et al. (2011) have reported 351 such cases in the last 122 years.
- Investigating circumstantial evidence may accelerate the process of discovery and description of potentially new species (Rossi 2011), as demonstrated by van Roosmalen et al. (1998, 2000, 2002).
- Some real animals may be mythified beyond recognition. For instance, while investigating bizarre legends about a monkey who sneezed on rainy days because water drops got into its nose, Geissmann et al. (2010) discovered and described a new species of the genus *Rhinopithecus* (*R. strykeri*) in Myanmar.

As for the criticism of eyewitness testimony, while it is true that this sort of evidence is the kind that is most often used in the majority of the pseudosciences (such as parapsychology, ufology, and several "alternative medicine" practices) and can be unreliable (see Polidoro 2006; Wiseman 2011), it is also true that a great deal of rare natural phenomena has initially been documented only due to eyewitness testimony (Paxton 2009). Eyewitness testimony is widely used in zoology both to assess the extinction of a species (e.g. Hume et al. 2004; Black et al. 2013) or its persistence (e.g. Boyd and Stanfield 1998) and to report new species that have yet to be formally described (e.g. Pitman et al. 1987). The main difference in cryptozoology is in the data elaboration that, differently from what happens in zoology, generally does not undergo any scientific control and refereeing. But it is interesting to note that peer-reviewed journals do not discard cryptozoological contributions a priori as long as they satisfy the criteria for scientific publication (Paxton 2011). Cryptozoology may be taken more seriously by the academic world if cryptozoologists start to apply the scientific method more consistently. However, if cryptozoology must rid itself of what public opinion deems its "symbolic animals" (Bigfoot, lake monsters, surviving dinosaurs, and so on) in order to become a scientific discipline proper, then we must ask ourselves whether—once these beings have been set aside—this field of study will become redundant with zoology. According to McKinney (2013), "each and every animal" currently known to science has been previously known only through vague description; yet this does not at all imply that the whole history of zoology should be included in cryptozoology. It is also worth noting that, according to Heuvelmans (1984), at least until the end of the eighteenth century, zoology had no need at all for cryptozoology because the systematic search for species of unknown status was the norm for naturalists of that time. Yet, strictly speaking, there appears to be no impediment to instituting cryptozoology as a science, as it is not redundant with zoology. First of all, McKinney (2013) seems to ignore that stating that "each and every animal" has been ethnoknown before its discovery and description is erroneous. Investigation to demonstrate the existence of a species is but one of the methods zoology may use to collect useful data and samples. For instance, one of the most commonly employed methods in field research uses area-specific surveys and traplines which randomly inventory species at a given location (Arment 2004). Furthermore, a new species may be unexpectedly collected purely by chance, as in the case of the megamouth shark (*Megachasma pelagios*), accidentally caught off the Hawaiian Islands in 1976 (Taylor et al. 1983). Therefore, it is entirely possible for a zoologist to encounter a new species, not previously ethnoknown, and not have any information about it before its official discovery. Arment (2004) defines cryptozoology as a "targeted-search methodology for zoological discovery"; yet even in this case one could ask what the differences are between this method and the zoological method. However, this problem ceases to exist if cryptozoology is considered a branch, or subdiscipline, of zoology. For instance, even recognized disciplines such as mammalogy, ichthyology, or herpetology, do not address problems that have not already been addressed by zoology—yet no one doubts their scientific value. Such disciplines may be considered specializations of zoology and may be further subdivided into additional subdisciplines (e.g. primatology and cetology for mammalogy ophidiology for herpetology, and so on). The point is that, if the methodology of any given discipline renders correct results, then its acceptance as a scientific subject is sanctioned by sheer conviction. For instance, only in recent years has bathrachology (the study of amphibians) been supported as a distinct discipline, yet this does not mean that before then zoology did not include the study of amphibians. Amphibians were, however, studied by herpetologists merely because of historical tradition, but in 1982 a group of French zoologists founded the first bathrachological society in the world in Paris and published *Alytes,* the first journal completely devoted to this new discipline. Mammalogy, ornithology, ichthyology, and bathrachology are specializations within zoology that are based on the precise taxonomic status of their subjects of study, while palaeontology—based on the remains and traces of extinct beings—is based on the temporal status of their subjects. Therefore, there is no reason cryptozoology should not be based on the cognitive status of the *taxa* and considered the branch of zoology that studies and researches *taxa* whose possible existence is initially based only on circumstantial and testimonial evidence. However, in my opinion, the only way to achieve this status would be if zoologists with a shared perception of this concept of "scientific cryptozoology" founded a new society and published a new journal devoted to the aforementioned topics. Given time, this would prevent anyone dabbling in "mysterious animals", such as lake monsters, chupacabras, or mothmen, from being called "cryptozoologists". Only in this way could and would cryptozoology avoid the often embarrassing "image problems" that have affected it since its birth as a field. As a matter of fact, a great deal of interesting information collected by professional zoologists risks falling into oblivion due to fear of it being associated with a pseudo-scientific discipline. For instance, in a paper dedicated to the possible recent survival of the pigmy hippopotamus and giant lemurs in Madagascar, deduced by the local people's description of animals called respectively "Kilopilopitsofy" and "Kidoky", Burney and Ramilisonina (1999) report that they have been reluctant to publish their work for fear of it being associated with cryptozoology.

Acknowledgments For their invaluable help, I wish to thank: Francesco Maria Angelici and Matt Bille for their advice, Marco Signore and Mauro Cella for their suggestions and translation, Dario Marcello Soldan, Charles Paxton, and Ulrich Magin for some papers I needed and was not able to find. I thank the two anonymous referees for their careful reading of my manuscript and their insightful comments and suggestions.

References

Arment C (2004) Cryptozoology: science & speculation. Coachwhip, Landisville

Binns R (1983) The Loch Ness mystery solved. Rigby, Adelaide

Black SA, Fellous A, Yamaguchi N, Roberts DL (2013) Examining the extinction of the barbary lion and its implications for felid conservation. PLoS One 8(4):e60174. doi:10.1371/journal.pone.0060174

Blancou L (1959) Géographie cynégétique du monde. Presses Universitaires de France, Paris

Bousfiled EL, LeBlond PH (1995) An account of *Cadborosaurus willsi*, new genus, new species, a large aquatic reptile from the Pacific coast of North America. Amphipacifica 1:1–25

Boyd IJ, Stanfield MP (1998) Circumstantial evidence for the presence of monk seals in the West Indies. Oryx 32:310–316

Burney DA, Ramilisonina (1999) The Kilopilopitsofy, Kidoky, and Bokyboky: accounts of strange animals from Belo-sur-Mer, Madagascar, and the megafaunal "extinction window". Am Anthropol 100:957–966

Campbell S (1996) The loch ness monster the evidence. BPC-AUP Aberdeen, London

Carroll RT (2003) The skeptic's dictionary: a collection of strange beliefs, amusing deceptions, and dangerous delusions. Wiley, New York

Clark J, Coleman L (1978) Creatures of the outer space. Warner Books, New York
Conway J, Koseman CM, Naish D (2013) Cryptozoologicon, vol 1. Irregular Books
Cozzuol MA, Clozato CL, Holanda EC, Rodrigues FH, Nienow S, de Thoisy B, Redondo RAF, Santos FR (2013) A new species of tapir from the Amazon. J Mammal 94:1331–1345
Donegan TM (2008) New species and subspecies descriptions do not and should not always require a dead type specimen. Zootaxa 1761:37–48
Dubois A, Nemésio A (2007) Does nomenclatural availability of nomina of new species or subspecies require the deposition of vouchers in collections? Zootaxa 1409:1–22
Geissmann T, Lwin G, Aung S, Naing Aung T, Aung ZM, Hla T, Grindley M, Momberg F (2010) A new species of snub-nosed monkey, Genus *Rhinopithecus* Milne-Edwards, 1872 (Primates, Colobianae), from Northern Kachin State, Northeastern Myanmar. Am J Primatol. doi:10.1002/ajp.20894
Giam X, Scheffers BR, Sodhi NS, Wilcove DS, Ceballos G, Ehrlich PR (2012) Reservoirs of richness: least disturbed tropical forests are centres of undescribed species diversity. Proc Biol Sci 279(1726):67–76. doi:10.1098/rspb.2011.0433
Greenwell R (1982) Formation of the society. ISC Newslet 1(1):1–3
Greenwell R (1983) Interview. ISC Newslet 2(1):1–5
Greenwell R (1984) Interview. The father of cryptozoology gives his views on many matters. ISC Newslet 3(3):1–6
Greenwell R (1985) A classificatory system for cryptozoology. Cryptozoology 4:1–14
Groves CP (1984) But how many large, terrestrial animal species remain to be discovered? Cryptozoology 3:111–115
Happel D (1983) Parataxa and hypothetical concepts—their irrelevance to cryptozoology. Cryptozoology 2:147–154
Hedgpeth JW (1968) Elusive specimens. Science 162(3855):787–788
Heuvelmans B (1955) Sur la piste des bêtes ignorées. Plon, Paris
Heuvelmans B (1958a) Dans le sillage des monstres marins: le kraken et le poulpe colossal. Plon, Paris
Heuvelmans B (1958b) On the Track of Unknown animals. Rupert Hart-Davis, London
Heuvelmans B (1958c) Tras la pista de los animales desconoscidos (I y II). Luis de Caralt, Barcelona
Heuvelmans B (1965) Le grand serpent-de-mer. Plon, Paris
Heuvelmans B (1968) In the wake of the sea-serpent. Hill & Wang, New York
Heuvelmans B (1969) Note préliminaire sur un spécimen conservé dans le glace, d'une forme encore inconnue d'Hominidé vivant: *Homo pongoides* (sp. Seu subsp. Nov.). Bullettin de l'Institut Royal des Sciences Naturelles de Belgique 45:1–24
Heuvelmans B (1978) Les derniers dragons d'Afrique. Plon, Paris
Heuvelmans B (1980) Les bêtes humaines d'Afrique. Plon, Paris
Heuvelmans B (1982) What is cryptozoology? Cryptozoology 1:1–12
Heuvelmans B (1983) How many animal species remain to be discovered? Cryptozoology 2:1–24
Heuvelmans B (1984) The birth and early history of cryptozoology. Cryptozoology 3:1–30
Heuvelmans B (1986) Annotated checklist of apparently unknown animals with which cryptozoology is concerned. Cryptozoology 5:1–26
Heuvelmans B (1987a) La criptozoologia: cosa è cosa non è. Abstracta 12:68–75
Heuvelmans B (1987b) La metamorfosi degli animali sconosciuti in bestie favolose. Abstracta 18:68–75
Heuvelmans B (1988) The sources and method of cryptozoological research. Cryptozoology 7:1–21
Heuvelmans B (1990) The metamorphosis of unknown animals into fabulous beasts and of fabulous beasts into known animals. Cryptozoology 9:1–12
Heuvelmans B (1997) Histoire de la Cryptozoologie—Seconde partie essor et officialisation de la Cryptozoologie. Criptozoologia 3:23–40
Heuvelmans B (2007) The natural history of hidden animals. Kegan Paul, London
Heuvelmans B, Porchnev BF (1974) L'homme de Néanderthal est toujours vivant. Plon, Paris

Hume J, Martill DM, Dewdney C (2004) Dutch diaries and the demise of the dodo. Nature. doi:10.1038/nature02688
Johnson DH (1959) On the track of unknown animals. Science 130(3384):1245–1246
Keel JA (1970) Strange creatures from time and space. Fawcett, Minnesota
Krantz GS (1999) Bigfoot Sasquatch evidence. Hancock House, Surrey
Krumbiegel I (1950) Von neuen und unentdeckten Tierarten. Kosmos, Stuttgart
LeBlond PH, Bousfiled EL (1995) Cadborosaurus survivor from the deep. Horsdal & Schubart, Victoria
Loxton D, Prothero DR (2013) Abominable science!: origins of the Yeti, Nessie, and other famous cryptids. Columbia University Press, New York
Mackal RP (1976) The monsters of loch ness. Swallow Press, Chicago
Mackal RP (1980) Searching for hidden animals. Garden City, New York
Mackal RP (1987) A living dinosaur? In search of Mokele-Mbembe. E.J. Brill, New York
Magin U (1996) St George without a dragon. Fortean Studies 4:223–234
May RM (1984) Science journals: cryptozoology. Nature 307:687
McKinney (2013) In response to Darren Naish: why cryptozoology is pseudoscience. Available at http://skeptoid.com/blog/2013/01/08/in-response-to-darren-naish-why-cryptozoology-is-pseudoscience/. Accessed 14 Jan 2015
Meurger M, Gagnon C (1998) Lake monster traditions: a Cross-cultural analysis. Fortean Tomes, London
Moore RC, Sylvester-Bradley PC (1957) Zoological nomenclature: proposed addition to the "Règles" of provisions recognizing and regulating the nomenclature of "Parataxa". J Paleontol 31:1180–1183
Morris R, Morris D (1966) Men and pandas. Hutchinson, London
Naish D (2001) Sea serpents, seals and coelacanths: an attempt at a holistic approach to the identity of large aquatic cryptids. In: Simmons I, Quin M (eds) Fortean studies, vol 7. John Brown, London, pp 75–94
Naish D (2007) Monster hunting? Well, no. No. Available at http://scienceblogs.com/tetrapodzoology/2007/10/10/monster-hunting-well-no-no/ . Accessed 14 Jan 2015
Oudemans AC (1892) The great sea-serpent. An historical and critical treatise. Luzac, London
Pauwels O, Chérot F (1997) Cryptoherpétologie et nomenclature: le problème et sa solution. Bulletin de la Société herpetologique de France 82–83: 41–49 [in French]
Paxton C (2002) In search of monster? A defence to cryptozoology. The Skeptic 15(3):10–14
Paxton C (2009) The plural of 'anecdote' can be 'data': statistical analysis of viewing distances in reports of unidentified large marine animals 1758-2000. J Zool 279:381–387
Paxton C (2011) Putting the "ology" into cryptozoology. Biofortean Notes 1:7–20
Pitman RL, Aguayo AL, Urbán RJ (1987) Observations of an unidentified beaked whale (*Mesoplodon* sp.) in the eastern tropical Pacific. Mar Mamm Sci 3:245–352
Polaszek A, Grubb P, Groves C, Ehardt CL, Butynski TM (2005) What constitutes a proper description: response. Science 309:2164–2166
Polidoro M (2006) Il sesto senso. Piemme, Casale Monferrato
Prothero DR (2007) Evolution: what the fossils say and why it matters. Columbia University Press, New York
Raynal M (1989) Cryptozoology: science or pseudoscienze? Cryptozoology 8:98–102
Reed IC (1959) On the track of unknown animals by Bernard Heuvelmans. Am Sci 47:378
Rossi L (2011) The role of circumstantial evidence in the discovery and description of new species of Primates since 2000 and conservational implication. Paper presented at the 2nd International congress problematic wildlife—conservation and management, Università della Tuscia, Genazzano, Rome 2–5, February 2011
Rossi L (2012) Criptozoologia—Animali misteriosi tra scienza e leggenda. Photocity, Pozzuoli
Saggese P (2009) *Cadborosaurus willsi*: indagine attributiva. In: Saggese P, Mosca M (eds) All'ombra dei falsi mostri. Torino, Ananke, pp 185–209
Sanderson IT (1948) There could be dinosaurs. Saturday Evening Post (January 3)
Scott P, Rines R (1976) Naming the Loch Ness monster. Nature 258:466–468

Sheffers BR, Yong DL, Harris JBC, Giam X, Sodhi NS (2011) The world's rediscovered species: back from the brink? PLoS One 6, e22531. doi:10.1371/journal.pone.0022531
Sheil D, Lawrance A (2004) Tropical biologists, local people and conservation: new opportunities for conservation. Trends Ecol Evol 19:634–638
Shermer M (1997) Why people believe weird things. W. H. Freeman, New York
Shermer M (2003) Show me the body. Sci Am 288:27
Shine A (2006) Loch Ness. Loch Ness Project, Drumnadrochit
Simpson GG (1984) Mammals and cryptozoology. Proc Am Philos Soc 128:1–19
Smith JLB (1953) The second coelacanth. Nature 171:99–101
Smith JLB (1956) Old fourlegs. The story of the coelacanth. Longmans, London
Taylor RT, Compagno LJV, Struhsaker PJ (1983) Megamouth a new species, genus and family of lamnoid sharks, Megachasma pelagios (Family Megachasmidae), from the Hawaiian Islands. Proc Cal Acad Sci 43(8):87–110
van Roosmalen MGM (2014) *Tapirus pygmaeus* Van Roosmalen & Van Hooft in Van Roosmalen, 2013 (Mammalia, Perissodactyla, Tapiridae): proposed confirmation of availability of the specific name and of the book in which this nominal species was proposed. Bull Zool Nomencl 71:84–87
van Roosmalen MGM, van Hooft P (2013) New species of living tapir, the dwarf tapir (*Mammalia: Tapiridae*) from the Brazilian Amazon. In: van Roosmalen MGM (ed) Barefoot through the Amazon—on the path of evolution. CreateSpace, North Charleston, pp 400–404
van Roosmalen MGM, van Roosmalen T (2002) The description of a new marmoset genus, *Callibella* (Callitrichinae, Primates), including its molecular phylogenetic status. Neotrop Primates 11:1–10
van Roosmalen MGM, van Roosmalen T, Mittermeier RA, Fonseca GAB (1998) A new and distinctive species of marmoset (Callitrichidae, Primates) from the lower Rio Aripuanã, state of Amazonas, central Brasilian Amazonia. Goeldiana Zoologia 22:1–27
van Roosmalen MGM, van Roosmalen T, Mittermeier RA, Rylands AB (2000) Two new species of marmoset, genus *Callithrix* Erxleben, 1777 (Callitrichidae, Primates), from the Tapajós/Madeira interfluvium, south central Amazonia, Brazil. Neotrop Primates 8:1–18
van Roosmalen MGM, van Roosmalen T, Mittermeier RA (2002) A taxonomic review of the titi monkeys, genus *Callicebus* Thomas, 1903, with the description of two new species, *Callicebus bernhardi and Callicebus stephennashi*. Neotrop Primates 10:1–52
van Valen LM (1983) Cryptozoology, paleontology, and evidence. Cryptozoology 2:155–157
Wakeham-Dawson A, Morris S, Tubbs P, Dalebout ML, Baker CS (2002) Type specimens: dead or alive? Bull Zool Nomencl 59(4):282–286
Wall JE (1983) Cryptoletters. ISC Newslet 2(2):10
Welton LJ, Siler CD, Bennett D, Diesmos A, Duya MR, Dugay R, Rico ELB, van Weerd M, Brown RM (2010) A spectacular new Philippine monitor lizard reveals a hidden biogeographic boundary and a novel flagship species for conservation. Biol Lett. doi:10.1098/rsbl.2010.0119
West R (2011) Pickled punks & girlie shows. Schiffer, Atglen
Wigmore JH (1935) Code of the rules of evidence in trials at law. Little, Brown, Boston
Wiseman R (2011) Paranormality: why we see what isn't there. Pan Macmillan, London
Woodley M (2011) Introducing Aequivotaxa: a new classificatory system for cryptozoology. Kraken: Archives de Cryptozoologie 3:63–85
Woodley MA, Naish D, Shanahan HP (2008) How many extant pinniped species remain to be described? Hist Biol 20:225–235

ERRATUM TO

Chapter 2
Mammal Species Extinction and Decline: Some Current and Past Case Studies of the Detrimental Influence of Man

Katarzyna Daleszczyk, Amy E. Eycott, and Jörg E. Tillmann

F.M. Angelici (ed.), *Problematic Wildlife*, DOI 10.1007/978-3-319-22246-2_2

DOI 10.1007/978-3-319-22246-2_27

The following table was not included in Chapter 2 of this book.

The updated online version of the original chapter can be found under http://dx.doi.org/10.1007/978-3-319-22246-2_2

Table 2.1 Some past and present examples of species suffering from various anthropogenic threats. If habitat loss/degradation/fragmentation was caused partially by agriculture or logging activities, it is indicated in appropriate column. 'Human population growth' includes encroachment and demand for land, development of settlements and infrastructure as roads or dams, mining, etc. 'Domestic animals' include competition, altering habitat, and diseases and parasites transmitted. 'Introduced species' include competition, predation, and exotic diseases. 'Other' include direct conflict with people, capture as pets/ for trade, war/ civil unrest, land-use preventing gene exchange, pollution, use of rodenticides, etc. Based on data from IUCN website http://www.iucnredlist.org

Species	Order	Present (past) IUCN category	General former range	Habitat loss/ degradation/ fragmentation	Agriculture	Logging/ illegal timber extraction	Human population growth	Domestic animals	Hunting, poaching, human persecution	Introduced species	Population fragmentation	Other
DECLINING												
Giant panda *Ailuropoda melanoleuca*	Carnivora	EN	China	X	X	X			X		X	
Southern river otter *Lontra provocax*	Carnivora	EN	Argentina, Chile	X	X		X		X			X
Tiger *Panthera tigris*	Carnivora	EN	Asia	X	X	X	X		X			X
Flat-headed cat *Prionailurus planiceps*	Carnivora	EN	Sumatra, Borneo, Malayan peninsula	X	X	X	X		X			X
Indian water buffalo *Bubalus arnee*	Cetartiodactyla	EN	Southern and south-eastern Asia	X	X		X	X	X			X
Przewalski's gazelle *Procapra przewalskii*	Cetartiodactyla	EN	Western China	X	X		X				X	
Golden-capped fruit bat *Acerodon jubatus*	Chiroptera	EN	Philippines	X		X	X		X			X
Ecuadorian sac-winged bat *Balantiopteryx infusca*	Chiroptera	EN	Colombia, Ecuador	X	X	X						
Northern bettong *Bettongia tropica*	Diprotodontia	EN	Australia (Queensland)	X							X	X
Asiatic wild ass *Equus hemonius*	Perissodactyla	EN	Central and south-western Asia	X	X		X	X	X		X	

Yucatán black howler monkey *Alouatta pigra*	Primates	EN	Belize, Guatemala, Mexico	X		X			X			X
Woolly spider monkey *Brachyteles arachnoides*	Primates	EN	Brazil	X	X	X	X		X			X
Eastern gorilla *Gorilla beringei*	Primates	EN	Central Africa	X	X	X	X	X	X	X		X
Asian elephant *Elephas maximus*	Proboscidea	EN	Western, southern and south-eastern Asia	X	X		X		X			X
Utah prairie dog *Cynomys parvidens*	Rodentia	EN	USA (Utah)	X	X		X	X	X			
AT RISK OF EXTINCTION												
Black-footed ferret *Mustela nigripes*	Carnivora	EN(EW)	Central North America	X	X					X		X
European mink *Mustela lutreola*	Carnivora	CR	Europe	X					X			X
European bison *Bison bonasus*	Cetartiodactyla	VU(EW)	Western, central and south-eastern Europe, Caucasus	X	X	X			X		X	X
Scimitar-horned oryx *Oryx dammah*	Cetartiodactyla	EW	Probably North Africa	X				X	X			
Saiga *Saiga tatarica*	Cetartiodactyla	CR	South-East Europe, Central Asia	X					X			X
Philippine bare-backed fruit bat *Dobsonia chapmani*	Chiroptera	CR	Philippines	X	X	X	X		X			X

(continued)

Table 2.1 (continued)

Species	Order	Present (past) IUCN category	General former range	Habitat loss/ degradation/ fragmentation	Agriculture	Logging/ illegal timber extraction	Human population growth	Domestic animals	Hunting, poaching, human persecution	Introduced species	Population fragmentation	Other
Talaud bear cuscus *Ailurops melanotis*	Diprotodontia	CR	Indonesia	X	X	X	X		X		X	
Mountain pygmy-possum *Burramys parvus*	Diprotodontia	CR	South-eastern Australia	X						X	X	X
Tenkile *Dendrolagus scottae*	Diprotodontia	CR	Papua New Guinea	X			X		X			
Riverine rabbit *Bunolagus monticularis*	Lagomorhpha	CR	South Africa	X		X		X	X			
Pygmy three-toed sloth *Bradypus pygmaeus*	Pilosa	CR	Panama	X		X			X			
Blond titi monkey *Callicebus barbarabrownae*	Primates	CR	Brazil	X	X	X	X				X	
Siau Island tarsier *Tarsius tumpara*	Primates	CR	Indonesia	X					X			
Mexican agouti *Dasyprocta mexicana*	Rodentia	CR	Mexico	X								
Vancouver Island marmot *Marmota vancouverensis*	Rodentia	CR	Canada (British Columbia)	X		X						
EXTINCT Schomburgk'sdeer *Rucervus schomburgki*	Cetartiodactyla	EX	Thailand	X	X				X			
Lesser mascarene flying-fox *Pteropus subniger*	Chiroptera	EX	Mauritius, Réunion	X		X			X			

Tasmanian tiger *Thylacinus cynocephalus*	Dasyuromorphia	EX	Australia	X					X	X		
Red-bellied gracile mouse opossum *Cryptonanus ignitus*	Didelphimorphia	EX	Argentina	X	X	X	X	X				
Crescent nailtail wallaby *Onychogalea lunata*	Diprotodontia	EX	Western and central Australia	X				X		X		
Atalaye nesophontes *Nesophontes hypomicrus*	Eulipotyphla	EX	Dominican Republic, Haiti	X		X				X		
Lesser bilby *Macrotis leucura*	Peramelemorphia	EX	Central Australia	X						X		X

Index

F.M. Angelici (ed.), *Problematic Wildlife*, DOI 10.1007/978-3-319-22246-2

B

I

J

T

U

Z

MIX
Papier aus verantwortungsvollen Quellen
Paper from responsible sources
FSC® C105338

If you have any concerns about our products,
you can contact us on
ProductSafety@springernature.com

In case Publisher is established outside the EU,
the EU authorized representative is:
Springer Nature Customer Service Center GmbH
Europaplatz 3, 69115 Heidelberg, Germany

Printed by Libri Plureos GmbH
in Hamburg, Germany